# Stereochemistry and Organic Reactions

Conformation, Configuration, Stereoelectronic Effects and Asymmetric Synthesis

# Stereochemistry and Organic Reactions

## Conformation, Configuration, Stereoelectronic Effects and Asymmetric Synthesis

**Dipak K. Mandal**
Formerly of Presidency College/University
Kolkata, India

ELSEVIER

ACADEMIC PRESS
An imprint of Elsevier

Academic Press is an imprint of Elsevier
125 London Wall, London EC2Y 5AS, United Kingdom
525 B Street, Suite 1650, San Diego, CA 92101, United States
50 Hampshire Street, 5th Floor, Cambridge, MA 02139, United States
The Boulevard, Langford Lane, Kidlington, Oxford OX5 1GB, United Kingdom



**Notices**
Knowledge and best practice in this field are constantly changing. As new research and experience broaden our understanding, changes in research methods, professional practices, or medical treatment may become necessary.

Practitioners and researchers must always rely on their own experience and knowledge in evaluating and using any information, methods, compounds, or experiments described herein. In using such information or methods they should be mindful of their own safety and the safety of others, including parties for whom they have a professional responsibility.

To the fullest extent of the law, neither the Publisher nor the authors, contributors, or editors, assume any liability for any injury and/or damage to persons or property as a matter of products liability, negligence or otherwise, or from any use or operation of any methods, products, instructions, or ideas contained in the material herein.

**Library of Congress Cataloging-in-Publication Data**
A catalog record for this book is available from the Library of Congress

**British Library Cataloguing-in-Publication Data**
A catalogue record for this book is available from the British Library

ISBN 978-0-12-824092-2

For information on all Academic Press publications
visit our website at https://www.elsevier.com/books-and-journals

*Publisher:* Susan Dennis
*Acquisitions Editor:* Emily McCloskey
*Editorial Project Manager:* Allison Hill
*Production Project Manager:* Paul Prasad Chandramohan
*Cover Designer:* Victoria Pearson

Typeset by SPi Global, India

# Contents

# Preface

Organic stereochemistry is a vast and important subject. It is an intrinsic part of all organic chemistry. Planning and executing stereoselective synthesis to prepare single enantiomers as well as single diastereomers is an exhilarating science that has registered enormous progress over the years, and still is in fast development. Understanding of the factors that determine the high levels of three-dimensional control in organic synthesis is an essential and fundamental topic for all students of organic chemistry. It is not surprising therefore that stereochemistry is covered in every undergraduate, advanced undergraduate and graduate course in organic chemistry. It is my involvement in teaching this subject to undergraduate and graduate students for more than 30 years that has inspired me to write this book. The purpose of this endeavour is entirely pedagogic, keeping in view that our students desperately crave understanding, not factual knowledge alone. Understanding requires a 'feel' for the set of principles that influence the stereochemistry of organic molecules and their reactions, which is in the heart of this book; the historical development has been generally given short shrift.

The book is primarily structured in three parts (Part I–III), and each part is factored into several chapters. An appendix is included in Part IV. Part I deals with the stereochemistry—conformation and configuration—of acyclic molecules (Chapters 1 and 2) and cyclic molecules (Chapter 3). The stereochemistry of organic reactions follows thereafter. Part II provides an introduction to perturbation molecular orbital theory for the origin of stereoelectronic effects (Chapter 4) and an introduction to the principles of stereoselectivity and hierarchical levels or generations of asymmetric synthesis (Chapter 5) as a background aid to follow the reaction stereochemistry covered in Part III. I hope this unique style would help enhance the pedagogy of this text. The stereochemistry of reactions with particular emphasis on diastereoselectivity and asymmetric synthesis are presented in seven chapters along the lines of mechanistic classes. Two elaborate chapters (Chapters 6 and 7) are devoted to ionic reactions, the most abundant class, two for pericyclic reactions (Chapters 8 and 9), and one each for transition metal-catalysed reactions (Chapter 10), radical reactions (Chapter 11) and photochemical reactions (Chapter 12).

The book is distinct from other contemporary textbooks in that it brings a holistic approach by providing, within the limits of space and time, separate chapters on the stereochemistry of reactions of all mechanistic types ranging from ionic, pericyclic, transition metal-catalysed to radical and photochemical. This would undoubtedly help reveal the unifying principles for stereochemical understanding and prediction. Another important difference between this book and others is the emphasis on the working of stereochemistry, specifically how to delineate the stereochemistry of products. Learning stereochemistry is hard work. I have found that students are often not quite comfortable to work stereochemistry for themselves. They need some more help. To address their concerns, I have always been looking for innovative approaches to stereochemical issues. My efforts have produced some simple and general stereochemical rules/guidelines, some of which have been published in three papers in the *Journal of Chemical Education*. These published (also some unpublished) rules/mnemonics have been used extensively in the relevant chapters as an aid to draw quickly and correctly the product stereochemistry in organic reactions.

As an element of learning and pedagogy, I have provided over 150 problems within the chapters, reinforcing the main themes in the text. I hope that students could test their learning immediately while reading through the chapters. These problem sets should be considered an integral part of the course. The detailed answers to the problems with references, wherever necessary, are given in Appendix in Part IV. I have provided a total of about 1,400 references to primary and review literature, with each chapter containing a reference list at its end. These references will enable the readers to go further into the subject.

The approach presented in this book is distinct and class tested. I have tried to find a level to suit everyone in his/her own way from the undergraduate to the research level. I hope this book will be of value and interest to the students, teachers and researchers of organic, biological and medicinal chemistry, and also to a wider circle of readers including biologists, pharmacologists, polymer chemists and chemists working in industry who might wish to have an overview of this highly fascinating field of chemistry.

I would like to thank the anonymous reviewers for helpful suggestions. I am grateful to the editorial members Emily M. McCloskey and Allison Hill, production manager Paul P. Chandramohan, and their colleagues at Elsevier for excellent support and cooperation. Finally, I thank my wife Tapasi and my daughter Sudipta for their continuous support and my son Tirtha for his active help in artwork and referencing. I could not finish without thanking my little granddaughter Adrita who has provided me the necessary fillip during the preparation of this book.

**Dipak K. Mandal**
Kolkata, India

# Abbreviations

| | |
|---|---|
| **Ac** | acetyl (MeCO) |
| **acac** | acetylacetonato ($MeCOCHCOMe^-$) |
| **AIBN** | azobisisobutyronitrile [$Me_2C(CN)N{=}N(CN)CMe_2$] |
| **Alpine-Borane** | *B*-isopinocamphenyl-9-borabicyclo[3.3.1]nonane |
| **AQN** | anthraquinone |
| **9-BBN** | 9-borabicyclo[3.3.1]nonane |
| **BINAL-H** | binaphthyllithium aluminium hydride |
| **BINAP** | 2,2′-bis(diphenylphosphino)-1,1′-binaphthalene |
| **BINOL** | 1,1′-bi-2-naphthol |
| **Bn** | benzyl ($PhCH_2$) |
| **Boc** | *tertiary*-butoxycarbonyl ($Me_3COC{=}O$) |
| **box** | bis-oxazoline |
| **bpy** | 2,2′-bipyridine |
| **Bs** | brosyl or *p*-bromobenzenesulphonyl ($4\text{-}BrC_6H_4SO_2$) |
| **Bu** | butyl [$Me(CH_2)_3$] |
| ***i*-Bu** | isobutyl ($Me_2CHCH_2$) |
| ***s*-Bu** | *secondary*-butyl [$MeCH_2C(Me)H$] |
| ***t*-Bu** | *tertiary*-butyl ($Me_3C$) |
| **Bz** | benzoyl (PhCO) |
| **CAN** | cerium(IV) ammonium nitrate [$Ce(NH_4)_2(NO_3)_6$] |
| **Cbz** | benzyloxycarbonyl ($PhCH_2OC{=}O$) |
| **CD** | circular dichroism |
| **CE** | Cotton effect |
| **CIR** | chiral inducing reagent |
| **cod** | 1,5-cyclooctadiene |
| ***m*CPBA** | *meta*-chloroperbenzoic acid ($3\text{-}ClC_6H_4CO_3H$) |
| **CSP** | chiral stationary phase |
| **Cy** | cyclohexyl ($C_6H_{11}$) |
| **DAIB** | 3-*exo*-(dimethylamino)isoborneol |
| **dba** | *trans*,*trans*-dibenzylideneacetone [$(PhCH{=}CH)_2CO$] |
| **de** | diastereomeric excess |
| **DET** | diethyl tartrate |
| **DHQ** | dihydroquinine |
| **DHQD** | dihydroquinidine |
| **DIBAL-H** | diisobutylaluminium hydride ($i\text{-}Bu_2AlH$) |
| **DIP-Cl** | diisopinocamphenylboron chloride ($Ipc_2BCl$) |

| | |
|---|---|
| **DIPT** | diisopropyl tartrate |
| **DMAD** | dimethyl acetylenedicarboxylate ($MeO_2CC{\equiv}CCO_2Me$) |
| **DME** | 1,2-dimethoxyethane ($MeOCH_2CH_2OMe$) |
| **DMF** | *N,N*-dimethylformamide ($Me_2NCH{=}O$) |
| **DMSO** | dimethylsulphoxide ($Me_2S{=}O$) |
| **dr** | diastereomer ratio |
| **DTBP** | 2,6-di-*t*-butylpyridine |
| **$e^-$** | electron |
| **ee** | enantiomeric excess |
| **ESR** | electron spin resonance |
| **Et** | ethyl ($MeCH_2$) |
| **FMO** | frontier molecular orbital |
| **fod** | 2,2-dimethyl-6,6,7,7,8,8,8-heptafluoro-3,5-octanedionate |
| **1G/2G/3G/4G** | first/second/third/fourth generation |
| **gb** | gauche-butane |
| **GC** | gas chromatography |
| **HOMO** | highest occupied molecular orbital |
| **HMPA** | hexamethylphosphoramide [$(Me_2N)_3P{=}O$] |
| **HPLC** | high performance liquid chromatography |
| **IND** | *N*-carboxy-indoline |
| **IR** | infrared |
| **ISC** | intersystem crossing |
| **$Ipc_2BH$** | diisopinocamphenylborane |
| **$Ipc_2BOTf$** | diisopinocamphenylboron triflate |
| **KHMDS** | potassium hexamethyldisilazide [$(Me_3Si)_2NK$] |
| **LCAO** | linear combination of atomic orbitals |
| **Lcp** | left circularly polarized |
| **LDA** | lithium diisopropylamide ($i$-$Pr_2NLi$) |
| **L-DOPA** | (*S*)-3′,4′-dihydroxyphenylalanine |
| **LiHMDS** | lithium hexamethyldisilazide [$(Me_3Si)_2NLi$] |
| **LiTMP** | lithium tetramethylpiperidide |
| **L-Selectride** | lithium tri(*sec*-butyl)borohydride [$Li(s\text{-}Bu)_3BH$] |
| **LUMO** | lowest unoccupied molecular orbital |
| **2,6-Lutidine** | 2,6-dimethylpyridine |
| **Me** | methyl ($CH_3$) |
| **Meerwein's salt** | trimethyloxonium tetrafluoroborate ($Me_3O^+BF_4^-$) |
| **Ms** | mesyl or methanesulphonyl ($MeSO_2$) |
| **$NAD^+$** | nicotinamide adenine dinucleotide |
| **NADH** | nicotinamide adenine dinucleotide hydride (reduced form of $NAD^+$) |
| **NaHMDS** | sodium hexamethyldisilazide [$(Me_3Si)_2NNa$] |
| **NBS** | *N*-bromosuccinimide |
| **NGP** | neighbouring-group participation |

| | |
|---|---|
| **NMO** | *N*-methylmorpholine-*N*-oxide |
| **NMR** | nuclear magnetic resonance |
| **op** | optical purity |
| **ORD** | optical rotatory dispersion |
| **Oxone** | potassium peroxymonosulphate ($2KHSO_5 \cdot KHSO_4 \cdot K_2SO_4$) |
| **Ph** | phenyl ($C_6H_5$) |
| **PHAL** | phthalazine |
| **PHOX** | phosphinooxazoline |
| **PMP** | 1,2,2,6,6-pentamethylpiperidine |
| **PPTS** | pyridinium *p*-toluenesulphonate |
| **Pr** | propyl [$Me(CH_2)_2$] |
| ***i*-Pr** | isopropyl ($Me_2CH$) |
| **2-Py** | 2-pyridinyl |
| **QP** | quinine hypophosphite |
| **QDP** | quinidine hypophosphite |
| **RAMP** | (*R*)-1-amino-2-(methoxymethyl)pyrrolidine |
| **Red-Al** | sodium bis(2-methoxyethoxy)aluminium hydride [$NaAlH_2(OCH_2CH_2OMe)_2$] |
| **Rcp** | right circularly polarized |
| **r.t.** | room temperature |
| **SAEP** | (*S*)-1-amino-2-(1-ethyl-1-methoxypropyl)pyrrolidine |
| **SAMP** | (*S*)-1-amino-2-(methoxymethyl)pyrrolidine |
| **SET** | single electron transfer |
| **SOMO** | singly occupied molecular orbital |
| **TADDOLs** | $\alpha,\alpha,\alpha',\alpha'$-tetraaryl-1,3-dioxolane-4,5-dimethanols |
| **TAPA** | α-(2,4,5,7-tetranitro-9-fluorenylideneaminooxy) propionic acid |
| **TEMPO** | 2,2,6,6-tetramethylpiperidine-1-oxyl |
| **TBDMS** | *tertiary*-butyldimethylsilyl (*t*-$BuMe_2Si$) |
| **TBDPS** | *tertiary*-butyldiphenylsilyl (*t*-$BuPh_2Si$) |
| **Tf** | triflyl or trifluoromethanesulphonyl ($CF_3SO_2$) |
| **TFA** | trifluoroacetic acid ($CF_3CO_2H$) |
| **TFAE** | 2,2,2-trifluoro-1-(9-anthryl) ethanol |
| **$TfO^-$** | triflate ($CF_3SO_3^-$) |
| **THF** | tetrahydrofuran |
| **TIPS** | triisopropylsilyl (*i*-$Pr_3Si$) |
| **TLC** | thin layer chromatography |
| **TMEDA** | *N,N,N′,N′*-tetramethylethylenediamine [$Me_2N(CH_2)_2NMe_2$] |
| **TMS** | trimethylsilyl ($Me_3Si$) |
| **TOT** | tri-*o*-thymotide |
| **Ts** | tosyl or *p*-toluenesulphonyl ($4\text{-}MeC_6H_4SO_2$) |
| **TS** | transition structure |
| **UV** | ultraviolet |

# Part I

# Stereochemistry of organic molecules

A molecule is a microscopic grouping of atoms linked together by bonds, and its structure is a concept at the molecular level, whereas a physical property refers to an assembly of molecules or compound at the macroscopic level. Part I comprises three chapters (Chapters 1–3) which deal with the concepts of symmetry, chirality, conformation, configuration and topicity in organic molecules, and methods based on physical properties to achieve resolution of racemates and to determine conformation, configuration and enantiomeric composition.

Chapter 1

# Acyclic molecules 1: Conformation and symmetry

Organic stereochemistry[1–4] is the chemistry of organic molecules in three-dimensional, and is an intrinsic part of all organic chemistry. Organic stereochemistry is usually factorized into stereochemistry of molecules (static aspects) and stereochemistry of reactions (dynamic aspects). Stereochemistry of molecules broadly refers to conformation and configuration. In this chapter, we will describe the conformation and symmetry of acyclic molecules.

## 1.1 Stereoisomerism in molecules

The constitution of a molecule is defined by the sequence of atoms and bonds building up the molecule. Molecules with the same molecular formula but with a difference in constitution (bonding connectivity) are called *constitutional isomers*. For example, 1-butene and 2-butene are two constitutional isomers of $C_4H_8$. The differences between the constitutional isomers are *topological*. Again, molecules with identical constitution can differ in their stereochemical structure, i.e. arrangement of atoms or groups in three-dimensional space. Such molecules with the same constitution but different stereochemical arrangements are called *stereoisomers*. Stereoisomers, in turn, may differ in conformation (*conformational isomers or conformers*) or configuration (*configurational isomers*). A configurational isomer may exist in several conformational isomers.

The stereoisomers (conformational or configurational) are further classified into *enantiomers* and *diastereomers*. Enantiomers are the stereoisomers that are mirror images of each other. Notably, the enantiomeric relationship can exist only between two stereoisomers. Since the two enantiomers do not differ in the distances between the constitutionally equivalent atoms but differ when one is reflected in a mirror plane, their difference is *topographical*. On the other hand, stereoisomers that are *not* mirror images of each other are called diastereomers. Thus, the stereoisomers that are not enantiomers are diastereomers. Evidently, the diastereomeric relationship is possible between two or more stereoisomers. As the distances between the constitutionally equivalent atoms of

Stereochemistry and Organic Reactions. https://doi.org/10.1016/B978-0-12-824092-2.00001-0

the diastereomers are different, their difference is *geometric*. For the classification of isomerism, see also Fig. 2.1 (Chapter 2).

## 1.2 Representation of tetrahedral molecules: Perspective (flying wedge, zigzag) and projection (Fischer, sawhorse, Newman) formulas

A tetrahedral $sp^3$ carbon is bonded to four ligands (atoms or groups). Of the four bonds, a *maximum of two bonds* can lie on a plane (say, the plane of the paper) when one bond (third) is above the plane and the other (fourth) is below the plane, as shown for CHFClBr **1.1** (Fig. 1.1). To draw the tetrahedral structure, the following convention is adopted in this text: a bold bond in wedged form indicates a front (up) ligand; a hashed bond in untapered form indicates a rear (down) ligand.

The representation **1.1** is called a flying wedge formula. A rotation around the C—H bond lying on the plane can place other three ligands (F, Cl, Br) out of the plane **1.2** (use a molecular model). In this representation, only one bond (C—H) is lying on the plane. Now, if the C—H bond is tilted backwards, a representation **1.3** is obtained when no bond is lying on the plane. This type of representation in which two horizontal bonds lie above the plane and two vertical bonds are below the plane can be transformed into a projection formula called the Fischer projection. The representations **1.1**–**1.3** indicate perspective formulas. The projections of the bonds in the perspective formula **1.3** on the plane of the paper give a Fischer projection **1.4**. Remember that horizontal and vertical bonds in a Fischer projection correspond to above the plane and below the plane bonds respectively. All four representations (**1.1**–**1.4**) are equivalent.

A given perspective or projection formula can be transformed into equivalent formula(s) by rotating in groups of three, or by making two (or even number) interchanges. Fig. 1.2A shows that a flying wedge formula of CHFClBr is converted into an equivalent formula by clockwise rotation of three ligands (F, Br, Cl) through 120 degrees around C—H bond or by making two interchanges as indicated. Similar procedures also apply to a Fischer projection, as shown in Fig. 1.2B. Note that keeping any ligand (say, H) fixed, the other three ligands (F, Cl, Br) are rotated. This procedure can be repeated to draw other equivalent Fischer projections. As shown in Fig. 1.2B, a Fischer projection can also be

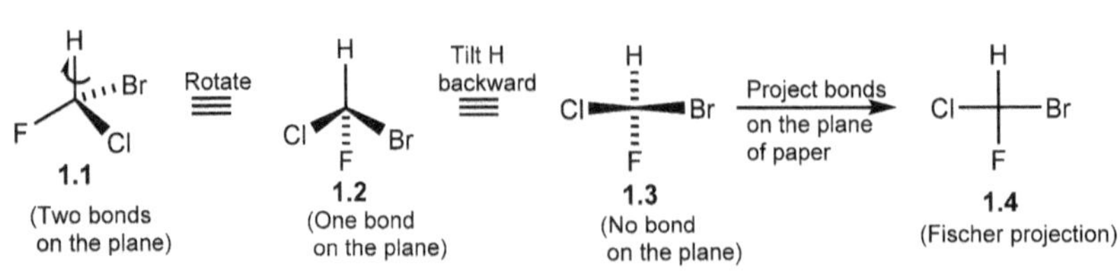


**FIG. 1.1** Representations of a tetrahedral molecule CHFClBr in perspective formulas and Fischer projection.

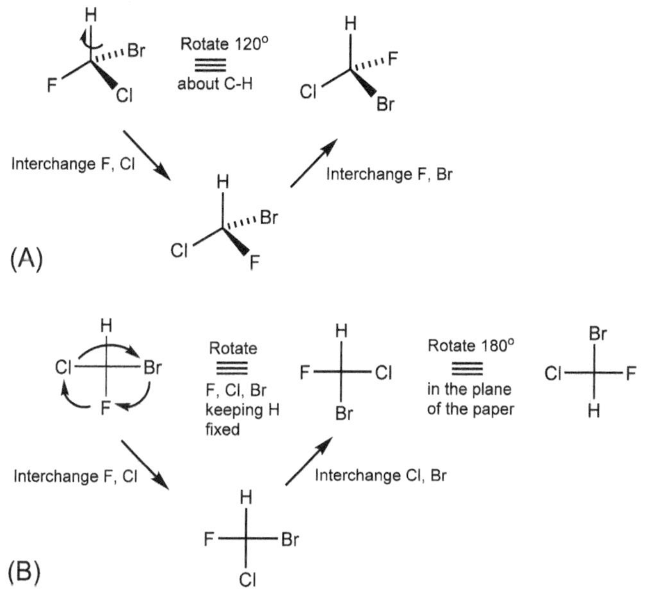


**FIG. 1.2** Methods for drawing (A) equivalent flying wedge formulas and (B) equivalent Fischer projections for a tetrahedral molecule CHFClBr.

rotated 180 degrees in the plane to obtain an equivalent Fischer projection. Fig. 1.2B also shows that two interchanges give an equivalent Fischer projection. Remember that any two ligands can be chosen for an act of exchange.

The representations of a C—C bond with the spatial disposition of associated ligands are needed in stereochemical descriptions. The simplest molecule with a C—C bond is ethane. For molecules with a number of C—C bonds, a stereochemical representation often needs to be drawn with respect to a particular C—C bond. As for example, consider 2-bromo-3-chlorobutane (MeCHBrCHClMe) in a staggered conformation and in an eclipsed conformation (see Section 1.3).

Fig. 1.3A shows perspective formulas (flying wedge and zigzag) and projection formulas (sawhorse and Newman) of a staggered conformation of 2-bromo-3-chlorobutane. The flying wedge formula is drawn viewing the C2—C3 bond *sideways* (use a molecular model) when the four-carbon chain lies in the plane of the paper. Note that, for each tetrahedral carbon C2 or C3, two bonds are on the plane, one bond is above the plane and the other is below the plane. If the C2—C3 bond is viewed in *oblique* fashion, one obtains a sawhorse projection formula, in which Br and Cl are on the same side of the C—C bond as in the equivalent flying wedge formula. A Newman projection formula is obtained when the C2—C3 bond is viewed *end-on*. The main carbon chain of the flying wedge formula can also be drawn as a horizontal *zigzag* chain with ligands

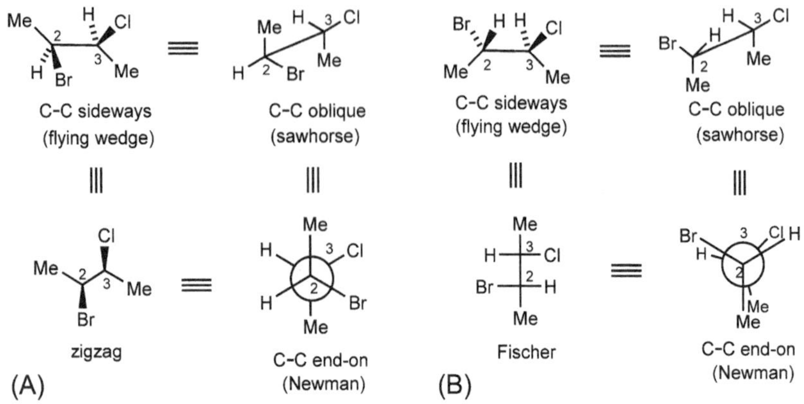


**FIG. 1.3** Representations of a C—C bond and associated ligands of 2-bromo-3-chlorobutane: (A) a staggered conformation in flying wedge, sawhorse, Newman and zigzag formulas, and (B) an eclipsed conformation in flying wedge, sawhorse, Newman and Fischer formulas.

(other than H) pointing up or down. This perspective zigzag formula (devised by Masamune) always denotes a staggered conformation. A zigzag formula does not represent an eclipsed conformation.

Fig. 1.3B shows a perspective formula (flying wedge) and projection formulas (sawhorse, Newman and Fischer) for an eclipsed conformation. The flying wedge, sawhorse and Newman formulas are drawn in a similar manner as described above. Note that, in the Newman projection, the ligands on the back carbon (C3) are artificially rotated a little so that they are not obscured completely. In the Fischer projection, horizontal and vertical ligands at the C2—C3 bond indicate above the plane and below the plane ligands respectively. In contrast to a zigzag formula, a Fischer projection always denotes an eclipsed conformation, and does not represent a staggered conformation.

## 1.3 Conformation

Conformation has been defined[5] as "the spatial arrangement of the atoms affording distinction between stereoisomers which can be interconverted by rotations about formally single bonds." The term has been extended to include inversion at trigonal pyramidal centres (see later Section 2.5.2). Conformational analysis refers to the study of the relative stability of conformations and relating conformation to the properties and reactivity of molecules. Since a C—C σ bond is cylindrically symmetrical, internal rotation at the C—C bond is not likely to affect the bond orbital. However, the rotation about the C—C bond is not absolutely free and there occurs an energy barrier to rotation. Conformations also arise as a result of rotation about C—N and C—O bonds.

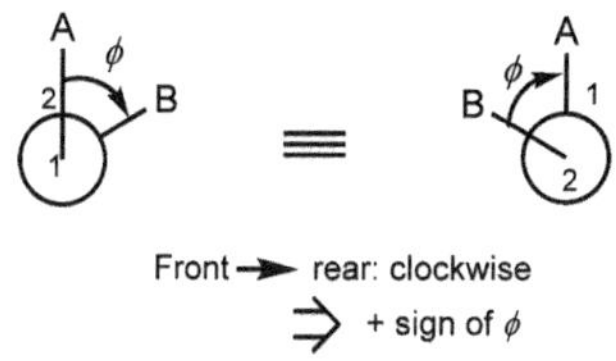


**FIG. 1.4** A positive torsion angle ($\phi$).

### 1.3.1 Torsion angles and torsional strain

In a nonlinear chain of four atoms A—C1—C2—B, the torsion angle ($\phi$) for the two bonds A—C1 and C2—B is defined as the angle between the projections of A—C1 and C2—B on a plane perpendicular to C1—C2 (Fig. 1.4). The torsion angle has both magnitude and a sign. Looking along the C—C bond in *either* direction (C1 → C2 or C2 → C1), if the turn from the front atom to the rear atom is clockwise, $\phi$ is positive and if the turn is anticlockwise $\phi$ is negative.

Internal rotation changes the torsion angle. Conformation may be described in terms of exact magnitude and sign of $\phi$. However, when exact values of $\phi$ are not known or relevant, conformation may be described using a range of torsion angles (±30 degrees), known as Klyne–Prelog classification[6] (Table 1.1).

The dihedral angle is a term similar to the torsion angle but unlike a torsion angle, the dihedral angle is an unsigned angle. Torsion angles of −120 and −60 degrees (see Table 1.1) correspond to dihedral angles of 240 and 300 degrees, respectively. The term dihedral angle will not generally be used in this text.

The torsional strain (or Pitzer strain) refers to the interaction that leads to the variation of potential energy as a function of torsion angle. The torsion angle

**TABLE 1.1 Klyne–Prelog specification of torsion angles.**

| Torsion angle ($\phi$) (degrees) | Designation |
|---|---|
| 0 ± 30 (−30 to +30) | synperiplanar (*sp*) |
| +60 ± 30 (+30 to +90) | +synclinal (+*sc*) |
| +120 ± 30 (+90 to +150) | +anticlinal (+*ac*) |
| 180 ± 30 (+150 to −150) | antiperiplanar (*ap*) |
| −120 ± 30 (−150 to −90) | −anticlinal (−*ac*) |
| −60 ± 30 (−90 to −30) | −synclinal (−*sc*) |

Synclinal ($\phi \approx 60$ degrees) is often called gauche ($g^+ \approx +60$ degrees, $g^- \approx -60$ degrees). Antiperiplanar ($\phi \approx 180$ degrees) is frequently called anti.

function is naturally a Fourier series. However, for a simple $C(sp^3)$—$C(sp^3)$ single bond, the torsional potential ($V_\phi$) appears to follow a simple cosine function as

$$V_\phi = \frac{1}{2} V_0 (1 + \cos 3\phi) \quad (1.1)$$

where $V_0$ is a constant. For a C—C—C—C chain, $V_0$ is 3.1 kcal mol$^{-1}$.[7] When $\phi = 0$ degrees for the eclipsed (synperiplanar) conformation, the Eq. (1.1) gives $V_\phi = V_0$, i.e. the torsional potential is at a maximum. The torsional potential is at a minimum ($V_\phi = 0$) when $\phi = 60$ degrees for the staggered (synclinal) conformation.

### 1.3.2 Nonbonded van der Waals interactions

At infinite distance, the energy of interaction of two nonbonded atoms is zero. As the two atoms approach each other, an attractive force (London or dispersion force) operates which is taken as $-ar^{-6}$ ($a$ is a constant and r is the internuclear distance). This leads to a lowering of the energy. At a still closer distance, a repulsive force due to closed-shell repulsion takes place which is expressed as $br^{-12}$ ($b$ is a constant). The overall nonbonded potential ($V_{nb}$) is then expressed by Eq. (1.2), called a Lennard-Jones potential.

$$V_{nb} = -ar^{-6} + br^{-12} \quad (1.2)$$

Fig. 1.5 shows a plot of $V_{nb}$ against r.

van der Waals radius refers to the distance between two nonbonded atoms at which there is no net attraction or repulsion. At this point, the distance (r′) between the two atoms is the sum of their van der Waals radii. Nonbonded interactions may be attractive or repulsive depending on the internuclear distance. When the distance between two specified atoms is less than the sum of their van der Waals radii, they repel each other. Such repulsion due to crowding is referred to as van der Waals strain or steric strain.

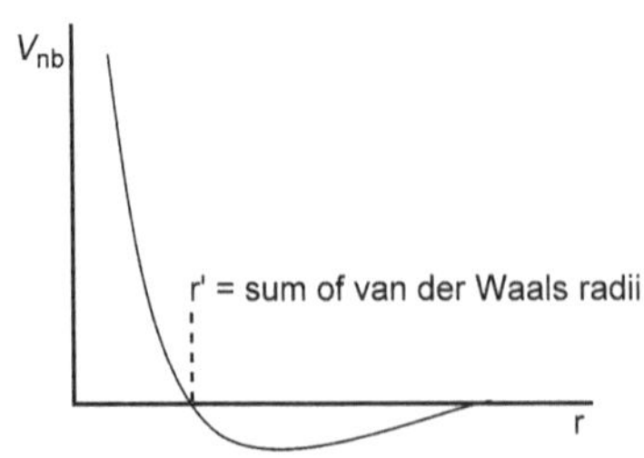


**FIG. 1.5** Nonbonded van der Waals interactions between two atoms.

The approximate values of van der Waals radii (in Å) of a few atoms or groups are given below.

| H | F | Cl | Br | I | $CH_3$ |
|---|---|---|---|---|---|
| 1.25 | 1.35 | 1.8 | 1.95 | 2.15 | 2.0 |

### 1.3.3 Molecular mechanics

Only a very brief introduction to molecular mechanics is presented here. An overview and a general account of the methods of molecular mechanics can be found elsewhere.[8,9]

Strain in a molecule arises from a nonideal molecular geometry and several strain factors contribute to the total strain energy of the molecule. These include bond length strain, bond angle strain, torsional strain and nonbonded interactions. The total strain energy ($V_{\text{strain}}$) is given by

$$V_{\text{strain}} = \sum V_r + \sum V_\theta + \sum V_\phi + \sum V_{\text{nb}} \tag{1.3}$$

where $V_r$ is bond length strain due to bond stretching or compression, $V_\theta$ is bond angle deformation strain, $V_\phi$ is torsional strain (Eq. 1.1) and $V_{\text{nb}}$ is nonbonded interactions (Eq. 1.2).

The bond length strain ($V_r$) is assumed to follow a Hooke's law expression as

$$V_r = \frac{1}{2}k_r(r - r_0)^2 \tag{1.4}$$

where r is the bond length at any given instant, $r_0$ is the mean or equilibrium bond length and $k_r$ is a constant for a type of bond. It is assumed that such constants can be transferred between molecules in molecular mechanics calculation. Typically, $V_r$ is 3.2 kcal mol$^{-1}$ if $r - r_0$ is only 10 pm (0.1 Å) for a C—C single bond. Thus a change of bond length is energetically very costly.

The bond angle bending is also, to a first approximation, a quadratic function as

$$V_\theta = \frac{1}{2}k_\theta(\theta - \theta_0)^2 \tag{1.5}$$

where $\theta - \theta_0$ is the change in bond angle from the equilibrium bond angle taken as 109°28′, and $k_\theta$ is the appropriate force constant. Typically, $k_\theta$ is 0.025 kcal mol$^{-1}$ degrees$^{-2}$ for CCC angle. Thus $V_\theta$ is only 0.3 kcal mol$^{-1}$ for a 5 degrees deformation in bond angle. The bond angle change may arise from the rehybridization of the pivotal atom through variations of s and p character of the orbitals involved in bonding.

It is of note that the list of terms in Eq. (1.3) can be enlarged to include intramolecular electrostatic (coulombic) interactions ($V_E$) for polar groups if present, solvation ($-V_S$) and cross terms with two parameters to improve $V_{strain}$. A molecular mechanics calculation begins with the calculation of strain energy of a trial structure obtained from a molecular model or from a similar molecule in the X-ray structural database. Computationally, the structure of a molecule with $n$ atoms is a vector of $3n$ Cartesian coordinates. By changing the coordinates and recomputing the energies, the energies of all geometries slightly deviated from the original trial structure can be explored. The energy as a function of the coordinates is minimized using a suitable computer algorithm and the exploration is repeated until a structure is found at an energy minimum so that any further deformation leads to a higher energy. The structure at the energy minimum is taken to be the predicted structure of the molecule. The method permits the calculation of both structure and energy.

The mathematical functions (force fields) that accurately model the various strain interactions for a wide variety of molecules have been developed. The term force field signifies that the array of atoms is in a field of interatomic forces. The method allows the strain energy of any particular conformation to be estimated. By calculating different energies of various conformations, it is possible to predict the lowest energy conformation.

### 1.3.4 Conformation of ethane

The simplest molecule with a single C—C bond is ethane. The internal rotation of one methyl group relative to the other changes the torsion angle and gives rise to conformations. In principle, infinite number of conformations is possible. The potential energy curve for rotation about the C—C bond in ethane shows that there are three degenerate minima and three degenerate maxima (Fig. 1.6).

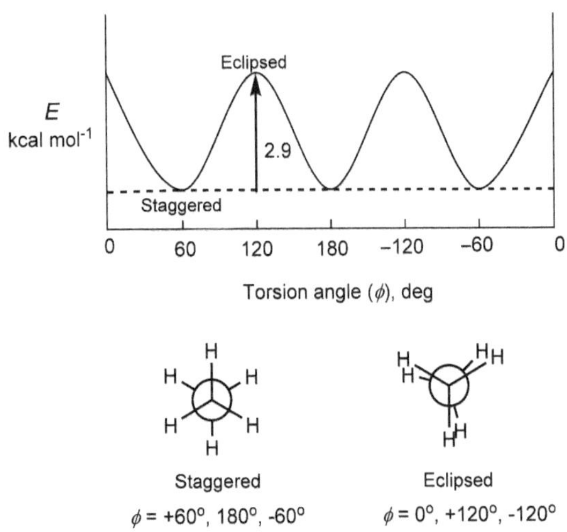


**FIG. 1.6** Potential energy as a function of torsion angle for ethane.

The three minima correspond to torsion angles $\phi = \pm 60$ degrees, 180 degrees and represent staggered conformations while the three maxima correspond to eclipsed conformations ($\phi = 0$ degrees, $\pm 120$ degrees). The conformations that are located at the energy minima are called conformational isomers or conformers. Therefore ethane has three indistinguishable staggered conformers. The energy difference between the eclipsed and the staggered conformation is $\sim 2.9\,\text{kcal mol}^{-1}$ which gives the energy barrier to internal rotation in ethane.[10] Thus, at any instant, any individual molecule is likely to be in the staggered conformation. The eclipsed conformation is regarded as a transition structure (TS) conformation for the interconversion between staggered conformations. The energy barrier is not so high; at room temperature a rapid interconversion of staggered conformers takes place through an eclipsed conformation.

#### 1.3.4.1 *Torsional barrier and kinetics*

An idea of the rate of rotation can be obtained from an estimate of approximate rate constant using transition state theory. The rate constant ($k$) for a process derived from transition state theory is given by

$$k = \frac{k_B T}{h} e^{-\Delta G^{\neq}/RT} \qquad (1.6)$$

where $k_B$ is Boltzmann constant and $\Delta G^{\neq}$ is the free energy barrier. Assuming entropy term to be insignificant for a conformational process, $\Delta G^{\neq}$ is taken to be the conformational energy barrier ($2.9\,\text{kcal mol}^{-1}$). Substituting this value in Eq. (1.6), the value of $k$ at 25°C (298 K) is obtained as

$$k = \frac{(1.38 \times 10^{-23})(298)}{(6.62 \times 10^{-34})} e^{(-2900 \times 4.18)/(8.31)(298)}\,\text{s}^{-1} = 4.6 \times 10^{10}\,\text{s}^{-1}$$

The rate of rotation at 25°C is thus extremely fast with an average time between crossings is to the order of $10^{-10}$ s.

The relative instability of the eclipsed conformation is attributed to torsional strain or Pitzer strain. Any deviation from the lowest energy staggered conformation is accompanied by torsional strain which reaches a maximum at the eclipsed conformation.

#### 1.3.4.2 *Origin of the torsional strain*

The origin of the rotational barrier (torsional barrier) has been much debated.[11,12] The steric repulsion in the eclipsed conformation is at most a minor factor since the H atoms of the two methyl groups are barely within the van der Waals distance. In orbital terms, the steric repulsion, if any, would arise from the filled orbital/filled orbital repulsion of the C—H bonds in eclipsed ethane. The main factor responsible for the torsional barrier is σ–σ* hyperconjugation (in frontier orbital terms, $\text{HOMO}_{\sigma}/\text{LUMO}_{\sigma*}$ interaction),

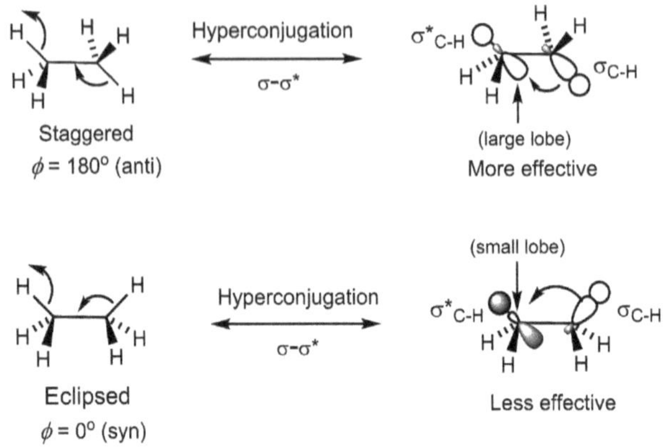


**FIG. 1.7** Origin of torsional barrier in ethane attributed to σ–σ* hyperconjugation.

which stabilizes the staggered conformation much more than the eclipsed conformation, and hence favours the former (Fig. 1.7).

Calculations[13] indicate that if the hyperconjugation effects are switched off, eclipsed conformation is preferred. Thus hyperconjugation, not the filled/filled repulsion, is the origin of the torsional barrier in ethane.[11,13] It should be noted that there are three anti H/H pairs in staggered ethane so that each σ–σ* hyperconjugation stabilizes it roughly by 1 kcal mol$^{-1}$. In other words, each H/H eclipsing raises the energy by ~1 kcal mol$^{-1}$.

### 1.3.4.3 *Conformation of propane and butane for rotation about C1-C2 bond*

In molecules of the type $H_3C—CH_2R$, the conformational situation is very similar to that of ethane. Thus propane (R=Me) and butane (R=Et) has three degenerate staggered conformers due to rotation about C1—C2 bond. There are also three degenerate eclipsed conformations. Fig. 1.8 shows one staggered and one eclipsed conformation of propane and butane. Due to R/H (R=Me, Et) eclipsing, the torsional barrier in propane or butane is higher than in ethane. The torsional barrier in propane is estimated to be ~3.4 kcal mol$^{-1}$.[14,15] Thus a

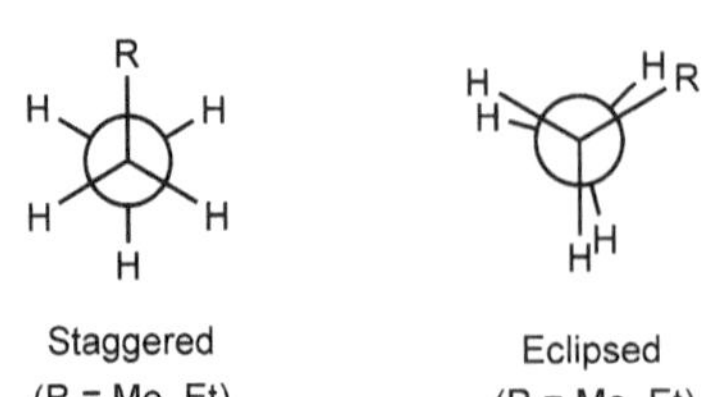


**FIG. 1.8** Staggered and eclipsed conformations of propane and butane due to rotation about C1—C2 bond.

Me/H eclipsing contributes ~1.4 kcal mol$^{-1}$ compared to H/H eclipsing (~1 kcal mol$^{-1}$) because of an additional Me/H steric (van der Waals) repulsion.

**Problem 1.1** The torsional barriers in ethyl halides ($CH_3CH_2X$, X=F, Cl, Br, I) are remarkably similar (3.3–3.5 kcal mol$^{-1}$). Explain.

### 1.3.5 Conformation of butane for rotation about C2—C3 bond

The eclipsed and staggered conformations and potential energy curve as a function of torsion angle for rotation about the central C2—C3 bond in butane is shown in Fig. 1.9A. Unlike ethane, the three staggered conformers of butane at potential energy minima are not equivalent. There are two degenerate minima corresponding to two gauche forms ($g^+$ and $g^-$) and one lowest minimum representing an anti form. The two gauche forms are chiral and enantiomeric while the anti form is achiral (see Figs 1.26B and 1.28B). Three conformers of butane thus comprise two gauche and one anti forms. The gauche butane suffers from repulsive steric (van der Waals) interaction between two $CH_3$ groups which

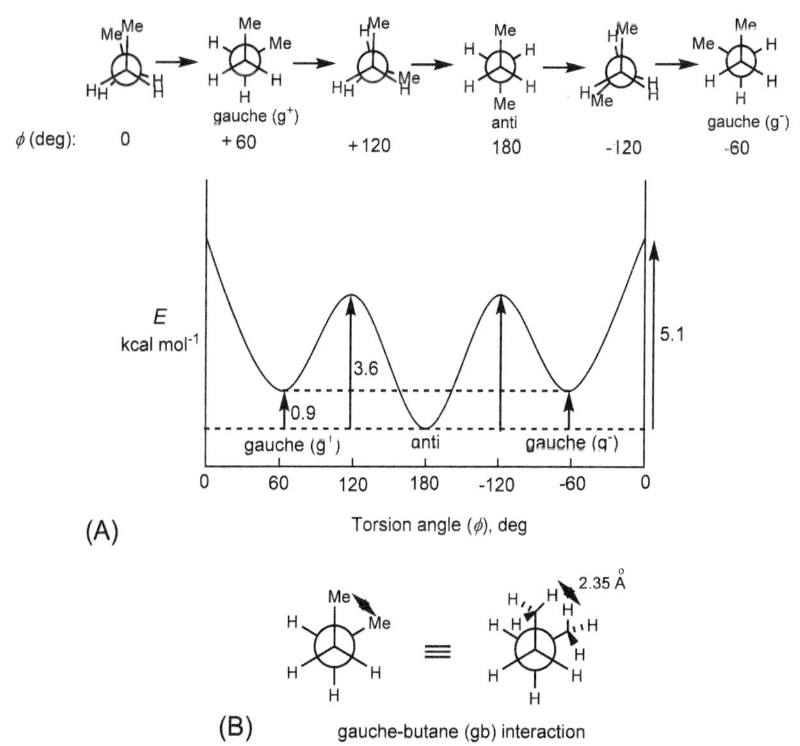


**FIG. 1.9** (A) Potential energy profile of butane as a function of torsion angle for rotation about C2—C3 bond and (B) gauchebutane interaction as H/H van der Waals repulsion.

raises its energy by 0.9 kcal mol$^{-1}$ relative to the anti butane which is devoid of such steric repulsion.[14,16] The relative stability of gauche and anti conformers is thus attributed to steric or van der Waals strain. This type of steric interaction between two $CH_3$ groups at a torsion angle $\phi = \pm 60$ degrees is referred to as *gauche–butane* (gb) interaction. Notably, the $CH_3/CH_3$ interaction in gauche butane implies a H/H van der Waals repulsion between two hydrogen atoms on the two $CH_3$ groups (the H···H distance is 2.35 Å which is less than the sum of their van der Waals radii) (Fig. 1.9B). In orbital terms, the steric strain refers to the filled orbital/filled orbital interaction (see Section 4.2.1). It is pertinent to mention that the torsion angle in the gauche conformer is somewhat larger than the 60-degree angle in a perfectly staggered conformation. This is because the van der Waals repulsion rotates the two methyl groups away from each other even at the expense of increasing torsional strain, and energy minimization gives a torsion angle of 65 degrees for the gauche conformation.

Of the three maxima, two degenerate maxima correspond to eclipsed conformations with torsion angles $\phi = \pm 120$ degrees involving twofold Me/H eclipsing, which are of lower energy than the other eclipsed conformation at $\phi = 0$ degrees involving a Me/Me eclipsing. The energy barrier for anti → gauche conversion[14] via the TS eclipsed conformations at $\phi = \pm 120$ degrees is 3.6 kcal mol$^{-1}$, whereas that for $g^- \rightarrow g^+$ conversion via the TS eclipsed conformation at $\phi = 0$ degrees is 4.2 (5.1 − 0.9) kcal mol$^{-1}$ (Fig. 1.9A). It should be noted that the height of the energy barrier between the two conformers, say, gauche and anti depends on whether it is measured from the side of the less stable conformer (gauche) or the more stable conformer (anti) (Fig. 1.9A).

**Problem 1.2** Using eclipsing interactions, estimate the increase in energy for the eclipsed conformation of butane at $\phi = \pm 120$ degrees relative to the anti conformer of butane.

**Problem 1.3** The energy difference between the eclipsed conformation at $\phi = 0$ degrees and the anti conformer of butane is 5.1 kcal mol$^{-1}$. Estimate the energetic cost of a $CH_3/CH_3$ eclipsing interaction.

### 1.3.5.1 Conformational equilibrium and thermodynamic parameters

The conformational equilibrium of gaseous butane is represented as

$$\text{anti} \rightleftharpoons \text{gauche} \qquad \Delta H^0 = 0.9 \text{ kcal mol}^{-1}$$

(Remember that gauche butane is a 1:1 mixture of $g^+$ and $g^-$ forms). The gauche–anti enthalpy difference is referred to as conformational enthalpy of butane, which is 0.9 kcal mol$^{-1}$.

The population of gauche and anti conformers can be estimated from the equilibrium constant ($K$) where

$$K = \frac{[\text{gauche}]}{[\text{anti}]} \quad \text{and} \quad \Delta G^0 = -RT\ln K$$

$\Delta G^0$ is given by $\Delta G^0 = \Delta H^0 - T\Delta S^0$. The entropy difference ($\Delta S^0$) between the two conformers arises mainly from entropy of symmetry and entropy of mixing. (The translational entropy is same; the rotational and vibrational entropies for the two conformers, though not the same, differ by a small margin.)

The entropy of symmetry ($S_{sym}$) is given by $-R \ln \sigma$ where $\sigma$ is the symmetry number. For both gauche and anti forms, $\sigma = 2$ (see Sections 1.5.1 and 1.5.2.2).

The entropy of mixing ($S_{mix}$) is given by $-R \sum x_i \ln x_i$ where $x_i$ is the mole fraction of the $i$th component. The gauche conformer has a $S_{mix}$ due to 1:1 mixing of $g^+$ and $g^-$ forms whereas there is no $S_{mix}$ for the achiral anti conformer.

Therefore,

$$\begin{aligned}\Delta S^0 &= S^0{}_{gauche} - S^0{}_{anti}\\ &= \left[-R\ln\sigma - R\sum {}^1/_2 \ln {}^1/_2\right] - [-R\ln\sigma]\\ &= \left[-R\ln 2 - R({}^1/_2\ln{}^1/_2 + {}^1/_2\ln{}^1/_2)\right] - [-R\ln 2]\\ &= -R\ln{}^1/_2 = R\ln 2 = 1.38\,\text{cal}\,\text{mol}^{-1}\text{K}^{-1}\end{aligned}$$

The value of $\Delta G^0$ at 25°C for anti $\rightleftharpoons$ gauche is estimated as

$$\begin{aligned}\Delta G^0 &= \Delta H^0 - T\Delta S^0\\ &= [900 - (298 \times 1.38)]\,\text{cal}\,\text{mol}^{-1} = 489\,\text{cal}\,\text{mol}^{-1}\end{aligned}$$

Now,

$$\frac{[\text{gauche}]}{[\text{anti}]} = K = e^{-\Delta G^0/RT} = e^{-489/(1.987\times 298)} = 0.43 = 30:70$$

Hence the conformational equilibrium of gaseous butane at room temperature comprises a population of 30% gauche (15% $g^+$ and 15% $g^-$) and 70% anti forms.

In general, populations in any two states (including conformational states) can be estimated from their free energy difference ($\Delta G^0$). Table 1.2 shows relative populations in two conformational states at 25°C as a function of their free energy difference ($\Delta G^0$).

**Problem 1.4** The gauche–anti enthalpy difference of butane in the liquid phase ($\Delta H^0 = 0.55\,\text{kcal}\,\text{mol}^{-1}$) is lower than that in the gaseous phase ($\Delta H^0 = 0.9\,\text{kcal}\,\text{mol}^{-1}$). Explain.

### 1.3.6 Conformation of 2,3-dimethylbutane

The conformational analysis of 2,3-dimethylbutane provides an interesting case. Here the gauche and anti conformers appear to have the same stability, and the three conformers (two gauche and one anti) are equally populated.

**TABLE 1.2** Relative populations in two conformational states as a function of $\Delta G^0$ at 298 K.

| $\Delta G^0$, kJ mol$^{-1}$ (kcal mol$^{-1}$)[a] | $K$ | More stable state (%) |
|---|---|---|
| 0 (0) | 1.0 | 50 |
| 1 (0.24) | 1.5 | 60 |
| 2 (0.48) | 2.2 | 69 |
| 3 (0.72) | 3.5 | 77 |
| 4 (0.96) | 5.0 | 83 |
| 5 (1.2) | 7.5 | 88 |
| 10 (2.4) | 57 | 98 |
| 15 (3.6) | 430 | 99.8 |
| 20 (4.8) | 3200 | 99.97 |

[a] *1 cal = 4.184 J.*

This rather unexpected result is ascribed to bond angle deformation and the consequent steric effect. In butane, Me—C—H bond angle is close to tetrahedral. In contrast, Me—C—Me angle in 2,3-dimethylbutane opens up to near 114 degrees.[17,18] This widening of bond angle (Thorpe–Ingold effect) leads to enhanced van der Waals repulsion in anti form but diminished steric strain in gauche form as shown in Fig. 1.10. Thus the anti form is destabilized and the gauche form is stabilized making them almost equienergetic. Similar considerations apply to other branched-chain alkanes of the type RR′CHCHRR′.[18,19]

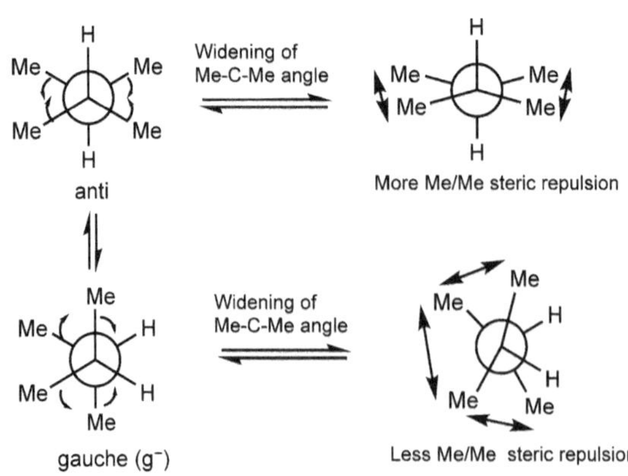


**FIG. 1.10** anti and gauche conformers of 2,3-dimethylbutane. Only one gauche conformer ($g^-$) is shown.

### 1.3.7 Conformation of 1,2-dihaloethanes and related molecules[20–22]

The molecules with polar substituents possess substantial dipoles; thus electrostatic effects by dipole–dipole interactions come into play (besides the steric effects) to govern the relative stability of the conformers. The magnitude of the dipole interactions will, however, be solvent dependent (see below).

#### *1.3.7.1 1,2-Dichloroethane and 1,2-dibromoethane*

The conformational equilibrium for 1,2-dichloroethane in the gas-phase is

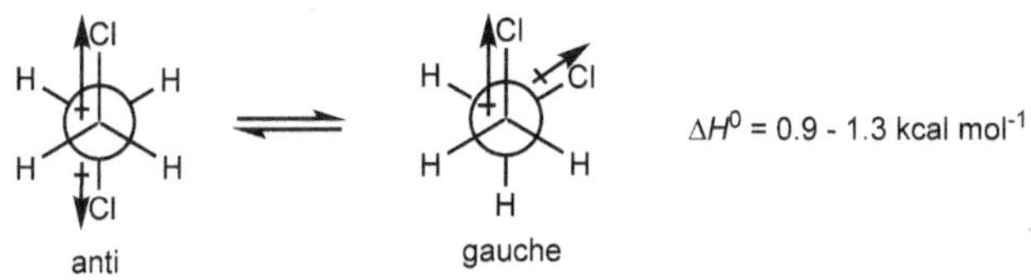


It has been observed that the anti conformer is favoured over the gauche conformer by 0.9–1.3 kcal mol$^{-1}$.[20] Here, besides the steric effects, the gauche conformer is further destabilized by dipole–dipole repulsion (two C—Cl dipoles in gauche are more or less aligned in contrast to anti where the two dipoles are opposed). The gauche conformer has a higher dipole moment than anti form in which the bond moments of two C—Cl dipoles cancel. With rise in temperature, the population of the gauche form will increase which will lead to an increase in dipole moment of 1,2-dichloroethane. Evidence for both gauche and anti conformers in the conformational equilibrium has been obtained from the IR spectrum of the compound. If the anti conformer were populated exclusively, 1,2-dichloroethane will have no dipole moment and hence would be weakly IR active. The presence of significant bands in the IR spectrum suggests a significant population of the gauche conformer in conformational equilibrium.

In polar solvents, the dipole–dipole repulsion will be less significant as the polar solvent molecules will surround the dipoles and thereby minimize the dipole–dipole repulsion. The population of the gauche conformer in a polar solvent will therefore increase. In other words, the gauche conformer with the higher dipole moment will benefit from the increased energy of solvation, resulting in an increase in the population of the gauche form. As the polarity of the solvent (which correlates with the dielectric constant ε) increases, the gauche form is less destabilized by dipole repulsion, and consequently gauche–anti enthalpy difference ($\Delta H^0$) decreases (Table 1.3).

The situation in 1,2-dibromoethane is similar.[21] The gauche–anti enthalpy difference is larger ($\Delta H^0 = 1.4$–$1.8$ kcal mol$^{-1}$) than that for 1,2-dichloroethane. Both steric and dipole–dipole repulsions contribute to destabilize the gauche conformer. As Br is larger than Cl, the steric effect is more significant in the case of 1,2-dibromoethane.

**TABLE 1.3 Solvent dependence of gauche–anti enthalpy difference ($\Delta H^0$) of 1,2-dichloroethane.**

| Solvent | ε | $\Delta H^0$ (kcal mol$^{-1}$) |
|---|---|---|
| Cyclohexane | 2.0 | 0.91 |
| Diethyl ether | 4.3 | 0.69 |
| Liquid (neat) | 10.1 | 0.31 |
| Acetone | 20.7 | 0.18 |

ε = dielectric constant.

### 1.3.7.2 *1,2-Difluoroethane*

The conformational equilibrium for 1,2-difluoroethane seems anomalous; the gauche form is favoured over the anti form by 0.6–0.9 kcal mol$^{-1}$ even in the gas phase notwithstanding the repulsive dipole–dipole interaction.[23]

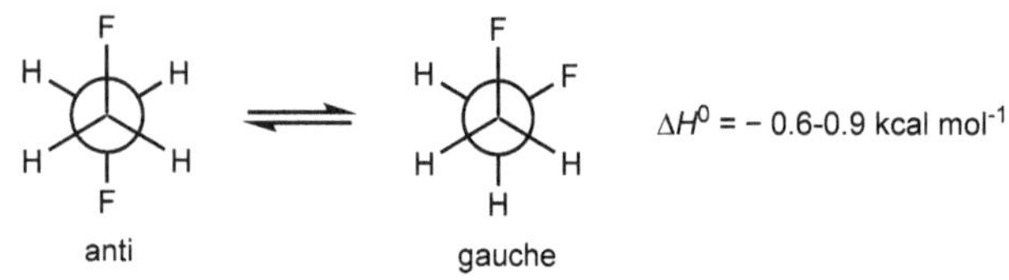


The reason for the preference of the gauche form is attributed to the relatively small van der Waals repulsion between two fluorine atoms (van der Waals radius = 1.5–1.6 Å) and a strong stereoelectronic effect by $\sigma_{C-H}$–$\sigma^*_{C-F}$ hyperconjugation[24] (Fig. 1.11). The stereoelectronic requirement for an effective hyperconjugation is the antiperiplanarity of the involved C—H and C—F bonds ($\phi$ = 180 degrees). In the anti conformer, no C—F bond is antiperiplanar to a C—H bond. In contrast, both C—F bonds in the gauche conformer can benefit from the favourable $\sigma_{C-H}$–$\sigma^*_{C-F}$ overlap (in frontier orbital terms,

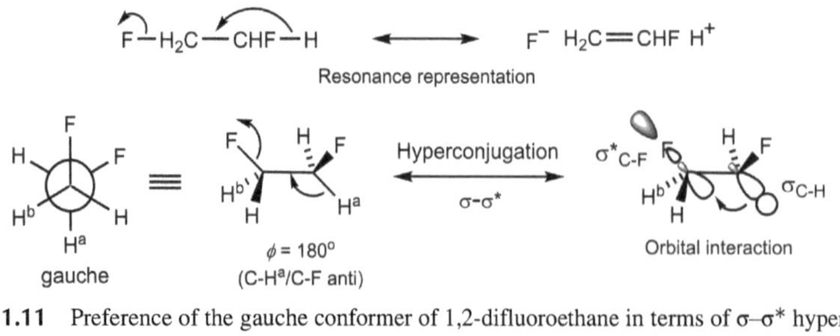


**FIG. 1.11** Preference of the gauche conformer of 1,2-difluoroethane in terms of σ–σ* hyperconjugation. The orbital interaction involving one C-H bond (labelled C—H$^a$) is shown only.

HOMO$_{\sigma}$/LUMO$_{\sigma*}$ interaction). As shown for the gauche conformer in Fig. 1.11, one C—F bond is anti to C—H$^a$ and the other C—F bond is anti to C—H$^b$. The resonance representation of the hyperconjugative effect involving a C—H bond, and the $\sigma_{C—H}$–$\sigma^*_{C—F}$ hyperconjugation with only C—H$^a$ bond are shown.

### 1.3.7.3 *1,2-Ethanediol*

Intramolecular H-bonding, wherever possible, can play a crucial role to govern the relative stability of conformers. For example, in ethylene glycol, the gauche conformer is preferred over the anti since the gauche form is stabilized by intramolecular H-bonding[25] (Fig. 1.12). The intramolecular H-bonding dominates the steric effect and determines the conformational equilibrium. Note that intramolecular H-bonding is not possible in the anti conformer since the two OH groups are far apart.

In general, 2-substituted ethanols ($HOCH_2CH_2X$ where X = OMe, F, Cl, or Br) have the preferred gauche conformation[26] due to intramolecular H-bonding. The most important examples of such effects are observed in biopolymers such as proteins, nucleic acids and polysaccharides.

**Problem 1.5** Explain the following:

**(a)** In 1-chloropropane, the gauche conformer is enthalpically preferred slightly over the anti conformer in the vapour state.
**(b)** Acetylcholine ($Me_3N^+CH_2CH_2OAc$) exists almost exclusively in the gauche conformation.
**(c)** In molecules of the type $CH_3OCH_2X$ (X = Cl, OR, OCOR), the gauche conformer is preferred.
**(d)** The IR spectra of racemic as well as *meso* isomers of butane-2,3-diol show two distinct O—H bands. For the racemic isomer, the band separation ($\Delta\nu$) is 49 cm$^{-1}$, whereas $\Delta\nu$ is 42 cm$^{-1}$ for the *meso* isomer.

## 1.3.8 Conformation of pentane as a function of two torsion angles

In pentane, rotations around two C—C bonds (C2—C3 and C3—C4) are possible, and the resulting conformations can be represented in terms of two torsion angles ($\phi_1$ and $\phi_2$) **1.5**.

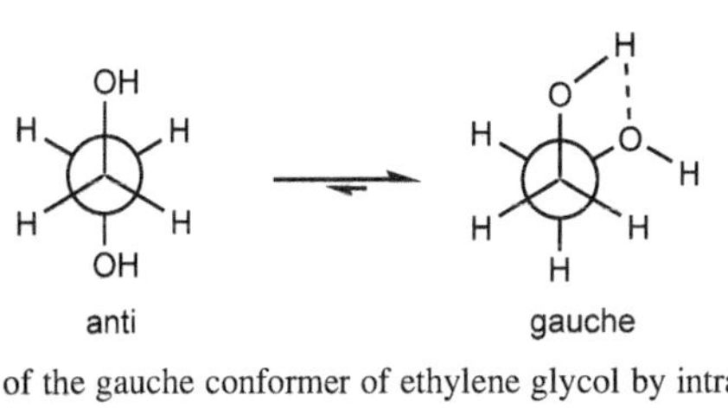


**FIG. 1.12** Preference of the gauche conformer of ethylene glycol by intramolecular H-bonding.

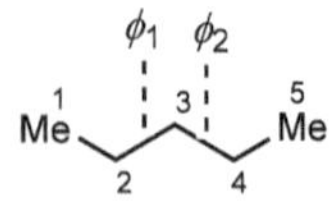


**1.5**

Since rotation around a single C—C bond (C2—C3) in butane leads to three staggered conformers [anti (a), $g^+$, $g^-$], rotations around two C—C bonds in pentane would give rise to $3^2 = 9$ staggered conformations as derived below.

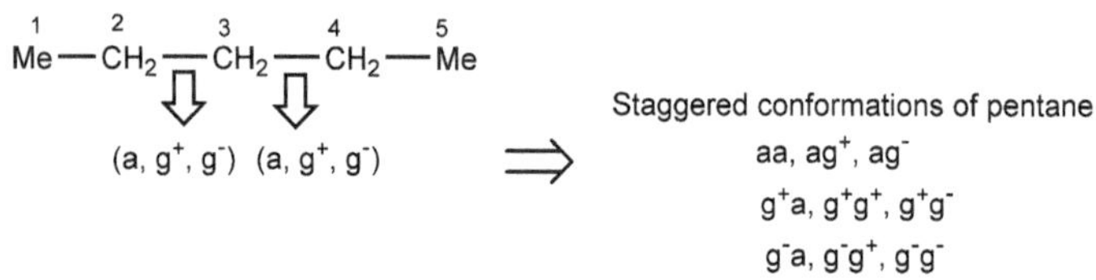


As pentane is a symmetrical molecule, the conformers $ag^+ \equiv g^+a$, $ag^- \equiv g^-a$ and $g^+g^- \equiv g^-g^+$. Thus, pentane has six distinct conformers: aa, $ag^+$, $ag^-$, $g^+g^+$, $g^+g^-$, $g^-g^-$. Again, since $g^+$ and $g^-$ are enantiomeric, $ag^+$ is enantiomeric with $ag^-$, and $g^+g^+$ is enantiomeric with $g^-g^-$. The enantiomers have the same energy. Therefore, pentane has four conformers of differing energies: anti–anti (aa), anti–gauche ($ag^+$) and gauche–gauche ($g^+g^+$, $g^+g^-$), which are depicted in Fig. 1.13A. Note that nine staggered conformers comprise a single anti–anti (a,a), four degenerate anti–gauche ($ag^+$, $g^+a$, $ag^-$, $g^-a$) and two degenerate pairs of gauche–gauche ($g^+g^+$, $g^-g^-$) and ($g^+g^-$, $g^-g^+$).

The aa conformer with the least steric strain has the lowest enthalpy. The enthalpy increases from aa to $ag^+$ conformer (one gauche interaction) to $g^+g^+$ conformer (two gauche interactions). For the estimation of their relative enthalpies, a Me/Et gauche interaction in pentane can be approximated to a gauche–butane (gb) interaction (0.9 kcal mol$^{-1}$). Between two gauche–gauche conformers ($g^+g^+$ and $g^+g^-$), the $g^+g^-$ conformer is further destabilized by a severe steric crowding of two methyl groups at relative 1,3 positions (Fig. 1.13B). This 1,3(Me/Me)-interaction (steric strain) is known as a double gauche pentane or *syn* pentane interaction, which is estimated to be 3.7 kcal mol$^{-1}$. (In cyclohexane systems, this interaction is called 1,3-diaxial (Me/Me) interaction, see Fig. 3.24A.) The $g^+g^-$ conformer is, therefore, a very high energy conformer and is effectively unpopulated at room temperature.[16,27] The energies of the four pentane conformers thus increase in the order: $aa < ag^+ < g^+g^+ \ll g^+g^-$. The energy profile of pentane in terms of two torsion angles ($\phi_1$ and $\phi_2$) has been described in the form of a contour diagram.[28]

It is of note that, in the conformational equilibrium, anti-gauche (ag) conformers have an entropic contribution due to mixing of four degenerate conformers ($ag^+$, $g^+a$, $ag^-$, $g^-a$). Similarly, the gauche–gauche enantiomeric pair ($g^+g^+$, $g^-g^-$) contributes entropy of mixing.

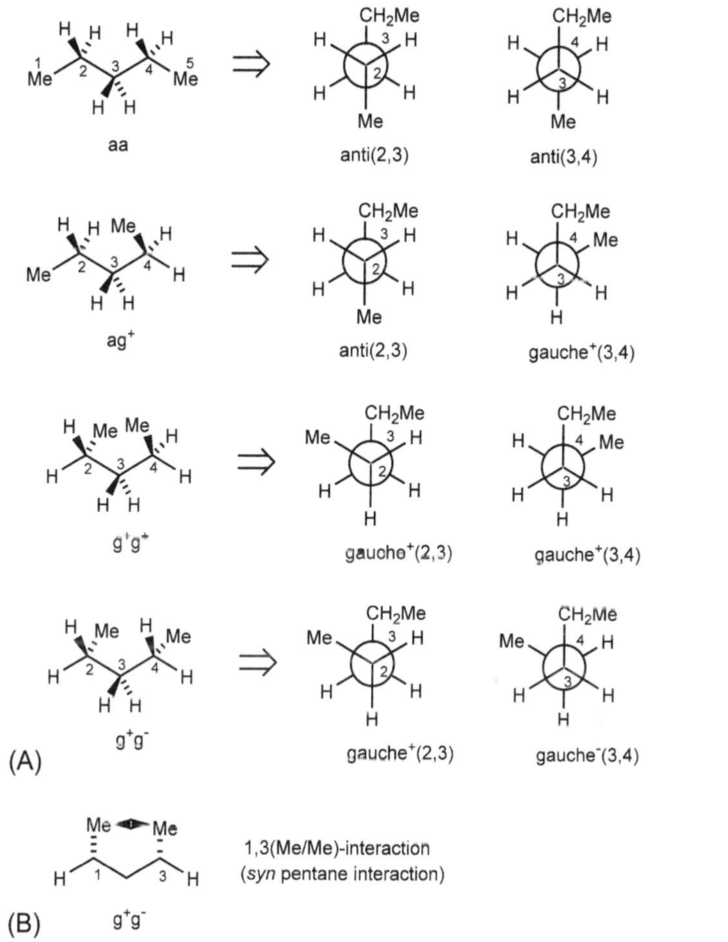


**FIG. 1.13** (A) Relative stability of four pentane conformers: aa $<$ ag$^+$ $<$ g$^+$g$^+$ $<$ g$^+$g$^{--}$. (B) Destabilization of g$^+$g$^-$ conformer by *syn* pentane interaction.

The principles of conformational analysis based on two torsion angles have also been applied to protein conformations (see Section 1.3.12).

**Problem 1.6** Draw all nine staggered conformers of pentane.

### 1.3.9 Conformation of alkenes

Unlike alkanes, alkenes exhibit unusual conformational properties. Consider the case of propene. The conformations arise as a result of rotation about C(sp$^3$)—C(sp$^2$) bond. Fig. 1.14 shows the eclipsed and staggered conformations of propene. In contrast to propane, the eclipsed conformation of propene

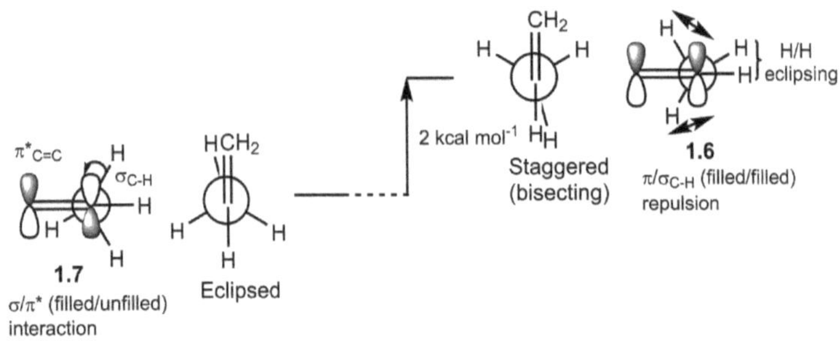


**FIG. 1.14** Conformations of propene.

($H_2C{=}C/C{-}H$ eclipsing) is more stable than the staggered or bisecting conformation ($H_2C{=}C/C{-}H$ staggered).[29,30] The energy barrier to rotation is about 2 kcal mol$^{-1}$.[14,30] As shown in **1.6**, the bisecting conformation is destabilized by both H/H eclipsing and filled/filled repulsive interaction of two C—H bonds with C=C π orbital, whereas the eclipsed conformation is stabilized by favourable filled/unfilled ($\sigma_{C-H}/\pi^*_{C=C}$) interaction **1.7**.[31,32] The detailed analysis of the barrier indicates that the rotation is coupled with vibrational processes.[33]

For1-butene, the eclipsed conformation is also more stable than the bisecting conformation. In this case, there are two eclipsed conformers, called the *cis* form ($C{=}CH_2/CH_3$ eclipsing) and the gauche form ($C{=}CH_2/H$ eclipsing) (Fig. 1.15A). The energy difference is however small, with the gauche conformer being slightly preferred by ~0.5 kcal mol$^{-1}$.[34] There are also two bisecting conformations (Fig. 1.15B). These are TS conformations for the interconversion of *cis* and gauche conformers. The energy barrier for gauche → *cis* conversion is 1.74 kcal mol$^{-1}$.

### 1.3.10 Conformation of carbonyl compounds

The conformational situation in acetaldehyde is similar to that of propene; the eclipsed conformation being more stable than the bisecting conformation (Fig. 1.16A). The energy barrier is 1.17 kcal mol$^{-1}$, which is less than that in propene.[35] This is probably due to less repulsive filled/filled ($\sigma_{C-H}/\pi_{C=O}$)

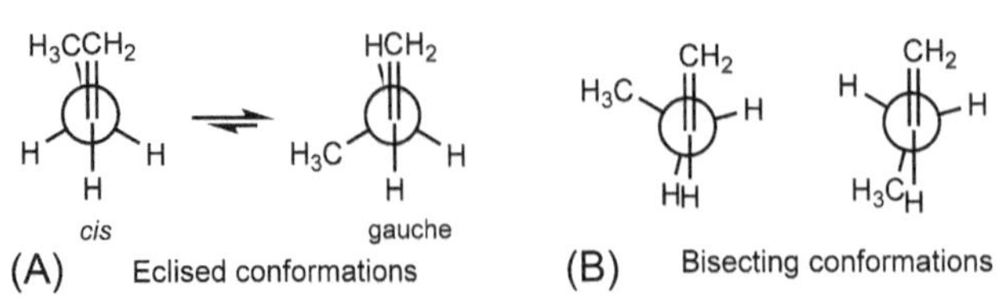


**FIG. 1.15** Conformations of 1-butene: (A) Preferred eclipsed conformers (*cis* and gauche); (B) Bisecting transition structure (TS) conformations.

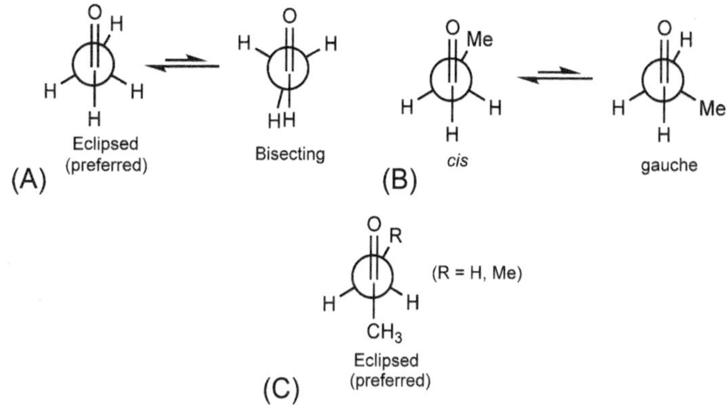


**FIG. 1.16** Conformations of (A) acetaldehyde, (B) propanal and (C) acetone and 2-butanone.

interaction in the bisecting conformation because of less π character of the C=O bond (C=O ↔ $C^+$ –$O^-$).

For propanal ($MeCH_2CHO$), the conformational preference is different from that of 1-butene (Fig. 1.16B). Here the *cis* conformation (O/Me eclipsing) is preferred over the gauche form by 0.7–1.2 kcal mol$^{-1}$.[36,37] The steric factor dominates in case of $Me_3CCH_2CHO$, making its gauche form preferred over the *cis* by 0.25 kcal mol$^{-1}$.[38]

The preferred conformation of acetone or 2-butanone resembles that of acetaldehyde, and the most populated conformation in each case is the eclipsed conformation with O/R eclipsing (R=H, Me)[39,40] (Fig. 1.16C).

The conformations of esters or amides arise from the rotation about C($sp^2$)—O or C($sp^2$)—N single bond. Due to resonance, the C—O single bond in ester or C—N single bond in amide has considerable double bond character as shown below.

The conjugation of nitrogen lone pair with carbonyl group is more effective and hence leads to more double bond character in C—N bond in amide compared to C—O bond in ester. The partial double bond character in the C—O or C—N bond increases the torsional barrier. For example, the barrier to rotation about C—O bond in methyl formate[41] is 10–13 kcal mol$^{-1}$, whereas the torsional barrier to rotation about C—N bond in *N*-methylformamide[42] is 20.6 kcal mol$^{-1}$.

The conformational situation in esters is shown in Fig. 1.17A. The esters can exist in *cis* (R/$CH_3$) or *trans* (R/$CH_3$) conformation; however, the *trans* conformation is preferred over the *cis* form. For methyl formate (R=H), the *trans*

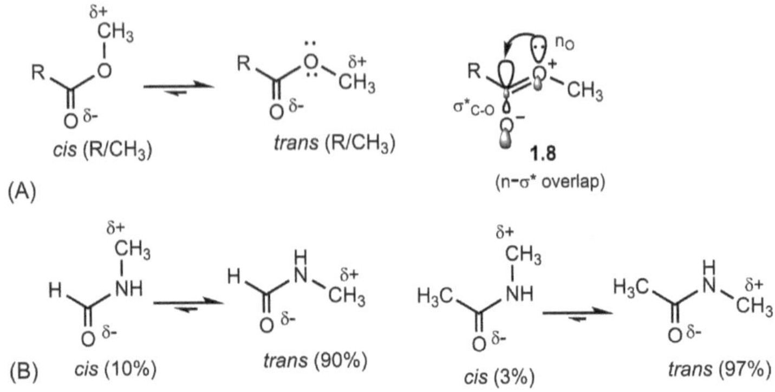


**FIG. 1.17** Conformations of (A) esters and (B) amides.

conformer is 5–6 kcal mol$^{-1}$ more stable than the *cis* form.[41] This rather large energy difference implies that methyl formate exists almost exclusively in the *trans* conformation (see Table 1.2). In general, the higher stability of the *trans* form is attributed to three factors: less R/$CH_3$ steric repulsion, electrostatic attraction between $O^{\delta-}$ and $H_3C^{\delta+}$, and favourable $n_O$–$\sigma^*_{C-O}$ interaction **1.8**.[43,44]

Similar to esters, amides can also exist in *cis* or *trans* conformation, and the *trans* form is more stable than the *cis* (Fig. 1.17B). For *N*-methylformamide, the *trans* form is preferred over the *cis* form by 1.4–1.6 kcal mol$^{-1}$ leading to a *trans*:*cis* = 9:1 in the conformational equilibrium.[42] The *trans* conformer of *N*-methylacetamide is highly preferred (~97%) with the *cis*-*trans* energy difference being 2.1–2.5 kcal mol$^{-1}$.[42] The preference of the *trans* form of amides is attributed to steric and electrostatic effects as described for esters; however, the n–σ* overlap does not occur for amides since the single lone pair on nitrogen is involved in resonance.

### 1.3.11 Conformation of conjugated dienes

The simplest conjugated diene is 1,3-butadiene. The molecule is planar and can exist in *s-cis* and *s-trans* conformations due to rotation about C($sp^2$)—C($sp^2$) single bond, as shown in Fig. 1.18. (Note that *s* in *s-cis* and in *s-trans* indicates $sp^2$—$sp^2$ single bond.)

The *s-trans* conformation is antiperiplanar ($\phi$ = 180 degrees), and is more stable than the *s-cis* conformation (synperiplanar, $\phi$ = 0 degrees) due to better

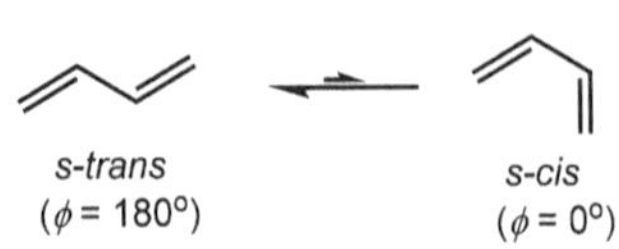


**FIG. 1.18** Conformations of butadiene.

orbital overlap and minimum steric interaction. The higher energy *s-cis* conformation can be isolated by a matrix isolation method.[38] The *s-trans* form is preferred over the *s-cis* by $2.9 \pm 0.4\,\text{kcal mol}^{-1}$ and exists almost exclusively (~99%) in the conformational equilibrium at room temperature. The energy barrier for *s-trans*→*s-cis* conversion is estimated to be $6.6$–$7.2\,\text{kcal mol}^{-1}$.[45] Like butadiene, α,β-unsaturated carbonyl compounds such as acrolein also exists largely in the sterically favoured *s-trans* conformation.[46]

### 1.3.12 Protein conformation: Ramachandran plot

Proteins are polymers comprised exclusively of α-amino acids. In the polymer chain, two successive amino acid units are held together by an amide bond called a peptide bond. Thus a protein has a polypeptide backbone. The peptide bonds in proteins are planar and predominantly *trans*. Two torsion angles $\phi$ and $\psi$ describe the rotation of the polypeptide chain around two bonds on both sides of the α-carbon ($C_\alpha$) **1.9**.

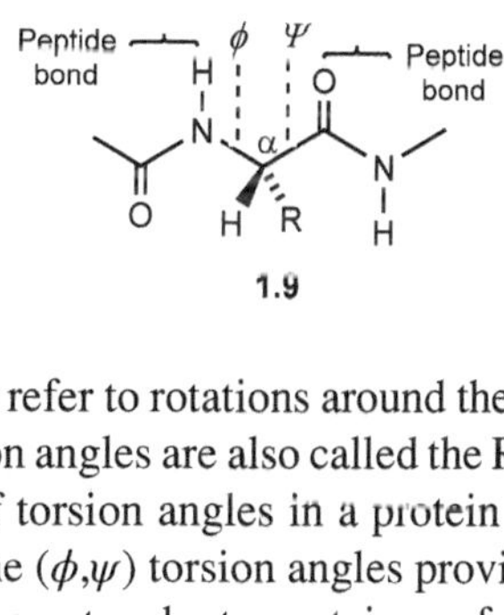


**1.9**

As shown in **1.9**, $\phi$ and $\psi$ refer to rotations around the N—$C_\alpha$ and $C_\alpha$—C bonds respectively. These torsion angles are also called the Ramachandran angles, and a plot (or distribution) of torsion angles in a protein structure is referred to as Ramachandran plot.[47] The ($\phi$,$\psi$) torsion angles provide the flexibility required for the polypeptide backbone to adopt a certain conformation or fold called the secondary structure elements of a protein. The plot also provides the excluded regions showing which rotations of polypeptide are not allowed due to steric hindrance.

A typical Ramachandran plot is shown in Fig. 1.19.

The horizontal and vertical axes show $\phi$ and $\psi$ values, respectively. Each dot on the plot shows the angles ($\phi$,$\psi$) for an amino acid. The restrictions on ($\phi$,$\psi$) angles in proteins to certain values are clearly seen in the plot. The regions with a high density of dots are called the allowed or low-energy regions. Some values of $\phi$ and $\psi$ are forbidden since the involved atoms will encounter a steric clash. These regions are empty or almost empty, and very few amino acid residues in proteins have their torsion angles in these regions. The plot shows that a particular type of protein conformation has its characteristic range of ($\phi$,$\psi$) angles, marked α for *α*-helix and β for *β*-sheet. The secondary structures in proteins thus arise from the repeating patterns of similar torsion angles for successive residues. Residues in *α*-helices cluster near $(\phi,\psi)=(-60,-45)$, while those

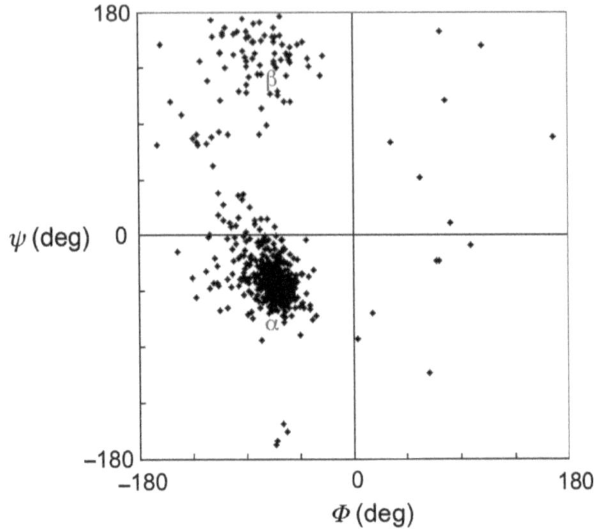


FIG. 1.19 Ramachandran plot showing the ($\phi$,$\psi$) signature for $\alpha$-helical and $\beta$-sheet conformations in proteins.

in $\beta$-sheet cluster around (−80,+150) in the upper left-hand quadrant of the Ramachandran plot. The plot demonstrates that ($\phi$,$\psi$) torsion angles belong mostly to two protein conformations.

The formation of $\alpha$-helices and $\beta$-sheets in proteins is illustrated with two examples in Fig. 1.20. As shown, myoglobin[48] is an $\alpha$-helical protein (Fig. 1.20A) whereas cellular retinoic acid-binding protein (CRABP)[49] is a predominantly $\beta$-sheet protein with a short region of $\alpha$-helix near the N-terminal

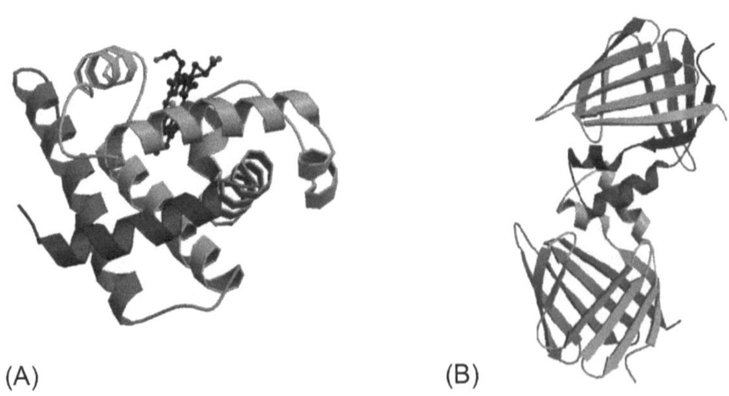


FIG. 1.20 Ribbon diagrams of (A) myoglobin (an $\alpha$-helical protein) [PDBid 1MBN], and (B) cellular retinoic acid-binding protein (a predominantly $\beta$-sheet protein) [PDBid 1CBI].

end (Fig. 1.20B). Hydrogen bonding plays a dominant role in stabilizing $\alpha$-helix and $\beta$-sheet conformations. In general, protein architecture involves the Linderstrøm-Lang hierarchy: primary → secondary → tertiary. The sequence of amino acids in a protein is its primary structure, local folding along the chain is called its secondary structure, and association of secondary structures via longer contacts gives its tertiary structure. Association of tertiary monomers via intersubunit contacts forms a quaternary structure for an oligomeric protein.

The questions of how and why a protein adopts a specific native conformation (tertiary structure) constitute the subject of protein folding problem. Generally, all the information needed to direct the folding of a protein is contained in its amino acid sequence.[50] Levinthal wondered how long it would take for a polypeptide chain to fold to its native conformation by a random search through the extensive conformational space available to it. For instance, a protein of 101 amino acid residues will have 100 peptide bonds, and therefore 100 ($\phi$,$\psi$) pairs of torsion angles. We have seen earlier that conformational analysis of pentane as a function of two torsion angles ($\phi$,$\psi$) gives three distinct and significant conformers (aa, ag, $g^+g^+$) (see Section 1.3.8). Thus, assuming three conformations per ($\phi$,$\psi$) pair, the total number of different conformations for the 101-residue protein $= 3^{100} = 5 \times 10^{47}$. If the protein is able to sample the conformations even at a very rapid rate of $10^{13}\,s^{-1}$, it would take $10^{27}$ years (larger than the age of the universe) to sample all possible conformations.[51] In reality, small proteins usually fold spontaneously within seconds, and even the large proteins fold within minutes. This difference between the theoretical calculation and the observed result is called Levinthal's paradox.[52] It has been concluded that proteins do not fold by a random search but there must exist specific folding pathways. During the evolution of the protein system with time, the formation of secondary structure elements through local cooperative interactions proceeds in parallel and the convergence of various structural features will increase with time until the native tertiary conformation is reached.

### 1.3.13 Summary

In general, rotation about single bond(s) can give rise to infinite number of conformations but only a few of these lie at the potential energy minima. The conformations that correspond to the energy minima are called conformational isomers or conformers. Thus all conformers are conformations but not vice versa. Thermodynamically, the conformational equilibrium comprises a mixture of conformers, the proportion of which depends on their relative stability or free energy. As regards kinetics, the energy barrier to interconversion of conformers is low enough that the conformers are readily interconvertible at room temperature, and hence are not isolable. However, the conformers are distinguishable on a short timescale by physical techniques such as spectroscopy and diffraction. A compound is a macroscopic assembly of molecules, and exists as a mixture of conformers.

Several factors are responsible for the relative stability of conformations. These include

**(a)** Torsional (or Pitzer) strain: Any deviation from the perfectly staggered conformation is accompanied by torsional strain, which reaches a maximum in the eclipsed conformation (see Fig. 1.6).

**(b)** van der Waals interactions: These can be repulsive or attractive depending on the internuclear distance between the nonbonded atoms. The repulsive nonbonded interaction is called steric strain (or van der Waals strain) that occurs due to crowding when the distance between two specified atoms (or groups) is less than the sum of their van der radii (see Figs 1.5, 1.9, 1.10, 1.13). The origin of the steric strain is the repulsive filled orbital/filled orbital interactions (see Section 4.2.1). The van der Waals attractive force (the London or dispersion force) between two atoms (or groups) comes into play at a distance equal to or slightly larger than the sum of their van der Waals radii (see Fig. 1.5, Problem 1.5a).

**(c)** Electrostatic interactions: These are dipole–dipole repulsions (see Section 1.3.7.1) or dipole–dipole attractions (see Fig. 1.17, Problem 1.5b). These interactions are subject to solvent effect (see Table 1.3).

**(d)** Hydrogen bonding: This is a stabilizing attractive force (see Section 1.3.7.3).

**(e)** Stereoelectronic effect: This refers mainly to stabilizing $\sigma$–$\sigma^*$, $\sigma$–$\pi^*$ and n–$\sigma^*$ hyperconjugation (see Figs. 1.7, 1.11, 1.14, 1.17). In frontier orbital terms, the hyperconjugative effect is a favourable interaction between a filled orbital (HOMO) and an unfilled orbital (LUMO). However, a filled orbital/filled orbital interaction is repulsive or destabilizing, such as $\sigma/\pi$ overlap (see Fig. 1.14).

**(f)** Statistical factor: This is an entropic contribution to free energy ($\Delta G^0$) (see Sections 1.3.5.1 and 1.3.8).

(For bond angle strain, see later Section 3.2)

## 1.4 Symmetry in molecules[53,54]

Symmetry plays an important role in understanding organic stereochemistry. For example, whether a molecule is chiral or achiral can be easily decided by its symmetry properties. A chiral molecule is one which is not superposable on its mirror image. The symmetry of a finite molecule is described in terms of *symmetry elements*. The operations performed to determine the presence of symmetry elements are called *symmetry operations*.

### 1.4.1 Symmetry elements and symmetry operations

A symmetry operation is an operation that converts the molecule into an *equivalent* spatial arrangement of atoms (i.e. an arrangement superposable with or

**TABLE 1.4 Specifying symmetry elements with respect to symmetry operations performed on the structure of a molecule or an object.**

| Symmetry element | Symmetry operation |
|---|---|
| Simple or proper axis of symmetry ($C_n$) | *Rotation* about the axis through 360 degrees/*n* |
| Plane of symmetry ($\sigma \equiv S_1$) | *Reflection* in the plane |
| Centre of symmetry ($i \equiv S_2$) | *Inversion* through the centre |
| Alternating or improper axis of symmetry ($S_n$) | *Rotation–reflection*: Rotation about the axis through 360 degrees/*n* followed by a reflection in a plane perpendicular to the rotation axis<br>or<br>*Reflection–rotation*: Reflection in a plane followed by rotation about a perpendicular axis (the reverse of *rotation–reflection*) |

indistinguishable from the original). A symmetry element can be a geometric line, plane or point with respect to which one or more symmetry operations are performed. The basic symmetry operations are rotations about an axis, reflections in a plane, and inversion through a centre. The symmetry elements are designated by the symbols due to Schönflies, which are also used to represent the corresponding symmetry operations.

There are four kinds of symmetry elements (Table 1.4). These are simple or proper axis of symmetry ($C_n$), plane of symmetry ($\sigma$), centre of symmetry or centre of inversion ($i$) and alternating or improper axis of symmetry ($S_n$). The subscript $n$ in the symbols $C_n$ and $S_n$ denotes the order or multiplicity of the axis such that $C_n$ denotes an $n$-fold simple axis and $S_n$ an $n$-fold alternating axis.

A molecule may possess a number of symmetry elements. The combination of those symmetry elements forms a symmetry group called a *point group* (since all the symmetry elements will intersect at a common point, not shifted by any of the symmetry operations). Note that a point group is different from the symmetry group known as space group in crystallography. A complete and nonredundant list of symmetry operations generated by the symmetry elements defining the point group satisfies the criteria for a mathematical group, and the problem of molecular symmetry can be solved by the group theory.

As per the requirements of the group theory, the identity ($E$) as a symmetry element is included in the list of symmetry elements (not shown in Table 1.4). The symmetry element identity ($E$) is universal and present in all objects and molecules, and the corresponding identity operation results in the conversion of the molecule into an arrangement *identical* (contrast with equivalent) with the original one. The identity operation is equivalent to $C_1$ or that resulting from a sequence of $C_n$, $\sigma$, $i$ or $S_n$ operations (see below).

### 1.4.2 Simple or proper axis of symmetry ($C_n$)

A simple or proper axis of symmetry ($C_n$) is an axis such that rotation of the molecule by 360 degrees/$n$ about the axis gives an equivalent spatial arrangement of atoms, i.e. an arrangement superposable with the original one. Notably, the rotation in symmetry operations implies a rigid rotation (rotation of the whole molecule) about an axis in contrast to rotation about a single bond for conformations in which the groups at one end of the single bond rotates with respect to those at the other end. The $C_n$ axis is illustrated with three examples in Fig. 1.21.

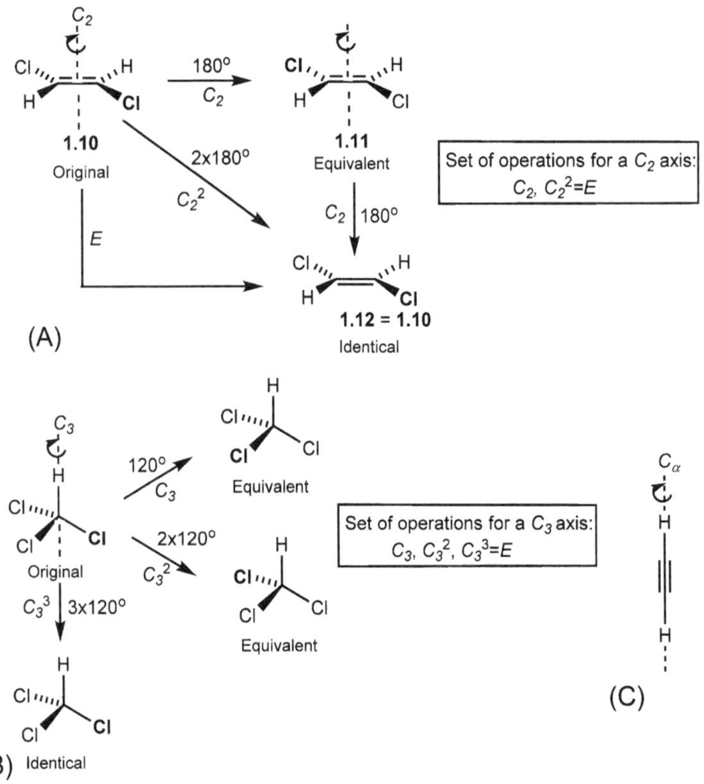


**FIG. 1.21** $C_n$ axis in molecules: (A) $C_2$ axis in *trans*-1,2-dichloroethene, (B) $C_3$ axis in chloroform and (C) $C_\infty$ axis in acetylene. In (A) and (B), one Cl in the original structure is highlighted in bold in order to understand whether the structure generated by the indicated operation is equivalent or identical with the original one.

Fig. 1.21A shows that *trans*-1,2-dichloroethene possesses a $C_2$ axis (twofold simple axis). The axis is designated $C_2$ since rotation of the original molecule **1.10** by an angle 180 degrees (360 degrees/2) produces an equivalent arrangement **1.11**. (A Cl atom is highlighted to understand whether the new spatial arrangement is equivalent or identical with the original one or otherwise.) Note the location of the axis in the molecule. Here $C_2$ passes through the centre of the C=C bond and is perpendicular to the plane of the molecule. If we assume the $C_2$ axis to be the z-axis in the Cartesian coordinate system, the effect of a $C_2$ operation is then to transpose a general point (x, y, z) to a point (−x, −y, z) as shown below.

$$[x, y, z] \xrightarrow{C_2(z)} [-x, -y, z]$$

A successive rotation of **1.11** by 180 degrees about the same axis (i.e. rotation of **1.10** by 2 × 180 degrees) gives **1.12** which is identical with the original. The symmetry operations for rotations by 180 degrees and 2 × 180 degrees about the axis are designated as $C_2$ and $C_2^2$ respectively. Since both the identity $E$ and $C_2^2$ operations give an arrangement identical with the original one, $C_2^2 = E$. Thus a $C_2$ axis generates one distinct operation $C_2$, besides $E$.

An example of a $C_3$ axis (threefold simple axis) in $CHCl_3$ is shown in Fig. 1.21B. The molecule is tetrahedral and the $C_3$ axis passes through the C—H bond. As shown, the $C_3$ axis generates the operations $C_3$ (rotation by 120 degrees), $C_3^2$ (rotation by 2 × 120 degrees) and $C_3^3$ (rotation by 3 × 120 degrees). Clearly, $C_3^3 = E$. Thus a $C_3$ axis generates two distinct operations $C_3$ and $C_3^2$, besides $E$.

For linear molecules, a $C_\infty$ axis exists since an equivalent arrangement is obtained by rotation through any angle about the axis. Fig. 1.21C shows a $C_\infty$ axis for acetylene, which passes through the linear chain H—C—C—H.

It should be noted that a $C_1$ axis is universal since rotation of any object or molecule by 360 degrees about an axis will give an arrangement identical with the original one, and therefore, $C_1 \equiv E$. If a molecule has no symmetry element except the identity $E \equiv C_1$, it is called *asymmetric*.

Since a $C_n$ operation is real in the sense that the original positions of atoms have shifted physically to new positions in the equivalent arrangement, it is called a proper rotation or a symmetry operation of the first kind.

### 1.4.3 Plane of symmetry ($\sigma_v$, $\sigma_h$, $\sigma_d$)

A plane of symmetry ($\sigma$) is a reflection plane that divides the molecule in such a way that the reflection of atoms in each half across the plane gives an arrangement superposable with the original one. In other words, a $\sigma$ plane is a plane that divides the molecule into two halves such that one half is the mirror image of the

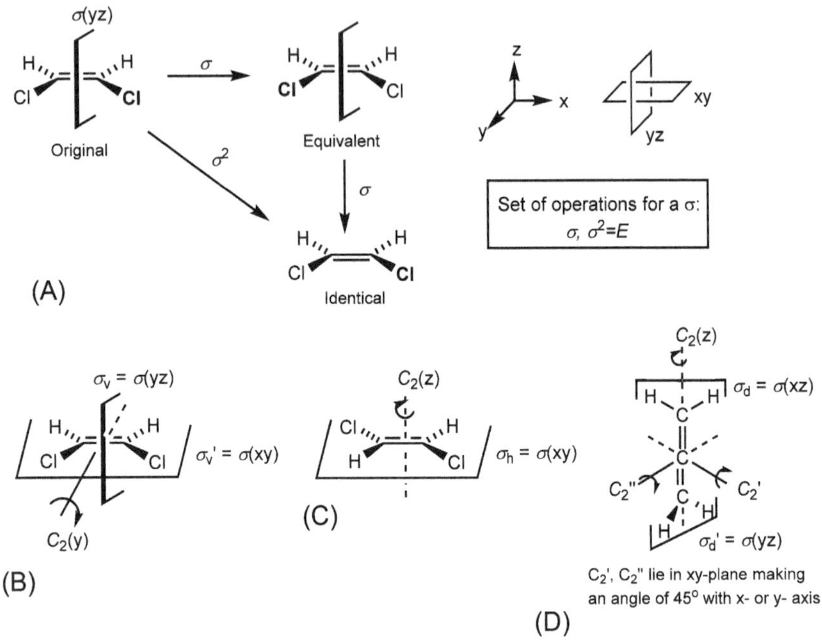


**FIG. 1.22** (A) Plane of symmetry ($\sigma$) and the corresponding symmetry operations in *cis*-1,2-dichloroethene, (B) two $\sigma_v$ planes ($\sigma_v$, $\sigma_v'$) in *cis*-1,2-dichloroethene, (C) a $\sigma_h$ plane in *trans*-1,2-dichloroethene and (D) two $\sigma_d$ planes ($\sigma_d$, $\sigma_d'$) in allene. In (A), one Cl in the original structure is highlighted in bold in order to understand whether the structure generated by the indicated operation is equivalent or identical with the original one.

other half. In contrast to the rotation operation which is real, a reflection operation is a virtual one, and the reflection is termed an improper operation or a symmetry operation of the second kind. A simple illustration of a plane of symmetry (σ) is shown in Fig. 1.22A. The σ plane is perpendicular to the plane of the molecule. The reflection of the two halves across the σ plane gives an arrangement indistinguishable from the original structure. Note that the two halves of the molecule are mirror images of each other. As shown, two successive reflections produce an arrangement identical with the original one, which implies $\sigma^2 = E$. Thus a $\sigma$ symmetry plane generates one distinct operation $\sigma$, besides $E$.

The position of the symmetry plane can be specified with respect to a Cartesian coordinate system. As shown in Fig. 1.22A, the symmetry plane is the $\sigma$(yz) plane when the plane of the molecule is xy-plane. In many instances, it is necessary to be more specific regarding the location of the $\sigma$ plane in relation to a simple axis ($C_n$) of the highest order, and various subscripts such as v (vertical), h (horizontal) or d (dihedral) are added to identify the position of the plane. A $\sigma_v$ plane contains the vertical axis which is the principal axis ($C_n$ axis of highest $n$). A $\sigma_h$ plane is a horizontal plane which is perpendicular to the vertical axis. A $\sigma_d$ plane contains the vertical axis and also bisects a pair of $C_2$ axes perpendicular to the vertical axis; thus a $\sigma_d$ is just a special case of $\sigma_v$. The presence of two $\sigma_v$ planes ($\sigma_v$, $\sigma_v'$) in *cis*-1,2-dichloroethene, one $\sigma_h$ plane in *trans*-1,2-dichloroethene, and two $\sigma_d$ planes ($\sigma_d$, $\sigma_d'$) in allene is illustrated in Fig. 1.22B–D respectively. In Fig. 1.22B, only one simple axis ($C_2$) lying in the plane of the molecule is present, which is therefore treated as a vertical or principal axis. In the Cartesian coordinate system, this is the $C_2$(y) axis. There are two $\sigma$ planes, $\sigma$(yz) and $\sigma$(xy) both containing the principal y-axis. Therefore both planes are $\sigma_v$ planes ($\sigma_v$, $\sigma_v'$). (Note carefully that the $\sigma$(xy) plane is a $\sigma_v$, not a $\sigma_h$.) In Fig. 1.22C, the principal axis is $C_2$(z) axis, and the $\sigma$ plane is the plane of the molecule, i.e. $\sigma$(xy) which is perpendicular to the principal z-axis. Therefore this symmetry plane is $\sigma_h$. In Fig. 1.22D, allene has three $C_2$ axes ($C_2$, $C_2'$, $C_2''$) mutually perpendicular to each other. Since all axes are $C_2$, the one passing through the most atoms is taken as the vertical or principal axis, which is the $C_2$ (z) axis. Note that $C_2'$ and $C_2''$ pass through the middle sp carbon (the origin of the coordinate system) and lie in the xy-plane. Each of the two axes makes an angle of 45 degrees with the x- or y-axis (use a molecular model). As shown, there are two vertical $\sigma$ planes [$\sigma$(xz), $\sigma$(yz)], each of which bisects the pair of $C_2$ axes perpendicular to the $C_2$(z) axis. These two $\sigma$ planes are therefore $\sigma_d$ planes ($\sigma_d$, $\sigma_d'$).

It is of note that a $\sigma$ operation changes only one coordinate of a general point (x, y, z) as

$$[x, y, z] \xrightarrow{\sigma(xz)} [x, -y, z] \qquad [x, y, z] \xrightarrow{\sigma(yz)} [-x, y, z] \qquad [x, y, z] \xrightarrow{\sigma(xy)} [x, y, -z]$$

### 1.4.4 Centre of symmetry or centre of inversion (*i*)

A centre of symmetry (*i*) is a specific point within a molecule through which it is possible to project every atom in a structure to an equivalent position on the opposite side of the centre. In other words, inversion of all atoms in the molecule through the point results in an arrangement indistinguishable from the original one. The point is also called a centre of inversion. In mathematical terms, a centre of symmetry will exist at the origin of the coordinate system (0, 0, 0) if, for every atom (x, y, z) in the molecule, there exists an equivalent atom at (−x, −y, −z). Molecules that contain a centre of symmetry are called centrosymmetric molecules. There can be only one centre of inversion in a molecule. The presence of a centre of symmetry in molecules is illustrated with three examples in Fig. 1.23. Fig. 1.23A shows that *trans*-1,2-dichloroethene has an inversion centre at the midpoint of the double bond. Inversion of all atoms through the centre gives an equivalent arrangement. Like a reflection operation, the inversion is also an improper operation. A successive inversion results in an

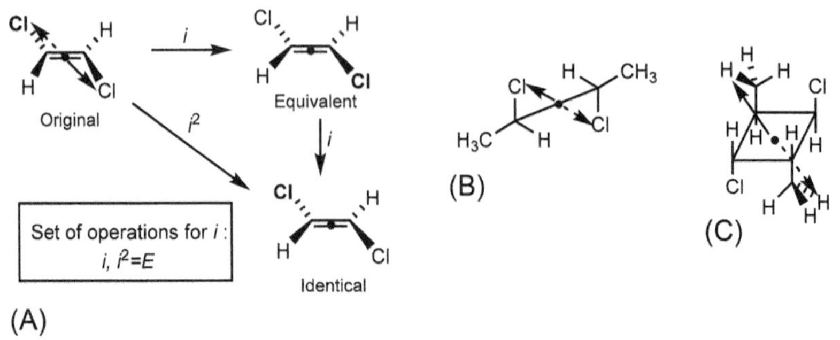


**FIG. 1.23** (A) Centre of symmetry ($i$) and the corresponding symmetry operations in *trans*-1,2-dichloroethene. Inversion centre in (B) anti conformer of *meso*-2,3-dichlorobutane and (C) a cyclobutane derivative. The centre of inversion is represented by a dot. In (A), one Cl in the original structure is highlighted in bold in order to understand whether the structure generated by the indicated operation is equivalent or identical with the original one.

identical structure, and therefore, $i^2 = E$. Thus a centre of symmetry generates one distinct operation $i$, besides $E$. The presence of an inversion centre in the anti conformer of *meso*-2,3-dichlorobutane, and in a cyclobutane derivative is shown in Fig. 1.23B and C respectively.

### 1.4.5 Alternating or improper axis of symmetry ($S_n$, $S_1 \equiv \sigma$, $S_2 \equiv i$)

An alternating axis of symmetry ($S_n$) is an axis such that rotation of the molecule by 360 degrees/$n$ about the axis followed by reflection in a plane perpendicular to it gives an arrangement superposable with the original one. The order of rotation and reflection is interchangeable. Note that rotation and reflection in $S_n$ constitute a single two-stage operation. As $S_n$ involves a reflection which is a virtual operation, it is also called an improper axis of symmetry.

A onefold alternating axis ($S_1$) involves rotation by 360 degrees followed by reflection in a perpendicular plane. Since rotation by 360 degrees gives an identical structure, $S_1$ operation is equivalent to performing only a $\sigma$ operation. Therefore, $S_1 \equiv \sigma$.

A two-fold alternating axis ($S_2$) involves a combination of 180-degree rotation ($C_2$ operation) and reflection ($\sigma$ operation) which will transform a general point (x, y, z) as given below. Note carefully that the $C_2$ and $\sigma$ operations in $S_2$ are not symmetry operations because each individual operation ($C_2$ or $\sigma$) does not produce an equivalent arrangement but a combined operation ($S_2$) does.

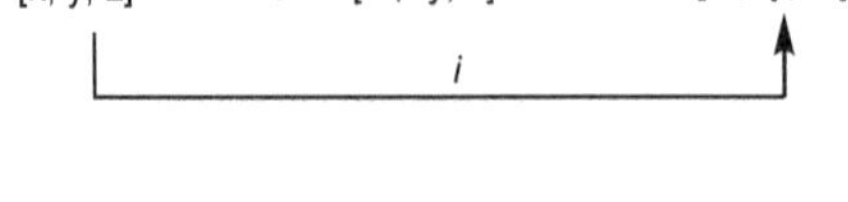

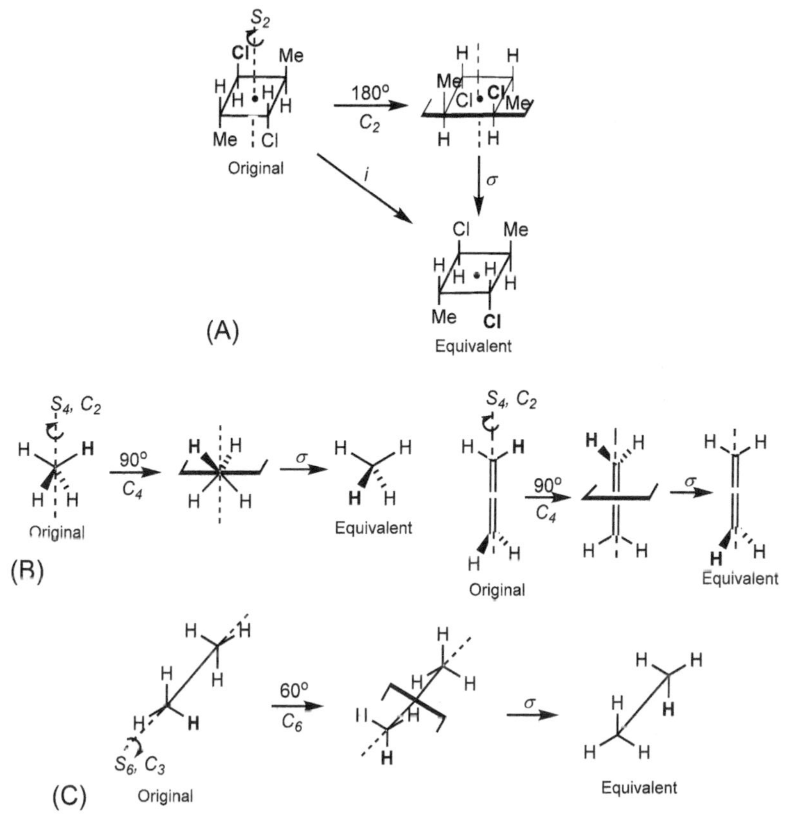


**FIG. 1.24** Molecules having alternating axis of symmetry ($S_n$): (A) equivalence of $S_2$ and $i$, (B) $S_4$ axis in methane and allene and (C) $S_6$ axis in the staggered ethane. One Cl in (A) and one H in (B) and (C) are highlighted in bold in order to understand whether the structure generated by the indicated operation is equivalent or identical with the original one.

The combined operation in $S_2$ is equivalent to the inversion operation through a centre ($i$). Therefore, $S_2 \equiv i$. The inversion centre is the point of intersection of the $C_2$ axis with the $\sigma$ plane. Notably, if a molecule has $C_2$ and $\sigma$ symmetry, it necessarily has an $S_2$ axis but not the vice versa. The equivalence of $S_2$ and $i$ is shown diagrammatically in Fig. 1.24A.

As $S_n$ axes include $\sigma \equiv S_1$ and $i \equiv S_2$, we will generally refer to $S_n$ axis when $n > 2$. Fig. 1.24B illustrates two examples for an $S_4$ axis. Methane has a regular tetrahedral structure. An axis passing through two opposite pairs of hydrogen atoms (as shown) is an $S_4$ axis. Evidently, there are two other similar $S_4$ axes (check). The three $S_4$ axes in methane coincide with the x, y and z axes. The structure of allene is like an extended tetrahedron, and an axis coinciding with the main molecular chain (C=C=C) is the $S_4$ axis.

The set of operations generated by an $S_4$ axis is $S_4$, $S_4^2$, $S_4^3$, $S_4^4$.

Of these, $S_4^2$ implies the sequence of operations $C_4$, $\sigma$, $C_4$, $\sigma$. (Here $C_4$ and $\sigma$ operations are not symmetry operations.) Since $\sigma^2 = E$, $S_4^2 = C_4^2 = C_2$.

Similarly, $S_4^4 = E$. In general, $S_n^n = E$ (when n = even).

Thus, **the set of operations generated by an $S_4$ axis = $S_4$, $C_2$, $S_4^3$, $E$.**

This set includes the operations $C_2$, $E$ generated by a $C_2$ symmetry axis. Therefore, *there must be a $C_2$ axis coincident with $S_4$.*

Fig. 1.24C shows that an $S_6$ axis is present in the staggered conformation of ethane. The C—C bond is the $S_6$ axis; rotation of the molecule by 60 degrees around the axis followed by reflection in a perpendicular plane gives an equivalent arrangement.

The set of operations generated by an $S_6$ axis is $S_6$, $S_6^2$, $S_6^3$, $S_6^4$, $S_6^5$, $S_6^6 = E$, where

$$S_6^2 = (C_6, \sigma, C_6, \sigma) = C_6^2 = C_3$$

$$S_6^3 = (C_6, \sigma, C_6, \sigma, C_6, \sigma) = (C_6^3, \sigma) = (C_2, \sigma) = i$$

$$S_6^4 = (C_6, \sigma, C_6, \sigma, C_6, \sigma, C_6, \sigma) = C_6^4 = C_3^2$$

Thus, **the set of operations generated by an $S_6$ axis = $S_6$, $C_3$, $i$, $C_3^2$, $S_6^5$, $E$.**

Since this set contains the operations $C_3$, $C_3^2$, $E$ generated by a $C_3$ symmetry axis, *the existence of the $S_6$ axis automatically requires that a $C_3$ axis (coincident with $S_6$) exists.*

## 1.5 Symmetry point groups

A complete and nonredundant set of symmetry operations generated by the symmetry elements present in a molecule constitutes a point group. A point group contains all the relevant symmetry information from which it would be possible to reconstruct the structure of the molecule. All the symmetry elements in a molecule may not be independent. For example, in *trans*-1,2-dichloroethene, the presence of a $C_2$ axis and a $\sigma_h$ plane (see Fig. 1.22C) requires that $S_2 \equiv i$ must be present. The existence of such relationships simplifies the symmetry considerations by what is known as the point group symbol. The symbol identifies the *minimum* number and type of symmetry elements needed to describe the symmetry properties of the molecule. The point group symbols are usually given in bold type (in contrast to symbols for symmetry elements or operations given in italics). A detailed discussion of the point groups is beyond the scope of this text; we will describe some common point groups for chiral and achiral molecules.

A simple procedure for classifying molecules into point groups is given in Table 1.5. The special point groups for linear and highly symmetrical molecules are not included in the table, which will be described very briefly later.

The total number of distinct operations that can be performed in a group is called the *order* (denoted by the symbol $h$) of the group. The order ($h$) of various

**TABLE 1.5 Point group classification, order and symmetry number.**

| Symmetry elements | Point group symbol | Order ($h$) | Symmetry number ($\sigma$) |
|---|---|---|---|
| Only $E$ ($C_1$) | $C_1$ | 1 | 1 |
| Only $\sigma$ | $C_s$ | 2 | 1 |
| Only $i$ | $C_i$ | 2 | 1 |
| Only $S_n$ ($n$=even) | $S_n$ | $n$ | $n/2$ |
| Only $C_n$ | $C_n$ | $n$ | $n$ |
| $C_n$ (no $\perp C_2$)+$\sigma_h$ | $C_{nh}$ | $2n$ | $n$ |
| $C_n$ (no $\perp C_2$)+$n\sigma_v$ | $C_{nv}$ | $2n$ | $n$ |
| $C_n$+$n \perp C_2$ | $D_n$ | $2n$ | $2n$ |
| $C_n$+$n \perp C_2$+$\sigma_h$ | $D_{nh}$ | $4n$ | $2n$ |
| $C_n$+$n \perp C_2$+$n\sigma_d$ | $D_{nd}$ | $4n$ | $2n$ |

$\perp C_2$ means a $C_2$ axis perpendicular to the principal $C_n$ axis.

point groups is also shown in Table 1.5. Symmetry reduces entropy, and the term *symmetry number* is employed in the calculation of entropy. The total number of distinct operations generated by all $C_n$ axes in a molecule is called the symmetry number (denoted by the symbol $\sigma$). Remember that $C_n^n = E$. Thus symmetry number denotes the total number of indistinguishable but nonredundant arrangements resulting from the rotations about all proper axes. Note that the same symbol $\sigma$ is used in two different contexts, symmetry plane and symmetry number. The symmetry number ($\sigma$) of various point groups is shown in Table 1.5.

Now we will describe briefly some point groups for chiral and achiral molecules. Chirality and achirality are purely geometrical properties. A molecule is chiral (handed from the Greek *cheir*, hand) if it is nonsuperposable on its mirror image.[55] A chiral molecule and its mirror image constitute an enantiomeric pair. To examine whether a molecule is chiral or achiral, one could build models of the molecule and its mirror image, and then try to superpose them. However, a simple alternative is to look for the symmetry properties of the molecule[56] (Fig. 1.25).

If the molecule lacks the symmetry elements $\sigma \equiv S_1$, $i \equiv S_2$ and $S_n$ ($n > 2$), the molecule is chiral. But a chiral molecule may possess one or more $C_n$ axes. If the chiral molecule possesses no symmetry elements other than $C_1 \equiv E$ (identity), it is called asymmetric. Chiral molecules possessing $C_n$ ($n > 1$) are not asymmetric (older term is dissymmetric). Thus all asymmetric molecules are chiral but all chiral molecules are not asymmetric.

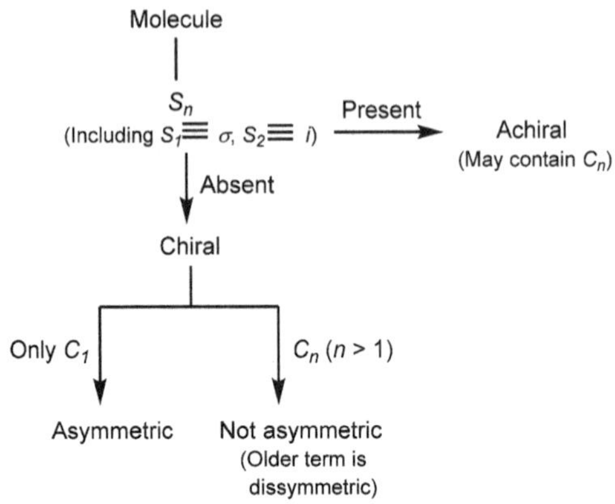


**FIG. 1.25** Chiral and achiral molecules based on symmetry properties.

## 1.5.1 Chiral point groups ($C_1$, $C_n$, $D_n$)

The chiral point groups are $\mathbf{C_1}$, $\mathbf{C_n}$ and $\mathbf{D_n}$ (see Table 1.5). Fig. 1.26 shows examples of molecules belonging to chiral point groups. All molecules that contain a single chiral centre (usually a tetrahedral carbon attached to four different ligands) belong to the point group $\mathbf{C_1}$. These molecules have no symmetry elements except $C_1 \equiv E$, and hence they are asymmetric. Fig. 1.26A provides an example. The order ($h$) of the point group $\mathbf{C_1}$ is 1, and the symmetry number ($\sigma$) is also 1.

In a chiral point group $\mathbf{C_n}$, the only symmetry element is $C_n$ ($n > 1$). The point group $\mathbf{C_2}$ is common, and two examples are shown in Fig. 1.26B. The only symmetry element present is $C_2$. In gauche butane, the $C_2$ axis passes through the midpoint of $C_2$—$C_3$ bond bisecting the torsion angle between the two methyl groups. In 1,3-dichloroallene, the $C_2$ axis passes diagonally through the central sp. carbon. The symmetry operations generated by $C_2$, order ($h$) and symmetry number ($\sigma$) for the point group $\mathbf{C_2}$ are given below.

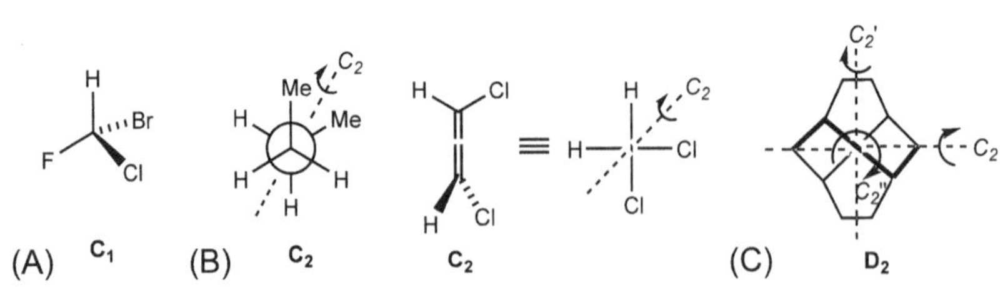


**FIG. 1.26** Examples illustrating chiral point groups: (A) $\mathbf{C_1}$, (B) $\mathbf{C_2}$ and (C) $\mathbf{D_2}$.

| Point group: $\mathbf{C_2}$ |
|---|
| Symmetry element: $C_2$ |
| Symmetry operations: $C_2$, $C_2^2=E \Rightarrow h=\sigma=2$ |

Since the total number of operations is 2, the order ($h$) is 2. The symmetry number ($\sigma$) is also 2. In general, for $\mathbf{C_n}$, $h=\sigma=n$. The point group $\mathbf{C_n}$ has been found for $n>2$, but is quite rare.

The chiral point group $\mathbf{D_n}$ (also called dihedral groups) are characterized by a principal $C_n$ axis and $n$ $C_2$ axes perpendicular to it (see Table 1.5). An example of the $\mathbf{D_2}$ point group (sometimes denoted by **V** for German *Vierergruppe*) is shown in Fig. 1.26C. The molecule twistane[57] has a principal $C_2$ axis and 2 $\perp C_2$ axes ($C_2'$ and $C_2''$). Each $C_2$ axis generates one operation $C_2$ besides $E$. As shown below, the total number of distinct symmetry operations generated by three $C_2$ axes is 4. The order ($h$) is 4, and the symmetry number ($\sigma$) is also 4. In general, for $\mathbf{D_n}$, $h=\sigma=2n$.

| Point group: $\mathbf{D_2}$ |
|---|
| Symmetry element: $C_2$, $C_2'$, $C_2''$ |
| Distinct symmetry operations: $E$, $C_2$, $C_2'$, $C_2'' \Rightarrow h=\sigma=4$ |

Chiral molecules with higher $\mathbf{D_3}$ symmetry are rare but known. Two examples are *trans-transoid-trans-transoid-trans*-perhydrotriphenylene[58] (see Problem 1.7k) and trishomocubane.[59]

## 1.5.2 Achiral point groups

A molecule is achiral if it is superposable on its mirror image. Thus an achiral molecule is identical with its mirror image. As regards symmetry, an achiral molecule *must* contain one or more of the following symmetry elements: $\sigma \equiv S_1$, $i \equiv S_2$ and $S_n$ ($n>2$), and it may also contain one or more proper axes $C_n$ (Fig. 1.25). Some achiral point groups are $\mathbf{C_s}$, $\mathbf{C_i}$, $\mathbf{S_n}$, $\mathbf{C_{nv}}$, $\mathbf{C_{nh}}$, $\mathbf{D_{nh}}$ and $\mathbf{D_{nd}}$ (see Table 1.5).

### *1.5.2.1 Point groups $C_s$, $C_i$ and $S_n$*

Fig. 1.27 shows examples of molecules belonging to $\mathbf{C_s}$, $\mathbf{C_i}$ and $\mathbf{S_n}$ point groups. The point group $\mathbf{C_s}$ has only a symmetry plane $\sigma$, and no $C_n$, as illustrated with chloroethylene and a cyclopropane derivative in Fig. 1.27A. The symmetry properties of the point group $\mathbf{C_s}$ are given below.

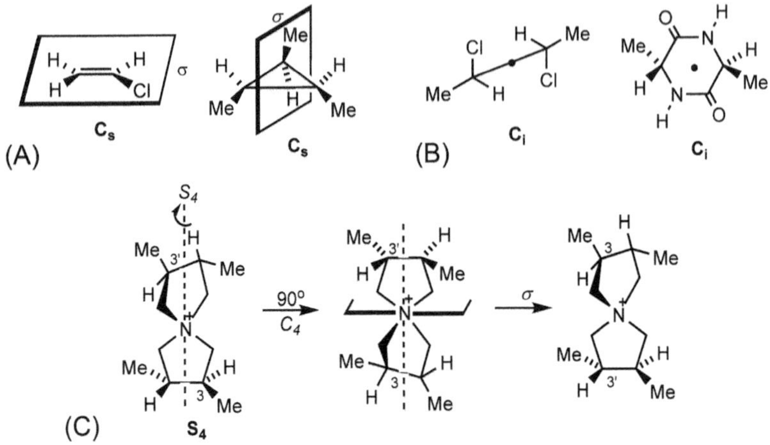


**FIG. 1.27** Examples of achiral point groups (A) $\mathbf{C_s}$, (B) $\mathbf{C_i}$ and (C) $\mathbf{S_4}$.

Point group: $\mathbf{C_s}$

Symmetry element: $\sigma$

Symmetry operations: $\sigma$, $\sigma^2 = E$ $\Rightarrow$ $h = 2$, $\sigma = 1$

The total number of operations is 2, and hence the order $h=2$. But the symmetry number $\sigma=1$ ($E \equiv C_1$).

The point group $\mathbf{C_i}$ possesses only a centre of symmetry $i$. Two examples are the anti conformer of *meso*-2,3-dichlorobutane and the *trans*-diketopiperazine, as shown in Fig. 1.27B. As $i \equiv S_2$, the point group $\mathbf{C_i} \equiv \mathbf{S_2}$. The symmetry characteristics of the point group $\mathbf{C_i}$ are given below.

Point group: $\mathbf{C_i}$

Symmetry element: $i$

Symmetry operations: $i$, $i^2 = E$ $\Rightarrow$ $h = 2$, $\sigma = 1$

The total number of operations is 2 and hence the order $h=2$. The symmetry number $\sigma=1$ ($E \equiv C_1$).

The point group $\mathbf{S_4}$ is illustrated with a spirane[60] in Fig. 1.27C (the counter ion $TsO^-$ is omitted). Only an $S_4$ axis is present in $\mathbf{S_4}$. *As a necessary consequence of the existence of the $S_4$ axis, the molecule must contain a $C_2$ axis coincident with the $S_4$ axis* (see Section 1.4.5, Fig. 1.24B). The set of nonredundant symmetry operations generated by the $S_4$ (see Section 1.4.5) and $C_2$ axes is given in the box below. Note that the symmetry element $C_2$ arising from the simple consequence of $S_4$ defining the point group is shown within bracket.

Point group: $\mathbf{S_4}$

| Symmetry elements | Symmetry operations | | Set of nonredundant operations: |
|---|---|---|---|
| $S_4$ | $E, S_4, C_2, S_4^3$ | ⇨ | $E, S_4, C_2, S_4^3$ |
| $(C_2)$ | $E, C_2$ | | ⇒ $h = 4$; $\sigma = 2$ $(E, C_2)$ |

The total number of distinct operations is 4, and hence the order $h=4$. The symmetry number $\sigma=2$ (operations $E$, $C_2$). In general, for $\mathbf{S_n}$ ($n$ even), $h=n$, $\sigma=n/2$ (see Table 1.5).

### 1.5.2.2 Point groups $C_{nv}$ and $C_{nh}$

The point group $\mathbf{C_{nv}}$ has a single $C_n$ axis and $n$ $\sigma_v$ planes whereas the point group $\mathbf{C_{nh}}$ has a single $C_n$ axis and a single $\sigma_h$ plane (see Table 1.5). In $\mathbf{C_{nv}}$ point groups, $\mathbf{C_{2v}}$ and $\mathbf{C_{3v}}$ are illustrated with examples in Fig. 1.28A. *cis*-1,2-Dichloroethene has a $C_2$ axis and 2 $\sigma_v$ planes, and therefore belongs to $\mathbf{C_{2v}}$. The *s-cis* butadiene has also $\mathbf{C_{2v}}$ symmetry (not shown). The symmetry operations generated by the symmetry elements in $\mathbf{C_{2v}}$ are given in the box below.

Point group: $\mathbf{C_{2v}}$

| Symmetry elements | Symmetry operations | | Set of nonredundant operations: |
|---|---|---|---|
| $C_2$ | $E, C_2$ | | $E, C_2, \sigma_v, \sigma_v'$ |
| $\sigma_v$ | $E, \sigma_v$ | ⇨ | $= E, C_2, 2\sigma_v$ |
| $\sigma_v'$ | $E, \sigma_v'$ | | ⇒ $h = 4$; $\sigma = 2$ $(E, C_2)$ |

Note that the operations $\sigma_v$, $\sigma_v'$ are indicated as $2\sigma_v$ in the set of operations. The total number of distinct operations is 4, and hence the order $h=4$. The symmetry number $\sigma=2$ (operations $E$, $C_2$).

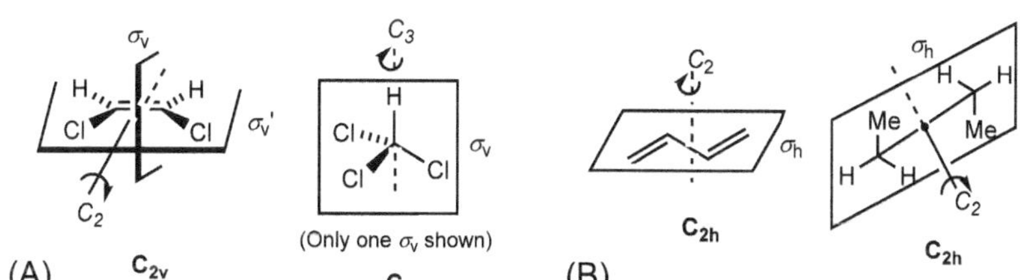


FIG. 1.28 Examples illustrating achiral point groups (A) $\mathbf{C_{2v}}$ and $\mathbf{C_{3v}}$ and (B) $\mathbf{C_{2h}}$.

$CHCl_3$ has $\mathbf{C_{3v}}$ symmetry having a $C_3$ axis and 3 $\sigma_v$ planes (only one of the 3 $\sigma_v$ planes is shown, each plane passing through C—H bond and a C—Cl bond). The symmetry operations generated by the symmetry elements in $\mathbf{C_{3v}}$ are listed below.

Point group: $\mathbf{C_{3v}}$

| Symmetry elements | Symmetry operations | Set of nonredundant operations: |
|---|---|---|
| $C_3$ | $E, C_3, C_3^2$ | $E, C_3, C_3^2, \sigma_v, \sigma_v', \sigma_v''$ |
| $\sigma_v, \sigma_v', \sigma_v''$ | $E, \sigma_v, \sigma_v', \sigma_v''$ | $= E, 2C_3, 3\sigma_v$ |
| | | $\Rightarrow h = 6;\ \sigma = 3\ (E, 2C_3)$ |

The order of $\mathbf{C_{3v}}$ point group $(h) = 6$, and the symmetry number $\sigma = 3$ (operations $E$, $2C_3$). In general, for $\mathbf{C_{nv}}$, $h = 2n$ and $\sigma = n$.

In $\mathbf{C_{nh}}$ point groups, the $\mathbf{C_{2h}}$ group is common and present in molecules such as *trans*-1,2-dichloroethene, *s-trans* butadiene, glyoxal and anti conformer of butane. Fig. 1.28 B illustrates the point group $\mathbf{C_{2h}}$ with *s-trans* butadiene and anti butane. In *s-trans* butadiene, a $C_2$ axis passes through the midpoint of the $C_2$—$C_3$ bond and is perpendicular to the plane of the molecule. The molecular plane serves as a single $\sigma_h$ plane. The anti conformer of butane has a $C_2$ axis bisecting the $C_2$—$C_3$ bond, and a $\sigma_h$ plane perpendicular to the $C_2$ axis and passing through the four-carbon chain. *The existence of both $C_2$ and $\sigma_h$ implies that an inversion centre i at the intersection of the axis and the plane must exist.* The symmetry operations generated by the complete set of symmetry elements are given in the box below. Note that the symmetry element *i* (arising from the simple consequence of the symmetry elements defining the point group) is shown within bracket.

Point group: $\mathbf{C_{2h}}$

| Symmetry elements | Symmetry operations | Set of nonredundant operations: |
|---|---|---|
| $C_2$ | $E, C_2$ | $E, C_2, \sigma_h, i$ |
| $\sigma_h$ | $E, \sigma_h$ | $\Rightarrow h = 4;\ \sigma = 2\ (E, C_2)$ |
| $(i)$ | $E, i$ | |

The order of the $\mathbf{C_{2h}}$ point group $(h) = 4$, and the symmetry number $\sigma = 2$. In general, for $\mathbf{C_{nh}}$, $h = 2n$, $\sigma = n$. It should be noted that anti butane belongs to the achiral point group $\mathbf{C_{2h}}$, whereas chiral gauche butane has $\mathbf{C_2}$ symmetry (see Fig. 1.26B); however, since they have the same symmetry number ($\sigma = 2$), they make equal contribution to entropy due to symmetry (see Section 1.3.5.1). The higher $\mathbf{C_{nh}}$ groups are rare but known.

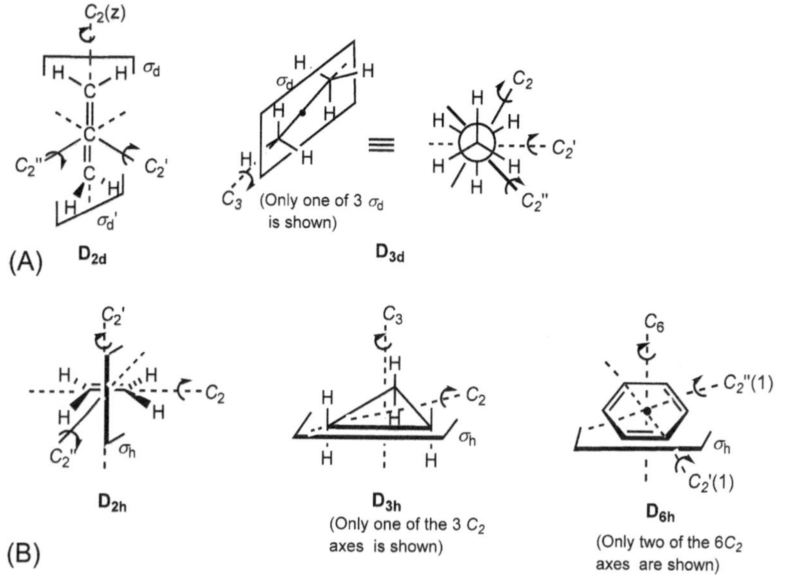


**FIG. 1.29** Examples illustrating achiral point groups (A) $\mathbf{D_{2d}}$ and $\mathbf{D_{3d}}$ and (B) $\mathbf{D_{2h}}$, $\mathbf{D_{3h}}$ and $\mathbf{D_{6h}}$.

### *1.5.2.3 Point groups $D_{nd}$ and $D_{nh}$*

The point group $\mathbf{D_{nd}}$ results when $n$ $\sigma_d$ planes are added to the ($C_n + n \perp C_2$) axes present in $\mathbf{D_n}$ (see Table 1.5). A representative example of the $\mathbf{D_{2d}}$ point group is allene (Fig. 1.29A). The $\mathbf{D_{2d}}$ group is characterized by a principal $C_2$ axis, 2 $C_2$ axes ($C_2'$ and $C_2''$) perpendicular to it, and 2 $\sigma_d$ planes ($\sigma_d$ and $\sigma_d'$) intersecting with the principal axis and bisecting the pair of $C_2$ axes (see also Fig. 1.22D). *The existence of these symmetry elements provides another improper axis $S_4$ coincident with the principal $C_2$ axis* (see Fig. 1.24B). The symmetry operations generated by the complete set of symmetry elements are included in the box given below. Note that the symmetry element $S_4$ arising from the simple consequence of the symmetry elements defining the point group $\mathbf{D_{2d}}$ is shown within bracket.

**Point group: $\mathbf{D_{2d}}$**

| Symmetry elements | Symmetry operations | | Set of nonredundant operations: |
|---|---|---|---|
| $C_2$ | $E, C_2$ | | $E, C_2, C_2', C_2'', S_4, S_4^3, \sigma_d, \sigma_d'$ |
| $C_2', C_2''$ | $E, C_2', C_2''$ | ⇨ | $= E, C_2, 2C_2', 2S_4, 2\sigma_d$ |
| $(S_4)$ | $E, S_4, C_2, S_4^3$ | | |
| $\sigma_d, \sigma_d'$ | $E, \sigma_d, \sigma_d'$ | | ⇒ $h = 8$; $\sigma = 4$ ($E, C_2, 2C_2'$) |

The order of $\mathbf{D_{2d}}$ point group ($h$) = 8, and the symmetry number $\sigma = 4$ (operations $E$, $C_2$, $2C_2'$).

The point group $\mathbf{D_{3d}}$ is illustrated with the staggered conformer of ethane (Fig. 1.29A). The molecule has a $C_3$ axis along the C—C bond, 3 $C_2$ axes ($C_2', C_2'', C_2'''$) perpendicular to it, and 3 $\sigma_d$ planes ($\sigma_d, \sigma_d', \sigma_d''$) passing through the pairs of antiperiplanar C—H bonds. The $C_2$ axes pass through the midpoint of the C—C bond bisecting the torsion angles as shown in the Newman projection (three $C_2$ axes are indicated by a simple line, a dashed line and a thick line). As a consequence of the presence of these symmetry elements, *there will be an $S_6$ axis coincident with $C_3$ axis and a centre of inversion i* in the $\mathbf{D_{3d}}$ group. The symmetry operations generated by the complete set of symmetry elements are given in the box below. For the operations generated by $S_6$, see Section 1.4.5.

Point group: $\mathbf{D_{3d}}$

| Symmetry elements | Symmetry operations |
|---|---|
| $C_3$ | $E, C_3, C_3^2$ |
| $C_2', C_2'', C_2'''$ | $E, C_2', C_2'', C_2'''$ |
| $(S_6)$ | $E, S_6, C_3, i, C_3^2, S_6^5$ |
| $(i)$ | $E, i$ |
| $\sigma_d, \sigma_d', \sigma_d''$ | $E, \sigma_d, \sigma_d', \sigma_d''$ |

⇨ Set of nonredundant operations:
$E, C_3, C_3^2, C_2', C_2'', C_2''', i, S_6, S_6^5, \sigma_d, \sigma_d', \sigma_d''$
$= E, 2C_3, 3C_2, i, 2S_6, 3\sigma_d$

⇒ $h = 12$; $\sigma = 6$ $(E, 2C_3, 3C_2)$

The order of the $\mathbf{D_{3d}}$ group ($h$) = 12, and the symmetry number $\sigma = 6$ (operations $E$, $2C_3$, $3C_2$). In general, for $\mathbf{D_{nd}}$, $h = 4n$, $\sigma = 2n$.

The $\mathbf{D_{nh}}$ point groups are characterized by a principal $C_n$ axis, $n$ $C_2$ axes perpendicular to it and a single $\sigma_h$ plane perpendicular to the principal axis (see Table 1.5). *As a necessary consequence of the existence of these symmetry elements, there will be n $\sigma_v$ planes and an $S_n$ axis in this group.* The complete set of symmetry elements in $\mathbf{D_{nh}}$ is, therefore, $C_n$, $n$ $C_2$, $\sigma_h$, ($n$ $\sigma_v$, $S_n$). Fig. 1.29B illustrates $\mathbf{D_{2h}}$, $\mathbf{D_{3h}}$ and $\mathbf{D_{6h}}$ point groups taking ethylene, cyclopropane and benzene respectively. The symmetry operations generated by the symmetry elements in $\mathbf{D_{2h}}$ are given below.

Point group: $\mathbf{D_{2h}}$

| Symmetry elements | Symmetry operations |
|---|---|
| $C_2$ | $E, C_2$ |
| $C_2', C_2''$ | $E, C_2', C_2''$ |
| $\sigma_h$ | $E, \sigma_h$ |
| $(\sigma_v, \sigma_v')$ | $E, \sigma_v, \sigma_v'$ |
| $(S_2 = i)$ | $E, i$ |

⇨ Set of nonredundant operations:
$E, C_2, C_2', C_2'', \sigma_h, \sigma_v, \sigma_v', i$
$= E, C_2, 2C_2', \sigma_h, 2\sigma_v, i$

⇒ $h = 8$; $\sigma = 4$ $(E, C_2, 2C_2')$

The order of the $\mathbf{D_{2h}}$ group $(h)=8$, and the symmetry number $\sigma=4$ (operations $E$, $C_2$, $2C_2'$).

The complete set of symmetry elements for cyclopropane ($\mathbf{D_{3h}}$) is $C_3$, $C_2$, $C_2'$, $C_2''$, $\sigma_h$, $(\sigma_v, \sigma_v', \sigma_v'', S_3)$. Each $C_2$ axis passes through a vertex and bisects the opposite bonds. The $\sigma_h$ plane is the plane of the ring. The $S_3$ axis generates 6 operations. In general, an $S_n$ axis ($n$ odd) generates $2n$ operations. The symmetry operations generated by the symmetry elements in $\mathbf{D_{3h}}$ are listed below.

Point group: $\mathbf{D_{3h}}$

| Symmetry elements | Symmetry operations |
|---|---|
| $C_3$ | $E, C_3, C_3^2$ |
| $C_2', C_2'', C_2'''$ | $E, C_2', C_2'', C_2'''$ |
| $\sigma_h$ | $E, \sigma_h$ |
| $(\sigma_v, \sigma_v', \sigma_v'')$ | $E, \sigma_v, \sigma_v', \sigma_v''$ |
| $(S_3)$ | $S_3, S_3^2=C_3^2, S_3^3=\sigma_h, S_3^4=C_3, S_3^5, S_3^6=E$ |

⇨ Set of nonredundant operations:
$E, C_3, C_3^2, C_2', C_2'', C_2''', \sigma_h, \sigma_v, \sigma_v', \sigma_v'', S_3, S_3^5$
$= E, 2C_3, 3C_2, \sigma_h, 3\sigma_v, 2S_3$

⇒ $h = 12$; $\sigma = 6$ $(E, 2C_3, 3C_2)$

The order of the $\mathbf{D_{3h}}$ group $(h)=12$, and the symmetry number $\sigma=6$ (operations $E$, $2C_3$, $3C_2$).

Benzene has $\mathbf{D_{6h}}$ symmetry. Besides the symmetry elements $C_6$, $6 \perp C_2$ and $\sigma_h$ defining the group, *there will necessarily be 6 $\sigma_v$ planes and an $S_6$ axis. Again, the presence of $S_6$ ensures the presence of $S_3$ and $i$*. Of six $C_2$ axes, three axes [$C_2'(1)$, $C_2'(2)$, $C_2'(4)$] pass through opposite atoms while the other three [$C_2''(1)$, $C_2''(2)$, $C_2''(4)$] bisect the opposite bonds. The $\sigma_h$ plane is the molecular plane. Of six $\sigma_v$ planes, three pass through opposite atoms and the other three bisect the opposite bonds. These vertical $\sigma$ planes comprise 3 $\sigma_v$ and 3 $\sigma_d$ planes. The complete set of symmetry elements in $\mathbf{D_{6h}}$ generates the operations as given in the box below.

Point group: $\mathbf{D_{6h}}$

| Symmetry elements | Symmetry operations |
|---|---|
| $C_6$ | $C_6, C_6^2=C_3, C_6^3=C_2, C_6^4=C_3^2, C_6^5, C_6^6=E$ |
| $C_2'(1), C_2'(2), C_2'(3)$ | $E, C_2'(1), C_2'(2), C_2'(3)$ |
| $C_2''(1), C_2''(2), C_2''(3)$ | $E, C_2''(1), C_2''(2), C_2''(3)$ |
| $\sigma_h$ | $E, \sigma_h$ |
| $(i)$ | $E, i$ |
| $(S_6)$ | $E, S_6, C_3, i, C_3^2, S_6^5$ |
| $(S_3)$ | $S_3, C_3^2, \sigma_h, C_3, S_3^5, E$ |
| $(\sigma_v, \sigma_v', \sigma_v''$ $\sigma_d, \sigma_d', \sigma_d'')$ | $E, \sigma_v, \sigma_v', \sigma_v''$ $E, \sigma_d, \sigma_d', \sigma_d''$ |

⇨ Set of nonredundant operations:
$E, C_6, C_6^5, C_3, C_3^2, C_2, 3C_2', 3C_2''$, $\sigma_h, i, S_6, S_6^5, S_3, S_3^5, 3\sigma_v, 3\sigma_d$

$= E, 2C_6, 2C_3, C_2, 3C_2', 3C_2'', \sigma_h, i, 2S_6, 2S_3, 3\sigma_v, 3\sigma_d$

⇒ $h = 24$; $\sigma = 12$ $(E, 2C_6, 2C_3, C_2, 3C_2', 3C_2'')$

The order ($h$) of the $\mathbf{D_{6h}}$ group is 24 and the symmetry number ($\sigma$) is 12 (operations $E$, $2C_6$, $2C_3$, $C_2$, $3C_2'$, $3C_2''$). In general, for $\mathbf{D_{nh}}$, $h=4n$, $\sigma=2n$.

### 1.5.3 Linear ($\mathbf{C_{\infty v}}$, $\mathbf{D_{\infty h}}$) and highly symmetrical point groups ($\mathbf{T_d}$, $\mathbf{O_h}$, $\mathbf{I_h}$)

In linear molecules, there is a $C_\infty$ axis along the linear chain of atoms, an infinite number of vertical $\sigma$ planes intersecting with it, and an infinite number of $C_2$ axes perpendicular to $C_\infty$ axis. If the molecules lack a horizontal plane $\sigma_h$, the point group is $\mathbf{C_{\infty v}}$. Otherwise, if the molecules contain a $\sigma_h$, the point group is $\mathbf{D_{\infty h}}$. These point groups are illustrated in Fig. 1.30A. HCN belongs to the point group $\mathbf{C_{\infty v}}$, whereas acetylene has $\mathbf{D_{\infty h}}$ symmetry.

The highly symmetrical achiral molecules possess multiple high-order rotation axes, i.e. two or more $C_n$ axes ($n>2$). These molecules resemble Platonic solids (as mentioned in Plato's dialogue *Timaeus*) such as the tetrahedron (4 faces), cube (6 faces), octahedron (8 faces), dodecahedron (12 faces) and icosahedron (20 faces). These five types of molecules fall into three point groups $\mathbf{T_d}$, $\mathbf{O_h}$ and $\mathbf{I_h}$.

The regular tetrahedral molecules such as $CH_4$ belong to the point group $\mathbf{T_d}$ (Fig. 1.30B). This symmetry is characterized by the symmetry elements 4 $C_3$, 3 $C_2$ and 6 $\sigma_d$. As shown, four C—H bonds are labelled 1, 2, 3 and 4. Each $C_3$ axis passes through a C—H bond (1, 2, 3 or 4) and the centre of the opposite face. Each $C_2$ axis bisects the opposite pairs of C—H bonds (1,2)/(3,4), (1,4)/(2,3) or (1,3)/(2,4), and each $\sigma_d$ plane passes through a pair of C—H bonds (1,2), (1,3), (1,4), (2,3), (2,4) or (3,4) and bisecting the other pair. There will be also 3 $S_4$ axes coincident with 3 $C_2$ axes (see also Fig. 1.24B). The full set of symmetry elements for the $\mathbf{T_d}$ point group is

$$4\,C_3, 3\,C_2, 3\,S_4, 6\,\sigma_d$$

It was shown earlier that besides $E$, each $C_3$ axis generates two operations ($C_3$, $C_3^2$), each $C_2$ axis generates one operation ($C_2$), each $S_4$ axis generates two distinct operations ($S_4$, $S_4^3$) and each $\sigma_d$ plane generates one operation ($\sigma_d$). Therefore, the complete set of distinct operations for the $\mathbf{T_d}$ point group is

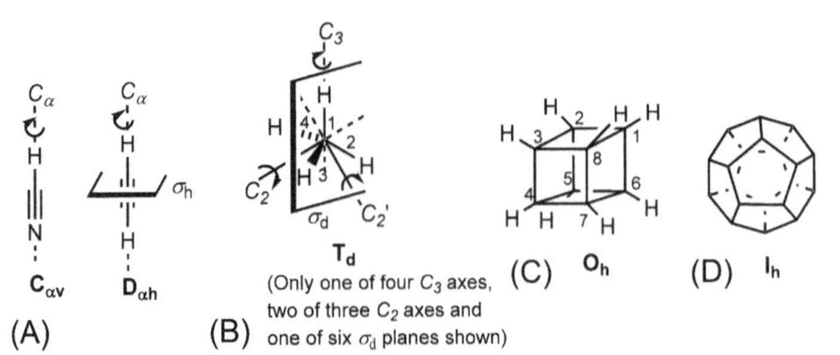


FIG. 1.30 Examples of achiral point groups (A) $\mathbf{C_{\infty v}}$ and $\mathbf{D_{\infty h}}$, (B) $\mathbf{T_d}$, (C) $\mathbf{O_h}$ and (D) $\mathbf{I_h}$.

$$E, 8C_3, 3C_2, 6S_4, 6\sigma_d$$

The order ($h$) of the $\mathbf{T_d}$ group is therefore 24, and the symmetry number ($\sigma$) is 12 (operations $E$, $8C_3$, $3C_2$).

The $\mathbf{O_h}$ symmetry is present in the cube and octahedron. This group is characterized by 3 $C_4$, 4 $C_3$ and 6 $C_2$ axes, and 9 $\sigma$ planes (3 $\sigma_h$ and 6 $\sigma_d$). There are also 3 $S_4$, 4 $S_6$ and $i$. An example of $\mathbf{O_h}$ symmetry is cubane[61] (Fig. 1.30C). The $\mathbf{I_h}$ symmetry is present in the dodecahedron and icosahedron. This group is characterized by 6 $C_5$, 10 $C_3$ and 15 $C_2$ axes, and 15 $\sigma$ planes. There are also 6 $S_{10}$ axes. The buckminsterfullerene ($C_{60}$),[62] an allotrope of carbon having 20 hexagonal and 12 pentagonal faces, is a closed-shell polygon and has $\mathbf{I_h}$ symmetry (Fig. 1.30D).

**Problem 1.7** Identify the point groups of the following molecules. Mention all symmetry elements and write the complete set of distinct symmetry operations. Determine the order ($h$) and the symmetry number ($\sigma$) for each.

(a) Alanine (b) *trans*-1,2-Dimethylcyclopropane (c) Pyridine (d) Naphthalene (e) Phenanthrene (f) Glyoxal (g) Adamantane (h) 1-Chloroadamantane (i) Eclipsed ethane (j) 1,4-Dichlorocubane (k) *trans-transoid-trans-transoid-trans*-Perhydrotriphenylene.

### 1.5.4 Desymmetrization

A point group of high symmetry can be reduced to a point group of lower symmetry by removal of symmetry element(s) in a process called desymmetrization. Desymmetrization can be performed by either distortion of the symmetrical framework or substitution on the framework.[63,64] Cyclohexane as a regular planar hexagon has $\mathbf{D_{6h}}$ symmetry but conformational analysis shows that the hexagon is actually distorted into a chair with $\mathbf{D_{3d}}$ symmetry (see Section 3.3.3), a subsymmetry of $\mathbf{D_{6h}}$. Desymmetrization by substitution

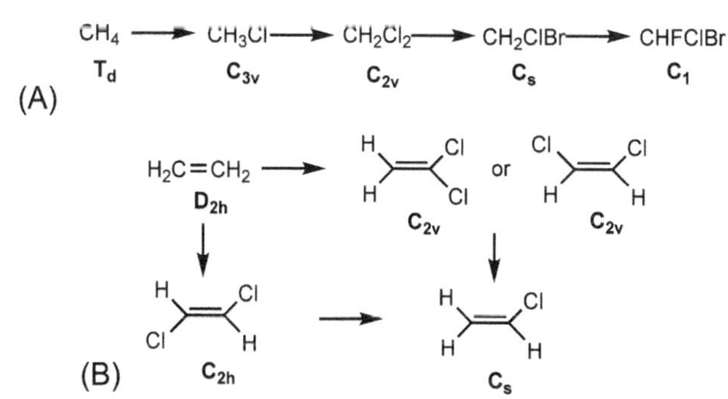


**FIG. 1.31** Desymmetrization by substitution on a symmetrical structure using (A) methane and (B) ethylene.

on a symmetrical structure is illustrated in Fig. 1.31. A single substitution on methane leads to desymmetrization of the highly symmetrical $CH_4$ (**$T_d$**) to $CH_3Cl$ (**$C_{3v}$**) (Fig. 1.31A). Successive substitution then leads to $CH_2Cl_2$ (**$C_{2v}$**), $CH_2ClBr$ (**$C_s$**), and finally to CHFClBr (**$C_1$**). Fig. 1.31B shows that ethylene (**$D_{2h}$**) can be desymmetrized to **$C_{2v}$** in 1,1-dichloroethene or *cis*-1,2-dichloroethene, whereas desymmetrization of ethylene to *trans*-1,2-dichloroethene leads to **$C_{2h}$**. Further desymmetrization of **$C_{2v}$** or **$C_{2h}$** leads to chloroethene (**$C_s$**).

### 1.5.5 Dynamic symmetry

The symmetry of molecules described above is determined on the basis of rigid structures. However, a molecule can have multiple equivalent conformers that are in rapid equilibrium. For example, there are three equivalent staggered conformations of ethane that are rapidly interconvertible on a time scale that is much faster than that of the experiment designed to determine the symmetry. Such molecules have *dynamic symmetries* and one can at best determine the *averaged symmetry* of the molecule.[65] Therefore, the symmetry determined earlier for the staggered ethane (see Fig. 1.29A) refers to the rigid point group **$D_{3d}$**; actually it will have an averaged symmetry. The nonrigid symmetry can also arise in other cases. For example, the cyclobutane derivative shown in Fig. 1.23C has an inversion centre. But rotation of the methyl groups about the C—C bond would result in several rapidly interconvertible rotamers of the molecule when the inversion centre *i* is present in a specific rotamer (as shown in Fig. 1.23C) with the point group **$C_i$**. The **$C_i$** symmetry thus refers to a rigid rotamer. In reality, the molecule has an averaged symmetry. The rigorous treatment[66] of nonrigid or averaged symmetry requires knowledge of permutation group theory which is beyond the scope of this text. However, we will consider briefly the nonrigid or dynamic symmetry in cyclic conformations in a nonmathematical or intuitive fashion later (see Section 3.5.2.1).

## References

1. Ramsay, O. B. *Stereochemistry*; Heyden & Son: Philadelphia, 1981.
2. Eliel, E. L.; Wilen, S. H.; Mander, L. N. *Stereochemistry of Organic Compounds*; Wiley: New York, 1994.
3. Robinson, M. J. T. *Organic Stereochemistry*; Oxford University Press: Oxford, 2000.
4. Mezey, P. G., Ed. *New Developments in Molecular Chirality*; Kluwer Academic: Boston, MA, 1991.
5. IUPAC. *Compendium of Chemical Terminology: Gold Book*; Version 2.3.3, 2014; p. 319.
6. Klyne, W.; Prelog, V. *Experientia* **1960**, *16*, 521.
7. Engler, E. M.; Andose, J. D.; Schleyer, P.v. R. *J. Am. Chem. Soc.* **1973**, *95*, 8005.
8. Clark, T. *A Handbook of Computational Chemistry*; Wiley-Interscience: New York, 1985.

9. Grant, G. H.; Richards, W. G. *Computational Chemistry*; Oxford University Press: Oxford, 1995.
10. Császár, A. G.; Allen, W. D.; Schaefer, H. F., III. *J. Chem. Phys.* **1998**, *108*, 9751.
11. Schreiner, P. R. *Angew. Chem. Int. Ed. Engl.* **2002**, *41*, 3579.
12. Mo, Y.; Gao, J. *Acc. Chem. Res.* **2007**, *40*, 113.
13. Pophristic, V.; Goodman, L. *Nature* **2001**, *411*, 565.
14. Lowe, J. P. *Prog. Phys. Org. Chem.* **1968**, *6*, 1.
15. Durig, J. R.; Craven, S. M.; Bragin, J. *J. Chem. Phys.* **1970**, *53*, 38.
16. Wiberg, K. B.; Murcko, M. A. *J. Am. Chem. Soc.* **1988**, *110*, 8029.
17. Bartell, L. S.; Boates, T. L. *J. Mol. Struct.* **1976**, *32*, 379.
18. Rüchardt, C.; Beckhaus, H.-D. *Angew. Chem. Int. Ed. Engl.* **1980**, *19*, 429.
19. Hellmann, G.; Hellmann, S.; Beckhaus, H.-D.; Rüchardt, C. *Chem. Ber.* **1982**, *115*, 3364.
20. Mizushima, S.-I. *Structure of Molecules and Internal Rotation*; Academic Press: New York, 1954.
21. Neu, J. T.; Gwinn, W. D. *J. Chem. Phys.* **1950**, *18*, 1642.
22. Morino, Y. *J. Mol. Struct.* **1985**, *126*, 1.
23. Abraham, R. J.; Kemp, R. H. *J. Chem. Soc. B* **1971**, 1240.
24. Allinger, N. L.; Hindmann, D.; Hönig, H. *J. Am. Chem. Soc.* **1977**, *99*, 3282.
25. Tichý, M. In *Advances in Organic Chemistry, Methods and Results*; Raphael, R. A., Taylor, E. C., Wynberg, H., Eds.; Vol. 5; Wiley: New York, 1965; p. 115.
26. Snyder, E. I. *J. Am. Chem. Soc.* **1966**, *88*, 1165.
27. Bushweller, C. H.; Whalon, M. R.; Laurenzi, B. J. *Tetrahedron Lett.* **1981**, *22*, 2945.
28. Dale, J. *Stereochemistry and Conformational Analysis*; Universitetsforlaget: Oslo, 1978; p. 98.
29. Herschbach, D. R.; Krisher, L. C. *J. Chem. Phys.* **1958**, *28*, 728.
30. Durig, J. R.; Guirgis, G. A.; Bell, S. *J. Phys. Chem.* **1989**, *93*, 3487.
31. Wiberg, K. B.; Martin, E. *J. Am. Chem. Soc.* **1985**, *107*, 5035.
32. Dorigo, A. E.; Pratt, D. W.; Houk, K. N. *J. Am. Chem. Soc.* **1987**, *109*, 6591.
33. Goodman, L.; Kundu, T.; Leszczynski, J. *J. Phys. Chem.* **1996**, *100*, 2770.
34. Durig, J. R.; Compton, D. A. C. *J. Phys. Chem.* **1980**, *84*, 773.
35. Kilb, R. W.; Lin, C. C.; Wilson, E. B. *J. Chem. Phys.* **1957**, *26*, 1695.
36. Abraham, R. J.; Pople, J. A. *Mol. Phys.* **1960**, *3*, 609.
37. Durig, J. R.; Compton, D. A. C.; McArver, A. Q. *J. Chem. Phys.* **1980**, *73*, 719.
38. Squillacote, M. E.; Sheridan, R. S.; Chapman, O. L.; Anet, F. A. L. *J. Am. Chem. Soc.* **1979**, *101*, 3657.
39. Karabatsos, G. J.; Fenoglio, D. J. *Top. Stereochem.* **1970**, *5*, 167.
40. Bowen, J. P.; Pathiaseril, A.; Profeta, S.; Allinger, N. L. *J. Org. Chem.* **1987**, *52*, 5162.
41. Allinger, N. L.; Chang, S. H. M. *Tetrahedron* **1977**, *33*, 1561.
42. Drakenberg, T.; Forsén, S. *J. Chem. Soc. D* **1971**, 1404.
43. Kirby, A. J. *The Anomeric Effect and Related Stereoelectronic Effects at Oxygen*; Springer: New York, 1983.
44. Wiberg, K. B.; Laidig, K. E. *J. Am. Chem. Soc.* **1987**, *109*, 5935.
45. Carreira, L. A. *J. Chem. Phys.* **1975**, *62*, 3851.
46. Cherniak, E. A.; Costain, C. C. *J. Chem. Phys.* **1966**, *45*, 104.
47. Ramachandran, G. N.; Ramakrishnan, C.; Sasisekharan, V. *J. Mol. Biol.* **1963**, *7*, 95.
48. Watson, H. C. *Prog. Stereochem.* **1969**, *4*, 299.
49. Thompson, J. R.; Bratt, J. M.; Banaszak, L. J. *J. Mol. Biol.* **1995**, *252*, 433.
50. Anfinsen, C. B. *Science* **1973**, *181*, 223.
51. Zwanzig, R.; Szabo, A.; Bagchi, B. *Proc. Natl. Acad. Sci. U. S. A.* **1992**, *89*, 20.

52. Levinthal, C. In *Mossbauer Spectroscopy in Biological Systems*; Degennes, P., Ed.; University of Illinois Press: Urbana, IL, 1969; p. 22.
53. Cotton, F. A. *Chemical Applications of Group Theory*, 2nd ed.; New York: Wiley, 1971 (chapter 3).
54. Ogden, J. S. *Introduction to Molecular Symmetry*; Oxford University Press: Oxford, 2001 (chapter 1).
55. Mislow, K. *Introduction to Stereochemistry*; Benjamin: New York, 1965; p. 52.
56. Cahn, R. S.; Ingold, C.; Prelog, V. *Angew. Chem. Int. Ed. Engl.* **1966**, *5*, 385.
57. Whitlock, H. W. *J. Am. Chem. Soc.* **1962**, *84*, 3412.
58.. Farina, M.; Audisio, G. *Tetrahedron* **1970**, *26*, 1827–1839.
59. Helmchen, G.; Staiger, G. *Angew. Chem. Int. Ed. Engl.* **1977**, *16*, 116.
60. McCasland, G. E.; Proskow, S. *J. Am. Chem. Soc.* **1955**, *77*, 4688. **1956**, *78*, 5646.
61. Barborak, J. C.; Watts, L.; Pettit, R. *J. Am. Chem. Soc.* **1966**, *88*, 1328.
62. Kroto, H. W.; Heath, J. R.; O'Brien, S. C.; Curl, R. F.; Smalley, R. E. *Nature* **1985**, *318*, 162.
63. Donaldson, J. D.; Ross, S. D. *Symmetry and Stereochemistry*; Wiley: New York, 1972.
64. Prelog, V. *Chem. Br.* **1968**, *4*, 382.
65. Bauder, A.; Meyer, R.; Günthard, H. H. *Mol. Phys.* **1974**, *28*, 1305.
66. Longuet-Higgins, H. C. *Mol. Phys.* **1963**, *6*, 445.

Chapter 2

# Acyclic molecules 2: Configuration and resolution

## 2.1 Distinction between configuration and conformation

In Chapter 1, we have seen that conformational isomers have a relatively low energy barrier for their interconversion, and hence they are not separable under normal conditions (25°C). In contrast, configurational isomers are isolable and exist at room temperature. This is because the barriers to interconversion of configurational isomers are quite high, usually involving breaking and forming of bonds. However, there are some unavoidable grey areas. For example, the barriers to rotation about C—C single bond in biphenyl systems can vary over a wide range (~2 to >40 kcal mol$^{-1}$) depending on the substitution pattern (see later Section 2.9.2.3). The rate of interconversion for any rotational barrier can be calculated using Eq. (1.6) (see Section 1.3.4.1). At 25°C, a torsional barrier of 20 kcal mol$^{-1}$ corresponds to an interconversion rate $(k) = 1.3 \times 10^{-2}\ s^{-1}$ $(t_{1/2} = 0.693/k = 53\ s)$ which indicates that the isomers cannot be separated by conventional methods, whereas a barrier of 25 kcal mol$^{-1}$ gives a value of $k = 2.9 \times 10^{-6}\ s^{-1}$ $(t_{1/2} \cong 66\ h)$, thereby allowing the separation of the isomers. Thus, at low barriers the biphenyl stereoisomers are conformational isomers whereas at high barriers (≥22.3 kcal mol$^{-1}$) the biphenyl stereoisomers represent configurational isomers. Isolability or separability of stereoisomers (based on the kinetics of interconversion) provides an important criterion to distinguish between conformation and configuration. It should be noted that the term isolability cannot be defined precisely since the temperature of the experiment as well as the lifetime required to consider a compound as an isolable entity vary. However, it has a thermodynamic foundation in the form of the phase rule which gives the number of isolable components ($C$) irrespective of the observable conditions. According to the phase rule, $F = C - P + 2$ where $F$ is the number of degrees of freedom of the system and $P$ is the number of phases.

*Configuration* has been defined[1] as 'the arrangements of atoms of a molecular entity in space that distinguishes stereoisomers, the isomerism between which is not due to conformation differences'. The configuration of a molecule will remain unchanged by rotation about single bonds, except the grey areas

Stereochemistry and Organic Reactions. https://doi.org/10.1016/B978-0-12-824092-2.00002-2

mentioned above. A configurational isomer may also possess several conformational isomers.

## 2.2 Classification of isomerism

Fig. 2.1 summarizes the types of isomerism possible for organic molecules (see also Section 1.1).

There is some confusion as to whether the term stereoisomers (in general, isomers) should be used in the *microscopic* sense at the molecule level or in the *macroscopic* sense at the compound level. We will use the term in both contexts with the following connotations. When we deal with the structure of a stereoisomer, the term refers to a molecule but when we deal with an observable property or separation of a stereoisomer, the term refers to a compound.

The configurational stereoisomerism in organic molecules arises from the presence of stereogenic elements, namely, stereogenic centre, stereogenic axis or stereogenic plane. Besides, helicity is also considered a stereogenic element.

## 2.3 Stereogenic centre, chiral centre, chiral molecule and enantiomers

Stereoisomerism due to the presence of stereogenic centre(s) [in short, stereocentre(s)][2,3] is most common. A stereocentre is usually a tetrahedral $sp^3$ atom (most commonly C) attached to four different ligands. For example, the carbon in CHFClBr is a stereocentre. The centre is called stereogenic because exchange of any two ligands generates a new stereoisomer (Fig. 2.2). Here the molecule is chiral (point group $\mathbf{C_1}$), and hence gives two nonidentical structures that are

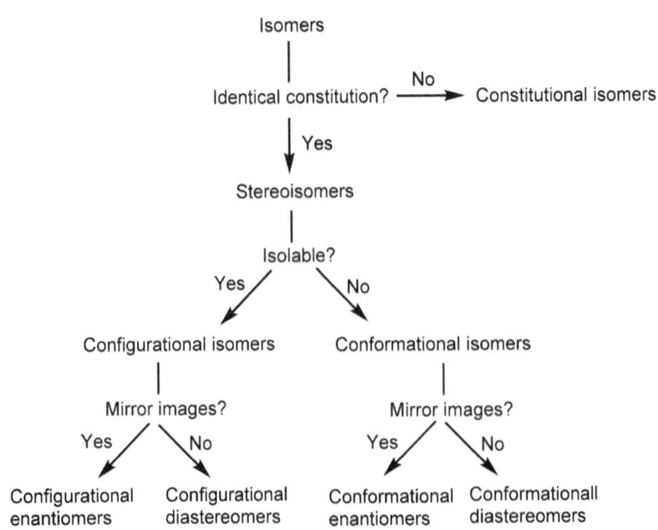


**FIG. 2.1** Types of isomerism for organic molecules.

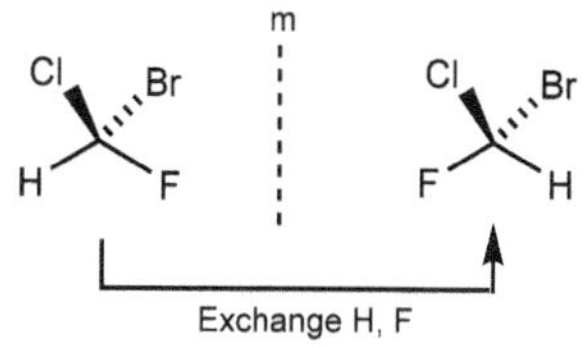


FIG. 2.2 Two stereoisomers (enantiomers) of CHFClBr due to the presence of a stereogenic centre.

mirror images of each other. Thus CHFClBr has two enantiomers (Fig. 2.2). Interconversion of the enantiomers requires bonds to be broken and reformed, and therefore involves a very high energy barrier. Thus the stereoisomers due to a stereocentre are isolable and belong to configurational isomers.

The stereogenic carbon in the above chiral molecule is chirotopic[3] as it resides in a chiral environment, and is therefore a chiral centre (see Section 2.8.2). It is worth noting that *all* atoms in a chiral molecule are chirotopic. An achiral molecule may also have some chirotopic atoms (see later Section 2.8.5). In general, all chiral centres are stereocentres but the reverse is not always true (see later Fig. 2.37). Since a stereocentre with four ligands differing in *constitution* is chirotopic, we will use the terms stereocentre and chiral centre interchangeably for an atom linked to four constitutionally different ligands.

Molecules containing a *single* chiral centre are always chiral (point group $C_1$) and lead to a pair of enantiomers. Notably, chiral molecule and chiral centre are distinct terms; molecules having two or more chiral centres may be achiral (see later Section 2.8.5). Some examples of molecules with a single chiral carbon are shown in Fig. 2.3. The chiral centre is labelled with an asterisk (*), and only one enantiomer is shown in each case.

The scaler properties such as interatomic distances and energies are same for the two enantiomers, and they have identical scaler physical properties (like melting point, boiling point, refractive index, solubility, heat of combustion, UV–Vis, IR and NMR spectra) in an achiral environment. Therefore, they cannot be separated by common and conventional methods of separation, that is crystallization, distillation or chromatography. But their interactions with chiral molecules differ, and they can be distinguished in a chiral environment (such as chiral solvents, chiral stationary phases in chromatography and biochemical reagents like enzymes and hormones). A simple way of differentiating the two enantiomers is however the determination of optical rotation which is a vectorial physical property (see Section 2.4.4).

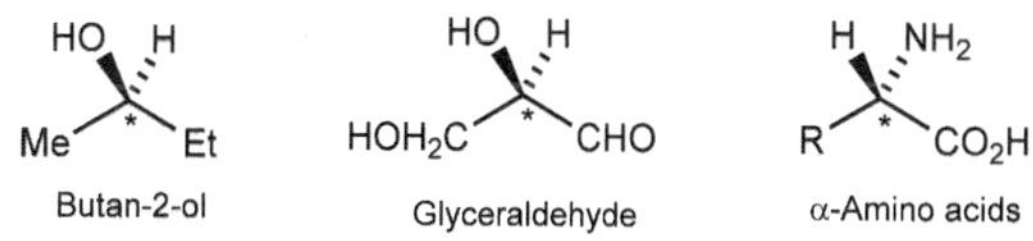


FIG. 2.3 Examples of molecules with a single chiral carbon.

Optical rotation denotes the rotation of the plane of polarization of plane-polarized light when it passes through a sample of chiral molecules. The angle of optical rotation ($\alpha$ in deg) is detected by a polarimeter. If the rotation is clockwise, it is said to be positive (+), and if anticlockwise it is negative (−). The enantiomer with (+) rotation is also called dextrorotatory, and the enantiomer with (−) rotation is said to be levorotatory. The two enantiomers have equal but opposite optical rotations, and each enantiomer is said to be optically active. An equimolar mixture of the two enantiomers is called a racemate, which will show no net optical rotation due to external compensation. Therefore a racemic (±) form is optically inactive. It should be noted that, if the plane of polarization is rotated by $\pm 180$ degrees $\times n$ ($n = 1, 2, 3, \ldots$), the new plane will coincide with the old plane and no distinction can be made between the rotations of $\alpha$ and $\alpha \pm 180n$ (see Problem 2.1).

## 2.4 Chirality and optical activity

Chirality[4] is a geometrical concept at the molecular or *microscopic* level while optical activity is a physical property of a compound at the *macroscopic* level. A sample of a compound is a macroscopic assembly of molecules homogeneous to ordinary methods of separation, that is crystallization, distillation or chromatography. (Note that 10 mg of a compound ($M_r$ 100) will contain $6 \times 10^{19}$ molecules.) A sample of a chiral substance used to detect optical activity might be *enantiomerically pure* (in which all molecules are chiral and of a single enantiomer) or *enantiomerically enriched* (wherein all molecules are chiral but with an excess of molecules of one enantiomer) or a *racemate* (in which all molecules are chiral but with equal number of molecules of the two enantiomers). It therefore follows that an enantiomerically pure (enantiopure) or enantiomerically enriched (enantioenriched) sample will show optical activity (the former will give higher optical rotation), whereas a racemic sample will be optically inactive, though all three samples of the substance consist of chiral molecules and no achiral molecules. Thus *chirality is the necessary and sufficient condition for the existence of enantiomers but not for optical activity; for a chiral compound to be optically active, there must be an excess of one enantiomer over the other.*

A racemate cannot be separated into the constituent enantiomers by the conventional methods used for ordinary mixtures. The process of separation of a racemate into enantiomers is called *resolution* which depends on diastereomeric interactions through the use of chiral reagents or solvents (see later Section 2.13).

### 2.4.1 Chiral conformations and optical inactivity

A compound may contain several conformers (chiral or achiral) which cannot be separated at room temperature. Consider the case of butane which is optically

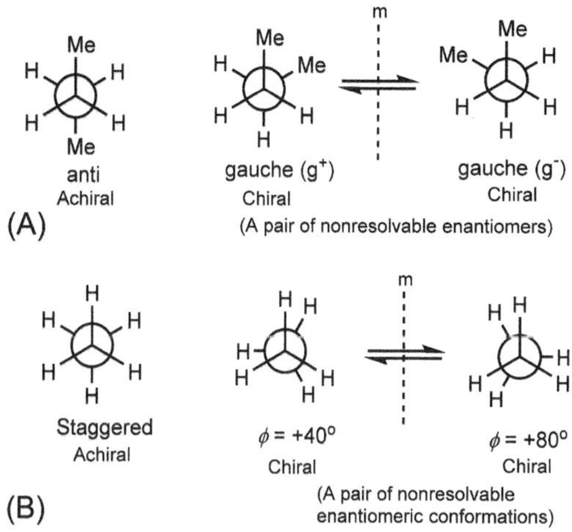


**FIG. 2.4** Optical inactivity of (A) butane and (B) ethane from conformational standpoint.

inactive. The simple reason for optical inactivity is that butane is regarded as an achiral compound with no chiral centres. But the conformational analysis of butane (see Section 1.3.5) shows that its conformational equilibrium comprises one achiral anti conformer and two chiral gauche conformers that are mirror images of each other (Fig. 2.4A). Two gauche forms are related as conformational enantiomers, whereas a gauche conformer ($g^+$ or $g^-$) is related to the anti as conformational diastereomers. The anti conformer is optically inactive since it is achiral. The gauche conformers, though chiral, interconvert rapidly at room temperature constituting a *nonresolvable* enantiomeric pair (or racemate) that is optically inactive. Thus, from the conformational standpoint, butane is macroscopically optically inactive.

Next, we consider the case of ethane having no chiral centres. The conformational analysis (see Section 1.3.4) shows that ethane has three degenerate staggered conformers and three degenerate eclipsed conformations with a low energy barrier (2.9 kcal mol$^{-1}$). The staggered conformer (point group $\mathbf{D_{3d}}$) and the eclipsed conformation (point group $\mathbf{D_{3h}}$) are both achiral. The conformational profile of ethane (see Fig. 1.6) consists of three indistinguishable segments, one of which is in the range of $\phi = 0$–120 degrees. Interestingly, the conformations on the two sides of the staggered conformer ($\phi = +60$ degrees) in the potential energy well, say, at $\phi = +40$ degrees and +80 degrees are chiral and enantiomeric (Fig. 2.4B). There are such innumerable chiral and enantiomeric conformations in the same potential well. The small energy differences between the chiral conformations and the lowest energy achiral staggered conformer imply that the chiral conformational states are significantly populated; however they exist as nonresolvable enantiomeric pairs or racemates. Thus, conformational arguments also suggest that ethane is optically inactive.

## 2.4.2 Optical rotation

The magnitude of optical rotation $\alpha$ (in deg) depends on the number of molecules of the sample being traversed by the plane-polarized light, and hence on the path length $l$ and the concentration of the solution $c$ (the density in the case of a pure liquid) at a specified wavelength of light $\lambda$ and temperature $T$. The measured optical rotation is usually normalized by a quantity called *specific rotation* $[\alpha]$ defined as

$$[\alpha] = \frac{\alpha}{lc} \tag{2.1}$$

where the observed rotation $\alpha$ is in deg., $l$ in dm (the length of a typical polarimeter cell is 10 cm = 1 dm), and $c$ in g/cm$^3$. The unit of $[\alpha]$ is given by

$$[\alpha] = \frac{\alpha}{lc} = \frac{\text{deg cm}^3}{10\,\text{cm g}} = 10^{-1}\,\text{deg cm}^2\text{g}^{-1}$$

But the unit of specific rotation is usually omitted; for example a specific rotation is denoted as +20, *not* +20 degrees. The wavelength of light and the temperature are usually indicated as subscript and superscript of $[\alpha]$, respectively. Thus $[\alpha]_D^{25}$ indicates the specific rotation using the wavelength of the emission line of a sodium lamp ($D$-line, 589 nm) at 25°C.

When optical activity is compared for analogous chiral compounds, it is usual to use the quantity called *molar rotation* $[\Phi]$ which depends on the number of moles of the substance traversed by the plane-polarized light. $[\Phi]$ is defined as

$$[\Phi] = \frac{[\alpha]M}{100} \tag{2.2}$$

where $M$ is the molar mass. The unit of $[\Phi]$ is.

$[\Phi] = \frac{[\alpha]M}{100} = 10^{-1}\,\text{degcm}^2\text{g}^{-1}\,\text{g}\left(10^{-2}\text{mol}\right)^{-1} = \text{deg cm}^2\ \text{dmol}^{-1}$ (dmol = decimole = $10^{-1}$mol).

Note that since the unit of $M$ is g mol$^{-1}$, $M$/100 implies g $(10^{-2}\text{mol})^{-1}$.

**Problem 2.1** How can you decide whether an observed rotation of a solution of an optically active compound is +50 degrees and not actually −130 degrees?

## 2.4.3 Enantiomeric excess

A sample of a chiral compound can be enantiomerically pure (enantiopure), enantiomerically enriched (enantioenriched) or a racemate. This refers to the enantiomeric composition of the chiral compound isolated as a natural product or obtained by synthesis. The enantiomeric excess (ee) is defined as the excess of one enantiomer over the other and is expressed in percentage. The ee (%) is given by

$$\text{ee}\,(\%) = (\text{mole fraction of one enantiomer} - \text{mole fraction of other enantiomer}) \times 100$$

For an enantiopure sample, ee = (1–0) × 100 = 100%, whereas for a racemate (1:1 mixture of enantiomers), ee = (½ – ½) × 100 = 0%. A chiral but nonracemic sample will be optically active (+ or −). The ee of a nonracemic (+)-sample is 50% implies that the sample contains 75% (+)-enantiomer and 25% (−)-enantiomer, that is 50% (+)-enantiomer and 50% racemate [25% (+) + 25% (−)].

A variety of methods (including NMR, chromatographic and kinetic methods) is available for the determination of enantiomeric excess of a chiral sample (see later Section 2.14). For the determination of ee by optical rotation, a sample of a pure enantiomer is required. It is assumed that ee is the same as the optical purity (op) which is given by

$$\text{op}\,(\%) = \frac{[\alpha]_{\text{sample}}}{[\alpha]_{\text{max}}} \times 100 \tag{2.3}$$

where $[\alpha]_{max}$ is the optical rotation of the pure enantiomer.

However, discrepancies in the measured values of op occur due to the inaccuracy of the $[\alpha]_{max}$ value used because the sample might contain traces of contaminants of high specific rotation. Further, a strong solute–solute interaction and intermolecular association would affect the results. This was first demonstrated by Horeau[5] showing a difference of op values of 2-ethyl-2-methylsuccinic acid [$HO_2CCH_2C(Et)(Me)CO_2H$] measured in nonpolar and polar solvents (the Horeau effect). Therefore, experimental values of op may not be always equal to ee.

**Problem 2.2** A sample of 2-butanol showed a specific rotation of +8.11. The specific rotation of pure (−)-2-butanol is 13.52. Determine the enantiomeric composition of the sample.

### 2.4.4 Origin of optical activity

Optical activity is a chiroptical property[6,7] that originates from the nondestructive interactions of chiral molecules with polarized light. Light is electromagnetic radiation that consists of two mutually perpendicular oscillating electric and magnetic fields. In ordinary light, the electric field vector oscillates in all directions perpendicular to the direction of propagation. The same is true for the magnetic field. Ordinary light is therefore isotropic or unpolarized. When ordinary light passes through a filter (polarizer), the electric field (**E**) oscillations of the emerging radiation can occur only in one plane, say, *xz* plane as shown in Fig. 2.5. The magnetic field oscillations perpendicular to the electric field vectors are then in the *yz* plane (not shown). Such light is anisotropic and is called plane polarized or linearly polarized light.

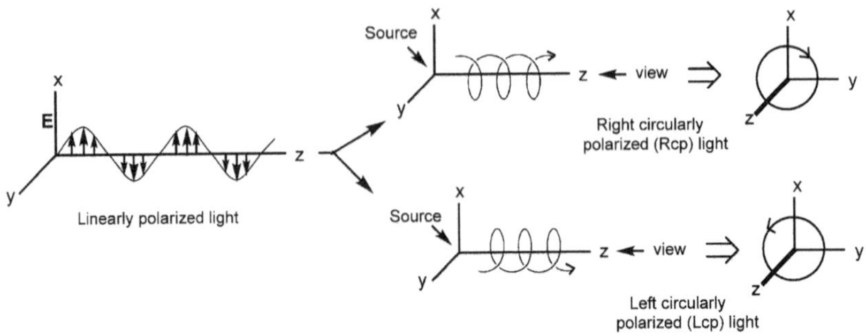


**FIG. 2.5** Linearly polarized light resolved into left and right circularly polarized lights.

When linearly polarized light is passed through an electrooptic modulator, it is resolved into two circularly polarized components called left circularly polarized (Lcp) and right circularly polarized (Rcp) lights (Fig. 2.5). The electric field vector describes a left-handed helix or a right-handed helix as viewed towards the light source. The projections of the helix are circles which lie on the xy plane as shown in Fig. 2.5. The magnitude of the vector remains constant but its direction changes in a helical manner. The Lcp and Rcp rays are related to each other as nonsuperposable mirror images and are therefore chiral. Theoretically, if the **E** vectors of two electromagnetic waves are one-quarter wavelength out of phase and perpendicular, the sum of the **E** vectors rotates so that its tip follows a helical path generating circularly polarized light. If Lcp and Rcp rays of equal intensity (amplitude) are superimposed, the result is linearly polarized light. Similarly, linearly polarized light can be resolved into L and R components.

In an isotropic medium of achiral molecules, the Lcp and Rcp lights travel at the same velocity. They are in-phase but in opposite senses (anticlockwise and clockwise). The resultant vector sum is the linearly polarized light emerging in the same original plane (Fig. 2.6A). We will now see what happens when the linearly polarized light passes through a sample of chiral molecules of an enantiomer. The interactions of chiral molecules with the chiral Lcp and Rcp rays will be diastereomeric in nature and will lead to refraction of Lcp and Rcp lights to different extents, that is unequal slowing down of the Lcp and Rcp beams.[8]

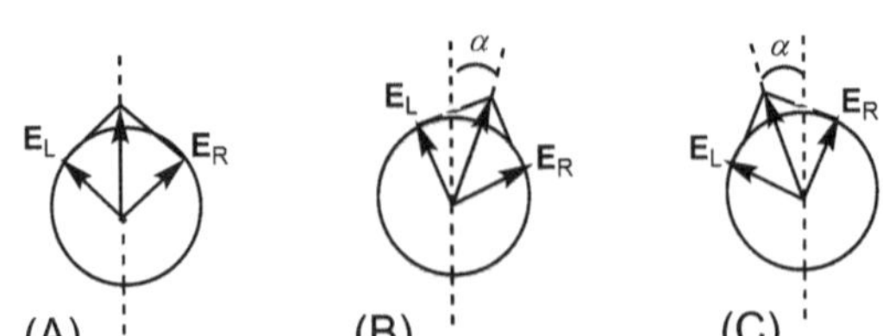


**FIG. 2.6** (A) No optical rotation ($n_L = n_R$). Optical rotation due to circular birefringence: (B) (+) rotation when $n_L > n_R$; (C) (−) rotation when $n_L < n_R$.

Let $n_L$ and $n_R$ be the refractive indices for Lcp and Rcp rays. If $n_L \neq n_R$ (anisotropic refraction), the chiral substance is said to exhibit *circular birefringence*. Consider the following cases:

Case I: $n_L > n_R$.

Since the refractive index $n = c_0/c$ where $c_0$ is the velocity of light in vacuum and $c$ is the velocity of light in the medium,

$$n_L > n_R \Rightarrow c_0/c_L > c_0/c_R \Rightarrow c_L < c_R.$$

Thus, the velocity of the Lcp beam is slower than that of the Rcp beam. Here the resultant linearly polarized light emerges in a plane which is rotated clockwise with respect to the incident beam (Fig. 2.6B). The optical rotation $\alpha$ is positive, and the chiral sample is optically active and dextrorotatory (+).

Case II: $n_L < n_R \Rightarrow c_0/c_L < c_0/c_R \Rightarrow c_L > c_R$.

Here the Lcp beam moves faster than the Rcp beam. The emergent linearly polarized light then appears anticlockwise rotated relative to the incident beam (Fig. 2.6C). The sample is levorotatory (−).

The angle of rotation $\alpha$ is related to the difference in refractive index $(n_L - n_R)$ and is given by the Fresnel equation:

$$\alpha - \frac{1800l\,(n_L - n_R)}{\lambda_0} \tag{2.4}$$

where $l$ is the path length (in dm) and $\lambda_0$ is the wavelength of light in vacuum (in cm). The difference in refractive index ($\Delta n = n_L - n_R$) can be calculated using the Fresnel equation. For example, 2-butanol exhibits an optical rotation $\alpha = 11.2$ degrees with a cell of path length 1 dm and light of wavelength 589 nm (sodium *D* line) at 20°C. The value of $\Delta n$ is obtained as

$$\Delta n = \alpha\lambda_0/1800l = 36.6 \times 10^{-9}$$

The refractive index ($n$) of neat 2-butanol is 1.3954, thus $\Delta n/n = 2.6 \times 10^{-6}\%$.- The value of $\Delta n$ is extremely small but since $\lambda_0$ is also very small ($589 \times 10^{-7}$ cm), $\alpha$ is appreciable.

### 2.4.5 Optical rotation and ligand polarizability: Brewster's rule

The sign and magnitude of optical rotation can be predicted by some empirical and semiempirical rules. One such rule is Brewster's rule[9] which predicts whether an enantiomer of a chiral compound will be dextrorotatory or levorotatory based on the differences in the polarizabilities of the atoms attached to the chiral centre (called atomic asymmetry). As the refractive index is related to the polarizability, the rotatory power of the chiral molecules also depends on polarizabilities of ligands, that is susceptibility to deformation by the interaction of their valence electrons with the electric fields of the radiation.[10] The polarizability of ligands can thus be derived from atomic refraction.[11] The polarizability order of some common ligands are

$$I > Br > SH > Cl > C{\equiv}C > CN > C{=}C > Ph > CO_2H > Me > NH_2 > OH > H > D > F$$

Notably, the atomic refractions of the attached atoms in C=C and C≡C are taken as half the value of the group refraction, and for groups such as CN, Ph, and $CO_2H$, a share of the group refraction is assigned to the attached carbon.

The Brewster's rule states that, assuming the polarizability order of four ligands attached to the chiral carbon as $a > b > c > d$, a compound will be dextrorotatory if $a \rightarrow b \rightarrow c$ is clockwise when viewed away from d, and levorotatory if the sequence is anticlockwise (cf. *R*/*S* descriptors, see Section 2.7.1.3). The rule is described using Fischer projections in Fig. 2.7A, when d is placed on the vertical line (top or bottom). If d is on the horizontal line, the opposite assignment is correct, that is (+) for anticlockwise and (−) for clockwise. Fig. 2.7A also shows that the enantiomer of isotopically labelled ethanol is predicted to be dextrorotatory in agreement with the experiment.[12]

If two ligands can interact by intramolecular hydrogen bonding (as in α-hydroxy acids or α-amino acids) forming a cyclic structure, then in addition to atomic asymmetry, there is a conformational dissymmetry contribution[13], as shown in Fig. 2.7B. The H-bonded cyclic structures **2.1** and **2.2** *as drawn* contribute (+) rotation and (−) rotation respectively, when polarizability order is $p > q$. The contributions due to atomic asymmetry and conformational dissymmetry may reinforce or oppose each other. As also shown in Fig. 2.7B,

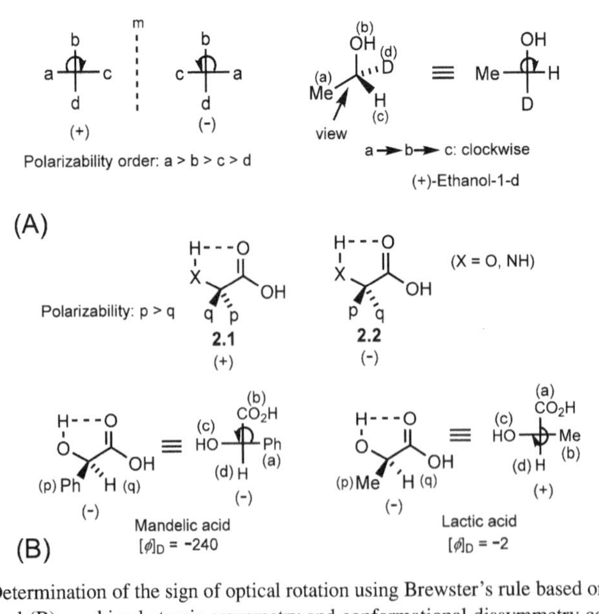


FIG. 2.7 Determination of the sign of optical rotation using Brewster's rule based on (A) atomic asymmetry and (B) combined atomic asymmetry and conformational dissymmetry contributions.

the two contributions reinforce for mandelic acid (the enantiomer is highly levorotatory in water, molar rotation $[\Phi]_D = -240$) and oppose for lactic acid (the enantiomer is weakly levorotatory in water, $[\Phi]_D = -2$). In the latter situation, the model may sometimes lead to ambiguous results, for example, the sodium salt of the lactic acid enantiomer is found to be dextrorotatory. Brewster also developed an empirical approach to determine the sign and magnitude of optical rotation for conformationally flexible compounds based on conformational dissymmetry model.[9,14]

**Problem 2.3** Using Brewster's rule, predict the sign of optical rotation of the product formed by treatment of (−)-1-chloro-1-phenylethane with $LiAlD_4$. Assign *R*/*S* descriptor to the reactant and the product.

## 2.5 Chiral centres other than carbon[15,16]

Besides carbon, a chiral centre can be a heteroatom such as N, Si, P or S. The chiral heteroatom is either tetracoordinate (tetrahedral) or tricoordinate (pyramidal) with a lone pair ligand.

### 2.5.1 Molecules with chiral tetracoordinate heteroatoms

Molecules with tetracoordinate chiral centres (N, Si, P and S) are illustrated in Fig. 2.8. Each compound has two separable enantiomers, one of which is shown. In the sulphone, two ligands at the chiral S atom differ only from oxygen isotopic substitution; still the compound shows a measurable optical rotation ($[\alpha]_D = -0.16$).

### 2.5.2 Molecules with chiral tricoordinate heteroatoms

Molecules with chiral pyramidal nitrogen, for example a tertiary amine (NMeEtPr) has two enantiomers (Fig. 2.9A). But these are readily interconvertible by pyramidal inversion and hence are not isolable.[17] As shown in Fig. 2.9A, the nitrogen inversion proceeds through a planar ($sp^2$ hybridized) TS. The barrier to inversion (about 8 kcal mol$^{-1}$) is low enough as not to permit the isolation of the enantiomers. Therefore, these stereoisomers are configurationally unstable and regarded as conformational isomers.

However, ring strain and electronegative substituent on nitrogen increases the barrier to inversion substantially so as to permit the isolation of the

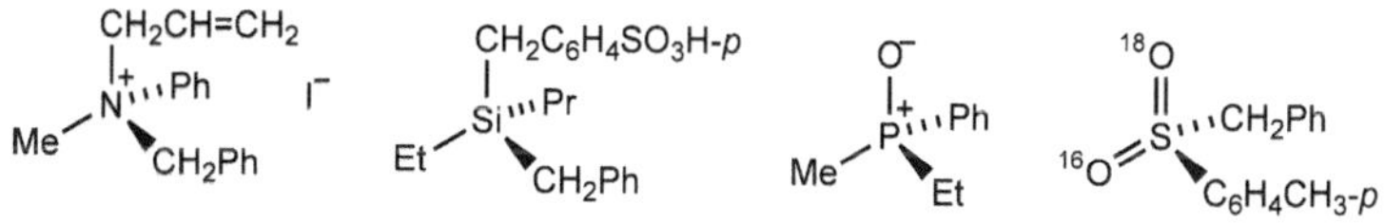


**FIG. 2.8** Molecules with tetracoordinate chiral centres (N, Si, P and S).

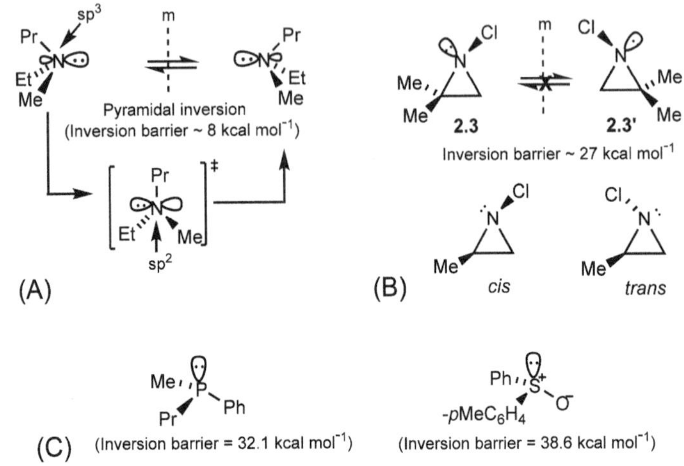


**FIG. 2.9** (A) Pyramidal inversion of a tertiary amine responsible for the interconversion of the two enantiomers. (B) No nitrogen inversion in the *N*-chloroaziridines forming configurationally stable stereoisomers. (C) Examples of configurationally stable enantiomers with tricoordinate P or S. The smaller lobe of heteroatom hybrid orbital is not shown in (B) and (C).

enantiomers. For example, the *N*-chloroaziridine **2.3** (Fig. 2.9B) is configurationally stable and does not undergo racemisation (a process of conversion of an enantiomer into a racemic mixture) at room temperature with an inversion barrier of about 27 kcal $mol^{-1}$.[17,18] Similarly, *N*-chloro-2-methylaziridine exists as isolable *cis* and *trans* isomers[19], also shown in Fig. 2.9B. For the aziridine ring, the change of hybridization at nitrogen from $sp^3$ to $sp^2$ (required in pyramidal inversion) is accompanied by a further increase in ring strain with the deviation of bond angle increasing from 49°28′ (109°28′–60°) to 60° (120°–60°). Furthermore, electronegative Cl substituent on the nitrogen would increase the s character of the lone pair orbital (since nitrogen uses a hybrid orbital with more p character to form the N—Cl bond). The increased s character would lower the energy of the lone pair orbital and thereby makes it more difficult to raise its energy to that of a p orbital in the TS. In other words, the lone pair on nitrogen will be more tightly held due to the presence of electron withdrawing Cl and hence less available for inversion. The two factors thus reinforce each other to significantly raise the barrier to inversion.

For the tricoordinate P or S (elements in the second period), the nonbonding orbital of lone pair has more s character (e.g. the lone pair in $PH_3$ occupies the 3s orbital). As a result, the inversion barrier for phosphines can be so high that the enantiomers can be separated.[20,21] For example, the energy barrier for the interconversion of the enantiomers of the chiral phosphine (Fig. 2.9C) is 32.1 kcal $mol^{-1}$. Chiral phosphines serve as useful chiral ligands in asymmetric synthesis. Compounds with tricoordinate S as the chiral centre also provide configurationally stable enantiomers. For example, the inversion barrier for the enantiomers of the chiral sulphoxide shown in Fig. 2.9C is 38.6 kcal $mol^{-1}$.[22]

**Problem 2.4** Chiral trialkylammonium salts undergo racemisation in aqueous solution. Explain.

## 2.6 Stereocentre configuration

The configuration at a stereocentre can be defined by the *relationships* among any three ligands: two ligands are connected by a *clockwise* (or *anticlockwise*) arrow through a tetrahedral angle when the third becomes a *front* (or *rear*) ligand.[23] Note that the geometry of the fourth ligand will be automatically fixed as front or rear in the tetrahedral space, that is if the third ligand is in front, the fourth ligand must be in rear. This definition of stereocentre configuration is represented in Fig. 2.10. The four ligands attached to the stereocentre are a, b, c and d, and any two of them can be chosen for the arrow relationship. In Fig. 2.10A, a → b is clockwise when c is in front; equivalently, c → d can be drawn clockwise when a is in front (Fig. 2.10B).

Notably, for a set of three ligands (a, b, c), both changes in relationships will indicate the same configuration (use a molecular model):

(a → b clockwise/c front) ≡ (a → b anticlockwise/c rear)
(a → b clockwise/c rear) ≡ (a → b anticlockwise/c front)

This is illustrated with an example shown below.

Any stereocentre has two opposite configurations. They display only one change in ligand relationships: (clockwise/front → anticlockwise/front) **or** (clockwise/front → clockwise/rear), as shown below. The configuration of

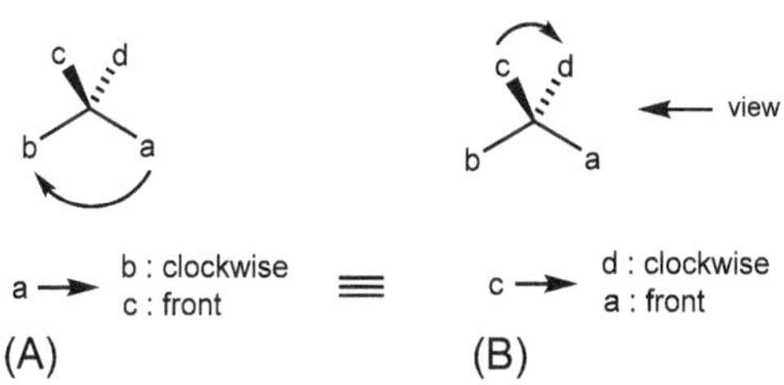


**FIG. 2.10** Configuration defined at a stereocentre in terms of relationships among three ligands; the geometry of the fourth ligand being automatically fixed in the tetrahedral space: (A) a and b are connected by a clockwise arrow when c is in front and (B) c and d are related by a clockwise arrow when a is in front. The configurations defined in (A) and (B) are equivalent.

the enantiomer on the left is defined as F→H clockwise/Cl front, while the enantiomer on the right as F→H anticlockwise/Cl front.

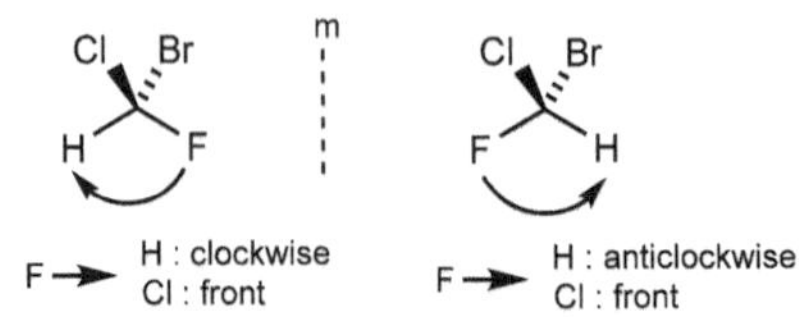


The number of changes in ligand relationships can also be used to delineate the inversion or retention of configuration at a stereocentre in reactions, as shown in Fig. 2.11A.[23] The reaction at a stereocentre usually involves the replacement of an old ligand with a new ligand, say a by a′. If the spatial disposition of the ligands a′, b, c and d in the product is the same as that of a, b, c and d in the reactant, there is retention and, if different, there is inversion. One change in ligand relationships implies inversion, whereas retention involves either no change or both changes.

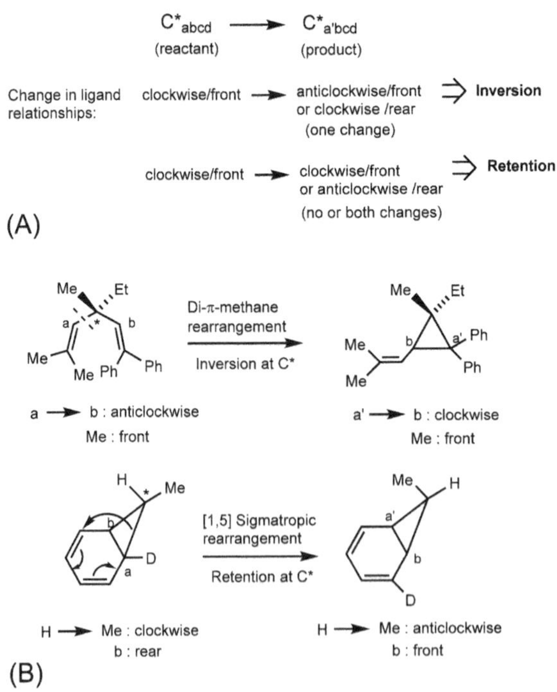


**FIG. 2.11** (A) Inversion or retention of configuration at a stereocentre based on the number of changes in ligand relationships from reactant to product and (B) examples showing delineation of product stereochemistry using the rules.

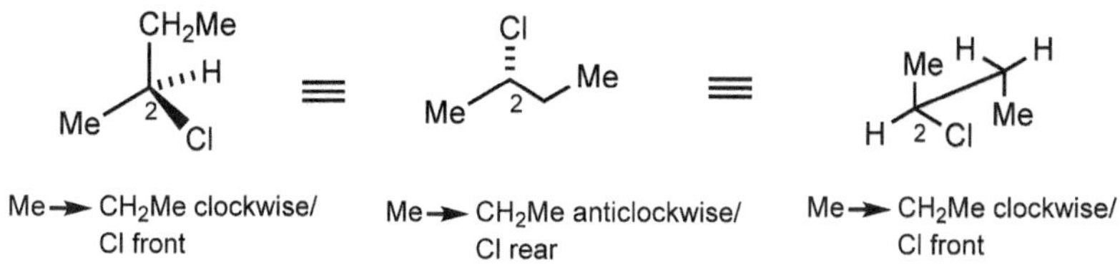


**FIG. 2.12** Representation of an enantiomer of 2-chlorobutane in equivalent flying wedge, zigzag and sawhorse formulas using the ligand relationships.

Fig. 2.11B illustrates the rules for delineating product stereochemistry as a result of inversion of configuration in a di-π-methane rearrangement and retention of configuration in a [1, 5] sigmatropic rearrangement.

Any stereocentre configuration can be depicted in several stereochemical formulas. For example, Fig. 2.12 shows that an enantiomer of 2-chlorobutane can be represented in equivalent flying wedge, zigzag and sawhorse formulas using the ligand relationships.

A stereocentre configuration in any stereochemical formula can also be transformed into an equivalent Fischer projection using a mnemonic[24] as shown in Fig. 2.13.

Thus the flying wedge formula in Fig. 2.12 can be transformed into a Fischer projection as.

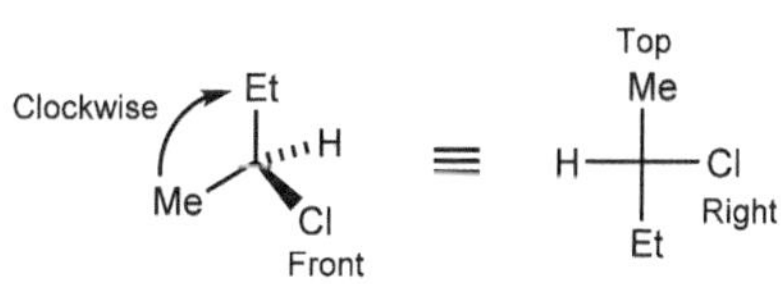


## 2.7 Absolute configuration and descriptors

Enantiomers have the same relative configuration but different absolute configuration. Thus, two enantiomers of a chiral compound with several stereocentres

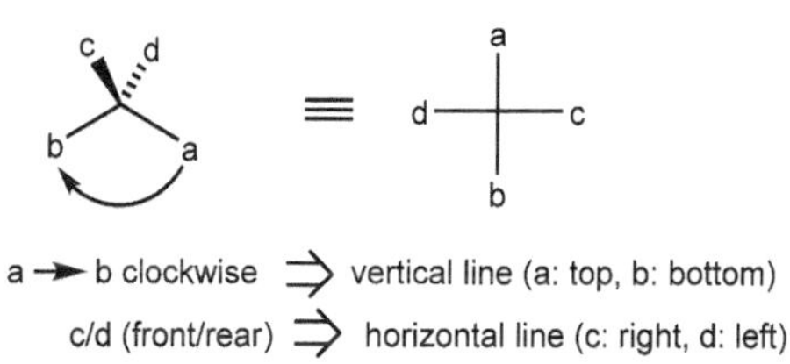


**FIG. 2.13** A mnemonic for transformation of a 3D formula into a Fischer projection.

*must have* opposite absolute configuration at *each* stereocentre. The absolute configuration is specified by a symbol (stereodescriptor). Two descriptors are needed to specify two possible configurations at a stereocentre.

### 2.7.1 The *R*/*S* system

The system of *R*/*S* descriptors is known as the Cahn-Ingold-Prelog (CIP) system[25,26], which is universally applicable to all stereocentres. Since a variety of ligands can be linked to stereocentres, the four ligands at a stereocentre are normalized on the basis of their priority (see below). The assignment of *R* or *S* to a stereocentre involves two steps:

- Establishing a priority order of the ligands using the CIP sequence rules
- Specifying a descriptor (*R* or *S*) using the CIP chirality rule

#### *2.7.1.1 Determination of ligand priority order*

The ligands attached to a stereocentre may differ in constitution or in configuration. The CIP sequence rules[25,26] for determining the priority order of ligands comprise a set of rules to be used *in turn*. The main features of the sequence rules are as follows.

1. **Atomic number**: Higher atomic number precedes lower atomic number.
2. **Isotopes**: Higher mass number precedes lower mass number (see Figs. 2.19 and 2.23C).
3. **Double bond configuration**: *cis* precedes *trans*. For the purpose of this rule, if the higher-priority ligand on the terminal alkene carbon and the chiral centre are on the same side of the double bond, the relationship is *cis*, and if they are on the opposite side, the relationship is *trans*. This is a modified rule by Prelog and Helmchen[26] (see Fig. 2.41B).
4. **Relative configuration and pseudoasymmetry**: *l* (like) precedes *u* (unlike). For two stereocentres, *l* implies *R*,*R* or *S*,*S* and *u* implies *R*,*S* or *S*,*R*. (see Fig. 2.38) For pseudoasymmetry, *r* precedes *s*.
5. **Enantiomorphic ligands**: *R* precedes *S*; *M* precedes *P* (see Figs. 2.37 and 2.38).

In the vast majority of organic molecules, the ligands differ only in constitution and their priority order can be determined by the **Sequence rule 1** based on atomic number: *higher atomic number precedes lower atomic number*. If the ligand atoms attached directly to the stereocentre are all different, for example in CHFClBr, the priority order is determined straightway (Br > Cl > F > H). However, when two or more atoms directly attached to the stereocentre are the same, the priority order is to be decided with the help of *subrules*. There are several subrules or schemes that have been developed, such as (i) tree-graph exploration[27]; (ii) use of complemented (duplicate or phantom) atoms for multiple bonds[25,27]; (iii) graphical flowchart scheme[28]; and (iv) ligand complementation when a ligand is polydentate or has a cyclic or bicyclic component[26,29]. The proper application of these subrules/schemes is not always simple, and it involves a lengthy and cumbersome procedure in many cases. Instead of those subrules, an alternative 'application rules' for the determination of ligand priority order is described here.[24]

### 2.7.1.2 *Mandal's approach to determining ligand priority*

This approach treats the ligands in their normal bonding connectivity. A ligand chain denotes a chain of first atom $A_1$ (directly linked to a stereocentre $C^*$) and outward atoms (second atom $A_2$, third atom $A_3$ and so on) as shown below. The first or an outward atom in a ligand chain is a carbon atom or a heteroatom.

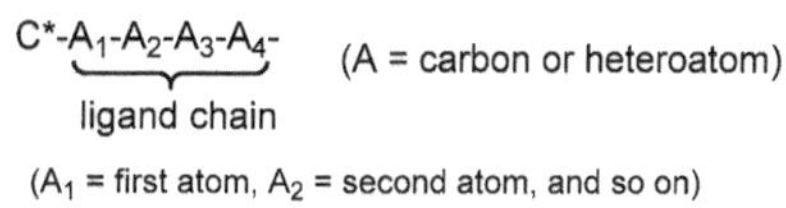


A decision on ligand priority is reached at the earliest possible atom in ligand chains based on the following 'application rules'.

**I.** C—X bonds: Count the number of C—O/C—N bonds at a similar carbon in ligands, taking a double bond as 2 bonds and a triple bond as 3 bonds. The rule is: *greater number of bonds gives higher priority*. (C—O bond gets precedence over C—N bond).

**II.** H atoms: Count the *least* number of H atoms attached to a similar atom in ligand chains. The rule is: *the fewer the number of H atoms, the higher is the priority*.

**III.** Chain length: If the rule of H atoms cannot provide a decision, then compare whether a ligand has the next outward atom (longer chain) or not (shorter chain). The rule is: *longer chain precedes shorter chain.*

These application rules are illustrated below with several sets of ligands for determining the priority order. The rules are used successively from I to II to III, wherever necessary.

First, consider a simple set of four ligands (OH, $CO_2H$, $CH_3$, H) attached to the chiral centre of lactic acid:

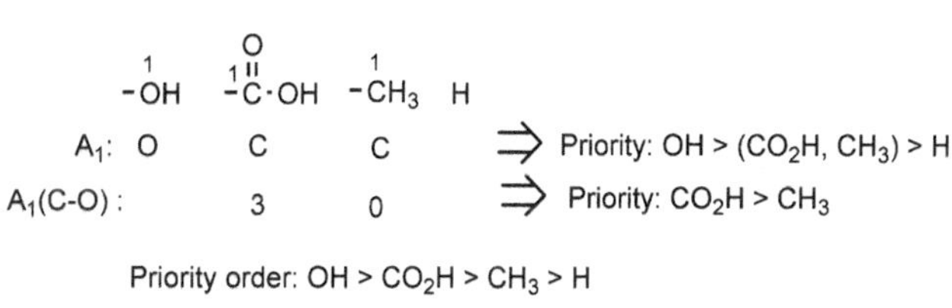

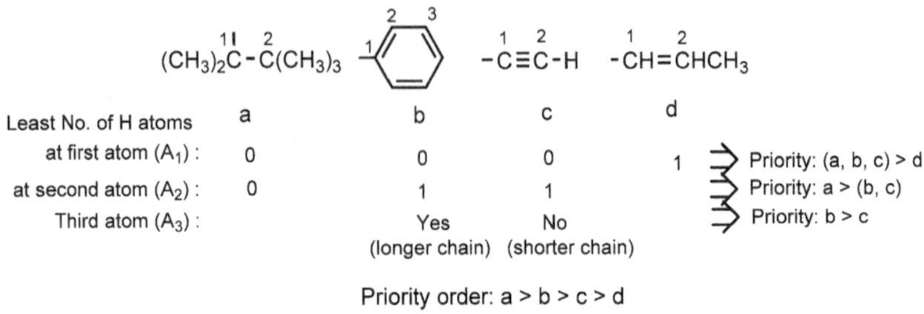


**FIG. 2.14** Ligand priority order using rule of H atoms (application rule II) and rule of chain length (application rule III).

Based on the atomic number of first atom ($A_1$) in ligands, the partial priority order is OH > ($CO_2H$, $CH_3$) > H. Note that H ligand has the lowest priority. Then the rule of C—X bonds (application rule I) gives the priority $CO_2H$ (number of C—O bonds = 3) > $CH_3$ (no C—O bond). Thus the overall priority order is OH > $CO_2H$ > $CH_3$ > H.

Fig. 2.14 shows a set of four ligands whose priority order is decided by the rule of H atoms (application rule II) and the rule of chain length (application rule III).

Another set of ligands is shown in Fig. 2.15, for which the priority order is determined by the application rules I and II.

(The above illustrations of the application rules are described in detail as an aid to understanding only; once understood, the ligand priority can be decided at a glance.)

If the ligand priority order can be obtained on the basis of atomic number of first/outward atom in ligand chains, no other rules need to apply as shown in Fig. 2.16. Note that, for ligands c and d, a third atom Br vs F (labelled 3′) decides the branch for the ligand chain and a fourth atom Br vs Cl (labelled 4) decides the relative priority as c > d.

If a lone pair ligand is present at a stereocentre, it has the least priority since it is regarded as an atom with atomic number zero.

In many cases, the ligand priority order can be simply determined on the basis of the rule of H atoms (application rule II). Three examples are shown in Fig. 2.17. Only one stereocentre (marked *) is considered in each case.

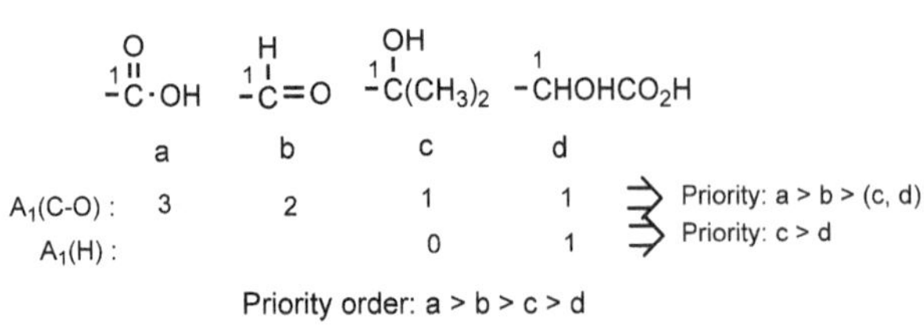


**FIG. 2.15** Ligand priority order using application rule I and application rule II.

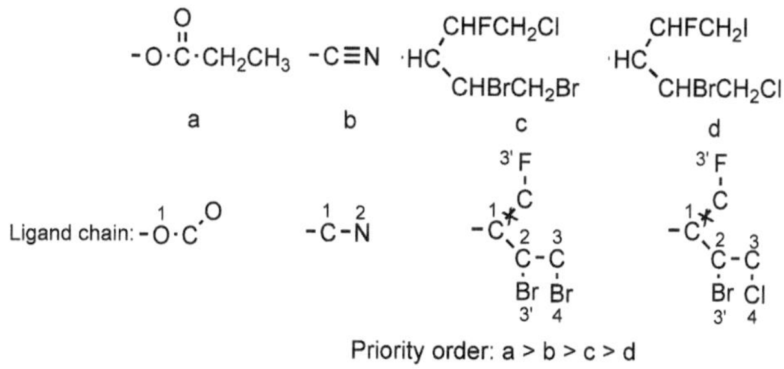

**FIG. 2.16** Ligand priority order from the atomic number of atoms in ligand chains.

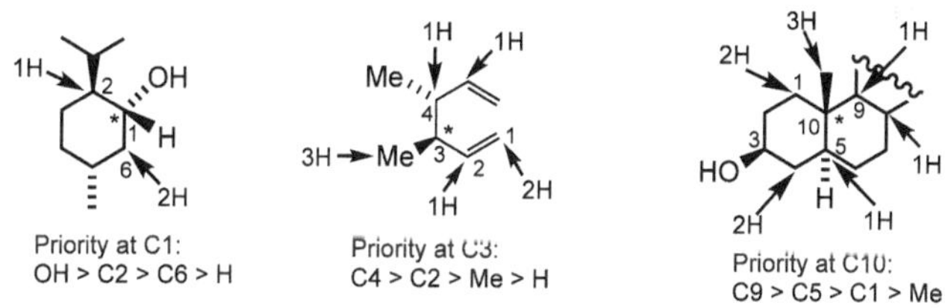

**FIG. 2.17** Ligand priority order using the rule of H atoms (application rule II) only.

Fig. 2.18 provides a rather complex example. Here the priority order is decided using all three application rules (I–III). Note that the chiral centre (C*) is not a member of the ligand chain in a and d.

It should be noted that H atoms in the application rule II includes hydrogen isotopes (D or T). If there is no decision, then the **Sequence rule 2** based on mass number, that is *higher mass number precedes lower mass number* will apply. This is illustrated using a set of three ligands (Pr, Et, $CD_2Me$) in Fig. 2.19.

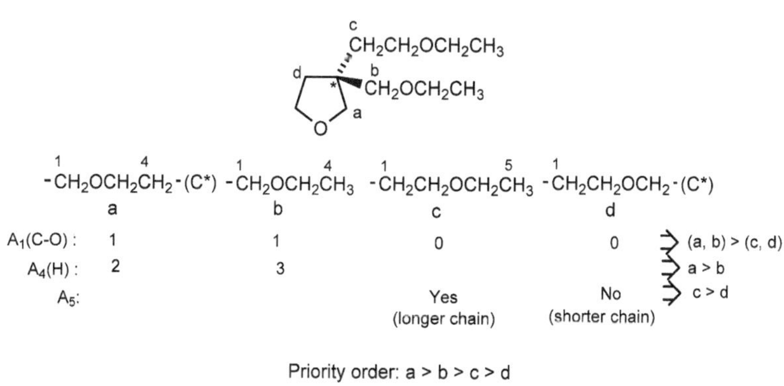

**FIG. 2.18** Ligand priority order using application rules I, II and III.

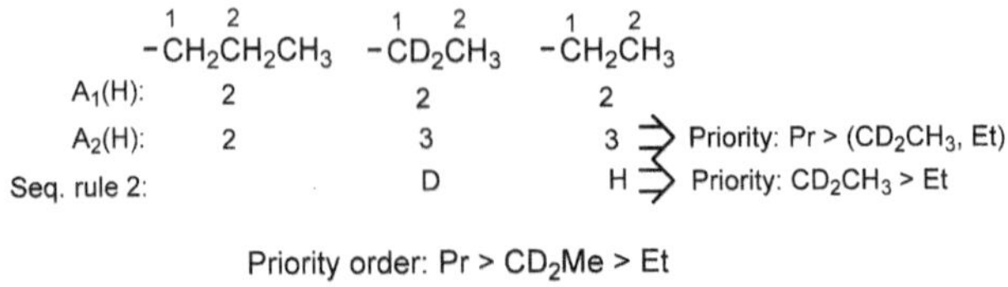


**FIG. 2.19** Ligand priority order using the application rule II and the sequence rule 2.

The stereocentres with ligands differing in configuration are much less common, and some select examples will be described later using the relevant sequence rules (see Sections 2.8.9 and 2.9.1.1).

### 2.7.1.3 *Assignment of* R/S *descriptor*

Once the priority order of four ligands attached to a stereocentre is established, the configurational descriptor *R* or *S* can be assigned using the **Chirality rule.**[25] Let the priority order of four ligands (a, b, c and d) attached to a stereocentre be a > b > c > d. According to the chirality rule, the absolute configuration (*R* or *S*) to the stereocentre is specified in terms of a model as described below.

The model requires that the molecule is viewed away from the lowest priority ligand (d). By the chirality rule, the configuration is *R* (from the Latin *rectus*, right) if a → b → c traces clockwise, and *S* (*sinister*, left) if a → b → c traces anticlockwise. The model resembles a steering wheel on which a, b, and c are placed while d is put on the shaft, the axis of the wheel being the bond between the chiral centre and d. Rotating the wheel clockwise or right corresponds to *R* while rotating the wheel anticlockwise or left corresponds to *S*. Fig 2.20A illustrates the assignment of *R*/*S* descriptors to the enantiomers of 2-butanol.

*Corollary*: If the molecule is viewed from the side of d; the opposite is true, that is the descriptor is *R* if the array a → b → c is anticlockwise, and *S* if a → b → c traces clockwise.

This is illustrated in Fig 2.20B with the natural enantiomer of the α-amino acid alanine whose configuration is assigned as *S*.

(The CIP chirality model and Brewster's model for optical rotation (see Section 2.4.5) are analogous; the former based on ligand priority gives the

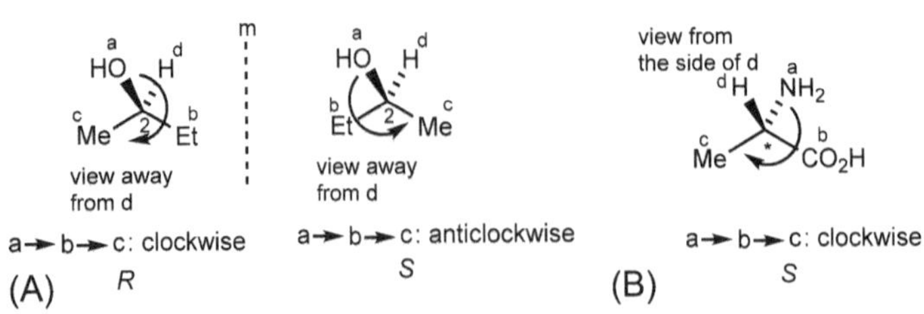


**FIG. 2.20** Assignment of *R*/*S* descriptor to (A) the enantiomers of 2-butanol and (B) an enantiomer of alanine.

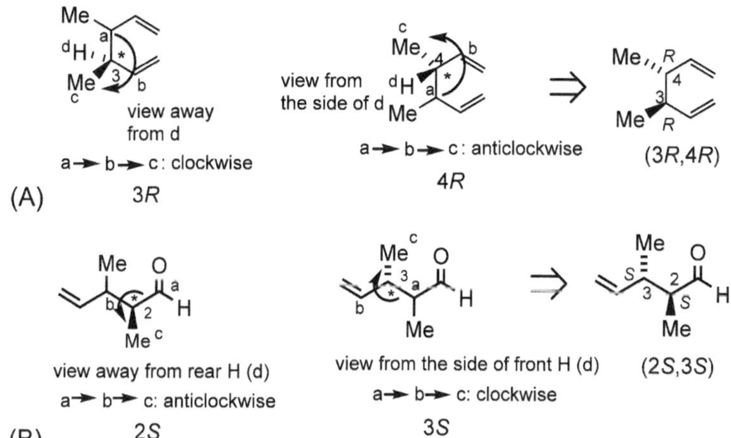


**FIG. 2.21** Assignment of *R*/*S* descriptors to an enantiomer of (A) 3,4-dimethyl-1,5-hexadiene and (B) 2,3-dimethyl-4-pentenal.

symbol for absolute configuration while the latter based on ligand polarizability predicts the sign of optical rotation.)

Fig. 2.21 illustrates the assignment of *R*/*S* descriptors to molecules containing two chiral centres. In Fig. 2.21A, the molecule is viewed away from the lowest priority ligand at C3 when the configuration is *R*, and from the side of the lowest priority ligand at C4 when the configuration is also *R*. The absolute configuration at C3 and C4 for the enantiomer of 3,4-dimethyl-1,5-hexadiene is thus determined to be (3*R*,4*R*). In Fig. 2.21B, the molecule is viewed away from the lowest priority ligand at C2 when the configuration is *S*, and from the side of the lowest priority ligand at C3 when the configuration is also *S*. Note that the lowest priority ligand (H) at the chiral centres is not shown in the zigzag formula. The descriptors at C2 and C3 determined for the enantiomer of 2,3-dimethyl-4-pentenal is (2*S*,3*S*).

If a stereocentre is represented in a Fischer projection, the assignment of *R*/*S* descriptor is straightforward and simple. If the lowest priority ligand is on the *vertical* line, the array of the remaining three in descending priority sequence gives the correct descriptor, *R* for clockwise and *S* for anticlockwise; but if the lowest priority ligand is on the *horizontal* line, the opposite assignment is correct, that is *R* for anticlockwise and *S* for clockwise. This is shown for a general case Cabcd (priority order: a > b > c > d) in Fig. 2.22A. Fig. 2.22B illustrates two examples using Fischer projections.

Conversely, for a stereoisomer with the descriptor *R* or *S*, equivalent Fischer projections or flying wedge formulas can be drawn as shown explicitly in Fig. 2.23. Similarly, stereoisomers with chiral P and S atoms are represented in Fig. 2.23B and Fig. 2.23C respectively.

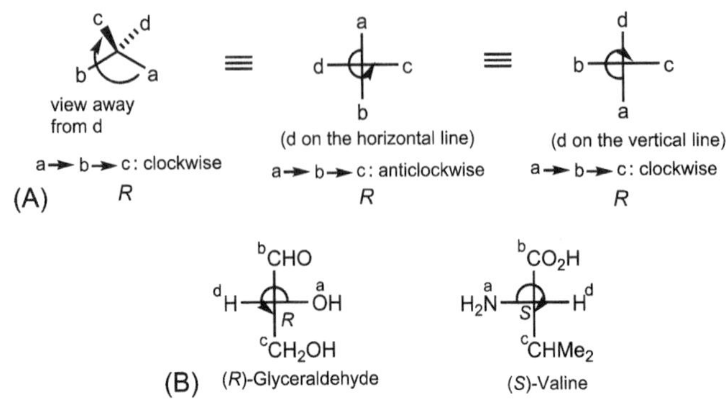


FIG. 2.22 (A) Assignment of *R*/*S* descriptor to a stereocentre in the Fischer projection for a general case and (B) two examples.

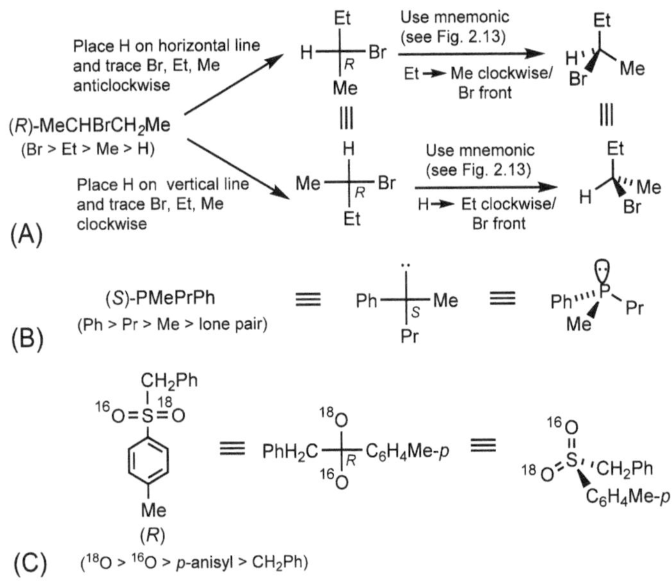


FIG. 2.23 Ficher Projections and flying wedge formulas for (A) (*R*)-2-bromobutane, (B) (*S*)-methylphenylpropylphosphine and (C) (*R*)-benzyl *p*-tolyl sulphone ($^{18}O$ labelled).

The stereocentres can also be represented in other projection or perspective formulas. However, the direct assignment of descriptors in those formulas is often difficult. Here a unifying procedure[24] via the Fischer projection (see the mnemonic in Fig. 2.13) is illustrated in Fig. 2.24.

In each example, there are two stereocentres of which one is assigned the descriptor. Note that the ligands are labelled a, b, c and d, with the priority a>b>c>d. Fig. 2.24A shows the sawhorse formula of a stereoisomer of

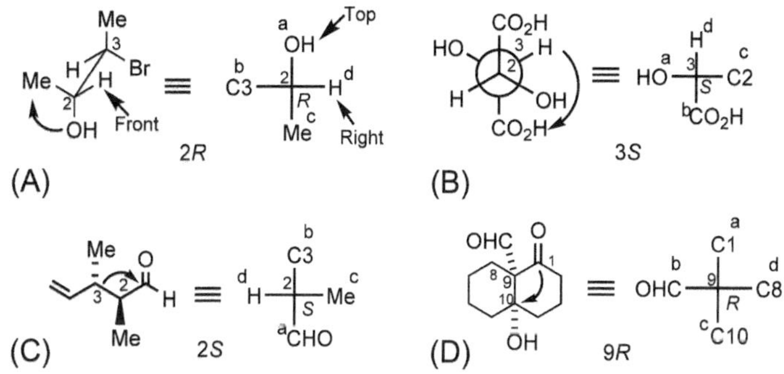


**FIG. 2.24** Assignment of *R*/*S* descriptor to a stereocentre in (A) sawhorse, (B) Newman, (C) zigzag and (D) planar cyclic representations via a Fischer projection drawn using the mnemonic in Fig. 2.13.

3-bromo-2-butanol. The stereocentre at C2 is transformed into a Fischer projection using the mnemonic, which gives the descriptor *R* at C2. The Newman projection of a stereoisomer of tartaric acid is shown in Fig. 2.24B. Transforming to a Fischer projection gives the descriptor *S* at C3 (back carbon). Fig. 2.24C shows the zigzag formula of a stereoisomer of 2,3-dimethyl-4-pentenal. The configuration at C2 is determined as *S* from the Fischer projection. In the planar cyclic formula of a decalone derivative (Fig. 2.24D), the stereocentre C9 has the configuration *R* as obtained from the Fischer projection.

Finally, we consider a chiral spirane shown in Fig. 2.25. The descriptors for the stereocentres C1 and C6 in the enantiomer **A** are assigned easily as 1*S* and 6*R*. The spirocarbon C5 is also a stereocente but the two rings attached to it are constitutionally same. Thus, to assign the descriptor at C5, one ring is *arbitrarily* given preference over the other.[30] Then, in the priority sequence a > b > c > d, the more substituted branch in the *preferred* ring is given the priority a and the less substituted the priority c, while in the *other* ring, the more substituted will get the priority b and the less substituted the priority d as shown in Fig. 2.25. The descriptor for C5 in **A** is thus assigned as 5*S* (a → b → c is anticlockwise, viewing away from d). Similarly, the enantiomer **A′** has the absolute configuration (1*R*,5*R*,6*S*).

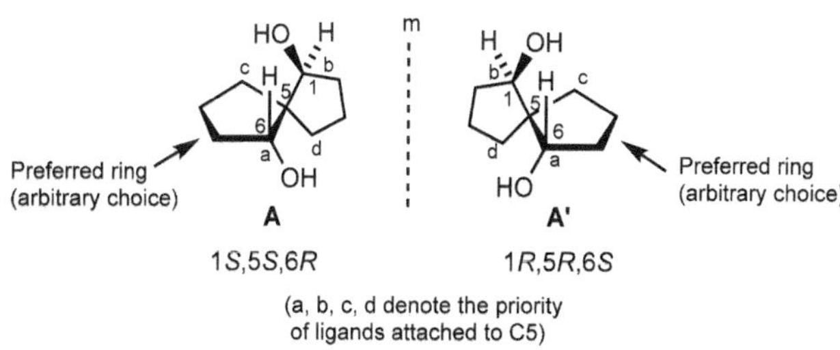


**FIG. 2.25** Assignment of *R*/*S* descriptors to a chiral spirane.

**Problem 2.5** Assign *R*/*S* descriptors to the following molecules:

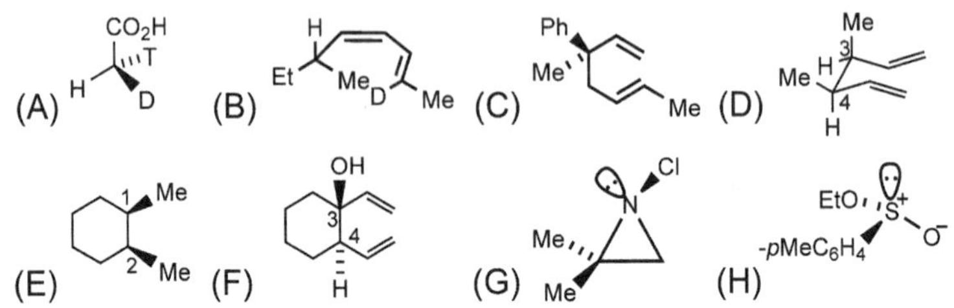


## 2.7.2 The D/L system

The D/L system[31] of configurational nomenclature is used nowadays only for α-amino acids and carbohydrates, and is defined in a Fischer projection. The vertical line denotes the main carbon chain with C1 (IUPAC numbering) at the top. The configuration is D if $NH_2$ (α-amino acids) or OH at the *highest numbered* chiral centre (carbohydrates) is on the horizontal right, and the configuration is L when $NH_2$ or OH is on the horizontal left. Fig. 2.26A illustrates the D/L configuration of α-amino acids taking alanine as an example. Only L-amino acids are the building blocks for proteins, and L ≡ *S*. An aldose monosaccharide is a polyhydroxy aldehyde. The simplest aldose is an aldotriose, glyceraldehyde. The D/L configuration of glyceraldehyde, threose (an aldotetrose),

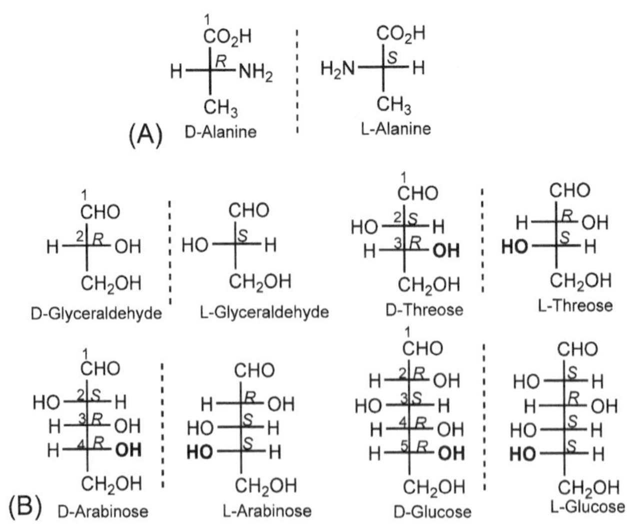


**FIG. 2.26** D/L configuration for α-amino acids and aldose monosaccharides. The OH at the highest numbered chiral centre that decides the configuration to be D or L is shown in bold in threose, arabinose and glucose.

arabinose (an aldopentose) and glucose (an aldohexose) is illustrated in Fig. 2.26B. The descriptor *R* or *S* is also assigned to each chiral centre.

### 2.7.3 Experimental methods for determining absolute configuration

Two enantiomers of a chiral compound have opposite optical rotations + and − (experimental quantity) and also opposite absolute configurations *R* and *S* at a stereocentre (theoretical descriptors). If one enantiomer (+) has *R* configuration, then the other enantiomer (−) must have *S* configuration. But the question is: which enantiomer has which absolute configuration? There was no simple method to solve this problem experimentally. In fact, in 1888, Fischer in his work on the stereochemistry of (+)-glucose *arbitrarily assumed* the absolute configuration of (+)-glyceraldehyde as D (≡ *R* in CIP notation), and thereby determined the D configuration of (+)-glucose. Thereafter, absolute configurations of many chiral compounds were determined by chemical correlation with (+)- or (−)-glyceraldehyde. It was not until 1951 when Bijvoet[32] determined the absolute configuration of (+)-tartaric acid to be L (≡ *R*,*R*) from the anomalous X-ray scattering of its sodium rubidium salt **2.4**. The chemical correlation of L-(+)-tartaric acid with (+)-glyceraldehyde confirmed that the absolute configuration of (+)-glyceraldehyde is D (Fig. 2.27). Fischer's assumption was thus found to be fortunately correct. As shown in Fig. 2.27, the integrity of the chiral centre C2 is maintained since no bond to C2 is affected by the chemical transformations. The absolute configuration (+)-glyceraldehyde is established by its oxidation with HgO to give (−)-glyceric acid as obtained from L-(+)-tartaric

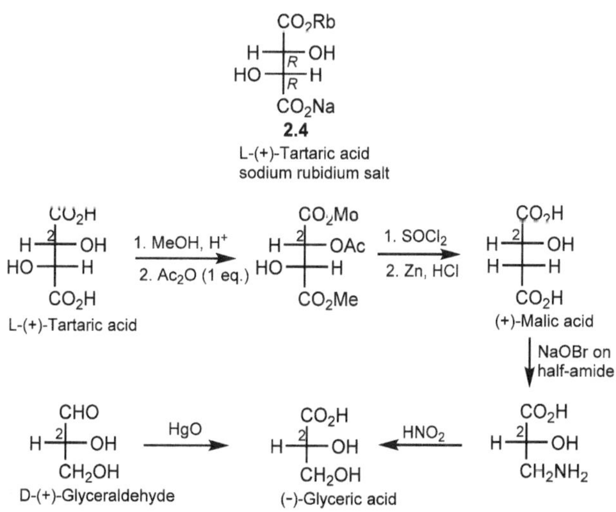


**FIG. 2.27** Chemical correlation of (+)-glyceraldehyde with L-(+)-tartaric acid. The absolute configuration of (+)-tartaric acid was determined by X-ray crystallography of its sodium rubidium salt **2.4**.

acid by the sequence of transformations. Note that acetylation of OH at C2 or C3 gives the same product (the two possible products being equivalent by rotation of 180 degrees in the plane of the paper). Further, optical rotation by itself cannot provide its absolute configuration since oxidation of (+)-glyceraldehyde gives (−)-glyceric acid although both have the same configuration.

X-ray crystallography is the most reliable general method for the determination of absolute configuration of a chiral compound provided that suitable crystals from enantiomerically pure sample are available. The determination is based on the anomalous X-ray scattering caused by a heavy atom in the crystal (zirconium was used as a heavy atom in the experiment of Bijvoet et al.[32]). The anomalous dispersion refers to the pairs of reflections that are equally intense for an achiral crystal become unequal for a chiral crystal and the signs of inequalities relate to the absolute configuration of the molecules.

The methods to determine absolute configuration also include chiroptical methods such as electronic and vibrational circular dichroism[33], vibrational and Raman optical activity[34] and the comparison of experimentally and theoretically computed chiroptical data[35,36]. The advantage of these methods is that they do not require chiral crystals. Indirect methods are also available which determine absolute configuration based on the known absolute configuration of the reference compound. Besides chemical correlation[37,38], these include NMR anisotropy[39] using chiral shift reagents and chiral solvents, and enzymatic methods.[38] The most useful method for determining the structure in solution is 2D NMR.[40]

## 2.8 Relative configuration and descriptors

If there are two or more stereocentres in a molecule, the configurations at the stereocentres with respect to each other denote the relative configuration of the molecule. The term relative configuration is also used in the context of reaction stereochemistry. This refers to retention or inversion of configuration at a stereocentre undergoing a change (see Fig. 2.11), which will be described later in Part III.

### 2.8.1 Molecules with two unlike stereocentres (AB type)

Consider a molecule with two unlike stereocentres, for example 3-bromo-2-butanol ($CH_3CHOHCHBrCH_3$). The two stereocentres C2 and C3 are unlike (AB type) as the set of ligands attached to C2 (OH, $CHBrCH_3$, $CH_3$, H) is different from the set at C3 (Br, $CHOHCH_3$, $CH_3$, H). Since each stereocentre has two possible configurations (*R* and *S*), the two stereocentres can generate four possible configurations giving four stereoisomers (**A**, **A′**, **B**, **B′**) as shown below.

$$H_3\overset{2}{C}HOH\overset{3}{C}HBrCH_3$$

| Stereocentre: A | B | Stereoisomer | |
|---|---|---|---|
| *R* | *R* | **A** | Enantiomers |
| *S* | *S* | **A'** | |
| *R* | *S* | **B** | Enantiomers |
| *S* | *R* | **B'** | |

(**A**, **A′**) and (**B**, **B′**) indicate two enantiomeric pairs. Note that *any pair of enantiomers has opposite configuration at both stereocentres*. Now, consider the relationship between a member of one enantiomeric pair with a member of the other pair. The relationship between **A** (*R*,*R*) and **B** (*R*,*S*) is diastereomeric since they differ in configuration at one stereocentre and hence are not mirror images of each other. Similarly, **A** and **B′** are diastereomeric, as also **A′** and **B**, and **A′** and **B′**. Thus, a particular stereoisomer can be diastereomeric with more than one stereoisomer but is enantiomeric with only its mirror image stereoisomer. Notably, diastereomeric relationship occurs between two stereoisomers which differ in configuration at one or more (but not all) stereocentres.

In general, to determine the possible stereoisomers of any compound containing multiple stereocentres, it is convenient to work out a set of distinct diastereomers (chiral/achiral) in which each chiral diastereomer would exist as a racemate that can be resolved into two enantiomers. Thus 3-bromo-2-butanol has two diastereomers, say, **A** and **B**, each of which is chiral. The two chiral diastereomers then provide two racemates, that is enantiomeric pairs (**A**, **A′**) and (**B**, **B′**), thereby accounting for all four stereoisomers of the compound. Fig. 2.28 shows the Fischer projections for the stereoisomers of 3-bromo-2-butanol indicating *R*/*S* descriptor to C2 and C3. (Remember that the *R*/*S* assignment to just one stereoisomer, say **A**, is needed using CIP rules, because the *R*/*S* descriptors to other stereoisomers will follow by an exchange of ligands at a stereocentre which changes *R* to *S* or *S* to *R*.)

In contrast to enantiomers, the interatomic distances in diastereomers are different. They have therefore different energies and physical properties, which allow for their easy separation and purification by conventional methods.

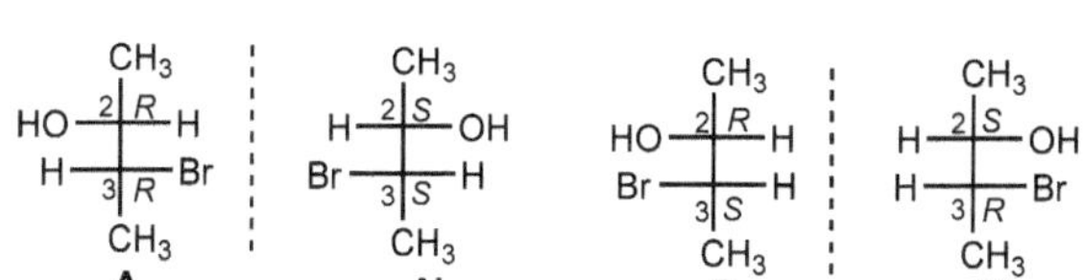


Priority at C2: OH > C3 > $CH_3$ > H

Priority at C3: Br > C2 > $CH_3$ > H

**FIG. 2.28** Fischer projections of the stereoisomers of 3-bromo-2-butanol.

### 2.8.2 Stereogenicity and chirotopicity

The tetrahedral centres in a chiral or achiral stereoisomer have two distinct attributes known as stereogenicity and chirotopicity.[3]

If exchange of any two ligands at a tetrahedral centre results in a different stereoisomer, the centre is stereogenic (see Fig. 2.2), otherwise it is nonstereogenic.

Chirotopicity is determined by local symmetry as to whether or not an atom lies on a symmetry plane ($\sigma$), centre of inversion ($i$) or $S_n$ ($n > 2$) axis. A chirotopic atom is any atom that resides in a local chiral environment. *All* atoms in a *chiral molecule* are necessarily chirotopic. On the other hand, an achirotopic atom lies on $\sigma$, $i$, or $S_n$ axis.

Three kinds of tetrahedral centres in stereoisomers are possible: (a) chirotopic and stereogenic, (b) chirotopic but nonstereogenic and (c) achirotopic but stereogenic.

In each stereisomer of 3-bromo-2-butanol (see Fig. 2.28), C2 and C3 are stereogenic and both reside in a chiral environment in the chiral molecule. Therefore, C2 and C3 are chirotopic and stereogenic. Such *chirotopic and stereogenic centres are called chiral centres.* Note that all atoms of 3-bromo-2-butanol are chirotopic.

For other kinds of $sp^3$ centres, see later Section 2.8.9.

### 2.8.3 *l* (like)/*u* (unlike) and *pref*/*parf* descriptors

There are many proposals by which the descriptors for relative configuration of diastereomers can be assigned, some of which are described in this and the following sections.

The *l* (like) and *u* (unlike) descriptors[41] for relative configuration are based on the CIP system. The two chiral centres are like for *R*,*R* or *S*,*S* and unlike for *R*, *S* or *S*,*R*. In this notation, the descriptor for the diastereomer **A** (2*R*,3*R*) of 3-bromo-2-butanol (see Fig. 2.28) is *l* and that for **B** (2*R*,3*S*) is *u*. The method also applies to molecules containing several chiral centres with reference to the configuration of the lowest numbered chiral centre. For example, D-(+)-glucose has the absolute configuration 2*R*,3*S*,4*R*,5*R* (see Fig. 2.26). The relationship between the reference chiral centre C2 (lowest numbered) and other chiral centres is *u* (2*R*,3*S*), *l* (2*R*,4*R*) and *l* (2*R*,5*R*). The same *l*/*u* relationship necessarily applies to L-(−)-glucose. Therefore the *l*/*u* descriptor for (+), (−) or racemic glucose is *ull*. The *l*/*u* descriptors for relative configuration of diastereomers are concise but need the *R*/*S* descriptors for the chiral centres to be determined by the CIP rules.

The *pref*/*parf* descriptors[42] for diastereomers are also based on the CIP rules. In this nomenclature, the ligands in a staggered conformation are viewed from the side of the bond joining the chiral centres. If the three ligands in the CIP sequence at each chiral centre appear in the clockwise (*Re*) or anticlockwise

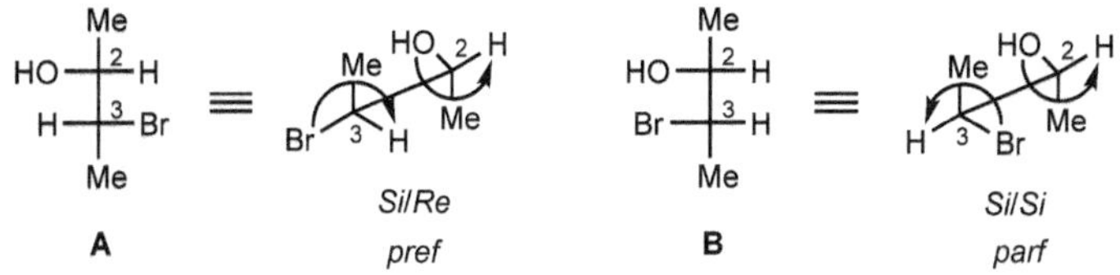


**FIG. 2.29** *pref/parf* Descriptors for the diastereomers of 3-bromo-2-butanol.

(*Si*) order, the relative configuration may then be specified as *Re/Re* (or *Si/Si*) for one diastereomer and *Re/Si* (or *Si/Re*) for the other. The latter configuration (*Re/Si* or *Si/Re*) is called *pref* (priority reflective) and the former (*Re/Re* or *Si/Si*) is denoted by *parf* (priority antireflective). Note that *Re* and *Si* are reflective like *R* and *S*. Thus, for 3-bromo-2-butanol, the diastereomer **A** is *pref* and **B** is *parf* (Fig. 2.29).

From the practical perspective of a synthetic chemist, the *l/u*, *pref/parf* and other descriptors based on the CIP rules are not particularly convenient and have not been generally embraced, and will not be described further in this text.

## 2.8.4 *erythro/threo* descriptors

For acyclic diastereomers particularly with two chiral centres, simpler ways have been devised to specify relative configuration without the need for specifying absolute configuration at the individual chiral centres using the CIP rules. These descriptors such as *erythro/threo* and *syn/anti* are called soft descriptors. These are simple to use but not unambiguous. The *erythro/threo* descriptors are based on the configuration of four-carbon sugars erythrose or threose in the Fischer projection (Fig. 2.30A). Consider a general molecule RCHXCHYR′ in which R—C—C—R′ is the main carbon chain, and X and Y are similar heteroatom containing ligands. If a diastereomer has two identical or similar

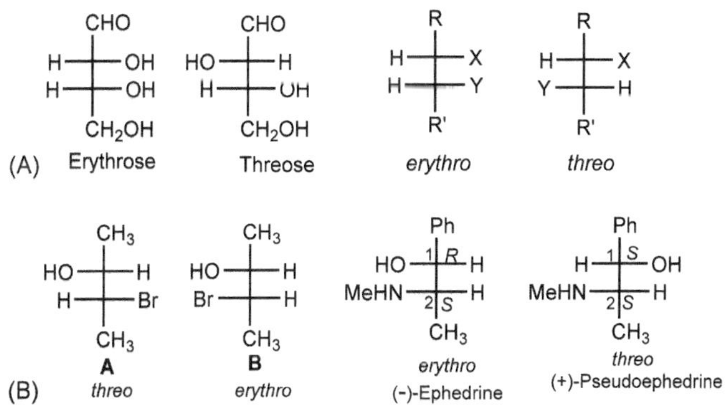


**FIG. 2.30** (A) *Erythro/threo* descriptors based on the sugars *erythrose* and threose and (B) illustrations of *erythro/threo* nomenclature.

ligands on the same side of the Fischer projection as in erythrose, it is called *erythro*, and if on the opposite sides as in threose, it is called *threo* (Fig. 2.30A). Thus, for 3-bromo-2-butanol, diastereomer **A** is *threo* and **B** is *erythro* (Fig. 2.30B). It should be noted that these descriptors apply to any enantiomer or racemic form of a diastereomer. Fig. 2.30B also shows two natural chiral diastereomers. The *erythro* diastereomer is ephedrine and the *threo* is pseudoephedrine. (−)-Ephedrine (1*R*,2*S*) and (+)-pseudoephedrine (1*S*,2*S*) are the natural enantiomers obtained from plants.

### 2.8.5 Molecules with two like stereocentres (AA type)

A representative example of molecules with two like stereocentres is tartaric acid ($HO_2CCHOHCHOHCO_2H$). The stereocentres C2 and C3 are like (AA type) as the set of ligands attached to the stereocentres is the same (OH, $CO_2H$, $CHOHCO_2H$ and H). Like 3-bromo-2-butanol, tartaric acid has also two diastereomers but here one diastereomer is chiral and the other is achiral as shown in Fig. 2.31. The chiral diastereomer is *threo* that exists as a resolvable enantiomeric pair (+, −). We have seen earlier that absolute configuration of (+)-tartaric acid is *R*,*R* (see Fig. 2.27). Therefore, (−)-tartaric acid is *S*,*S*. It is of note that C2 and C3 in the enantiomers are stereogenic and chirotopic. The *erythro* diastereomer is achiral due to the presence of a plane of symmetry ($\sigma$). (The Fischer projection of the *erythro* isomer is also superposable with its mirror image by rotation of 180 degrees in plane). The absolute configuration of the *erythro* isomer is *R*,*S* which indicates that optical rotation contributions from the two like chiral centres in the molecule will be of same magnitude but opposite in sign, and thereby cancel each other. The achiral isomer is thus said to be optically inactive by internal compensation (contrast with a racemate which is optically inactive by external compensation). The optically inactive achiral diastereomer is called *meso*. The stereoisomers of tartaric acid are usually referred to as +, − and *meso*. Notably, C2 and C3 in the *meso* isomer are also stereogenic and chirotopic (since they do not lie on the symmetry plane but reside in a local chiral environment) and are chiral centres. Thus the presence of two (or more) chiral centres in a molecule does not imply that the molecule is necessarily chiral; it can well be achiral.

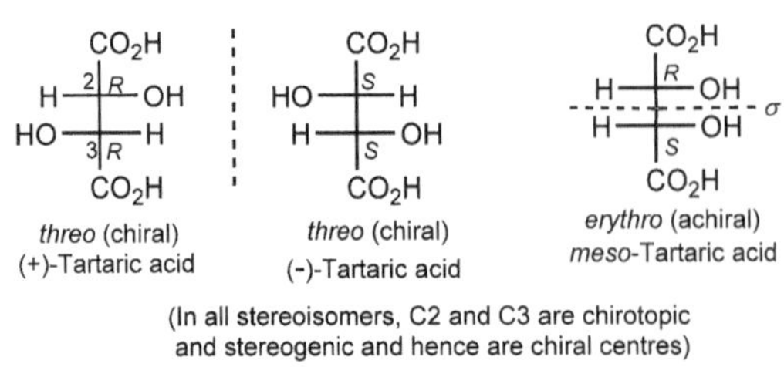


**FIG. 2.31** Stereoisomers of tartaric acid containing two like stereocentres (chiral centres).

### 2.8.6 Diastereomer conformation and *erythro/threo* or *syn/anti* descriptors

A Fischer projection of a diastereomer always represents an eclipsed conformation (see Fig. 1.3B). Thus the Fischer projections of the diastereomers shown earlier in Figs. 2.28, 2.30 and 2.31 denote eclipsed conformations corresponding to energy maxima. Rotation about the single bond transforms an eclipsed conformation (torsion angle $\phi = 0$ degrees) to a staggered conformer ($\phi = +60$ degrees, $-60$ degrees or 180 degrees) for a diastereomer. Fig. 2.32A illustrates a case with *threo* and *erythro* diastereomers of RCHXCHXR′. From the eclipsed conformation ($\phi = 0$ degree) to the staggered conformer ($\phi = 180$ degrees), the ligands (X's) on opposite sides of the main chain (R—C—C—R′) plane go to the same side and vice versa. The staggered sawhorse formulas can be transformed to zigzag formulas to which *syn/anti* descriptors[43] are defined: two substituents oriented in the same direction (*syn*) or in the opposite directions (*anti*). As shown, *threo* corresponds to *syn*, and *erythro* (or *meso* for R = R′) corresponds to *anti*. Between *syn* and *anti*, *anti* is sterically more stable than *syn* or in other words, *erythro* is more stable

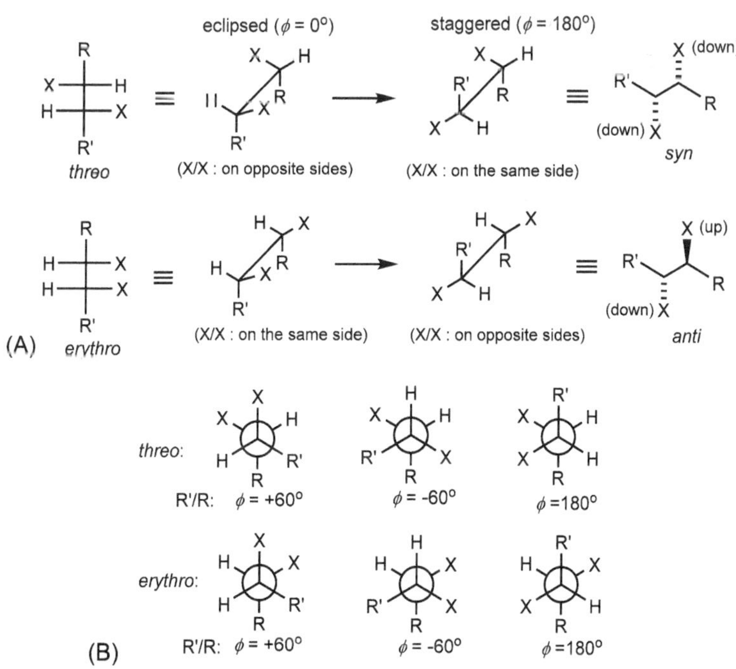


**FIG. 2.32** (A) Correlation of *threo/erythro* with *syn/anti* descriptors for RCHXCHXR′ and (B) Newman projections for all three conformers of *threo* or *erythro* diastereomer.

than *threo*. Notably, *syn*/*anti* descriptors apply only to the staggered zigzag formula and not to any eclipsed formula.

The Newman projections of all three staggered conformers ($\phi = \pm$ 60 degrees, 180 degrees for R′/R) for *threo* or *erythro* diastereomer are shown in Fig. 2.32B. It should be noted that, of three similar pairs (R/R′, X/X, H/H), two pairs are gauche and the remaining pair is anti in *threo*, while all three pairs are either gauche or anti in *erythro*.

Fig. 2.33 shows the assignment of *syn*/*anti* descriptors (cf. Fig. 2.32A) for molecules containing unlike and like chiral centres. Note carefully how the zigzag main chain is drawn so that you do not end up in drawing a wrong enantiomer! The *meso*-tartaric acid is *anti* with a centre of inversion (*i*), and is achiral.

The translation of eclipsed wedge formulas for *threo*/*erythro* isomers to staggered zigzag formulas with *syn*/*anti* descriptors is shown in Fig. 2.34. The main carbon chain (1-2-3-4) which is eclipsed in the wedge formula becomes staggered in the zigzag formula by rotation about 2-3 bond as shown.

**Problem 2.6** Draw a correct zigzag formula for the following enantiomer and designate by the *syn*/*anti* descriptor.

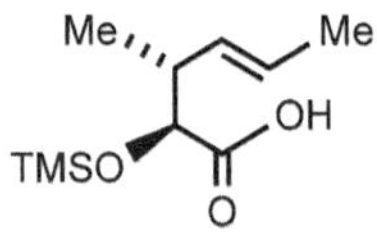


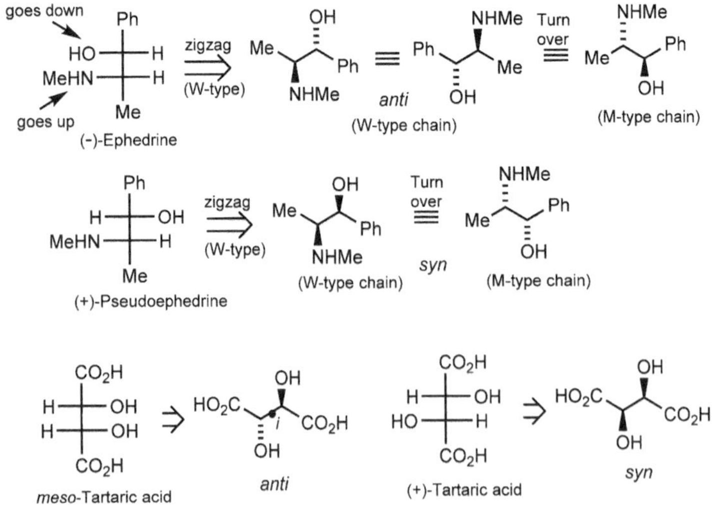


**FIG. 2.33** Assignment of *syn*/*anti* descriptors (cf. Fig. 2.32A) for molecules containing two unlike and two like chiral centres.

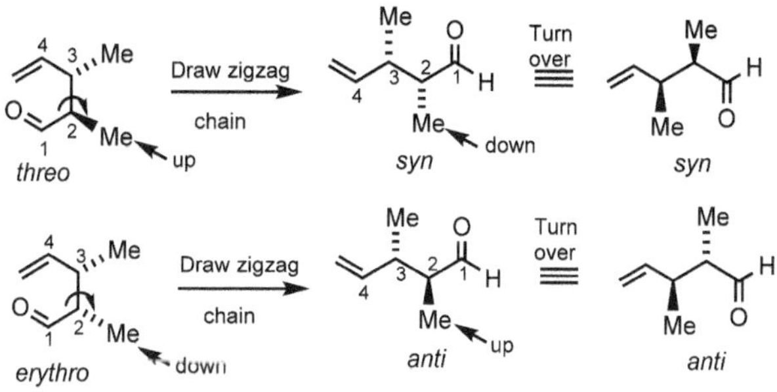


**FIG. 2.34** Translation of eclipsed flying wedge formulas to staggered zigzag formulas and the consequent change in the descriptor: *threo* to *syn*, *erythro* to *anti*.

## 2.8.7 Molecules with three and four unlike stereocentres (ABC and ABCD types): Aldopentose and aldohexose families

In general, a molecule with $n$ unlike stereocentres has $2^{n-1}$ diastereomers, each of which is chiral and exists as an enantiomeric pair. Therefore, the total number of stereoisomers is $2^n$. Here all stereocentres are chiral centres.

To illustrate, we will consider examples from the family of sugars. The aldopentose family ($HOCH_2CHOHCHOHCHOHCHO$) contains three chiral centres C2, C3 and C4 that are unlike (ABC type) and the aldohexose family ($HOCH_2CHOHCHOHCHOHCHOHCHO$) contains four unlike chiral centres C2, C3, C4 and C5 (ABCD type). Therefore, the aldopentose has $2^{3-1}=4$ chiral diastereomers and the aldohexose has $2^{4-1}=8$ chiral diastereomers. Fig. 2.35A shows the D-family of aldopentoses. (The L-family constitutes the mirror image enantiomers.) Aldopentoses can be obtained by stepping up aldotetroses. As shown, D-erythrose gives D-ribose and D-arabinose, whereas D-threose gives D-xylose and D-lyxose. The stereochemistry of sugars is usually drawn in eclipsed Fischer projections but the more realistic representation is the staggered zigzag formula. In the D-family, the highest numbered stereocentre (C4 for aldopentose) has *R* configuration (see Section 2.7.2). As shown in Fig. 2.35A, a Fischer projection can be translated into a zigzag formula first by drawing 4*R* (OH down) and then making *threo* → *syn* and *erythro* → *anti* correspondence. In case of D-ribose, 4,3-*erythro* and 3,2-*erythro* lead to 4,3-*anti* and 3,2-*anti*. For D-arabinose, 4,3-*erythro* and 3,2-*threo* lead to 4,3-*anti* and 3,2-*syn*. Similarly, the zigzag formulas of D-xylose and D-lyxose are drawn. The above procedure serves only as an aid to drawing a zigzag formula. But as descriptors, *syn*/*anti* are assigned to the aldoses with respect to the reference centre C2. Thus ribose (+, − or racemate) is 2,3-*anti*,2,4-*syn*; arabinose is 2,3-*syn*,2,4-*anti*, and so on (Fig. 2.35A). Remember that the relative descriptors *syn*/*anti* apply to any enantiomer or racemate of a diastereomer.

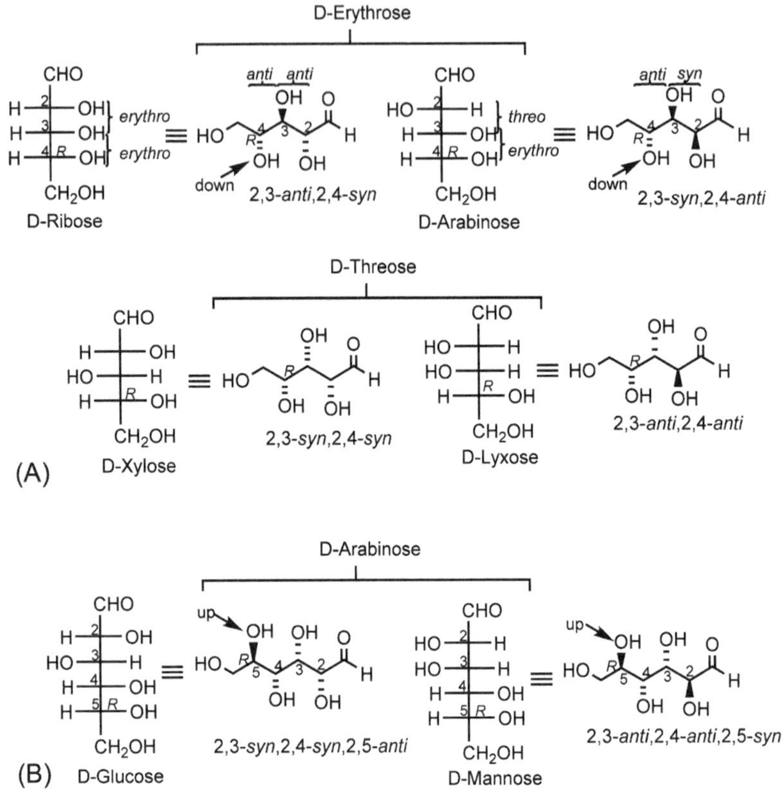


**FIG. 2.35** Fischer projections and zigzag formulas for (A) D-family of aldopentoses and (B) D-glucose and D-mannose in the D-aldohexose family.

The D-family of aldohexoses comprising eight diastereomers can be obtained by stepping up D-aldopentoses. For example, D-glucose and D-mannose are obtained by stepping up D-arabinose as shown in Fig. 2.35B. The staggered zigzag formulas of D-glucose and D-mannose are drawn in a similar manner as described for D-aldopentoses with *R* configuration at C5 (OH up). In terms of *syn*/*anti* descriptors, glucose (+, − or racemate) is 2,3-*syn*,2,4-*syn*,2,5-*anti*, and mannose is 2,3-*anti*,2,4-*anti*,2,5-*syn*.

The other six diastereomers of the D-aldohexose family can also be obtained by the stepping up procedure: D-ribose gives D-allose and D-altrose, D-xylose gives D-gulose and D-idose, and D-lyxose gives D-galactose and D-talose. All D-aldohexoses are dextrorotatory (+), except D-gulose and D-idose which are levorotatory (−). It may be easy to remember the sequence in the aldohexose family allose, altrose, glucose, mannose, gulose, idose, galactose, talose using the mnemonic '*all altr*uists *gl*adly *ma*ke *gu*m *i*n *gal*lon *ta*nks'.

**Problem 2.7** Draw Fischer projections and zigzag formulas for D-aldohexoses except D-glucose and D-mannose. Designate each aldohexose by *syn*/*anti* descriptors.

## 2.8.8 Constitutionally symmetrical molecules with even number of stereocentres (ABBA type)

Molecules of the type R(*CHX*)$_n$R with $n$ stereocentres are constitutionally symmetrical. Polyhydroxy dicarboxylic acids [$HO_2C(CHOH)_nCO_2H$] are suitable examples of this type. With $n$ = even, the simplest compound is tartaric acid ($n$ = 2), which has been described earlier (see Section 2.8.5). The next member ($n$ = 4) is 2,3,4,5-tetrahydroxyadipic acid (or hexaric acid). It has four stereocentres (C2, C3, C4, C5) where C2 = C5 and C3 = C4. The molecule thus represents ABBA type and the possible diastereomers are shown in Fig. 2.36A.

There are four chiral diastereomers (**A**–**D**), each existing as an enantiomeric pair, and two achiral or *meso* diastereomers (**E**, **F**) having a plane of symmetry ($\sigma$). The total number of stereoisomers is therefore 10. A representative chiral diastereomer and two *meso* isomers are shown in Fig. 2.36B. Remember that other chiral diastereomers can be easily drawn by an exchange of ligands at a stereocentre for a change in descriptor.

In general, in molecules of the type R(*CHX*)$_n$R ($n$ = even), the number of chiral diastereomers $= 2^{n-2}$ (hence the number of optically active isomers $= 2^{n-1}$), the number of *meso* isomers $= 2^{(n-2)/2}$, and the total number of stereoisomers $= 2^{n-1} + 2^{(n-2)/2}$.

$HO_2C\overset{2}{C}HOH\overset{3}{C}HOH\overset{4}{C}HOH\overset{5}{C}HOHCO_2H$

| Stereocentre: | A | B | B | A | Diastereomer |
|---|---|---|---|---|---|
| | *R* | *R* | *R* | *R* | **A** Chiral |
| | *R* | *R* | *R* | *S* | **B** Chiral |
| | *R* | *R* | *S* | *R* | **C** Chiral |
| | *R* | *S* | *S* | *R* | **D** Chiral |
| | *R* | *R* | *S* | *S* | **E** *meso* |
| | *R* | *S* | *R* | *S* | **F** *meso* |

(A)

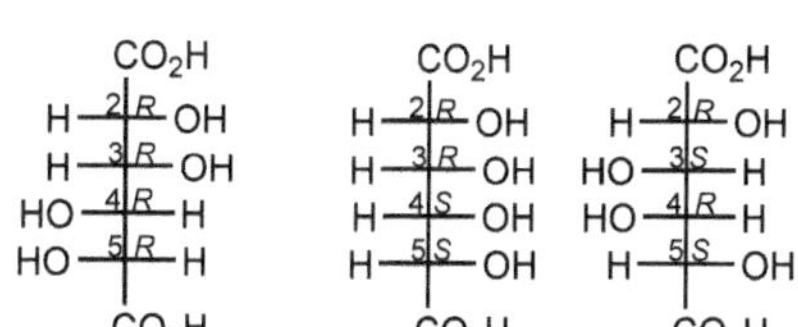


(B) **A** **E** **F**

**FIG. 2.36** (A) Chiral and *meso* diastereomers of 2,3,4,5-tetrahydroxyadipic acid (ABBA type) and (B) Fischer projections of a representative chiral diastereomer (one enantiomer shown) and two *meso* isomers.

### 2.8.9 Constitutionally symmetrical molecules with odd number of stereocentres (ABA and ABCBA types)

Polyhydroxy dicarboxylic acids also provide suitable examples for constitutionally symmetrical molecules of the type R($CHX$)$_n$R for $n$ = odd. The simplest compound is 2,3,4-trihydroxyglutaric acid ($n$ = 3). The molecule contains three stereocentres (C2, C3, C4) where C2 = C4), and belongs to ABA type. The possible diastereomers are shown in Fig. 2.37. The diastereomer **A** is chiral and exists as two enantiomers **A** (2*R*,4*R*) and **A′** (2*S*,4*S*). In **A** (or **A′**), C3 is not a stereocentre as it bears two ligands that are identical in constitution and in configuration. (It is easy to see that an exchange of H and OH at C3 in either **A** or **A′** gives the same stereoisomer). But C3 resides in a chiral environment in the chiral molecule and hence is chirotopic (see Section 2.8.2). Thus *C3 in* ***A*** *or* ***A′*** *is chirotopic but nonstereogenic.*

The diastereomer **B** has a plane of symmetry passing through C3 and is achiral (*meso*). (Note that an achiral diastereomer is called *meso* if the diastereomeric set contains at least one chiral diastereomer.) Here C3 is a stereocentre; it bears four different ligands of which two have the same constitution but different configuration—(*R*) $CHOHCO_2H$ and (*S*) $CHOHCO_2H$. But, since C3 resides in a local achiral environment (lies on the symmetry plane), it is achirotopic (see Section 2.8.2). Thus *C3 in* ***B*** *is achirotopic but stereogenic.* It is not called a chiral centre, although a stereocentre. Such an achirotopic but stereogenic centre having two enantiomorphic ligands is called a

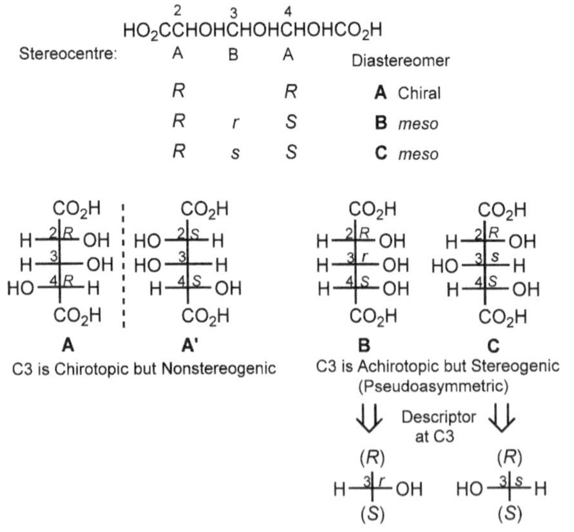


**FIG. 2.37** Stereoisomers of 2,3,4-trihydroxyglutaric acid (ABA type). The *r*/*s* descriptor at the pseudoasymmetric centre C3 in the *meso* isomers **B** and **C** is based on the priority order OH > (*R*) $CHOHCO_2H$ > (*S*) $CHOHCO_2H$ > H.

*pseudoasymmetric* centre. Since C3 is stereogenic, an exchange of two ligands (H, OH) at C3 gives another diastereomer **C** which is also *meso*. In **D**, *C3 is also achirotopic but stereogenic*. Thus 2,3,4-trihydroxyglutaric acid has four stereoisomers comprising an enantiomeric pair (**A**, **A′**) and two *meso* forms (**B** and **C**) as shown in Fig. 2.37. The descriptor at the pseudoasymmetric centre C3 in **B** or **C** is denoted in lower case as *r* or *s*.[25] The ligand priority order at C3 is determined by invoking the **Sequence rule 5** (see Section 2.7.1.1), that is $R > S$. The overall priority order is therefore OH > ($R$) $CHOHCO_2H$ > ($S$) $CHOHCO_2H$ > H. Thus, the descriptor at C3 is *r* in **B** and *s* in **C**.

The next member in the odd series is 2,3,4,5,6-pentahydroxypimelic acid ($n=5$). The molecule contains five stereocentres (C2, C3, C4, C5, C6) where C2 = C6 and C3 = C5, and thus belongs to ABCBA type. The possible diastereomers are shown in Fig. 2.38. There are six chiral diastereomers **A–F** (each with an enantiomeric pair) and four *meso* forms **G–J**. The total number of stereoisomers is therefore 16.

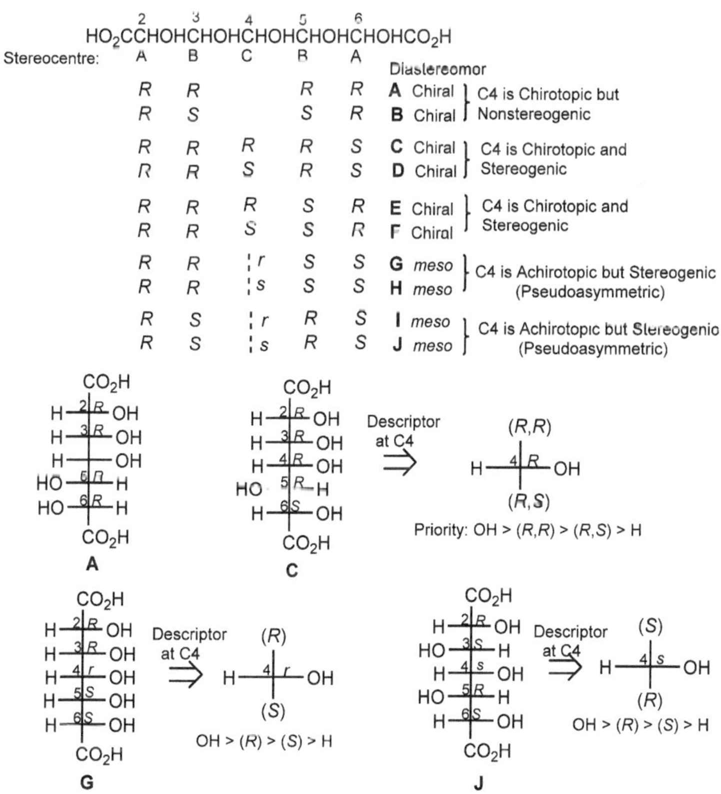


**FIG. 2.38** Diastereomers of 2,3,4,5,6-pentahydroxypimelic acid (ABCBA type) and Fischer projections of representative chiral (**A**, **C**) and *meso* (**G**, **J**) isomers with the assignment of *R*/*S* or *r*/*s* descriptor at C4.

In **A** and **B**, C4 is chirotopic but nonstereogenic.

In **C** and **D**, C4 is chirotopic and stereogenic (**C** and **D** are related by an exchange of H and OH at C4). Similarly, C4 is chirotopic and stereogenic in **E** and **F**. The *R*/*S* descriptor at C4 in **C–F** can be determined by invoking the **Sequence rule 4** (see Section 2.7.1.1), i.e. *l* (like: *R*,*R* or *S*,*S*) > *u* (unlike: *R*,*S* or *S*,*R*). Remember that the sequence rule 4 has precedence over the sequence rule 5 (*R* > *S*). The assignment of the descriptor at C4 is illustrated with **C** in Fig. 2.38.

In *meso* isomers **G** and **H**, C4 is achirotopic (lies on the symmetry plane) but stereogenic. Therefore, C4 is pseudoasymmetric in **G** and **H**. Similarly, C4 is pseudoasymmetric in **I** and **J**. The *r*/*s* descriptor at C4 in **G** or **H** cannot be determined by invoking the Sequence rule 4 (*l* > *u*) because the two enantiomorphic ligands (*R*,*R*/*S*,*S*) are like (Fig. 2.38). Similarly, in **I** or **J**, the sequence rule 4 (*l* > *u*) is not relevant as the two enantiomorphic ligands (*R*,*S*/*S*,*R*) are both unlike. Therefore, in turn, the **Sequence rule 5** (*R* > *S*) decides the priority order, as shown in Fig. 2.38. Note that Fischer projections of only representative chiral diastereomers **A** and **C,** and *meso* isomers **G** and **I** are depicted in Fig. 2.38.

In general, in molecules of the type $R(CHX)_nR$, ($n$ = odd), the number of chiral diastereomers = $2^{n-2} - 2^{(n-3)/2}$ (hence the number of optically active isomers = $2^{n-1} - 2^{(n-1)/2}$), the number of *meso* isomers = $2^{(n-1)/2}$, and the total number of stereoisomers = $2^{n-1}$.

## 2.9 Stereogenic axis

A stereogenic axis (or stereoaxis) is a line connecting two atoms, each of which is linked to a pair of unlike ligands. Unlike all four different ligands at a stereocentre, the two ligand pairs at the two ends of a stereoaxis may be same pairwise (a, b)/(a, b) [a ≠ b] or different (a, b)/(c, d) [a ≠ b, c ≠ d].

A stereoaxis may be achiral or chiral. If the two ligand pairs are coplanar, the stereoaxis is achiral (the plane of the molecule is a symmetry plane) as shown in Fig. 2.39A. The double-bond systems such as alkenes may possess an achiral axis and exhibit diastereomerism. On the other hand, if the two ligand pairs are noncoplanar or lie in orthogonal planes, the stereoaxis becomes chiral (Fig. 2.39B). The molecules such as allenes, biphenyls and related systems may possess a chiral axis[44] and thereby exhibit enantiomerism.

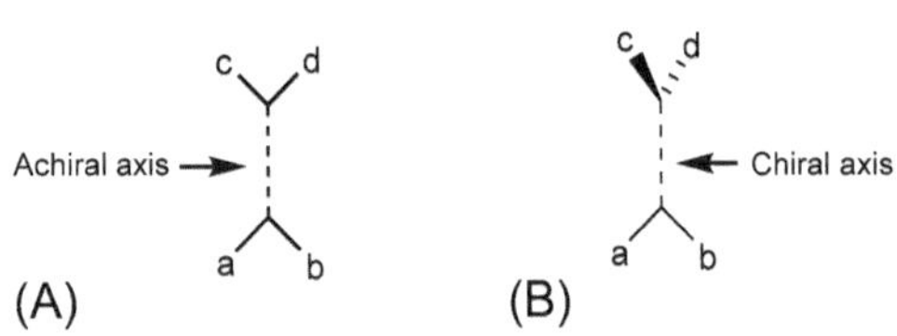


**FIG. 2.39** (A) Achiral stereoaxis and (B) chiral stereoaxis. a, b, c and d denote ligands; a ≠ b, c ≠ d.

### 2.9.1 Achiral stereoaxis: *E*/*Z* descriptors

Historically, stereoisomerism in alkenes has been depicted by the descriptors *cis* (two identical ligands are on the same side of the double bond) and *trans* (two identical ligands are on the opposite sides of the double bond). For example, 2-butene has *cis* and *trans* isomers. The energy barrier to *cis–trans* isomerization involving the breaking of a π bond is quite high (62 kcal $mol^{-1}$)[45] and they exist as configurational stereoisomers. Since the *cis* and *trans* isomers are not mirror images of each other, they are related as diastereomers. The *cis*/*trans* nomenclature has limitations but is still acceptable. A general method of nomenclature is based on the CIP system and is denoted by *Z* and *E* descriptors.[46] If two higher-priority (or lower-priority) ligands lie on the same side of the double bond, the descriptor is *Z* (from the German *zusammen*, together), and if they are on the opposite sides, the descriptor is *E* (*entgegen*, opposite) as shown in Fig. 2.40A. The *E*/*Z* descriptors are illustrated with an alkene **2.5** and an oxime **2.6**. Usually an *E* isomer is more stable than a *Z* isomer for steric reasons; however, a *Z* isomer can be highly stabilized due to chelation through H-bonding as shown for the enol form of a 1,3-diketone (acetylacetone) **2.7**. For cumulenes [Cab(=C)$_n$=Ccd] in which total number of double bonds is *odd*, the terminal ligand pairs are coplanar and therefore such cumulenes would exhibit *E*/*Z* diastereomerism.[47]

Each double bond stereoaxis in a diene or polyene can have *E*/*Z* stereochemistry. In general, a diene with two stereoaxes has four diastereomers (*E*,*E*), (*E*,*Z*), (*Z*,*E*) and (*Z*,*Z*). If terminal ligands in a diene are the same, (*E*,*Z*) ≡ (*Z*,*E*) and there will be three diastereomers. Further, a diene diastereomer can exist in *s-cis* and *s-trans* conformations, with the *s-trans* conformer being sterically more stable. Similarly, in a polyene diastereomer, each diene unit can be *s-cis* or *s-trans*. These are illustrated with (2*E*,4*Z*)-2,4-hexadienoic acid and (2*E*,4*Z*,6*E*-2,4,6-octatriene in Fig. 2.40B.

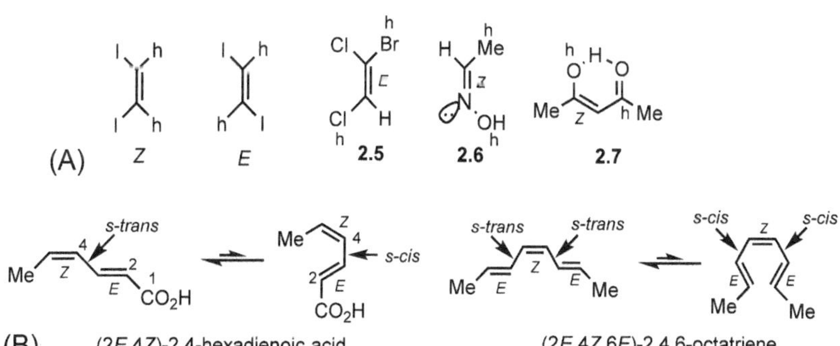


**FIG. 2.40** (A) *E*/*Z* descriptors with examples and (B) *E*/*Z* descriptors in *s-cis* and *s-trans* conformations of a diene and a triene. h and l denote higher-priority and lower-priority ligands, respectively.

#### 2.9.1.1 *Chiral alkenes and dienes:* E,Z/R,S *descriptors*

A diastereomeric alkene will be chiral if it contains a chiral centre. The *E* diastereomer would then exist as an enantiomeric pair (*E*,*R*) and (*E*,*S*). Similarly, the Z diastereomer has (Z,*R*) and (Z,*S*) enantiomers. These are depicted in Fig. 2.41A.

Fig. 2.41B illustrates the descriptors for two chiral dienes **2.8** and **2.9**. In each case, the chiral centre bears an (*E*) double bond and a (*Z*) double bond, besides two other ligands (Me, H). The *R*/*S* descriptor at the chiral centre is determined using the **Sequence rule 3** (see Section 2.7.1.1), that is *cis* precedes *trans*. In terms of the sequence rule, *cis*/*trans* relationship is considered between the *ligand containing the chiral centre* (marked *) and the *higher priority ligand (h) on the terminal alkene carbon*. Thus, in **2.8**, the (*E*) double bond corresponds to *trans* and the (*Z*) double bond *cis*. In contrast, the (*E*) and (*Z*) double bonds in **2.9** correspond to *cis* and *trans*, respectively. The descriptor thus assigned at the chiral centre in **2.8** is *R*, and in **2.9**, it is *S*.

### 2.9.2 Axial chirality

Molecules such as allenes, spiranes, alkylidenecycloalkanes and biphenyls may possess a chiral axis[44] (see Fig. 2.39B). The two ligand pairs at the two ends of the chiral axis are either pairwise same or different, but the two ligands in a pair must be unlike.

#### 2.9.2.1 *Allenes and Lowe's rule*

Fig. 2.42 shows two chiral allenes **2.10** and **2.11**, in which C=C=C is the chiral axis. In **2.10**, a pair of unlike ligands ($CO_2H$/H) is linked at both ends of the axis while in **2.11**, the two ligand pairs (H/Cl) and ($CMe_3$/Me) are different. Each chiral allene exists as an enantiomeric pair. The two enantiomers (**2.10**, **2.10′**) [or (**2.11**, **2.11′**)] are distinguished by their opposite sign of optical

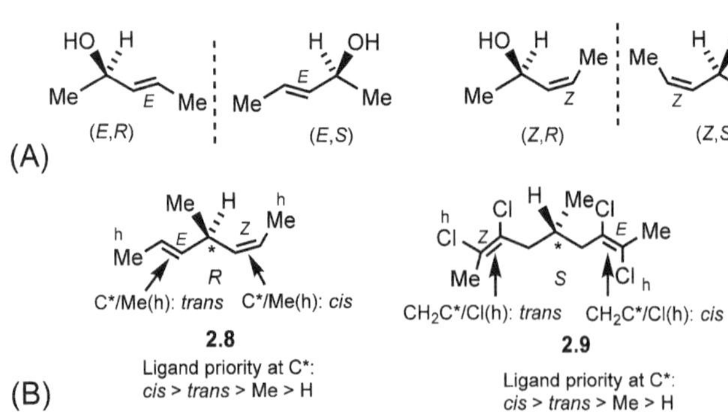


**FIG. 2.41** (A) Stereoisomers of a chiral alkene and (B) assignment of *R*/*S* descriptor to chiral dienes.

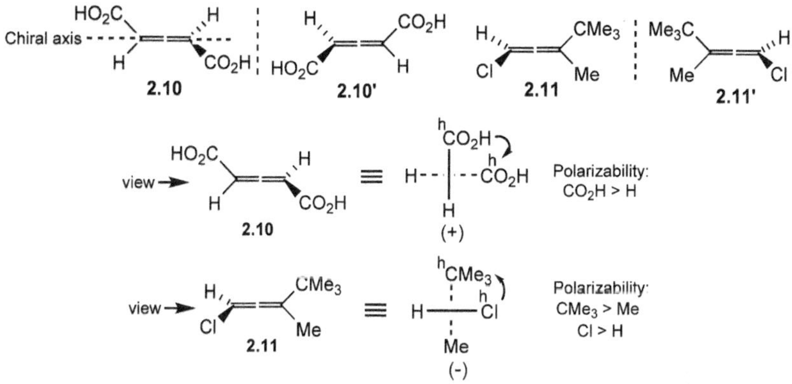


**FIG. 2.42** Chiral allenes and the application of the Lowe's rule.

rotation. Like Brewster's rule (see Section 2.4.5), there is also an empirical rule called Lowe's rule[48] that can predict the sign of optical rotation (at the sodium *D* line) of an allene based on the pattern of polarizability of the ligands about the chiral axis. For the purpose of the rule, the allene is viewed along the chiral axis and is so projected that the front ligands are on a vertical or horizontal line when the rear ligands are on a perpendicular line. If the turn from the front ligand of higher polarizability (h) to the rear ligand of higher polarizability (h) makes a clockwise screw pattern, the allene is dextrorotatory (+), and if the turn is anti-clockwise, it is levorotatory (−). The rule is illustrated with **2.10** and **2.11** which are found to be dextrorotatory and levorotatory respectively (Fig. 2.42).

## 2.9.2.2 *Spiranes and alkylidene cycloalkanes*

In allenes, the high energy barrier to rotation about double bonds ensures that the two enantiomers do not interconvert, and hence are separable. If two double bonds in allene are replaced by two rings, a spirane results, and if one double bond is replaced by a ring, an alkylidenecycloalkane is formed. Fig. 2.43 shows examples of a chiral spirane **2.12** and a chiral alkylidenecyclohexane **2.13**, each with an enantiomeric pair (an exchange of two ligands at any end of the axis

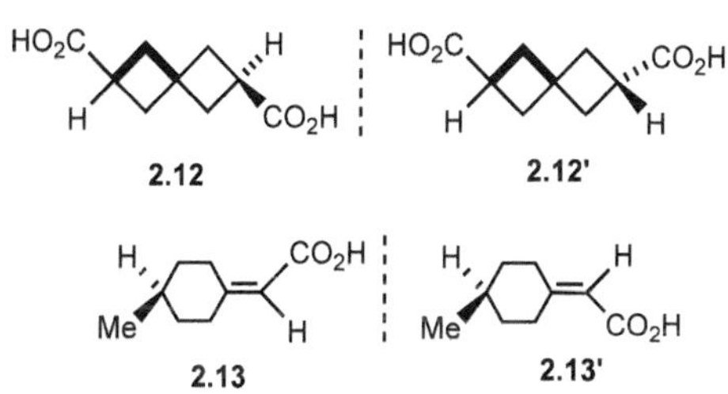


**FIG. 2.43** Axially chiral spirane and alkylidenecyclohexane.

gives the opposite enantiomer). The chiral axis is maintained due to the rigidity of the molecular framework in spiranes, and a combination of rigid framework and hindered rotation about a double bond in alkylidenecycloalkanes. Notably, the barrier to racemisation in allenes, spiranes or alkylidenecycloalkanes is independent of the size of the substituents.

### 2.9.2.3 Biphenyls: Atropisomerism

In the biphenyl derivative **2.14**, the barrier to rotation about the central C(sp$^2$)—C(sp$^2$) single bond (pivotal bond) determines whether two axially chiral enantiomers would exist (Fig. 2.44A). Here the size of the barrier depends on the steric demands of *ortho* substituents in each phenyl ring. Sufficient steric repulsion between *ortho* substituents in the planar TS is necessary for the enantiomers to be resolved. This type of stereoisomerim (enantiomerism) due to hindered rotation about a single bond is called atropisomerism[49] (atrop from the Greek *a*, not and *tropos*, turn; it implies that molecules do not turn around the molecular axis). The biphenyl enantiomers are thus called atropisomers. The atropisomerism was first discovered in the case of 6,6′-dinitro-2,-2′-diphenic acid (**2.14**: a=$NO_2$, b=$CO_2H$).[50] In general, for atropisomerism in biphenyl systems, two conditions are to be met:

**(i)** Each phenyl ring must lack a vertical plane of symmetry so that the molecule is chiral.

**(ii)** The *ortho* substituents must have a large size so that the steric repulsion leads to a barrier to rotation that permits the separation of enantiomers.

The biphenyl derivative **2.14** or **2.14′** is chiral (point group $\mathbf{C_2}$). The $C_2$ axis bisects the C—C bond and makes an angle of 45 degrees with the aromatic planes. The planar TS can be *cisoid* when two a's are on the same side or *transoid* when two a's are on the opposite side (Fig. 2.44A). Racemisation is more likely to occur through the lower energy *transoid* TS. The energy profile for rotation in biphenyl derivatives may show an energy minimum at 90 degrees (when the two rings are orthogonal) or a double minimum around 90 degrees

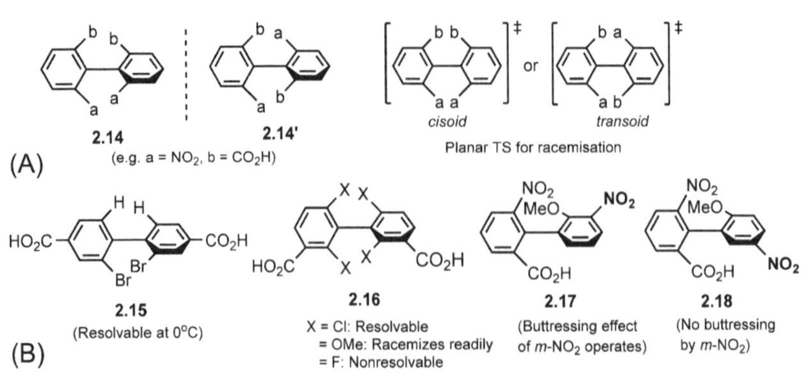


**FIG. 2.44** (A) Atroisomerism in axially chiral biphenyl systems and (B) examples.

(when the inter-ring torsional angle deviates from 90 degrees) depending on the steric repulsions between *ortho* substituents.[51]

The energy barrier required for the existence of atropisomerism falls into grey areas; the required free energy barrier is $\geq$22.3 kcal mol$^{-1}$ ($t_{1/2}$ ~ 2000 s) at 300 K (see also Section 2.1). An arbitrarily value of a half-life ($t_{1/2}$) of at least 1000 s is defined as the condition for the atropisomers to be isolated.[52]

The apparent order of steric hindrance by different *ortho* substituents is.

$$Br > Me > Cl > NO_2 > CO_2H > OMe > F > H > D$$

which roughly parallel to their van der Waals radii. D has a smaller van der Waals radius than H since it has a lower zero point energy.

Biphenyl isomerism has been extensively described elsewhere.[51,53] Here we will consider only a few examples (Fig. 2.44B). Even one bulky *ortho* substituent such as Br in each ring **2.15** can allow the separation of enantiomers but at a lower temperature of 0°C. Depending on the nature of the X substituent, the biphenyl derivative **2.16** is resolvable, or can racemize readily, or is nonresolvable.

A *meta* substituent can also increase the barrier to racemisation, for example, **2.17** racemizes at a much slower rate than **2.18**. This is attributed to a buttressing effect of the *meta* substituent ($NO_2$) in **2.17** that tends to prevent the outward bending of the *ortho* substituent (OMe) in the planar TS necessary to slip past readily the *ortho* substituents of the other ring. In **2.18**, the *m*-$NO_2$ group is not adjacent to the *ortho* OMe and hence the buttressing effect is not operative. The buttressing effect of some groups follows the order: $NO_2 > Br > Cl > Me$ which however does not correspond to the order of their van der Waals radii. The atropisomerism about C(sp$^2$)—C(sp$^3$) or C(sp$^3$)—C (sp$^3$) single bond is also known.[54,55]

**Problem 2.8** Which of the following biphenyl derivatives (**X** and **Y**) is resolvable? Explain.

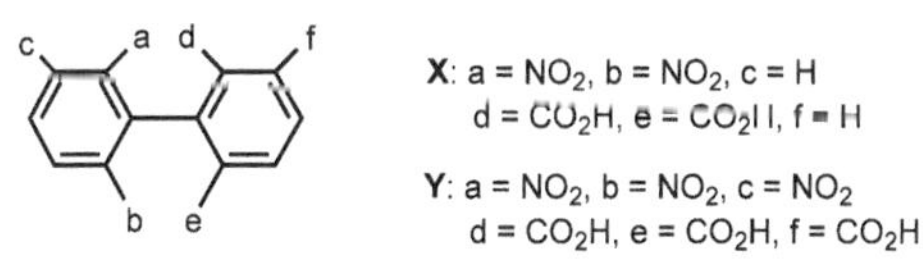


**Problem 2.9** Complete the following scheme and predict whether **A**, **B**, **C** or **D** is resolvable or not.

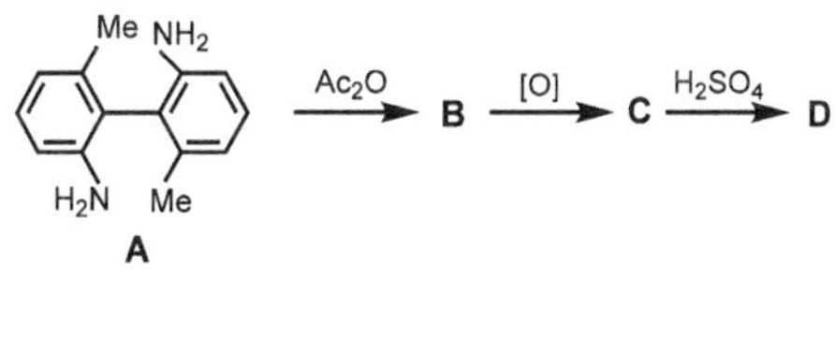

## 2.9.3 Chiral axis descriptors (*R*/*S* or *M*/*P*)

The absolute configuration of axially chiral molecules is also specified by *R*/*S* descriptors. But an additional sequence rule, [25] that is *near groups precede far groups* is required to determine the ligand priority order. This is illustrated with a chiral allene and a biphenyl in Fig. 2.45. Note that the molecule can be viewed along the chiral axis from either side.

Another system involves the assignment of *M* (minus) or *P* (plus) descriptor by viewing the molecule as a helix[26]. If the turn from the front higher-priority ligand to the rear higher-priority ligand is clockwise, the descriptor is *P*, and if anticlockwise, the descriptor is *M*. As shown in Fig. 2.45, axial-$R \equiv M$ and axial-$S \equiv P$.

### 2.9.3.1 *Mandal's approach to determining chiral axis configuration*

A simple and straightforward approach[56] to assigning a chiral axis descriptor (*R* or *S*) is based on a directional relationship between two h→l arrows at the two ends of the chiral axis, where h and l denote higher-priority and lower-priority ligand respectively (Fig. 2.46). The arrows are drawn through the bond angle in two orthogonal planes as shown.

The chiral axis configuration is specified as Con (when both arrows are clockwise or anticlockwise) or Dis (when one arrow is clockwise and the other anticlockwise). Con and Dis correspond, respectively, to the descriptors *R* and *S* for a chiral axis.

The below mnemonic can be used to draw a proper h→l arrow in the perpendicular plane:

h (front) ⇒ h→ l is clockwise
h (rear) ⇒ h→ l is anticlockwise

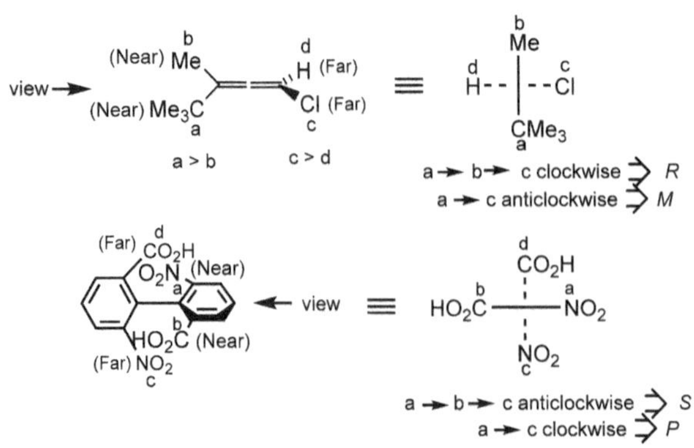


FIG. 2.45 *R*/*S* and *M*/*P* descriptors for chiral allene and biphenyl systems.

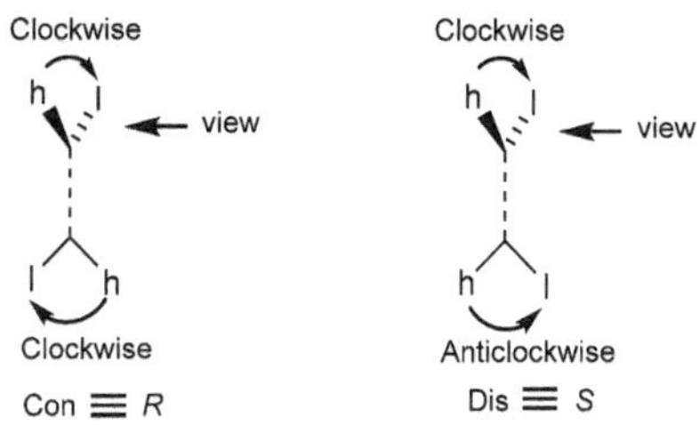


**FIG. 2.46** Mandal's approach to assigning *R*/*S* descriptor to a chiral axis.

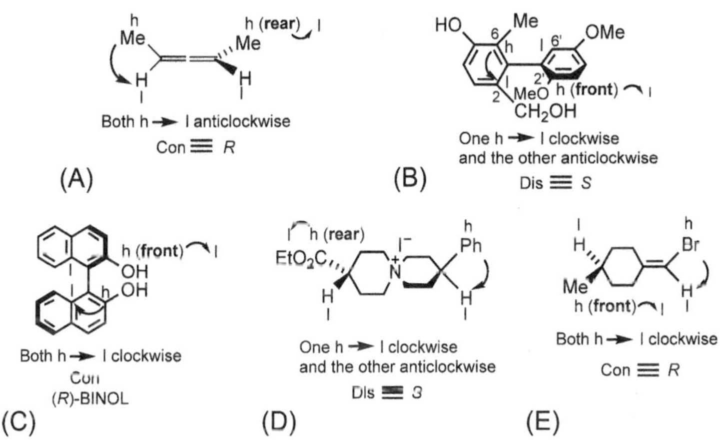


**FIG. 2.47** Assignment of *R*/*S* descriptor to (A) an allene, (B) a biphenyl, (C) a 1,1′-bi-2-naphthol, (D) a spirane and (E) an alkylidenecyclohexane. Two h → l arrows being clockwise or anticlockwise indicate Con ≡ *R* while one arrow being clockwise and the other anticlockwise denote Dis ≡ *S*. See also the mnemonic.

The procedure is illustrated with five examples in Fig. 2.47. Note that h → l arrow in the perpendicular plane is indicated as per the above mnemonic. For the biphenyl derivative (Fig. 2.47B), the ligands are explored from the centre to outward. Thus the ligand pair in the plane of the paper comprises a C2 branch (l) and a C6 branch (h), and the ligands in the perpendicular plane are a C2′ branch (h) and a C6′ branch (l). (The arrows are drawn explicitly only as an aid to understanding; the descriptors can be assigned at a glance.)

**Problem 2.10** Adamantane-2,6-dicarboxylic acid (shown below) is chiral due to the presence of a chiral axis. Identify the chiral axis and assign its configuration by *R*/*S* descriptor.

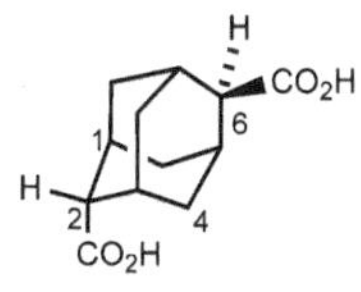

## 2.10 Planar chirality

Planar chirality[25,57] cannot be defined simply. A chiral plane in a molecule contains as many atoms as possible but the chirality is due to the presence of one or more ligands outside the chiral plane. Molecules with chiral planes are relatively few; some examples are cyclophanes, paracyclophanes and *trans*-cycloalkenes.[57] Fig. 2.48A shows a chiral cyclophane[58] as a pair of enantiomers. This type of compound was previously called *ansa* compound (from the Latin *ansa*, handle). The plane of the aromatic ring is the chiral plane. For the cyclophane ($n=8$), the benzene ring cannot turn through the alicyclic ring as the methylene chain is too short and the two enantiomers are separable. But if $n=10$, the benzene ring can turn freely leading to a nonresolvable racemate. Fig. 2.48A also shows a chiral [2.2]paracyclophane[59] that is resolvable. Here the benzene ring bearing $CO_2H$ is the chiral plane.

Fig. 2.48B depicts the chiral (*E*)-cyclooctene that can be resolved into two enantiomers.[57,60] In this case, the chiral plane is the plane of the double bond. Since the methylene chain in the (*E*)-cyclooctene is very short, it is forced out of the plane and the molecule exhibits planar chirality. The molecule belongs to the $\mathbf{C_2}$ point group, having only a $C_2$ axis passing through the centre of the double bond and bisecting 5–6 bond. In larger (*E*)-cycloalkenes such as (*E*)-cyclodecene, the double bond can swivel freely through the polymethylene bridge and the compound is nonresolvable.[61]

### 2.10.1 Chiral plane descriptors (*R*/*S* or *P*/*M*)

The chiral plane configuration is also specified by *R*/*S* or *P*/*M* descriptors.[25,26] For assigning the *R*/*S* configuration, the chiral plane is viewed from the *pilot atom* (the out-of-plane atom closest to the atom of highest priority in the chiral plane) as shown in Fig. 2.49. If the adjacent three atoms trace a→b→c clockwise on the chiral plane, the configuration is *R*, and if the array is anticlockwise, the configuration is *S*. The assignment of the *R*/*S* descriptor is illustrated with an enantiomer of a cyclophane, a [2.2]paracyclophane and (*E*)-cyclooctene

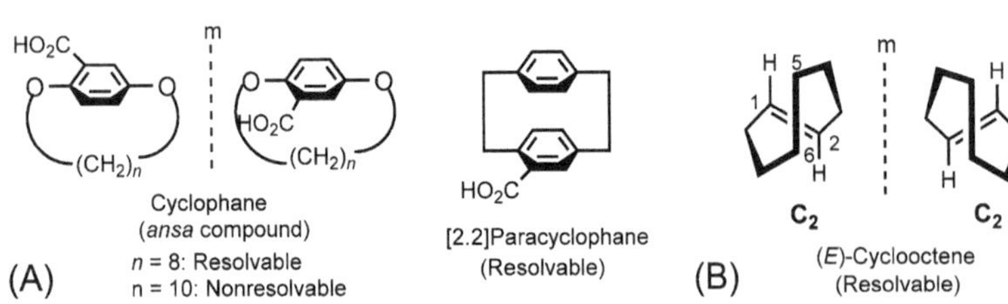


**FIG. 2.48** Molecules with planar chirality: (A) a cyclophane (*ansa* compound) and a [2.2]paracyclophane with an aromatic ring as the chiral plane and (B) (*E*)-cyclooctene with a double bond as the chiral plane.

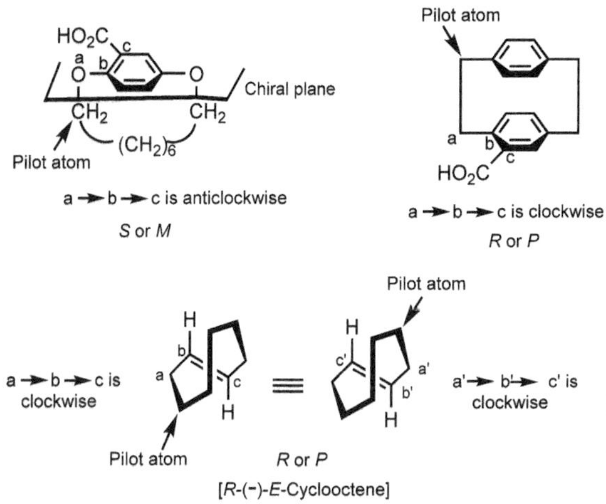


**FIG. 2.49** Assignment of the *R*/*S* or *P*/*M* descriptor to an enantiomer of a cyclophane, a [2.2]paracyclophane and (*E*)-cyclooctene.

(Fig. 2.49). For (*E*)-cyclooctene, there are two equivalent pilot atoms. As seen from either pilot atom, the configuration is *R*.

The chiral plane configuration can also be assigned in terms of *P*/*M* descriptors in the sense of a right-twisted or left-twisted helix. As also shown in Fig. 2.49, if the orientation of c in the three-atom chain from the pilot atom is clockwise, the descriptor is *P*, and if the orientation is anticlockwise, the descriptor is *M*. It is seen that planar-$R \equiv P$ and planar-$S \equiv M$. Note that this correlation is opposite to that in axial chirality, where axial-$R \equiv M$ and axial-$S \equiv P$. The absolute configuration of (−)-(*E*)-cyclooctene has been determined experimentally and found to be *R* or *P*.[62]

## 2.11 Molecules that are inherently chiral

Some molecules have an inherently chiral framework and no substituents are needed to make the molecule chiral. The most important examples are molecules with helical structures such as helicenes.[63,64] The helical chirality is specified by *P*/*M* descriptors. If a helix spirals downward clockwise (right handed) along the helical axis, the helicity is denoted by *P* (plus) while if a helix spirals downward anticlockwise (left handed), the helicity is indicated as *M* (minus) (Fig. 2.50A).

Fig. 2.50B shows an example of a helicene, namely, hexahelicene ([6]helicene). If the molecule were planar, the terminal rings would tend to occupy the same space causing a severe steric clash. The molecule is therefore twisted to give helical structures to avoid the severe steric repulsion between the atoms of the terminal rings. Note that relative to a central part of the molecule, some rings

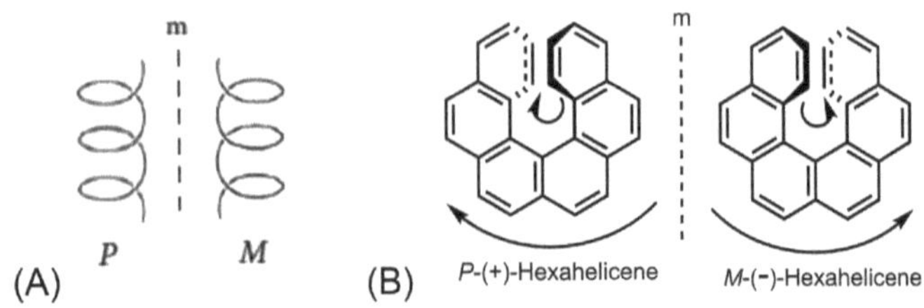


**FIG. 2.50** (A) *P*-helicity and *M*-helicity. (B) Two enantiomers (*P* and *M*) of hexahelicene or [6] helicene. (+) and (−) indicate sign of optical rotation.

are gradually bent up on one side and the others bent down on the other side. X-ray crystallography of hexahelicene[65] demonstrates that the interplanar angle between the terminal rings is 58.5 degrees. The helical structures are chiral and give a pair of enantiomers of hexahelicene. The chirality of the helicene is said to be due to molecular overcrowding. The two enantiomers of hexahelicene are assigned as *P* and *M*.

Hexahelicene has a very large specific rotation $[\alpha]_D=3640$ ($CHCl_3$)][66] as the aromatic chromophoric system itself being chiral. Inherently chiral chromophores follow the helicity rule[67] which states that *P*-helicity leads to a (+) rotation while *M*-helicity gives a (−) rotation. This prediction for hexahelicene is in conformity with the experimental observation.[68]

It may be mentioned that a polypeptide derived from natural L-amino acids often coil to form an α-helix with *P*-helicity (see Fig. 1.20A).

## 2.12 Topicity of ligands and faces

If two ligands are identical in isolation (when separated from the molecule), they are called *homomorphic* ligands[69] (from the Greek *homos*, same and *morphe*, form). Topicity (from the Greek *tropos*, place) refers to the phenomenon when two homomorphic ligands may be distinguishable when part of a molecule. If the two places they occupy are equivalent, the ligands are homotopic, and if nonequivalent, the ligands are called heterotopic (or stereoheterotopic).[70] Whether two ligands in a molecule are homotopic or heterotopic can be determined by *substitution* or *symmetry* criterion.[70,71]

By the substitution criterion, if either ligand is replaced by a *new* ligand (different from others attached to the same atom) and the two resulting structures are superposable or equivalent, the ligands are homotopic but if the structures are nonequivalent, they are heterotopic. Heterotopic ligands can be either enantiotopic (when the two resulting structures are enantiomeric) or diastereotopic (when the two structures are diastereomeric).

By the symmetry criterion, two ligands are homotopic if they are exchangeable by a $C_n$ operation, enantiotopic when exchanged by $\sigma$, $i$ or $S_n$ operation, and diastereotopic when not related by $C_n$ and $S_n$ (including $\sigma$ and $i$).

The concept of topicity is also applicable to the faces of a trigonal ($sp^2$) atom that may be attacked by an incoming reagent. Depending on the environment created by the attached ligands, the two faces of the trigonal atom can be either homotopic or heterotopic. The homotopic or heterotopic faces of a trigonal atom can also be determined by *addition* or *symmetry* criterion[70,71] (see Sections 2.12.4 and 2.12.5).

### 2.12.1 Homotopic ligands

To describe topicity, we will consider mainly hydrogen atoms as homomorphic ligands; however, other homomorphic ligands can also be chosen.

Fig. 2.51 illustrates the presence of two homotopic hydrogen ligands (labelled $H_a$ and $H_b$) in dichloromethane, (*R,R*)-(+)-tartaric acid and *trans*-1,2-dichlorocyclopropane respectively using both substitution and symmetry criteria. In each case, replacement of $H_a$ or $H_b$ by a new ligand generates the same structure, and the two H's are exchangeable by a $C_2$ operation. The two H atoms at the chiral centres of *trans*-1,2-dichlorocyclopropane are also homotopic (check). α-Cyclodextrin (point group $\mathbf{C_6}$)[72] has six homotopic glucose units that are related by a $C_6$ axis perpendicular to the average plane of the molecule and passing through its centre.

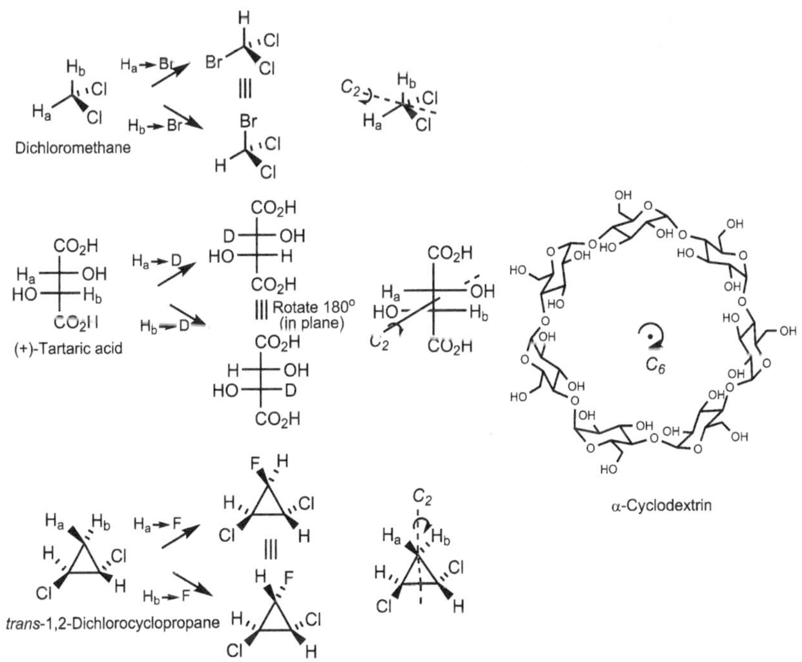


**FIG. 2.51** Examples of homotopic ligands.

### 2.12.2 Enantiotopic and diastereotopic ligands

Fig. 2.52A illustrates enantiotopic H ligands determined by either substitution or symmetry criterion. In propionic acid, the H atoms of the methylene group are enantiotopic as the replacement of either H by OH gives two enantiomers of lactic acid (*R* and *S*). The two H atoms of the methylene group in 1-methylallene are also enantiotopic as shown by substitution with Cl. The two H ligands in *meso*-tartaric acid and [2.2]paracyclophane cannot be exchanged by a $C_n$ operation but are related by a plane of symmetry $\sigma$, and are therefore enantiotopic.

A chiral molecule with a single chiral centre belongs to the point group $\mathbf{C_1}$, and is devoid of $C_n$ ($n > 1$) and $S_n$ axes. Two homomorphic ligands in such molecules will then be diastereotopic. For example, two H atoms of the methylene group in (*R*)-2-chlorobutane cannot be exchanged by $C_n$ ($n > 1$) or $S_n$ operation, and are diastereotopic (Fig. 2.52B). The diastereotopicity is also examined by substitution with Br which leads to *threo* (*R*,*R*) and *erythro* (*S*,*R*) diastereomers. The two H atoms of the methylene group in styrene are also diastereotopic as the substitution by Cl leads to (*Z*) and (*E*) diastereomers as shown in Fig. 2.52B.

A molecule can possess both enantiotopic and diastereotopic ligands. For example, in *cis*-1,2-dichlorocyclopropane (Fig. 2.52C), $H_a$ and $H_b$ are enantiotopic as they are related by a symmetry plane ($\sigma$). In contrast, the ligands $H_c$ and $H_d$ lie on the $\sigma$ plane but not related by $\sigma$ and hence are diastereotopic (verify also by substitution criterion).

The H ligands in cyclobutanone provide an interesting case (Fig. 2.52D). $H_a$ and $H_b$ are related by $\sigma_1$ symmetry plane (the plane of the molecule) and are enantiotopic. Similarly, $H_c$ and $H_d$ are enantiotopic. Further, $H_a$ and $H_c$ are enantiotopic as are $H_b$ and $H_d$ since they are related by the perpendicular $\sigma_2$ plane passing through C3 and C=O. Thus *$H_a$ is enantiotopic with $H_b$ and $H_c$*, and *$H_d$ is enantiotopic with $H_b$ and $H_c$*. The molecule also possesses an in-plane $C_2$ axis passing through C3 and C=O. $H_a$ and $H_d$ are interchangeable by the $C_2$ operation and are homotopic. Similarly, $H_b$ and $H_c$ are homotopic.

**Problem 2.11** Two $CH_2OH$ ligands in D-mannitol, D-galactitol and D-glucitol are homotopic, enantiotopic and diastereotopic respectively. Explain using substitution ($CH_2OH$ by CHO) or symmetry criterion.

### 2.12.3 Prostereogenic elements, prochirality and *pro-R*/*pro-S* descriptors

A tetrahedral centre of the type Caabc where a, a denote enantiotopic or diastereotopic ligands is a *prostereogenic centre* since the replacement of either homomorphic ligand by a new ligand a′ makes the centre stereogenic (Caa′bc).[69] If the generated stereocentre is a chiral centre, the prostereogenic centre is called a *prochiral centre*.[3,73] For example, the two H atoms at C2 in propionic acid are enantiotopic and the replacement of either H by OH generates a chiral centre (see Fig. 2.52A). Here C2 is a prochiral centre and propionic acid itself with

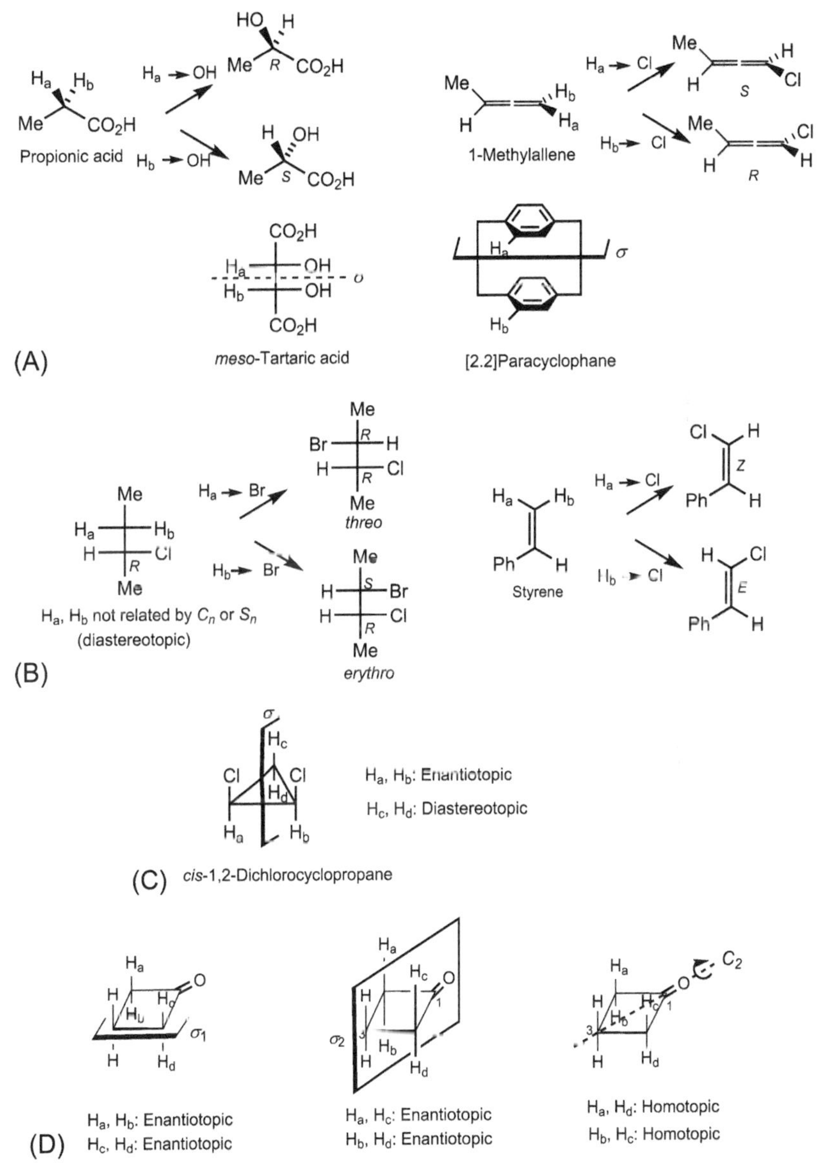


**FIG. 2.52** (A) Enantiotopic H ligands in propionic acid, 1-methylallene, *meso*-tartaric acid and [2.2]paracyclophane; (B) diastereotopic H ligands in 2-chlorobutane and styrene; (C) enantiotopic and diastereotopic H ligands in *cis*-1,2-dichlorocyclopropane and (D) sets of enantiotopic and homotopic H ligands in cyclobutanone.

no chiral centre is said to be prochiral. *A prochiral molecule* contains only enantiotopic ligands. In (*R*)-2-chlorobutane, the two H atoms at C3 are diastereotopic and C3 is a prochiral centre (see Fig. 2.52B) but the molecule cannot be called prochiral as it is already chiral with a chiral centre at C2.

Besides the prostereogenic centre, there are other prostereogenic elements, i.e. prostereogenic axis or plane. For example, 1-methylallene (see Fig. 2.52A) with two enantiotopic H ligands has a prostereogenic C=C=C axis which is also prochiral. Thus 1-methylallene is a prochiral molecule. Similarly, [2.2] paracyclophane is prochiral with a prochiral benzene plane (see Fig. 2.52A). A prostereogenic axis is not always a prochiral axis. For example, the C=C axis in styrene with two diastereotopic H atoms is a prostereogenic axis but it is not prochiral as the substitution of either H does not generate a chiral axis (see Fig. 2.52B).

Two homomorhic ligands at a prostereogenic centre or axis have the same priority under the CIP rules. Therefore, to specify descriptors[2,73] for the enantiotopic/diastereotopic ligands, one of them is *hypothetically* assigned a higher priority than the other without disturbing the priorities of the other ligands. Thus, for the two H ligands ($H_a$, $H_b$), $H_a$ is raised to a higher priority ($H_a > H_b$) for assigning the descriptor to $H_a$. Similarly, to determine the descriptor of $H_b$, $H_b$ is given a higher priority ($H_b > H_a$). The descriptor is *pro-R* if the hypothetical configuration at the prostereogenic centre or axis is *R*, and *pro-S* if *S*. Evidently, if $H_a$ is *pro-R*, $H_b$ will be *pro-S* and vice versa. A *pro-R* (or *pro-S*) H ligand is often indicated as $H_R$ (or $H_S$).

The assignment of descriptors to enantiotopic ligands is illustrated in Fig. 2.53A. The descriptor of $H_a$ in propionic acid is assigned using the priority sequence $CO_2H > Me > H_a > H_b$ which leads to the hypothetical configuration *R*, and hence $H_a$ is *pro-R* ($H_R$). Alternatively, $H_a$ can be replaced by D when the priority order is $CO_2H > Me > D > H_b$. The descriptor of $H_b$ is obtained using the priority order $CO_2H > Me > H_b > H_a$ or by replacing $H_b$ with D. Note that since $H_a$ is *pro-R* ($H_R$), $H_b$ must be *pro-S* ($H_S$). In 1-methylallene, $H_a$ is *pro-S* ($H_S$) using the priority order Me > H, $H_a > H_b$. Consequently, $H_b$ will be $H_R$. The descriptors for the enantiotopic ligands in other molecules are determined similarly.

Fig. 2.53B depicts the assignment of descriptors to the diastereotopic ligands in a similar manner as shown for the enantiotopic ligands. In (*R*)-2-chlorobutane, $H_a$ is *pro-R* ($H_R$) using the priority sequence C2 > Me > $H_a > H_b$. Hence $H_b$ will be *pro-S* ($H_S$). The descriptors for diastereotopic ligands are sometimes denoted by a double indexing system[74] when a second subscript refers to the absolute configuration of the chiral centre already present. By this convention, $H_a$ is $H_{RR}$ (the second subscript being the *R* descriptor at the chiral centre C2). Similarly, $H_b$ is $H_{SR}$. The double indexing system has the advantage that it distinguishes diastereotopicity from enantiotopicity.

In case of diastereotopic ligands of an alkene with a prostereogenic axis, *pro-E*/*pro-Z* descriptors are used. As shown in Fig. 2.53B, the descriptor of $H_a$ in styrene is *pro-Z*, and $H_b$ is *pro-E*.

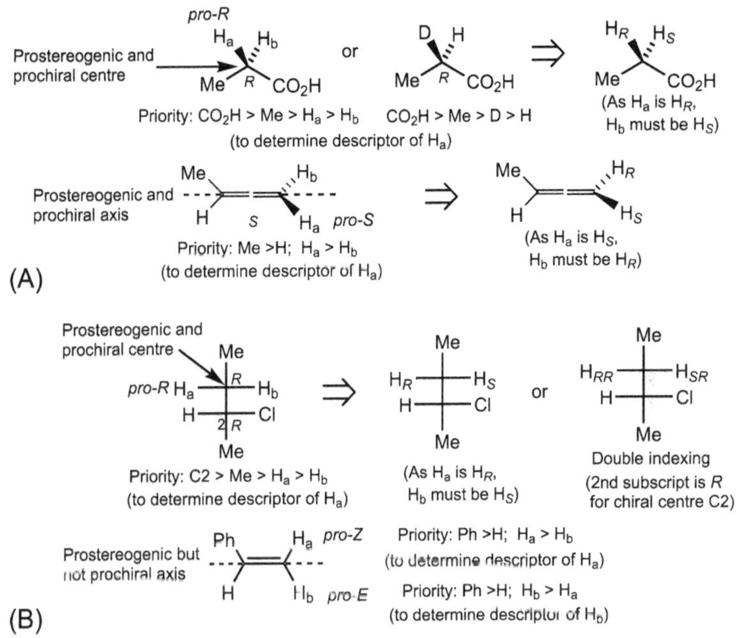


**FIG. 2.53** (A) *pro-R/pro-S* Descriptors of enantiotopic ligands at a prochiral centre and a prochiral axis, (B) *pro-R/pro-S* descriptors of diastereotopic ligands at a prochiral centre and *pro-E/pro-Z* descriptors at a prostereogenic but not prochiral axis. *Pro-R* H ligand is also denoted as $H_R$ and *pro-S* H as $H_S$. In double indexing, *R/S* configuration at the chiral centre is added as a second subscript in the descriptor.

For *Re/Si* descriptors to enantiotopic ligands, see Fig. 2.57.

**Problem 2.12** Citric acid [$HO_2CCH_2C(OH)(CO_2H)CH_2CO_2H$] has three prochiral centres and contains both enantiotopic and diastereotopic sets of H atoms. Explain.

Also assign proper descriptors to the H atoms at C2 and C4, and to $CH_2CO_2H$ ligands at C3.

**Problem 2.13** In (2*R*,4*S*)-2,4-dihydroxyglutaric acid, C3 is prostereogenic but proachirotopic whereas in the half-ester of the acid, C3 is prostereogenic and prochirotopic. Explain.

## 2.12.4 Homotopic faces

The π-faces of a trigonal atom (usually carbonyl or alkene carbon) can be homotopic or heterotopic. By the *addition* criterion, the faces are homotopic if addition of a reagent to either face generates the same product. Fig. 2.54A shows two carbonyl faces to be homotopic as the addition of $CN^-$ to the front face or the rear face gives the same product. By the *symmetry* criterion, they are homotopic

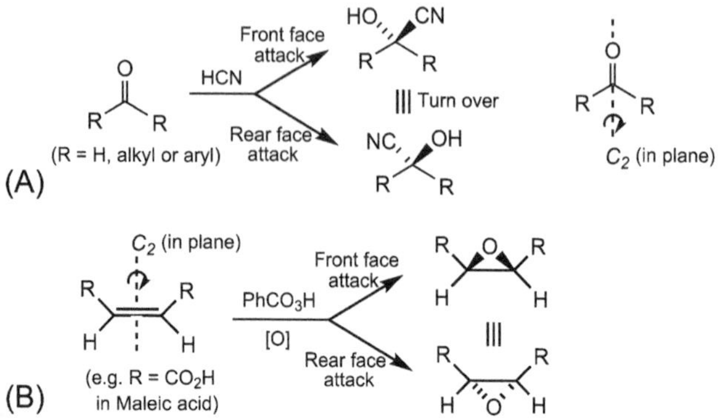


**FIG. 2.54** Homotopic faces of (A) a carbonyl compound and (B) an alkene.

since the two faces are exchangeable by a $C_2$ axis. In general, if the two ligands attached to the carbonyl carbon are identical, the faces are homotopic.

Fig. 2.54B depicts two alkene faces that are homotopic. Addition of oxygen on the front and rear faces with perbenzoic acid gives the same epoxide. The two faces are also related by a $C_2$ axis.

## 2.12.5 Enantiotopic and diastereotopic faces and *Re/Si* descriptors

If the carbonyl group is attached to two unlike ligands, addition to the trigonal carbon generates a chiral centre. Here the carbonyl group is prochiral and the molecule is said to be two-dimensionally chiral.[75] For a prochiral carbonyl group, the two carbonyl faces can be enantiotopic or diastereotopic. If the addition to the two carbonyl faces gives enantiomeric products, the faces are enantiotopic, and if the products are diastereomeric, the faces are diastereotopic.

The two enantiotopic or diastereotopic carbonyl faces are designated by *Re*/*Si* descriptors.[75,76] A carbonyl face is a plane defined by three ligands with priority order $a > b > c$. If the sequence $a \rightarrow b \rightarrow c$ is clockwise, it is the *Re* face, and if it is anticlockwise, it is the *Si* face. This is illustrated with two carbonyl faces of benzaldehyde in Fig. 2.55. The descriptor for the front face is easy to assign: the sequence $O \rightarrow Ph \rightarrow H$ is anticlockwise, and hence *Si*. A direct view of the rear face is difficult. If the molecule is turned upside down, the original rear face will now be the front face which is *Re*. Notably, the two faces must have opposite descriptors; easy assignment of a descriptor to the front face automatically gives the opposite descriptor to the rear face which may be difficult to observe directly.

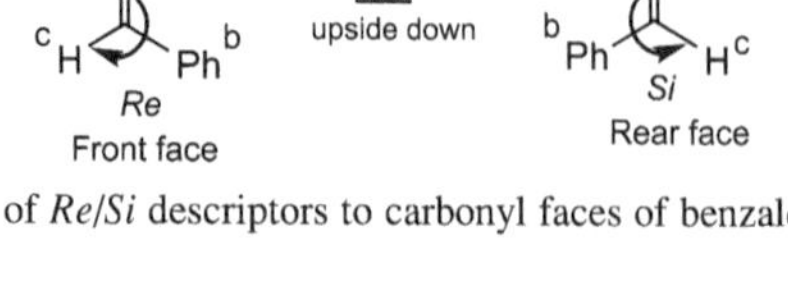


**FIG. 2.55** Assignment of *Re*/*Si* descriptors to carbonyl faces of benzaldehyde.

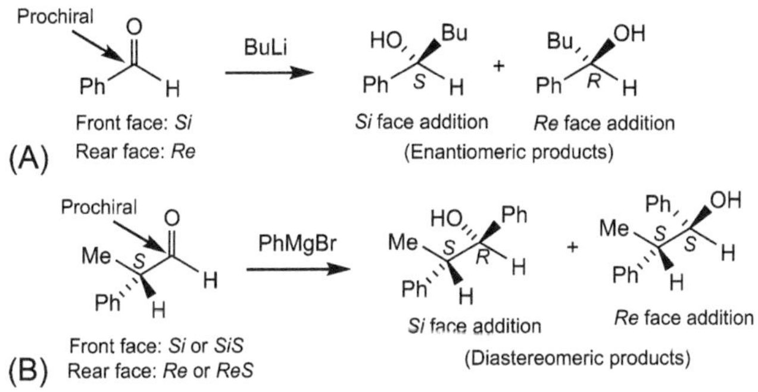


**FIG. 2.56** Formation of a new chiral centre from the *Re* face and the *Si* face addition to (A) enantiotopic carbonyl faces and (B) diastereotopic carbonyl faces.

Fig. 2.56 illustrates enantiotopic and diastereotopic carbonyl faces showing products resulting from attack on the *Re* face and the *Si* face. The two prochiral faces of achiral benzaldehyde are enantiotopic as the products are enantiomeric (Fig. 2.56A). The enantiotopic relationship also follows from the symmetry criterion: the two faces are not interchangeable by a $C_2$ operation but are related by a symmetry plane $\sigma$ (the plane of the molecule). In general, carbonyl *Re*/*Si* faces are enantiotopic for a prochiral carbonyl group in an achiral molecule.

For a chiral carbonyl substrate with a chiral centre, the faces are diastereotopic (Fig. 2.56B). The front face is *Si* or *SiS* (using double indexing, cf. Fig. 2.53B) and the rear face is *Re* or *ReS*. Nucleophilic attack on the front (*Si*) face or the rear (*Re*) face creates a new chiral centre (*R* or *S*).

Like carbonyl faces, the two faces of an alkene double bond can also be enantiotopic or diastereotopic. The *Re*/*Si* descriptors to the two alkene faces of α,β-unsaturated carbonyl compounds are illustrated in Fig. 2.57. Fig. 2.57A shows the descriptors in respect of a trigonal α-carbon for the two enantiotopic faces of an acrylic acid ester. (If the ester group contains a chiral centre, the two faces will be diastereotopic.) The upper face or lower face addition is indicated by a dashed line. The upper face can be observed directly and its descriptor in respect of α-carbon is easily assigned as α-*Si*. The lower face is however difficult to observe directly but it holds a mirror image

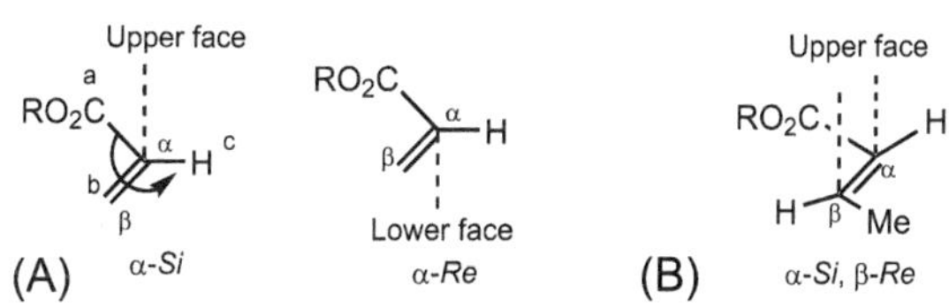


**FIG. 2.57** *Re*/*Si* Descriptors to (A) upper and lower alkene faces of an acrylic acid ester with respect to α-carbon and (B) upper alkene face of a crotonic acid ester with respect to both α- and β-carbons.

relationship to the upper face and its descriptor must be opposite, that is α-*Re*. The *Re*/*Si* descriptors also apply to both trigonal alkene carbons (α and β) as shown in the case of crotonic acid ester (Fig. 2.57B). The upper face is α-*Si*, β-*Re*.

**Problem 2.14** The alkene protons of the diester of maleic acid with (−)-menthol appear as a singlet (2H, s) in $^1$H NMR but the signal is split and a double doublet (dd) is observed for its epoxide. Explain.

### 2.12.6 *Re*/*Si* descriptors to enantiotopic ligands

Like enantiotopic faces, *Re*/*Si* descriptors can be assigned to enantiotopic ligands as an alternative to *pro-R*/*pro-S* descriptors. This follows from a view of prochirality based on chirality in two-dimension.[75] To specify *Re* or *Si* descriptor to a homomorphic ligand (say, $H_a$), the molecule is viewed from $H_a$ when the other three ligands (a, b and c with the priority order $a > b > c$) are defined in a plane. If the sequence $a \rightarrow b \rightarrow c$ is clockwise, the descriptor is *Re*, and if anticlockwise, the descriptor is *Si*. Fig. 2.58A shows the descriptors for the enantiotopic ligands $H_a$ and $H_b$ in propionic acid as *Re* and *Si*, respectively. Here *Re* corresponds to *pro-R* and *Si* to *pro-S* (see Fig. 2.53A). However, this correspondence is not general. It is valid if the homomorphic ligand in the set of three ligands is of lowest priority (c) (Fig. 2.58A) or of highest priority (a). In the case of intermediate priority, the correspondence is not valid, as shown for an enantiotopic ligand ($Me_a$) in 2-bromopropane when *Re* corresponds to *pro-S* (Fig. 2.58B).

### 2.12.7 Topicity and enzyme-catalysed reactions

In enzyme-catalysed reactions, one of the two enantiotopic or diastereotopic ligands/faces of a substrate is involved specifically. To illustrate, we will consider a classical experiment of enzyme-catalysed reduction of acetaldehyde to ethanol in the presence of the enzyme yeast alcohol dehydrogenase and NADH, the reduced form of the coenzyme nicotinamide adenine dinucleotide ($NAD^+$).[77,78] The reducing agent NADH has a prochiral centre bearing two diastereotopic hydrogens ($H_R$ and $H_S$) as shown in Fig. 2.59. The carbonyl group in acetaldehyde is also prochiral and has two enantiotopic (*Re*/*Si*) faces. In the

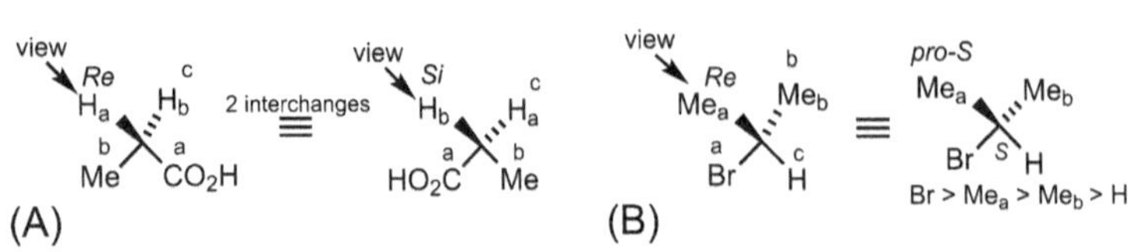


**FIG. 2.58** *Re*/*Si* Descriptors to the enantiotopic ligands in (A) propionic acid and (B) 2-bromopropane.

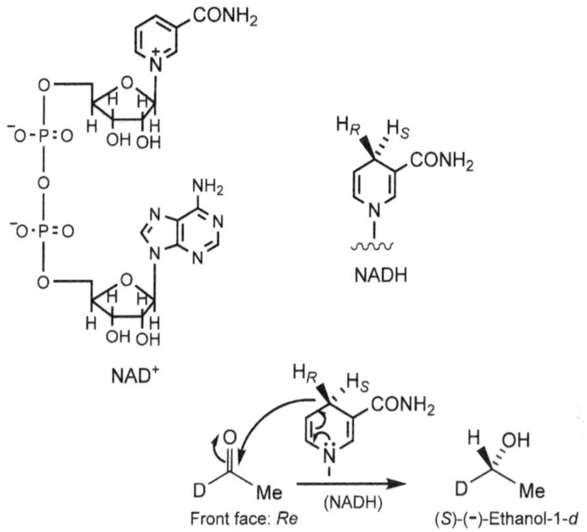


FIG. 2.59 Enzyme-catalysed reduction of deuterated acetaldehyde (MeCDO) to ethanol-1-*d* in the presence of yeast alcohol dehydrogenase and NADH.

reduction, NADH acts as a hydride transfer agent like $NaBH_4$ and one of the two hydrogens ($H_R$ or $H_S$) is transferred to one of the two faces (*Re* or *Si*) of the aldehyde. The question is: which hydrogen is transferred to which face? To address this issue, a deuterated acetaldehyde (MeCD=O) has been employed in the experiment. The product is then ethanol-1-*d* which is levorotatory ($[\alpha]_D = -0.3$), the absolute configuration of which was established earlier as *S*. The formation of *S*-(−)-ethanol-1-*d* from MeCDO confirms that $H_R$ of NADH is transferred to the *Re* face of the aldehyde (Fig. 2.59). This is attributed to a lock and key model in which the aldehyde fits into the active site of the enzyme in such a way as to allow the attachment of only $H_R$ to the *Re* face.[79] It may be mentioned that in the reverse enzyme catalysed oxidation, it is the $H_R$ of ethanol that is abstracted by the oxidizing agent $NAD^+$.

## 2.13 Resolution

### 2.13.1 Introduction

Obtaining a chiral compound as single enantiomers is extremely important in biological recognition and response. For example, the (*S*) enantiomer of the analgesic naproxen functions as an active drug whereas the (*R*) enantiomer is essentially inactive. There are broadly three ways that can give single enantiomers of chiral compounds.

(a) **Chiral pool**: A chiral pool is a collection of naturally occurring enantiomerically pure compounds. This includes amino acids and their derivatives and carbohydrates. Since most of the chiral molecules from plants and animals are synthesized by enzyme-catalysed reactions, most of them are available as single enantiomers. Nowadays, the chiral pool has been expanded by the inclusion of semi-synthetic and synthetic enantiomeric products.[80]

**(b)** **Resolution**: If a chiral compound is synthesized from achiral or chiral but racemic starting materials, the product is always obtained as a racemate (1:1 mixture of two enantiomers). The process of separation of a racemate into the component enantiomers is called resolution.[81] Thus, resolution can be used to obtain single enantiomers.

**(c)** **Asymmetric synthesis**: Asymmetric synthesis is the synthesis of an enantiomer (see Section 5.3). This is the preferred method to obtain a single enantiomer of a variety of chiral compounds, wherever this is possible.

In this section, we will describe briefly the principles and methods of resolution.[82] Enantiomers have identical physical properties (such as melting point, boiling point, solubility etc.) unless placed in a chiral environment. Thus enantiomers cannot be separated by conventional methods of separation. On the other hand, diastereomers have different physical properties and can be separated by the methods used for ordinary mixtures. Two enantiomers of a chiral compound interact differently with an enantiomer of other chiral compounds including chiral solvents and chiral adsorbents in chromatography. Such interactions are diastereomeric. The principles of resolution methods are based on the diastereomeric interactions.

Resolution thus requires an enantiomerically pure chiral substance as a resolving agent that is available from the chiral pool or can be obtained by a simple derivatization of the chiral pool compound. One advantage of resolution is that both enantiomers of a compound are obtained by using one chiral substance from the chiral pool. However, the *maximum* yield of one enantiomer is 50%. Again, if only one enantiomer is useful, the other enantiomer is wasted. One practical problem for resolution is the difficulty in separation of the diastereomeric derivatives or complexes and to determine how well a separation has taken place. Polyfunctional or aromatic resolving agents are better than monofunctional or aliphatic ones because multiple interaction sites can cause more efficient separation of diastereomers based on their solubility differences or elution behaviour.[83,84]

### 2.13.2 Resolution via diastereomeric derivatives

If the diastereomeric derivatives of a racemic sample can be obtained as salts by acid–base reactions, resolution is relatively simple. Salts are easily prepared and the constituent acids and bases can be easily recovered from them. Furthermore, salts are usually crystalline and can be separated by crystallization. The packing of ions in diastereomeric salts leads to difference in lattice energies and hence

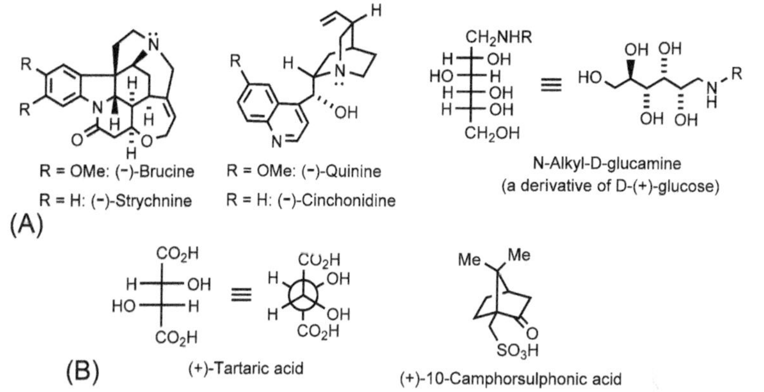


**FIG. 2.60** (A) Resolving agents for acids and (B) resolving agents for bases.

their solubilities. The less soluble salts are easier to purify and thus one pure enantiomer is obtained easily. In order to resolve racemic acids, an enantiomerically pure chiral base from the chiral pool or a simple derivative of that base is employed. Some examples of these resolving bases[85] are (−)-brucine, (−)-strychnine, (−)-quinine, (−)-cinchonidine and *N*-alkyl-D-glucamine (a derivative of D (+)-glucose) (Fig. 2.60A). Acids such as (+)-tartaric acid and (+)-10-camphorsulphonic acid are often used as resolving agents[85] for the resolution of racemic bases (Fig. 2.60B). As 10-camphorsulphonic acid is a stronger acid than tartaric acid, it is generally used for making salts of weak racemic bases.

A general scheme for resolution of a racemic acid using (−)-brucine as a resolving agent is shown in Fig. 2.61A. Note that since (+)-acid and (−)-acid have opposite configurations, the two salts must be diastereomeric. The diastereomeric salts have different solubilities and are separated by careful crystallization. The separated salts are then treated with HCl when the enantiomers are precipitated from the separate acidic solutions. The resolving agent (−)-brucine remains in the solution as its salt from which (−)-brucine can be recovered by treatment with NaOH and reused. The alkaloid brucine has a rigid heptacyclic skeleton which imparts low solubility on the carboxylate salts and brucine is widely used as an effective basic resolving agent.[86]

Fig. 2.61B shows a specific example of resolution of the analgesic drug naproxen[87] (a carboxylic acid). Of the two enantiomers of naproxen, only the (*S*) enantiomer is active as a drug while the (*R*) enantiomer is essentially inactive. An effective resolving agent for racemic naproxen is *N*-octyl-D-glucamine[88] (see Fig. 2.60A). The diastereomeric salt of (*S*)-naproxen with (+)-*N*-octyl-D-glucamine crystallizes out while the other diastereomeric salt of (*R*)-naproxen remains in solution as a soluble salt. The crystals are filtered off and treated with HCl to release (*S*)-naproxen. Thus naproxen is obtained easily as its active (*S*) enantiomer. The resolving base *N*-octyl-D-glucamine can also be recycled.

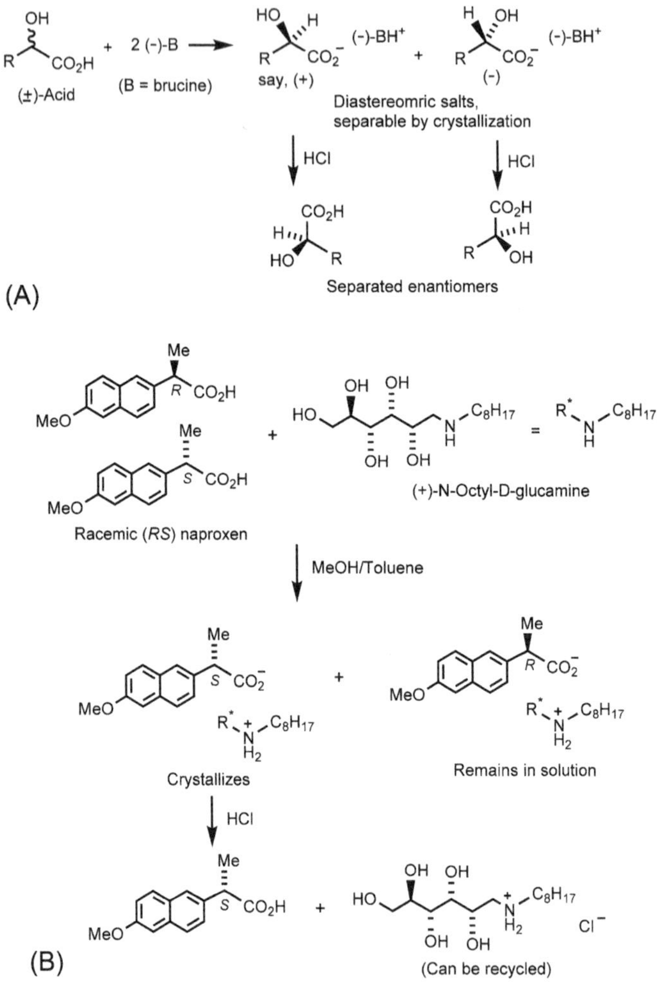


**FIG. 2.61** (A) A general scheme for resolution of a racemic acid via formation of diastereomeric salts and (B) resolution of a racemic analgesic drug naproxen using *N*-octyl-D-glucamine as a resolving chiral base.

For the resolution of a neutral racemate, one strategy is to convert the neutral compound to its acidic derivative and then resolve using a chiral base such as (−)-brucine. A racemic alcohol 2-butanol can be converted to the half-ester of a dibasic acid using a cyclic anhydride such as phthalic anhydride. The racemic acid-ester is then resolved using (−)-brucine, and the separated enantiomeric acid-esters are either hydrolyzed or reduced with $LiAlH_4$ to the enantiomeric alcohols (Fig. 2.62A). The phthalate method[85,89] was a popular choice in the earlier years but the efficiency of chromatographic separation has largely replaced this method. The racemic alcohols can be converted into their diastereomeric esters with the semi-synthetic or synthetic acidic resolving agents such

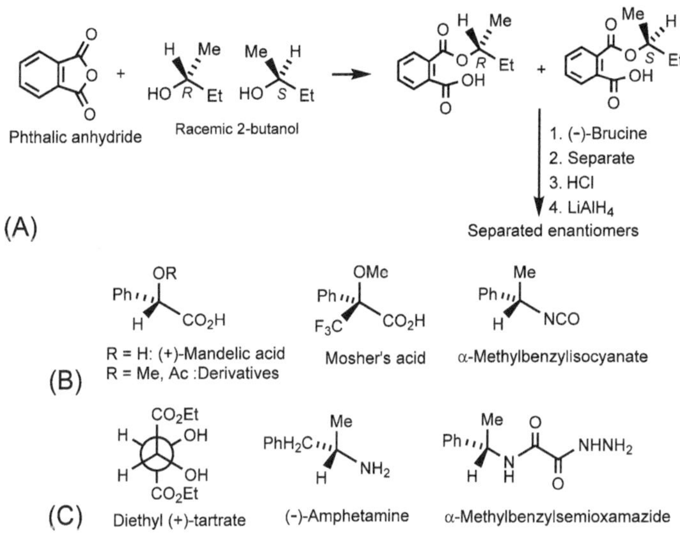


FIG. 2.62 (A) Resolution of a racemic alcohol by the phthalate method. (B) Resolving agents for alcohols. (C) Resolving agents for aldehydes and ketones.

as (+)-mandelic acid (or its derivatives)[90], and Mosher' acid[91] (Fig. 2.62B). The diastereomeric esters are then separated by chromatography. Isocyanates as resolving agents have also been employed to resolve racemic alcohols which are converted into diastereomeric urethane esters. An example of such resolving agent is α-methylbenzylisocyanate[92] (Fig. 2.62B). Racemic aldehydes or ketones can be resolved via formation of diastereomeric cyclic acetals[93] using diethyl (+)-tartrate as a resolving agent (Fig. 2.62C). Other carbonyl derivatizing agents such as (−)-amphetamine[94] (which forms Schiff's bases) and α-methylbenzylsemioxamazide[95], an analogue of semicarbazide, have also found application in the resolution of racemic aldehydes and ketones (Fig. 2.62C).

## 2.13.3 Resolution via complexes and inclusion compounds

Resolution of chiral alkenes such as (*E*)-cyclooctene (see Fig. 2.48B) can be performed via diastereomeric complexes using a chiral square planar platinum complex **2.19** (Fig. 2.63).[96] Diastereomeric complexes result from the

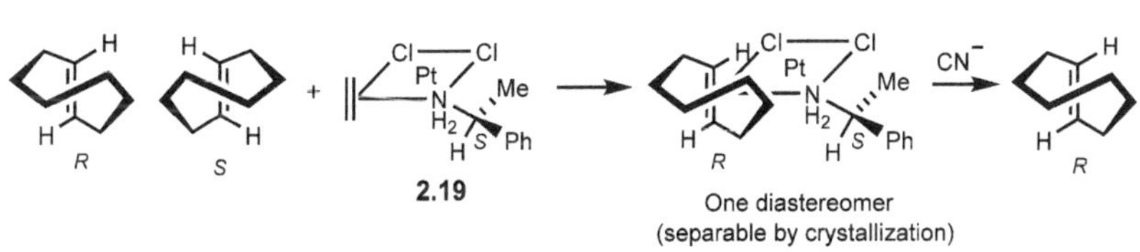


FIG. 2.63 Resolution of (*E*)-cyclooctene via diastereomeric platinum complexes (only one diastereomeric complex is shown).

replacement of ethylene in **2.19** by the enantiomers of (*E*)-cyclooctene (only one diastereomeric complex is shown in Fig. 2.63). The C=C π bond in (*E*)-cyclooctene is twisted and therefore coordinates more strongly with platinum than ethylene does. The diastereomeric complexes are separable by crystallization at low temperature. The enantiomers of (*E*)-cyclooctene are obtained from the separated complexes by displacement with $CN^-$.

Resolution can also be achieved through formation of diastereomeric π- or charge-transfer complexes using a chiral tetranitro aromatic compound, TAPA [α-(2,4,5,7-tetranitro-9-fluorenylideneaminooxy)propionic acid][97,98] as a resolving agent. These complexes can be separated by crystallization. In some cases, one diastereomeric π-complex crystallizes and the other remains in solution as dissociated. A remarkable resolution of racemic hexahelicene (see Fig. 2.50) has been achieved[99,100] using (−)-TAPA. In this case (−)-hexahelicene crystallizes whereas (+)-hexahelicene forms a relatively stable π-complex that remains in solution (Fig. 2.64).

Resolution of racemic samples that are inert or difficult to resolve can be carried out via formation of lattice inclusion compounds. Some substances crystallize in such a fashion that a hole or channel is created in the lattice. The efficient packing in the crystal lattice can occur when other guest molecules fill up the holes or channels. The formation of these inclusion compounds depends primarily on the steric fit between the host and guest molecules. A useful host employed for resolution of racemic 2-halobutanes by lattice inclusion is chiral tri-*o*-thymotide (TOT)[101] shown in Fig. 2.65A. (+)-TOT crystallizes preferentially with (*S*)-2-bromobutane, thereby effecting partial resolution (ee = 32–45%). (−)-Brucine also forms lattice inclusion compounds with a variety of small molecules.[102] Fig. 2.65B shows that (−)-brucine crystallizes with (+) enantiomer of an acetylenic alcohol from which (+)-alcohol is recovered in a pure form (ee 100%).[103] The (−) enantiomer is also recovered from the solution with an ee 93%. Examples of racemic haloalkanes partly resolved through lattice inclusion compounds with (−)-brucine are shown in Fig. 2.65C.

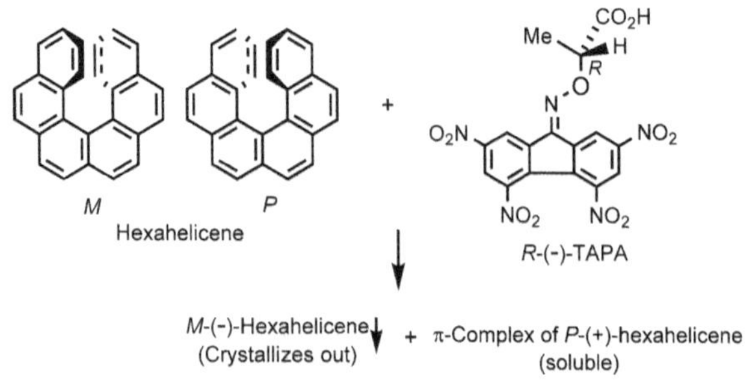


**FIG. 2.64** Resolution of racemic hexahelicene using (−)-TAPA as a resolving π-acid.

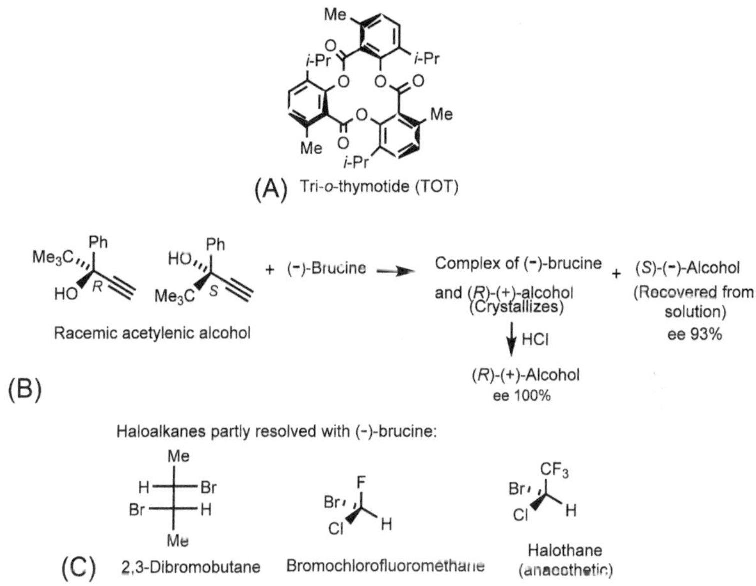


**FIG. 2.65** Resolution through formation of lattice inclusion compounds: (A) tri-*o*-thymotide (TOT) as a host for resolving 2-halobutanes; (B) resolution of a racemic acetylenic alcohol using (−)-brucine; (C) examples of racemic haloalkanes partly resolved using (−)-brucine (in each case one enantiomer is drawn).

### 2.13.4 Resolution by chromatography using chiral stationary phase

Since the interactions of enantiomeric molecules with the surface of a chiral adsorbent (stationary phase) are diastereomeric, the chiral stationary phase (CSP) binds one enantiomer more tightly than the other. This binding increases the retention time of that enantiomer and thereby permits chromatographic separation of the enantiomers. In general, the differences in interactions are small and the chromatographic passes are repeated for a preparative resolution. A simple chiral adsorbent is powdered lactose **2.20** (a disaccharide of (+)-glucose and (+)-galactose) which was first used for resolution of Tröger's base **2.21**[104] (Fig. 2.66). The interaction is mediated mainly by hydrogen bonding. The most widely used chiral adsorbents are based on natural polymers such as cellulose. For example, a cellulose derivative **2.22** was employed to resolve racemic carboxylic acids such as mandelic acid ($PhCHOHCO_2H$).[105] Synthetic chiral polymers such as polyacrylamides **2.23** are also used as a stationary phase.[106] Chiral polyacrylamide is usually prepared by polymerization of the acryl derivative of enantiopure amino acid ester as shown in Fig. 2.66. Racemic chlorthalidone **2.24** has been resolved on a chiral polyacrylamide.[107]

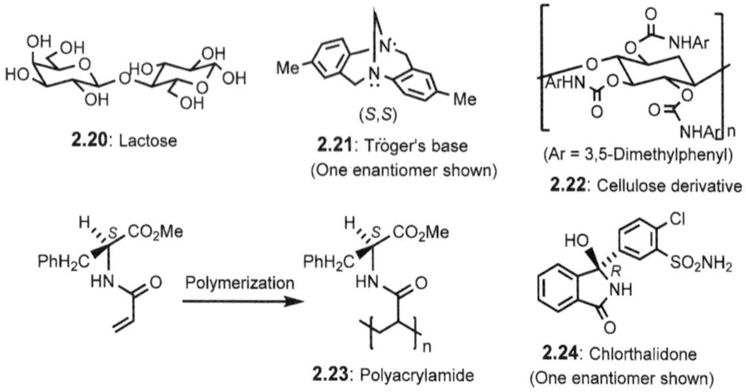


**FIG. 2.66** Chiral stationary phases based on lactose **2.20**, cellulose derivative **2.22** and polyacrylaide **2.23** for chromatographic resolutions and structures of Tröger's base **2.21** (resolved on lactose) and chlorthalidone **2.24** (resolved on a chiral polyacrylamide).

Synthetic resolving agents such as TAPA (see Fig. 2.64) can be covalently bonded to an achiral silica gel support and the resulting chiral phase is employed effectively for the resolution of chiral hydrocarbons such as helicenes.[108,109]

A very useful technique for resolution is based on high-performance liquid chromatography (HPLC) using a CSP developed by Pirkle[110] (Fig. 2.67). The CSP **2.25** is prepared by covalent attachment of an enantiopure chiral agent 2,2,2-trifluoro-1-(9-anthryl)-ethanol (TFAE) to mercaptopropyl silica. The racemic sample to be resolved is converted to a 3,5-dinitrophenyl derivative. A variety of chiral substrates such as amines, alcohols and amino acid esters are converted to 3,5-dinitrophenyl derivatives **2.26** by treatment with

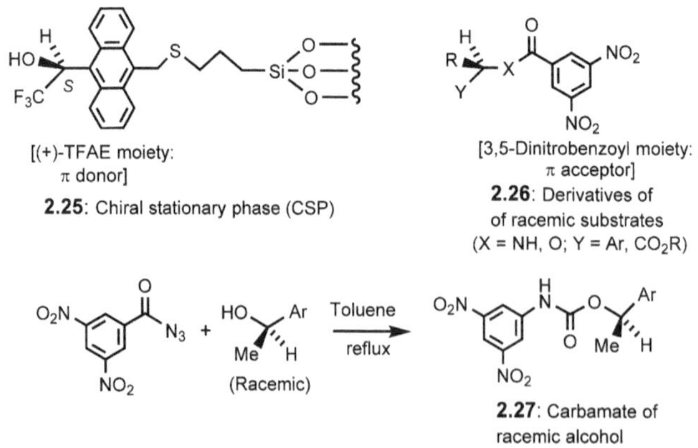


**FIG. 2.67** Chiral stationary phase with a π base **2.25** and a complementary π acid derivative of racemic substrates **2.26**. Preparation of carbamate derivative of a racemic alcohol **2.27**.

3,5-dinitrobenzoyl chloride for their resolution by this method. An alcohol can also be converted to a carbamate **2.27** via an isocyanate intermediate from the Curtius rearrangement of 3,5-dinitrobenzoyl azide. The binding of a chiral enantiomer of the substrate to the CSP occurs largely through π-π or charge-transfer interaction between the electron-deficient 3,5-dinitrophenyl ring (a π acceptor or π acid) and the electron-rich anthracene ring (a π donor or π base) of TFAE moiety.[111]

## 2.13.5 Resolution using achiral phase chromatography

Resolution with both achiral phases in chromatography has been developed recently with the racemic sample premixed with a chiral inducing reagent (CIR).[112] The mixture is subjected to thin layer chromatography (TLC) on plain silica gel plates, and the chromatograms are developed with the mobile phase containing no chiral additive. The separated enantiomers are isolated and the resolution ($R_S$) is estimated by dividing the distance between two spots by the sum of two spot radii. A value of $R_S \geq 1.50$ is taken as indicative of complete resolution.[113] For example, complete resolution has been achieved for (±)-selenomethionine with (−)-quinine as a CIR and for the β blocker (±)-orciprenaline using L-(+)-glutamic acid as a CIR (Fig. 2.68).

The resolution has been explained in terms of formation of diastereomeric associates (via noncovalent interactions) which have different chromatographic mobilities in achiral phases. Theoretical calculations support the formation of diastereomers and their elution order. It has been shown that stronger interactions in the heterochiral enantiomeric associate [(−)-(+)] diastereomer[112,114] (from (±)-orciprenaline and (+)-Glu) contribute to the enhanced hydrophobicity and hence increased retention time (lower $R_F$, 0.33) in comparison to the other [(+)-(+)] diastereomer (higher $R_F$, 0.53).

For kinetic resolution, see later Section 5.3.4.2.

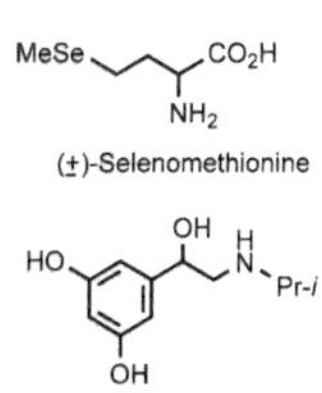


| CIR | $R_F$ (+) | $R_F$ (-) | $R_S$ |
|---|---|---|---|
| (-)-Quinine | 0.26 | 0.13 | 1.69 |
| L-(+)-Glu | 0.53 | 0.33 | 2.1 |

**FIG. 2.68** Resolution using achiral phase thin layer chromatography (TLC) on plain silica gel plates using the mobile phase ($MeCN-MeOH-CH_2Cl_2-H_2O$). CIR = chiral inducing reagent, $R_F$ = retention factor, $R_S$ = resolution (calculated by dividing the distance between two spots by the sum of two spot radii).

## 2.14 Determination of enantiomeric composition

Enantiomeric composition of a chiral sample is commonly expressed as enantiomeric excess (ee), an excess of one enantiomer over the other (see Section 2.4.3). Several methods are employed to determine ee. A method based on a chiroptical property (optical rotation) has been described earlier (see Section 2.4.3). Methods based on the separate measurement of each enantiomer or its derivative, in a mixture are superior. These methods include: (a) NMR spectroscopy with a mixture of diastereomeric derivatives (b) Chromatography using a chiral stationary or mobile phase (c) NMR in a chiral environment using a chiral solvent or chiral shift reagent and (d) Method based on enzyme specificity.

### 2.14.1 NMR methods based on forming diastereomeric derivatives

The enantiomers of a chiral but nonracemic sample are converted quantitatively into their diastereomeric derivatives using an enantiopure chiral derivatizing agent. The diastereomeric mixture is then analysed by NMR.[115,116] NMR is an achiral probe but since diastereomers differ in an achiral environment, they would give distinguishable and measurable signals, the intensity ratio of which would correlate with the ratio of enantiomers in the original sample. Fig. 2.69 shows that a nonracemic amine is converted into the diastereomeric (*R*,*R*) and (*R*,*S*) amides using acid chloride of (*R*)-Mosher's acid (see Fig. 2.62B) as a chiral derivatizing agent. Due to the presence of a $CF_3$ group in the amide derivatives, the diastereomer analysis is possible by $^{19}$F NMR which simplifies the analysis due to smaller number of peaks and better signal separation than $^{1}$H NMR.[117]

A major problem in this method is that the derivatizing agent must be enantiomerically pure. If not, the minor component, the (*S*) reagent would lead to (*S*,*S*) and (*S*,*R*) derivatives which will be indistinguishable in NMR from the enantiomeric (*R*,*R*) and (*R*,*S*) derivatives, and the estimated enantiomeric purity will be less than the true value. Furthermore, the reaction must be quantitative otherwise the rate difference between the enantiomers might influence the estimate of ee.

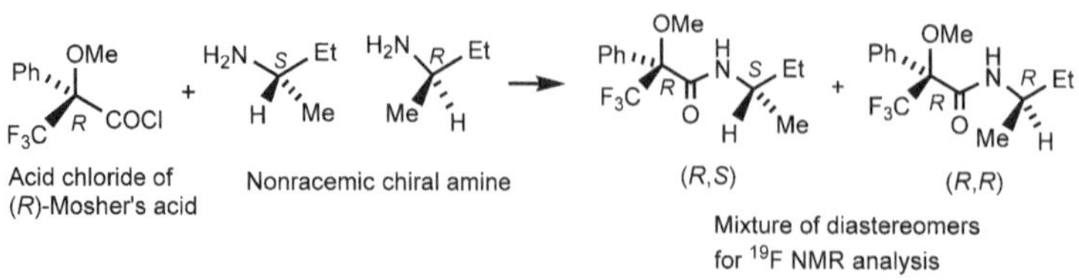


**FIG. 2.69** Formation of diastereomeric derivatives of a nonracemic amine with the acid chloride of (*R*)-Mosher's acid) for $^{19}$F NMR analysis to determine the enantiomeric composition.

## 2.14.2 Chromatographic methods based on diastereomeric interactions

The direct analysis of enantiomers without derivatization is possible using a CSP in efficient analytical columns in gas chromatography (GC) or HPLC.[118,119] In the case of HPLC, a chiral mobile phase can also be used in combination with an achiral stationary phase.

The interactions of enantiomers of a nonracemic sample with the chiral stationary phase are diastereomeric in nature. The enantiomer that binds more strongly with a larger number of stabilizing interactions appears at longer elution time and the less tightly bound enantiomer elutes early. It is possible to perform the chromatography of the same sample on a CSP of opposite enantiomer when the retention times are reversed. Fig. 2.70 shows an enzyme-catalysed asymmetric synthesis of aspartate. The product is converted into a volatile derivative before analysis by gas chromatography on the stationary phase Chirasil-Val (a chiral polysiloxane) **2.28**[120].A mass spectrometer is used as a selective detector. The enzymatic reaction is found to be highly enantioselective. In the presence of enzyme L-aspartase, the fumarate is predominantly converted into L-aspartate by the attack of $NH_3$ at an alkene carbon on the *Si* face. The enantiomeric composition from GC analysis shows that L-aspartate is almost exclusive (99.99%) with the minor D-aspartate being as little as 0.01%. This shows the remarkable sensitivity of the GC method with a sensitivity of 0.01% in detecting a minor component.

The use of a chiral mobile phase together with an achiral stationary phase in HPLC has been demonstrated in the determination of enantiomeric excess of the

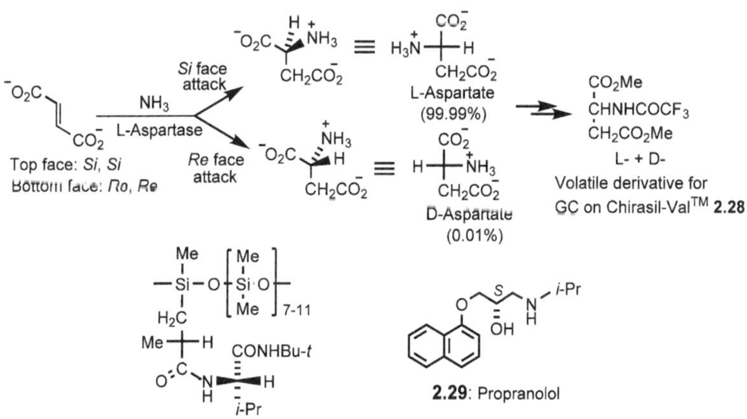


**2.28**: Chirasil-Val™

**FIG. 2.70** Determination of enantiomeric composition of aspartate from an enzyme-catalysed asymmetric synthesis by gas chromatography on a chiral stationary phase Chirasil-Val **2.28**. **2.29**: Propanolol, a β-blocking drug, whose enantiomeric composition has been determined by HPLC using a chiral mobile phase.

β-blocking drug propranolol **2.29** (Fig. 2.70). The chiral mobile phase is set up by the addition of (+)-10-camphorsulphonic acid (see Fig. 2.60B) to a solvent of low polarity. Propranolol **2.29** is a chiral amine and forms diastereomeric ion pairs with (+)-10-camphorsulphonic acid. The diastereomeric ion pairs migrate differently leading to their separation.[121]

### 2.14.3 NMR methods based on chiral solvents or chiral shift reagents

The enantiomers of a chiral substance may exhibit different NMR chemical shifts when the measurement is performed in a chiral nonracemic solvent. The specific nuclei of solutes that were enantiotopic in an achiral environment become diastereotopic by their interaction with a chiral nonracemic environment and may thus exhibit nonequivalence or *anisochrony* in NMR.[122,123] This is a much simpler and most direct method of enantiomer differentiation. Instead of using expensive chiral solvents, a nonracemic chiral solvating agent (chiral additive) may be added to the NMR sample in an achiral solvent. Several chiral solvating agents have been employed for the determination of ee.[116,124] The most commonly used is enantiopure 2,2,2-trifluoro-1-(9-anthryl)-ethanol (TFAE) **2.30**[125](Fig. 2.71) (see also Fig. 2.67) which can form diastereomeric complexes with the enantiomers of nonracemic solute. These diastereomeric complexes have different chemical shifts, and the integration of the NMR peaks gives the ratio of enantiomers. This method however depends on the mobile equilibrium between the complexed and free solute molecules, which are very fast on the NMR timescale. The complexation by TFAE may involve H-bonding by OH group, π-π interaction through the anthracene moiety and the interaction through the polar $CF_3$ group. This model is known as a *three point interaction model*.[70,126] The effect of these interactions also limits the relative orientations of solute–solvent pairs and thereby maximizes the diastereomeric differences in the complexes and minimizes the averaging out the chemical shifts of the groups in solute molecules. Furthermore, the ring current of the aromatic ring of TFAE can induce effective differences in proton chemical shifts of groups in the shielding or deshielding region in the diastereomeric complexes. The determination of enantiomer composition using TFAE as a chiral solvating agent is applicable to a wide variety of functionalized solutes

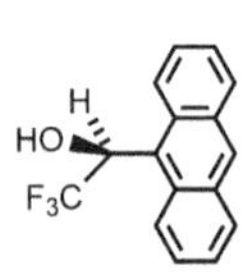


**2.30**

(*S*)-(+)-TFAE

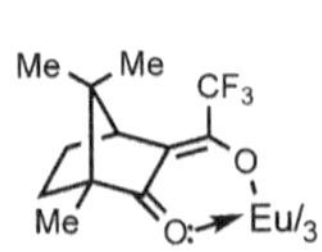


**2.31**

tris(3-Trifluoroacetyl-*d*-camphorato)-europium(III)

**FIG. 2.71** A chiral solvating agent, 2,2,2-trifluoro-1-(9-anthryl)-ethanol (TFAE) **2.30** and a chiral shift reagent, tris(3-trifluoroacetyl-*d*-camphorato)-europium(III) **2.31**.

including alcohols, ethers, ketones, carboxylic acids, lactones, amines, amino acids and sulphoxides.

The chiral shift reagents are generally paramagnetic chiral β-diketone complexes of lanthanides such as Eu or Pr.[127] A typical chiral shift reagent is tris(3-trifluoroacetyl-*d*-camphorato)-europium(III) **2.31** (Fig. 2.71). The reagent can bind rapidly and reversibly to a range of compounds such as alcohols, ethers, ketones, esters, amines and sulphoxides by increasing the coordination number of the metal from 6 to 7. Addition of chiral shift reagents to NMR sample of nonracemic solutes leads to diastereomeric complexes. Here the NMR spectra are usually simpler as the peaks are well separated in the presence of the para magnetic metal.[128] The chiral shift reagents are thus more advantageous than chiral solvating agents for the determination of enantiomeric composition. However, as with chiral solvating agents, this method also depends on the mobile equilibrium between the free and bound states.

**Remarks**: In the methods based on chiral solvating agents or chiral shift reagents, it is not essential that the reagent be enantiomerically pure (ee 100%). The enantiomeric purity of the reagent influences the separation of peaks, and so long the peaks are sufficiently separated for analysis of the enantiomeric composition of the solute which depends on the area or integration of the peaks, a nonracemic reagent that gives a reasonable peak separation will suffice. For similar reasons, a nonracemic chiral stationary phase (see Section 2.14.2) need not necessarily be enantiomerically pure, in contrast to the enantiopure chiral derivatizing agent required to form diastereomeric derivatives (see Section 2.14.1).

## 2.14.4 Enzymatic methods

Enzymes react specifically with one of the enantiomers of a chiral substrate. This specificity of enzymes can be exploited to determine the ee of a nonracemic sample.[129] There are also pairs of enzymes that are specific to the two enantiomers separately. If both enzymes are employed to determine the enantiomeric purity of a sample, one analysis can put a check on the other since the combined results ought to be ee = 100%. Fig. 2.72 shows that the enantiomeric purity of a sample of (*R*)-lactic acid can be determined with the enzyme (*S*)-lactate dehydrogenase in the presence of a stoichiometric amount of the cofactor $NAD^+$ (see Fig. 2.59). The reaction is an enzymatic oxidation or dehydrogenation. Since the enzyme is specific to (*S*)-lactic acid, the minor (*S*)

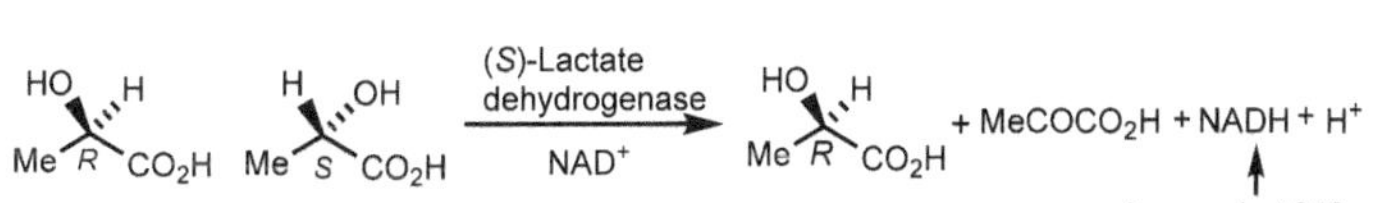


**FIG. 2.72** Determination of enantiomeric purity of (*R*)-lactic acid by enzymatic oxidation in the presence of the enzyme (*S*)-lactate dehydrogenase and the cofactor $NAD^+$.

enantiomer is converted to pyruvic acid when $NAD^+$ is reduced to NADH. The major (*R*) enantiomer would remain unreacted. The formation of NADH is assayed by UV spectroscopy at 340nm to estimate ee.[130] It is necessary that the reaction goes to completion; otherwise, the results have to be corrected with the equilibrium constant measured under the same conditions.

## References


1. IUPAC. *Compendium of Chemical Terminology: Gold Book*; Version 2.3.3, 2014; p. 318.
2. Hirschmann, H.; Hanson, K. R. *J. Organomet. Chem.* **1971**, *36*, 3293.
3. Mislow, K.; Siegel, J. *J. Am. Chem. Soc.* **1984**, *106*, 3319.
4. Mislow, K. *Introduction to Stereochemistry*; Benjamin: New York, 1965; p. 52.
5. Horeau, A. *Tetrahedron Lett.* **1969**, 3121.
6. Kelvin, L. *Baltimore Lectures on Molecular Dynamics and the Wave Theory of Light*; Clay: London, 1904.
7. Weiss, U. *Experientia* **1968**, *24*, 1088.
8. Mislow, K. *Introduction to Stereochemistry*; Benjamin: New York, 1965; p. 54.
9. Brewster, J. W.; Am, J. *Chem. Soc.* **1959**, *81*, 5475. 5483, 5493.
10. Brewster, J. W. *Top. Stereochem.* **1967**, *2*, 1.
11. Vogel, A. I. *J. Chem. Soc.* **1948**, 1833.
12. Klyne, W.; Buckingham, J. *Atlas of Stereochemistry: Absolute Configurations of Organic Molecules*, 2nd ed, Vol. 1; Oxford University Press: New York, 1978; p. 211.
13. Eliel, E. L.; Wilen, S. H.; Mander, L. N. *Stereochemistry of Organic Compounds*; Wiley: New York, 1994; p. 1085.
14. Brewster, J. W. *Tetrahedron* **1961**, *13*, 106.
15. Gilman, H., Ed. *Organic Chemistry: An Advanced Treatise*; 2nd ed.; New York: Wiley, 1943 (chapter 4).
16. Finar, I. L. *Organic Chemistry*, 5th ed, Vol. Vol. 2; Longman: London, 1975 (chapter 6).
17. Lehn, J. M. *Top. Curr. Chem.* **1970**, *15*, 311.
18. Jennings, W. B.; Boyd, D. R. In *Cyclic Organonitrogen Stereodynamics*; Lambert, J. B., Takeuchi, Y., Eds.; VCH Publishers: New York, 1992; p. 105.
19. Brois, S. J.; Am, J. *Chem. Soc.*, 90; 1968; p. 508. 1680.
20. Horner, L. *Pure Appl. Chem.* **1964**, *9*, 225.
21. Gallagher, M. J.; Jenkins, I. D. *Top. Stereochem.* **1968**, *3*, 1.
22. Rayner, D. R.; Gordon, A. J.; Mislow, K. *J. Am. Chem. Soc.* **1968**, *90*, 4854.
23. Mandal, D. K. *J. Chem. Educ.* **2007**, *84*, 274.
24. Mandal, D. K. *J. Chem. Educ.* **2000**, *77*, 866.
25. Cahn, R. S.; Ingold, C.; Prelog, V. *Angew. Chem. Int. Ed. Engl.* **1966**, *5*, 385.
26. Prelog, V.; Helmchen, G. *Angew. Chem. Int. Ed. Engl.* **1982**, *21*, 567.
27. Eliel, E. L.; Wilen, S. H.; Mander, L. N. *Stereochemistry of Organic Compounds*; Wiley: New York, 1994; p. 103.
28. Starkey, R. *J. Chem. Educ.* **1995**, *72*, 315.
29. Eliel, E. L. *J. Chem. Educ.* **1985**, *62*, 223.
30. Eliel, E. L.; Wilen, S. H.; Mander, L. N. *Stereochemistry of Organic Compounds*; Wiley: New York, 1994; p. 1138.
31. Eliel, E. L. *Stereochemistry of Carbon Compounds*, McGraw-Hill: New York, 1962; p. 22, 88.
32. Bijvoet, J. M.; Peerdeman, A. F.; van Bommel, A. J. *Nature* **1951**, *168*, 271.

33. Nafie, L. A.; Diem, M. *Acc. Chem. Res.* **1979**, *12*, 296.
34. Barron, L. D. *Acc. Chem. Res.* **1980**, *13*, 90.
35. Wood, W. W.; Fickett, W.; Kirkwood, J. G. *J. Chem. Phys.* **1952**, *20*, 561.
36. Moffitt, W. *J. Chem. Phys.* **1956**, *25*, 467.
37. Klyne, W.; Buckingham, J. *Atlas of Stereochemistry*, 2nd ed, Vols. I and II; Chapman & Hall: London, 1978.
38. Fiaud, J. C. In *Stereochemistry, Fundamentals and Methods*; Kagan, H. B., Ed.; Vol. 3; Georg Thieme Publishers: Stuttgart, Germany, 1977.
39. Mayo, B. C. *Chem. Soc. Rev.* **1973**, *2*, 49.
40. Neuhaus, D.; Williamson, M. P. *The Nuclear Overhauser Effect in Structural and Conformational Analysis*; VCH Publishers: New York, 1989.
41. Seebach, D.; Prelog, V. *Angew. Chem. Int. Ed. Engl.* **1982**, *21*, 654.
42. Carey, F. A.; Kuehne, M. E. *J. Organomet. Chem.* **1982**, *47*, 3811.
43. Masamune, S.; Ali, S. A.; Snitman, D. L.; Garvey, D. S. *Angew. Chem. Int. Ed. Engl.* **1980**, *19*, 557.
44. Cahn, R. S.; Ingold, C.; Prelog, V. *Experientia* **1956**, *12*, 81.
45. Rabinovitch, B. S.; Michel, K.-W. *J. Am. Chem. Soc.* **1959**, *81*, 5065.
46. Blackwood, J. E.; Gladys, C. L.; Loening, K. L.; Petrarca, A. E.; Rush, J. E. *J. Am. Chem. Soc.* **1968**, *90*, 509.
47. Bertsch, K.; Karich, G.; Jochims, J. C. *Chem Ber.* **1977**, *110*, 3304.
48. Lowe, G. *Chem. Commun.* **1965**, 411.
49. Kuhn, R. In *Stereochemie*; Freudenberg, K., Ed.; Franz Deutike: Leipzig, Germany, 1933; p. 803.
50. Christie, G. H.; Kenner, J. H. *J. Chem. Soc.* **1922**, *121*, 614.
51. Eliel, E. L.; Wilen, S. H.; Mander, L. N. *Stereochemistry of Organic Compounds*; Wiley: New York, 1994; p. 1142.
52. Ōki, M. *Top. Stereochem.* **1983**, *14*, 1.
53. Krow, G. *Top. Stereochem.* **1970**, *5*, 31.
54. Chandross, E. A.; Sheley, C. F. *J. Am. Chem. Soc.* **1968**, *90*, 4345.
55. Ōki, M. *The Chemistry of Rotational Isomers*; Springer: New York, 1993.
56. Mandal, D. K. *Bull. Chem. Soc. Jpn.* **2002**, *75*, 365.
57. Schlögl, K. *Top. Curr. Chem.* **1984**, *125*, 27.
58. Lüttringhaus, A.; Gralheer, H. *Justus Liebigs Ann. Chem.* **1942**, *550*, 67. **1947**, *557*, 108, 112.
59. Cram, D. J.; Steinberg, H. *J. Am. Chem. Soc.* **1951**, *73*, 5691.
60. Nakazaki, M.; Yamamoto, K.; Naemura, K. *Top. Curr. Chem.* **1984**, *125*, 1.
61. Binsch, G.; Roberts, J. D. *J. Am. Chem. Soc.* **1965**, *87*, 5157.
62. Cope, A. C.; Mehta, A. S. *J. Am. Chem. Soc.* **1964**, *86*, 5626.
63. Martin, R. H. *Angew. Chem. Int. Ed. Engl.* **1974**, *13*, 649.
64. Laarhoven, W. H.; Prinsen, W. J. C. *Top. Curr. Chem.* **1984**, *125*, 63.
65. DeRango, C.; Tsoucaris, G.; Declerq, J. P.; Germain, G.; Putzeys, J. P. *Cryst. Struct. Commun.* **1973**, *2*, 189.
66. Newman, M. S.; Lednicer, D. *J. Am. Chem. Soc.* **1956**, *78*, 4765.
67. Eliel, E. L.; Wilen, S. H.; Mander, L. N. *Stereochemistry of Organic Compounds*; Wiley: New York, 1994; p. 1038.
68. Lightner, D. A.; Hefelfinger, D. T.; Powers, T. W.; Frank, G. W.; Trueblood, K. N. *J. Am. Chem. Soc.* **1972**, *94*, 3492.
69. Hirschmann, H.; Hanson, K. R. *Tetrahedron* **1974**, *30*, 3649.
70. Mislow, K.; Raban, M. *Top. Stereochem.* **1967**, *1*, 1.

71. Eliel, E. L.; Wilen, S. H.; Mander, L. N. *Stereochemistry of Organic Compounds*; Wiley: New York, 1994 (chapter 8).
72. Hybl, A.; Rundle, R. E.; Williams, D. E. *J. Am. Chem. Soc.* **1965**, *87*, 2779.
73. Hanson, K. R. *J. Am. Chem. Soc.* **1966**, *88*, 2731.
74. Retey, J.; Robinson, J. A. *Stereospecificity in Organic Chemistry and Enzymology*; Weinheim: Verlag Chemie, 1982.
75. Prelog, V.; Helmchen, G. *Helv. Chim. Acta* **1972**, *55*, 2581.
76. Hanson, K. R. *Annu. Rev. Biochem.* **1976**, *45*, 307.
77. Loewus, F. A.; Westheimer, F. H.; Vennesland, B. *J. Am. Chem. Soc.* **1953**, *75*, 5018.
78. Levy, H. R.; Loewus, F. A.; Vennesland, B. *J. Am. Chem. Soc.* **1957**, *79*, 2949.
79. Vennesland, B. *Top. Curr. Chem.* **1974**, *48*, 39.
80. ten Hoeve, W.; Wynberg, H. *J. Organomet. Chem.* **1985**, *50*, 4508.
81. IUPAC. In *Compendium of Chemical Terminology the "Gold Book"*; McNaught, A. D., Wilikinson, W., Eds.; 2nd ed.; Blackwell Scientific Publications: Oxford, 1997.
82. Eliel, E. L.; Wilen, S. H.; Mander, L. N. *Stereochemistry of Organic Compounds*; Wiley: New York, 1994; p. 297.
83. Wilen, S. H. *Top. Stereochem.* **1971**, *6*, 107.
84. Pirkle, W. H.; Simmons, K. A. *J. Organomet. Chem.* **1983**, *48*, 2520.
85. Wilen, S. H. In *Tables of Resolving Agents and Optical Resolutions*; Eliel, E. L., Ed.; University of Notre Dame Press: Notre Dame, Indiana, 1972.
86. Jacques, J.; Collet, A.; Wilen, S.; Enantiomers, H. *Racemates and Resolutions*; Wiley: New York, 1981; p. 254.
87. Harrington, P. J.; Lodewijk, E. *Org. Process. Res. Dev.* **1997**, *1*, 72.
88. Holton, P. G. *Chem. Abstr.* **1981**, *95*, 61852n.
89. Klyashchitskii, B. A.; Shvets, V. I. *Russ. Chem. Rev.* **1972**, *41*, 592.
90. Whitesell, J. K.; Reynolds, D. *J. Organomet. Chem.* **1983**, *48*, 3548.
91. Koreeda, M.; Yoshihara, M. *J. Chem. Soc. Chem. Commun.* **1981**, 974.
92. Donaldson, R. E.; Saddler, J. C.; Byrn, S.; McKenzie, A. T.; Fuchs, P. L. *J. Organomet. Chem.* **1983**, *48*, 2167.
93. Fessner, W.-D.; Prinzbach, H. *Tetrahedron* **1986**, *42*, 1797.
94. Arcamone, F.; Bernardi, L.; Patelli, B.; Di Marco, A. *Chem. Abstr.* **1976**, *85*, 142918.
95. Tökés, A. L. *Liebigs Ann. Chem.* **1987**, 1007.
96. Cope, A. C.; Ganellin, C. R.; Johnson, H. W., Jr. *J. Am. Chem. Soc.* **1962**, *84*, 3191.
97. Block, P., Jr.; Newman, M. S. *Org. Synth.* **1968**, *48*, 120.
98. Jacques, J.; Collet, A.; Wilen, S. H. *Enantiomers, Racemates and Resolutions*; Wiley: New York, 1981; p. 273.
99. Newman, M. S.; Lutz, W. B.; Lednicer, D. *J. Am. Chem. Soc.* **1955**, *77*, 3420.
100. Wynberg, H.; Lammertsma, K. *J. Am. Chem. Soc.* **1973**, *95*, 7913.
101. Arad-Yellin, R.; Green, B. S.; Knossow, M.; Tsoucaris, G. *J. Am. Chem. Soc.* **1983**, *105*, 4561.
102. Worsch, D.; Vögtle, F. *Top. Curr. Chem.* **1987**, *140*, 21.
103. Toda, F.; Tanaka, K.; Ueda, H. *Tetrahedron Lett.* **1981**, *22*, 4669.
104. Prelog, V.; Wieland, P. *Helv. Chim. Acta* **1944**, *27*, 1127.
105. Okamoto, Y.; Aburatani, R.; Kaida, Y.; Hatada, K. *Chem. Lett.* **1988**, 1125.
106. Blaschke, G. In *Chromatographic Chiral Separations*; Zief, M., Crane, L. J., Eds.; Marcel Dekker: New York, 1988 (chapter 7).
107. Blaschke, G.; Markgraf, H. *Chem. Ber.* **1980**, *113*, 2031.
108. Klemm, L. H.; Reed, D. *J. Chromatogr.* **1960**, *3*, 364.
109. Numan, H.; Helder, R.; Wynberg, H. *Recl. Trav. Chim. Pays-Bas* **1976**, *95*, 211.

110. Pirkle, W. H.; Finn, J. In *Asymmetric Synthesis*; Morrison, J. D., Ed.; Vol. 1; Academic Press: New York, 1983 (chapter 6).
111. Pirkle, W. H.; House, D. W. *J. Organomet. Chem.* **1979**, *44*, 1957.
112. Bhusan, R.; Nagar, H.; Martens, J. *RSC Adv.* **2015**, *5*, 28316.
113. Sherma, J.; Fried, B., Eds. *Handbook of TLC*; Marcel Dekker: New York, 1990; p. 50.
114. Martens, J.; Bhusan, R. *Helv. Chim. Acta* **2014**, *97*, 161.
115. Gaudemer, A. In *Stereochemistry, Fundamentals and Methods*; Kagan, H. B., Ed.; Vol. 3; Georg Thieme Publishers: Stuttgart, Germany, 1977; p. 117.
116. Parker, D. *Chem. Rev.* **1991**, *91*, 1441.
117. Guerrier, L.; Royer, J.; Grierson, D. S.; Husson, H.-P. *J. Am. Chem. Soc.* **1983**, *105*, 7754.
118. Lochmüller, C. H.; Souter, R. W. *J. Chromatogr.* **1975**, *113*, 283.
119. Gil-Av, E. *J. Chromatogr. Libr.* **1985**, *32*, 111.
120. Frank, H.; Nicholson, G. J.; Baeyer, E. *J. Chromatogr. Sci.* **1977**, *15*, 174.
121. Pettersson, C.; Schill, G. *J. Chromatogr.* **1981**, *204*, 179.
122. Raban, M.; Mislow, K. *Tetrahedron Lett.* **1965**, 4249.
123. Pirkle, W. H.; Am, J. *Chem. Soc.* **1837**, *1966*, 88.
124. Pirkle, W. H.; Hoover, D. J. *Top. Stereochem.* **1982**, *13*, 263.
125. Pirkle, W. H.; Beare, S. D. *J. Am. Chem. Soc.* **1969**, *91*, 5150.
126. Ogston, A. G. *Nature* **1948**, *162*, 963.
127. Sullivan, G. R. *Top. Stereochem.* **1978**, *10*, 287.
128. Fraser, R. R. In *Asymmetric Synthesis*; Morrison, J. D., Ed.; Vol. 1; Academic Press: New York, 1983 (chapter 9).
129. Bergmeyer, H. U., Ed. *Methods of Enzymatic Analysis*, Vols. VI–VIII,; 3rd ed.; VCH: Weinheim, Germany, 1984–1985.
130. Wong, C. H.; Whitesides, G. M. *J. Am. Chem. Soc.* **1981**, *103*, 4890.

Chapter 3

# Cyclic molecules: Configuration and conformation

## 3.1 Configuration of cyclic molecules

Like acyclic molecules, the absolute configuration at a stereocentre in cyclic molecules is also specified by *R*/*S* descriptors.[1] The relative configuration for cyclic diastereomers is usually designated by the descriptor: *cis* (when two substituents are on the same side of the ring) or *trans* (when two substituents are on opposite sides of the ring). The stereoisomerism in cyclic molecules due to the presence of stereocentres is similar to that in acyclic molecules; however, some additional features can arise from the rigidity of the cyclic framework.

### 3.1.1 Cyclic molecules with two unlike/like stereocentres and a systematic *cis*/*trans* nomenclature

Consider, for example, 1-ethyl-2-methylcyclohexane that possesses two unlike stereocentres (cf. Fig. 2.28). The compound has two diastereomers (*cis* and *trans*), each of which is chiral and exists as a resolvable enantiomeric pair (Fig. 3.1A). The absolute configuration at a stereocentre of each stereoisomer is assigned as *R* or *S* (see Fig. 2.20). If the two stereocentres are like as in 1,2-dimethylcyclohexane, one diastereomer (*cis*) is achiral with a symmetry plane while the *trans* diastereomer is chiral and resolvable (Fig. 3.1B). The achiral *cis* diastereomer is also called *meso* (cf. stereoisomers of tartaric acid, Fig. 2.31). However, in many cyclic systems, the number of stereoisomers could be less than expected because of the enhanced symmetry and rigidity of the cyclic framework (see later).

When the ligands at the two stereocentres of cyclic diastereomers are all different, the assignment of *cis* (*c*)/*trans* (*t*) descriptors is made with respect to a reference (*r*) group (Fig. 3.1C).[2] If the cyclic compound is named with a suffix such as carboxylic acid or carboxaldehyde, the reference group is $CO_2H$ or CHO. For example, the compound **3.1** is *c*-2-hydroxy-2-phenyl-*r*-1-cyclopropanecarboxylic acid. (*cis* precedes *trans*; as *cis* OH ensures Ph to be *trans*, the symbol *t* for Ph is omitted.) In other cases of cyclic compounds,

Stereochemistry and Organic Reactions. https://doi.org/10.1016/B978-0-12-824092-2.00003-4

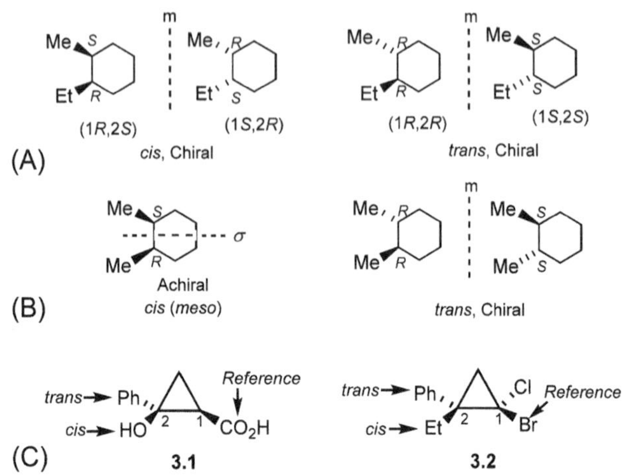


**FIG. 3.1** Stereoisomers of cyclic molecules with (A) two unlike stereocentres and (B) two like stereocentres. (C) A systematic *cis* (*c*)/*trans* (*t*) nomenclature for cyclic diastereomers.

the reference group is taken to be the group of highest CIP priority. Thus the compound **3.2** is *r*-1-bromo-1-chloro-*c*-2-ethyl-2-phenylcyclopropane.

### 3.1.2 Cyclic molecules containing a pseudoasymmetric centre

A centre that is achirotopic but stereogenic is called pseudoasymmetric (see Section 2.8.9). 2,3-Dichlorocyclopropanecarboxylic acid contains two like chiral centres (C2 and C3) and a pseudoasymmetric centre C1 (Fig. 3.2). Three diastereomers (**A**–**C**) are possible, of which two are *meso* (**A** and **B**) and the other (**C**) is chiral and racemic. In the *meso* isomers, C1 is achirotopic (lies on the symmetry plane) but stereogenic (exchange of $CO_2H$ and H at C1 gives a stereoisomer) and hence pseudoasymmetric. In the chiral diastereomer (**C**), C1 is chirotopic (resides in the chiral molecule) but nonstereogenic (exchange of $CO_2H$ and H at C1 does not give a stereoisomer). According to *cis*/*trans* nomenclature,[2] **A** is *c*-2,*c*-3-dichloro-*r*-1-cyclopropanecarboxylic acid, **B** is *t*-2,*t*-3-dichloro-*r*-1-cyclopropanecarboxylic acid and **C** is *c*-2,*t*-3-dichloro-*r*-1-cyclopropanecarboxylic acid (ring numbering follows the rule *cis* precedes *trans*).

### 3.1.3 Cyclic molecules with three and six like stereocentres

1,2,3-Trimethylcyclopropane and 1,3,5-trimethylcyclohexane possess three like stereocentres and their stereoisomerism is similar (Fig. 3.3A). Due to high symmetry of these cyclic molecules, degeneracies occur which lead to a fewer number of stereoisomers. The possible diastereomers can be derived from *cis* (or *trans*) relationship of the methyl substituents as shown in Fig. 3.3A. Three

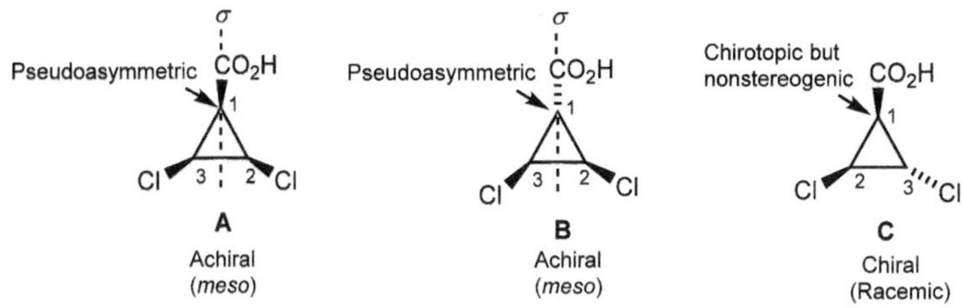


**FIG. 3.2** Diastereomers of a cyclopropane with two like stereocentres and a pseudoasymmetric centre. **A** is *c*-2,*c*-3-dichloro-*r*-1-cyclopropanecarboxylic acid, **B** is *t*-2,*t*-3-dichloro-*r*-1-cyclopropanecarboxylic acid and **C** is *c*-2,*t*-3-dichloro-*r*-1-cyclopropanecarboxylic acid.

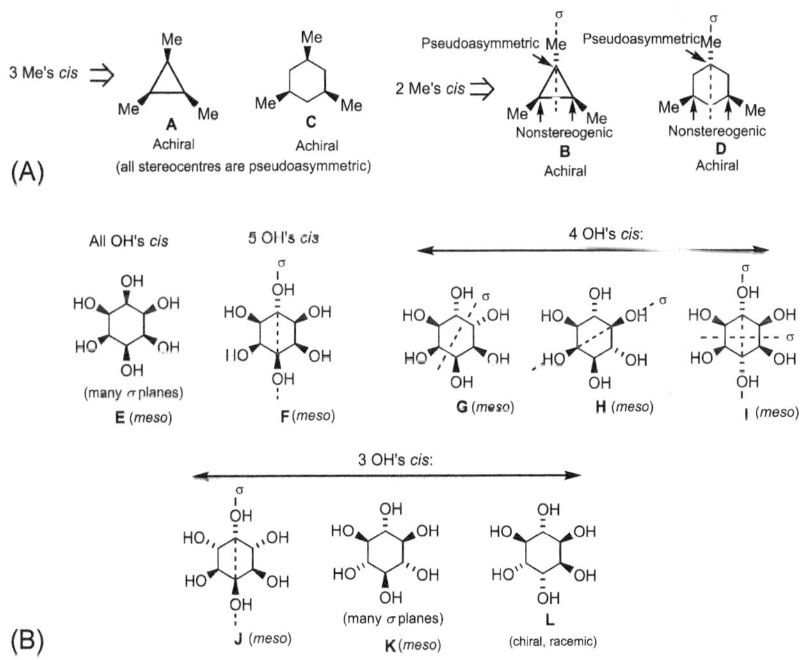


**FIG. 3.3** Diastereomerism in cyclic molecules with (A) three like stereocentres and (B) six like stereocentres.

*cis* Me's give **A** for the cyclopropane derivative (or **C** for the cyclohexane derivative), whereas two *cis* Me's lead to **B** (or **D**). Each diastereomer is achiral due to the presence of at least one vertical plane of symmetry. All three stereocentres in **A** (or **C**) are achirotopic but stereogenic (each lies on a symmetry plane and exchange of two ligands at any ring carbon generates a stereoisomer) and therefore pseudoasymmetric, whereas in **B** (or **D**), the ring carbon lying in the symmetry plane is pseudoasymmetric and the other two ring carbons are nonstereogenic. Since there is no chiral diastereomer in the diastereomeric set, the two achiral diastereomers of 1,2,3-trimethylcyclopropane or 1,2,3-trimethylcyclohexane are *not* called *meso* isomers.

Fig. 3.3B shows eight possible diastereomers (**E–L**) for the naturally occurring inositols[3] (hexahydroxycyclohexanes) with six like stereocentres. Only one of them (**L**) is chiral and resolvable into enantiomers. The other seven diastereomers (**E–K**) have at least one plane of symmetry and are *meso*. Similarly, hexachlorocyclohexane has eight possible diastereomers, of which the γ-isomer (similar to **I** in Fig. 3.3B) is a powerful insecticide known as gammexane.

**Problem 3.1** 2,4-Diphenylcyclobutane-1,3-dicarboxylic acid (truxillic acid) has five diastereomers, all of which are achiral. One of the diastereomers, the α-form, possesses an inversion centre (*i*). Draw the diastereomers.

### 3.1.4 Assignment of *r*/*s* descriptors to pseudoasymmetric centres

1,3-Disubstituted cyclobutanes or 1,4-disubstituted cyclohexanes have two achiral diastereomers (*cis* and *trans*), each possessing two pseudoasymmetric centres. Consider, for example, *cis*- and *trans*-1,3-cyclobutanediol in which C1 and C3 are pseudoasymmetric. The assignment of *r*/*s* descriptors[4] to C1 and C3 of the *cis* isomer is illustrated in Fig. 3.4. The determination of priority order of ligands attached to the pseudoasymmetric centre C1 or C3 is rather complex and is based on artificially transforming the didentate ring ligand into two monodentate ligands as shown in Fig. 3.4. For the didentate ring at C1, one monodentate ligand is obtained by artificially cleaving 1–2 bond and the second results from the artificial cleavage of 1–4 bond. Each monodentate branch has a duplicate pseudoasymmetric atom (C1) at the end (shown within bracket). The relative priority of the two monodentate ligands is determined on the basis of *R*/*S* configuration of the stereogenic centre in the two branches using the sequence rule 5, i.e. $R > S$ (see Section 2.7.1.1). The descriptor *r* or *s* to C1 then follows from the priority order of the four ligands: $OH > R > S > H$. As shown in Fig. 3.4, the descriptor assigned to C1 is *s*. The descriptor to C3 is similarly assigned as *s*. Thus *cis*-1,3-cyclobutanediol is (1*s*,3*s*). Using the similar procedure, the descriptors for the *trans* isomer will be (1*r*,3*r*). It is interesting to note that one exchange of ligands (OH, H) at C1 or C3 for *cis* → *trans* transformation changes descriptors at both pseudoasymmetric centres C1 and C3 (1*s*,3*s* → 1*r*,3*r*).

**Problem 3.2** Assign *r*/*s* descriptors to the following molecules: (a) *cis*-1-ethyl-4-methylcyclohexane (b) *trans*-1,4-cyclohexanediol.

### 3.1.5 Cyclic molecules with a bridged framework

All possible configurations of a cyclic molecule are not sterically accessible for a rigid framework, which leads to a smaller number of available stereoisomers. This is illustrated with two examples in Fig. 3.5A. Camphor **3.3** has two unlike

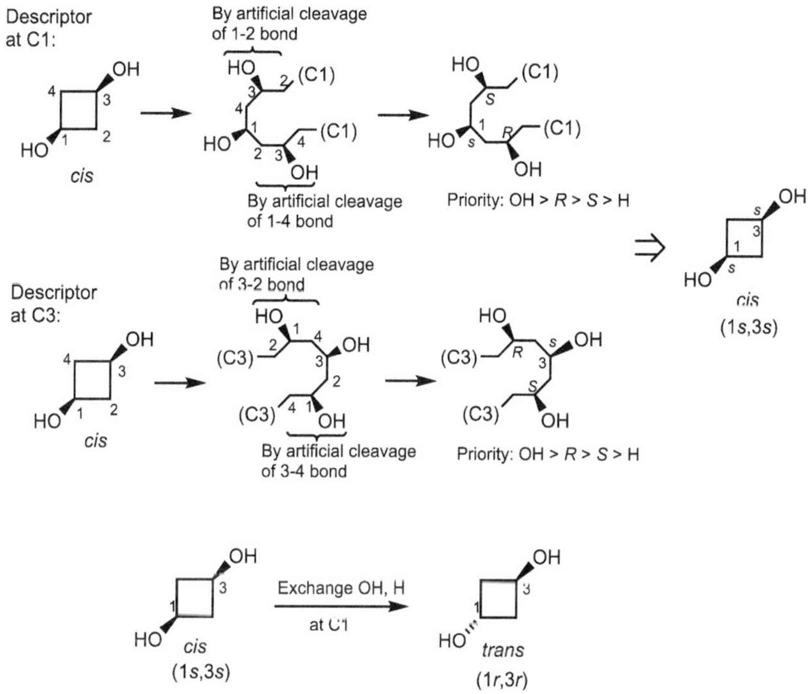


**FIG. 3.4** Assignment of *r*/*s* descriptors to the pseudoasymmetric centres of *cis*-1,3-cyclobutanediol. The *cis*→*trans* transformation changes (1*s*,3*s*) → (1*r*,3*r*).

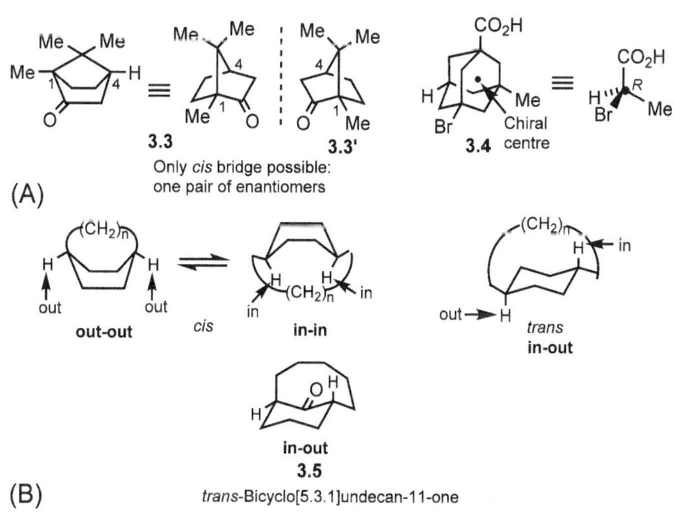


**FIG. 3.5** (A) Bridged cyclic systems with two or more chiral centres but having only one enantiomeric pair: camphor **3.3** and an adamantoid **3.4**. (B) out–out, in–in and in–out isomers of a general bicyclic system, and an example of a *trans* (in–out) compound **3.5**.

chiral centres at bridgehead carbons C1 and C4. Yet camphor exists as a single diastereomer which is chiral and resolvable into two enantiomers (**3.3**, **3.3′**). One out of two possible diastereomers is obtained because the configurations at C1 and C4 cannot be varied independently. The small one-carbon bridge must be *cis* with respect to six-membered ring; thus Me and H at the bridgehead carbons C1 and C4 can only be *cis*. The *trans* one-carbon bridge is sterically impossible.

Adamantane is a highly symmetrical molecule like $CH_4$ and belongs to the point group $\mathbf{T_d}$. If the four bridgehead ligands of adamantane are different as in **3.4** (Fig. 3.5A), the molecule becomes chiral (point group $\mathbf{C_1}$). There are four unlike chiral centres at the four bridgehead carbons but the adamantane derivative **3.4** exists only as a single pair of enantiomers. This is attributed to a chiral centre (represented by a dot) with a tetrahedral arrangement of four bridgehead ligands. Interestingly, this chiral centre is not any real atom but an empty centre in the molecule. This example demonstrates that a chiral centre can lie in an empty space of a molecule. Clearly, the adamantoid **3.4** represents the *R* enantiomer. Exchange of any pair of ligands gives the *S* enantiomer.

In contrast to camphor, if the bridge is large enough, the restriction for a *cis* bridge doesn't apply. This is illustrated in Fig. 3.5B, when both *cis* and *trans* bridges are possible.[5,6] The *cis* configuration can be out–out configuration based on the location of bridgehead hydrogens or other groups. This out–out isomer is drawn in a boat conformation as in camphor, which can convert to an in–in isomer depending on the strain present in it. The *trans* configuration refers to an in–out (or out–in) isomer when the cyclohexane ring can assume the preferred chair conformation. An example of an in–out bicyclic compound with a five-carbon bridge is *trans*-bicyclo[5.3.1]undecan-11-one **3.5** that has been synthesized.[7] This appears to be the smallest known example of in–out bicyclic system.

## 3.2 Cyclic conformation: Introduction and Baeyer strain theory

In order to correlate the stability of monocyclic compounds as a function of ring size,[8] Baeyer in 1885 introduced the concept of angle strain (also known as Baeyer strain). The Baeyer strain refers to the deviation of bond angle of cycloalkanes from the normal tetrahedral angle of 109°28′. This is defined in Fig. 3.6A. Baeyer assumed that the rings were planar and regular polygons. Based on the assumption, the geometric bond angle $\alpha$ (not the interorbital angle) for each ring was calculated and thereby strain was determined. Note that total angle deviation was divided equally between the two ring bonds and therefore one-half of the magnitude of total deviation was the absolute value of Baeyer strain. The strains calculated for different ring sizes are shown in Fig. 3.6B.

It is seen that strain decreases rapidly from 3-ring to 5-ring and then increases appreciably to 6-ring and thereafter continues to increase for the larger

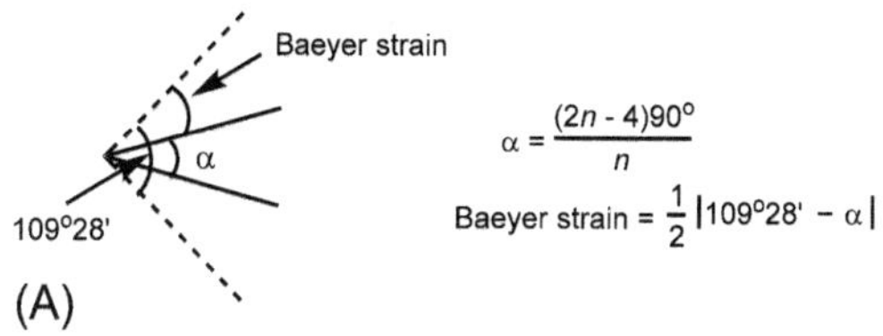


(A)

| Ring size, $n$ | 3 | 4 | 5 | 6 | 7 | 8 | 15 |
|---|---|---|---|---|---|---|---|
| Baeyer strain | 24°44' | 9°44' | 0°44' | 5°16' | 9°33' | 12°46' | 23°16' |

(B)

**FIG. 3.6** (A) Baeyer strain defined and (B) Baeyer strain calculated for different ring sizes.

rings. Thus stability of ring increases from cyclopropane to cyclopentane. Then, as one proceeds to cyclohexane and larger rings, the ring stability decreases progressively. The question is: how did Baeyer's theory agree with experiments? Experimentally, a measure of stability of rings, $(CH_2)_n$ ($n=3, 4, 5\ldots$) can be obtained from thermochemical data such as heat of combustion $[(CH_2)_n + 1.5nO_2 \rightarrow nCO_2 + nH_2O + \text{heat}]$. Table 3.1 shows the experimental

**TABLE 3.1** Ring strain for cycloalkanes from heats of combustion.

| Ring size, $n$ | Strain per $CH_2$ (kcal mol$^{-1}$) | Classification |
|---|---|---|
| 3 | 9.17 | Small |
| 4 | 6.58 | |
| 5 | 1.24 | Common |
| 6 | 0.02 | |
| 7 | 0.89 | |
| 8 | 1.21 | Medium |
| 9 | 1.40 | |
| 10 | 1.24 | |
| 11 | 1.02 | |
| 12 | 0.34 | Large |
| 13 | 0.40 | |
| 14 | 0.14 | |
| 15 | 0.13 | |
| 16 | 0.12 | |

values of strain per $CH_2$ (in kcal mol$^{-1}$) for different ring sizes. The strain per $CH_2$ is determined as follows.

$$\text{Strain per } CH_2 \left(\text{kcal mol}^{-1}\right)$$

$$= \text{Observed value (heat of combustion}/n) - \text{Reference value } (157.44)$$

The reference value for a $CH_2$ group is taken as the difference of heat of combustion between a straight-chain alkane $CH_3(CH_2)_xCH_3$ and its next higher homologue $CH_3(CH_2)_{x+1}CH_3$ ($x \geq 5$). This value is nearly constant at 157.44 kcal mol$^{-1}$. Table 3.1 shows that highest experimental strain is in cyclopropane; strain then decreases rapidly to cyclobutane to cyclopentane. These trends are consistent with the Baeyer's prediction. But the strain reaches a minimum for cyclohexane (not cyclopentane as suggested by Baeyer's theory) which is essentially stainless. Thereafter, the strain increases and there is a substantial strain in 8–11 rings. Then the strain drops and remains almost constant with a very small value at 14-ring and above. Thus Bayer's theory failed completely for cyclohexane and larger rings. The next question is: What was wrong in Baeyer's theory? It is the Baeyer's assumption that the rings are planar is not correct. Later, in 1892, Sachse recognized that cyclohexane ring is puckered (nonplanar) and is virtually free of angle strain.[9]

From the experimental strain data in Table 3.1, rings can be classified into four types[10]: (a) highly strained small rings (3-, 4-), (b) common or normal rings (5-, 6-, 7-) with relatively small strain (the 6-ring being nearly strainless), (c) medium rings (8- to 11-) with somewhat more strain and (d) large rings (12-onwards) with little or negligible strain. There are several strain factors that can affect the stability of a ring. Broadly, these are angle strain, torsional strain, steric (or van der Waals) strain and dipole–dipole repulsion (in case of polar substituents). When a kind of strain is introduced in a molecule, the strain energy is minimized by a distribution and compromise among several strain factors. In small rings, angle strain is the main factor. Among common rings, cyclohexane is essentially strain-free, whereas in cyclopentane, there is a compromise between angle strain and torsional strain. In medium rings, unfavourable torsional strain and van der Waals repulsion between internal hydrogens, known as transannular strain, contribute significantly. The large rings have very low strain energies.

The common rings are found in a vast majority of natural and synthetic compounds and cyclohexane ring is by far the most important.

## 3.3 Conformation of cyclohexane

### 3.3.1 Chair conformation

From the heats of combustion data, cyclohexane is found to be essentially strain-free (see Table 3.1). It is possible to construct nonplanar (puckered)

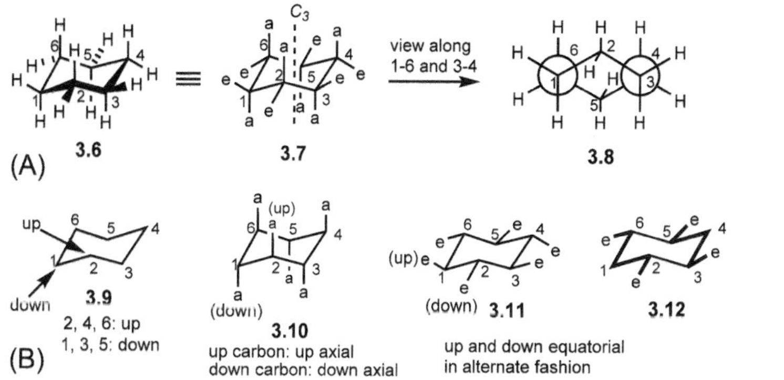


**FIG. 3.7** (A) Chair conformation of cyclohexane and (B) points to draw a chair conformation correctly.

models[9] of cyclohexane that are free of angle strain (when all bond angles are tetrahedral). An important model is chair cyclohexane **3.6**[11] (looks like a chair) shown in Fig. 3.7A. The structure **3.6** is a perspective formula of cyclohexane molecule where C1 and C4 are in plane, C2 and C3 are in front, and C5 and C6 are in rear. The chair cyclohexane has a $C_3$ axis that passes through the centre of the molecule and is perpendicular to the mean plane of the ring. Each carbon has two C—H bonds, one bond is parallel to the $C_3$ axis called axial (a) bond while the other lies roughly on the equatorial plane of the ring (cf. equator of the earth) and is called equatorial (e) bond. Thus a chair cyclohexane has six axial and six equatorial bonds **3.7**. Note carefully the spatial orientations of the equatorial and axial C—H bonds (cf. **3.7** with **3.6**). If viewed along 1–6 and 3–4 bonds, **3.7** represents two staggered sawhorse structures linked at C2 and C5, which is also shown by an equivalent Newman projection **3.8**. Clearly, all bonds in chair cyclohexane are fully staggered and the molecule is thus free of torsional strain. Further, there is no axial–axial, axial–equatorial or equatorial–equatorial van der Waals repulsion between the hydrogens. Thus chair cyclohexane is free of angle strain, torsional strain and steric (vander Waals) strain and becomes the lowest energy conformation (see Section 3.3.4).

It is important to draw the conformations of cyclohexane with reasonable accuracy for understanding stereochemistry. A few points should be noted as regards the chair conformation of cyclohexane (Fig. 3.7B).

**(a)** Of six ring carbons, three carbons are relatively up and three down **3.9**.
**(b)** Up carbons have up axial bonds and down carbons have down axial bonds **3.10**.
**(c)** The equatorial bonds also slope up and down alternately **3.11**. The spatial orientation of the equatorial bonds at C6, C2 gives an M-shape while at C5, C3 makes a W-shape as highlighted in bold in **3.12**.

**(d)** If an axial bond is up, the equatorial bond at the same carbon must be down and vice versa (**3.10**, **3.11**).

**(e)** The relative *cis* or *trans* stereochemistry of two substituents indicates their up/down relationship: *cis* when both up or both down, and *trans* when one is up and the other down. The up and down bonds are also denoted as β and α, respectively (cf. steroid nomenclature, Section 3.11.4).

## 3.3.2 Boat and twist-boat conformations

If one end of a chair is flipped up as shown in Fig. 3.8, a boat conformation **3.13** is formed which is also free of angle strain. But the boat form suffers from a considerable torsional strain due to eclipsing of four pairs of C—H bonds at C2—C3 and C5—C6. Further, there is a severe steric strain involving two flagpole hydrogens at C1 and C4 (the distance is 1.83 Å which is much less than the sum of their van der Waals radii, 2.5 Å). The bonds at C1 or C4 are called bowsprit and flagpole, and those at C2—C3 and C5—C6 are designated as pseudoaxial (a′) and pseudoequatorial (e′). An alternative drawing of boat conformation **3.14** and the equivalent Newman projection **3.15** are also shown, which depict clearly the eclipsing strain and van der Waals repulsion between flagpole hydrogens. Evidently, the boat conformation is much less stable than the chair conformation, and we will see later that boat conformation is a transition structure (TS) conformation corresponding to an energy maximum.

The boat conformation is however more flexible than the chair. If the two side C—C bonds (C2—C3 and C5—C6) twist relative to each other, the eclipsing interactions and flagpole repulsions are relieved to some extent creating a different conformation called a twist-boat conformation **3.16** as shown in Fig. 3.9. The twist-boat conformation is more stable than the boat conformation, and the twist-boat form corresponds to a local energy minimum (see later). The types of C—H bonds in **3.16** are indicated as twist-axial (ta), twist-equatorial (te) and isoclinal (ic).

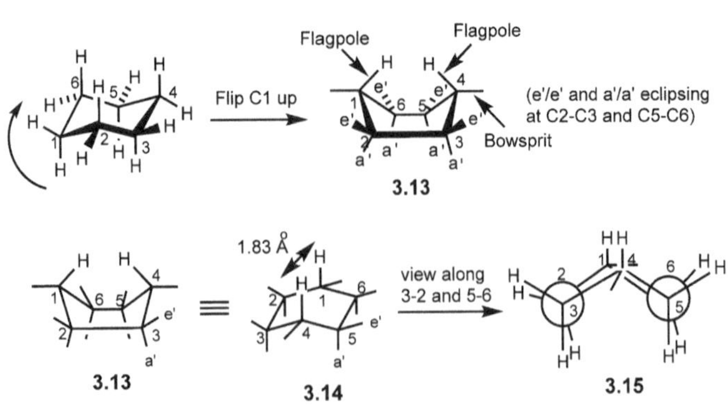


**FIG. 3.8** Boat conformation of cyclohexane.

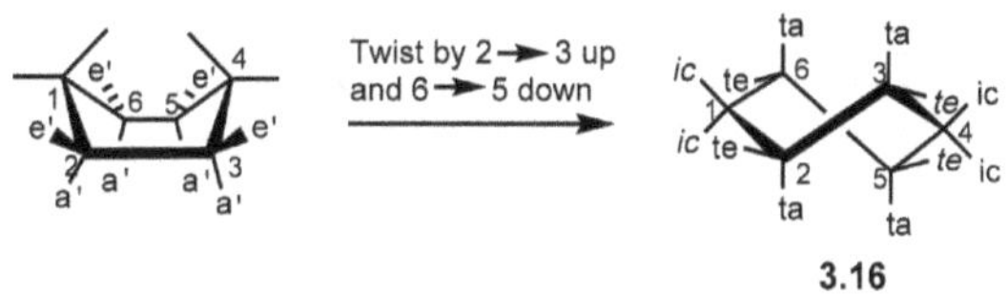


FIG. 3.9 Twist-boat conformation of cyclohexane.

### 3.3.3 Symmetry of chair, boat and twist-boat conformations

The chair conformation has a principal $C_3$ axis, 3 perpendicular $C_2$ axes bisecting pairs of opposite C—C bonds, and 3 $\sigma_d$ planes passing through diagonally opposite ring carbons and bisecting two $C_2$ axes (Fig. 3.10: **3.17**). (Only two $C_2$ axes and one $\sigma_d$ plane are shown.) Therefore chair cyclohexane belongs to the achiral point group $\mathbf{D_{3d}}$ (see Table 1.5 for point group classification). The chair cyclohexane also has necessarily an $S_6$ axis (coincident with $C_3$) and a centre of inversion *i* for the $\mathbf{D_{3d}}$ point group (see Section 1.5.2.3). The symmetry number ($\sigma$) is 6 (operations $E$, $2C_3$, $3C_2$). The axial H atoms are homotopic since they are related by $C_n$ ($C_3$, $C_2$) axes. For the same reason, the equatorial H atoms are also homotopic. But the set of axial H's is diastereotopic with the set of equatorial H's as the two types are not related by $C_n$ ($C_3$, $C_2$) or by $S_n$ ($S_6$, $\sigma$ and *i*).

The boat conformation is also achiral belonging to the point group $\mathbf{C_{2v}}$ (Fig. 3.10: **3.18**) due to the presence of a $C_2$ axis and 2 $\sigma_v$ planes, one passing through two opposite carbons with flagpole bonds and the other bisecting two side C—C bonds (not shown). The symmetry number ($\sigma$) is 2 (operations $E$, $C_2$).

In contrast to the chair and the boat conformation, the twist-boat conformation is chiral and belongs to the point group $\mathbf{D_2}$ having 3 mutually perpendicular $C_2$ axes (Fig. 3.10: **3.19**) (cf. Fig. 1.26C). The symmetry number ($\sigma$) is 4 (operations $E$, $C_2$, $C_2'$, $C_2''$).

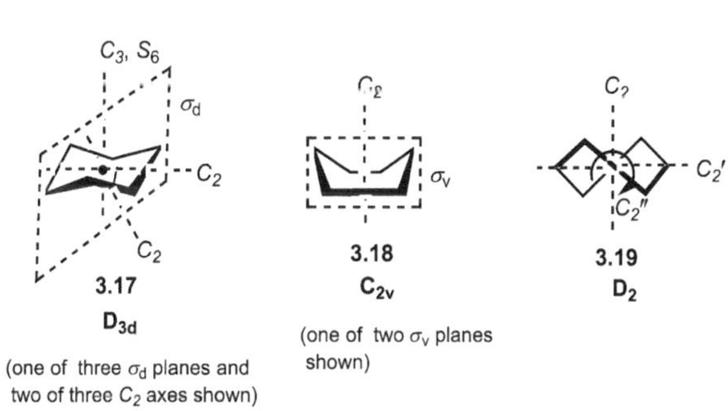


FIG. 3.10 Symmetry and point groups for chair, boat and twist-boat conformations of cyclohexane. In **3.17**, only one $\sigma_d$ out of 3 $\sigma_d$ planes and 2 $C_2$ out of 3 $C_2$ axes are shown. In **3.18**, one $\sigma_v$ is shown and the other $\sigma_v$ bisecting two side C—C bonds is not shown.

### 3.3.4 Chair–chair ring inversion

If one end (C1) of a chair is flipped down and the other end (C4) is flipped up, another chair results as shown in Fig. 3.11A (try yourself with the help of a model). The chair–chair interconversion is called ring inversion. Note that the flipping is equivalent to rotating bonds and not breaking them and therefore gives rise to conformations. The ring inversion of cyclohexane is not a one-step process[12] and calculations suggest a twist-boat intermediate.[13] The inversion process is delineated in Fig. 3.11B. The arrows indicate the direction of the movement of the individual carbons to give the next conformation. A characteristic feature of this pathway is that a $C_2$ axis (bisecting 2–3 and 5–6 bonds) is maintained throughout the process: chair → half-chair → twist-boat → half-chair → chair. As such, this is called a $C_2$ pathway of ring inversion. In the half-chair conformation four carbons are in plane, one carbon is above the plane

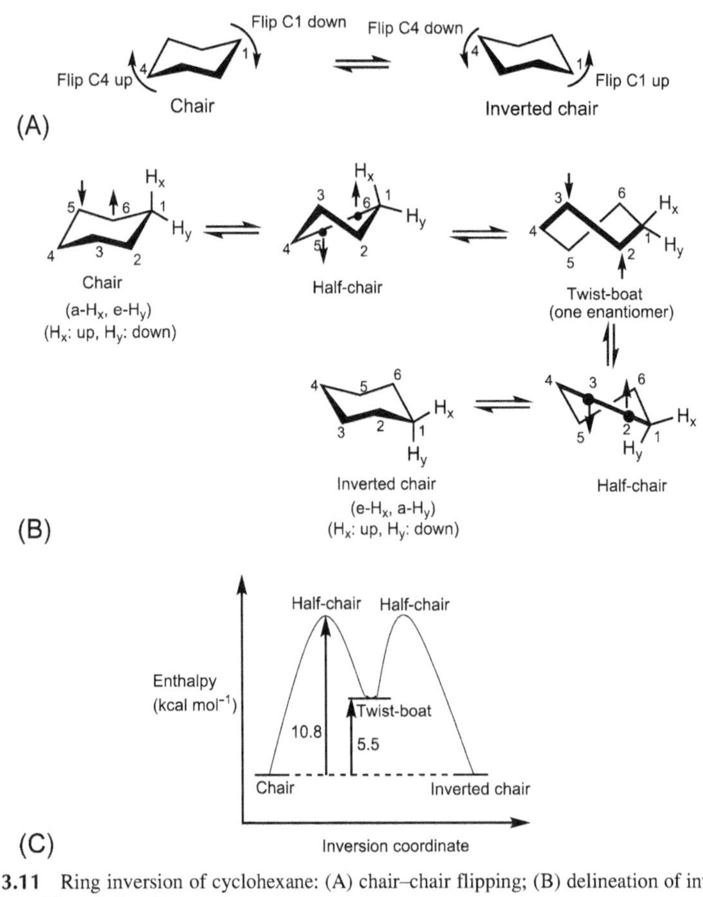


**FIG. 3.11** Ring inversion of cyclohexane: (A) chair–chair flipping; (B) delineation of inversion pathway; (C) enthalpy diagram.

and the other below the plane. The half-chair conformation has a $C_2$ axis only and therefore belongs to the chiral point group $\mathbf{C_2}$. In the ring inversion, a twist-boat enantiomer can convert to the chair or the inverted chair through the half-chair conformation (Fig. 3.11B). Ring inversion through the other twist-boat enantiomer then involves the other enantiomeric half-chair.

In the ring inversion, all axial hydrogens become equatorial and vice versa but up ($\beta$) hydrogen remains up ($\beta$) and down ($\alpha$) hydrogen remains down ($\alpha$) (Fig. 3.11B). Since the axial set and the equatorial set are diastereotopic, the ring inversion is also called topomerization.[14]

An enthalpy diagram for the ring inversion of cyclohexane is shown in Fig. 3.11C. The chair to twist-boat conversion involves a half-chair TS with $\Delta H^{\neq} = 10.8$ kcal mol$^{-1}$. Since the chair and twist-boat conformations correspond to energy minima, they are conformational isomers or conformers. The chair conformer is enthapically more stable than the twist-boat conformer by 5.5 kcal mol$^{-1}$.[13,15]

The conformational equilibrium of cyclohexane is represented as

$$\text{Chair} \rightleftharpoons \text{Twist-boat} \qquad \Delta H^0 = 5.5 \text{ kcal mol}^{-1}$$

The population of chair and twist-boat conformers at 25°C can be estimated in a similar manner as described for butane (see Section 1.3.5.1). Relative to the achiral chair, the chiral twist-boat has a greater flexibility with a lower symmetry number ($\sigma$) and entropy of mixing for the two enantiomeric twist-boat forms. The high enthalpy ($\Delta H^0 = 5.5$ kcal mol$^{-1}$) of twist-boat conformer is thus opposed by a favourable entropy ($\Delta S^0 = 4.9$ cal K$^{-1}$ mol$^{-1}$).[16] The free energy difference ($\Delta G^0$) for the equilibrium is then given by

$$\Delta G^0 = \Delta H^0 - T\Delta S^0 = 4\,\text{kcal mol}^{-1}$$

$$\text{Therefore}\ \frac{[\text{Twist}-\text{boat}]}{[\text{Chair}]} = K = e^{-\Delta G^0/RT} = e^{-4000/(1.987\times 298)} = 10^{-3} = 1:1000$$

The population of twist-boat conformer is thus estimated to be 0.1%. Other estimations with somewhat different values of $\Delta H^0$ and $\Delta S^0$ also suggest that the twist-boat is not more than 0.4%. Thus cyclohexane exists almost exclusively in the chair conformation at room temperature.

The two enantiomeric twist-boat conformers are depicted in Fig. 3.12A. The twist-boat form is flexible and a little energy (1–1.5 kcal mol$^{-1}$) is needed to distort a twist-boat to the boat conformation.[13,15] As shown in Fig. 3.12A, the boat conformation can twist in two alternative fashions to form the enantiomeric twist-boats. However, the twist-boat conformers are rapidly interconvertible via a boat TS and form a nonresolvable enantiomeric pair. Thus cyclohexane as an equilibrium mixture of achiral chair and chiral twist-boat conformers is optically inactive. It is pertinent to mention that the boat

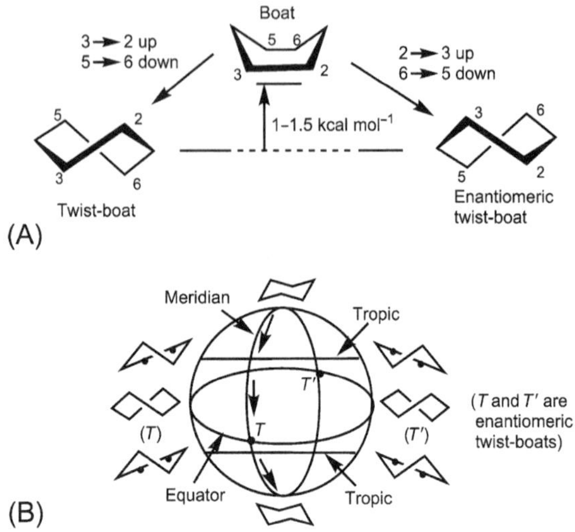


**FIG. 3.12** (A) Interconversion of two enantiomeric twist-boat conformers via a boat conformation. (B) A conformational map of cyclohexane depicting ring inversion along any meridian on a spherical surface.

conformation as TS for the twist–twist interconversion is not a part of the ring inversion (see Fig. 3.11C).

It is possible to depict the conformational map of cyclohexane on a sphere (cf. earth), as shown in Fig. 3.12B. The ring inversion is effected along the meridians. The intersections of the meridians with the equator represent twist-boat forms, and the half-chair conformations lie on the tropics. The rapid twist-twist interconversion ($T \Leftrightarrow T'$) occurs through a boat; however, the boat conformation is not a part of chair-chair ring inversion.

### 3.3.5 NMR: Dynamics of conformational changes

The axial and equatorial protons in a chair cyclohexane are diastereotopic (hence in different environments) and two separate chemical shift signals are expected in $^1$H NMR. However, whether different signals can be observed depends on the rate of site exchange between axial and equatorial protons as a result of chair–chair ring inversion. The faster the ring inversion, the faster is the rate at which a proton moves between the two sites. Broadly, at slow site exchange, two separate signals ($\delta_a$ and $\delta_e$) can be observed but at fast site exchange, the observed spectrum shows a single averaged signal ($\delta_{av}$) with chemical shift $(\delta_a+\delta_e)/2$ (Fig. 3.13). At intermediate rate of exchange, two peaks broaden and can coalesce to a single wide band. The rate of ring inversion is a function of temperature and analysis of temperature dependent NMR can provide information on the dynamics of the conformational processes.[17–19] At room temperature, the ring inversion of cyclohexane is fast ($k = 10^4$–$10^5\ s^{-1}$) and $^1$H NMR spectrum of cyclohexane-$d_{11}$ shows a single averaged

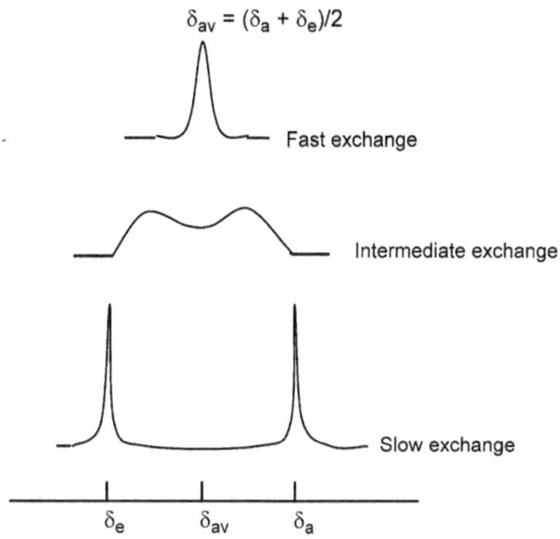


**FIG. 3.13** Schematic $^1$H NMR spectra for the site exchange of axial and equatorial protons in a chair cyclohexane at slow, intermediate and fast rates.

resonance signal ($\delta$ = 1.40 ppm). At −90°C, the rate is greatly slowed down ($k \sim 1\ s^{-1}$), and two separate peaks are observed with a chemical shift difference $\Delta\delta = 0.48$ ppm, with the axial protons at upfield. On raising temperature to −60°C ($k \sim 60\ s^{-1}$), a broadened spectrum is observed on a 60-MHz instrument. With further rise in temperature to −20°C ($k \sim 3 \times 10^3\ s^{-1}$), the spectrum shows an averaged signal which is almost the same as that at room temperature.

Whether an NMR spectrometer can show signals for the individual sites depends on the 'speed' of the NMR spectrometer (like the shutter speed in a camera) which is given by

$$v = \frac{\pi}{\sqrt{2}} \Delta\delta\nu_o \tag{3.1}$$

where $v$ is the NMR speed, $\Delta\delta$ is the chemical shift difference for the two sites in ppm and $\nu_o$ is the spectrometer frequency in MHz. For example, if two signals are separated by 0.5 ppm, $v = 4.4 \times 10^2\ s^{-1}$ on a 400-MHz spectrometer, whereas $v = 66\ s^{-1}$ on a 60-MHz spectrometer. If the rate of site exchange is appreciably slower than the 'speed' of NMR machine ($k \ll v$), the spectrum will show two separate signals and if the exchange rate is much faster ($k \gg v$), an averaged signal will be observed. As mentioned above, a broadened spectrum is observed at −60°C on a 60-MHz machine ($k \sim 60\ s^{-1}$, $v = 66\ s^{-1}$); however, this band might be split into separate signals on a 400-MHz instrument ($k \sim 60\ s^{-1}$, $v = 4.4 \times 10^2\ s^{-1}$).

### 3.3.6 Trapping of twist-boat conformer by matrix isolation method

The population of the twist-boat conformer at room temperature is ∼0.1% but increases at high temperature ($T\Delta S^0$ will favour the twist form). The chair and twist-boat conformers can be trapped in a frozen argon matrix when a mixture of cyclohexane vapour and argon heated to 800°C is rapidly cooled to −253°C.[20] The composition of the conformers in the solid matrix has been determined by IR spectroscopy which shows the presence of ~25% twist conformer. Further, the rate of disappearance of the twist form on a slight rise in temperature gives the activation parameter ($\Delta G^{\neq}$) for the twist to chair conversion to be 5.3 kcal mol$^{-1}$ (cf. Fig. 3.11C).

### 3.3.7 A closer look at chair cyclohexane

Electron diffraction studies[21,22] of cyclohexane in the gas phase reveal that the molecule is a slightly flattened chair with C—C—C bond angles of 111.4 degrees,[22] which is somewhat greater than the ideal tetrahedral angle. (The ring will be completely flat if the bond angles were 120 degrees.) The torsion angles are 54.9 degrees, being less than the optimum 60 degrees for a perfectly staggered arrangement in the ideal chair. Due to these deviations, the axial C—H bonds are not perfectly parallel and lean outwards from the $C_3$ axis by about 7 degrees. It is instructive to compare these bond angles and torsion angles of cyclohexane with those of the acyclic propane as a reference system. The C—C—C bond angle in propane is 112.4 degrees (the repulsion between two C—C bonds is more than that between two C—H bonds) with the torsion angle near 60 degrees. In acyclic propane, bond angles and torsion angle can vary independently which is not possible in cyclohexane due to ring constraints. If bond angle ($\alpha$) increasers, the adjacent torsion angle ($\phi$) decreases according to the equation[23]

$$\cos \phi = -\frac{\cos \alpha}{1 + \cos \alpha} \tag{3.2}$$

The equation satisfies the condition that torsion angle ($\phi$) is 60 degrees for a bond angle ($\alpha$) of 109.5 degrees. If bond angle increases by 1 degrees ($\alpha$=110.5 degrees), $\phi$ decreases to 57.4 degrees. Thus if the bond angle of chair cyclohexane is 112.4 degrees as for propane, torsion angle $\phi$ will be 52 degrees (a considerable deviation from the ideal 60 degrees), and the molecule would then suffer from an appreciable torsional strain. The total strain is however minimized by a compromise between angle strain and torsional strain with the bond angle of 111.4 degrees (instead of 112.4 degrees) when the torsion angle is 54.9 degrees (these observed values are consistent with the equation). Thus with some angle and torsional strain, the chair cyclohexane is not fully strain-free. We have seen earlier that heats of combustion data (see Table 3.1) indicate that cyclohexane has negligible strain based on straight-chain alkanes as standards. However, these alkanes are mixtures of anti and gauche forms; the difference in heats of combustion of two successive alkanes then incorporates some strain energies in the reference value. If successive alkanes with all-anti conformers can be chosen as standards,[24] the reference value will be slightly lower than

157.44 kcal $mol^{-1}$ and hence the strain energy of cyclohexane will be greater than 0.12 (6 × 0.02) kcal $mol^{-1}$ (see Table 3.1). With these standards, the strain energy of cyclohexane is estimated to be 1.31 kcal $mol^{-1}$ which vindicates that chair cyclohexane is not entirely strain-free.

## 3.4 Conformation of monosubstituted cyclohexanes: Conformational free energy (or *A* Value)

Like cyclohexane, a monosubstituted cyclohexane also exists mainly in the chair conformation. There are two chair conformers in equilibrium with the substituent (X) occupying axial (a) and equatorial (e) positions (Fig. 3.14).[25] The conformers (a-X and e-X) are diastereomeric in relationship, and their equilibrium population depends on their relative stability (free energy difference). In general, a-X conformer is destabilized by nonbonded steric repulsions between axial X and axial hydrogens on C3 and C5 on the same side of the ring. These steric interactions are referred to as *syn*-diaxial interactions or 1,3-diaxial interactions (as the axial substituent and an axial hydrogen on C3 or C5 are at relative 1,3 positions). In contrast, e-X conformer does not suffer from any steric repulsion between the equatorial X and any hydrogen on ring carbons. Thus e-X conformer is more stable than a-X conformer.

For the equilibrium shown in Fig. 3.14, $\Delta G^0_{a \rightarrow e}$ is negative ($G^0_e < G^0_a$), and $-\Delta G^0 = RT \ln K = RT \ln ([e\text{-}X]/[a\text{-}X])$. The free energy difference between the conformers, $-\Delta G^0$ is referred to as the conformational free energy or *A* value of the substituent on the cyclohexane ring.[26]

Table 3.2 shows conformational free energies ($-\Delta G^0$ values) or *A* values of some common substituents.[25,27,28] A positive value of $-\Delta G^0$ for a substituent implies that the substituent is favoured in the equatorial position. The larger the *A* value, the stronger is the preference for the equatorial position.

### 3.4.1 Methyl-, ethyl- and isopropylcyclohexane

In methyl cyclohexane, the a-Me conformer is destabilized by two 1,3-diaxial (Me/H) interactions relative to the e-Me conformer (cf. Fig. 3.14). Therefore

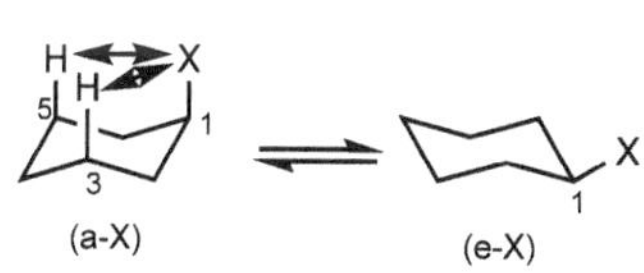


$$\Delta G^0 = G^0_e - G^0_a = -\,ve$$

$$-\Delta G^0 = RT \ln K = RT \ln \frac{[e\text{-}X]}{[a\text{-}X]}$$

**FIG. 3.14** Conformational free energy ($-\Delta G^0$) or *A* value of a substituent on cyclohexane defined. The arrow (↔) indicates *syn*-diaxial or 1,3-diaxial interaction in the axial conformer.

**TABLE 3.2** Conformational free energies ($-\Delta G^0$) or *A* values for some common substituents in the cyclohexane system.

| Substituent | $-\Delta G^0$ (*A* value) ($kcal\,mol^{-1}$) | Substituent | $-\Delta G^0$ (*A* value) ($kcal\,mol^{-1}$) |
|---|---|---|---|
| $CH_3$ | 1.74 | C≡CH | 0.46 |
| $CH_2CH_3$ | 1.79 | C≡N | 0.2 |
| $CH(CH_3)_2$ | 2.21 | | |
| $C(CH_3)_3$ | 4.7 | $OCH_3$ | 0.63 |
| $Si(CH_3)_3$ | 2.5 | $OCOCH_3$ | 0.68 |
| | | OTs | 0.5 |
| $CH{=}CH_2$ | 1.49 | OH (aprotic solvent) | 0.6 |
| $C_6H_5$ | 2.87 | OH (protic solvent) | 0.95 |
| | | $NH_2$ (protic solvent) | 1.7 |
| $COCH_3$ | 1.21 | F | 0.25 |
| $CO_2CH_3$ | 1.25 | Cl | 0.53 |
| $CO_2H$ | 1.4 | Br | 0.48 |
| $CO_2^-$ | 2.0 | I | 0.47 |

e-Me is the preferred conformation. A 1,3-diaxial (Me/H) interaction actually involves the steric repulsion between two nonbonded hydrogens (a-H and H of Me) that are 2.34 Å apart, which is less than the sum of their van der Waals radii (2.5 Å) (Fig. 3.15A). In order to determine the energy difference between the a-Me and the e-Me conformers, it is necessary to estimate the interaction energy of a 1,3-diaxial (Me/H) interaction. As shown in Fig. 3.15B, a 1,3-diaxial (Me/H) interaction is equivalent to a gauche–butane (gb) interaction between the Me group and a ring $CH_2$ group at C3 while the same groups are anti in the e-Me conformer. Note that C—C bonds making the torsion angle $\phi = 60$ degrees (gauche) and $\phi = 180$ degrees (anti) are highlighted in bold. Similarly, the other 1,3-diaxial (Me/H) interaction in the a-Me conformer is equivalent to a gb interaction between the Me group and a ring $CH_2$ group at C5 (not shown).

Thus for the a-Me conformer,

$$\text{interaction energy of two 1,3-diaxial (Me/H) interactions}$$
$$= \text{two gauche-butane (gb) interactions} = 2 \times 0.9\,\text{kcal mol}^{-1} = 1.8\,\text{kcal mol}^{-1}$$

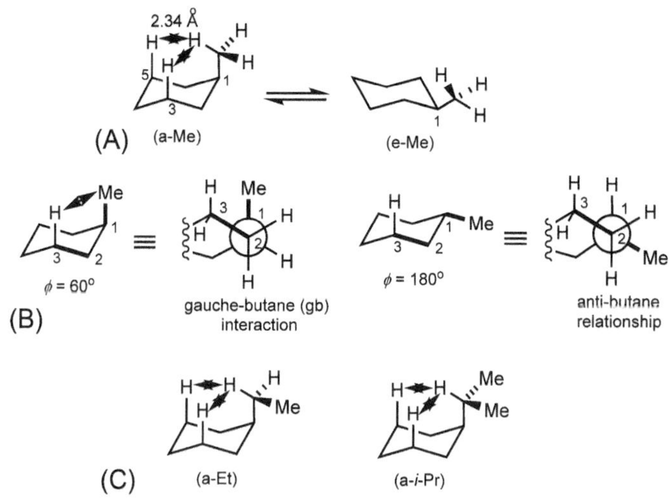


**FIG. 3.15** (A) Conformation of methylcyclohexane showing the steric origin of 1,3-diaxial interaction; (B) estimation of interaction energy of a 1,3-diaxial (Me/H) interaction based on gauche–butane (gb) interaction and (C) 1,3-diaxial interactions in the axial conformer of ethylcyclohexane and isopropylcyclohexane.

For the e-Me conformer, there is no steric interaction with Me in antibutane relationship.

The conformational equilibrium for methylcyclohexane is then represented as

$$\text{a-Me} \rightleftharpoons \text{e-Me} \qquad \Delta H^0 = -1.8 \text{ kcal mol}^{-1}$$

The enthalpy difference between the conformers, $-\Delta H^0$, is called conformational enthalpy or steric enthalpy of the substituent. Since, for chair–chair ring inversion, $\Delta S^0 \sim 0$, $-\Delta G^0 = 1.8$ kcal mol$^{-1}$. The conformational free energy or *A* value of a Me substituent is thus estimated to be 1.8 kcal mol$^{-1}$ which agrees well with the value given in Table 3.2. This free energy difference gives >90% e-Me conformer at equilibrium (cf. Table 1.2).

Et is a larger substituent than Me but it has almost the same $-\Delta G^0$ or *A* value as Me, and the *A* value for much bulkier isopropyl (*i*-Pr) is only about 0.4 kcal mol$^{-1}$ higher (see Table 3.2). This is mainly due to similar 1,3-diaxial interactions as for methylcyclohexane, involving one H of Et (or *i*-Pr) and *syn*-axial H's (Fig. 3.15C). If a Me group of Et (or *i*-Pr) is turned into the ring, the axial conformer will be of very high energy due to severe 1,3-diaxial interactions between Me of Et (or *i*-Pr) and axial H's (cf. *tert*-butyl substituent, Section 3.4.2) and therefore Me group of Et or *i*-Pr will turn outside.

### *3.4.1.1 Conformational enthalpy and conformational entropy of methyl, ethyl and isopropyl substituents*

The thermodynamic parameters for Me, Et and *i*-Pr substituents reveal that conformational enthalpy ($-\Delta H^0$) decreases and conformational entropy ($\Delta S^0$) increases along the series[29]:

$$-\Delta H^0\ (\text{kcal mol}^{-1}) : \text{Me}\ (1.75) > \text{Et}\ (1.60) > i-\text{Pr}\ (1.52)$$

$$\Delta S^0\ (\text{cal mol}^{-1}\ \text{K}^{-1}) : \text{Me}\ (-0.03) < \text{Et}\ (0.64) < i-\text{Pr}\ (2.31)$$

Qualitatively, this can be explained on the basis of rotamers possible due to rotation about a single bond between the ring carbon and the substituent (Et or *i*-Pr). The sterically favourable rotamer(s) results when an H atom of Et or *i*-Pr is turned into the ring as in the case of Me substituent. The significant rotamers for the a-Et and the e-Et conformers are depicted in Fig. 3.16A. There are two significant rotamers ($a_1$ and $a_2$) for the a-Et conformer and three rotamers ($e_1$, $e_2$ and $e_3$) for the e-Et conformer. The additional steric interaction (besides 1,3-diaxial interactions) in each rotamer arises from the gauche–butane (gb) interaction involving Me (of Et) and a ring $CH_2$. As shown, $a_1$ and $a_2$ each has 1 gb while $e_1$ and $e_2$ each has 1 gb and $e_3$ has 2 gb. The enthalpic contribution to the axial or the equatorial conformer is then a weighted average of gb in the rotamers. An enthalpy diagram of ethylcyclohexane vis-á-vis methylcyclohexane (having no additional gb) is shown in Fig. 3.16B. Since the weighted average

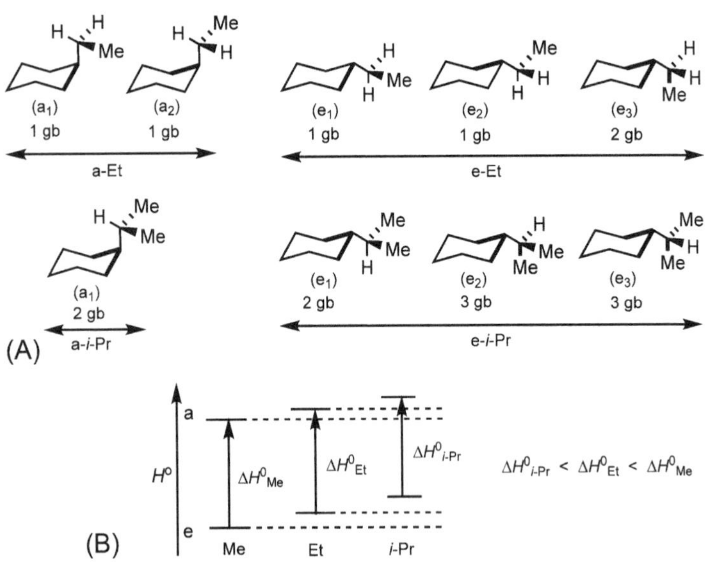


**FIG. 3.16** (A) Rotamers of axial and equatorial conformers of ethylcyclohexane and isopropylcyclohexane and (B) enthalpy diagram for methyl-, ethyl- and isopropylcyclohexane. gb = gauche–butane.

gb for the e-Et conformer is more than that of the a-Et conformer, the enthalpy of the former is raised more than the latter. Therefore $\Delta H^0_{Et} < \Delta H^0_{Me}$. In the case of *i*-Pr substituent, there is a single rotamer ($a_1$) for the a-*i*-Pr conformer with 2 gb and three rotamers ($e_1$, $e_2$ and $e_3$) are possible for the e-*i*-Pr conformer with 2, 3 and 3 gb (Fig. 3.16A). It therefore follows that $\Delta H^0_{i\text{-}Pr} < \Delta H^0_{Et}$ (Fig. 3.16B).

The entropic contribution of rotamers to the axial and the equatorial conformer is provided by their entropy of mixing. The entropy of mixing for the e-Et conformer (three rotamers) is greater than that for the a-Et conformer (two rotamers). Therefore $\Delta S^0_{Et} > \Delta S^0_{Me}$ for a $\rightleftharpoons$ e equilibrium. Similarly, entropy of mixing for the e-*i*-Pr conformer (three rotamers) is much greater than that for the a-*i*-Pr conformer (one rotamer), which leads to $\Delta S^0_{i\text{-}Pr} > \Delta S^0_{Et}$. The somewhat higher value of $-\Delta G^0$ of *i*-Pr relative to Me can be attributed at least in part to an appreciable contribution of $-T\Delta S^0$ to $-\Delta G^0$.

### 3.4.2 *tertiary*-Butyl- and trimethylsilylcyclohexane

Table 3.2 shows that *tert*-butyl group has a high $-\Delta G^0$ or $A$ value (4.7 kcal mol$^{-1}$). This is because the axial conformer is highly destabilized by severe 1,3-diaxial interactions as one Me of *tert*-butyl must turn into the ring (Fig. 3.17). Consequently, a *t*-Bu group holds the conformational equilibrium to the side in which it is equatorial; however, it does not stop the process of ring inversion.[26,30] The system is not conformationally locked but conformationally biased or anancomeric[31] (from the Greek *anankein*, fix by fate or law). It is noteworthy that the conformational equilibrium with a *t*-Bu substituent is comparable to chair $\rightleftharpoons$ twist-boat equilibrium ($\Delta G^0 = 4$ kcal mol$^{-1}$, see Section 3.3.4). Interestingly, when *tert*-butyl ($CMe_3$) is replaced by trimethylsilyl ($SiMe_3$), the $A$ value drops appreciably to about half, i.e. 2.5 kcal mol$^{-1}$ (see Table 3.2). As C—Si bond is longer than C—C, the inwards Me of $SiMe_3$ in the axial conformer is more distant from the *syn*-axial hydrogens (Fig. 3.17). Consequently, 1,3-diaxial ($SiMe_3$/H) interactions will be less compared to 1,3-diaxial ($CMe_3$/H) interactions. Further, $SiMe_3$ with a longer C—Si bond would benefit more from outwards bending due to flattening of cyclohexane chair (see Section 3.3.7).

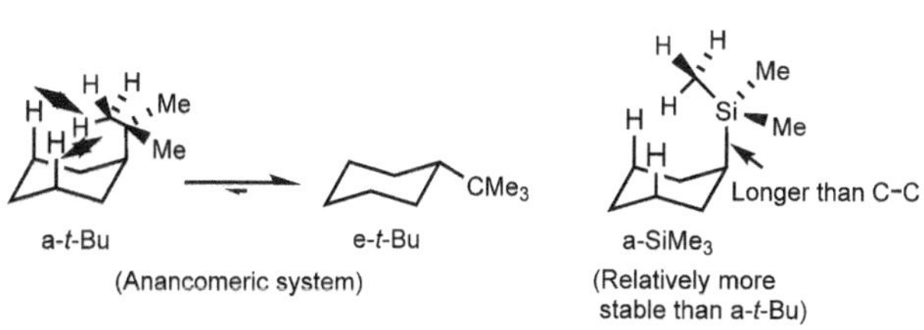


**FIG. 3.17** Anancomeric equilibrium for *t*-butylcyclohexane and axial conformer of trimethylsilylcyclohexane.

### 3.4.3 Phenyl- and vinylcyclohexane

Phenyl and vinyl are planar substituents. The conformational equilibrium of phenylcyclohexane is shown in Fig. 3.18A. A rotamer ($a_1$) of the axial Ph conformer in which Ph lies on the vertical plane bisecting the ring (bisector plane) suffers from severe 1,3-diaxial interactions resembling those for *t*-Bu substituent (see Fig. 3.17). Therefore the phenyl ring is rotated to a plane perpendicular to the bisector plane. In this rotamer ($a_2$), the 1,3-diaxial strain is relieved but a new steric strain develops due to crowding of *ortho*-H atoms of phenyl ring with the adjacent equatorial H atoms of cyclohexane ring which is less severe. Thus $a_2$ is a sterically more favourable rotamer of the a-Ph conformer. Ring inversion of $a_2$ produces a rotamer ($e_1$) of the e-Ph conformer in which the steric strain persists. But this rotamer can give a rotamer ($e_2$) with Ph in the bisector plane when there is no steric strain. Thus $e_2$ is the favoured rotamer of the e-Ph conformer. The conformation equilibrium of phenylcyclohexane is then established as $a_2 \rightleftharpoons e_2$. Clearly, the e-Ph conformer is the preferred conformer of phenylcyclohexane. The larger *A* value of Ph (2.87 kcal $mol^{-1}$) relative to Me (1.74 kcal $mol^{-1}$) presumably arises from two e-H/*o*-H interactions in the a-Ph conformer.[32]

Like phenylcyclohexane, axial vinyl conformer of vinylcyclohexane has two rotamers: $a_1$ with vinyl group in the bisector plane and $a_2$ with vinyl in the perpendicular plane (Fig. 3.18B). The steric interactions in $a_1$ resemble those for Me substituent (*A* value 1.74 kcal $mol^{-1}$). In $a_2$, interaction energy for e-H/vinyl H interaction is expected to be about one-half of that in the a-Ph conformer having two e-H/*o*-H interactions and is ~1.4 kcal $mol^{-1}$ (½ × *A* value of Ph). The steric enthalpy of the a-vinyl conformer is then a weighted average of interaction energies in $a_1$ and $a_2$, which is ~1.5–1.6 kcal $mol^{-1}$.

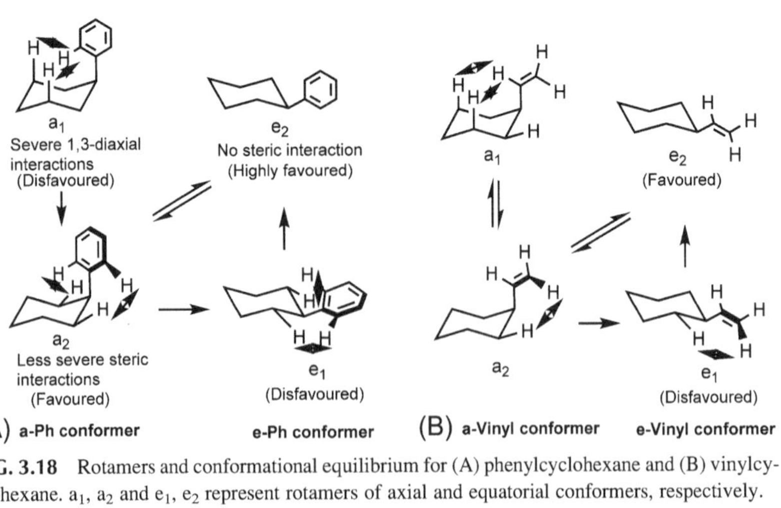


**FIG. 3.18** Rotamers and conformational equilibrium for (A) phenylcyclohexane and (B) vinylcyclohexane. $a_1$, $a_2$ and $e_1$, $e_2$ represent rotamers of axial and equatorial conformers, respectively.

This is slightly opposed by the favourable entropy of mixing of $a_1$ and $a_2$. On the other hand, the favoured rotamer ($e_2$) of the e-vinyl conformer has no steric strain, and it is the preferred conformer of vinylcyclohexane. Thus the conformational free energy ($-\Delta G^0$) of a vinyl substituent (1.49 kcal $mol^{-1}$) is found to be about one-half of that of a phenyl substituent (2.87 kcal $mol^{-1}$).

### 3.4.4 Cyclohexanes with linear, planar carbonyl and heteroatom containing substituents

If the substituent is linear such as C≡CH or CN, the conformational free energies ($-\Delta G^0$) or *A* values are quite small (see Table 3.2). This is because 1,3-diaxial interactions in the axial conformer are much less since no atom is turned into the ring. Only the sp. carbon with associated π electrons can possibly interact with *syn*-axial hydrogens.

For the planar carbonyl substituents (COR and $CO_2R$), the $-\Delta G^0$ or *A* values fall in between tetrahedral and linear substituents and are somewhat smaller than planar vinyl substituent (see Table 3.2). The $-\Delta G^0$ value of ionic species ($CO_2^-$) is greater than that of neutral species ($CO_2H$). This can be attributed to solvent effect. Both $CO_2^-$ and $CO_2H$ in the axial position suffer from steric hindrance to solvation by syn-axial hydrogens and are less solvated. But since the equatorial $CO_2^-$ is more strongly solvated and stabilized than equatorial $CO_2H$, the difference in free energy ($-\Delta G^0$ value) will be greater for $CO_2^-$.

For substituents with an $sp^3$ heteroatom directly linked to the ring such as OR, OCOR, OTs, the conformational free energies are much smaller than those for substituents with $sp^3$ carbon (see Table 3.2). Though these substituents are large in size by themselves, the group attached to $sp^3$ oxygen is turned outside the ring with a lone pair pointing inwards in the favoured rotamer of the axial conformer (Fig. 3.19). The 1,3-diaxial interactions then involve an oxygen lone pair and are quite small. Table 3.2 also shows that the $-\Delta G^0$ value for OH in a protic solvent (isopropanol) is higher than that in an aprotic solvent (cyclohexane). This is due to solvation effect. In a protic solvent, axial OH suffers from steric hindrance to solvation by hydrogen bonding but no steric hindrance operates for solvation of equatorial OH. Therefore equatorial OH is more solvated and stabilized leading to a larger free energy difference ($-\Delta G^0$) compared to

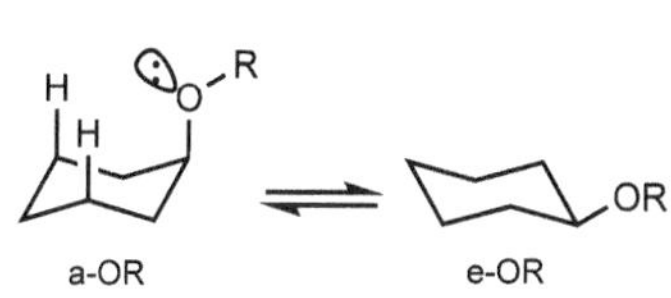


**FIG. 3.19** Conformational equilibrium of cyclohexane bearing OR (R=H, Me, COMe, Ts) substituents.

that in an aprotic solvent in which the solvent effect is minimal. Similar solvent effects operate for the $NH_2$ substituent.

For halogen substituents, $-\Delta G^0$ value is least for F which is smallest in size. However, $-\Delta G^0$ values are similar for Cl, Br and I. Though size increases from $Cl \rightarrow Br \rightarrow I$, the C-halogen bond distance also increases along the series. Therefore, the increase in 1,3-diaxial interactions due to increase in size will be offset, at least in part, due to increasing distances. Other factors also contribute to offset the size effect. A longer C-halogen axial bond would benefit more from outwards bending due to flattening of cyclohexane chair. Also, a larger halogen atom is more polarizable and an attractive part of van der Waals force would tend to stabilize the axial conformer.

The slight increase in $-\Delta G^0$ value for Cl (0.53 kcal $mol^{-1}$) relative to Br (0.48 kcal $mol^{-1}$) is however attributed to vibrational and rotational entropy that stabilizes the e-Cl conformer ($\Delta S^0 = 0.32$ cal $K^{-1}$ $mol^{-1}$) more than the e-Br conformer ($\Delta S^0 = 0.06$ cal $K^{-1}$ $mol^{-1}$).[19,33]

## 3.4.5 Experimental determination of conformational free energy ($-\Delta G^0$) or *A* value

NMR spectroscopy is a valuable tool for thermodynamic and kinetic studies of conformational inversion.[18,34–36] Chemical methods involving diastereomer equilibration[37] are also used to determine the conformational free energy or *A* value of substituents.

### *3.4.5.1 NMR method*

A substituent on a cyclohexane ring does not greatly change the rate of ring inversion. Therefore the conditions for slow exchange of protons between axial and equatorial sites for substituted cyclohexanes can also be achieved at low temperatures when ring inversion is slow on NMR timescale (see Section 3.3.5). We will describe briefly three techniques for the determination of conformational free energy ($-\Delta G^0$) of substituents using NMR.

#### $^1$H NMR: Chemical shifts and peak areas

Consider the case of iodocyclohexane (Fig. 3.20A). At −80°C, the $^1$H NMR spectrum of iodocyclohexane shows two distinct signals.[38] An upfield signal shows a multiplet (a triplet of triplets) with coupling constants $J = 3.5$ Hz and 12 Hz. This pattern is characteristic of a proton in the axial position ($H_a$ in the e-I conformer in Fig. 3.20A). The coupling constant $^3J$ (coupling through *three* bonds in H—C—C—H) varies as a function of torsion angle $\phi$ (known as Karplus equations[39]). Coupling is largest (about 10–13 Hz) when $\phi = 180$ degrees for diaxial protons, whereas diequatorial or axial/equatorial protons have $^3J$ values around 2–5 Hz when $\phi = 60$ degrees. In the e-I conformer, $H_a$ has two identical axial neighbours and two identical equatorial

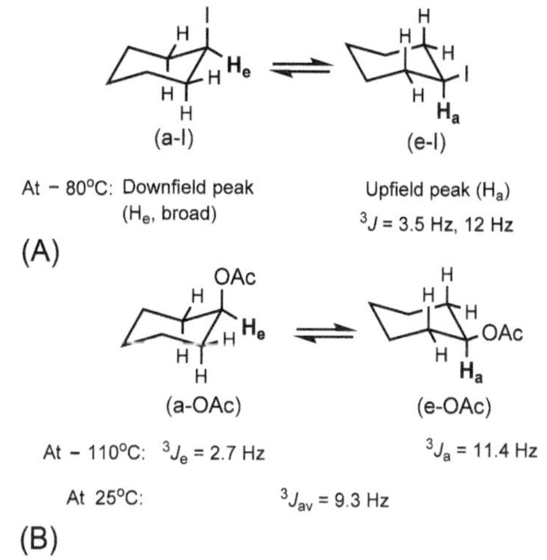


**FIG. 3.20** $^{1}$H NMR data for peaks and coupling constants at low temperatures with respect to (A) iodocyclohexane and (B) acetoxyxyclohexane for determining conformational free energy ($-\Delta G^0$) or $A$ value of the respective substituent.

neighbours. Thus $H_a$ appears as a triplet of triplets with two axial–axial couplings and two axial–equatorial couplings. The NMR spectrum of iodocyclohexane also shows a broad downfield peak which is characteristic of a proton in the equatorial position ($H_e$ in the a-I conformer in Fig. 3.20A). Here two equatorial–equatorial couplings and two equatorial–axial couplings are possible but no splitting is observed.

The relative areas of the two peaks are estimated as $H_a$:$H_e$ = 3.4:1, which gives[38]

$$-\Delta G^0 = RT \ \ln(3.4) = 1.987 \times 193 \ \ln(3.4)\,\text{cal mol}^{-1} = 0.47\,\text{kcal mol}^{-1}$$

Note that the conformational free energy ($-\Delta G^0$) of I is measured at −80°C; however, this value is often used at room temperature (see Table 3.2) assuming $\Delta S^0 = 0$ and $\Delta H^0$ being independent of temperature. But these assumptions are not always valid and the extrapolation of low temperature measurement of $-\Delta G^0$ to room temperature might not be reasonable.

## $^{1}$H NMR: Coupling constants and equilibrium

Unlike chemical shifts of axial and equatorial protons ($\delta_a$ and $\delta_e$) which vary with temperature, coupling constants are practically independent of temperature. Thus if $^3J_a$ and $^3J_e$ (coupling constant for axial and equatorial protons) are known from low temperature measurements, the equilibrium constant ($K$) can be determined from a weighted average coupling constant $^3J_{av}$ for the two rapidly interconverting conformers at room temperature.[40] Using the Winstein–Holness equation, $K$ is given by

$$K = \left({}^3J_{\text{av}} - {}^3J_{\text{e}}\right)/\left({}^3J_{\text{a}} - {}^3J_{\text{av}}\right) \tag{3.3}$$

Consider the case of acetoxycyclohexane (Fig. 3.20B). The e-OAc conformer and the a-OAc conformer are identifiable by $^1$H NMR at low temperature (−110°C) which shows two distinct peaks with the characteristic $^3J$ values. A sample of acetoxycyclohexane with substitution of H by D at C3, C4 and C5 was used to obtain sharp signals by deuterium decoupling. The largest coupling constant of 11.4 Hz denotes $^3J_a$ for the axial proton $H_a$ in the e-OAc conformer, whereas the smaller $J$ of 2.7 Hz gives $^3J_e$ for the equatorial proton $H_e$ in the a-OAc conformer (Fig. 3.20B). The NMR spectrum at room temperature (25°C) shows a weighted average single peak with $^3J_{av}$ of 9.3 Hz. Thus we get

$$K = \left({}^3J_{\text{av}} - {}^3J_{\text{e}}\right)/\left({}^3J_{\text{a}} - {}^3J_{\text{av}}\right) = (9.3 - 2.7)/(11.4 - 9.3) = 3.14$$

$$\text{Hence,} -\Delta G^0 = RT\ \ln(3.14) = 0.68\,\text{kcal mol}^{-1}$$

The conformational free energy ($-\Delta G^0$) of OAc at room temperature is thus estimated as 0.68 kcal mol$^{-1}$ (cf. Table 3.2).

### $^{13}$C NMR: Chemical shifts and peak areas

In $^1$H NMR, the signals for the two conformers are not always well resolved. $^{13}$C NMR is better for the determination of $-\Delta G^0$ through direct measurement of the intensity of the signals or by the use of $^1$H—$^{13}$C coupling constants.[41] For example, the conformational equilibrium for methylcyclohexane can be measured at −100°C using $^{13}$C NMR which gives separate $^{13}$C signals for equatorial Me and axial Me. The relative areas of the two peaks are in the ratio 160:1 in favour of the e-Me conformer, which gives $-\Delta G^0$ of Me = 1.74 kcal mol$^{-1}$ (see Table 3.2).

## *3.4.5.2 Chemical method*

A chemical method for the determination of $-\Delta G^0$ of OH involves the equilibration of two diastereomers in which OH is almost exclusively axial in one diastereomer and equatorial in the other.[42] Fig. 3.21 shows that *cis*- and *trans*-4-*t*-butylcyclohexanol can be equilibrated using a nickel catalyst in refluxing benzene (80°C). The *cis* isomer (in which both substituents are up) exists in the conformation in which *t*-butyl is equatorial, and hence OH is axial. The *trans* isomer (one substituent up and the other down) then exists in the conformation in which both *t*-butyl and OH are equatorial. Thus the free energy change for the equilibration reaction is equal to the free energy difference between the equatorial OH and the axial OH. Analysis of the equilibrium mixture gives 28% *cis* and 72% *trans* isomers. Therefore $K = [trans]/[cis] = 2.57$, which gives $-\Delta G^0 = 0.66$ kcal mol$^{-1}$ for the OH substituent in an aprotic solvent (see Table 3.2).

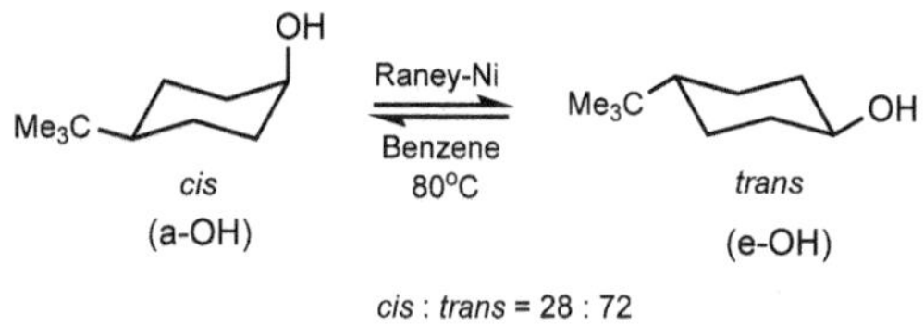


**FIG. 3.21** Determination of conformational free energy ($-\Delta G^0$) of OH in an aprotic solvent by chemical equilibration of *cis*- and *trans*-4-*t*-butylcyclohexanols.

### 3.4.5.3 *Isolation of conformers*

Like cyclohexane itself, the conformational inversion of substituted cyclohexanes is also very rapid at room temperature ($\sim 10^5\ s^{-1}$). The first-order rate constants ($k$) and half-life ($t_{1/2} = 0.693/k$) can be measured at low temperatures using NMR. The half-lives for conformational inversion of chlorocyclohexane at room temperature and at low temperatures are given below.[43]

| Temperature (°C) | Half-life ($t_{1/2}$) |
|---|---|
| 25 | $1.3 \times 10^{-5}$ s |
| −60 | $2.5 \times 10^{-2}$ s |
| −120 | 23 min |
| −160 | 22 years |

At low temperatures around −160°C (when $t_{1/2}$ = 22 years), the predominant e-Cl conformer will persist without conversion to the axial conformer. Crystallization of chlorocyclohexane under this condition gives only the crystals of equatorial chlorocyclohexane.[43] This is confirmed by the dissolution of these crystals at −150°C and measuring $^1$H NMR which shows the characteristic signal of the axial proton *H*—C—Cl ($\delta$ 3.8) of the e-Cl conformer. If the temperature is raised to −115°C, axial–equatorial equilibrium will be re-established giving two peaks at $\delta$ 3.8 and $\delta$ 4.5 for the axial proton in the e-Cl conformer and the equatorial proton in the a-Cl conformer, respectively.

## 3.5 Conformation of disubstituted cyclohexanes

The conformational free energy ($-\Delta G^0$) or $A$ value of a substituent indicates its preference for the equatorial position. Thus for disubstituted cyclohexanes, if the substituents are *not* mutually interacting, the equilibrium will favour the side in which the substituent with higher $A$ value is equatorial. The $A$ values are roughly additive for determining the equilibrium population.[44,45] The additivity might fail if the substituents interact with each other.[30,32]

Disubstituted cyclohexanes are of 1,1-, 1,2-, 1,3- or 1,4-type. 1,1-Disubstituted cyclohexanes do not exhibit configurational isomerism. In contrast, 1,2-, 1,3- or 1,4-disubstituted cyclohexanes can exist as *cis* and *trans* configurational isomers. A *cis* diastereomer has either both substituents up or both down, whereas a *trans* diastereomer has one substituent up and the other down.

### 3.5.1 1,1-Disubstituted cyclohexanes

1,1-Disubstituted cyclohexanes can exist in two interconvertible conformers if the substituents are different. Consider the case of 1-methylcyclohexanol (Fig. 3.22A). Since Me has higher *A* value than OH, the preferred conformer is expected to be (e-Me, a-OH) conformer, which has indeed been observed in DMSO (major:minor = 7:3). Note that the two conformers are diastereomeric. However, the population ratio corresponds to a free energy difference of 0.5 kcal mol$^{-1}$ which is much smaller than the expected difference of *A* values: $A(\text{Me}) - A(\text{OH}) = 1.74 - 0.6 = 1.14\,\text{kcal mol}^{-1}$. This possibly arises due to some levelling effect.

Fig. 3.22B shows the conformational analysis of 1-methyl-1-phenylcyclohexane.[32,46] Since Ph has higher *A* value than Me, the (e-Ph, a-Me) conformer is expected to be preferred over the other (a-Ph, e-Me) conformer by 1.13 kcal mol$^{-1}$. But the observed result is the opposite: (a-Ph, e-Me) conformer is preferred by 0.32 kcal mol$^{-1}$ indicating that additivity breaks down.[47] Ph itself prefers to be equatorial in the bisector plane (see Fig. 3.18A, rotamer $e_2$). But in the (e-Ph, a-Me) conformer, Ph in the bisector plane interacts with

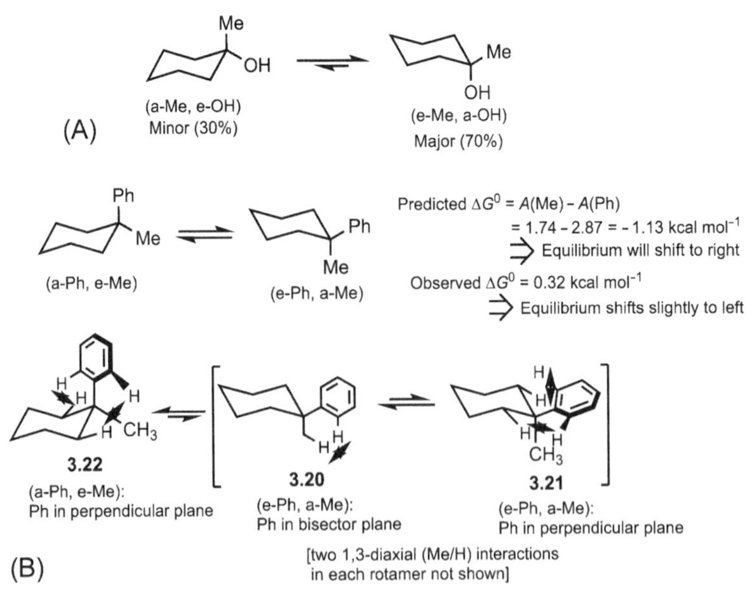


**FIG. 3.22** Conformational analyses of (A) 1-methylcyclohexanol and (B) 1-methyl-1-phenylcyclohexane.

Me sterically **3.20**, in addition to 1,3-diaxial interactions for the axial Me (not shown). If Ph goes to the perpendicular plane **3.21**, more severe e-H/*o*-H interactions come into play (cf. Fig. 3.18A, rotamer $e_1$). Therefore **3.20** is the favoured rotamer of the (e-Ph, a-Me) conformer. In the (a-Ph, e-Me) conformer, there is no additional steric interaction between Ph and Me besides the normal interactions destabilizing the axial Ph **3.22** (cf. Fig. 3.18A, rotamer $a_2$). The net effect is that the (a-Ph, e-Me) conformer is slightly preferred.

### 3.5.2 1,2-Dimethylcyclohexanc (*cis* and *trans*)

The chair–chair equilibrium for *cis*-1,2-dimethylcyclohexane is shown in Fig. 3.23A. (Start with a chair with both Me's up and down, then the other possible chair is obtained by ring inversion.) Note that *up axial* becomes *up equatorial* and vice versa, and the absolute configuration remains unchanged (1*R*,2*S* → 1*R*,2*S*) in ring inversion. In the both chair conformers (**X** and **X′**), one Me is equatorial and the other Me is axial. The question is whether the two chairs are identical or not? Since the chair **X** is chiral (point group $\mathbf{C_1}$), there exists an enantiomeric chair but this is identical with the inverted chair **X′** (Fig. 3.23A). The two enantiomeric conformers **X** and **X′** are rapidly interconvertible at room temperature and constitute a nonresolvable racemate, and hence *cis*-1,2-dimethylcyclohexane is optically inactive. The planar representation also indicates that the *cis* isomer is optically inactive, although for a different reason: it has a symmetry plane (σ) and is achiral.

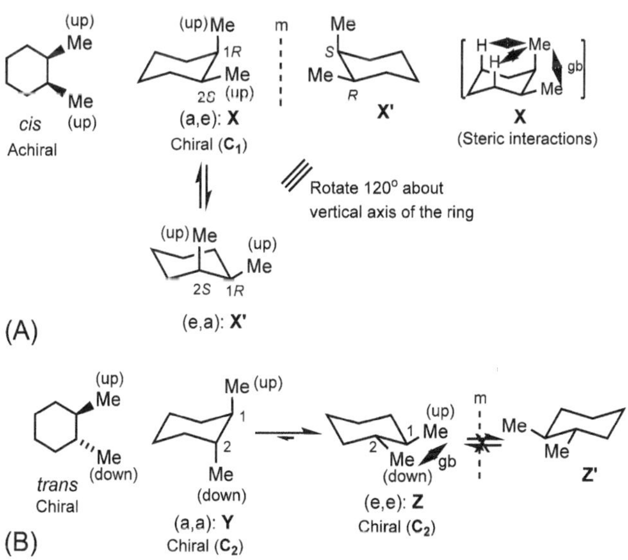


**FIG. 3.23** Conformational analysis of (A) *cis*-1,2-dimethylcyclohexane and (B) *trans*-1,2-dimethylcyclohexane.

The steric energy, $\Delta H^0_X$ of the conformer **X** can be estimated from the *A* value of Me (1.74 kcal mol$^{-1}$) and the steric interaction between a-Me and e-Me (a-Me↔e-Me) which resembles a butane–gauche (bg) interaction (0.9 kcal mol$^{-1}$). Since the enantiomeric conformers **X** and **X′** have the same energy, we get

$$\Delta H^0{}_X = \Delta H^0{}_{X'} = 1.74 + 0.9 = 2.64\,\text{kcal mol}^{-1}$$

The steric enthalpy of the *cis* isomer, $\Delta H^0_{cis}$, will be the weighted average of enthalpies of two conformers and is given by

$$\Delta H^0{}_{cis} = \tfrac{1}{2}\left(\Delta H^0{}_X + \Delta H^0{}_{X'}\right) = 2.64\,\text{kcal mol}^{-1}$$

In general, the entropy term, $\Delta S^0$, can be estimated primarily from three contributions: (i) entropy due to symmetry number ($-$R ln $\sigma$), (ii) entropy of mixing of the two conformers [$-$R ($x_1$ ln $x_1 + x_2$ ln $x_2$); $x_1$ and $x_2$ are the mole fractions] and (iii) entropy due to a racemic mixture (R ln 2, being the entropy of mixing when $x_1 = x_2 = \tfrac{1}{2}$). For the *cis* isomer (**C₁**), the symmetry number ($\sigma$) = 1, and since 1:1 mixture of the two conformers gives a racemate, entropy of mixing denotes the entropy due to a racemate. Therefore $\Delta S^0_{cis}$ is estimated as

$$\Delta S^0{}_{cis} = -\text{R ln } 1 + \text{R ln } 2 = 1.38\,\text{cal K}^{-1}\,\text{mol}^{-1}$$

Fig. 3.23B shows the conformational equilibrium for *trans*-1,2-dimethylcyclohexane. The conformer **Y** is diaxial (a,a) while the conformer **Z** is diequatorial (e,e). The steric enthalpy of each conformer is obtained as

$$\Delta H^0{}_Y = 2 \times A(\text{Me}) = 3.48\,\text{kcal mol}^{-1}$$

$$\Delta H^0{}_Z = \text{e} - \text{Me} \leftrightarrow \text{e} - \text{Me} = 1\,\text{gb} = 0.9\,\text{kcal mol}^{-1}$$

Therefore the (e,e) conformer **Z** is enthalpically more stable than the (a,a) conformer **Y** by (3.48–0.9) = 2.58 kcal mol$^{-1}$. If this energy difference is assumed to be the free energy difference, the population ratio **Z**:**Y** will be nearly 99:1 (cf. Table 1.2). Thus **Z** exists almost exclusively in the conformational equilibrium. The steric enthalpy of the *trans* isomer ($\Delta H^0_{trans}$) is given by

$$\Delta H^0{}_{trans} = 0.99\left(\Delta H^0{}_Z\right) + 0.01\left(\Delta H^0{}_Y\right) = 0.93\,\text{kcal mol}^{-1}$$

Since **Z** possesses only a $C_2$ axis bisecting the bond between the two stereocentres, it is chiral and belongs to the point group $\mathbf{C_2}$. There exists an enantiomer **Z′** but unlike in the case of the *cis* isomer, **Z′** cannot be obtained by ring inversion of **Z**. Thus the two enantiomers **Z** and **Z′** constitute a racemate that is resolvable, and the *trans* isomer can exist in optically active forms. The same conclusion is obtained from the planar representation, which is chiral. The entropy term, $\Delta S^0_{trans}$, is estimated as

$$\begin{aligned}\Delta S^0{}_{trans} &= -\text{R ln}\sigma - \text{R}(x_1\ \ln x_1 + x_2\ \ln x_2) + \text{R ln}2 \\ &= -\text{R ln}2 - \text{R}(0.99\ \ln 0.99 + 0.01\ \ln 0.01) + \text{R ln}2 = 0.11\,\text{cal K}^{-1}\text{mol}^{-1}\end{aligned}$$

The relative stability of *cis*- and *trans*-1,2-dimethylcyclohexanes can now be estimated from the *cis* ⇌ *trans* equilibrium:

$$\Delta\Delta H^0 = \Delta H^0{}_{trans} - \Delta H^0{}_{cis} = 0.93 - 2.64 = -1.71\,\text{kcal mol}^{-1}$$

$$\Delta\Delta S^0 = \Delta S^0{}_{trans} - \Delta S^0{}_{cis} = 0.11 - 1.38 = -1.27\,\text{cal K}^{-1}\,\text{mol}^{-1}$$

Thus the enthalpy ($\Delta H^0_{trans} < \Delta H^0_{cis}$) favours the *trans* isomer by 1.71 kcal mol$^{-1}$, whereas the entropy ($\Delta S^0_{cis} > \Delta S^0_{trans}$) favours the *cis* isomer by 1.27 cal K$^{-1}$ mol$^{-1}$. The enthalpy and entropy factors act in opposition, and the free energy difference, $\Delta\Delta G^0$ at 25°C, is given by

$$\Delta\Delta G^0 = \Delta\Delta H^0 - T(\Delta\Delta S^0) = [-1710 + (298 \times 1.27)]\,\text{cal mol}^{-1}$$
$$= -1.33\,\text{kcal mol}^{-1}$$

Therefore the *trans* is more stable than the *cis* by 1.33 kcal mol$^{-1}$.

Experimentally, the *cis* ⇌ *trans* equilibration can be performed by reversible dehydrogenation over a Pd catalyst near 300°C. The free energy change for the isomerization[48] was found to be −1.46 kcal mol$^{-1}$, which agrees fairly well with the estimated value of −1.33 kcal mol$^{-1}$.

### 3.5.2.1 *Symmetry and timescales*

The chair conformers of *cis*-1,2-dimethylcyclohexane have $\mathbf{C_1}$ symmetry but they are rapidly interconvertible by ring inversion on a timescale that is much faster than the timescale of the experiment at room temperature. At a very low temperature, say, at −90°C, the interconversion is slow enough on NMR timescale that the $^1$H-decoupled $^{13}$C NMR spectrum of the *cis* isomer shows eight signals for eight nonequivalent carbons of one particular conformer with $\mathbf{C_1}$ symmetry. However, the spectrum at 25°C shows only four signals for the eight carbons. This is because the interconversion of conformers at 25°C is fast on NMR timescale so that pairs of nuclei are exchanging environments, thereby averaging chemical shifts. The symmetry of the *cis* isomer at room temperature then refers to an averaged structure resulting from the dynamic or nonrigid symmetry (see also Section 1.5.5). The averaged structure seems to be the planar form having $\mathbf{C_s}$ symmetry with only a symmetry plane.[49] Notably, the averaged symmetry ($\mathbf{C_s}$) is higher than the symmetry of any contributing conformer ($\mathbf{C_1}$). In general, the averaged or dynamic symmetry of a substituted cyclohexane that undergoes rapid conformational inversion on the experimental timescale can be predicted from the planar cyclohexane ring. For example, the point group of chair cyclohexane is $\mathbf{D_{3d}}$ but the averaged symmetry of cyclohexane will be the symmetry of planar cyclohexane ($\mathbf{D_{6h}}$), though the high-energy planar form is not even a TS for ring inversion. A rigorous treatment of nonrigid symmetry using permutation group theory actually shows that the permutation group for the freely inverting cyclohexane ring and the point group $\mathbf{D_{6h}}$ are isomorphous.

### 3.5.3 1,3-Dimethylcyclohexane (*cis* and *trans*)

The conformational equilibrium for *cis*-1,3-dimethylcyclohexane is shown in Fig. 3.24A. Both chair conformers, the diaxial (a,a) and the diequatorial (e, e), are achiral due to the presence of a symmetry plane passing through C2 and C4 and belong to the point group $\mathbf{C_s}$ (symmetry number $\sigma = 1$). The planar representation is also achiral. Thus the *cis* isomer is optically inactive. The steric enthalpy of the (a,a) conformation is contributed by two 1,3-diaxial (Me/H) interactions (*A* value for Me) and one 1,3-diaxial (Me/Me) interaction. The interaction energy for the 1,3-diaxial (Me/Me) interaction is quite high and estimated to be 3.7 kcal mol$^{-1}$ (see Problem 3.3).[50] (In the acyclic system, this is referred to as *syn* pentane interaction, see Fig. 1.13B). Thus

$$\Delta H^0{}_{a,a} = A(\text{Me}) + 1,3 - \text{diaxial Me/Me} = 1.74 + 3.7 = 5.44\,\text{kcal mol}^{-1}$$

In contrast, the (e,e) conformer has no steric interaction involving Me's, and $\Delta H^0_{e,e} = 0$. The difference in enthalpies between the two conformers is quite large (5.44 kcal mol$^{-1}$). As such, the conformational equilibrium is biased to the e,e conformer, i.e. the system is anancomeric (cf. *t*-butylcyclohexane, Section 3.4.2). Thus the *cis* isomer exists practically as a single conformer with $\Delta H^0_{cis} \cong \Delta H^0_{(e,e)} = 0$. The entropy term, $\Delta S^0_{cis} = -\text{R} \ln \sigma = 0$. (Note that entropy of mixing is negligible and no racemate is possible.)

*trans*-1,3-Dimethylcyclohexane exists as a single chair conformation **X** since the two chair conformations (a,e) and (e,a) are superposable or identical as shown in Fig. 3.24B. (This is in contrast to the case of (a,e) and (e,a) conformers of *cis*-1,2-dimethylcyclohexane.) **X** is chiral (point group $\mathbf{C_1}$) and the enantiomeric **X′** is not interconvertible with **X**. The planar form is also

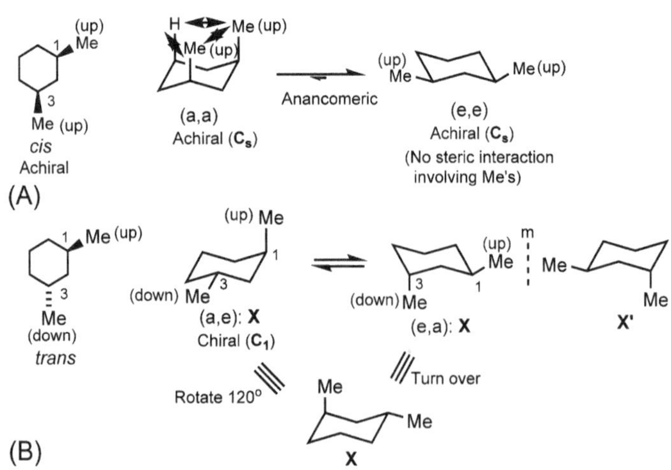


FIG. 3.24 Conformational analysis of (A) *cis*-1,3-dimethylcyclohexane and (B) *trans*-1, 3-dimethylcyclohexane.

chiral. The *trans* isomer thus exists as a resolvable racemate and in optically active forms. The steric enthalpy of the *trans* isomer with two 1,3-diaxial (Me/H) interactions is given by

$$\Delta H^0{}_{trans} = A(\text{Me}) = 1.74\,\text{kcal mol}^{-1}$$

The entropy term, $\Delta S^0_{trans} = -\text{R ln}\,\sigma + \text{R ln}\,2 = 0 + 1.38 = 1.38\,\text{cal K}^{-1}\,\text{mol}^{-1}$ (there is no entropy of mixing for a single conformer)

For the *cis* ⇌ *trans* equilibrium,

$$\Delta\Delta H^0 = \Delta H^0{}_{trans} - \Delta H^0{}_{cis} = 1.74 - 0 = 1.74\,\text{kcal mol}^{-1}$$

$$\Delta\Delta S^0 = \Delta S^0{}_{trans} - \Delta S^0{}_{cis} = 1.38 - 0 = 1.38\,\text{cal K}^{-1}\,\text{mol}^{-1}$$

Thus the enthalpy ($\Delta H^0_{cis} < \Delta H^0_{trans}$) favours the *cis* isomer by $1.74\,\text{kcal mol}^{-1}$ while the entropy ($\Delta S^0_{trans} > \Delta S^0_{cis}$) favours the *trans* isomer by $1.38\,\text{cal K}^{-1}\,\text{mol}^{-1}$. At 25°C,

$$\Delta\Delta G^0 = \Delta\Delta H^0 - T\left(\Delta\Delta S^0\right) = [1740 - (298 \times 1.38)]\,\text{cal mol}^{-1} = 1.33\,\text{kcal mol}^{-1}$$

Therefore the *cis* isomer has a lower free energy and is more stable than the *trans* isomer by $1.33\,\text{kcal mol}^{-1}$. This result agrees well with the experimental value[48] of $1.47\,\text{kcal mol}^{-1}$ in the liquid phase. Interestingly, the planar forms seem to suggest the opposite, i.e. the *cis* will be less stable than the *trans* due to steric reasons, which is not true.

**Problem 3.3** Equilibration studies of *cis*- and *trans*-1,1,3,5-tetramethylcyclohexane show that the *cis* isomer is favoured at equilibrium and is enthalpically more stable than the *trans* isomer by $3.7\,\text{kcal mol}^{-1}$. Estimate the interaction energy of a 1,3-diaxial (Me/Me) interaction.

### 3.5.4 1,4-Dimethylcyclohexane (*cis* and *trans*)

The chair conformations of both *cis*- and *trans*-1,4-dimethylcyclohexanes are achiral with a symmetry plane passing through C1 and C4 (Fig. 3.25). The planar forms are also achiral. The C1 and C4 centres are pseudoasymmetric (see Section 3.1.4). The *cis* isomer exists as a single chair conformation **X** since the (a,e) and (e,a) conformers are superposable. It belongs to the point group $\mathbf{C_s}$ (symmetry number $\sigma = 1$). The steric enthalpy of the *cis* isomer with two 1,3-diaxial (Me/H) interactions is given by

$$\Delta H^0{}_{cis} = A(\text{Me}) = 1.74\,\text{kcal mol}^{-1}$$

The entropy term, $\Delta S^0_{cis} = -\text{R ln}\,\sigma = 0$ (no entropy of mixing; no racemate).

The *trans* isomer has two possible chair conformations: the diaxial (a,a) and the diequatorial (e,e). The steric enthalpy of the (a,a) conformer, $\Delta H^0_{(a,a)} = 2 \times A$ $(\text{Me}) = 3.48\,\text{kcal mol}^{-1}$, whereas $\Delta H^0_{(e,e)} = 0$. The energy difference

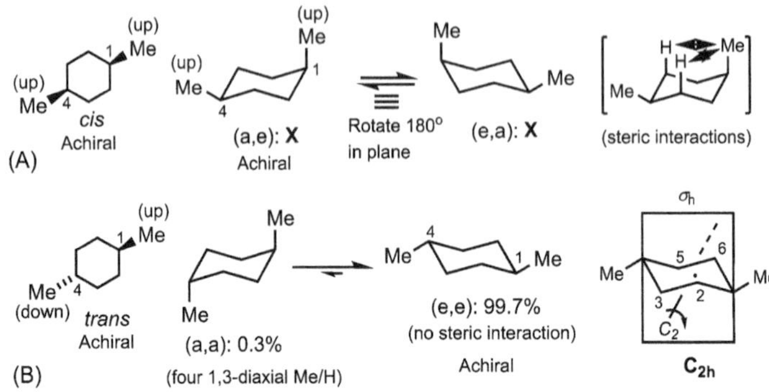


FIG. 3.25 Conformational analysis of (A) *cis*-1,4-dimethylcyclohexane and (B) *trans*-1, 4-dimethylcyclohexane.

(3.48 kcal mol$^{-1}$) gives roughly the population ratio (e,e:a,a) = 99.7:0.3. Therefore the steric enthalpy of the *trans* isomer is

$$\Delta H^0{}_{trans} = 0.997\left(\Delta H^0{}_{e,e}\right) + 0.003\left(\Delta H^0{}_{a,a}\right) = 0.01\ \text{kcal mol}^{-1}$$

The (e,e) conformer belongs to the point group $\mathbf{C_{2h}}$ (a $C_2$ axis bisecting 2–3 and 5–6 bonds and a $\sigma_h$ plane perpendicular to $C_2$). The entropy term is contributed by symmetry number and entropy of mixing and is given by

$$\begin{aligned}\Delta S^0{}_{trans} &= -\text{R}\ \ln 2 - \text{R}\left(0.997\ \ln 0.997 + 0.003\ \ln 0.003\right)\\ &= -1.34\ \text{cal K}^{-1}\ \text{mol}^{-1}\end{aligned}$$

For the *cis* ⇌ *trans* equilibrium,

$$\Delta\Delta H^0 = \Delta H^0{}_{trans} - \Delta H^0{}_{cis} = 0.01 - 1.74 = -1.73\ \text{kcal mol}^{-1}$$

$$\Delta\Delta S^0 = \Delta S^0{}_{trans} - \Delta S^0{}_{cis} = -1.34 - 0 = -1.34\ \text{cal K}^{-1}\ \text{mol}^{-1}$$

Thus the enthalpy favours the *trans* isomer and the entropy favours the *cis*. The free energy difference at 25°C is

$$\begin{aligned}\Delta\Delta G^0 &= \Delta\Delta H^0 - T\left(\Delta\Delta S^0\right) = [-1730 + (298 \times 1.34)]\ \text{cal mol}^{-1}\\ &= -1.33\ \text{kcal mol}^{-1}\end{aligned}$$

Therefore the *trans* isomer has a lower free energy and is more stable than the *cis* isomer by 1.33 kcal mol$^{-1}$. This estimated value agrees well with the experimental value[48] of −1.43 kcal mol$^{-1}$ in the liquid phase.

**Problem 3.4** The free energy change associated with conformational inversion of *cis*-1-isopropyl-4-methylcyclohexane as measured by $^{13}$C NMR at −100°C is 0.48 kcal mol$^{-1}$. Estimate the conformational free energy (*A* value) of *i*-Pr substituent.

### 3.5.5 Disubstituted cyclohexanes with bulky substituents

A bulky substituent such as *tert*-butyl is conformationally biased to be equatorial. If *t*-Bu is forced to occupy the axial position in a chair conformation, twist-boat conformation may compete. This is because the conformational free energy or *A* value of *t*-Bu (4.7 kcal mol$^{-1}$) is close to the free energy difference between the chair and twist-boat forms (4 kcal mol$^{-1}$) (see Section 3.3.4). In *cis*-1,3-di-*t*-butylcyclohexane, both *t*-Bu's are equatorial but the *trans* isomer must have one axial *t*-Bu in the chair conformation (Fig. 3.26A). Since axial *t*-Bu is highly unstable, the chair conformer will equilibrate with the twist-boat in which the same *t*-Bu is equatorial (te). Thus *trans*-1,3-di-*t*-butylcyclohexane exists as an equilibrium mixture of chair and twist-boat conformers, and the twist form is found to be slightly preferred.[48,51,52] Similarly, *cis*-1,4-di-*t*-butylcyclohexane exists as the chair-twist equilibrium mixture[53,54] (Fig. 3.26B).

An axial bulky substituent in the chair conformation can be further destabilized by an additional steric interaction with a vicinal equatorial substituent, as shown for *cis*-2-isopropylcyclohexanol in Fig. 3.27. In the preferred rotamer, H of axial *i* Pr is turned into the ring while two Me's are pointed outside. But the presence of an adjacent equatorial OH causes a severe (OH/Me) *syn* pentane-like interaction in the (a-*i*-Pr,e-OH) conformer (cf. Fig. 1.13B). In comparison, there is no additional strain in the (e-*i*-Pr,a-OH) conformer. Therefore the

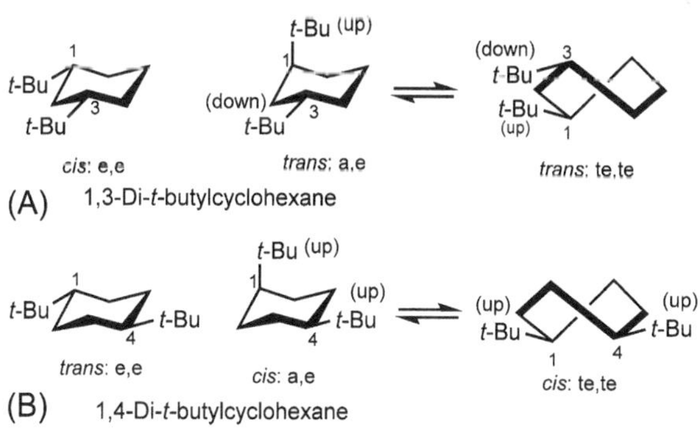


**FIG. 3.26** Conformational analysis of (A) *cis*- and *trans*-1,3-di-*t*-butylcyclohexane and (B) *cis*- and *trans*-1,4-di-*t*-butylcyclohexane.

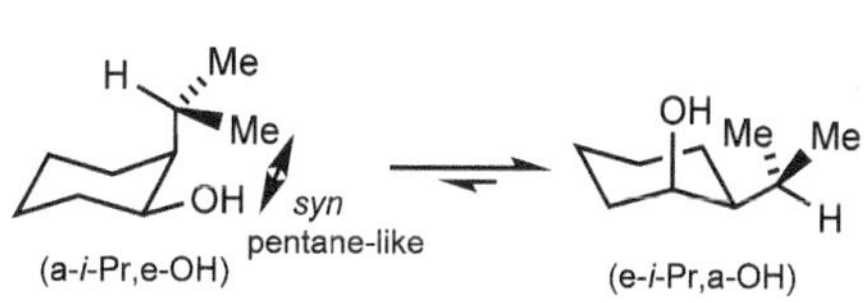


**FIG. 3.27** Vicinal nonadditivity in the case of *cis*-2-isopropylcyclohexanol.

equilibrium favours more strongly the (e-*i*-Pr,a-OH) conformer than expected from the additivity of *A* values of *i*-Pr and OH.[55] This is a case of vicinal nonadditivity.

### 3.5.6 Disubstituted cyclohexanes with polar substituents

If two vicinal substituents are polar, the relative stability of conformations will depend on dipole–dipole repulsion in addition to steric factor.[56] For example, *trans*-1,2-dibromocyclohexane in benzene exists as ~1:1 mixture of (a,a) and (e,e) conformers (Fig. 3.28A) in contrast to *trans*-1,2-dimethylcyclohexane that exists almost exclusively in the (e,e) conformer when only steric factor is operative (see Fig. 3.23B). The dipole–dipole repulsion is minimal in the (a,a) conformer as the two C—Br dipoles oppose each other while the (e,e) conformer is destabilized by dipole–dipole repulsion when two C—Br dipoles are aligned at a torsion angle of 60 degrees. Thus dipole–dipole repulsion tends to favour the (a,a) but steric factor tends to favour the (e,e). Two factors act in opposition and a balance between them gives almost equal amounts of (e,e) and (a,a) conformers. In polar solvents, the sterically favoured (e,e) form predominates as the solvent molecules surround the Br's by electrostatic attraction and thereby minimize the dipole repulsion. But in the gaseous state, the (a,a) conformer is highly predominant (95%) presumably because the dipole–dipole repulsion outweighs the steric effect.

Fig. 3.28B shows the (a,a) ⇌ (e,e) equilibria for *trans*-1,2-, *cis*-1,3- and *trans*-1,4-dichlorocyclohexanes. The (a,a) conformers of *trans*-1,2- and *trans*-1,4 have similar steric strain with four 1,3-diaxial (Cl/H) interactions but the (e,e) conformer of *trans*-1,2 is more destabilized by dipole–dipole repulsion and gauche interaction between two Cl's. Therefore *trans*-1,2 has higher (a,a) content than *trans*-1,4. On the other hand, (a,a) conformer of *cis*-1,3 suffers from a severe 1,3-diaxial (Cl/Cl) interaction (5.5 kcal mol$^{-1}$) as well

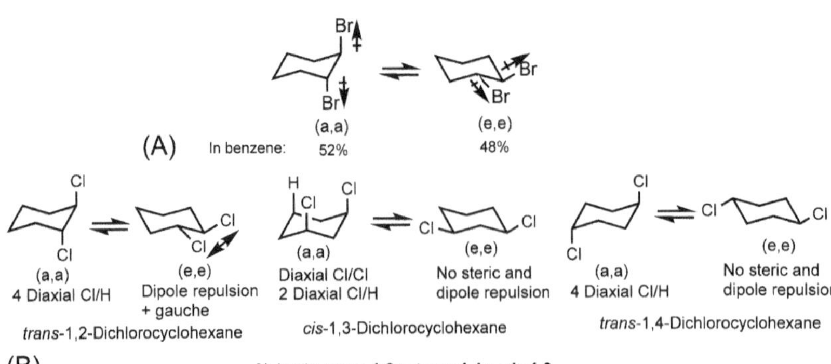


**FIG. 3.28** (A) Conformation of *trans*-1,2-dibromocyclohexane showing dipole repulsion and (B) relative (a,a) content in *trans*-1,2-, *cis*-1,3- and *trans*-1,4-dichlorocyclohexanes.

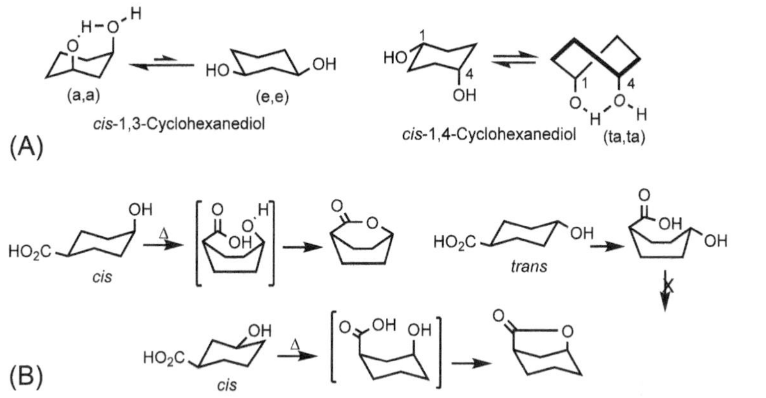


**FIG. 3.29** (A) Conformational analysis of *cis*-1,3- and *cis*-1,4-cyclohexanediols and (B) lactonization of *cis*-4-hydroxycyclohexanecarboxylic acid through boat conformation and of *cis*-3-hydroxycyclohexanecarboxylic acid through chair conformation.

as two 1,3-diaxial (Cl/H) interactions and therefore has the lowest (a,a) content. The % (a,a) content thus decreases in the order: *trans*-1,2-dichlorocyclohexane > *trans*-1,4-dichlorocyclohexane > *cis*-1,3-dichlorocyclohexane.

The conformational equilibrium of *cis*-1,3-cyclohexanediol favours the (a,a) conformer rather than the (e,e) conformer. This is due to intramolecular hydrogen bonding (from IR) that can stabilize the (a,a) but not the (e,e) (Fig. 3.29A). Intramolecular H-bonding is not possible between two *cis* OH's that are far apart at 1,4 positions in the chair conformation but can occur in the twist-boat leading to a chair-twist equilibrium for *cis*-1,4-cyclohexanediol[57] (Fig. 3.29A).

Fig. 3.29B shows that *cis*-4-hydroxycyclohexanecarboxylic acid forms a lactone on heating but the *trans* isomer does not.[56] The substituents OH and $CO_2H$ are far apart in the chair conformation for both *cis* and *trans* isomers. However, the *cis* isomer can assume a boat conformation in which OH and $CO_2H$ can come closer for lactonization to take place. But for the *trans* isomer, they cannot come within bonding distance even in the boat form and the reaction does not occur. A lactone is also formed on heating *cis*-3-hydroxycyclohexanecarboxylic acid (Fig. 3.29B). Here lactonization takes place through the chair but via a higher energy (a,a) conformation in which OH and $CO_2H$ are close to react. Although (a,a) content is very small (0.6% at room temperature), rate of lactonization is fast due to entropic effects.

## 3.6 Conformation of polysubstituted cyclohexanes

We will describe here conformations of only a few polysubstituted cyclohexanes. If three substituents are placed on different ring carbons, the possible constitutional isomers are 1,2,3-, 1,2,4- and 1,3,5-trisubstituted cyclohexanes. 1,2,3-Trimethylcyclohexane has three diastereomers, one chiral and two *meso*

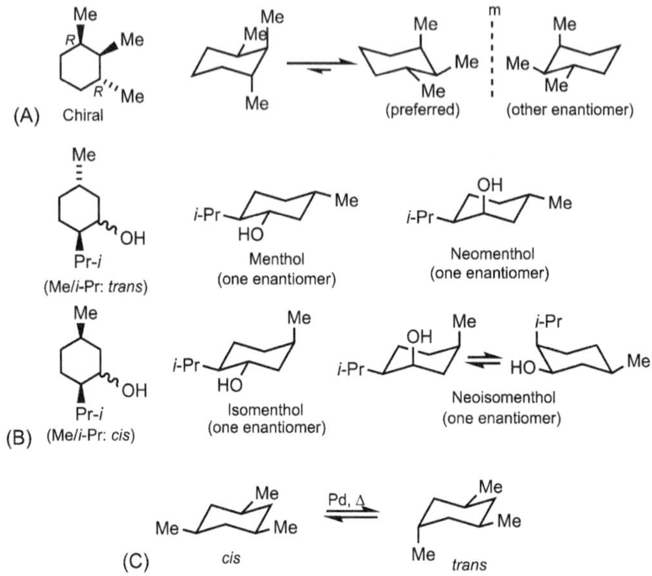


**FIG. 3.30** (A) Conformational analysis of chiral 1,2,3-trimethylcyclohexane, (B) preferred conformations of four diastereomeric menthols and (C) equilibration of *cis*- and *trans*-1,3,5-trimethylcyclohexane.

(cf. 2,3,4-trihydroxyglutaric acid in Fig. 2.37). The preferred conformation of a diastereomer is that in which maximum number of Me's can occupy equatorial positions, as shown for the chiral 1,2,3-trimethylcyclohexane in Fig. 3.30A. Note that C2 is chirotopic but nonstereogenic.

Menthols are diastereomers of a 1,2,4-trisubstituted cyclohexane, 2-isopropyl-5-methylcyclohexanol that contains three unlike chiral centres. Therefore four chiral diastereomers are possible, each with a (±)-pair (cf. aldopentoses in Fig. 2.35A). These are menthol, neomenthol, isomenthol and neoisomenthol. Fig. 3.30B shows the preferred conformations of all menthols in which *i*-Pr with the highest *A* value is equatorial. But in the case of neoisomenthol, 1,3-diaxial OH/Me interaction (2.4 kcal mol$^{-1}$)[58] is similar to the *A* value of *i*-Pr, and both conformers with the equatorial *i*-Pr and the axial *i*-Pr will populate substantially in the conformational equilibrium. Note that only one enantiomer of each diastereomer is shown. Based on steric interactions in the conformers, the stability order is: menthol > neomenthol > isomenthol > neoisomenthol. Experimentally, equilibration of menthol and neomenthol in the presence of aluminium isopropoxide can be used to determine their relative stability. The equilibration with Al(O-*i*-Pr)$_3$ occurs through reversible oxidation of the alcohols. Similarly, isomenthol and neoisomenthol can be equilibrated in the presence of Al(O-*i*-Pr)$_3$.

1,3,5-Trimethylcyclohexane has only two achiral diastereomers (see Fig. 3.3A). They are called *cis* and *trans* (but not *meso* since no chiral diastereomer exists) and can be equilibrated through reversible dehydrogenation over a Pd catalyst (Fig. 3.30C). The equilibrium favours the more stable *cis* isomer.

The nongeminal tetrasubstituted cyclohexanes are of the types 1,2,3,4-, 1,2,3,5- and 1,2,4,5-. For a tetrasubstituted case, consider 1,2,4,5-tetramethylcyclohexane. It has five diastereomers (**A**–**E**) that can be equilibrated over Pd catalyst at 300°C[59] (Fig. 3.31A). The position of equilibrium corresponds to their relative stability **A** > **B** > **C** > **D** > **E** in agreement with the calculated $\Delta H^0$ values. The barrier to ring inversion for the least stable **E** isomer is found to be significantly higher (12.4 kcal mol$^{-1}$) than that for cyclohexane (10.8 kcal mol$^{-1}$).

The ring inversion barrier can be as high as 17.3 kcal mol$^{-1}$ for all-*cis* hexamethylcyclohexane **X** (Fig. 3.31B). This is presumably because the vicinal Me's need to pass by each other through a crowded environment in all possible inversion pathways.[59] In the case of inositols, all eight diastereomers (see Fig. 3.3B, Problem 3.6) are known.[3] But for hexachlorocyclohexanes, the all-*cis* isomer **Y** (Fig. 3.31B) is too unstable and has not been synthesized.[60] This is presumably because of three severe 1,3-diaxial Cl/Cl interactions[50] (3 × 5.5 kcal mol$^{-1}$) besides dipole repulsions and gauche interactions. The γ isomer of hexachlorocyclohexane is an insecticide and its conformation is (a, a,a,e,e,e), as shown in Fig. 3.31B.

**Problem 3.5** The free energy difference ($\Delta G^0$) between the conformers of 1,1,2-trimethylcyclohexane is found to be 1.53 kcal mol$^{-1}$ which shows vicinal nonadditivity. Explain.

**Problem 3.6** Draw conformations for all diastereomeric inositols.

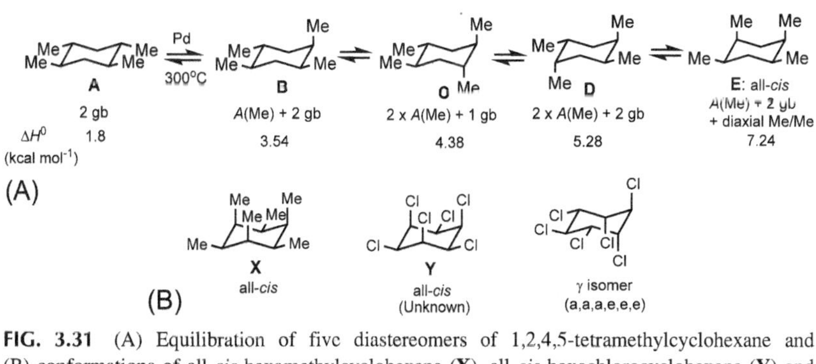


**FIG. 3.31** (A) Equilibration of five diastereomers of 1,2,4,5-tetramethylcyclohexane and (B) conformations of all-*cis*-hexamethylcyclohexane (**X**), all-*cis*-hexachlorocyclohexane (**Y**) and γ isomer of hexachlorocyclohexane. gb = gauche–butane.

## 3.7 Conformation of six-membered saturated heterocycles[61,62]

The replacement of one or more carbon atoms of cyclohexane ring by heteroatoms such as O, N and S gives six-membered saturated heterocycles. The monohetero system with O, N and S is called oxane (tetrahydropyran), piperidine and thiane, respectively. Like cyclohexane, they exist in the preferred chair conformation. An alkyl substituent in saturated heterocycles also prefers to be equatorial but the conformational free energies ($-\Delta G^0$) or *A* values may differ from one heterocycle to another, and also from cyclohexane. For 2-substituted heterocycles, the $-\Delta G^0$ values of Me are shown with reference to methylcyclohexane in Fig. 3.32A. The $-\Delta G^0$ values for 2-methyloxane[63] and 2-methylpiperidine[64] are larger than for methylcyclohexane, whereas the value for 2-methylthiane[65] is smaller. The larger *A* value for the oxygen or nitrogen heterocycle is attributed to the shorter C—O bond (1.43 Å) or C—N bond (1.47 Å) compared to C—C bond (1.53 Å). A shorter bond distance brings the axial Me closer to synaxial hydrogens and thereby increases 1,3-diaxial (Me/H) interactions. Consequently, the *A* value increases from cyclohexane to piperidine to oxane. Since the C—S bond in thiane is larger than C—C, the *A* value decreases.

For 3-substituted heterocycles such as 3-methyloxane,[63] the *A* value is found to be smaller than for methylcyclohexane (Fig. 3.32B). This is because one 1,3-diaxial (Me/H) interaction in cyclohexane system is replaced by a less strong (Me/lone pair) interaction in the oxane. For 4-methyloxane[63] (Fig. 3.32B), there are two 1,3-diaxial (Me/H) interactions similar to those in cyclohexane system. The small difference in the *A* values might result from some variations in the ring shapes.

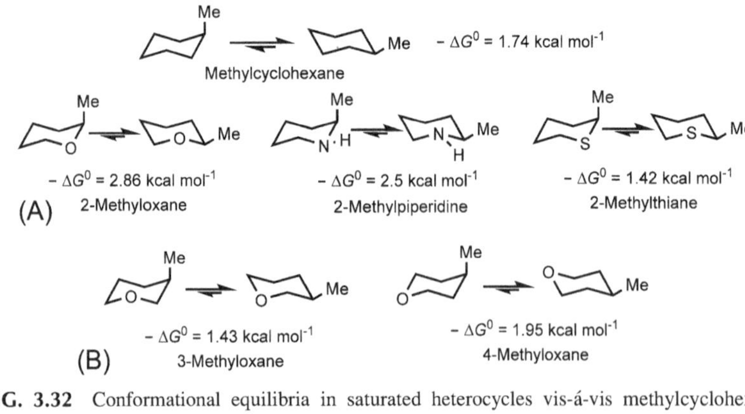


**FIG. 3.32** Conformational equilibria in saturated heterocycles vis-á-vis methylcyclohexane: (A) 2-methyloxane, 2-methylpiperidine and 2-methylthiane and (B) 3-methyloxane and 4-methyloxane.

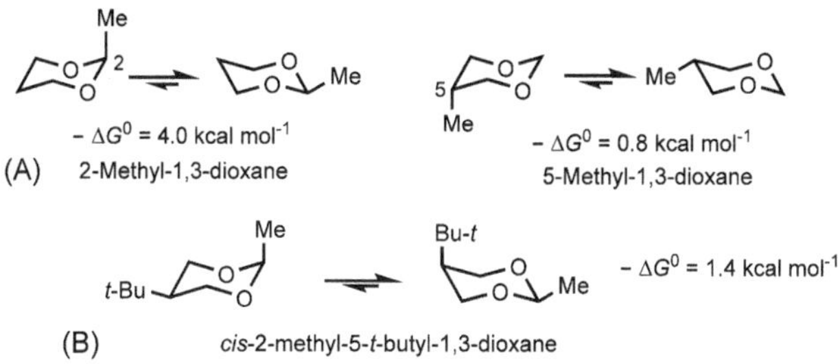


FIG. 3.33 Conformational equilibria for (A) 2-methyl-1,3-dioxane and 5-methyl-1,3-dioxane; (B) *cis*-2-methyl-5-*t*-butyl-1,3-dioxane.

Fig. 3.33A shows the $-\Delta G^0$ values for 2-methyl and 5-methyl derivatives of 1,3-dioxane.[66] For 2-methyl-1,3-dioxane, the $A$ value of Me (4.0 kcal mol$^{-1}$) is greater than that for 2-methyloxane (2.86 kcal mol$^{-1}$) since the axial Me in the former is more strongly destabilized by 1,3-diaxial (Me/H) interactions due to shorter C—O bonds. For 5-methyl-1,3-dioxane, the $A$ value is quite small because the two 1,3-diaxial (Me/lone pair) interactions are much less severe. An interesting case is *cis*-2-methyl-5-*t*-butyl-1,3-dioxane (Fig. 3.33B). NMR spectrum shows that Me is equatorial and *t*-Bu is axial in the preferred conformation.[67,68] This conformer is favoured over the other by 1.4 kcal mol$^{-1}$. The axial preference of *t*-Bu in this case is clearly due to a weaker 1,3-diaxial (*t*-Bu/lone pair) interactions compared to a strong destabilizing 1,3-diaxial (Me/H) interactions in the other conformer (cf. 2-methyl-1,3-dioxane).

**Problem 3.7** In a concentrated solution (2.8 M) in $CD_2Cl_2$, *cis*-3-hydroxythiane oxide exists preferentially in the diequatorial conformation ($-\Delta G^0 = 1.0$ kcal mol$^{-1}$) but in a dilute solution (2.3 m*M*), the equilibrium favours the diaxial conformer ($\Delta G^0 > 1.3$ kcal mol$^{-1}$). Explain.

## 3.8 The anomeric effect[69–71]

In 2-alkyl-substituted tetrahydropyrans, the alkyl group (R) prefers to be equatorial for steric reasons. In contrast, if the 2-substituent is an electronegative or electron withdrawing group (X), there is a definite preference for the X to be axial (Fig. 3.34). This unexpected preference is called the anomeric effect.

The anomeric effect was recognized first in carbohydrate chemistry. Since the 2-position of tetrahydropyran (oxane) ring in glucose is anomeric carbon (C1), such effect has been termed the anomeric effect. This acts against the steric preference of the substituent X and a measure of the anomeric effect is expressed by a parameter $E_{An}$ given by[72]

$$E_{An} = RT \ln\frac{[a-X]}{[e-X]} + A_X = \Delta G^0_{oxane} - \Delta G^0_{cyclohexane} \quad (3.4)$$

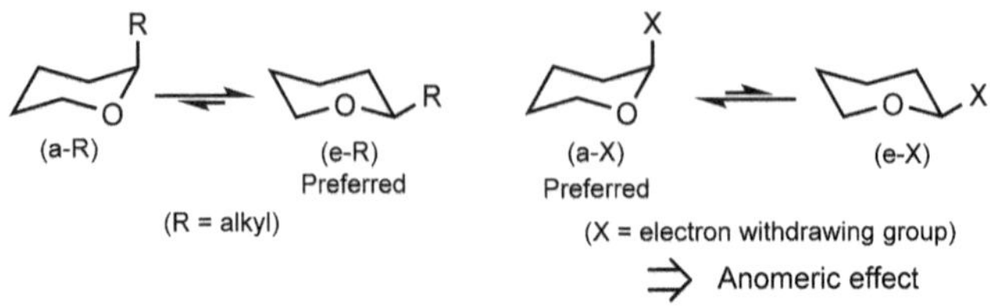


**FIG. 3.34** Anomeric effect in 2-X-substituted tetrahydropyran system (X = electron withdrawing substituent).

**TABLE 3.3 Anomeric effect ($E_{An}$) of some substituents in tetrahydropyran (oxane) system.**

| | Anomeric effect, $E_{An}$ (kcal mol$^{-1}$) | |
|---|---|---|
| **Substituent (X)** | *Calculated* | *Observed* |
| OH | 1.5 | 0.61 |
| OMe | 1.7 | 0.75 |
| Cl | 3.5 | 2.12 |
| Me | −1.0 | −1.12 |
| $NH_2$ | −2.1 | −0.33 |

Usually $A_X$ is taken as the $A$ value for cyclohexane system; however, in some cases this is corrected for oxane system. The parameter $E_{An}$ denotes the preference of a substituent for the axial position. In an anomeric effect, $E_{An}$ is positive. If $E_{An}$ is negative, it is called a reverse anomeric effect. Table 3.3 shows the calculated and experimentally determined anomeric effects for some common substituents in tetrahydropyran system.[73] Electron withdrawing substituents (OH, OMe and Cl) exhibit (positive) anomeric effect, whereas electron donating Me and $NH_2$ give negative or the reverse anomeric effect. The largest anomeric effect is given by Cl.

### 3.8.1 Origin of the anomeric effect

The generally accepted explanation of the anomeric effect is stereoelectronic, in terms of $n$-$\sigma^*$ hyperconjugation.[23,73] The delocalization of a nonbonding electron pair on oxygen ($n_O$) with the vacant $\sigma^*$ orbital of the adjacent C—X bond ($\sigma^*_{C-X}$) is optimal when the interacting orbitals are antiperiplanar. In the axial conformer, C—O—C—X fragment is gauche (highlighted in bold bonds) so that $n_O$ and $\sigma^*_{C-X}$ are antiperiplanar (Fig. 3.35A). Consequently, an effective delocalization leads to a thermodynamic stabilization of the axial conformation.

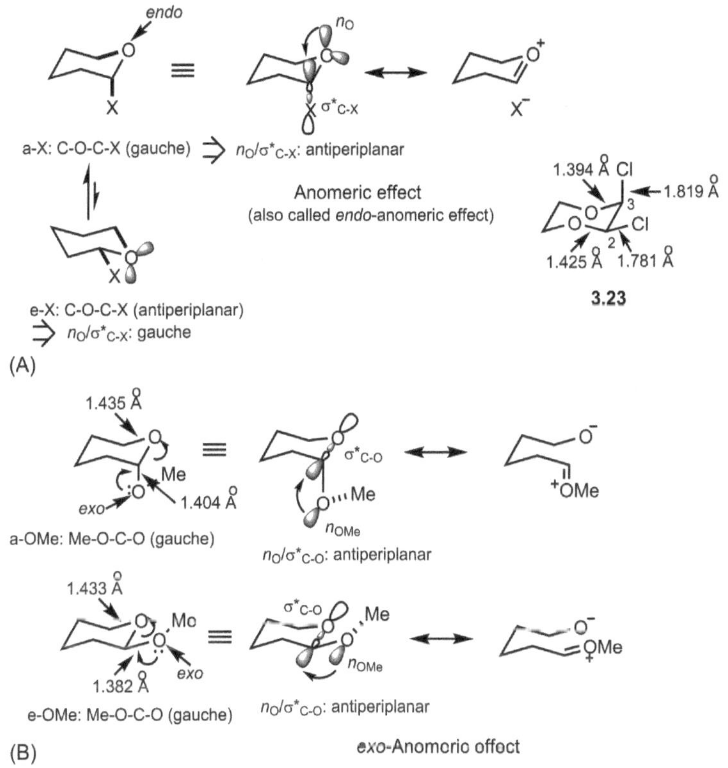


**FIG. 3.35** (A) Explanation of the anomeric effect (also called *endo*-anomeric effect) based on $n$–$\sigma^*$ negative hyperconjugation in tetrahydropyran system. (B) *exo*-Anomeric effect possible in both axial and equatorial conformers of 2-methoxytetrahydropyran. (smaller lobe of lone pair orbital on *endo* or *exo* O not shown)

But in the equatorial conformer, C—O—C—X fragment is antiperiplanar; C—X bond is then oriented gauche to the lone pairs on oxygen and no effective interactions can occur. Thus the axial conformer is more stable and hence is the preferred conformation.

As shown in Fig. 3.35A, the $n$–$\sigma^*$ orbital interaction is also represented as no-bond resonance or hyperconjugation. This type of hyperconjugation is called negative hyperconjugation because a filled orbital is delocalized into a $\sigma^*$ antibonding orbital. The filled orbital may be an $n$ orbital or a π orbital; thus $n$–$\sigma^*$ or π–$\sigma^*$ interaction denotes a negative hyperconjugation. (In positive hyperconjugation, a filled σ orbital is delocalized into a vacant p or π* orbital indicating σ–p or σ–π* interaction.) In frontier orbital terms, the $n$–$\sigma^*$ interaction is called $HOMO_n/LUMO_{\sigma^*}$ interaction since lone pairs on oxygen are HOMOs and the antibonding $\sigma^*$ orbital of the C—X bond is LUMO. Note that this frontier orbital interaction refers to the ground state stability of the molecule, not its reactivity.

The resonance representation in Fig. 3.35A shows that *n*–$\sigma$* negative hyperconjugation leads to a shortening of the C—O bond distance (with double bond character) and a lengthening of the exocyclic C—X bond (with less single bond character). This has indeed been confirmed by calculation and X-ray crystallography. The structural data from the X-ray crystal structure of *cis*-2,3-dichloro-1,4-dioxane **3.23**[74] show that the axial C—Cl bond is appreciably longer (1.819 Å) than the equatorial C—Cl bond (1.781 Å). Further, the C3—O bond is considerably shortened (1.394 Å) compared to the C2—O bond (1.425 Å).

The anomeric effect shown in Fig. 3.35A involves the lone pair on ring O (*endo* O) and is sometimes called *endo*-anomeric effect. Another kind of anomeric effect is possible in both axial and equatorial conformers in which the lone pair on *exo* O is delocalized into the vacant $\sigma$* orbital of adjacent ring C—O bond. This type of anomeric effect is called *exo*-anomeric effect.[75,76] This is illustrated with 2-methoxytetrahydropyran in Fig. 3.35B.[77] Here Me—O—C—O fragment can be gauche in a rotamer of both axial and equatorial conformers. Thus a lone pair on *exo*-oxygen ($n_{OMe}$) is antiperiplanar with the $\sigma$* orbital of the ring C—O bond ($\sigma^*_{C-O}$) in each conformer. A shortening of C—OMe bond distance (1.382 Å) in the equatorial conformer with reference to the usual C—O bond distance (1.433 Å) is indicative of an *exo*-anomeric effect. Such shortening of C—OMe bond distance also takes place in the axial conformer but C—OMe bond distance is longer (1.404 Å) than that of the equatorial conformer because an *endo*-anomeric effect (Fig. 3.35A) is operative in the axial conformer.

Another interpretation of the *endo*-anomeric effect is based on electrostatic interactions.[78] In the equatorial conformer, the C—X dipole and the resultant dipole from two oxygen lone pairs are nearly aligned. Therefore electrostatic repulsion between them would lead to the destabilization of the equatorial conformation (Fig. 3.36). But in the axial conformer, the two dipoles are in different spatial directions and there is no significant dipole–dipole repulsion. Thus the axial conformation is preferred.

The dipole repulsion between the oxygen lone pairs (which resemble rabbit ears) and the C-X dipole is known as 'rabbit ear effect'. This interpretation has the merit in accounting for the solvent effect[71] and is also supported by theoretical calculations.[79] The conformational equilibria of 2-methoxytetrahydropyran in different solvents show that the proportion of axial conformer decreases on increasing the polarity of the solvent as given below.

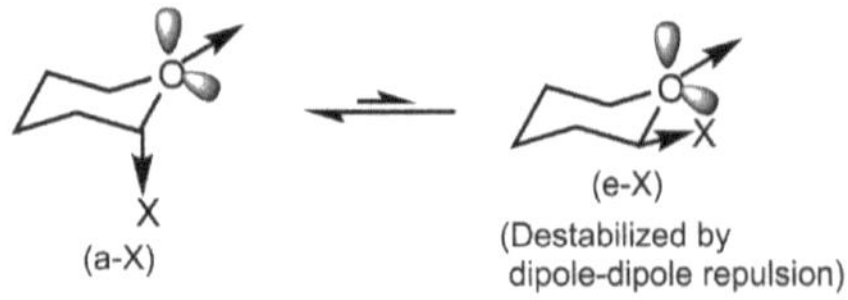


**FIG. 3.36** Interpretation of the *endo*-anomeric effect based on electrostatic interactions (rabbit ear effect).

| **Solvent:** | $CCl_4$ (ε 2.2) | Acetone (ε 20.7) | Acetonitrile (ε 37.5) | Water (ε 78.5) |
|---|---|---|---|---|
| **Axial (%):** | 83 | 72 | 68 | 52 |

The anomeric effect is less pronounced in more polar solvents because the dipole–dipole repulsion is weakened by solvation spheres.

**Problem 3.8** Predict the preferred conformation of *trans*-2,5-dichloro-1, 4-dioxane.

## 3.8.2 Anomeric effects in glucose and its derivatives

The configuration of D-glucose was shown earlier as an open-chain structure (see Fig. 2.35B); however, this structure is present negligibly in aqueous solution because of spontaneous intramolecular hemiacetal formation to give a cyclic pyranose (tetrahydropyran) structure (Fig. 3.37). The hemiacetal formation creates a new stereocentre at C1, and two diastereomers are formed. These diastereomers differing in configuration at only C1 (previously carbonyl carbon) are called anomers, and C1 is called anomeric carbon. The two anomers of D-glucose are α-D-glucopyranose and β-D-glucopyranose. Both planar representations (Haworth formulas) and preferred chair conformations of α and β anomers are shown in Fig. 3.37. It may be mentioned that any pair of diastereomeric sugars that differ in configuration at only one chiral centre are known as epimers; α- and β-anomers are therefore C1 epimers.

In the preferred chair conformations, all substituents are equatorial except OH at the anomeric carbon (C1) which is down axial in the α-anomer and up equatorial in the β-anomer. Note that in the D-series of sugars, α indicates a down (below the plane) bond and β an up (above the plane) bond. The sugar derivatives formed from the reaction of anomeric OH are called glycosides

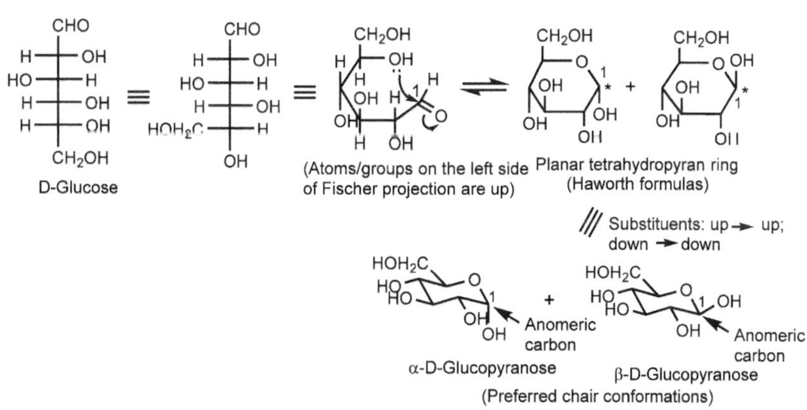


**FIG. 3.37** Pyranose structures of α-D-glucose and β-D-glucose (Howarth formulas and preferred chair conformations).

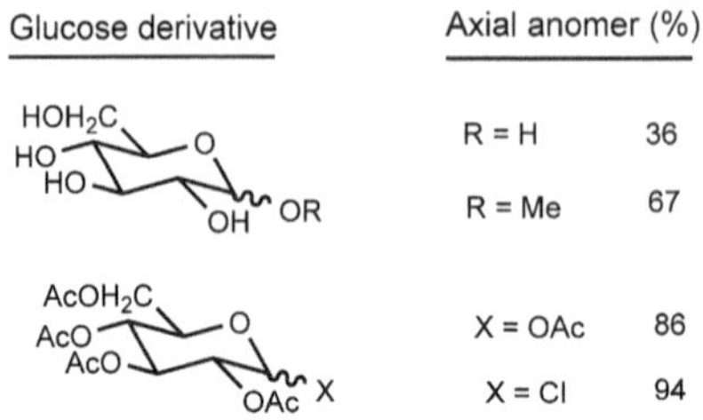


**FIG. 3.38** Anomeric effects in glucose and its derivatives.

and the anomeric OH is referred to as glycosidic OH. For example, if the anomeric OH of α-D-glucose is derivatized to OMe, the resulting glucose derivative is called methyl α-D-glucoside.

Glucose and its derivatives exhibit anomeric effects as shown in Fig. 3.38.[80]

For the glucose derivatives, the axial anomer (α) predominates over the equatorial anomer (β) in the diastereomeric equilibrium. The proportion of the axial isomer increases progressively from 67% to 94% as the substituent at the anomeric carbon becomes more electron withdrawing or electronegative. This tendency for the axial anomer to predominate over the equatorial anomer has been termed by Lemieux as the anomeric effect.[81] The parent glucose in aqueous solution seems to be an exception with the axial:equatorial ratio = 36:64. However, this amount of the axial isomer is significantly higher than that expected from the conformational free energy of OH in polar solvent ($-\Delta G^0 = 0.95\,\text{kcal mol}^{-1}$). The $-\Delta G^0$ value gives the equilibrium constant $K = 5$ with axial:equatorial ratio = 17:83. Further, the $-\Delta G^0$ value of OH will be even more in tetrahydropyran (oxane) system, and the expected axial isomer will then be less than 17%. Thus the greater axial content (36%) than expected (<17%) in aqueous solution of glucose indicates that anomeric effect operates; however, this is less pronounced in a highly polar solvent water.

The α- and β-anomers of D-glucose can be obtained pure by crystallization under different conditions: crystallization from aqueous solution below 50°C gives α-D-glucose (specific rotation $[\alpha]_D^{20} = +112.2$) while crystallization by vacuum evaporation of saturated aqueous solution at 115°C produces β-D-glucose ($[\alpha]_D^{20} = +19.2$). When α-D-glucose is dissolved in water, it can convert to the β-anomer via the open-chain structure, and finally equilibrium is established between the two anomers (α = 36%, β = 64%) with $[\alpha]_D^{20} = +52.7$. The same equilibrium mixture can also be obtained starting from the β-anomer. This phenomenon of gradual change of initial specific rotation of either anomer to the equilibrium value of +52.7 is called mutarotation.

## 3.9 Conformation of cyclohexyl systems with a double bond

The cyclohexyl systems with a double bond include cyclohexanones, alkylidenecyclohexanes and cyclohexenes.

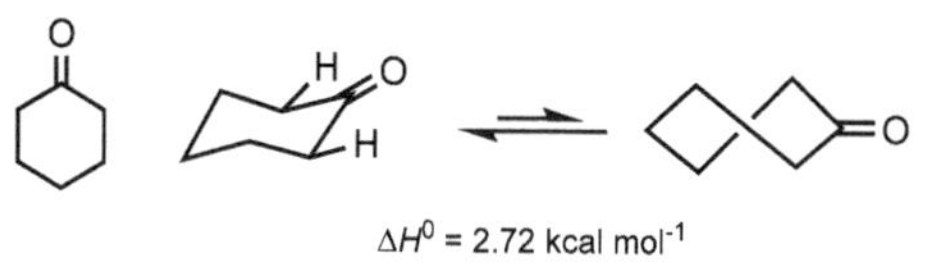

FIG. 3.39 Conformational equilibrium for cyclohexanone.

### 3.9.1 Cyclohexanones

Replacing one $CH_2$ of cyclohexane by C=O gives cyclohexanone but does not alter its conformational preference.[82] The chair conformation is the more stable conformation of cyclohexanone in the chair-twist equilibrium (Fig. 3.39).

However, the cyclohexanone chair is appreciably flattened at the carbonyl end and is destabilized to some extent by eclipsing interactions of C=O with the adjacent equatorial H's. On the other hand, the twist-boat form is slightly stabilized by some relief from carbonyl eclipsing and flagpole-type interactions. As a result, the energy difference between the twist-boat and chair conformers is less (2.72 kcal $mol^{-1}$)[83] than that for cyclohexane (5.5 kcal $mol^{-1}$). The energy barrier for ring inversion of cyclohexanone is also considerably smaller (4.0 kcal $mol^{-1}$)[83] since the barrier to rotation about C—C(—O) is lower than that for C—C in alkanes (see Section 1.3.10). In the net, cyclohexanone is relatively more strained than cyclohexane by about 3 kcal $mol^{-1}$ due to the presence of some angle strain and torsional strain.

#### 3.9.1.1 2-Alkylketone effect

The conformational free energy ($-\Delta G^0$) of substituents in 2-substituted cyclohexanones can be determined from the equilibration studies of the diastereomeric substrates through facile epimerization via an enol or enolate intermediate in the presence of acid or base. For example, equilibration of *cis*- and *trans*-4-*t*-butyl-2-alkylcyclohexanone in the presence of a base gives the $-\Delta G^0$ values of 2-alkyl substituents (R = Me, Et, *i*-Pr and *t*-Bu) (Fig. 3.40A).[84] Interestingly, the trend in $-\Delta G^0$ values from Me → Et → *i*-Pr is opposite to that for cyclohexane system. This trend of progressively lower $-\Delta G^0$ values signifies increasing accumulation of the axial substituent (R) and is known as 2-alkylketone effect. Note that the net equilibrium is still favourable to the side of the equatorial substituent. The $-\Delta G^0$ value for 2-Me is similar to that in cyclohexane. For 2-Et or 2-*i*-Pr substituent, there is relatively less steric strain in the axial conformation (*trans* isomer) **3.24** than in cyclohexane system. For 2-*i*-Pr, the equatorial conformation (*cis* isomer) is further destabilized sterically by Me/O interaction **3.25**. The 2-alkylketone effect is thus attributed to less steric hindrance in the axial conformation and additional steric interaction in the equatorial conformation in the case of 2-*i*-Pr. With 2-*t*-Bu substituent, the *trans* isomer with the axial *t*-Bu is unstable and exists predominantly in the twist-boat. On the other hand, the equatorial *t*-Bu

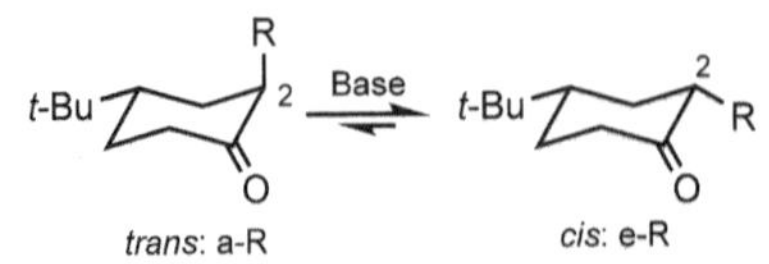


| 2-R | $-\Delta G^0$ (kcal mol$^{-1}$) | *A* value (cyclohexane) |
|---|---|---|
| Me | 1.56 | 1.74 |
| Et | 1.09 | 1.79 |
| *i*-Pr | 0.59 | 2.21 |
| *t*-Bu | 1.62 | 4.7 |

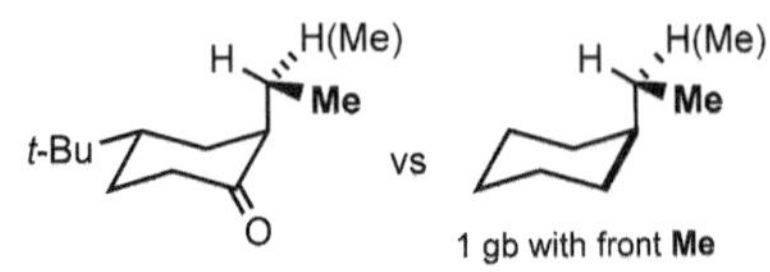


3.24

(A) No gb with front **Me**

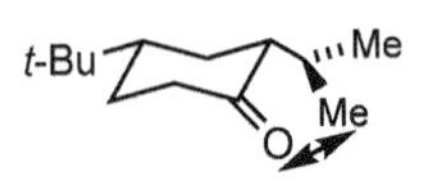


3.25

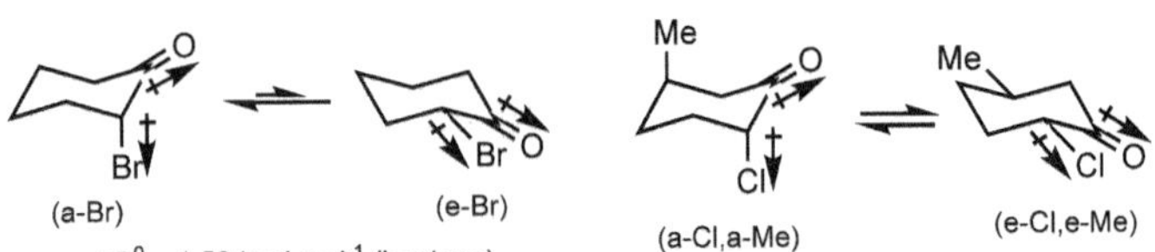


(B) 2-Bromocyclohexanone

*trans*-2-Chloro-5-methylcyclohexanone

**FIG. 3.40** (A) 2-Alkylketone effect: conformational free energies ($-\Delta G^0$) of 2-alkyl substituents (Me, Et, *i*-Pr and *t*-Bu) in cyclohexanone system and (B) conformational equilibria for cyclohexanones with polar substituents at 2-position.

(*cis*) is destabilized by eclipsing with carbonyl oxygen. This leads to an appreciably smaller $-\Delta G^0$ value compared to the *A* value of *t*-Bu in cyclohexane system.

If the 2-substituent is polar as in 2-bromocyclohexanone,[85] the equatorial conformation is highly destabilized by the electrostatic repulsions between the C—Br and C=O dipoles (Fig. 3.40B). There is also steric crowding between Br and O in the equatorial conformation. Thus the equatorial conformation is unfavourable by both electrostatic and steric factors.[86] In the axial conformation, the two dipoles are nearly orthogonal and essentially noninteracting though axial Br is sterically unfavourable. Again, the a-Br conformation could be stabilized by $\pi_{C=O}$–$\sigma^*_{C-Br}$ interaction (cf. anomeric effect).[87] In the net, the axial conformer is preferred ($\Delta G^0 = 1.03\,\text{kcal mol}^{-1}$ in cyclohexane). In the case of *trans*-2-chloro-5-methylcyclohexanone (Fig. 3.40B), the (a-Cl, a-Me) conformation is sterically unfavourable but the electrostatic repulsion between the C—Cl and C=O dipoles is maximal in the (e-Cl, e-Me) conformer. In a nonpolar solvent, electrostatic repulsion dominates and the diaxial conformer is preferred while in a polar solvent such as methanol, the electric dipoles are strongly solvated and the steric factor favours the diequatorial conformer.

### 3.9.1.2 3-Alkylketone effect

In a 3-substituted cyclohexanone, the axial substituent is impeded by one 1,3-diaxial interaction in contrast to two in cyclohexane system. Consequently,

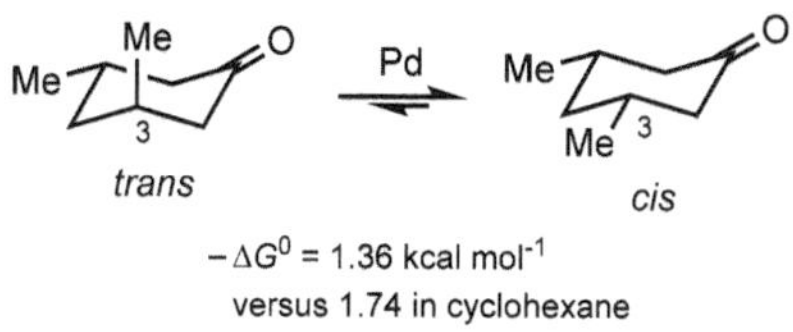


**FIG. 3.41** 3-Alkylketone effect in cyclohexanone system.

the energy difference between the axial and equatorial conformers is less. This is shown for 3-Me substituent from equilibrium studies in Fig. 3.41. The equilibration between *cis*- and *trans*-3,5-dimethylcyclohexanone is achieved through reversible catalytic dehydrogenation over Pd catalyst at 220°C. The $-\Delta G^0$ value is 1.36 kcal mol$^{-1}$, whereas the *A* value for Me in cyclohexane is 1.74 kcal mol$^{-1}$.[88] Note that the *cis* (e-Me) isomer is still major but to a less extent. The tendency for 3-alkyl substituent in a cyclohexanone to enhance axial content in conformational equilibrium is called 3-alkylketone effect.

### *3.9.1.3 4-Alkylketone effect*

An axial 4-alkyl substituent in a cyclohexanone is far from C=O but encounters two 1,3-diaxial interactions with *syn*-axial H's at C2 and C6. However, relative to cyclohexane, the axial H's at C2 and C6 in cyclohexanone are leaned outwards causing some relief of 1,3-diaxial strain (Fig. 3.42A). This feature leads to 4-alkylketone effect in cyclohexanone system.[30] Fig. 3.42B shows that rate of $CrO_3$ oxidation of 4,4-dimethylcyclohexanol is greater than that of cyclohexa nol. This is due to 4-methylketone effect causing some relief of 1,3-diaxial strain in the former case. In the latter case, no 1,3-diaxial strain is present in either alcohol substrate or ketone product. The 4-methylketone effect is also responsible for the faster rate of oxidation of a triterpene system compared to a steroidal system, also shown in Fig. 3.42B. Fig. 3.42C shows that the equilibrium constant for cyanohydrin formation with 4,4-dimethylcyclohexanone is smaller than with cyclohexanone. This is because 4-methylketone effect in the case of substituted cyclohexanone leads to more 1,3-diaxial strain in the cyanohydrin product. In comparison, the substrate or the product in the unsubstituted case has no 1,3-diaxial strain. Thus 4-methylketone effect influences both kinetics (Fig. 3.42B) and thermodynamics (Fig. 3.42C) of cyclohexanone reactions.

## 3.9.2 Alkylidenecyclohexanes

The simplest alkylidenecyclohexane is methylenecyclohexane. It has an exocyclic C=C and it resembles cyclohexanone. Like cyclohexanone, the preferred conformation of methylenecyclohexane is chair; however, the barrier to inversion is higher (~8 kcal mol$^{-1}$) due to the greater barrier to rotation about C—C(=$CH_2$) compared to C—C(=O).[89] The conformational free energies of substituents ($-\Delta G^0$) at C3 of methylenecyclohexane are shown in

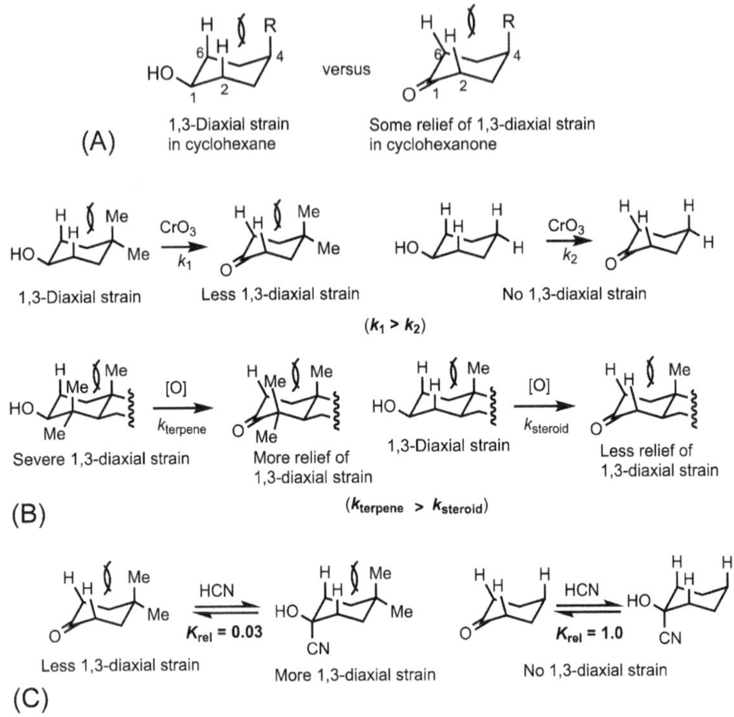


**FIG. 3.42** (A) 4-Alkylketone effect in cyclohexanone system. Role of 4-methylketone effect in (B) relative rates of $CrO_3$ oxidation and (C) relative equilibrium constants for cyanohydrin formation.

Fig. 3.43. The $-\Delta G^0$ value of Me is about one-half of the *A* value in cyclohexane.[90] This is presumably due to the decrease in 1,3-diaxial Me/H interaction (one instead of two) in the axial conformer (a-R). Interestingly, the $-\Delta G^0$ values of polar substituents (OMe, OAc) in a nonpolar solvent tend to be larger than or similar to the A values in cyclohexane.[91] A possible explanation is that the repulsive electrostatic interaction between the trigonal ring carbon C1 (which is more electronegative than sp$^3$ C) and the axial polar substituent tends to offset the benefit from the decreased 1,3-diaxial interactions.

For the conformational preference of a substituent at 2-position (allylic position) of methylenecyclohexane, see allylic ($A^{1,3}$) strain (Section 3.9.4.1).

### 3.9.3 Cyclohexenes

Unlike methylenecyclohexane, cyclohexene has an endocyclic C=C and its preferred conformation is half-chair as determined by electron diffraction.[92] Fig. 3.44A shows the half-chair conformation of cyclohexene in two perspectives **3.26** (as viewed through the plane of the double bond) and **3.27** (as viewed from

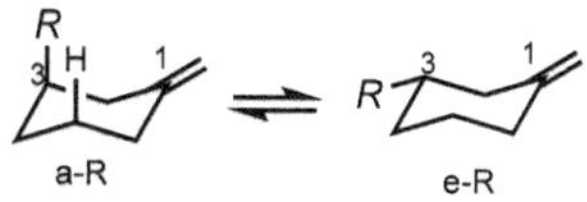


| 3-R | $-\Delta G^0$ (kcal mol$^{-1}$) | *A* value (cyclohexane) |
|---|---|---|
| Me | 0.8 | 1.74 |
| OMe | 0.8 | 0.63 |
| OAc | 0.6 | 0.68 |

**FIG. 3.43** Conformational free energies of substituents at C3 of methylenecyclohexane.

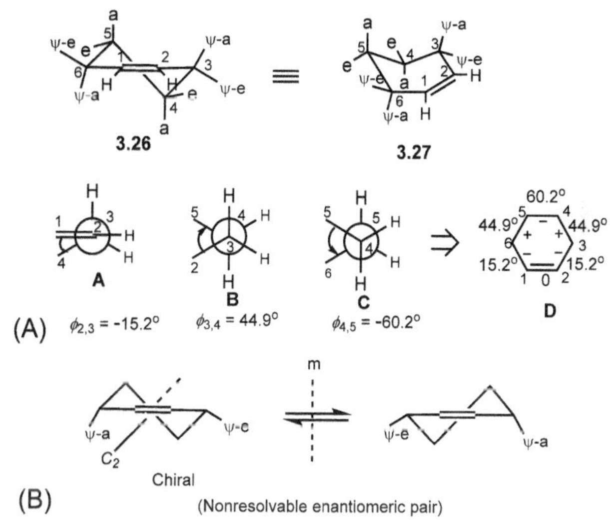


**FIG. 3.44** (A) Half-chair conformation of cyclohexene in two perspectives and its torsion angle picture and (B) half-chair inversion of cyclohexene producing a nonresolvable (±)-pair.

above the plane) in which four ring carbons (C1, C2, C3 and C6) lie in the plane of the double bond in conformity with $sp^2$ geometry and two ring carbons (C5 and C4) are above and below the plane. There are three different torsion angles: $\phi_{2,3}=\phi_{6,1}=-15.2$ degrees, $\phi_{3,4}=\phi_{5,6}=+44.9$ degrees and $\phi_{4,5}=-60.2$ degrees. These are depicted in Newman projections (**A–C**) that lead to the torsion angle picture **D** (Fig. 3.44A).[93] The value of torsion angle $\phi_{4,5}$ (60.2 degrees) indicates normal axial (a) and equatorial (e) bonds at C4 and C5. In contrast, since the value of $\phi_{2,3}$ is not 0 degrees but 15.2 degrees, and $\phi_{3,4}$ is not 60 degrees but 44.9 degrees, the two bonds at C3 are not exactly axial or exactly equatorial and are called pseudoaxial (ψ–a) and pseudoequatorial (ψ–e). For the similar reasons, C6 has ψ–a and ψ–e bonds.

The half-chair conformation of cyclohexene possesses only a $C_2$ axis bisecting the double bond and is chiral (point group $\mathbf{C_2}$). The two enantiomeric half-chairs are readily interconvertible by flipping via a boat TS, and constitute a nonresolvable (±)-pair (Fig. 3.44B). Thus cyclohexene is optically inactive. This conforms to the conclusion obtained from the planar cyclohexene structure which is achiral. The barrier to half-chair inversion is determined as 5.3 kcal mol$^{-1}$ from low-temperature NMR of cyclohexene-$d_6$.[94]

For 4-substituted cyclohexenes, the equatorial conformation is more stable than the axial, and the conformational free energy ($-\Delta G^0$) is about one-half the value in cyclohexane.[95] For example, $-\Delta G^0$ value for 4-Me is 0.86 kcal mol$^{-1}$. This is due to one 1,3-diaxial Me/H interaction in the a-Me conformer (check). In 3-substituted cyclohexenes (substituents at an allylic position), an alkyl group (R) prefers to be pseudoequatorial (ψ–e), whereas an electronegative substituent such as OMe, OAc, Cl or Br tends to be predominantly pseudoaxial (ψ–a) as a result of stabilization by π–σ* interaction (cf. anomeric effect).[96]

For the allylic ($A^{1,2}$) strain in cyclohexene system, see Section 3.9.4.2.

### 3.9.4 Allylic 1,3 ($A^{1,3}$) and allylic 1,2 ($A^{1,2}$) strains

The allylic strain is a kind of steric strain and refers to the nonbonded interaction between an allylic substituent (allylic position is numbered 1) and a substituent at 2-position or 3-position of the allylic moiety. This is illustrated with a general allylic system in Fig. 3.45. X and Y are carbon atoms or heteroatoms (N, O). The allylic systems include alkenes, imines, immonium ions and nitrones.

The $A^{1,3}$ strain occurs particularly in alkylidenecyclohexanes and related imine-enamine systems,[91] whereas the $A^{1,2}$ strain manifests in cyclohexenes and related systems.[97] In both cases, the allylic strain tends to disfavour the allylic substituent in the equatorial or pseudoequatorial position (see below).

#### *3.9.4.1 $A^{1,3}$ strain in alkylidenecyclohexanes and related systems*

Fig. 3.46A illustrates the $A^{1,3}$ strain in an alkylidenecyclohexane. As shown, the equatorial Me is destabilized so much by the $A^{1,3}$ strain that the conformational equilibrium favours the side in which Me is axial. In this case, the $A^{1,3}$ strain dominates over two 1,3-diaxial (Me/H) interactions.

Here we will also examine the role of $A^{1,3}$ strains in reactions and synthesis.[98] Fig. 3.46B shows that *cis*-4-*t*-butyl-2-methylcyclohexanone is converted to the *trans* isomer by treatment with pyrrolidine followed by mild hydrolysis. Because of the $A^{1,3}$ strain, *cis* immonium ion does not form by condensation with pyrrolidine. Instead, epimerization of the *cis* ketone to the *trans* ketone can occur in the presence of base (pyrrolodine) but the equilibrium favours the more stable *cis* isomer. However, a small amount of the *trans* isomer formed can condense with pyrrolidine to give the favoured *trans* immonium ion with no

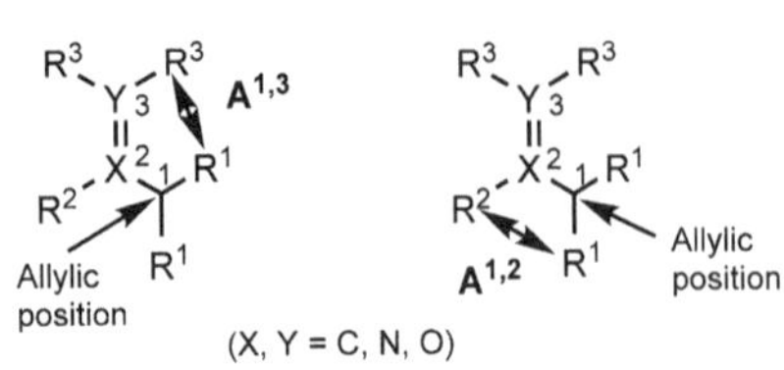


FIG. 3.45 $A^{1,3}$ and $A^{1,2}$ strains in a general allylic system.

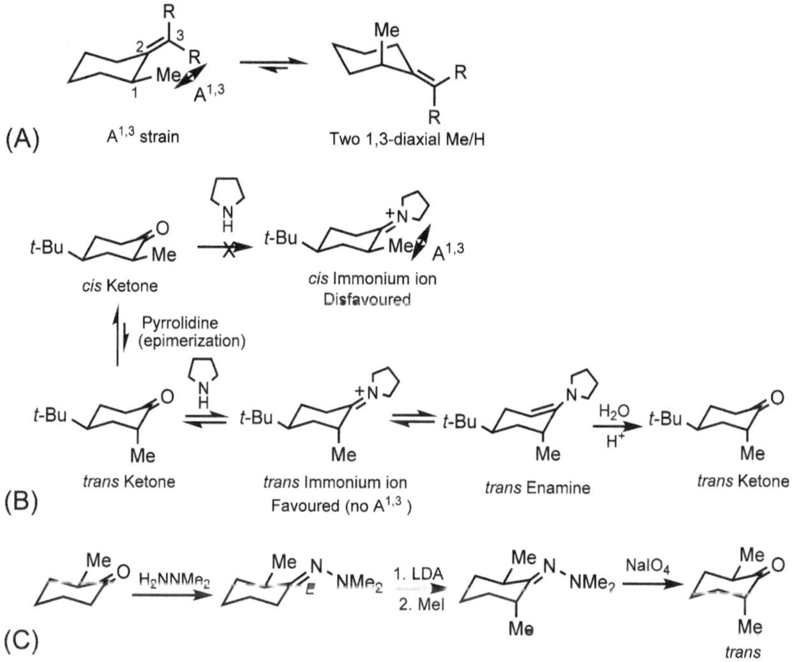


**FIG. 3.46** (A) $A^{1,3}$ strain in alkylidenecyclohexanes. Applications of $A^{1,3}$ strains in (B) conversion of *cis*-4-*t*-butyl-2-methylcyclohexanone to *trans*-4-*t*-butyl-2-methylcyclohexanone and (C) synthesis of *trans*-2,6-dimethylcyclohexanone from 2-methylcyclohexanone.

$A^{1,3}$ strain, which then tautomerizes to the *trans* enamine. Finally, the *trans* ketone product is obtained after hydrolysis.

Fig. 3.46C depicts the synthesis of *trans* 2,6-dimethylcyclohexanone from 2-methylcyclohexanone.[99] Treatment of 2-methylcyclohexanone (predominantly equatorial) with *N*,*N*-dimethylhydrazine produces the sterically favourable (*E*) hydrazone avoiding the $A^{1,3}$ strain. Alkylation of the hydrazone in the presence of LDA and MeI gives *trans* alkylation since the incoming Me goes to the axial position because an $A^{1,3}$ strain will develop for the equatorial attack. The less hindered α-proton will be abstracted under kinetic enolate condition with LDA. Finally, cleavage of the *trans* methylated hydrazone with $NaIO_4$ gives the product.

**Problem 3.9** Predict the preferred conformation of the cyclic amide shown below. Compare with *cis*-1,3-dimethylcyclohexane.

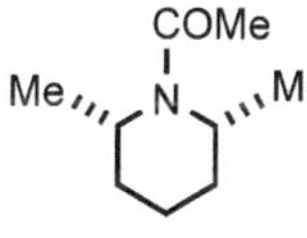

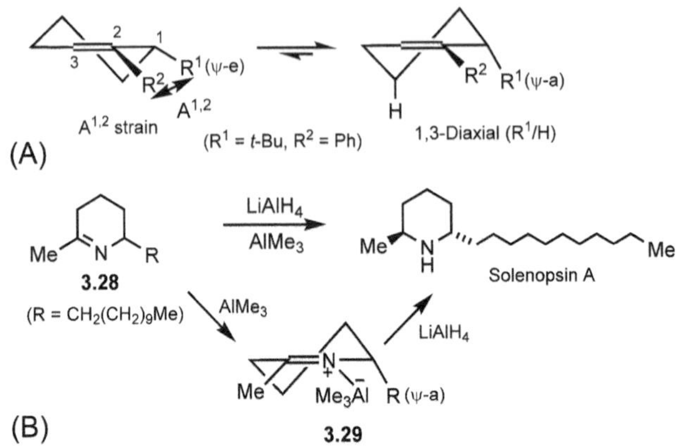


**FIG. 3.47** (A) $A^{1,2}$ strain in a cyclohexene. (B) Application of the $A^{1,2}$ strain in a synthesis of solenopsin A.

#### 3.9.4.2 *$A^{1,2}$ strain in cyclohexenes and related systems*

Fig. 3.47A illustrates the $A^{1,2}$ strain in a cyclohexene.[97] The torsion angle between the allylic substituent ($R^1$) in the pseudoequatorial (ψ–e) position and the 2-substituent ($R^2$) is considerably smaller than 60 degrees and the $A^{1,2}$ strain comes into play when the substituents are bulky. For example, if $R^1$ = *t*-Bu and $R^2$ = Ph, the $A^{1,2}$ strain dominates over 1,3-diaxial (*t*-Bu/H) interaction, favouring the pseudoaxial (ψ–a) conformation. It should be noted that $R^1$ moves away from the 2-substituent ($R^2$) when it goes to the pseudoaxial position, thereby causing relief of the $A^{1,2}$ strain (use a model).

An application of the $A^{1,2}$ strain in a synthesis of an alkaloid solenopsin A is shown in Fig. 3.47B.[100] The reduction of the cyclic imine **3.28** with $LiAlH_4$ in the presence of Lewis acid $AlMe_3$ gives *trans*-2-methyl-6-undecylpiperidine (solenopsin A). Due to complexation of $AlMe_3$ with the imine nitrogen, a large $A^{1,2}$ strain forces the bulky undecyl (R) substituent into the pseudoaxial (ψ–a) position in the complex **3.29**. The reduction of **3.29** with $LiAlH_4$ then leads to the *trans* product. It may be mentioned that in the absence of Lewis acid, $LiAlH_4$ reduction produces the more stable *cis* piperidine (cf. relative stability of *cis*- and *trans*-1,3-dimethylcyclohexane, see Section 3.5.3).

## 3.10 Monocyclic systems other than cyclohexane

### 3.10.1 Cyclopropane

A three-membered cyclopropane ring is necessarily planar. Therefore cyclopropane has a major angle strain (see Section 3.2) and also possesses a considerable torsional strain due to six pairs of eclipsing H atoms (three pairs above the ring plane and three below the ring plane). Thus both angle and torsional strains contribute to the high strain energy of cyclopropane. It appears that $sp^3$ carbon

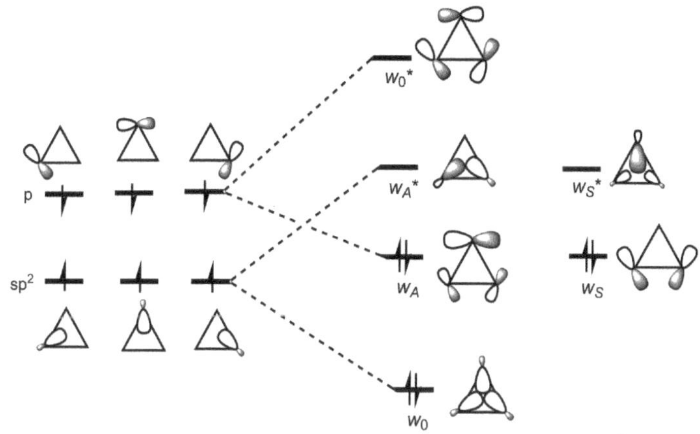


**FIG. 3.48** The *Walsh* orbitals of cyclopropane.

orbitals cannot overlap in end-on fashion required for normal C—C σ bonding, and lead to bent bonds when the interorbital angle in cyclopropane is appreciably larger than the internuclear or geometric angle of 60 degrees.[101]

### 3.10.1.1 Bonding in cyclopropane: Walsh model

The C—H stretching frequency in cyclopropane is 3080 $cm^{-1}$, which indicates a C($sp^2$)—H bond, not C($sp^3$)—H. The *Walsh* model[102] of cyclopropane is based on three $CH_2$ units with $sp^2$ carbons. Each carbon atom then provides one $sp^2$ hybrid orbital and a p orbital for C—C bonding. The six *Walsh* molecular orbitals of cyclopropane are shown in Fig. 3.48. The symmetry-adapted linear combination of three $sp^2$ hybrids gives the bonding *Walsh* orbital $w_0$ and two degenerate antibonding *Walsh* orbitals $w_A$* and $w_S$*. The symmetry-adapted linear combination of three p orbitals on the other hand yields a degenerate pair of bonding *Walsh* orbitals $w_A$ and $w_S$ and an antibonding *Walsh* orbital $w_0$*. The bonding *Walsh* orbitals $w_A$ and $w_S$ are the degenerate HOMOs. This model can thus explain the double bond-like behaviour of cyclopropane ring.

### 3.10.1.2 Cyclopropylcarbinyl cation

The C—C bonds in cyclopropyl rings can act as powerful donors and thus greatly stabilize a carbocation with a cyclopropyl ring in α position. To illustrate, consider the relative rates of $S_N1$solvolysis of the tertiary substrates in Fig. 3.49.[103] Interestingly, replacement of a phenyl group by a cyclopropyl group leads to a large increase in the reaction rate (about 500-fold). This means that a cyclopropyl ring is able to stabilize a carbocation intermediate much better than a conjugating phenyl ring.

The pronounced stabilization of cyclopropylcarbinyl cations has been explained in terms of the frontier orbital interaction of a filled *Walsh*-type

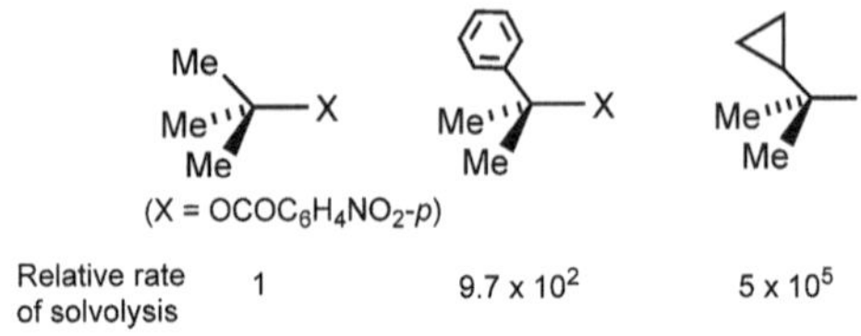


**FIG. 3.49** Relative rates of solvolysis of tertiary substrates in 80% aqueous acetone at 25°C.

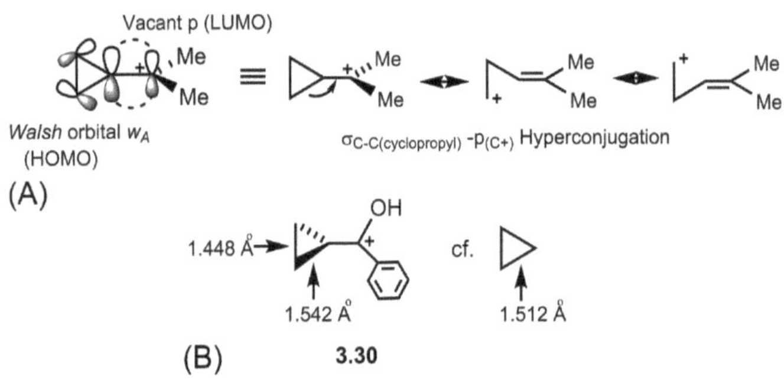


**FIG. 3.50** (A) Frontier orbital interaction or $\sigma_{C—C(cyclopropyl)}$–$p_{(C+)}$ hyperconjugation for stabilization of the cyclopropylcarbinyl cation in the bisected conformation; (B) experimental evidence in favour of the frontier orbital or hyperconjugative interaction.

cyclopropyl σ orbital (HOMO) with the vacant p orbital (LUMO) at the carbocation centre (Fig. 3.50A). This type of interaction is also described as σ–p hyperconjugation ($\sigma_{C—C(cyclopropyl)} \rightarrow p_{C(+)}$). The cyclopropyl ring possesses two degenerate high-energy HOMOs ($w_A$ and $w_S$). However, interaction of $w_A$ as the HOMO in the bisected conformation (when the plane of the carbocation bisects the ring) is most effective because the carbocation p orbital can overlap very efficiently with the adjacent p orbital in $w_A$ having a large coefficient as shown in Fig. 3.50A.[104] (Note that this frontier orbital interaction refers to the ground state stability of a molecule or intermediate, not its reactivity).

The cyclopropylcarbinyl cation can be generated in super acid and characterized by $^1$H NMR. A rather high torsional barrier of about 13.7 kcal mol$^{-1}$ for rotation about the C—C bond between the cyclopropyl ring and the cationic centre is consistent with the σ–p hyperconjugation. The best evidence for the σ–p interaction in the bisected conformation is demonstrated by X-ray diffraction of crystalline cyclopropylcarbinyl cations such as **3.30** (Fig. 3.50B).[105] Here the cyclopropyl ring adopts the bisected conformation with an angle of nearly 90 degrees with respect to the carbocation plane. The bond lengths in the cyclopropyl moiety show that the vicinal bonds are larger (1.542 Å) and the distal bond is shorter (1.448 Å) than in free cyclopropane (1.512 Å). These results support the electron donation from the cyclopropyl σ orbital to the vacant p orbital of the carbocation as shown in the frontier orbital interaction or the σ–p hyperconjugation.

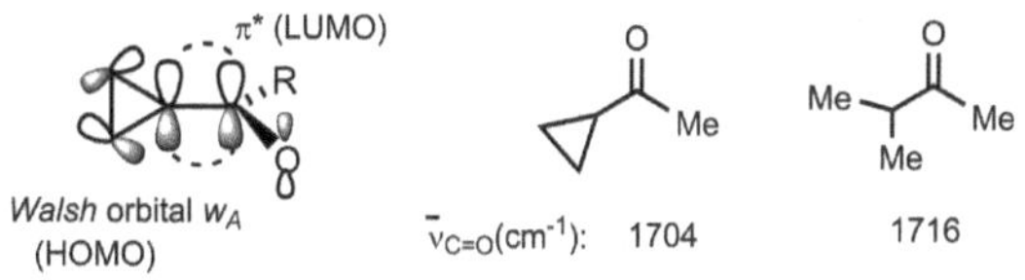


**FIG. 3.51** Frontier orbital interaction or $\sigma_{C-C(cyclopropyl)}$–$\pi^*_{C=O}$ hyperconjugation and IR stretching frequencies for cyclopropyl ketones.

The hyperconjugation of cyclopropyl groups with electron withdrawing π substituents such as C=O is also possible and again the σ–π* interaction is very effective in the bisected conformation[106] (Fig. 3.51). The electron donation from the Walsh-type cyclopropyl σ orbital to the antibonding $\pi^*_{C=O}$ causes a decrease in the bond order of carbonyl group which will lead to a shift of the C=O stretching frequency to lower wavenumbers. Thus the C=O stretching frequency for cyclopropyl methyl ketone is 1704 cm$^{-1}$ compared to 1716 cm$^{-1}$ for isopropyl methyl ketone.

**Problem 3.10** Predict the product of the following reaction:

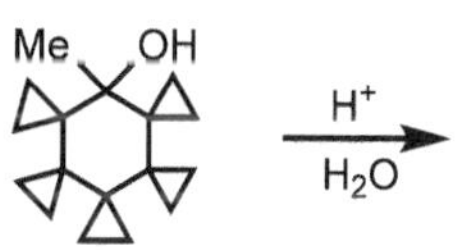


## 3.10.2 Cyclobutane

Planar cyclobutane has high torsional strain from eight pairs of eclipsing H atoms (four pairs above the ring plane and four below the ring plane). The puckering of the ring can reduce the torsional strain though at the expense of some angle strain. As a compromise, the cyclobutane molecule assumes a wing-shaped conformation with a puckering angle of 28 degrees (Fig. 3.52).[107] The angle between the two planes defined by 1–2–3 and 3–4–1 is then 180° − 28° = 152°. The wing-shaped cyclobutane has axial (a) and equatorial

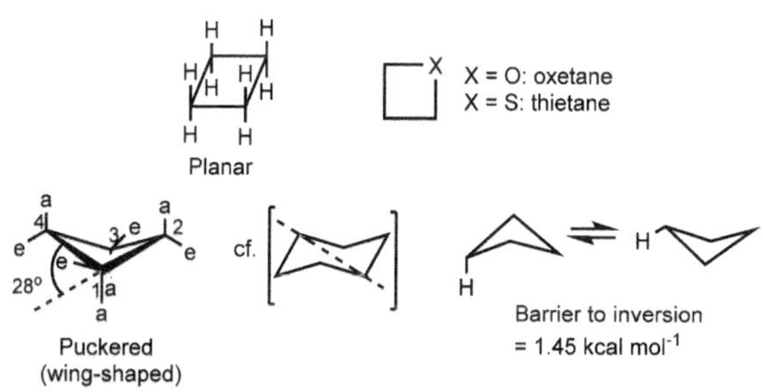


**FIG. 3.52** Conformations of cyclobutane.

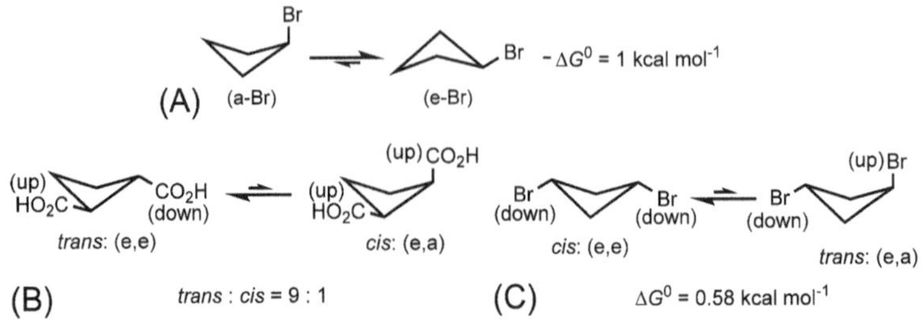


**FIG. 3.53** (A) Conformational equilibrium for bromocyclobutane, (B) equilibration of *cis*- and *trans*-1,2-cyclobutanedicarboxylic acids and (C) equilibration of *cis*- and *trans*-1,3-dibromocyclobutanes.

(e) bonds similar to those in cyclohexane. (Note that the wing-shaped conformation may be easily drawn based on chair cyclohexane.) The ring inversion takes place via a planar TS and the barrier to ring inversion is only 1.45 kcal mol$^{-1}$.[107] The inversion barrier is further diminished in the case of heteracyclobutanes due to a decrease in torsional strain in the planar structure. Thietane has an inversion barrier of 0.75 kcal mol$^{-1}$ while oxetane exists practically in the planar form with an inversion barrier of as little as 0.1 kcal mol$^{-1}$.

#### *3.10.2.1 Substituted cyclobutanes*

Like cyclohexane, the substituents in cyclobutane are preferably equatorial for steric reasons. For example, bromocyclobutane exists preferentially in the equatorial conformation and is more stable than the axial conformation by about 1 kcal mol$^{-1}$ (Fig. 3.53A).[108] For 1,2-disubstituted cyclobutanes, the *trans* diastereomer is preferentially diequatorial (e,e), whereas the *cis* isomer is equatorial–axial (e,a). Thus the *trans*-1,2 isomer will be more stable than the *cis*-1,2. Experimentally, equilibration of *cis*- and *trans*-1,2-cyclobutanedicarboxylic acids gives *trans*:*cis*=9:1 (Fig. 3.53B). In the case of 1,3-disubstituted cyclobutanes, for example, in 1,3-dibromocyclobutane, the *cis* (e,e) is more stable than the *trans* (e,a) by 0.58 kcal mol$^{-1}$ (Fig. 3.53C). Dipole moment data suggest that the *trans*-1,3 ring is more flattened.

### 3.10.3 Cyclopentane

Planar cyclopentane with bond angles close to tetrahedral angle is almost free of angle strain but is destabilized by a large torsional strain due to H/H eclipsing (10 pairs, 10 kcal mol$^{-1}$). The total strain is however minimized by puckering of the ring. The conformational analysis of cyclopentane is however not so simple. Two lowest energy conformations of cyclopentane are envelope and half-chair (Fig. 3.54). In the envelope conformation, four carbons are in a plane and the remaining one is above or below the plane. It has $\mathbf{C_s}$ symmetry with one symmetry plane and is achiral. On the other hand, the half-chair conformation has

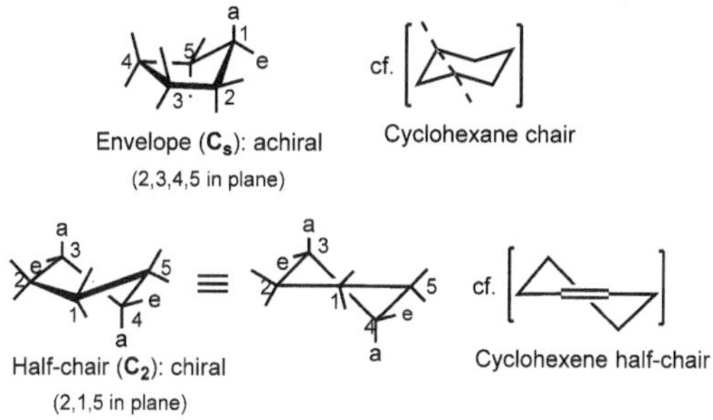


**FIG. 3.54** Envelope and half-chair conformations of cyclopentane.

three carbons in a plane, one carbon above the plane and the other below the plane. It has $\mathbf{C_2}$ symmetry with only a $C_2$ axis and is chiral. (Note that envelope and half-chair conformations may be drawn based on chair cyclohexane and half-chair cyclohexene, respectively.)

The energy difference between envelope and half-chair conformers is quite small, the envelope being favoured by only about 0.5 kcal mol$^{-1}$.[109] Their interconversion involves a very low energy barrier and does not proceed through high energy planar structure. Besides envelope and half-chairs, there exists a continuum of intermediate chiral conformations ($\mathbf{C_1}$). The cyclopentane molecule thus exists in a rapid conformational flux. The five carbon atoms of cyclopentane oscillate perpendicular to the average plane of the ring like a wave passing round the ring. This motion apparently causes the molecule to appear as if a single out-of-plane atom is rotated (not in reality) round the ring and is called *pseudorotation*.[110] This process is so rapid that it is considered a vibration with almost no change in energy. In the full pseudorotation circuit, envelope → half-chair → envelope → half-chair and so on, there are 10 envelope conformations with each of 5 carbon atoms being either up or down. There are also 10 half-chairs with each pair of adjacent carbons (1,2; 2,3; 3,4; 4,5; 5,1) being up/down or down/up. The up/down and down/up conformations for any two adjacent carbons constitute a pair of enantiomeric half-chairs; thus 10 half-chairs comprise 5 enantiomeric pairs. Since a total of 20 envelope/half-chair conformations are involved in the course of movement of the puckering by 360 degrees in one pseudorotation circuit, a half-chair from the envelope or vice versa will occur at 18-degree intervals.[111] Thus cyclopentane exists as a single family with pseudorotating envelope and half-chair conformers. This is in contrast to cyclohexane which has two families, chair and twist-boat, when each family consists of just one conformer.

### *3.10.3.1 Substituted cyclopentanes and cyclopentanone*

Although cyclopentane itself doesn't have a single highly favoured conformation, a substituent on the cyclopentane ring can upset the balance of strains

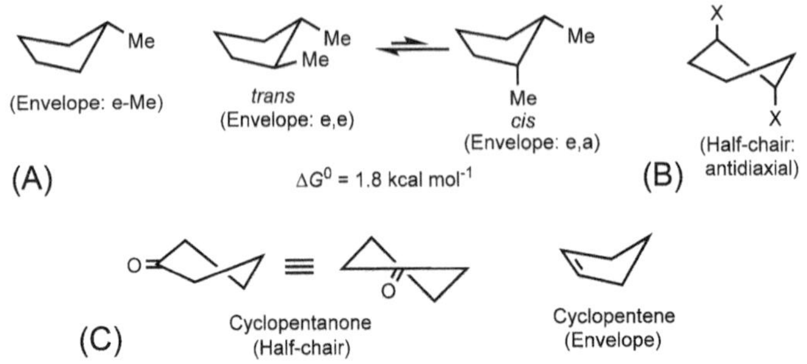


**FIG. 3.55** Preferred conformations of (A) methylcyclopentane, and *cis*- and *trans*-1,2-dimethylcyclopentane; (B) *trans*-1,2-dihalocyclopentane and (C) cyclopentanone and cyclopentene.

thereby favouring either the envelope or the half-chair. In monosubstituted cyclopentanes such as methylcyclopentane, the envelope conformation is preferred by 0.9 kcal mol$^{-1}$ when the substituent occupies the sterically favourable equatorial position at the tip or flap of the envelope (Fig. 3.55A).[111] In the case of 1,2-disubstituted cyclopentanes, the *trans* diastereomer (envelope e,e) is more stable than the *cis* isomer (envelope e,a); the energy difference is 1.8 kcal mol$^{-1}$ when both substituents are Me (Fig. 3.55A). Here the *cis* isomer (envelope e,a) is mainly destabilized by Me/Me steric interaction which is stronger than usual, since the two Me's are at a torsion angle significantly lower than 60 degrees.[112] For *trans*-1,2-dihalocyclopentanes, the preferred conformation is presumably the half-chair when the two halogens are anti and diaxial (Fig. 3.55B). This preference is consistent with the low dipole moment of *trans*-1,2-dihalocyclopentanes. For 1,3-disubstitutedcyclopentanes, no clear-cut stability order of *cis* and *trans* isomers has been observed.

For cyclopentanone with one sp$^2$ ring carbon, the half-chair conformation is preferred over the envelope by 2.4 kcal mol$^{-1}$.[111] Replacement of $CH_2$ by C=O reduces the torsional strain most in the half-chair with the carbonyl carbon at the middle of three-carbon plane on the $C_2$ axis (Fig. 3.55C). Here the barrier to pseudorotation (1.15 kcal mol$^{-1}$) is significantly higher than that for cyclopentane but this is still much lower compared to the barrier for ring inversion in cyclohexane. For cyclopentene with two sp$^2$ carbons, the double bond is probably at the bottom of the envelope to minimize the torsional strain (Fig. 3.55C). But the increase in angle strain offsets the decrease in torsional strain and the strain in cyclopentene is almost the same as that in cyclopentane.[113]

### *3.10.3.2 I-Strain*

The total strain energy in ring systems is often referred to as I-strain (internal strain), which is contributed by angle strain, torsional strain and van der Waals or steric strain in the case of small and common rings. The change in hybridization of a ring carbon from sp$^3$ to sp$^2$ or vice versa leads to a change

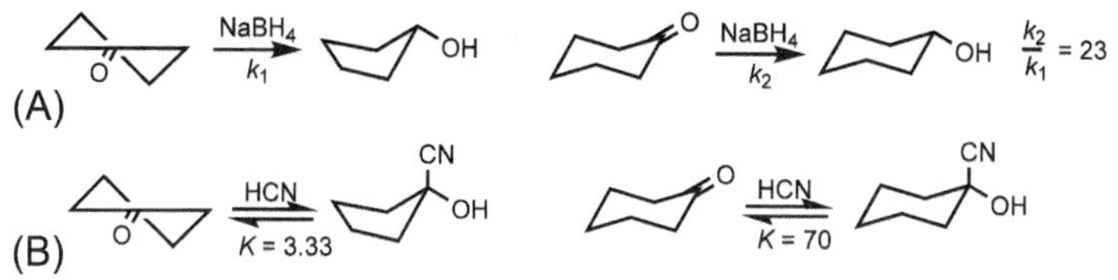


**FIG. 3.56** Nucleophilic additions to cyclopentanone and cyclohexanone: Effects of I-strain on (A) relative rates and (B) equilibrium constants.

in I-strain.[114] I-Strain affects both rates and equilibria of reactions. Nuclcophilic addition to a cyclopentanone causes $sp^2 \rightarrow sp^3$ change and is accompanied by increase in I-strain primarily due to an increase in torsional strain. There is however a minor relief in angle strain. But nucleophilic addition to a cyclohexanone leads to a decrease in I-strain mainly due to relief of torsional strain. Therefore addition reactions of cyclopentanone are less favourable that those of cyclohexanone both kinetically and thermodynamically because the hybridization change is associated with both TS and product. For example, cyclohexanone is reduced by $NaBH_4$ 23 times faster than cyclopentanone does (Fig. 3.56A).[115] The equilibrium constant for addition of HCN to cyclohexanone (70) is also greater compared to a value of 3.33 in the case of cyclopentanone (Fig. 3.56B).[116] On the other hand, reactions involving $sp^3 \rightarrow sp^2$ change will be more favourable for cyclopentanone in comparison to cyclohexanone.[117] For example, $S_N1$ solvolysis of cyclopentyl tosylate is 14 times faster than that of cyclohexyl tosylate.

**Problem 3.11** Explain the following observations:

**(a)** The $S_N1$ hydrolysis of 1-chloro-1-methylcyclopropane is slower than that of 1-chloro-1-methylcyclobutane.
**(b)** The hydrate formation for acetone in water is almost negligible, whereas cyclopropanone is hydrated to a significant extent.

### 3.10.4 Cycloheptane

As the ring size increases, the number of possible conformers increases due to more flexibility of the molecule.[118] For cycloheptane, there are two families of conformations and each family comprises two members: Family I—chair/twist-chair; Family II—boat/twist-boat (Fig. 3.57).[112,119] (Contrast with cyclohexane which has also two families but each family is single-membered—chair and twist-boat.) Two members of a cycloheptane family interconvert readily by pseudorotation but the two families are separated by a substantial barrier of about 8.5 kcal $mol^{-1}$ with the chair/twist-chair (Family I) having lower energy. Within each family, the twist conformer is more stable than the other which suffers from the severe eclipsing at the flat end (Fig. 3.57). Thus the most stable conformation of cycloheptane is twist-chair, which is preferred over the chair by

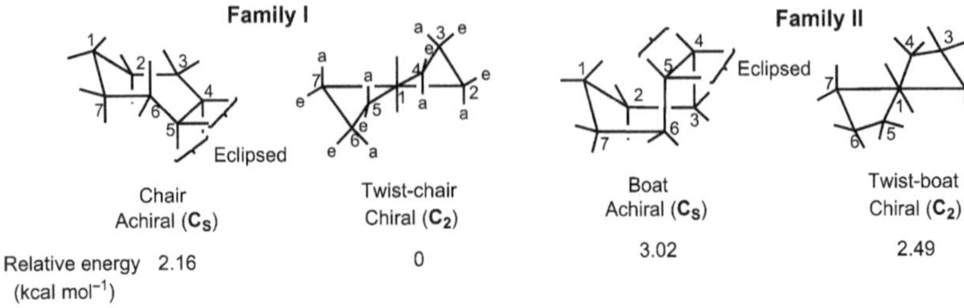


**FIG. 3.57** Conformations of cycloheptane (two families): Family I: chair/twist-chair and Family II: boat/twist-boat.

2.16 kcal mol$^{-1}$.[112] In boat/twist-boat (Family II), the twist-boat is more stable than the boat by 0.53 kcal mol$^{-1}$.[112] Between the twist-chair and the twist-boat, the energy difference is 2.49 kcal mol$^{-1}$. The twist conformer in each family has $\mathbf{C_2}$ symmetry and is chiral. In Family **I**, two enantiomeric twist-chairs interconvert by pseudorotation via the achiral chair ($\mathbf{C_s}$) while the enantiomeric twist-boats in Family **II** are interconvertible through the boat ($\mathbf{C_s}$).

### 3.10.4.1 *Cycloheptene and cycloheptatriene*

The preferred conformation of cycloheptene is chair with the double bond at the flat end (Fig. 3.58A),[120] which causes relief of torsional strain (cf. cycloheptane chair in Fig. 3.57). Interestingly, cycloheptatriene exists in boat conformation,[121] which can undergo a conformational inversion[122] with a barrier of 6.1 kcal mol$^{-1}$ when an axial substituent at the tetrahedral carbon becomes equatorial and vice versa (Fig. 3.58B).

**Problem 3.12** Predict the preferred conformation of (a) 7-methylcycloheptatriene and (b) 1-methyl-7-*t*-butylcycloheptatriene.

## 3.10.5 Medium rings

Conventionally, medium-sized rings are 8- to 11-membered (see Section 3.2, Table 3.1). The medium rings[118,123] are very flexible and have many conformers that are difficult to study in detail. Calculations indicate that such rings cannot attain favourable bond angles, torsion angles and nonbonded distances simultaneously. One unusual conformational feature in the medium rings is the presence of a kind of steric strain called *transannular strain.*[124]

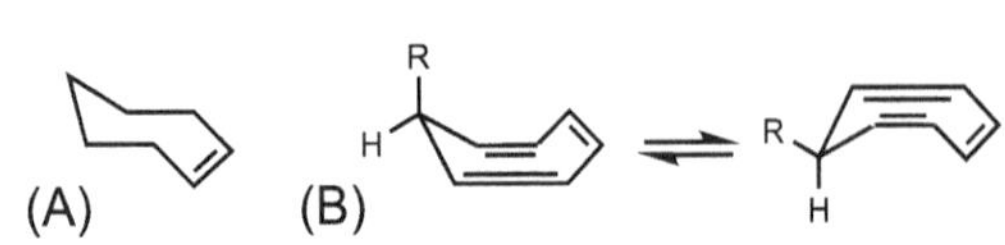


**FIG. 3.58** (A) Preferred chair conformation of cycloheptene and (B) conformational inversion of cycloheptatriene.

The conformations of these rings stipulate that some hydrogens are directed inside the rings which interact sterically across the ring leading to the transannular strain. This strain is a major contributor to I-strain in medium rings. The transannular interactions increase with substitution which leads to an increase in the barriers to pseudorotation as well as ring inversion. For example, the barrier to ring inversion for unsubstituted cyclononane (9-membered ring) is 6 kcal mol$^{-1}$ which increases to 20 kcal mol$^{-1}$ for 1,1,4,4-tetramethylcyclononane. These features of the medium rings are in contrast to ring inversions in cyclohexane system when the energy barriers increase only moderately with the increasing substitution. In reactions, the changes in hybridization and the resulting changes in transannular interactions play dominant roles. Reactions that reduce the number of ligands at a carbon (e.g. $S_N1$ reactions) are highly favoured. On the other hand, reactions that increase the number of ligands such as nucleophilic additions to ketones are disfavoured.

We will describe here briefly the conformations for cyclooctane and cyclodecane systems as representative members of the medium rings.

### 3.10.5.1 Cyclooctane, cyclooctanone, cyclooctene and cyclooctatetraene

Cyclooctane has at least 10 symmetrical conformations distributed in 4 families.[125] The lowest energy conformation of cyclooctane is boat-chair (BC) (also called chair–boat), determined by X-ray diffraction of its derivatives. The BC conformation exhibits a transannular interaction between two pseudoaxial H atoms at C3 and C7 (Fig. 3.59A). It has $\mathbf{C_s}$ symmetry and is achiral.

A substituent on cyclooctane ring generally prefers a sterically favourable equatorial (e) or pseudoequatorial (e′) position depending on the position of the ring carbon (Fig. 3.59B).[126] In cyclooctane, two bonds at C1 have a chair cyclohexane environment. So a substituent (Me) at C1 is preferably equatorial which is more stable than the axial Me conformer by 1.8 kcal mol$^{-1}$. But for a

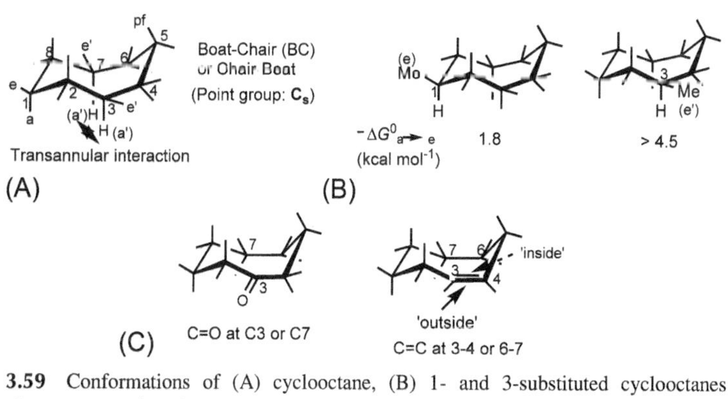


**FIG. 3.59** Conformations of (A) cyclooctane, (B) 1- and 3-substituted cyclooctanes and (C) cyclooctanone and cyclooctene.

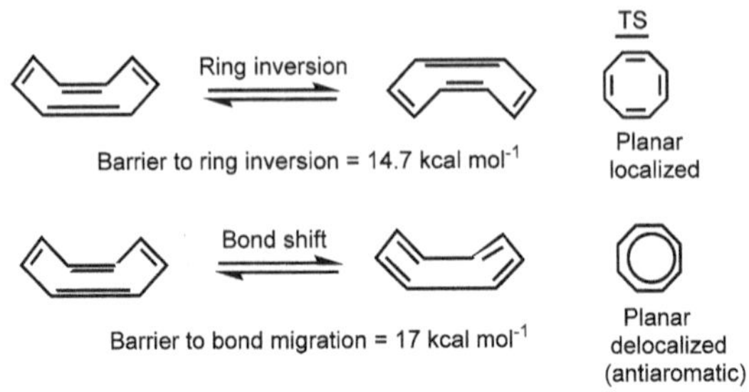


**FIG. 3.60** Ring inversion and double bond shift in cyclooctatetraene.

substituent (Me) at C3, the energy difference between the pseudoequatorial and pseudoaxial conformations is greater than 4.5 kcal mol$^{-1}$ as the pseudoaxial Me will be severely destabilized by a transannular interaction with the pseudoaxial H at C7. It may be mentioned that at C5, a substituent (Me) held by a flagpole-type bond of boat cyclohexane will be maximally destabilized to the extent of 6.1 kcal mol$^{-1}$. However, a substituent at C4 or C6 has almost equal preference for the two bonds (draw). As such, germinal disubstitution in cyclooctane is highly preferred at C4 or C6.

For cyclooctanone, C=O is preferentially at C3 or C7 due to relief of transannular strain and for similar reasons, the double bond in cyclooctene is between C3 and C4 or C6 and C7 (Fig. 3.59C).[126] It is of note that two faces of the double bond in cyclooctene indicate inside/outside faces rather than the usual top/bottom faces.

A planar delocalized cyclooctatetraene is predicted to be antiaromatic. But the actual cyclooctatetraene molecule is nonaromatic with four localized double bonds. The most stable conformation of cyclooctatetraene as determined by X-ray diffraction is tub-shaped (Fig. 3.60).[127] In this conformation, cyclooctatetraene can undergo ring inversion and bond migration. The ring inversion presumably occurs via a planar TS with localized double bonds and the barrier to inversion is estimated as 14.7 kcal mol$^{-1}$.[128] On the other hand, double bond shift probably takes place through a planar antiaromatic TS with an energy barrier of about 17 kcal mol$^{-1}$. This result implies that the planar delocalized structure (antiaromatic) is destabilized by 2.3 kcal mol$^{-1}$ relative to the planar localized structure.[129] This has been corroborated by theoretical calculations.

### *3.10.5.2 Cyclodecane and its derivatives*

The lowest energy conformation of cyclodecane and most of its derivatives is boat–chair–boat (BCB) as determined by low-temperature NMR[130] and X-ray diffraction.[131] The BCB conformation of cyclodecane is shown in Fig. 3.61. This conformation exhibits the transannular interactions most clearly, which is a characteristic feature of medium-sized rings. The BCB conformation is achiral with $\mathbf{C_{2h}}$ symmetry ($C_2$ axis bisects 4–5 and 9–10 bonds and

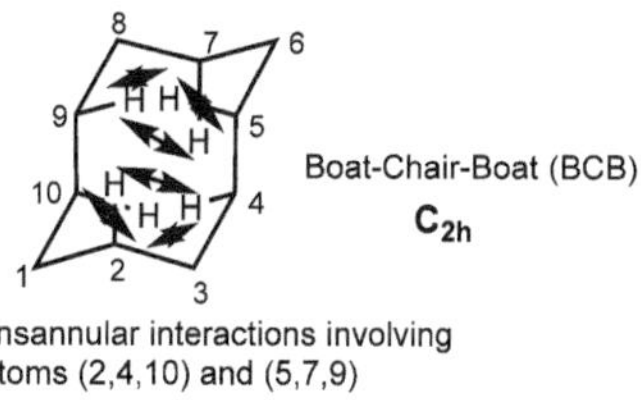


**FIG. 3.61** Conformation of cyclodecane.

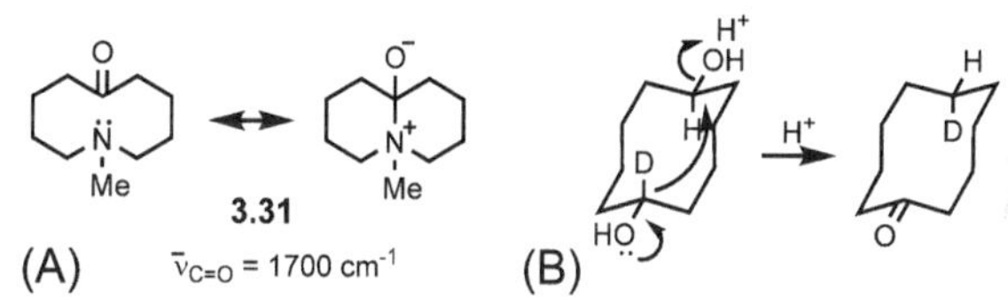


**FIG. 3.62** (A) Transannular resonance interaction in a cyclodecane derivative and (B) transannular pinacol rearrangement in a cyclodecane system (product from hydride shift not shown).

a perpendicular $\sigma_h$ passes through C2 and C7) and fits into a diamond lattice. As shown in Fig. 3.61, six H atoms are directed inside the ring when two separate sets of three H atoms (2,4,10) and (5,7,9) are involved in transannular interactions. Substitutents on a cyclodecane ring can lie only on the periphery as it is not possible to accommodate substituents larger than hydrogen in the inside positions due to steric reasons.

Transannular interactions may be responsible for physical and chemical behaviours of cyclodecane derivatives.[132] For example, a transannular resonance interaction in the cyclodecane derivative **3.31** might account for its high dipole moment and the shift of carbonyl stretching frequency to lower wavenumber (Fig. 3.62A). A transannular pinacol rearrangement involving a transannular shift of deuterium (or hydrogen) is illustrated in Fig. 3.62B. A transannular alkyl shift does not take place.

### 3.10.6 Large rings: Rotaxane and catenane

Large rings are 12-membered onwards (see Section 3.2, Table 3.1). These rings have too many conformers to study in detail, and only limited information is available. Importantly, large rings can allow new forms of topology to arise. A linear chain can be threaded through a large ring and then capped with bulky end groups to form a rotaxane (from the Latin *rota*, wheel and *axis*, axle).[133] A schematic diagram of rotaxane is shown in Fig. 3.63A. An example of rotaxane in which both the chain and the ring are based on polyethylene glycols is depicted in Fig. 3.63B.[134] A mixture of chains was employed when *n* is about eight on average. This glycol-based system is generally more efficient than threading

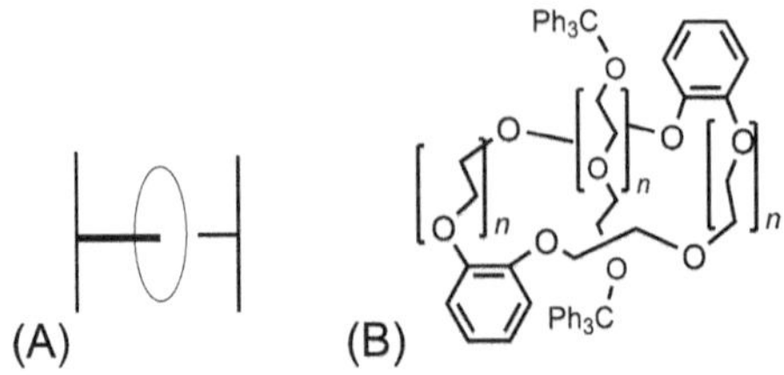


FIG. 3.63 (A) A schematic diagram of rotaxane and (B) an example of rotaxane based on polyethyelene glycols.

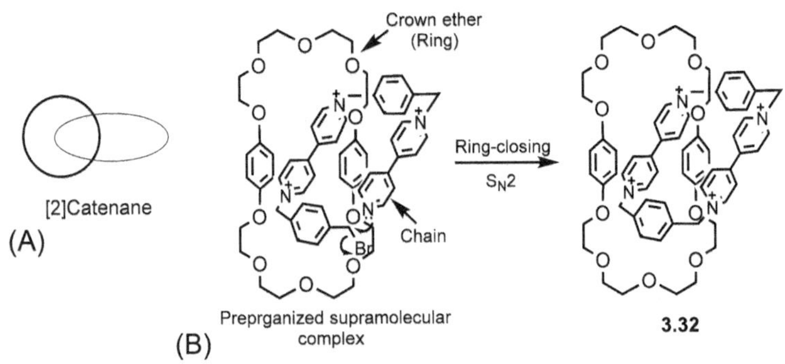


FIG. 3.64 (A) A schematic diagram of [2]catenane and (B) an example of catenane synthesis from a preorganized supramolecular complex.

of a polymethylene chain through a polymethylene ring because the polymethylene conformations usually have closely packed parallel segments.

In another type of topology, a chain can be threaded through a large ring and then closed to a second ring to form a *catena*ne (from the Latin *catena*, chain).[135] Fig. 3.64A shows a schematic diagram of [2]catenane. An example of [2]catenane synthesis is illustrated in Fig. 3.64B. The catenane **3.32** results from the ring-closing of a preorganized supramolecular complex formed by strong coulombic interactions between the chain of bis-bipyridinium salts and the crown ether bis(*para*-phenylene)-34-crown-10.[136] Ring-closing of the chain is achieved by nucleophilic substitution of benzyl bromide moiety by pyridine nitrogen. Catenanes also include biological macromolecules such as chained DNA molecules.

There are further types of molecular architectures such as trefoil knots (when a chain threading through a ring closes on itself) and Möbius strips (when the ends of a long chain are connected after a twist).[137]

## 3.11 Fused ring systems

The most common rings present in fused polycyclic systems[138] are 6-membered rings. We will describe here briefly the conformations of the fused systems decalins (6–6 bicyclics) and perhydrophenanthrenes/perhydroanthracenes (6–6–6 tricyclics). The naturally occurring steroids such as cholesterol are among the most important fused systems which provided the stimulation for the development of conformational analysis in the area of polycyclic ring systems. The steroid nucleus is a perhydrophenanthrene ring fused to a cyclopentane ring.

### 3.11.1 Decalins

Decalin (decahydronaphthalene or bicyclo[4.4.0]decane) with *cis* or *trans* ring junction can exist as a two-chair conformation (Fig. 3.65A), as determined by electron diffraction.[139] Note that *cis*-fusion and *trans*-fusion are analogous to *cis*- and *trans*-1,2-dimethylcyclohexane respectively. The chair–chair conformation of the *cis*-decalin possesses $\mathbf{C_2}$ symmetry and is chiral. The $C_2$ axis passes through the midpoint of 9–10 bond and bisects the torsion angle between 9-H and 10-H. On the other hand, the chair–chair *trans*-decalin is achiral. It has $\mathbf{C_{2h}}$ symmetry with a $C_2$ axis bisecting 2–3, 9–10 and 6–7 bonds and a perpendicular $\sigma_h$ plane passing through ring junction H atoms.

The chiral *cis*-decalin exists as a nonresolvable pair of enantiomers (**X** and **X′**) because they are rapidly interconvertible by ring inversion at room temperature (Fig. 3.65B). Hence, *cis*-decalin is optically inactive, as also judged from the planar structure which is achiral. Note that ring inversion requires inversion

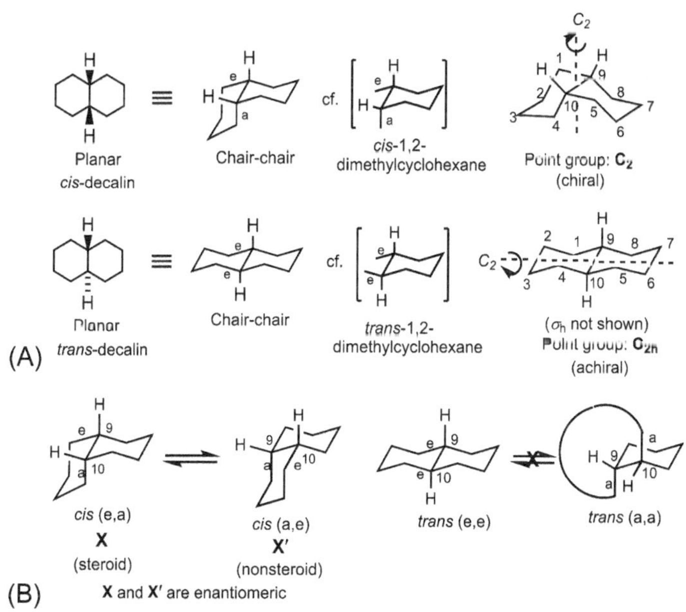


**FIG. 3.65** (A) Conformation and symmetry of *cis*- and *trans*-decalins and (B) all-chair ring inversion possible for *cis*-decalin but not for *trans*-decalin.

of both chairs when 9e,10a bonds in **X** invert to 9a,10e bonds in **X′** (use a molecular model). In analogy with the A/B rings in 5β steroids (see Section 3.11.4), **X** is called a steroid conformation while **X′** is nonsteroid. The barrier to ring inversion for *cis*-decalin[140] is 12.3 kcal $mol^{-1}$, which is somewhat higher than that for cyclohexane.

In contrast to *cis*-decalin, *trans*-decalin has a rigid chair–chair framework, and ring inversion to another chair–chair conformation is not possible since a, a fusion required by the inversion is geometrically unattainable (Fig. 3.65B). However, chair–chair to chair–boat conversion is possible, if needed.

### 3.11.1.1 *Relative stability of cis- and trans-decalins*

In *trans*-decalin, since either chair is fused to the other through equatorial bonds, no *interring or extraannular* gauche interactions (interaction of $CH_2$ group of A ring with $CH_2$ group of B ring) are involved. But in *cis*-decalin, there is one axial $CH_2$ residue with respect to each ring (C8 with respect to A ring, C4 with respect to B ring) as shown in Fig. 3.66. The steric interactions are depicted in terms of 1,3-diaxial interactions as well as their equivalent interring gauche–butane (gb) interactions. Note that the 1,3-diaxial interaction $C^4H_2$/8-H or $C^8H_2$/4-H represents the same C4 ↔ C8 gb interaction. In the net, *cis*-decalin has three gb interactions ($3 \times 0.9 = 2.7$ kcal $mol^{-1}$).

The difference in enthalpy ($\Delta\Delta H^0$) between the *cis* and *trans* isomers is

$$\Delta\Delta H^0 = \Delta H^0{}_{cis} - \Delta H^0{}_{trans} = 2.7 - 0 = 2.7\,\text{kcal mol}^{-1}$$

Thus enthalpy disfavours the *cis* isomer by 2.7 kcal $mol^{-1}$. This is in excellent agreement with the experimental value of 2.7 kcal $mol^{-1}$ in the liquid state determined from heats of isomerization or heats of combustion data.[141,142]

The entropy term ($\Delta S^0$) for *cis*- and *trans*-decalins can be estimated as follows. The chiral *cis* isomer has entropy of symmetry ($-R \ln \sigma$) as well as entropy of mixing due to a racemate ($-R \ln 2$), whereas the achiral *trans* isomer

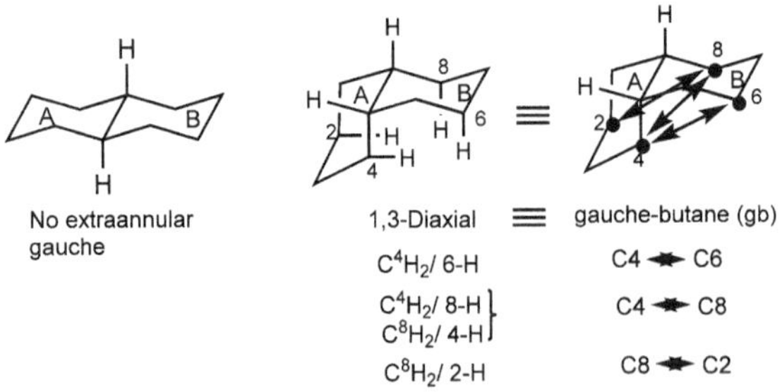


**FIG. 3.66** Interring or extraannular gauche interactions in *cis*- and *trans*-decalins. gb = gauche–butane.

has only entropy of symmetry ($-R\ \ln\ \sigma$). The symmetry number $\sigma = 2$ for both $\mathbf{C_2}$ (*cis*) and $\mathbf{C_{2h}}$ (*trans*). Therefore the entropy difference between the *cis* and *trans* isomers is

$$\begin{aligned}\Delta\Delta S^0 &= \Delta S^0{}_{cis} - \Delta S^0{}_{trans} = (-R\ \ln 2 + R\ \ln 2) - (-R\ \ln 2) = R\ \ln 2 \\ &= 1.38\,\text{cal}\,\text{K}^{-1}\,\text{mol}^{-1}.\end{aligned}$$

However, the experimental value is about $0.6\,\text{cal}\,\text{K}^{-1}\,\text{mol}^{-1}$, which is significantly lower than the estimated value.[141,143] This suggests some more ordering in the *cis* isomer, and *cis*-decalin is probably not as flexible as it seems to be with respect to the *trans* isomer. Entropy actually favours the *cis* isomer by a small margin of $0.6\,\text{cal}\,\text{K}^{-1}\,\text{mol}^{-1}$.

The free energy difference between the *cis* and *trans* isomers at 25°C is given by

$$\Delta\Delta G^0 = \Delta\Delta H^0 - T(\Delta\Delta S^0) = [2700 - (298 \times 0.6)]\,\text{cal}\,\text{mol}^{-1} = 2.52\,\text{kcal}\,\text{mol}^{-1}.$$

Thus *trans*-decalin is more stable than *cis*-decalin by $2.52\,\text{kcal}\,\text{mol}^{-1}$.

### 3.11.1.2 Angular methyl substitution

A decalin-type A/B rings in the steroid nucleus bears an angular methyl substituent. The angular methyl substituent affects the relative stability of *cis*- and *trans*-decalin frameworks. Introduction of angular Me in *trans*-decalin leads to four 1,3-diaxial (Me/H) interactions equivalent to four gb (Fig. 3.67). But in *cis*-decalin, angular Me introduces two 1,3-diaxial (Me/H) interactions because Me is axial with respect to B ring only (Fig. 3.67). Since *cis*-decalin has already three interring gb, *cis*-9-methyldecalin possesses a total of five gb. As a result, *cis*-9-methyldecalin will be enthalpically less stable than *trans*-9-methyldecalin by $0.9\,\text{kcal}\,\text{mol}^{-1}$ (one gb). The experimental value is $0.55 \pm 0.28\,\text{kcal}\,\text{mol}^{-1}$ (from isomerization equilibrium[144]) or $1.39 \pm 0.64\,\text{kcal}\,\text{mol}^{-1}$ (from heats of combustion[145]). It may be noted that *trans*-9-methyldecalin has $\mathbf{C_s}$ symmetry with one $\sigma$ plane passing through the ring junction, whereas *cis*-9-methyldecalin belongs to the point group $\mathbf{C_1}$

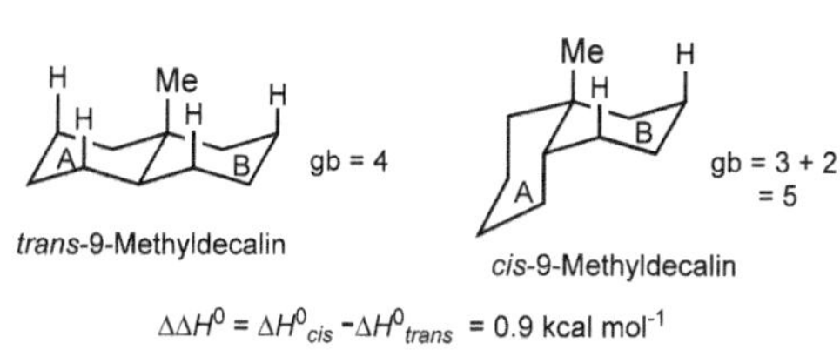


**FIG. 3.67** Effect of angular methyl substituent on the relative enthalpy of *cis*- and *trans*-decalins. gb = gauche–butane.

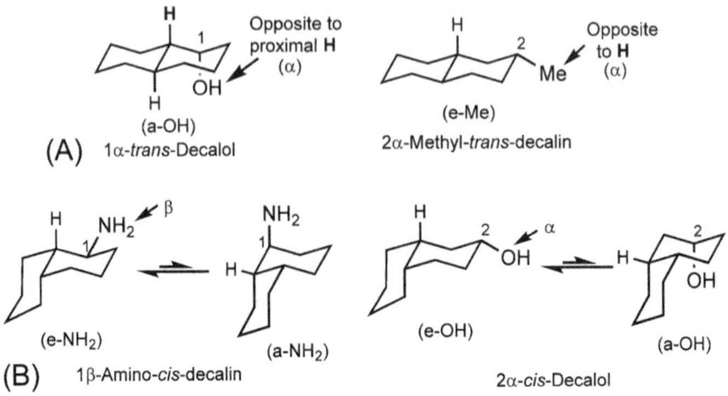


**FIG. 3.68** (A) Conformations of 1- and 2-substituted *trans*-decalin and (B) conformations of 1- and 2-substituted *cis*-decalin.

### 3.11.1.3 1- and 2-Substituted decalins

Since *trans*-decalin has a rigid chair–chair framework, a substituent occupies well-defined equatorial or axial position depending on the position of substitution (C1 or C2). As such, in 1α-hydroxy-*trans*-decalin (1α-*trans*-decalol), OH is axial, whereas 2α-methyl-*trans*-decalin exists in the equatorial Me conformation (Fig. 3.68A). The descriptors α and β refer to the stereochemical nomenclature in steroids. A substituent on the same side as the *proximal* angular H (or Me for steroids) is called β and that on the opposite side is denoted as α. In substituted *cis*-decalins, ring inversion takes place and the conformational equilibrium favours the side in which the substituent is equatorial, as illustrated with 1β-amino-*cis*-decalin and 2α-*cis*-decalol in Fig. 3.68B.

**Problem 3.13** Both *cis*- and *trans*-1-decalones are chiral and resolvable. Explain using conformational analysis.

**Problem 3.14** Predict the preferred conformation of *cis*-decahydroquinoline.

## 3.11.2 Perhydrophenanthrenes

Perhydrophenanthrene ring is a nonlinearly fused 6–6–6 tricyclic system which is present in steroids as A, B and C rings. A/B or B/C ring fusion represents a decalin moiety and is designated by *cis*/*trans* descriptors as for decalins. The stereochemistry of A/C linking is denoted by *cisoid*/*transoid* descriptors at the *direct bond* (Fig. 3.69). Note that A and C rings are not fused together. The stereochemical notations are illustrated with a perhydrophenanthrene isomer **3.33**.

Perhydrophenanthrenes have four stereocentres (ABBA type; see Section 2.8.8). Therefore six diastereomers (four chiral/racemic and two *meso*) are possible. The six perhydrophenanthrene diastereomers with their descriptors are shown in Fig. 3.70. The relative stability of the diastereomers was

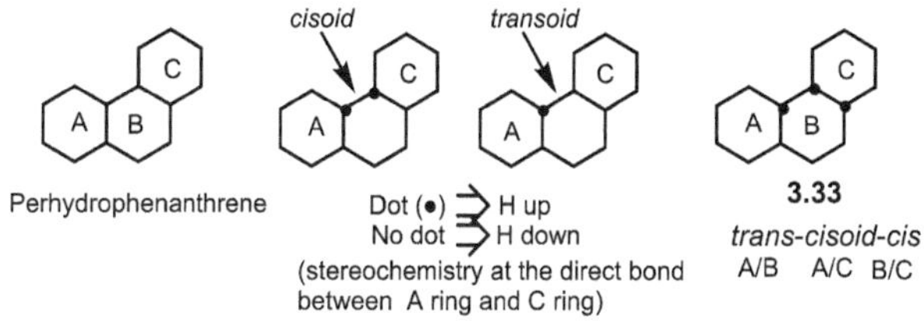


**FIG. 3.69** Stereochemical descriptors for perhydrophenanthrene system.

determined by Johnson[146] from conformational analysis based on the number of gauche–butane (gb) interactions (using 0.8 kcal mol$^{-1}$ for each) and/or other strains. The diastereomers (**A**–**F**) denote their increasing order of enthalpies. In the diastereomers, a *trans*-decalin (chair–chair) moiety contributes no gb, whereas a *cis*-decalin (chair–chair) contributes 3 gb (see Fig. 3.66). Further, additional strains may arise from C4 ↔ C5 gb, 1,3-diaxial interaction involving ring residues or ring conformation other than chair. The isomer **F** is least stable as it suffers from a severe 1,3-diaxial ($CH_2$/$CH_2$) strain which will be greater than 1,3-diaxial (Me/Me) interaction (3.7 kcal mol$^{-1}$) due to rigidity of the ring fusion. The middle ring in **E** is a twist-boat since the chair conformation of the middle B ring would require e,e fusion with the A ring but an impossible a,a fusion with the C ring. The twist-boat conformation of the middle ring in **E** can allow e,e-type fusion with the terminal rings, and the twist structure has been confirmed by X-ray crystallography.[147] **D** and **F** would exist as a mixture of two conformers since ring inversion is possible for a *cis*–*cis* isomer. Between the two conformers **D** and **D′**, the latter is slightly more stable as it lacks the C4 ↔ C5 gb. The isomer **F** is achiral in the planar form but has a chiral conformation. Ring inversion of **F** gives the enantiomeric **F′** and leads to a nonresolvable enantiomeric pair.

Fig. 3.70 also shows the relative enthalpies (kcal mol$^{-1}$) based on force field calculations[148]: **A** (0), **B** (2.44), **C** (2.57), **D** (4.01), **E** (7.03), **F** (9.01). These trends agree fairly well with the experimental free energies (kcal mol$^{-1}$) from the equilibrium data[149] but with a discrepancy of **E** having higher free energy than **F**: **A** (0), **B** (2.25), **C** (2.66), **D** (4.60), **E** (8.98), **F** (7.43).

### 3.11.3 Perhydroanthracenes

Perhydroanthracene ring is a linearly fused 6–6–6 tricyclic system and contains four equivalent stereocentres (AA/AA). Five perhydroanthracene diastereomers are possible comprising two chiral/racemic and three *meso* isomers (Fig. 3.71). The relative stability of the diastereomers was predicted by Johnson[146] based on the number of gauche–butane (gb) interactions and other strains (if any) due to 1,3-diaxial ($CH_2$/$CH_2$) interaction or boat conformation. The diastereomers (**A**–**E**) indicate their increasing order of enthalpies. The instability of

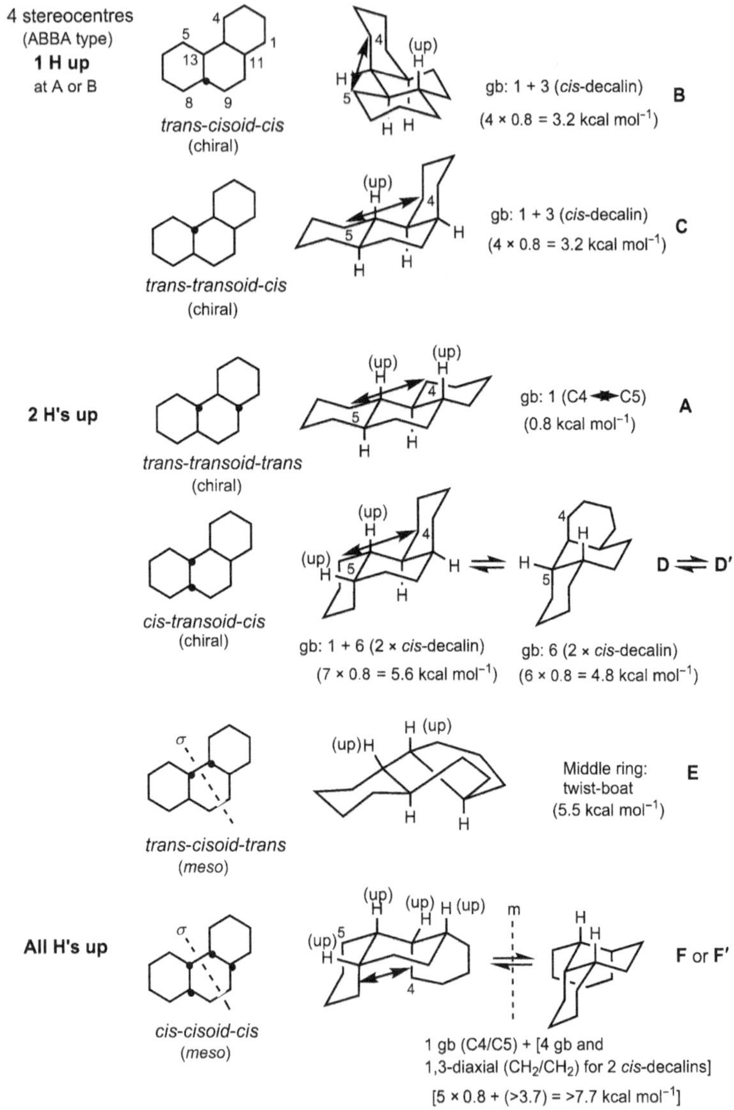


Relative enthalpy/free energy (kcal mol$^{-1}$)

| | A | B | C | D | E | F |
|---|---|---|---|---|---|---|
| Calculated enthalpy: (molecular mechanics) | 0 | 2.44 | 2.57 | 4.01 | 7.03 | 9.01 |
| Experimental free energy: (equilibrium data) | 0 | 2.25 | 2.66 | 4.60 | 8.98 | 7.43 |

**FIG. 3.70** Conformations and relative stabilities of perhydrophenanthrene diastereomers. gb = gauche–butane.

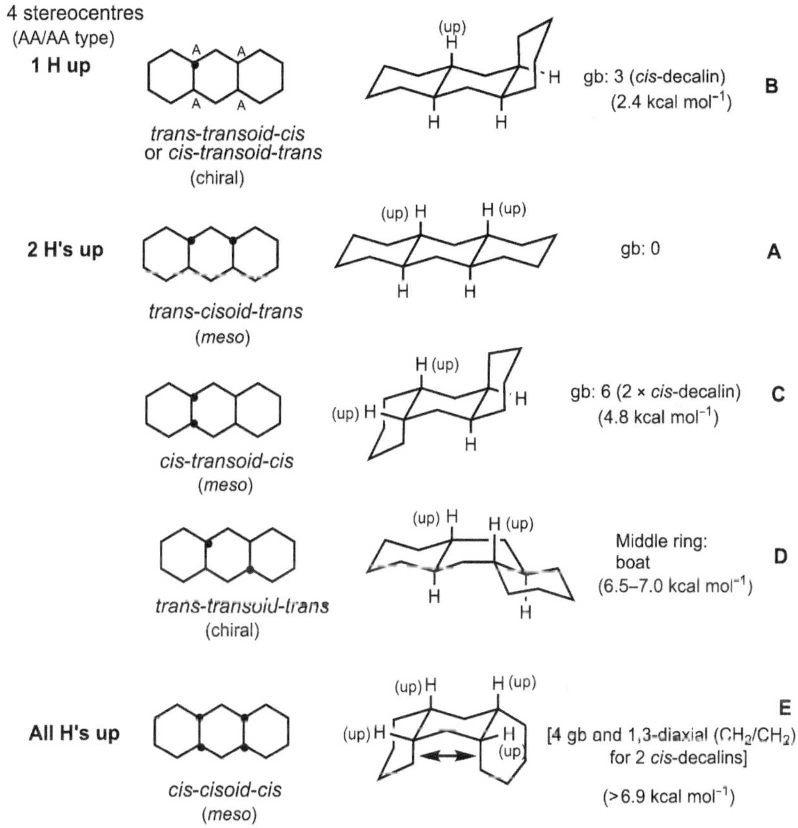


Relative enthalpy
(kcal mol$^{-1}$)

| | A | B | C | D | E |
|---|---|---|---|---|---|
| Calculated: | 0 | 2.62 | 5.56 | 5.86 | 8.13 |
| Experimental: | 0 | 2.76 | 5.58 | 4.15 | 8.74 |

FIG. 3.71 Conformations and relative enthalpies of perhydroanthracene diastereomers.

the diastereomer **D** arises from the enforced boat conformation of the middle ring. The isomer **E** is presumably least stable as it suffers from a severe 1,3-diaxial ($CH_2$/$CH_2$) interaction. Another conformer of **E** (not shown) with the middle ring in boat also contributes to a minor extent. The *cis–cis* conformers **C** and **E** can undergo ring inversion but give identical conformers. Both **C** (point group $\mathbf{C_i}$) and **E** (point group $\mathbf{C_s}$) are achiral.

Fig. 3.71 also shows the calculated and experimental relative enthalpies (kcal mol$^{-1}$) of the diastereomers.[150] The experimental enthalpies show a slightly lower value for **D** relative to **C**.

### 3.11.4 Steroids

The steroid nucleus has four fused rings, A/B/C (perhydrophenanthrene ring) and D (cyclopentane ring) **3.34** (Fig. 3.72A). In naturally occurring steroids, the perhydrophenanthrene unit is (*cis/trans*)-*transoid-trans*, which is *trans*-fused with the cyclopentane ring **3.35**. (Note that the C/D rings represent a *trans*-hydrindane system.) There are angular Me groups at C10 and C13. In steroids, the positions on the same side as the angular Me are specified as $\beta$ and those on the opposite side are specified as $\alpha$. Thus steroids with the A/B *cis* are called 5$\beta$ series and those with the A/B *trans* are called 5$\alpha$ series. As shown in Fig. 3.72A, androstane is 5$\beta$, whereas cholestane is 5$\alpha$.

Fig. 3.72B shows the conformations of 5$\alpha$ and 5$\beta$ series[11] taking 5$\alpha$-cholestane and 5$\beta$-cholestane, respectively. Since angular Me's are up, all up equatorial or up axial are $\beta$ while all down equatorial or down axial are $\alpha$ as indicated at C3, for instance.

## 3.12 Determination of configuration and conformation using chiroptical methods[151,152]

Chiroptical methods include optical rotatory dispersion (ORD) and circular dichroism (CD) techniques. Here we will describe briefly the theory and applications of ORD and CD for determining configuration or conformation of chiral molecules. A detailed discussion on the combined use of experimental and computational methods can be found elsewhere.[152]

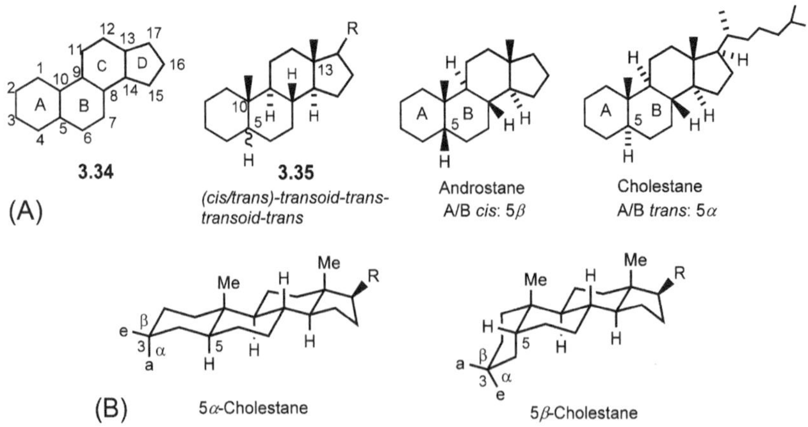


**FIG. 3.72** (A) Steroid numbering and descriptors ($\alpha$ and $\beta$) and (B) conformations of 5$\alpha$- and 5$\beta$-steroids.

### 3.12.1 Optical rotatory dispersion

The variation of refractive index with wavelength of light is called optical dispersion. Similarly, the variation of optical rotation (which is related to the refractive index; see Section 2.4.4) with wavelength is termed optical rotatory dispersion (ORD). Thus a plot of specific rotation $[\alpha]$ or molar rotation $[\Phi]$ as a function of wavelength ($\lambda$) gives an ORD curve. The ORD curve in the transparent spectral region shows a monotonic increase in optical rotation with decrease of wavelength and is known as plain curve. The monotonic increase in rotation exhibits an anomaly in the absorption region of chiral compounds, called anomalous rotatory dispersion.[153,154] This anomaly is known as the Cotton effect (CE).[155] Fig. 3.73 shows typical anomalous ORD curves exhibiting a positive CE (Fig. 3.73A) and a negative CE (Fig. 3.73B). An extremum in the anomalous ORD curve is a maximum (peak) or a minimum (trough). The CE is positive if the peak is at a longer wavelength ($\lambda$) than the trough when rotation first increases as $\lambda$ decreases (right to left on the horizontal axis). Conversely, the CE is negative if the trough is at a longer $\lambda$ (or the peak is at a shorter $\lambda$) when rotation first decreases as $\lambda$ decreases.[156] The point at which $[\Phi]=0$ is called the crossover point. For example, (+)-camphor (**3.3** in Fig. 3.5A) gives an anomalous ORD curve with a positive CE when the crossover point is at 294 nm close to its UV $\lambda_{max}$ (292 nm).[154]

The ORD technique is particularly suitable for chiral compounds containing a carbonyl chromophore. The forbidden $n \rightarrow \pi^*$ transition at 280–300 nm is readily accessible experimentally, and the weak absorption (low $\varepsilon$) allows enough light to be transmitted even at or near absorption $\lambda_{max}$ to permit the measurement of ORD.[157] The absolute configuration or the preferred conformation can be determined from the sign of the CE using empirical or semiempirical rules such as the axial haloketone rule and the octant rule.

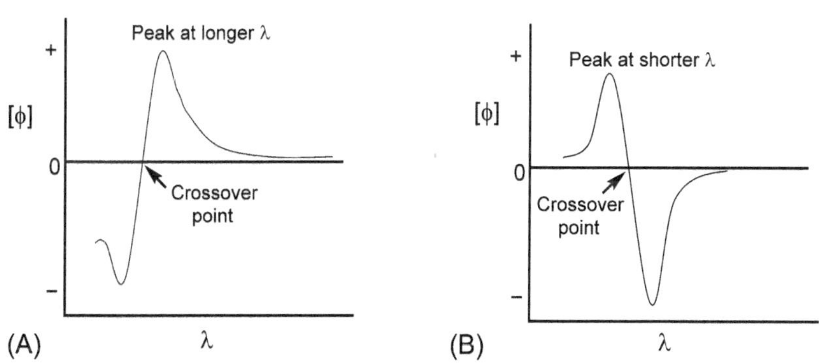


**FIG. 3.73** Typical anomalous ORD curves showing (A) a positive CE and (B) a negative CE.

### 3.12.1.1 The axial haloketone rule

The sign of the CE depends principally on the conformation of α-halogen (Cl, Br, I) of α-halocyclohexanones. The equatorial α-halogen on either side of C=O has little effect on the sign of the CE. But, if the α-halogen is axial, the sign of the CE may change according to the axial haloketone rule.[158] For the purpose of the rule, the molecule is to be viewed along O=C axis when carbonyl carbon is placed at the head (up carbon) of chair or boat cyclohexanone. If the halogen is on the left of the observer, the CE is predicted to be negative but if the halogen is on the right, the CE is positive (Fig. 3.74A).

*Corollary*: If the carbonyl carbon is a down carbon (not 'head'), the rule is reversed, i.e. the halogen on the left gives a positive CE while the halogen on the right gives a negative CE (Fig. 3.74B).

An application of the axial haloketone rule for the determination of absolute configuration of (−)-*trans*-1-decalone is shown in Fig. 3.75A.[159] The enantiomer will have one of the two possible configurations **A** and **A′**. Bromination of the compound produces (+)-*trans*-2-bromo-1-decalone in which Br is axial (spectral evidence by IR and UV). The ORD of the 2-bromo derivative shows a strong positive CE. According to the axial haloketone rule, 2-bromo derivative of **A** will show a positive CE, whereas that of **A′** will show a negative CE (Fig. 3.75A). Therefore the absolute configuration of (−)-*trans*-1-decalone is **A**.

Fig. 3.75B shows an application of the axial haloketone rule for determination of the preferred conformation of the A ring of a steroid.[160] 2*α*-Methyl-5*α*-cholestan-3-one on kinetic bromination gives 2-bromo-2-methyl derivative **3.36** which exhibits a negative CE. The bromine in **3.36** has been shown to be axial spectroscopically. The bromo derivative has two possible diastereomers **3.37** and **3.38**. The axial haloketone rule predicts that the normally expected **3.37** will show a positive CE which is opposite to the experimental result. The other diastereomer **3.38** has the equatorial Br in chair conformation but the derivative possesses an axial Br. If the A ring of **3.38** assumes a boat conformation, Br will be axial-like and a negative CE is then predicted by the haloketone rule in agreement with the experiment. Hence, the A ring in the steroid derivative is a boat rather than a chair. Notably, the boat

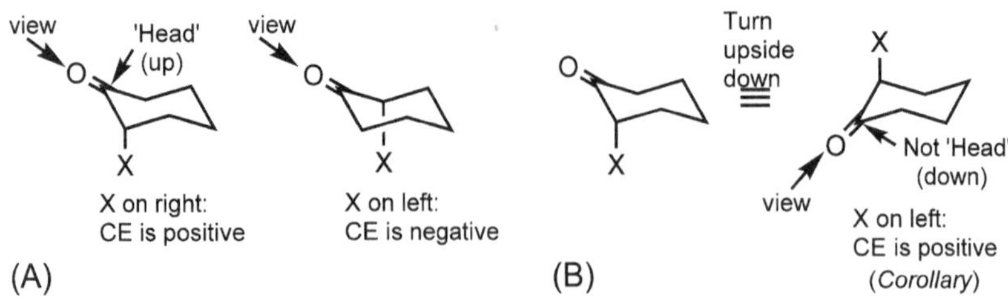


**FIG. 3.74** (A) The axial haloketone rule. (B) Corollary to the haloketone rule.

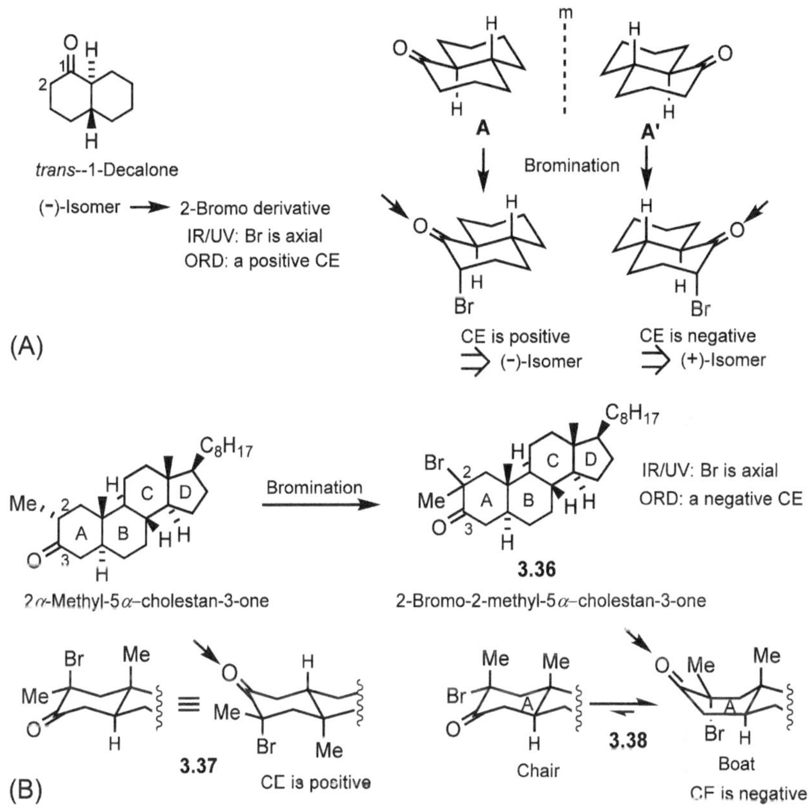


**FIG. 3.75** Application of the axial haloketone rule for the determination of (A) absolute configuration of *trans* 1-decalone and (B) preferred boat conformation of the A ring in a steroid derivative.

conformation of the A ring is preferred because the chair conformation is severely destabilized by 1,3-diaxial (Me/Me) interaction as well as unfavourable dipole–dipole repulsion between C—O and c Br. Mechanistically, bromination of 2$\alpha$-methyl-5$\alpha$-chlestan-3-one takes place preferentially by a parallel axial-like attack from the $\alpha$-face through a twist-boat TS leading to **3.38** (boat). The axial attack by Br to form the diastereomer **3.37** is highly unfavourable because of the severe 1,3-diaxial (Br/angular Me) strain.

**Problem 3.15** *R*-(+)-3-Methylcyclohexanone shows a positive CE. On chlorination, the compound produces a 2-chloro-5-methyl derivative that exhibits a negative CE in octane but a positive CE in methanol. The Cl has been shown to be axial spectroscopically. Determine the stereochemistry of the derivative and explain the solvent effect.

### 3.12.1.2 The octant rule

The Octant rule[161] uses three Cartesian planes (xy, xz, yz) to divide the space around C═O in a cyclohexanone into 8 sectors (called octants) with the origin at the midpoint of the C═O bond (Fig. 3.76). The molecule is viewed along the O═C axis (z-axis) with carbonyl carbon placed at the 'head' of the chair. The xy plane (plane A) bisects the C═O bond and divides the space into two regions containing four front octants and four rear octants. Rarely, a part of the cyclohexanone ring and the substituents fall into front octants. Therefore only rear octants **3.39** are drawn in Fig. 3.76. A vertical xz plane (plane *V*) passing through C═O and 4-e, 4-a bonds bisects the ring while a horizontal plane (plane H) passes through C═O and 2-e, 6-e bonds.

The Octant rule provides the contributions of substituents in different octants towards the sign of the CE.

**(a)** Substituents that lie on the Cartesian planes make no or negligible contribution to the CE. Thus 4-e, 4-a substituents on plane V and 2-e, 6-e substituents on plane H are ignored.

**(b)** For the four rear octants **3.39**, substituents in the top left (TL) and bottom right (BR) contribute to the positive sign of the CE while those in the top right (TR) and bottom left (BL) contribute to the negative sign. (For the front octants, the contributions to the sign are reversed.) It is worthwhile to mention that the contribution of a substituent to the sign of the CE is the sign of the product of its coordinates: TL = (+x)(−y)(−z) = +; TR = (+x)(+y)(−z) = −; BR = (−x)(+y)(−z) = +; BL = (−x)(−y)(−z) = −. The substituents that are more polarizable than H conform to the octant rule and are called consignate and those that are less polarizable than H such as F do not conform to the rule and are called dissignate.[162]

**(c)** Substituents close to the carbonyl chromophore, 2-a and 6-a, contribute more strongly. As a result, the contribution of highly polarizable axial halogen at 2- or 6-position becomes dominant which then effectively controls the sign of the CE. This is the axial haloketone rule.

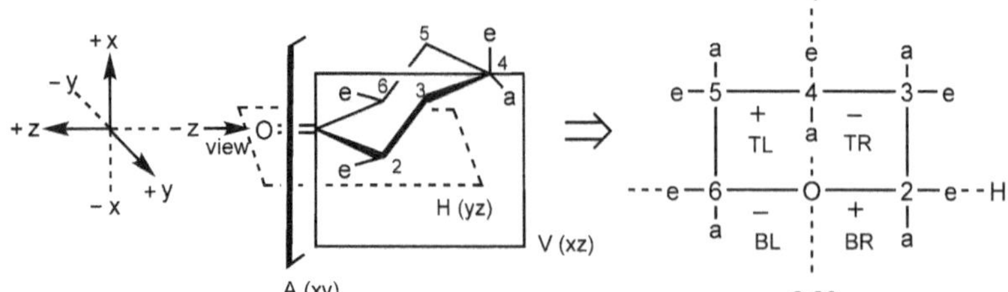


**3.39**
**Rear octants**:
-z (All), +x (Top), -x (Bottom)
+y (Right), -y (Left)

**FIG. 3.76** The Octant diagram for chair cyclohexanone. *BL*, bottom left; *BR*, bottom right; *TL*, top left; *TR*, top right.

The applications[157,163] of the Octant rule are illustrated with two examples in Fig. 3.77. In the first case (Fig. 3.77A), the preferred conformation of *R*-(+)-3-methylcyclohexanone is established using the Octant rule projections for e-Me and a-Me conformations.[164] The compound exhibits a positive CE which matches with the prediction for the e-Me conformer but not with the a-Me conformer which will show a negative CE. Note that the contribution of Me decides the sign of the CE. The preference for the e-Me conformation is in conformity with conformational analysis.

In the second example (Fig. 3.77B), the absolute configuration of (+)-*trans*-10-methyl-2-decalone is determined using the Octant rule projections for its two enantiomers **A** and **A′**.[165] The (+) enantiomer exhibits a positive CE which conforms to the prediction for the **A** enantiomer by the Octant rule. The other enantiomer **A′** will show a negative CE. Note that the contributions of larger number of atoms/groups in an octant decide the sign of the CE.

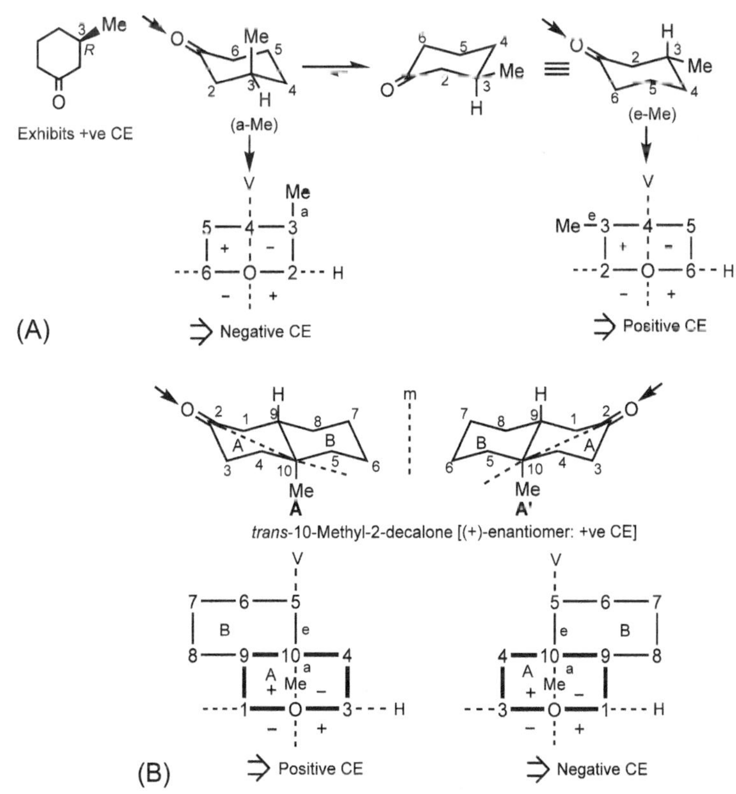


**FIG. 3.77** Application of the Octant rule for the determination of (A) preferred conformation of *R*-(+)-3-methylcyclohexanone; and (B) absolute configuration of (+)-*trans*-10-methyl-2-decalone.

**Problem 3.16** Using the Octant rule, predict the sign of the CE for the steroidal and nonsteroidal conformations of any one enantiomer of *cis*-10-methyl-2-decalone.

**Problem 3.17** Draw the Octant rule projections for 5$\alpha$-cholestan-1-one and 5$\alpha$-cholestan-3-one. Predict the sign of the CE in each case.

## 3.12.2 Circular dichroism

Anisotropic refraction of left circularly polarized (Lcp) and right circularly polarized (Rcp) lights gives optical rotation (see Section 2.4.4). But *anisotropic absorption* of Lcp and Rcp lights by chiral molecules leads to circular dichroism (CD).[166] If the chiral sample is optically active (one enantiomer in excess), then in the range of wavelengths in which absorption occurs,

$$\varepsilon_L \neq \varepsilon_R \ (\varepsilon = \text{molar absorption coefficient})$$

The difference, $\Delta\varepsilon = \varepsilon_L - \varepsilon_R$ is called the CD. The anisotropic absorption is also a Cotton effect (CE) that occurs in the spectral region for UV/visible absorption.

A plot of $\Delta\varepsilon$ as a function of wavelength ($\lambda$) in the region of the absorption band gives a CD spectrum. The sign of the CD is defined by the sign of $\Delta\varepsilon$. Thus $\varepsilon_L > \varepsilon_R$ denotes a positive CD curve (Fig. 3.78A), and $\varepsilon_L < \varepsilon_R$ indicates a negative CD curve (Fig. 3.78B).[154] For example, 5$\alpha$-cholestan-3-one gives a positive CD, whereas 5$\alpha$-cholestan-1-one gives a negative CD.

The CD spectra are simpler to interpret and CD has been essentially replacing ORD as the principal chiroptical method in the study of chiral molecules.[167]

### 3.12.2.1 *Theory*[168,169]

If a Lcp and a Rcp wave of equal intensity (amplitude) are superposed, the result is a linearly polarized light because at any point in space the $\mathbf{E}_L$ and $\mathbf{E}_R$ vectors will sum as shown in Fig. 3.79A.

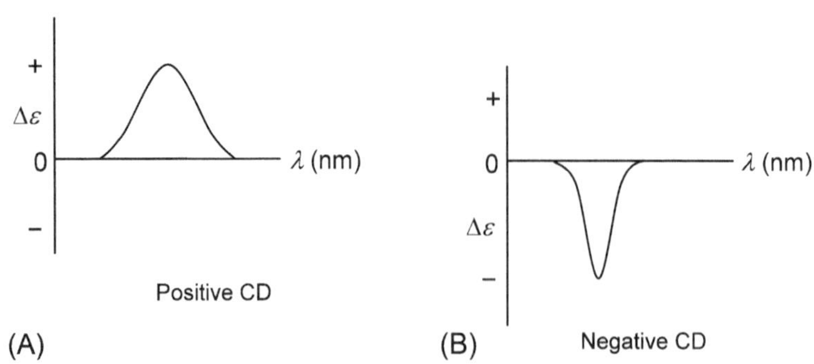


**FIG. 3.78** CD curves showing (A) a positive CD and (B) a negative CD.

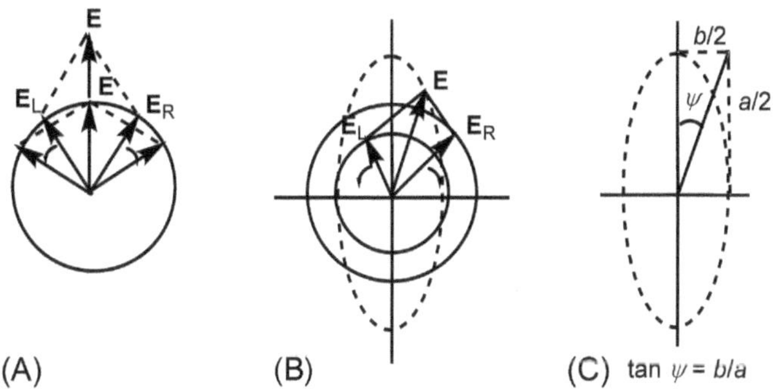


**FIG. 3.79** (A) Combination of left circularly polarized (Lcp) and right circularly polarized (Rcp) lights of equal intensity ($\mathbf{E}_L = \mathbf{E}_R$) producing linearly polarized light; (B) circular dichroism ($\varepsilon_L > \varepsilon_R$): combination of Lcp and Rcp lights of different intensity ($\mathbf{E}_L < \mathbf{E}_R$) generating elliptically polarized light and (C) ellipticity angle ($\psi$) defined. ($\mathbf{E}_L$ and $\mathbf{E}_R$ are the electric field vectors for Lcp and Rcp lights, respectively, and **E** is the resultant electric vector in each case.)

Now consider a CD when $\varepsilon_L > \varepsilon_R$. Since the absorption of Lcp light ($\varepsilon_L$) is greater, its intensity or amplitude ($\mathbf{E}_L$) will diminish more. Therefore $\mathbf{E}_L < \mathbf{E}_R$. Both $\mathbf{E}_L$ and $\mathbf{E}_R$ will however be smaller than $\mathbf{E}_0$ for the incident light. Fig. 3.79B shows that the tip of the resultant **E** vector traces an elliptical path and such light is said to be elliptically polarized A parameter called the ellipticity ($\psi$) is often used to describe an elliptical polarization. This is the ellipticity angle whose tangent, by geometry, is the ratio of the minor axis ($b$) and the major axis ($a$) of the ellipse, that is $\tan \psi = b/a$ or $\psi = \tan^{-1}(b/a)$ (Fig. 3.79C). $\psi$ is usually measured in mdegrees.

By analogy with specific rotation ($[\alpha]$) and molar rotation ($\Phi$) (see Section 2.4.2), specific ellipticity ($[\psi]$) and molar ellipticity ($[\theta]$) are defined as

$$[\psi] = \frac{\psi}{lc} \tag{3.5}$$

where $l$ is in dm and $c$ in $\mathrm{g\,cm^{-3}}$. Unit of $[\psi]$ is $10^{-1}\,\mathrm{deg\,cm^2\,g^{-1}}$.

$$[\theta] = \frac{[\psi]M}{100} \tag{3.6}$$

where $M$ is in $\mathrm{g\,mol^{-1}}$. Unit of $[\theta]$ is $\mathrm{deg\,cm^2\,dmol^{-1}}$.

Since the ellipticity $\psi$ is small, $\tan \psi \approx \psi$, which is proportional to $\Delta A$ as

$$\psi = 2.303\,\Delta A\,(180/4\pi) = 33\,\Delta A\ \text{degrees} \tag{3.7}$$

where $\Delta A = A_L - A_R$

Since $\Delta A = \Delta\varepsilon\ c'\ l'$ ($c'$ is the concentration in $\mathrm{mol\,L^{-1}}$ and $l'$ is the path length in cm), the molar ellipticity $[\theta]$ is given by

$$[\theta] = 3300\,\Delta\varepsilon \tag{3.8}$$

Experimentally, it is usual to measure $\Delta A$ or $\Delta\varepsilon$ (if the concentration is known) but for historical reasons, the ellipticity $\psi$ or molar ellipticity $[\theta]$ is plotted in a CD spectrum. For biological macromolecules such as proteins, mean residue ellipticity instead of molar ellipticity is often used when $M$ is taken as the mean residue molecular mass.

### 3.12.2.2 Exciton chirality

If two chromophores lie in close proximity to form a chiral array, the interaction between them can lead to the splitting of the CE to distinct couplets in the CD spectrum. This is called exciton coupling.[170] The term 'exciton' coined by Davydov refers to the nondegenerate excited states of a polychromophoric system. The exciton coupling has been successfully utilized for determining the absolute configuration of glycols through their dibenzoate derivatives.[171] Fig. 3.80A illustrates a case. The CD spectrum of bis(*p*-dimethylaminobenzoate) ester of (−)-spiro[4.4]nonane-1,6-diol **3.40** in ethanol shows two CEs for exciton splitting: a positive CE at a shorter wavelength of 295 nm and a negative CE at a longer wavelength of 321 nm with a separation of 26 nm.[172] This relationship between the two CEs denotes negative chirality,[171] which is also consistent with the dibenzoate chirality rule[173] (Fig. 3.80B). It may be mentioned that the two CEs are centred at 307 nm near the UV $\lambda_{max}$ 310.5 nm (UV spectrum not shown). The relative configuration of the stereocentres C1 and C6 in the glycol ester **3.40** is known from $^1$H NMR to be 1,5-*cis*-5,6-*trans*.[174] Hence its absolute configuration is deduced from the negative chirality as 1*S*,5*S*,6*R* (cf. Fig. 2.25).

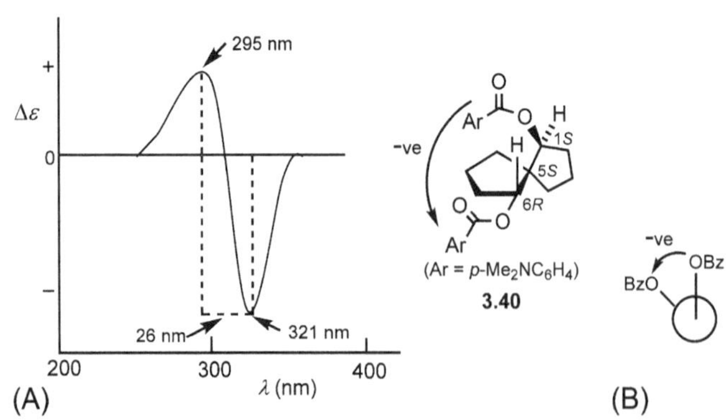


**FIG. 3.80** (A) Determination of absolute configuration of bis(*p*-dimethylaminobenzoate) ester of (−)-spiro[4.4]nonane-1,6-diol using exciton splitting in the CD spectrum. (B) The dibenzoate chirality rule (−ve = negative chirality).

### 3.12.2.3 Helicity rule

Inherently chiral chromophores follow the helicity rule which states that *P*-helicity (a right-handed helix) leads to a positive CE while *M*-helicity (a left-handed helix) gives rise to a negative CE. For helical molecules such as hexahelicene (see Fig. 2.50), electron movement during excitation has the character of a helical motion. The direction of electron transition moment is identical for either enantiomer of hexahelicene but the magnetic transition moment direction is reversed as to whether the helix is *P* or *M*. The parallel magnetic transition moments for *P*-hexahelicene lead to a positive CE, whereas the antiparallel magnetic transition moments for *M*-hexahelicene produce a negative CE.[175] The positive or negative CE for hexahelicene refers to a band at about 325 nm which corresponds to the high intensity π–π* transition between HOMO and LUMO.[176]

### 3.12.2.4 Determination of protein conformation

CD is particularly useful for the determination of protein conformation and conformational changes. The principal secondary conformations of proteins

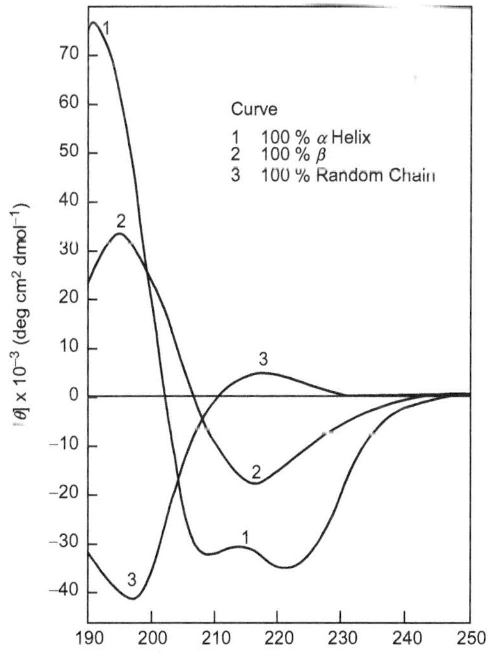


**FIG. 3.81** CD spectra of poly(L-lysine) in α-helical (curve 1), β-sheet (curve 2) and random coil (curve 3) conformations. *(Reproduced with permission from Greenfield, N.; Fasman, G. D.* Biochemistry ***1969***, 8, *4108. Copyright © 1969 American Chemical Society.)*

are $\alpha$-helix and $\beta$-sheet (see Section 1.3.12). Besides $\alpha$-helix and $\beta$-sheet conformations, proteins usually contain the random coil. A model polypeptide has only a single conformation and its structure is known from X-ray scattering. The principal polypeptide models are three forms of poly-L-lysine—$\alpha$-helix, $\beta$-sheet and random coil, whose characteristic CD spectra are shown in Fig. 3.81.[177] Based on these standard curves, the conformation of a protein can be determined from its observed CD spectrum. For example, myoglobin gives a CD spectrum that indicates a predominantly $\alpha$-helix conformation.[177] An estimate from the CD analysis shows that myoglobin contains 68.3% $\alpha$-helix, 4.7% $\beta$-sheet and 27% random coil. The analysis is based on the construction from the standard curves in Fig. 3.81 a weighted sum which is the same as the observed curve, assuming that no other conformations exist.[177]

## References

1. Cahn, R. S.; Ingold, C.; Prelog, V. *Angew. Chem. Int. Ed. Engl.* **1966**, *5*, 385.
2. Cross, L. C.; Klyne, W.; Collators. *Pure Appl. Chem.* **1976**, *45*, 11.
3. Angyal, S. J. *Q. Rev. Chem. Soc.* **1957**, *11*, 212.
4. Prelog, V.; Helmchen, G. *Angew. Chem. Int. Ed. Engl.* **1982**, *21*, 567.
5. Park, C. H.; Simmons, H. E. *J. Am. Chem. Soc.* **1972**, *94*, 7184.
6. Alder, R. W. *Acc. Chem. Res.* **1983**, *16*, 321.
7. Winkler, J. D.; Hey, J. P.; Williard, P. G. *J. Am. Chem. Soc.* **1986**, *108*, 6425.
8. Eliel, E. L.; Wilen, S. H.; Mander, L. N. *Stereochemistry of Organic Compounds*; Wiley: New York, 1994; p. 675.
9. Sachse, H. *Z. Phys. Chem.* **1892**, *10*, 203.
10. Brown, H. C.; Fletcher, R. S.; Johannesen, R. B. *J. Am. Chem. Soc.* **1951**, *73*, 212 (footnote 21).
11. Barton, D. H. R. *Experientia* **1950**, *6*, 316.
12. Anderson, J. E. *Top. Curr. Chem.* **1974**, *45*, 139.
13. Kellie, G. M.; Riddell, F. G. *Top. Stereochem.* **1974**, *8*, 225.
14. Binsch, G.; Eliel, E. L.; Kessler, H. *Angew. Chem. Int. Ed. Engl.* **1971**, *10*, 570.
15. Allinger, N. L. *J. Am. Chem. Soc.* **1977**, *99*, 8127.
16. Allinger, N. L.; Freiberg, L. A. *J. Am. Chem. Soc.* **1960**, *82*, 2393.
17. Anet, F. A. L.; Anet, R. In *Dynamic Nuclear Magnetic Resonance Spectroscopy*; Jackman, L. M., Cotton, F. A., Eds.; Academic Press: New York, 1975.
18. Sandström, J. *Dynamic NMR Spectroscopy*; Academic Press: New York, 1982.
19. Höfner, D.; Lesko, S. A.; Binsch, G. *Org. Magn. Reson.* **1978**, *11*, 179.
20. Squillacote, M.; Sheridan, R. S.; Chapman, G. L.; Anet, F. A. L. *J. Am. Chem. Soc.* **1975**, *97*, 3244.
21. Davis, M.; Hassel, O. *Acta Chem. Scand.* **1963**, *17*, 1181.
22. Geise, H. J.; Buys, H. R.; Mijlhoff, F. C. *J. Mol. Struct.* **1971**, *9*, 447.
23. Romers, C.; Altona, C.; Buys, H. R.; Havinga, E. *Top. Stereochem.* **1969**, *4*, 39.
24. Schleyer, P.v. R.; Williams, J. E.; Blanchard, K. R. *J. Am. Chem. Soc.* **1970**, *92*, 2377.
25. Jensen, F. R.; Bushweller, C. H. *Adv. Alicycl. Chem.* **1971**, *3*, 139.
26. Winstein, S.; Holness, N. J. *J. Am. Chem. Soc.* **1955**, *77*, 5562.
27. Hirsch, J. A. *Top. Stereochem.* **1967**, *1*, 199.
28. Eliel, E. L.; Wilen, S. H.; Mander, L. N. *Stereochemistry of Organic Compounds*; Wiley: New York, 1994; p. 690.

29. Booth, H.; Everett, J. R. *J. Chem. Soc. Perkin. Trans. 2* **1980**, 255.
30. Eliel, E. L.; Allinger, N. L.; Angyal, S. J.; Morrison, G. A. *Conformational Analysis*; Wiley: New York, 1965.
31. Antenunis, M. J. O. In *Conformational Analysis, Scope and Present Limitations*; Chiurdoglu, G., Ed.; Academic Press: New York, 1971; p. 32.
32. Allinger, N. L.; Tribble, M. T. *Tetrahedron Lett.* **1971**, 3259.
33. Reisse, J. In *Conformational Analysis, Scope and Present Limitations*; Chiurdoglu, G., Ed.; Academic Press: New York, 1971; p. 219ff.
34. Binsch, G. *Top. Stereochem.* **1968**, *3*, 97.
35. Oki, M. *Applications of Dynamic NMR in Organic Chemistry*; VCH Publishers: Deerfield Beach, FL, 1985.
36. Takeuchi, Y.; Marchand, A. P., Eds. *Applications of NMR Spectroscopy in Stereochemistry and Conformational Analysis*; VCH Publishers: Deerfield Beach, FL, 1986.
37. Eliel, E. L.; Schroeter, S. H. *J. Am. Chem. Soc.* **1965**, *87*, 5031.
38. Jensen, F. R.; Bushweller, C. H.; Beck, B. H. *J. Am. Chem. Soc.* **1969**, *91*, 334.
39. Karplus, M. *J. Chem. Phys.* **1959**, *30*, 11. *J. Am. Chem. Soc.* **1963**, *85*, 2870.
40. Eliel, E. L. *Chem. Ind.* **1959**, 568.
41. Anet, F. A. L.; Bradley, C. H.; Buchanan, G. W. *J. Am. Chem. Soc.* **1971**, *93*, 258.
42. Eliel, E. L.; Ro, R. S. *J. Am. Chem. Soc.* **1957**, *79*, 5992.
43. Jensen, F. R.; Bushweller, C. H. *J. Am. Chem. Soc.* **1969**, *91*, 3223.
44. Eliel, E. L.; Kandasamy, D. *J. Org. Chem.* **1976**, *44*, 3899.
45. Manoharan, M.; Eliel, E. L.; Carroll, F. I. *Tetrahedron Lett.* **1983**, *24*, 1855.
46. Hodgson, D. J.; Rychlewska, U.; Eliel, E. L.; Manoharan, M.; Knox, D. E.; Olefirowicz, E. M. *J. Org. Chem.* **1985**, *50*, 4838.
47. Eliel, E. L.; Manoharan, M. *J. Org. Chem.* **1981**, *46*, 1959.
48. Allinger, N. L.; Szkrybalo, W.; Van-Catledge, F. A. *J. Org. Chem.* **1968**, *33*, 784.
49. Leonard, J. E.; Hammond, G. S.; Simmons, H. E. *J. Am. Chem. Soc.* **1975**, *97*, 5052.
50. Corey, E. J.; Feiner, N. F. *J. Org. Chem.* **1980**, *45*, 765.
51. Remijnse, J. D.; van Bekkum, H.; Wepster, B. M. *Recl. Trav. Chim. Pays-Bas* **1974**, *93*, 93.
52. Loomes, D. J.; Robinson, M. J. T. *Tetrahedron* **1977**, *33*, 1149.
53. van de Graaf, B.; van Bekkum, H.; van Koningsveld, H.; et al. *Recl. Trav. Chim. Pays-Bas* **1974**, *93*, 135.
54. Schubert, W. K.; Southern, W. J.; Schäfer, L. *J. Mol. Struct.* **1973**, *16*, 403.
55. Stolow, R. D. *J. Am. Chem. Soc.* **1964**, *86*, 2170.
56. Eliel, E. L. *Stereochemistry of Carbon Compounds*; McGraw-Hill: New York, 1962.
57. Stolow, R. D. *J. Am. Chem. Soc.* **1961**, *83*, 2592.
58. Eliel, E. L.; Haubenstock, H. *J. Org. Chem.* **1961**, *26*, 3504.
59. Werner, H.; Mann, G.; Mühlstädt, M.; et al. *Tetrahedron Lett.* **1970**, *11*, 3563.
60. de la Mare, P. B. D.; Hall, D.; Pavitt, N. *J. Comput. Chem.* **1983**, *4*, 114.
61. Riddell, F. G. *The Conformational Analysis of Heterocyclic Compounds*; Academic Press: New York, 1980.
62. Delpuech, J.-J. In *Cyclic Organic Stereodynamics*; Lambert, J. B., Takeuchi, Y., Eds.; VCH Publishers: New York, 1992.
63. Eliel, E. L.; Hargrave, K. D.; Pietrusiewicz, K. M.; Manoharan, M. *J. Am. Chem. Soc.* **1982**, *104*, 3635.
64. Eliel, E. L.; Kandasamy, D.; Yen, C.-y.; Hargrave, K. D. *J. Am. Chem. Soc.* **1980**, *102*, 3698.
65. Willer, R. L.; Eliel, E. L. *J. Am. Chem. Soc.* **1977**, *99*, 1925.
66. Eliel, E. L. *Angew. Chem. Int. Ed. Engl.* **1972**, *11*, 739.

67. Eliel, E. L.; Knoeber, M. C. *J. Am. Chem. Soc.* **1968**, *90*, 3444.
68. Jones, A. J.; Eliel, E. L.; Grant, D. M.; Knoeber, M. C.; Bailey, W. F. *J. Am. Chem. Soc.* **1971**, *93*, 4772.
69. Kirby, A. J. *The Anomeric Effect and Related Stereoelectronic Effects at Oxygen*; Springer: New York, 1983.
70. Thatcher, G. R. J., Ed. *ACS Symposium Series 539*; *The Anomeric Effect and Associated Stereoelectronic Effects*; American Chemical Society: Washington, DC, 1993.
71. Juaristi, E.; Cuevas, G. *The Anomeric Effect*; CRC Press: Boca Raton, 1995.
72. Gleiter, R.; Haberhauer, G. *Aromaticity and Other Conjugation Effects*; Wiley-VCH: Weinheim, 2012; p. 331.
73. Salzner, U.; Schleyer, P.v. R. *J. Org. Chem.* **1994**, *59*, 2138.
74. Altona, C.; Knobler, C.; Romers, C. *Acta Crystallogr.* **1963**, *16*, 1217. Altona, C.; Romers, C. *Acta Crystallogr.* **1963**, *16*, 1225.
75. Lemieux, R. U.; Koto, S.; Voisin, D. In *Anomeric Effect, Origin and Consequences*; Szarek, W. A., Horton, D., Eds.; ACS Symposium Series 87; American Chemical Society: Washington, DC, 1979; p. 17.
76. Box, V. G. S. *Heterocycles* **1990**, *31*, 1151.
77. Jeffrey, G. A. In *Anomeric Effect, Origin and Consequences*; Szarek, W. A., Horton, D., Eds.; ACS Symposium Series 87; American Chemical Society: Washington, DC, 1979; p. 50ff.
78. Edward, J. T. *Chem. Ind.* **1955**, 1102.
79. Mo, Y. *Nat. Chem.* **2010**, *2*, 666.
80. Kirby, A. J. *Stereoelectronic Effects*; Oxford University Press: New York, 1996; p. 14.
81. Lemieux, R. U.; Chü, N. J. *Abstract Papers American Chemical Society 133rd Meeting*; 1958; p. 31N.
82. Dillen, J.; Geise, H. J. *J. Mol. Struct.* **1980**, *69*, 137.
83. Anet, F. A. L.; Chmurny, G. N.; Krane, J. *J. Am. Chem. Soc.* **1973**, *95*, 4423.
84. Eliel, E. L.; Allinger, N. L.; Angyal, S. J.; Morrison, G. A. *Conformational Analysis*; Wiley: New York, 1965. p. 114, 465.
85. Pan, Y.-H.; Stothers, J. B. *Can. J. Chem.* **1967**, *45*, 2943.
86. Meyer, A. Y.; Allinger, N. L.; Yuh, Y. *Isr. J. Chem.* **1980**, *20*, 57.
87. Corey, E. J.; Burke, H. J. *J. Am. Chem. Soc.* **1955**, *77*, 5418.
88. Allinger, N. L.; Freiberg, L. A. *J. Am. Chem. Soc.* **1962**, *84*, 2201.
89. Jensen, F. R.; Beck, B. H. *J. Am. Chem. Soc.* **1968**, *90*, 1066.
90. Lambert, J. B.; Clikeman, R. R. *J. Am. Chem. Soc.* **1976**, *98*, 4203.
91. Lambert, J. B. In *Conformational Analysis of Cyclohexenes, Cyclohexadienes and Related Hydroaromatic Compounds*; Rabideau, P. W., Ed.; VCH Publishers: New York, 1989; p. 47.
92. Chiang, J. F.; Bauer, S. H. *J. Am. Chem. Soc.* **1969**, *91*, 1898.
93. Auf der Heyde, W.; Lüttke, W. *Chem. Ber.* **1978**, *111*, 2384.
94. Anet, A. F. L.; Haq, M. Z. *J. Am. Chem. Soc.* **1965**, *87*, 3147.
95. Lambert, J. B.; Clikeman, R. R.; Taba, K. M.; Marko, D. E.; Bosch, R. J.; Xue, L. *Acc. Chem. Res.* **1987**, *20*, 454.
96. Lessard, J.; Tan, P. V. M.; Martino, R.; Saunders, J. K. *Can. J. Chem.* **1977**, *55*. 1015, 1017.
97. Johnson, F. *Chem. Rev.* **1968**, *68*, 375.
98. Johnson, F.; Whitehead, A. *Tetrahedron Lett.* **1964**, 3825.
99. Corey, E. J.; Enders, D. *Tetrahedron Lett.* **1976**, 3.
100. Maruoka, K.; Miyazaki, T.; Ando, M.; et al. *J. Am. Chem. Soc.* **1983**, *105*, 2831.
101. Liebman, J. F.; Greenberg, A. *Chem. Rev.* **1989**, *89*, 1215.
102. Walsh, A. D. *Nature* **1947**, *159*, 712. *Trans. Faraday Soc.* **1949**, *45*, 179.

103. Brown, H. C.; Peters, E. N. *J. Am. Chem. Soc.* **1973**, *95*, 2400.
104. Hoffmann, R. *J. Chem. Phys.* **1964**, *40*, 2480.
105. Childs, R. F.; Faggiani, R.; Lock, C. J. L.; Mahendran, M.; Zweep, S. D. *J. Am. Chem. Soc.* **1986**, *108*, 1692.
106. Allen, F. H. *Acta Crystallogr. Sect. B* **1980**, *81*, B36.
107. Egawa, T.; Fukuyama, T.; Yamamoto, S.; et al. *J. Chem. Phys.* **1987**, *86*, 6018.
108. Durig, J. R.; Little, T. S.; Lee, M. J. *J. Raman Spectrosc.* **1989**, *20*, 757.
109. Pitzer, K. S.; Donath, W. E. *J. Am. Chem. Soc.* **1959**, *81*, 3213.
110. Kilpatrick, J. E.; Pitzer, K. S.; Spitzer, R. *J. Am. Chem. Soc.* **1947**, *69*, 2483.
111. Fuchs, B. *Top. Stereochem.* **1978**, *10*, 1.
112. Hendrickson, J. B. *J. Am. Chem. Soc.* **1961**, *83*, 4537.
113. Cox, J. D.; Pilcher, G. *Thermochemistry of Organic and Organometallic Compounds*; Academic Press: New York, 1970.
114. Brown, H. C.; Fletcher, R. S.; Johannesen, R. B. *J. Am. Chem. Soc.* **1951**, *73*, 212.
115. Brown, H. C.; Ichikawa, K. *Tetrahedron* **1957**, *1*, 221.
116. Prelog, V.; Kobelt, M. *Helv. Chim. Acta* **1949**, *32*, 1187.
117. Heck, R.; Prelog, V. *Helv. Chim. Acta* **1955**, *38*, 1541.
118. Sicher, J. *Prog. Stereochem.* **1962**, *3*, 202.
119. Bocian, D. F.; Strauss, H. L. *J. Am. Chem. Soc.* **1977**, *99*. 2866, 2876.
120. Dale, J. *Stereochemistry and Conformational Analysis*; Verlag Chemie: New York, 1978; p. 194.
121. Traetteberg, M. *J. Am. Chem. Soc.* **1964**, *86*, 4265.
122. Anet, A. F. L. *J. Am. Chem. Soc.* **1964**, *86*, 458.
123. Dale, J. *Top. Stereochem.* **1976**, *9*, 199.
124. Prelog, V. In *Perspectives in Organic Chemistry*; Todd, A. R., Ed.; Interscience: New York, 1956; p. 96ff.
125. Anet, A. F. L. *Top. Curr. Chem.* **1974**, *45*, 169.
126. Still, W. C.; Galynker, I. *Tetrahedron* **1981**, *37*, 3981.
127. Zhou, Z.; Parr, R. G. *J. Am. Chem. Soc.* **1989**, *111*, 7371.
128. Anet, A. F. L.; Bourn, A. J. R.; Lin, Y. S. *J. Am. Chem. Soc.* **1964**, *86*, 3576.
129. Paquette, L. A. *Acc. Chem. Res.* **1993**, *26*, 57.
130. Pawar, D. M.; Smith, S. V.; Mark, H. L.; Odom, R. M.; Noe, E. A. *J. Am. Chem. Soc.* **1998**, *120*, 10715.
131. Dunitz, J. D. In *Perspectives in Structural Chemistry*; Dunitz, J. D., Ibers, J. A., Eds.; Vol. II; Wiley: New York, 1968; p. 1.
132. Prelog, V.; Schenker, K. *Helv. Chim. Acta* **1952**, *35*, 2044.
133. Schill, G.; Zollenkopf, H. *Justus Liebigs Ann. Chem.* **1969**, *721*, 53.
134. Agam, G.; Gravier, D.; Zilkha, A. *J. Am. Chem. Soc.* **1976**, *98*, 5206.
135. Wasserman, E. *J. Am. Chem. Soc.* **1960**, *82*, 4433. *Sci. Am.* **1962**, *207*, No. 5, 94.
136. Ashton, P. R.; Goodnow, T. T.; Kaifer, A. E.; et al. *Angew. Chem. Int. Ed. Engl.* **1989**, *28*, 1396.
137. Walba, D. M.; Richards, R. M.; Haltiwanger, R. C. *J. Am. Chem. Soc.* **1982**, *104*, 3219.
138. Liebman, J. F.; Greenberg, A. *Chem. Rev.* **1976**, *76*, 311.
139. Van den Enden, L.; Geise, H. J.; Spelbos, A. *J. Mol. Struct.* **1978**, *44*, 177.
140. Dalling, D. K.; Grant, D. M.; Johnson, L. F. *J. Am. Chem. Soc.* **1971**, *93*, 3678.
141. Allinger, N. L.; Coke, J. L. *J. Am. Chem. Soc.* **1959**, *81*, 4080.
142. Speros, D. M.; Rossini, F. D. *J. Phys. Chem.* **1960**, *64*, 1723.
143. Schucker, R. C. *J. Chem. Eng. Data* **1981**, *26*, 239.
144. Allinger, N. L.; Coke, J. L. *J. Org. Chem.* **1961**, *26*, 2096.

145. Dauben, W. G.; Rohr, O.; Labbauf, A.; Rossini, F. D. *J. Phys. Chem.* **1960**, *64*, 283.
146. Johnson, W. S. *J. Am. Chem. Soc.* **1953**, *75*, 1498.
147. Allinger, N. L.; Honig, H.; Burkert, U.; Asolnai, L.; Huttner, G. *Tetrahedron* **1984**, *40*, 3449.
148. Allinger, N. L.; Gorden, B. J.; Tyminski, I. J.; Wuesthoff, M. T. *J. Org. Chem.* **1971**, *36*, 739.
149. Honig, H.; Allinger, N. L. *J. Org. Chem.* **1985**, *50*, 4630.
150. Allinger, N. L.; Wuesthoff, M. T. *J. Org. Chem.* **1971**, *36*, 2051.
151. Eliel, E. L.; Wilen, S. H.; Mander, L. N. *Stereochemistry of Organic Compounds*; Wiley: New York, 1994 (chapter 13).
152. Zhu, H.-J. In *Organic Stereochemistry: Experimental and Computational Methods*; Wiley-VCH: Weinheim, 2015 (chapters 3–6).
153. Mislow, K. *Introduction to Stereochemistry*; Benjamin: New York, 1965; p. 85.
154. Crabbé, P. *ORD and CD in Chemistry and Biochemistry: An Introduction*; Academic Press: New York, 1972.
155. Cotton, A. *Compt. Rend.* **1895**, *120*. 989, 1044.
156. Djerassi, C.; Klyne, W. *Proc. Chem. Soc.* **1957**, 55. London.
157. Klyne, W.; Kirk, D. N. In *Fundamental Aspects and Recent Developments in Optical Rotatory Dispersion and Circular Dichroism*; Ciardelli, F., Salvadori, P., Eds.; Heyden: London, 1973 (chapter 3.1).
158. Djerassi, C.; Klyne, W. *J. Am. Chem. Soc.* **1957**, *79*, 1506.
159. Djerassi, C.; Staunton, J. *J. Am. Chem. Soc.* **1961**, *83*, 736.
160. Djerassi, C.; Finch, N.; Mauli, R. *J. Am. Chem. Soc.* **1959**, *81*, 4997.
161. Moffitt, W.; Woodward, R. B.; Moscowitz, A.; Klyne, W.; Djerassi, C. *J. Am. Chem. Soc.* **1961**, *83*, 4013.
162. Klyne, W.; Kirk, D. N. *Tetrahedron Lett.* **1973**, 1483.
163. Kirk, D. N. *Tetrahedron* **1986**, *42*, 777.
164. Charney, E. *The Molecular Basis of Optical Activity, Optical Rotatory Dispersion and Circular Dichroism*; Wiley: New York, 1979; p. 176.
165. Lightner, D. A.; Gurst, J. E. *Organic Conformational Analysis and Stereochemistry from Circular Dichroism Spectroscopy*; Wiley-VCH: New York, 2000; p. 189.
166. Schellman, J. A. *Chem. Rev.* **1975**, *75*, 323.
167. Scopes, P. M. *Fortschr. Chem. Org. Naturst.* **1975**, *32*, 167.
168. Velluz, L.; Legrand, M.; Grosjean, M. *Optical Circular Dichroism*; Verlag Chemie: Weinheim, 1965.
169. Sears, D.; Beychok, S. In *Physical Principles and Techniques of Protein Chemistry*; Leach, S. J., Ed.; Vol. C; Academic Press: New York, 1973; p. 446.
170. Davydov, A. S. *Theory of Molecular Excitons* (Kasha M.; Oppenheimer Jr. M., translators); McGraw-Hill: New York, 1962.
171. Harada, N.; Nakanishi, K. *J. Am. Chem. Soc.* **1969**, *91*, 3989.
172. Harada, N.; Nakanishi, K. *Circular Dichroic Spectroscopy—Exciton Coupling in Organic Stereochemistry*; University Science Books: Mill Valley, CA, 1983; p. 150.
173. Harada, N.; Ohashi, M.; Nakanishi, K. *J. Am. Chem. Soc.* **1968**, *90*, 7349.
174. Harada, N.; Ochiai, N.; Takada, K.; Uda, H. *J. Chem. Soc. Chem. Commun.* **1977**, 495.
175. Eliel, E. L.; Wilen, S. H.; Mander, L. N. *Stereochemistry of Organic Compounds*; Wiley: New York, 1994; p. 1012.
176. Mason, S. F. *Molecular Optical Activity and the Chiral Discriminations*; Cambridge University Press: Cambridge, 1982; p. 69.
177. Greenfield, N.; Fasman, G. D. *Biochemistry* **1969**, *8*, 4108.

Part II

# Introduction to stereoelectronic effects and stereoselective reactions

Part II contains two chapters (Chapters 4 and 5) which provide an introduction to molecular orbital perturbation theory for the genesis of stereoelectronic effects, and to the general principles of stereospecificity, stereoselectivity and hierarchical designs of asymmetric synthesis, as a background aid to follow the reaction stereochemistry presented in Part III.

Chapter 4

# Perturbation theory and organic reactions

## 4.1 Molecular orbitals (MOs)

A molecular orbital (MO) is described by a wave function $\psi$ which can be expressed as a linear combination of atomic orbitals (LCAOs). Thus, the linear combination of two atomic orbitals $\phi_1$ and $\phi_2$ gives two molecular orbitals $\psi_1$ and $\psi_2$ which are expressed as

$$\psi_1 = c_1\phi_1 + c_2\phi_2$$

$$\psi_2 = c_1\phi_1 - c_2\phi_2$$

where $c_1$ and $c_2$ are the mixing coefficients which denote the relative contributions of the AOs $\phi_1$ and $\phi_2$ to an MO. The coefficients may be positive, negative or even zero.

### 4.1.1 C—H and C—C σ MOs

The interaction of a hybrid orbital (say, $sp^3$) of carbon with the 1s orbital of hydrogen forms a C—H bond. The two-orbital interaction gives the bonding $\sigma_{CH}$ and the antibonding $\sigma^*_{CH}$ as shown in Fig. 4.1A. It is of note that energy of $sp^3_C$ (−16.1 eV) is somewhat lower than that of $1s_H$ (−13.6 eV). A C—C σ bond is formed by the end-on overlap of two hybrid orbitals, one from each carbon (Fig. 4.1B). The bonding σ MO has no node, whereas the antibonding σ*

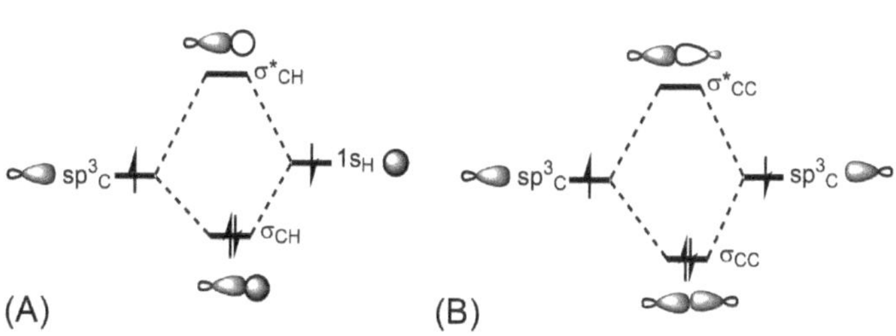


**FIG. 4.1** Bonding and antibonding MOs for (A) C—H bond and (B) C—C bond.

Stereochemistry and Organic Reactions. https://doi.org/10.1016/B978-0-12-824092-2.00004-6

MO has one node. In terms of the frontier orbitals, σ is the highest occupied molecular orbital (HOMO) and σ* is the lowest unoccupied molecular orbital (LUMO).

### 4.1.2 Alkene and carbonyl π MOs

The simplest alkene is ethylene in which two p orbitals are conjugated to each other to form a π bond. Using Hückel theory[1,2], the MO wavefunctions and their energies are as follows.[3]

$$\psi_1 = 0.707\phi_1 + 0.707\phi_2;\ E(\psi_1) = \alpha + \beta$$

$$\psi_2 = 0.707\phi_1 - 0.707\phi_2;\ E(\psi_2) = \alpha - \beta$$

Here the two coefficients have the same size for both $\psi_1$ and $\psi_2$. $\alpha$ is the Coulomb integral which denotes the energy of an electron in an isolated p orbital, and $\beta$ is the resonance integral which represents the energy of interaction between two p orbitals on adjacent atoms. Both $\alpha$ and $\beta$ are negative quantities. Thus, $\psi_1$ having lower energy than the energy of a p orbital ($\alpha$) is the bonding MO (π), whereas $\psi_2$ with energy higher than $\alpha$ is the antibonding MO (π*). The sketches of π (HOMO) and π* (LUMO) are shown below. π has no node while π* has one node.

The symmetry properties or phase relationships of carbonyl π MOs are similar to those of alkenes; however, the energy and coefficient patterns differ. The energy of a p orbital on oxygen (−15.9 eV) is much lower than that on carbon (−10.7 eV). This would lead to lowering of energy of both $\pi_{CO}$ and $\pi^*_{CO}$ compared with those for alkene (Fig. 4.2). The lower energy $p_O$ would contribute more to the lower energy $\pi_{CO}$ and the higher energy $p_C$ would contribute more to the higher energy $\pi^*_{CO}$. As a result, there will be a larger coefficient on

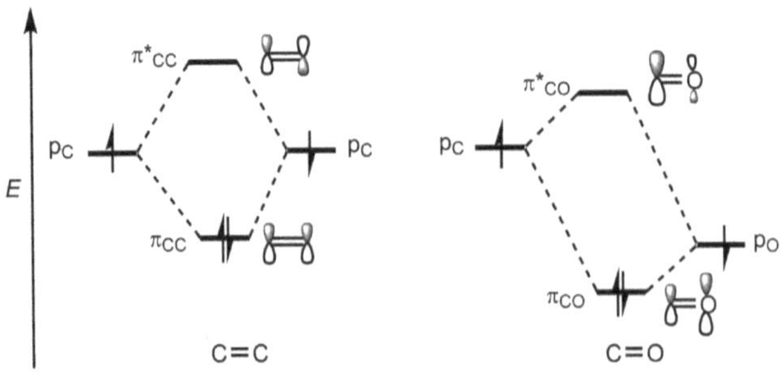


**FIG. 4.2** Energy diagram of π MOs of carbonyl group vis-a-vis alkene π MOs.

oxygen in $\pi_{CO}$ and a larger coefficient on carbon in $\pi^*_{CO}$. $\pi_{CO}$ is HOMO and $\pi^*_{CO}$ is LUMO, the polarization of HOMO and LUMO being opposite.

### 4.1.3 π MOs for allyl system

The allyl system (cation, radical or anion) represents a conjugated system of three p orbitals, and there are three π MOs. The sketches of the MOs and their energies are shown in Fig. 4.3.[3] For $\psi_2$, a node passes through the middle carbon C2. The energy of $\psi_2$ is α, which is the same as the energy of a p orbital and hence $\psi_2$ is a nonbonding MO while $\psi_1$ is bonding and $\psi_3$ is antibonding.

The frontier orbitals of an allyl system depend on whether it is a cation, a radical or an anion. The number of π electrons in allyl cation, radical and anion is 2, 3 and 4, respectively. Therefore, the frontier orbitals of allyl systems are:

| Allyl cation ($\psi_1^2$) | | Allyl radical ($\psi_1^2\psi_2^1$) | Allyl anion ($\psi_1^2\psi_2^2$) | |
|---|---|---|---|---|
| *HOMO* | *LUMO* | *SOMO* | *HOMO* | *LUMO* |
| $\psi_1$ | $\psi_2$ | $\psi_2$ | $\psi_2$ | $\psi_3$ |

Note that the frontier orbital for an allyl radical is $\psi_2$ which is a singly occupied molecular orbital (SOMO).

### 4.1.4 π MOs for conjugated diene

The simplest conjugated diene is butadiene. The four π MOs of butadiene in *s trans* conformation and the energy diagram are shown in Fig. 4.4.[3] $\psi_1$ and $\psi_2$ are bonding MOs, whereas $\psi_3$ and $\psi_4$ are antibonding MOs. The ground state π electron configuration of butadiene is $\psi_1^2\psi_2^2$. Thus, $\psi_2$ is HOMO and $\psi_3$ is LUMO.

The HOMO/LUMO energies in *s-trans* and *s-cis* conformations of butadiene are not the same. In *s-cis* conformation, the HOMO energy is raised and the LUMO energy is lowered relative to those in the *s-trans* conformation

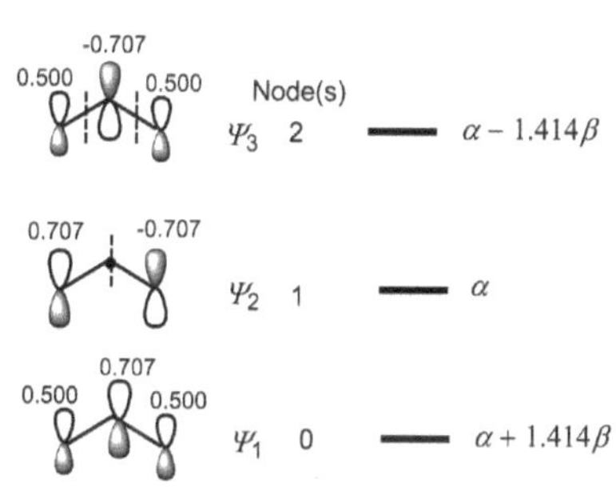


**FIG. 4.3** Sketches and energies of the π MOs of an allyl system.

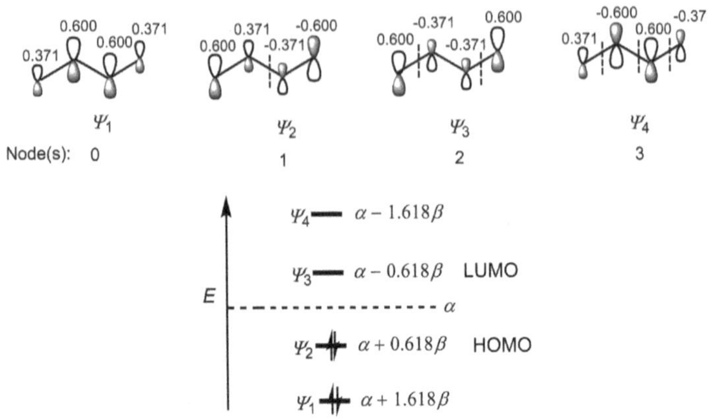


**FIG. 4.4** π MOs and the energy diagram of butadiene.

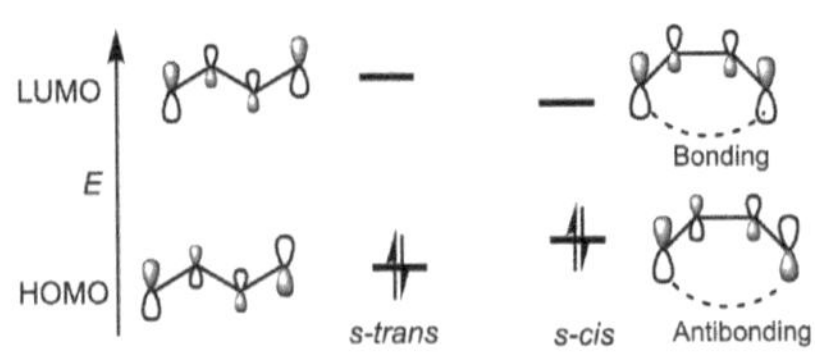


**FIG. 4.5** Relative HOMO/LUMO energies in *s-trans* and *s-cis* conformations of butadiene.

(Fig. 4.5). Unlike the *s-trans*, the *s-cis* conformation has a possible interaction between the two terminal p orbitals. An antibonding interaction raises the HOMO energy while a bonding interaction lowers the LUMO energy in *s-cis* conformation.

The lower HOMO/LUMO energy gap in *s-cis* conformation leads to UV absorption at a longer wavelength (253 nm) for a homoannular diene locked in *s-cis* conformation compared to a $\lambda_{max}$ of 215 nm for an acyclic or a heteroannular diene existing predominantly or exclusively in *s-trans* conformation.

**Problem 4.1** For a linear conjugated system, an MO $\psi_j$ has $(j-1)$ nodes. Using nodal properties, sketch the frontier orbitals of 1,3,5-hexatriene ignoring the size of coefficients.

## 4.2 Perturbation theory

A general approach to rationalizing chemical reactivity can be obtained in perturbation molecular orbital theory. Using this theory, Klopman[4] and Salem[5] derived a general perturbation expression, known as the Salem–Klopman equation.

### 4.2.1 The Salem–Klopman equation

In general, three major forces operate as two reacting molecules approach each other. The Salem–Klopman equation is expressed as a sum of three terms for the three forces.

The energy ($\Delta E$) gained or lost due to interactions of orbitals of one molecule (or part of the molecule) with those of another molecule (or its part) is given by

$$\Delta E = \underbrace{-\sum_{ab}(q_a+q_b)\beta_{ab}S_{ab}}_{\text{(1st term)}} + \underbrace{\sum_{k<l}\frac{Q_kQ_l}{\varepsilon R_{kl}}}_{\text{(2nd term)}} \underbrace{+\sum_{r}^{\text{occ.}}\sum_{s}^{\text{unocc.}} - \sum_{s}^{\text{occ.}}\sum_{r}^{\text{unocc.}} \frac{2\left(\sum_{ab}c_{ra}c_{sb}\beta_{ab}\right)^2}{E_r - E_s}}_{\text{(3rd term)}} \tag{4.1}$$

where $q_a$, $q_b$ = electron populations in atomic orbitals a and b; $\beta$, $S$ = resonance and overlap integrals; $Q_k$, $Q_l$ = total charges on atoms k and l; $\varepsilon$ = local dielectric constant; $R_{kl}$ = distance between the atoms k and l; $c_{ra}$ = coefficient of atomic orbital a in molecular orbital r; $c_{sb}$ = coefficient of atomic orbital b in molecular orbital s; $E_r$, $E_s$ = energy of molecular orbitals r and s.

The 1st term is positive as $\beta$ is a negative energy quantity. Therefore, the interactions represented by the 1st term are energy raising which would lead to destabilization of the interacting system. The repulsive 1st term actually describes the interactions of the occupied (filled) orbitals of one molecule with the occupied (filled) orbitals of the other, and generally represents the steric effect in molecules and in molecular interactions.

The 2nd term of Eq. (4.1) denotes Coulombic repulsion between like charges or attraction between unlike charges, and thereby contributes to the destabilization or stabilization of the system.

The 3rd term of Eq. (4.1) is negative since $E_{occ} < E_{unocc}$ (for the '+' summation, $E_r < E_s$ and for the '−' summation, $E_r > E_s$). This is therefore energy lowering and would lead to stabilization of the interacting system. This term represents the interactions of all occupied (filled) orbitals of one molecule with all unoccupied (unfilled) orbitals of the other, and is a donor–acceptor orbital interaction term. Since the interaction energy of two orbitals is indirectly proportional to their energy difference ($E_r - E_s$), the dominant interaction occurs between orbitals with smallest energy separation. These orbitals are the frontier orbitals (HOMOs and LUMOs). Thus, the orbital interactions in the 3rd term can be approximated to frontier orbital interactions. For the HOMO/LUMO interaction, the frontier orbital expression ($\Delta E_{FMO}$) can be obtained by simplifying this term, and is given by

$$\Delta E_{FMO} = -\frac{2\left(\sum_{ab}c_{HOa}c_{LUb}\beta_{ab}\right)^2}{E_{LU} - E_{HO}} \tag{4.2}$$

Clearly, $\Delta E_{FMO}$ is negative as $E_{LU} > E_{HO}$.

### 4.2.2 Stereoelectronic effects

The donor–acceptor orbital interaction represented by the 3rd term of the Salem–Klopman equation (Eq. 4.1) describes the stereoelectronic effects which lead to the stabilization of the interacting or reacting systems. The interaction is especially important with the frontier orbitals (Eq. 4.2). Three rules can be derived for the stereoelectronic effects.[6]

**Rule 1 The interacting orbitals must have suitable symmetry for a stabilizing interaction.**

To illustrate, consider an $S_N2$ reaction between a nucleophile ($Nu^-$) and an alkyl halide (R—X). The nucleophile can approach from the back side or the front side (Fig. 4.6). The crucial frontier orbital interaction is between the LUMO of the electrophile (alkyl halide) and the HOMO of the nucleophile.[7] The LUMO of an alkyl halide is the $\sigma^*$ orbital of the C—X bond and the HOMO of the nucleophile is a nonbonding orbital (n) containing lone pair electrons (typically an $sp^3$ hybrid orbital). Both orbitals have an axis of symmetry. During a back-side attack (Fig. 4.6A), an overlap between the orbitals occurs along the common axis, which gives a favourable bonding interaction. If the approach is from the front-side (Fig. 4.6B), the two orbitals show different symmetries and the interaction is both bonding and antibonding at the same time and therefore not favoured.

**Rule 2 The more efficient the overlap of the interacting orbitals, the stronger is the resulting interaction and stabilization.**

Eq. (4.2) shows that the frontier orbital term ($\Delta E_{FMO}$) is greater if the numerator $2(\sum_{ab} c_{HOa} c_{LUb} \beta_{ab})^2$ is larger. Qualitatively, this means that an efficient overlap is possible if the orbital coefficients (size of the orbital lobes) are larger. This can be also illustrated with the $S_N2$ reaction shown in Fig. 4.6. The antibonding $\sigma^*$ orbital has a large lobe on carbon directed away from the C—X bond.[8] The back-side approach involving the large lobe of $\sigma^*$ thus provides the most efficient HOMO/LUMO overlap. In contrast, since the density of the $\sigma^*$ orbital is small between the carbon and X, the overlap in the front-side approach becomes still less favourable.

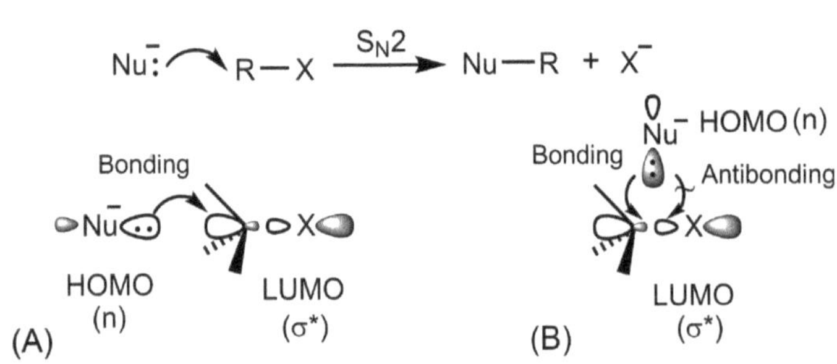


**FIG. 4.6** Frontier orbital interaction in an $S_N2$ reaction between a nucleophile ($Nu^-$) and an alkyl halide (RX): (A) back-side approach and (B) front-side approach.

**Rule 3 The interacting orbitals must be close in energy for a strong stabilizing interaction.**

Eq. (4.2) indicates that $\Delta E_{FMO}$ is greater if the HOMO/LUMO energy difference is smaller. In general, HOMO/LUMO energy separation decreases if HOMO energy is raised and/or LUMO energy is lowered. Consider the relative rates of the $S_N2$ reactions of an alkyl halide with the nucleophiles $Cl^-$ and $I^-$. The HOMO energy of $I^-$ (−8.31 eV) is greater than that of $Cl^-$ (−9.94 eV)[4] as the lone pair on iodine resides in a higher energy 5p orbital. Taking the calculated LUMO energy of an alkyl halide of about −5 eV, the HOMO/LUMO separations for the two reactions are

$$E_{\text{LUMO(alkyl halide)}} - E_{\text{HOMO(chloride)}} = -5 - (-9.94)\,\text{eV} = 4.94\,\text{eV}$$

$$E_{\text{LUMO(alkyl halide)}} - E_{\text{HOMO(iodide)}} = -5 - (-8.31)\,\text{eV} = 3.31\,\text{eV}$$

Therefore, the reaction of the alkyl halide with $I^-$ involving a smaller frontier orbital energy separation will be faster. Experimentally, the reaction rate for $I^-$ is 100 times greater than that for $Cl^-$.

### 4.2.3 Remarks

The frontier orbital approach deals with the reactant frontier orbitals. Thus, this is an initial interaction and this trend is assumed to be continued to the transition structure (TS). The theory is successful because the reactions in general have early TS, that is the TS is reactant-like and not product like. However, the prediction from the frontier orbital analysis depends upon how early the TS is.

Nucleophiles and electrophiles can also be classified as hard or soft in terms of the principle of hard and soft acids and bases.[7,9] Hard nucleophiles are generally negatively charged and have a relatively low-energy HOMO, whereas soft nucleophiles do not necessarily have a negative charge and possess high-energy HOMOs. Likewise, hard electrophiles are generally positively charged and have a high-energy LUMO while soft electrophiles do not necessarily have a positive charge and possess low-energy LUMOs. Other things being equal, hard nucleophiles react preferentially with hard electrophiles, and soft nucleophiles with soft electrophiles.

A soft–soft interaction is orbital controlled with a relatively small HOMO/LUMO energy separation. It represents the donor–acceptor orbital interaction (stereoelectronic effect) represented by the 3rd term of Eq. (4.1) or by Eq. (4.2). In contrast, a hard–hard interaction has a relatively large HOMO/LUMO energy separation and orbital interaction is not important. It refers to the electrostatic effects represented by the 2nd term of Eq. (4.1). The hard–hard interaction is favoured because of a large Coulombic attraction which is not intrinsically directional.

When the difference of reactivity and selectivity is of interest, the 1st term of the Salem–Klopman equation is often ignored. This is because if a molecule is attacked at two possible sites or if there are two possible orientations, the 1st term will be nearly the same for the two cases assuming that the substituents on the reactant sites are of constant size. However, in some cases all three terms of the Salem–Klopman equation play their part to determine the stereoelectronic effects (see Section 4.3.4).

It may be mentioned that there are other factors not involved in the Salem–Klopman equation which affect chemical reactivity. These include strain in the σ framework, solvent effects and the factors affecting entropy of activation.

## 4.3 Ionic reactions[10]

In this section, we will describe briefly the stereoelectronic effects in some fundamental ionic reactions.

### 4.3.1 Nucleophilic substitution at a saturated carbon

In the $S_N2$ reaction, the stereoelectronic factor is the back-side approach of the nucleophile (see Fig. 4.6A). The linear arrangement of the nucleophile, central carbon and leaving group is the essential feature, which results in the inversion of configuration (Fig. 4.7). As the TS is approached, the central carbon undergoes rehybridization and the orbital at the substitution assumes p character. The geometry of the TS is trigonal bipyramidal when the three 'spectator' substituents on the central carbon lie in a plane and there is partial bonding to the incoming nucleophile and the leaving group. In the product, $sp^3$ hybridization is re-established with inversion of configuration. There is smooth transition from the orbital system of reactants to that of the products maintaining maximum overlap throughout. The $S_N2$ reaction is stereospecific since one enantiomer of a chiral substrate gives one enantiomer of a chiral product.

In the $S_N1$ reaction, the intermediate carbocation is planar and both back-side and front-side attacks by the nucleophile are possible leading to inversion and retention of configuration, respectively. The extent of inversion and retention depends on a variety of factors including the stability of carbocation and the reaction conditions. If there are equal amounts of inversion and retention, the result is racemization, that is any one enantiomer of the starting material will give a racemic product.

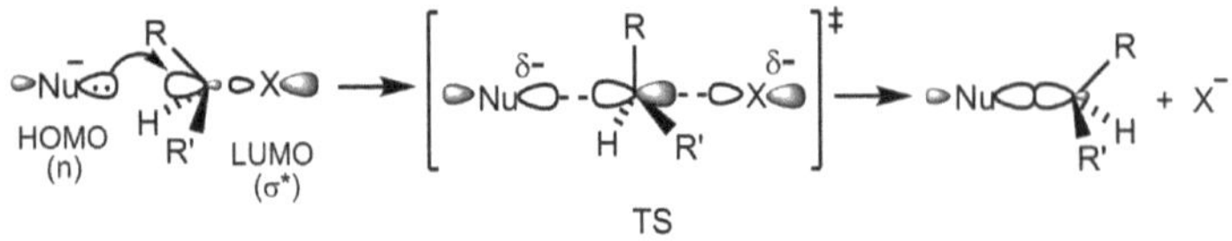


**FIG. 4.7** An $S_N2$ reaction leading to inversion of configuration.

### 4.3.2 Electrophilic substitution at a saturated carbon

In an electrophilic substitution ($S_E2$), the nucleophilic substrate is usually an organometallic reagent such as alkyllithium. The frontier orbitals are the LUMO of the electrophile and the HOMO of the nucleophile (σ orbital of C—M(Li) bond). Here, in contrast to $S_N2$ reactions, the overlaps between the HOMO and the LUMO (modelled by a vacant p orbital) are bonding for both the back-side and the front-side approach (Fig. 4.8).[7] Thus, an $S_E2$ reaction can take place with inversion of configuration in some cases, and retention of configuration in others. However, retention of configuration is the more usual pattern because in this case the electrophilic attack takes place at the site of the highest electron density of the σ orbital of the nucleophile.

### 4.3.3 Elimination reactions

Elimination reactions are typified by the elimination of HL from two adjacent tetrahedral centres to form alkenes (L = leaving group). As regards stereochemistry, the E2 elimination (involving concerted loss of proton and leaving group) is by far the most important. Here the HOMO is the σ orbital of the C—H bond and the LUMO is the σ* orbital of the C—L bond. There are two possible conformations (antiperiplanar and synperiplanar) in which the frontier orbitals are coplanar for efficient π overlap (Fig. 4.9A). The antiperiplanar conformation is more stable than synperiplanar since the former is staggered, whereas the latter is a high-energy eclipsed conformation. The difference in energy between the two conformations is still present in the TSs to some extent. Furthermore, only in the antiperiplanar conformation the orbitals are truly parallel for best overlap (check with models). Hence the preferred conformation for the E2 elimination is antiperiplanar,[11] and anti E2 elimination is favoured since it is faster than syn E2 elimination. Notably, there are three possible staggered conformers for an acyclic system, but only the conformer(s) in which the two departing groups (H and L) are antiperiplanar can participate in E2 elimination. In the cyclohexane system, the antiperiplanar conformation represents the chair diaxial conformation (Fig. 4.9B). Note that between the diaxial (a,a) and diequatorial (e,e)

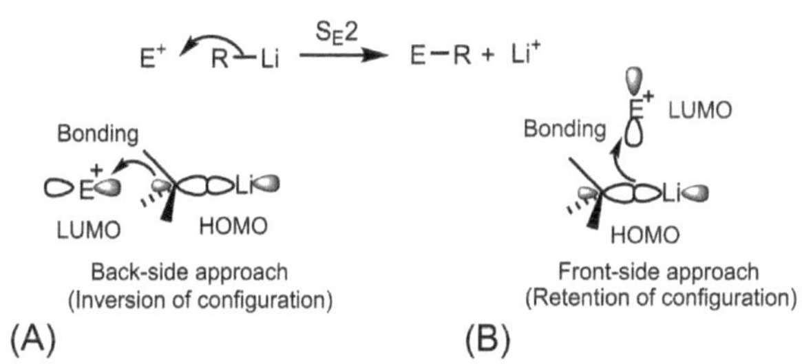


**FIG. 4.8** Frontier orbital interactions in an $S_E2$ reaction between an electrophile ($E^+$) and an alkyllithium: (A) back-side approach and (B) front-side approach.

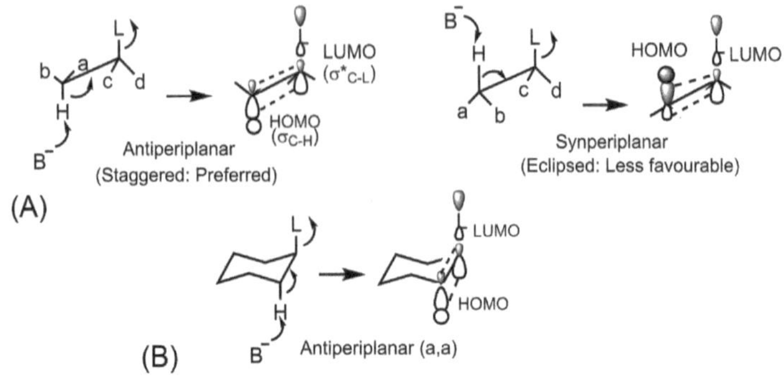


**FIG. 4.9** Stereoelectronic effects in E2 eliminations: (A) acyclic substrate and (B) cyclic substrate.

conformers of a cyclohexane derivative, only the (a,a) conformation provides the antiperiplanarity of the two breaking σ bonds.

Another extreme mechanism for eliminations is a stepwise E1cb mechanism. The first step which is not rate determining involves the loss of a proton to give an intermediate carbanion. The elimination of the leaving group L from the carbanion is rate determining, and the stereoelectronic factor is similar to that of E2 elimination; the carbanion nonbonding orbital is antiperiplanar to the C—L bond.

### 4.3.4 Nucleophilic addition to C=O

Nucleophilic attack to the carbonyl carbon can take place from above or below the plane of the carbonyl double bond. But the angle of attack of the nucleophile is *not* 90°, as might be imagined from the orthogonal approach along the axis of carbon $p_z$ orbital (Fig. 4.10A). Bürgi and Dunitz deduced from X-ray crystal structures that the trajectory of the nucleophile makes an angle close to 107 degrees with the C=O bond vector, which is called the Bürgi–Dunitz angle[12] (Fig. 4.10A). This angle is maintained from the initial attack to the end of the addition.

When the nucleophile on its trajectory is at a large distance (>2.5 Å) from the carbonyl carbon with little bonding developed, the angle of the nucleophile with the C–alkyl bond is less than 90° as expected. As the reaction proceeds, the carbonyl carbon changes its hybridization from $sp^2$ to $sp^3$. Thus in the TS, the angle of the nucleophile with the C–alkyl bond is not less than 90 degrees but an obtuse angle (Fig. 4.10A). The Bürgi–Dunitz angle is a somewhat larger angle.

The Bürgi–Dunitz trajectory of the nucleophilic attack to C=O can be rationalized not only with the frontier orbital term (Eq. 4.2) but the 1st term and the 2nd term of the Salem–Klopman equation (Eq. 4.1) also play their part (Fig. 4.10B). The HOMO is the nonbonding lone pair orbital of the nucleophile

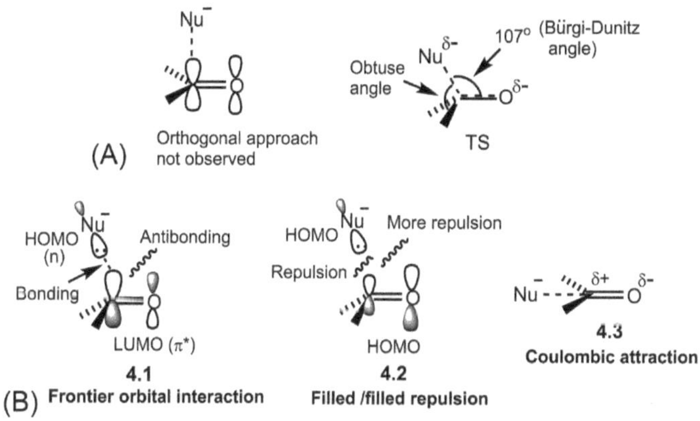


**FIG. 4.10** Nucleophilic attack to C=O: (A) the Bürgi-Dunitz angle; (B) rationalization of the Bürgi–Dunitz trajectory using the Salem–Klopman equation.

and the LUMO is $\pi^*_{C=O}$ as shown in the frontier orbital interactions **4.1**. The interaction of the HOMO with the carbonyl carbon in the LUMO having a larger coefficient is highly efficient but at the same time has an antibonding repulsion from the oxygen atom. There is also repulsive filled orbital/filled orbital interactions **4.2** (1st term in Eq. 4.1) between the HOMO of the nucleophile and the HOMO of the carbonyl group which will push the nucleophile away from the oxygen atom having a larger coefficient or more electron density. (Note that the filled/filled repulsion is not related to whether we draw the lobes of HOMOs facing each other with the same sign or opposite sign.) Furthermore, the Colombic forces (2nd term in Eq. 4.1) will tend to direct the nucleophile to approach from the opposite side of the electronegative oxygen bearing a partial negative charge **4.3**. All these factors act in concert to define the Bürgi–Dunitz trajectory for the nucleophilic attack on C=O. Calculations indicate that the filled/filled repulsion **4.2** is the most important of the three factors.[13]

### 4.3.5 Electrophilic addition to C=C

Electrophilic attack to C=C may lead to an open cation or a bridged cation depending on the nature of the electrophile.[14] The LUMO of an electrophile such as a proton or a cationic carbon is a vacant nonbonding orbital (usually an atomic orbital s or p). For heteroatom electrophiles such as the halogens and peracids, the LUMO is a vacant σ* orbital. The HOMO of the nucleophile (alkene) is the π orbital. Since the π orbital has no node, the electrons from HOMO are likely to move towards the central part of the intermolecular region. Therefore, an efficient HOMO/LUMO overlap requires an orthogonal and inside approach of the electrophile as shown in Fig. 4.11A.[15]

For a nonbridging electrophile like a proton, an open carbocation is finally formed as the hybridization of the attacked carbon changes from $sp^2$ to $sp^3$. For

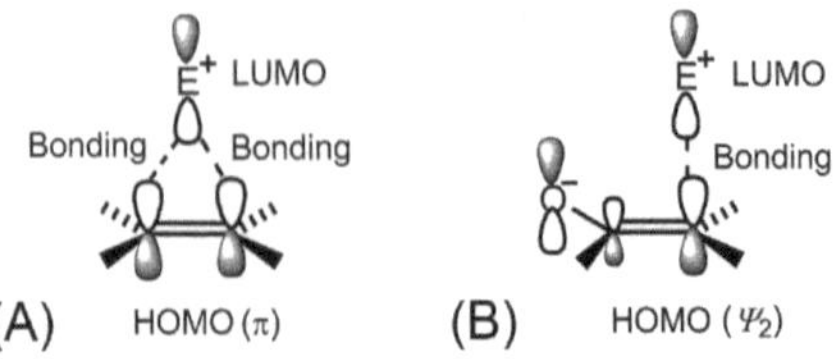


**FIG. 4.11** Frontier orbital interaction for electrophilic addition to (A) C=C and (B) enolate anion.

bridging electrophiles, the electrophilic atom bears a lone pair so that the bridging bonds can lead to a bridged intermediate or product.

Unlike in the case of a simple C=C, electrophilic attack on the enolate double bond involves a more or less perpendicular approach to the terminal carbon (Fig. 4.11B).[16] The HOMO of the enolate resembles the HOMO of an allyl anion ($\psi_2$) (see Fig. 4.3) but with a small orbital lobe at the middle carbon. An efficient HOMO/LUMO overlap involves the terminal carbon of the double bond with a larger coefficient. A more detailed analysis predicts that the electrophile should approach the enolate along a trajectory somewhat displaced from the 'perpendicular', due to the antibonding interaction between the orbital on the electrophile and the p orbital on oxygen in Fig. 4.11B.[17]

### 4.3.6 Fragmentation reactions

Fragmentation reactions take place when a strong electron donor (X) interacts with a good leaving group (L) three carbons away (Fig. 4.12A). The reaction, called Grob fragmentation, is controlled by stereochemistry.[18] The fragmenting C—C bond needs to be antiperiplanar to the leaving group for an efficient $HOMO_{\sigma(C-C)}/LUMO_{\sigma^*(C-L)}$ overlap (cf. E2 elimination) as shown in Fig. 4.12B. The required antiperiplanar bonds are highlighted. The direction of the lone pair orbital on the donor atom is not usually fixed but for a good

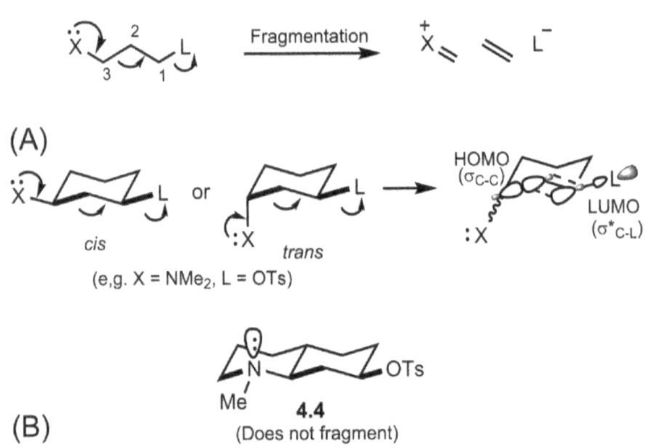


**FIG. 4.12** Stereoelectronic effects in fragmentation reactions.

overlap, it must get antiperiplanar to the breaking C—C bond. If that could not happen as in **4.4**, fragmentation does not occur. Note that in **4.4**, a ring C—N bond (in bold) is antiperiplanar, not the nitrogen lone pair.

For rearrangement reactions involving 1,2-shifts, see Section 4.4.7, Fig. 4.24A.

### 4.3.7 Intramolecular reactions: Baldwin's rules

Intramolecular reactions or cyclizations are in general more facile than intermolecular ones because of a favourable entropy factor (the reactive groups are held in close proximity). But all cyclizations are not successful even though on paper they appear to be fine. Baldwin identified the problem and produced a set of rules as to which cyclizations are favoured and which are disfavoured.[19,20] These are actually not rules which have to be obeyed but are guidelines based on experimental observations and stereochemical reasonings. Exceptions to Baldwin's rules are known; most well-defined exceptions include pericyclic reactions such as electrocyclic ring closure.

In Baldwin's rules, the reactions are classified by three parameters: the ring size ($n$), the hybridization of the atom attacked to form a new bond (*tet* for $sp^3$, *trig* for $sp^2$, and *dig* for sp) and the position of the breaking bond with respect to the new ring (*exo* if the breaking bond is outside the new ring, and *endo* if inside) (Fig. 4.13). When the nucleophile is a double bond, the process can be simultaneously specified at both electrophilic and nucleophilic carbons **4.5**.

The Baldwin's rules are summarized in Table 4.1.

For *exo-tet* cyclizations, there are no stereoelectronic problems as the nucleophile lone pair can overlap easily with the $\sigma^*_{C-Y}$ irrespective of ring size. Similarly, *exo-trig* cyclizations are successful as there are no geometric constraints for the overlap between lone pair orbital and $\pi^*_{C=Y}$.

The difficulty may arise with *endo-trig* cyclizations. For example, the cyclization of the enolate **4.6** proceeds easily in high yields by 5-*exo-trig* C-alkylation, whereas the enolate **4.7** does not undergo 5-*endo-trig* C-alkylation, instead 5-*exo-tet* O-alkylation takes place (Fig. 4.14A).[21] According to Baldwin's rules, 5-*endo-trig* reaction (see Table 4.1) is disfavoured. This is because of the geometric constraints that make 5-*endo-trig* reaction difficult. The

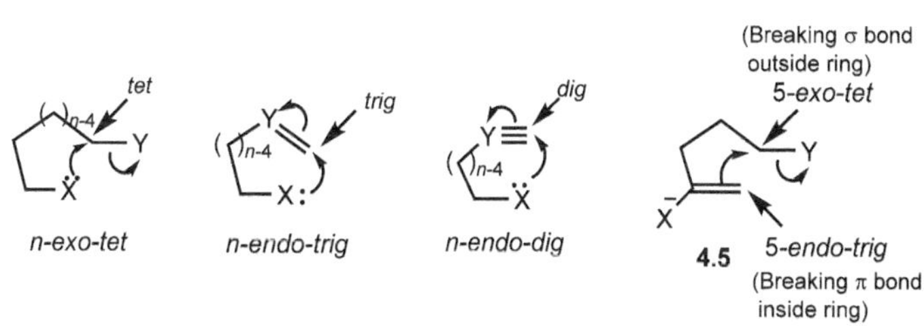

4.5

**FIG. 4.13** Classification of reactions for Baldwin's rules.

**TABLE 4.1 Baldwin's rules.**

| Reaction ($n=3$–7) | Favoured | Disfavoured |
|---|---|---|
| *n-exo-tet* | All | – |
| *n-endo-tet* | – | All |
| *n-exo-trig* | All | – |
| *n-endo-trig* | $n\geq6$ | $n<6$ |
| *n-exo-dig* | $n\geq5$ | $n<5$ |
| *n-ensdo-dig* | All | – |

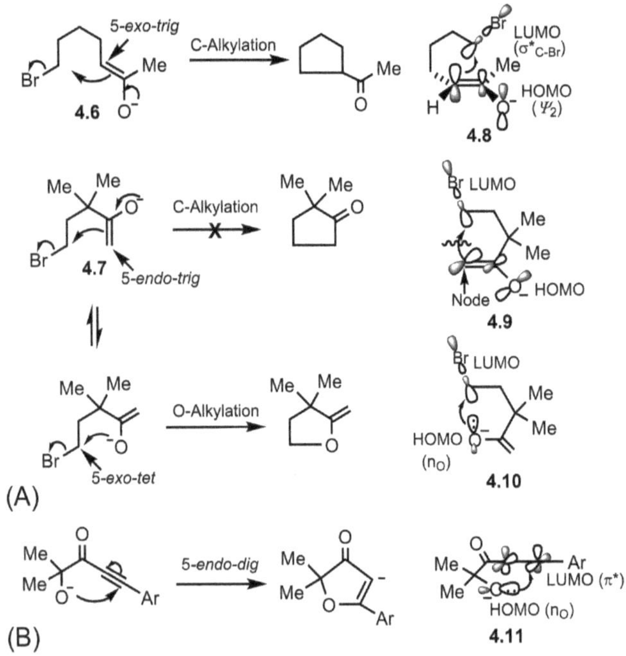


**FIG. 4.14** (A) Cyclization of enolates, Baldwin's rules and frontier orbital interactions; (B) a 5-*endo-dig* process and its frontier orbital picture.

stereoelectronic requirement for enolate alkylation is the orthogonal approach of the electrophile LUMO from above (or below) the π plane of the enolate HOMO (see Fig. 4.11B). In the case of 5-*exo trig* process, the σ* orbital of the C—Br bond as the LUMO can easily approach from above the plane of the π system because of a relatively long three-carbon alkyl chain joining the electrophilic and nucleophilic carbons **4.8**. In contrast, for the 5-*endo-trig* reaction, such attack of $\sigma^*_{C\text{-}Br}$ orbital from one face of the π plane of the enolate is impossible due to a shorter two-carbon alkyl chain; the σ* orbital can only approach to the node of the π system **4.9**. (If the alkyl chain is made longer even by one carbon, the process will be favourable as a 6-*endo-trig*.) In an alternative pathway, the enolate **4.7** undergoes O-alkylation by a favoured 5-*exo-tet* process when the lone pair on the enolate oxygen can easily approach to the $\sigma^*_{C—Br}$ orbital **4.10**.

In contrast to the 5-*endo-trig*, the 5-*endo-dig* process is favoured (see Table 4.1). Fig. 4.14B illustrates a 5-*endo-dig* cyclization. Apparently, the linear alkyne chain seems to place the nucleophilic oxygen lone pair far away from the electrophilic alkyne carbon. However, the alkyne has two orthogonal π* orbitals, one of which can lie in the plane of the new ring. This π* orbital as the LUMO can easily approach to the lone pair on oxygen (HOMO) in the 5-*endo-dig* cyclization **4.11**.

**Problem 4.2** Using Baldwin's rules, predict whether the following cyclizations are favoured or not.

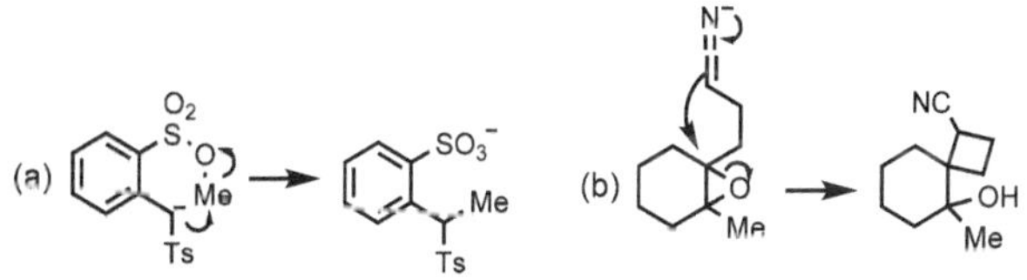


## 4.4 Pericyclic reactions[22,23]

### 4.4.1 Diels–Alder cycloadditions

The Diels–Alder reaction is a concerted cycloaddition of a diene ($_\pi4$) with an alkene or alkyne dienophile ($_\pi2$). The most favourable geometry of the reaction involves a parallel approach of the diene and the dienophile which permits easy end-on overlap for the developing σ bonds at the termini. In this approach, both diene and dienophile components are suprafacial (the new bonds develop on the same face of each π component), that is the Diels–Alder reaction is a $[_\pi4_s + _\pi2_s]$ cycloaddition.

#### *4.4.1.1 Frontier orbital analysis*

The frontier orbital interactions for the prototype Diels–Alder reaction is shown in Fig. 4.15A. The interactions at both ends are bonding for either $HOMO_{diene}$/$LUMO_{dienophile}$ or $LUMO_{diene}$/$HOMO_{dienophile}$ interaction, and the reaction is favoured. The Diels–Alder reaction becomes facile if appropriate substituents are put on the diene and the dienophile. The normal Diels–Alder cycloaddition involves electron-deficient (acceptor) dienophiles which interact with

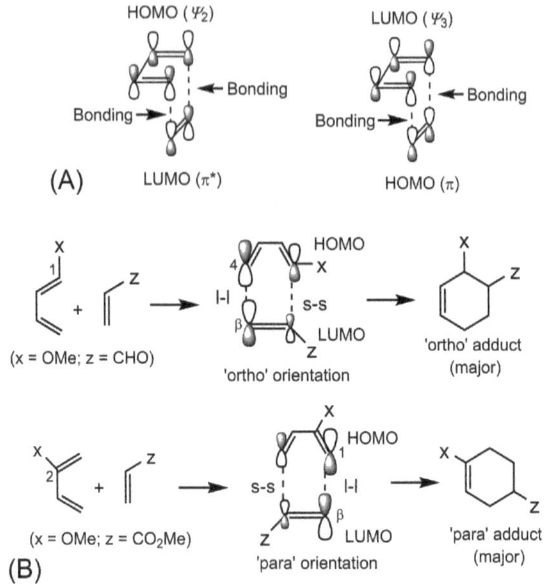


**FIG. 4.15** (A) Frontier orbital interactions for the prototype Diels–Alder reaction; (B) Frontier orbital pictures of regioselectivity. x = electron donating substituent, z = electron withdrawing substituent; l–l = large-large, s–s = small-small.

electron-rich (donor) dienes bearing conjugating (c-type) or electron donating (x-type) substituents. A c- or x-type substituent on a diene raises its HOMO energy, whereas an electron withdrawing (z-type) substituent on a dienophile lowers its LUMO energy. Thus, the HOMO/LUMO energy separation for the $HOMO_{diene}/LUMO_{dienophile}$ pair is smaller, and the normal Diels–Alder reactions are $HOMO_{diene}/LUMO_{dienophile}$ controlled. If the electron demands of the diene and the dienophile are reversed using an electron-deficient diene and electron-rich dienophile, the reaction is called inverse electron demand Diels–Alder (IEDDA) reaction, in which case the dominant interaction is $LUMO_{diene}/HOMO_{dienophile}$ interaction.

The regioselectivity of the Diels–Alder reaction between an unsymmetrical diene and an unsymmetrical dienophile generally follows the 'ortho/para' rule: 1-substituted dienes give preferentially the 'ortho' isomer while 2-substituted dienes give the 'para' isomer. The regioselectivity arises from the preferred coefficient interactions: large–large + small–small, as illustrated in Fig. 4.15B. Note the effect of substituents on the coefficient pattern: an electron withdrawing z substituent leads to a larger coefficient at the β-carbon in the dienophile LUMO while an electron donating x substituent gives a larger coefficient at C4 (1-substituted diene) or C1 (2-substituted diene) in the diene HOMO.[24,25] The large–large interaction provides the *leading bond* in the TS which is formed to a greater extent than the bond formed by small–small interaction. The reaction is therefore asynchronous.

**Problem 4.3** Using frontier orbital interactions, predict the regioselectivity of the following Diels-Alder reaction:

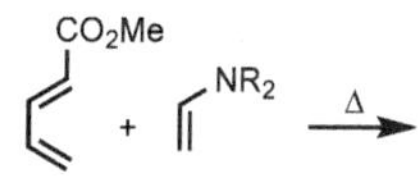


### *4.4.1.2 Diradicaloid model*

Femotosecond real-time studies of a retro Diels–Alder reaction of norbornene into cyclopentadiene and ethylene reveal that both concerted and nonconcerted trajectories are possible.[26] Since most Diels–Alder reactions are asynchronous, the asynchronous TS can be represented as a combination of two limiting structures: a diradical and its corresponding zwitterion. This is called the diradicaloid model.[27–30] The 'ortho/para' regioselectivity of the Diels–Alder reaction can be explained by the diradicaloid model, as shown in Fig. 4.16.[31] For the 1-substituted diene, both diradical and zwitterionic structures of the TS are stabilized by the substituent effects in the 'ortho' than in the 'meta' orientation, whereas the TS for the 'para' orientation is more stabilized than the 'meta' for the 2-substituted diene.

## 4.4.2 1,3-Dipolar cycloadditions

Like Diels–Alder, a 1,3-dipolar cycloaddition reaction is also a $[_{\pi}4_s + _{\pi}2_s]$ cyclo addition. The two components are a 1,3-dipole ($_{\pi}4$) and an alkene or alkyne

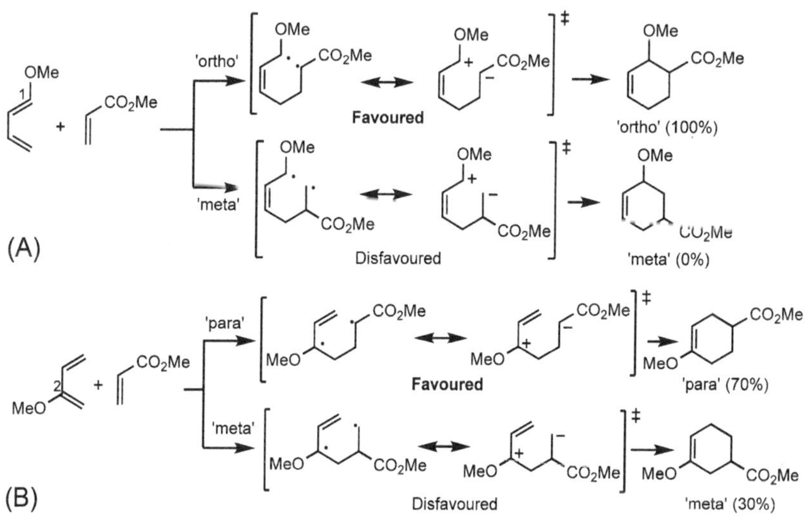


**FIG. 4.16** Diradicaloid TS model for regioselectivity of the Diels-Alder reaction using (A) 1-substituted diene and (B) 2-substituted diene.

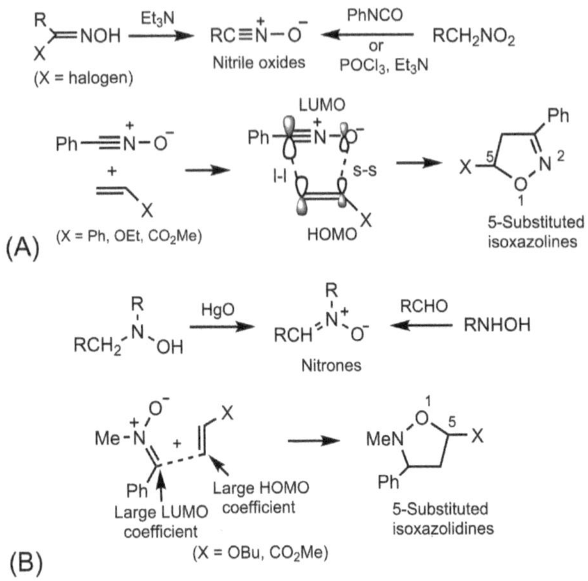


**FIG. 4.17** Regioselectivity of (A) nitrile oxide cycloadditions and (B) nitrone cycloadditions. Methods of generation of nitrile oxides and nitrones are also shown in (A) and (B), respectively.

dipolarophile ($_{\pi}2$). The regioselectivity of most 1,3-dipolar cycloadditions can be rationalized using frontier orbital theory, as illustrated with nitrile oxide cycloadditions (Fig. 4.17A) and nitrone cycloadditions (Fig. 4.17B). Nitrile oxides can be generated in situ by dehydrohalogenation of α-halooximes or by dehydration of primary nitroalkanes, whereas nitrones can be prepared by oxidation of N,N-disubstituted hydroxylamines or by condensation of *N*-alkylhydroxylamines with aldehydes or ketones. Both nitrile oxides and nitrones have low energy LUMOs[32,33], and their cycloadditions are often $LUMO_{dipole}/HOMO_{dipolarophile}$ controlled. The reactions are regioselective and give predominantly the 5-substituted products arising from the preferred large–large + small–small interactions. Note the coefficient pattern in the dipole LUMO and the dipolarophile HOMO.[32,33] If the dipolarophile is highly electron-deficient in nitrone cycloaddition, the regioselectivity is controlled by the $HOMO_{dipole}/LUMO_{dipolarophile}$ interaction and is reversed giving predominantly a 4-substituted isoxazolidine.

**Problem 4.4** Account for the regioselectivity in the following nitrone cycloaddition:

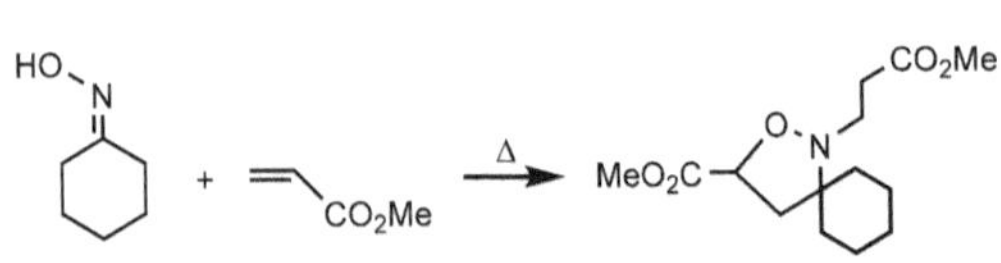

### 4.4.3 Ketene cycloadditions

A ketene cycloaddition is a [2+2] cycloaddition of a ketene with an alkene (ketenophile). The [2+2] cycloaddition is symmetry forbidden in the geometrically favourable supra/supra pathway but is possible by the supra/antara mode when the ketene and the alkene approach in orthogonal fashion (Fig. 4.18A). Here the alkene acts as a $_{\pi}2_s$ component and the ketene as a $_{\pi}2_a$ component (the new bonds develop on the opposite faces of the π component). The favourable frontier orbital interaction is $LUMO_{ketene}/HOMO_{alkene}$ interaction. The crossed TS is further stabilized by bonding secondary orbital interactions (wavy lines) between the lobes of the alkene HOMO and the orthogonal p orbital of C=O (π*) in the ketene LUMO.[34]

Ketene cycloadditions are also regioselective (Fig. 4.18B).[35] The regioselectivity arises from the leading bond between a large LUMO coefficient on the carbonyl carbon of ketene and a large HOMO coefficient on the alkene β-carbon with respect to a stronger electron donating OPr group.

**Problem 4.5** Predict the regioselectivity of the following ketene cycloaddition:

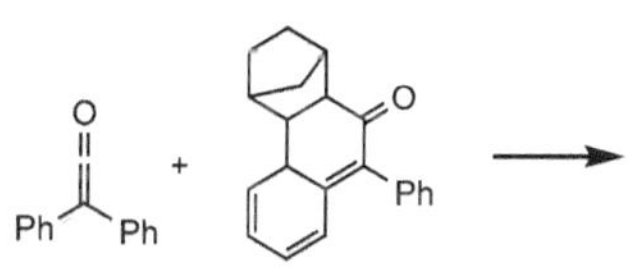


### 4.4.4 Carbene cycloadditions

Carbene cycloadditions lead to cyclopropanation of alkenes. Besides simple singlet carbenes such as dichlorocarbene ($:CCl_2$), carbenes are often associated with a metal, called carbenoid such as Simmons–Smith reagent ($I\text{-}CH_2\text{-}ZnI$).

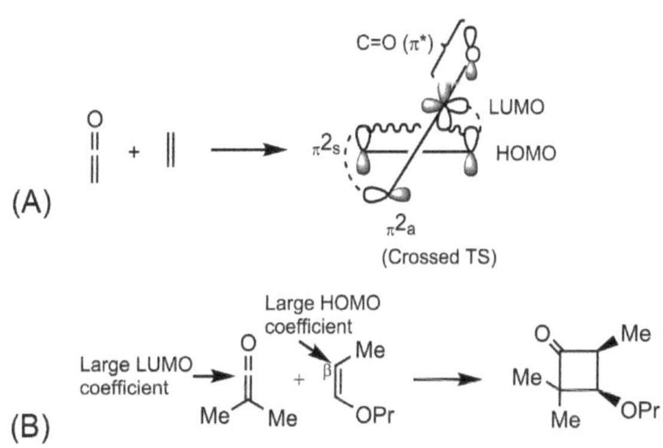


**FIG. 4.18** (A) FMO interactions for ketene cycloadditions (wavy lines indicate secondary orbital interactions); (B) regioselectivity of a ketene cycloaddition.

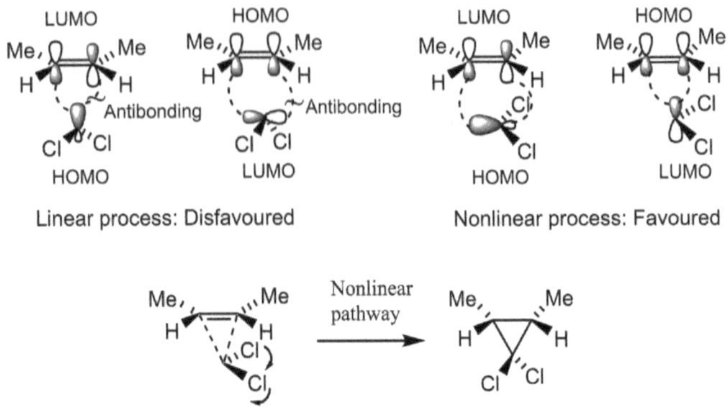


**FIG. 4.19** FMO interactions for carbene cycloadditions, and the nonlinear pathway leading to the product.

In carbene cycloadditions with alkenes, carbenes acts as a two-electron one-atom component ($_{\omega}2$), and such cycloadditions are called cheletropic reactions.[34] In the linear cheletropic process, the line of approach of carbene is the same as the axis of the lone pair, whereas in the nonlinear process, the line of approach is orthogonal to the axis of the lone pair (Fig. 4.19). The alkene component is suprafacial. The HOMO of the carbene is a filled $sp^2$ orbital and the LUMO is a vacant p orbital. The frontier orbital interactions show that the linear process is disfavoured but the nonlinear process is favoured. As the carbene carbon becomes nearly tetrahedral in the cyclopropane product, the Cl substituents in the nonlinear pathway would have to swing round 90 degrees as the product is reached.

### 4.4.5 Ene reactions

In the Alder ene reaction, an ene component (comprising a π bond and a σ bond) replaces the diene component in the Diels–Alder reaction. The ene component is usually an alkene bearing an allylic hydrogen, and the enophile is a π system similar to the dienophile. The ene reaction is a group transfer reaction which involves transfer of an allylic hydrogen atom to an enophile and gives a single product. Ene reactions are possible with heteroenophile or heteroene components. Such reactions include carbonyl ene reaction with the carbonyl group as enophile, and Conia reaction and metallo-ene reactions with enol and allylic metal reagent as ene component, respectively.

The frontier orbital picture of the ene reaction as $HOMO_{ene}/LUMO_{enophile}$ interaction is shown in Fig. 4.20A.[36] The HOMO of the ene component originates from the linear combination of σ (C—H) and π (C=C) and has one node similar to diene HOMO. An effective head-on orbital overlap for the C—H σ bond transfer gives an envelope TS, though the cyclic TS includes six atoms (five carbons and one hydrogen). The head-on overlap requires the three-atom chain (C···H···C) to lie on a single line of the envelope ring.

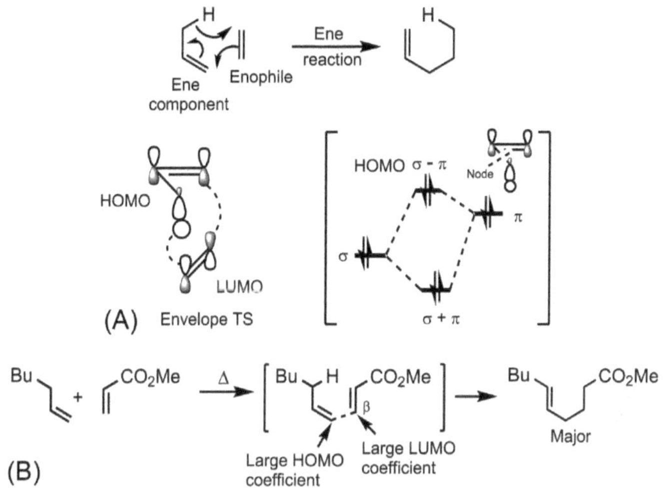


**FIG. 4.20** (A) Ene reaction and its frontier orbital picture (HOMO of the ene component generated by a linear combination is also shown); (B) regioselectivity of the ene reaction.

The regioselectivity of the ene reaction arises from the large–large coefficient interaction (leading bond), as illustrated in Fig. 4.20B.[37]

**Problem 4.6** Predict the regioselectivity of the following ene reaction:

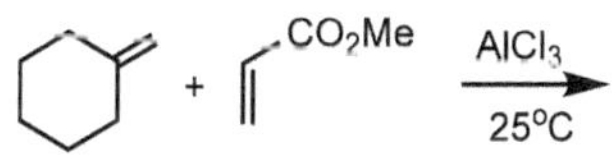


### 4.4.5.1 *Retro-ene reactions*

In retro-ene reactions, a π bond is formed at the expense of a σ bond (Fig. 4.21). The reaction is partly driven by gain in entropy (two species formed from one species). The entropic contribution becomes more significant at higher temperature ($\Delta G^0 = \Delta H^0 - T\Delta S^0$), as in pyrolytic eliminations. The substrates in retro-ene reactions can also be 1,2-dipoles when two σ bonds and a lone pair are

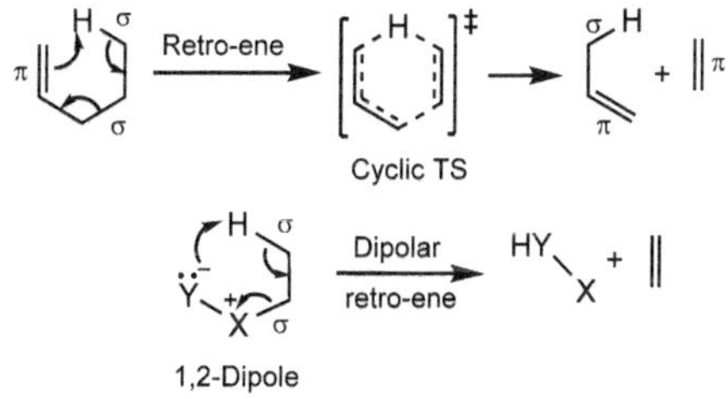


**FIG. 4.21** Retro-ene and 1,2-dipolar retro-ene reactions.

involved in the 6-electron cyclic TS (Fig. 4.21). These 1,2-dipolar retro-ene reactions are favoured by entropy, charge neutralization and breaking of a weak σ bond (C—X$^+$), and take place at a much lower temperature.

### 4.4.6 Electrocyclic reactions

An electrocyclic reaction denotes a ring closing process in which a new σ bond is formed between the ends of a conjugated π system, or the reverse ring opening process. For an electrocyclic ring closing, the two parallel p orbitals at the ends of a conjugated π system must be rotated to allow their overlap in the σ sense (end-on fashion). On the other hand, electrocyclic ring opening requires rotation of the orbital lobes of the breaking σ bond to give two parallel terminal p orbitals of the resulting π system. The rotations can occur in the same direction called conrotatory (in short, con) motion, or in opposite directions called disrotatory (in short, dis) motion. Electrocyclic reactions can be performed under thermal or photochemical condition, and the relevant selection rules for electrocyclic processes are given below.[38] Note that the rules for photochemical reactions are opposite to those for thermal reactions.

| Total no. of electrons | Thermal | Photochemical |
|---|---|---|
| $4n$ | Con | Dis |
| $4n+2$ | Dis | Con |

The frontier orbital interactions for the thermally allowed 4-electron ring opening of cyclobutene and photochemically allowed 4-electron ring closing of butadiene are shown in Fig. 4.22A. The former involves HOMO/LUMO interactions while the latter involves a singly occupied excited state LUMO

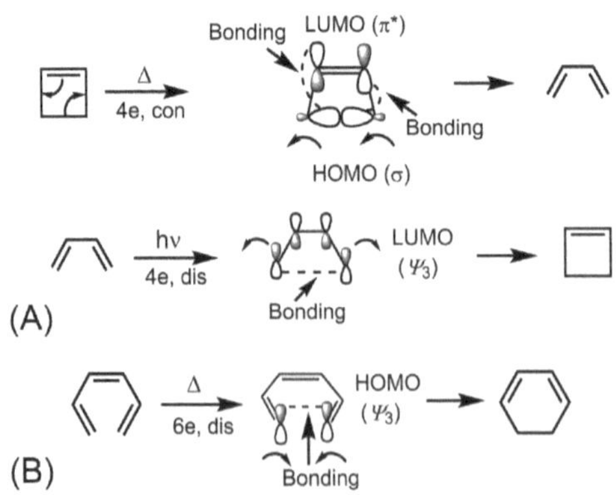


FIG. 4.22 FMO interactions: (A) thermal ring opening of cyclobutene and photochemical ring closing of butadiene and (B) thermal ring closing of hexatriene.

(see Section 4.6) for a single $_{\pi}4$ (butadiene) component. Fig. 4.22B shows the frontier orbital picture for the thermally allowed 6-electron ring closing of 1,3,5-hexatriene using HOMO of hexatriene.

### 4.4.7 Sigmatropic rearrangements

A sigmatropic rearrangement involves the migration of a σ bond (adjacent to one or more π components) from one position to another in a molecule with concomitant reorganization of the π component(s) and is designated by its order [*i*,*j*]. The migrating σ bond is labelled as bond 1–1 and the numbering is continued through π component(s) to each end of the new σ bond. The numbers (*i*,*j*) at the two ends of the new σ bond gives the order. In sigmatropic rearrangements, if the migrating group is hydrogen, the only stereochemical consideration is whether migration of hydrogen occurs on the same face of the π component, called the suprafacial shift or on the opposite face of the π component, termed the antarafacial shift. However, for the carbon shift, an additional stereochemical consideration is needed as to the relative change of configuration (retention or inversion) of the migrating stereocentre. The FMO analysis of sigmatropic shifts can be performed in terms of either the HOMO/LUMO interaction or the SOMO/SOMO interaction. In the latter, the molecule is divided *artificially* into two interacting radical components. The FMO interactions for the 6-electron [1,5] suprafacial H shift as a SOMO/SOMO interaction and the 8-electron [1,7] antarafacial H shift as a HOMO/LUMO interaction are shown in Fig. 4.23A. Fig. 4.23B shows the FMO picture for the 4-electron suprafacial [1,3]C shift with inversion of configuration.

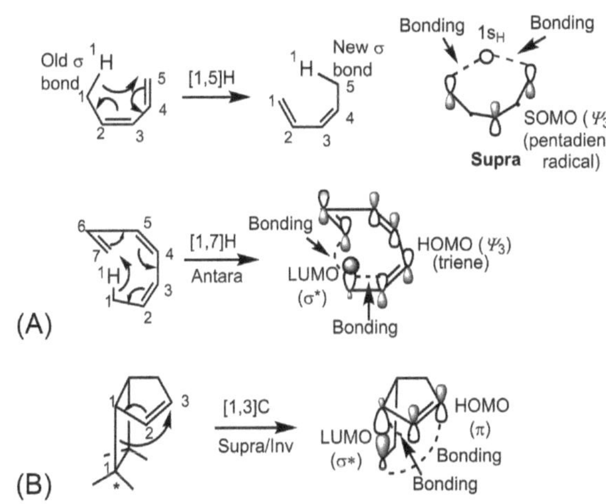


**FIG. 4.23** FMO interactions in sigmatropic rearrangements: (A) [1, 5] Suprafacial H shift and [1, 7] antarafacial H shift; (B) Suprafacial [1,3]C shift with inversion.

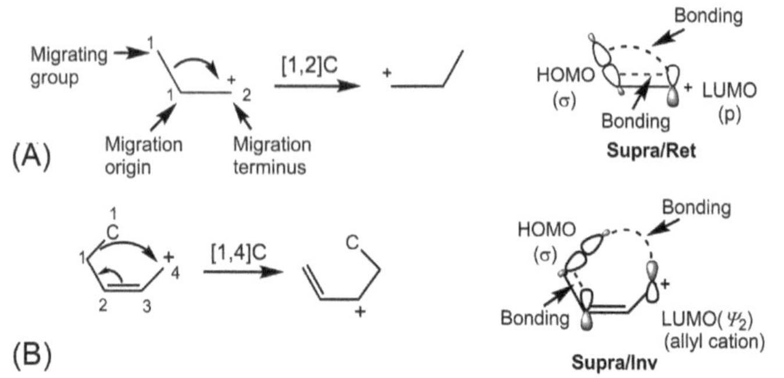


**FIG. 4.24** FMO interactions: (A) suprafacial [1,2]C shift with retention and (B) suprafacial [1,4]C shift with inversion.

The [1,2]C shifts in *cations* are most common rearrangements. A schematic representation of the [1,2]C shift, and its frontier orbital description for the supra/retention pathway are shown in Fig. 4.24A. Like the [1,3]C shift, the 4-electron [1,4]C shift in allyl cations also proceed via the supra/inversion pathway (Fig. 4.24B).

## 4.4.8 Cope/Claisen rearrangements

The Cope rearrangement is a 6-electron [3,3] sigmatropic rearrangement in which a C—C (1–1) σ bond migrates and a new C—C (3–3) σ bond is formed with concomitant reorganization of the π bonds (Fig. 4.25A). Replacing the migrating C—C bond by a C—O bond gives the Claisen rearrangement. The frontier orbital analysis of the Cope rearrangement in terms of two artificial allyl units is shown in Fig. 4.25B. The SOMO ($\psi_2$)/SOMO ($\psi_2$) interactions in both chair TS and boat TS are favourable. In each case, the two allyl components are suprafacial.

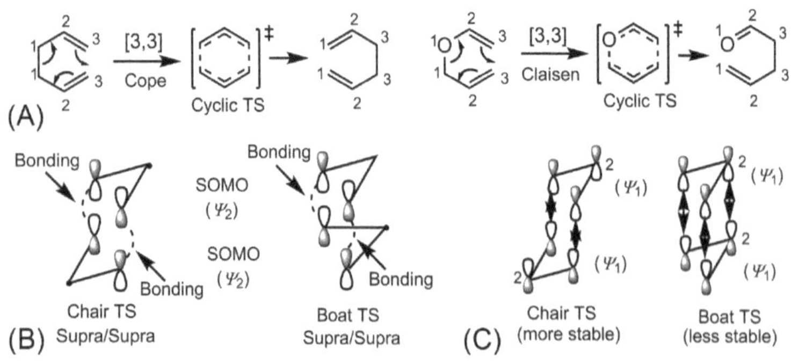


**FIG. 4.25** (A) [3,3] Cope and Claisen rearrangements; (B) FMO analysis for the Cope rearrangement involving a chair TS and a boat TS and (C) relative stability of chair and boat TSs (the repulsive filled/filled interactions are shown by double-headed arrows).

When both chair and boat TSs in the Cope or the Claisen rearrangement are geometrically accessible, the reaction preferentially takes the chair route because the chair TS is more stable than the boat TS. This is primarily attributed to the repulsive interaction of the filled orbital ($\psi_1$) of one allyl component with the filled orbital ($\psi_1$) of the other (the first term of the Salem–Klopman equation), which will destabilize the boat TS more than the chair due to an additional repulsion between the two p orbitals on C2 carbons in the boat TS (Fig. 4.25C).

### 4.4.9 Wittig rearrangements

A Wittig rearrangement is a 6-electron [2,3] sigmatropic rearrangement of carbanion of allyl ether (alloxycarbanion). The frontier orbital picture shows an allowed SOMO/SOMO interaction between the allyl unit and the O—C fragment under supra/supra mode (Fig. 4.26). The geometry of the five-centre TS has an envelope conformation.

## 4.5 Radical reactions[39,40]

Radicals are generally neutral species and soft. They react mostly with uncharged molecules. Thus, the Coulombic forces in their interactions are usually small. Frontier orbital interactions seem to play a dominant role in radical reactions.

### 4.5.1 Electrophilic and nucleophilic radicals: Orbital interaction diagrams

In contrast to electrophiles and nucleophiles in ionic reactions, radicals can be stabilized by both electron donating (x-type) and electron withdrawing (z-type) substituents. A radical centre adjacent to an x-substituent (e.g. R, OR) is nucleophilic whereas a radical centre adjacent to a z-substituent (e.g. COR, CN) is electrophilic. The frontier orbital interaction between the SOMO (p) of a radical and the HOMO (lone pair orbital) of an x-substituent is stabilizing, and creates a new higher energy SOMO for the nucleophilic radical (Fig. 4.27A). As two electrons are in a lower energy orbital and only one electron has gone up in energy, the net result is stabilizing. On the other hand, an interaction between

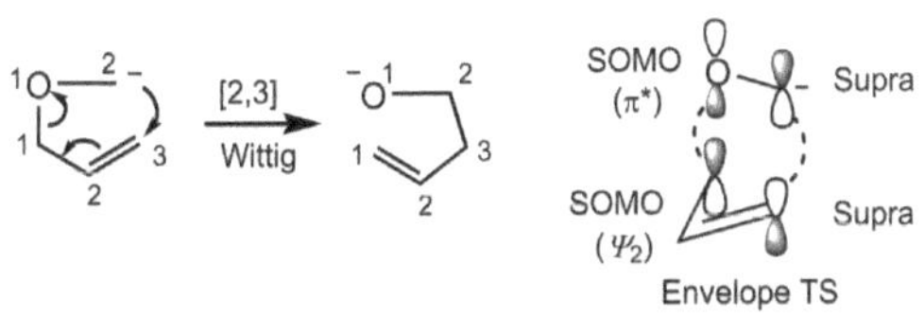


FIG. 4.26 FMO analysis for the Wittig rearrangement in the envelope TS.

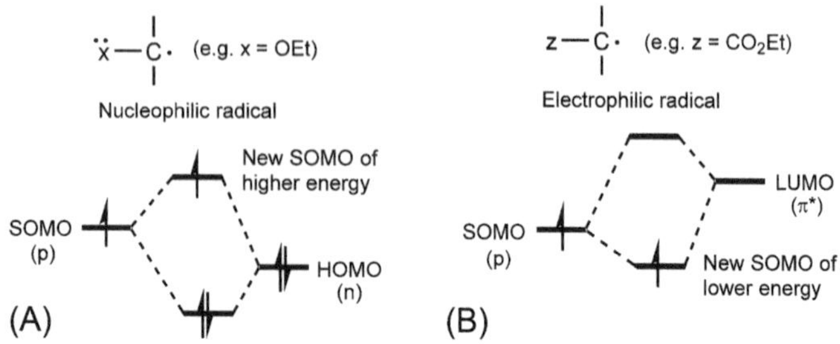


**FIG. 4.27** Frontier orbital interaction diagram: (A) nucleophilic radical and (B) electrophilic radical. $x$ = electron donating substituent; $z$ = electron withdrawing substituent.

the SOMO (p) and the LUMO (π*) of a z-substituent is also stabilizing, and creates a new lower energy SOMO for the electrophilic radical (Fig. 4.27B). (Here the frontier orbital interaction refers to the ground state stability of the radical, not its reactivity.)

In general, a nucleophilc radical has a high-energy SOMO and an electrophilic radical has a low-energy SOMO.

## 4.5.2 Reactivity and regioselectivity in radical reactions: Frontier orbital effects

As for ionic and pericyclic reactions, the reactivity of radicals is also primarily decided by the frontier orbital energy separations. Nucleophilic radicals (high-energy SOMO) will react preferentially with electron-deficient alkenes (low energy LUMO) because the SOMO/LUMO separation is smaller than the SOMO/HOMO separation (Fig. 4.28A). Similarly, electrophilic radicals (low-energy SOMO) will react preferably with electron-rich alkenes (high energy HOMO) since the SOMO/HOMO separation is smaller than the SOMO/LUMO separation (Fig. 4.28B).

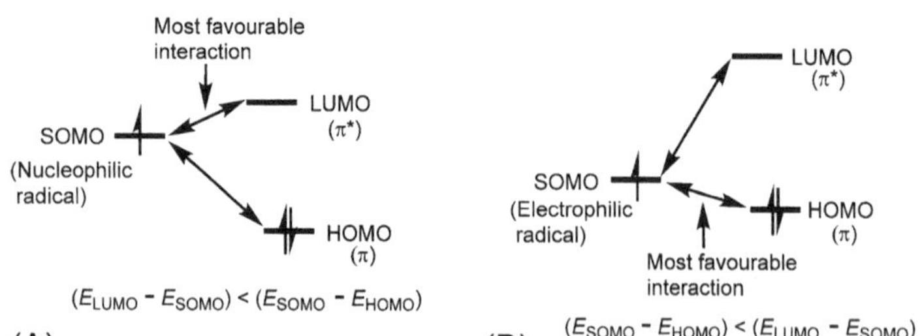


**FIG. 4.28** Frontier orbital effects: (A) reactivity of a nucleophilic radical with an electron-deficient π system (low-energy LUMO) and (B) reactivity of an electrophilic radical with an electron-rich π system (high-energy HOMO).

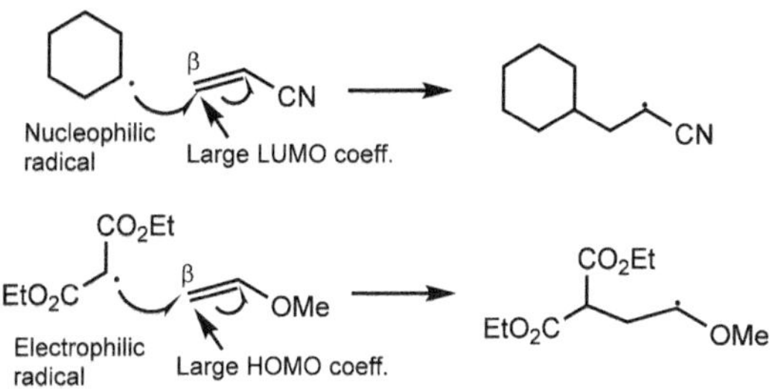


**FIG. 4.29** Frontier orbital effects on the regioselectivity of radical reactions.

The relative rates of radical additions also conform to the appropriate frontier orbital separations. The lower the energy of the LUMO of z-substituted alkenes, the faster will be the addition of a nucleophilic radical to it. On the other hand, an electrophilic radical will add more rapidly the higher the energy of the HOMO of x-substituted alkenes.[41]

Radicals add to the less substituted end of an alkene to give the more stable radical. This regioselectivity is generally consistent with the frontier orbital effects of HOMO/LUMO coefficients (Fig. 4.29).[42] As shown, a nucleophilic radical adds regioselectively to the β end of the z-substituted alkene (z = CN) having a larger LUMO coefficient. A similar pattern is observed with an electrophilic radical that adds to an x-substituted alkene (x = OMe) having a larger HOMO coefficient at β carbon.

### 4.5.3 Cyclization of radicals and regioselectivity

The cyclization of 5-hexenyl radicals is common in radical cyclizations. The cyclization usually shows a preference for the 5-*exo-trig* process to form a 5-membered ring over the 6-*endo-trig* cyclization to a 6-membered ring, even though the former gives the less stable primary radical (Fig. 4.30A). The regioselectivity is quite high (98:2). According to Baldwin's rules (Table 4.1), although the 5-*exo-trig* is favoured, the 6-*endo-trig* is also not disfavoured. Here the reason for the *exo* preference has been attributed to stereoelectronic effects. MO calculations suggest that the preferred direction of radical attack is from an angle of about 70° with respect to the plane of the double bond **4.17**.[43,44] This stereoelectronic requirement leads to a more strained *endo* cyclization **4.19** compared to an optimal orbital alignment in the *exo* cyclization **4.18**. There is also a coefficient interaction in favour of the *exo* closure. Since the radical is nucleophilic, the interaction will be influenced more by the LUMO of the alkene, and the radical attack will occur preferentially on the substituted terminus which has the higher coefficient in the LUMO **4.18**. Hence the formation of 5-membered ring is preferred.

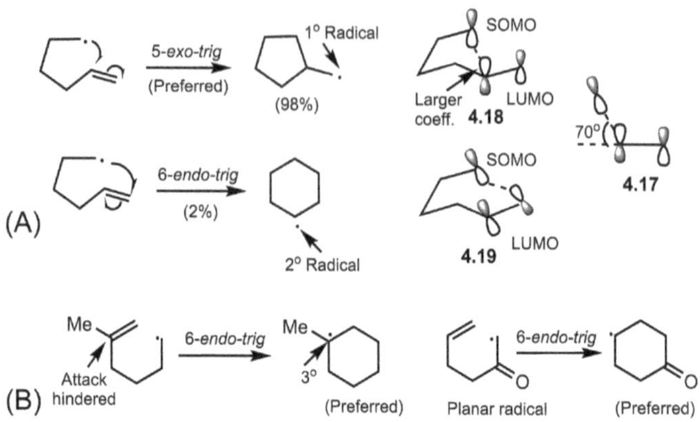


**FIG. 4.30** (A) Stereoelectronic effects in the preferred 5-*exo-trig* cyclization of 5-hexenyl radicals and (B) two examples where 6-*endo* cyclizations are preferred.

The relative rates of cyclization are in the order: 5-*exo* > 6-*endo* ~ 6-*exo* > 7-*endo*.[45] However, exceptions are observed if steric or other electronic effects are involved. For example, a 5-alkyl substituent promotes the 6-*endo-trig* pathway because the steric effect slows down the attack at the more substituted carbon in the *exo* closure (Fig. 4.30B).[46] The formation of a much more stabilized 3° radical also favours the *endo* pathway. The *endo*:*exo* selectivity is 1.6:1. The α-keto radicals also prefer the 6-*endo-trig* pathway (Fig. 4.30B). The conjugation of the radical centre with C=O imparts a planarity to the radical that favours the *endo* route.[47]

**Problem 4.7** Explain the regioselectivity observed in the following radical cyclizations:

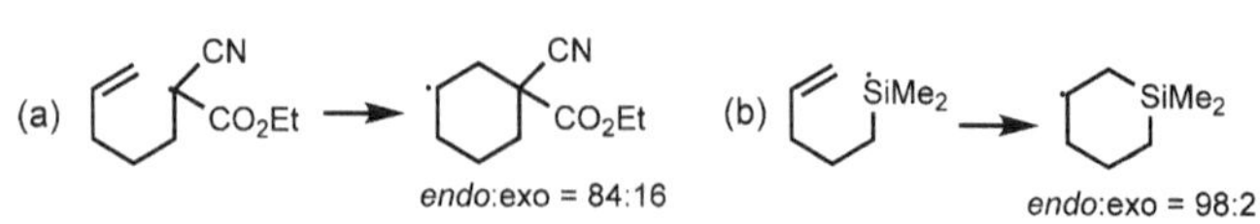


## 4.6 Photochemical reactions[48]

A photochemical reaction is usually a first excited state process. The promotion of one electron from a bonding orbital to an antibonding orbital creates two singly occupied orbitals in the excited state which can serve as excited state frontier orbitals in photochemical reactions. This is illustrated with a $_{\pi}2$ system in Fig. 4.31. A useful convention is to designate the singly occupied excited state orbitals of a molecule by its former ground state HOMO/LUMO labels, as shown. This convention will be used in this text.

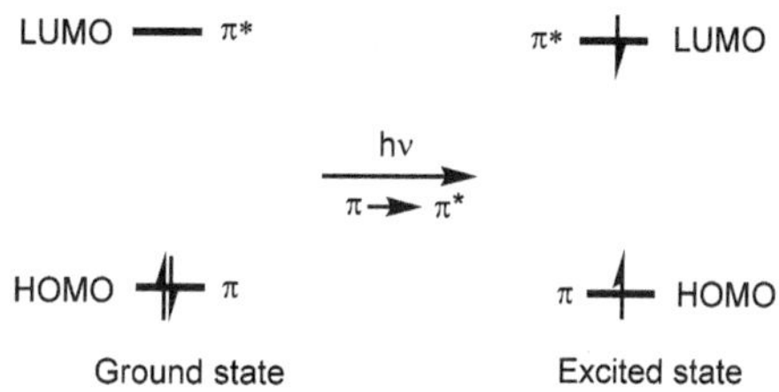


**FIG. 4.31** Frontier orbitals in the excited state of a $_{\pi}2$ system.

### 4.6.1 Photochemical [2+2] cycloadditions

A photochemical [2+2] cycloaddition involves the interaction of an excited state molecule with a ground state molecule since the probability of two excited state molecules encountering one another is quite small. The most important interactions are π/π (HOMO/HOMO) and π*/π* (LUMO/LUMO), each of which is energy lowering (Fig. 4.32). Both HOMO/HOMO and LUMO/LUMO interactions are bonding and the $[_{\pi}2_s+_{\pi}2_s]$ cycloaddition is thus favourable photochemically.

HOMO/HOMO and LUMO/LUMO interactions are presumably strong as the interacting orbitals are very close in energy. This may also lead to the formation of an exciplex (**exci**ted com**plex**) or excimer (**exci**ted di**mer**) as an intermediate and promote a stepwise photochemical pathway. The concerted bonding can result only from a singlet excited state. The triplet excited state obtained by intersystem crossing (ISC) or by sensitization would involve a spin inversion barrier to concerted addition, and hence the triplet state reactions would follow a stepwise mechanism.

### 4.6.2 Regioselectivity of photocycloadditions

The regioselectivity of photocycloadditions can be determined by the leading bond from the large–large interaction in appropriate frontier orbitals. The

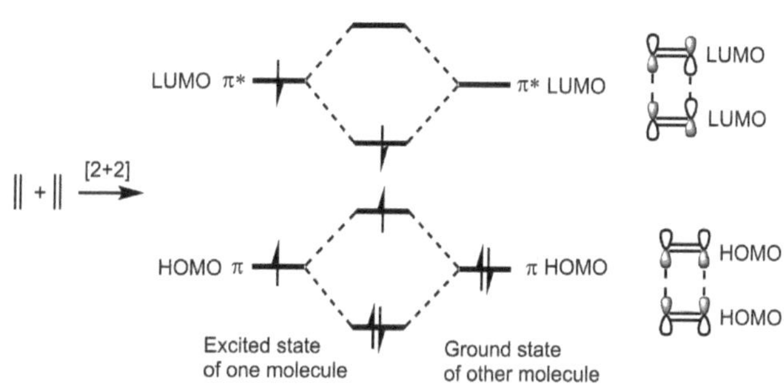


**FIG. 4.32** (A) Frontier orbital energy diagram for the photochemical [2+2] cycloaddition; (B) Frontier orbital interactions for the photochemical $[_{\pi}2_s+_{\pi}2_s]$ cycloaddition.

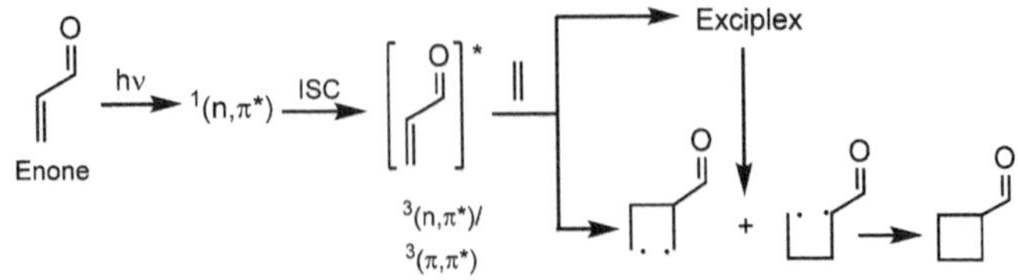


FIG. 4.33 Mechanism of triplet state enone photocycloaddition.

arguments based on coefficients seem to work for both pericyclic and stepwise photocycloadditions. Photocycloadditions of α,β-unsaturated carbonyl compounds (enones) with alkenes are stepwise, and involve triplet excited state of enones formed by ISC of the initial (n,π*) singlet.[49,50] The triplet excited state may be (n,π*) and/or (π,π*) depending on the substrate and the conditions employed. The cycloaddition may involve an exciplex and/or a diradical intermediate as shown in Fig. 4.33.

These cycloadditions can give regioisomeric products which are often termed as head-to-head and head-to-tail isomers. Consider a photocycloaddition of cyclohexenone with isobutene (Fig. 4.34). The reaction is regioselective and the major product is a head-to-tail adduct. The enone is an electron-deficient alkene, for which the HOMO is slightly but the LUMO is strongly polarized. The regioselectivity would therefore be governed more by the LUMO/LUMO interaction. As shown in Fig. 4.34, the large–large interaction leads to a head-to-tail regioisomer. Note that the LUMO of isobutene (Me is an electron donating substituent) has a large coefficient at the α-carbon.

**Problem 4.8** Account for the regioselectivity observed in the following Paterno-Büchi reaction:

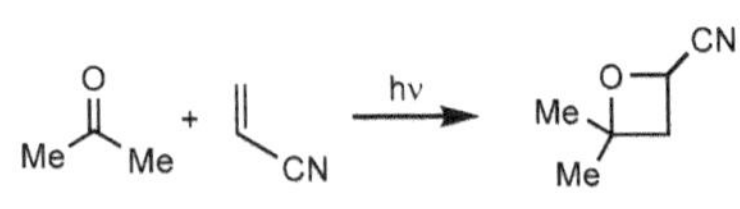


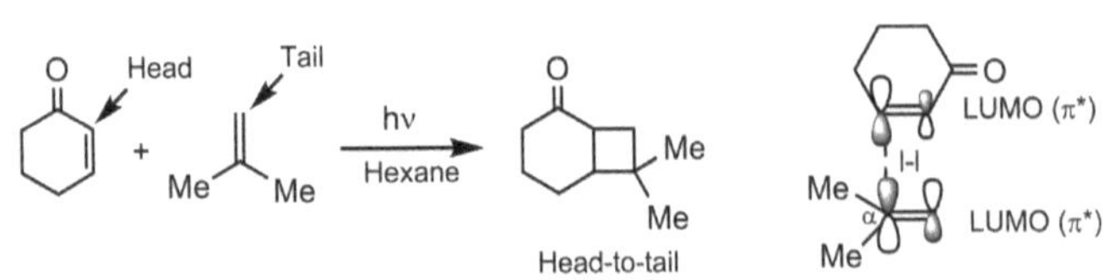


FIG. 4.34 Regioselectivity in the photocycloaddition of cyclohexenone with isobutene.

## References


1. Streitwieser, A., Jr. *Molecular Orbital Theory for Organic Chemists*; Wiley: New York, 1961.
2. Berson, J. A. *Angew. Chem. Int. Ed. Engl.* **1996**, *35*, 2751.
3. Mandal, D. K. *Pericyclic Chemistry: Orbital Mechanisms and Stereochemistry*; Elsevier: Cambridge, MA, 2018 (chapter 1).
4. Klopman, G. *J. Am. Chem. Soc.* **1968**, *90*, 223.
5. Salem, L. *J. Am. Chem. Soc.* **1968**, *90*, 543.
6. Kirby, A. J. *Stereoelectronic Effects*; Oxford University Press: New York, 1996 (chapter 2).
7. Klopman, G., Ed. *Chemical Reactivity and Reaction Paths*; New York: Wiley, 1974.
8. Salem, L. *Electrons in Chemical Reactions: First Principles*; Wiley: New York, 1982; p. 164.
9. Hudson, R. F.; Klopman, G. *Tetrahedron Lett.* **1967**, 1103.
10. Fleming, I. *Molecular Orbitals and Organic Chemical Reactions*; Wiley: Chichester, 2009. Student Edition; Reference Edition, 2010 (chapters 4 and 5).
11. Saunders, W. H., Jr.; Cockerill, A. F. *Mechanisms of Elimination Reactions*; Wiley: New York, 1973.
12. Cieplak, A. S.; Bürgi, H. B.; Dunitz, J. D. *Structure Correlation*, Vol. 1; VCH: Weinheim, 1994; p. 205.
13. Stone, A. J.; Erskine, R. W. *J. Am. Chem. Soc.* **1980**, *102*, 7185.
14. Houk, K. N.; Paddon-Row, M. N.; Rondan, N. G.; Wu, Y.-D.; Brown, F. K.; Spellmeyer, D. C.; Metz, J. T.; Li, Y.; Longarich, R. J. *Science* **1986**, *231*, 1108.
15. Anh, N. T.; Elkaim, L.; Thanh, B. T.; Maurel, F.; Flament, J. P. *Bull. Soc. Chim. Fr.* **1992**, *129*, 468.
16. Chopard, P. A.; Hudson, R. F.; Klopman, G. *J. Chem. Soc.* **1965**, 1379.
17. Agami, C.; Levisalles, J.; Lo Cicero, B. *Tetrahedron* **1979**, *35*, 961.
18. Grob, C. A. *Angew. Chem. Int. Ed. Engl.* **1969**, *8*, 535.
19. Baldwin, J. E. *J. Chem. Soc. Chem. Commun.* **1976**, 734.
20. Baldwin, J. E.; Thomas, R. C.; Kruse, L. I.; Silberman, L. *J. Org. Chem.* **1977**, *42*, 3846.
21. Baldwin, J. E.; Kruse, L. I. *J. Chem. Soc. Chem. Commun.* **1977**, 233.
22. Fleming, I. *Molecular Orbitals and Organic Chemical Reactions*; Wiley: Chichester, 2009. Student Edition; Reference Edition, 2010 (chapter 6).
23. Mandal, D. K. *Pericyclic Chemistry: Orbital Mechanisms and Stereochemistry*; Elsevier: Cambridge, MA, 2018 (chapters 3 and 4).
24. Houk, K. N. *J. Am. Chem. Soc.* **1973**, *95*, 4092.
25. Houk, K. N. *Acc. Chem. Res.* **1975**, *8*, 361.
26. Horn, B. A.; Herek, J. L.; Zewail, A. H. *J. Am. Chem. Soc.* **1996**, *118*, 8755.
27. Dewar, M. J. S.; Olivella, S.; Stewart, J. J. P. *J. Am. Chem. Soc.* **1986**, *108*, 5771.
28. Bernard-Henriet, C.; Chanon, M. *Tetrahedron Lett.* **1996**, *37*, 2417.
29. Ess, D. H.; Hayden, A. E.; Klärner, F.-G.; Houk, K. N. *J. Org. Chem.* **2008**, *73*, 7586.
30. Vogel, P.; Houk, K. N. *Organic Chemistry: Theory, Reactivity and Mechanisms in Modern Synthesis*; Wiley-VCH: Weinheim, 2019; p. 392 (chapter 5).
31. Vogel, P.; Houk, K. N. *Organic Chemistry: Theory, Reactivity and Mechanisms in Modern Synthesis*; Wiley-VCH: Weinheim, 2019; p. 399 (chapter 5).
32. Houk, K. N.; Sims, J.; Duke, R. E., Jr.; Strozier, R. W.; George, J. K. *J. Am. Chem. Soc.* **1973**, *95*, 7287.
33. Houk, K. N.; Sims, J.; Watts, C. R.; Luskus, L. J. *J. Am. Chem. Soc.* **1973**, *95*, 7301.

34. Woodward, R. B.; Hoffmann, R. *Angew. Chem. Int. Ed. Engl.* **1969**, *8*, 781.
35. Huisgen, R.; Feiler, L. A.; Binsch, G. *Chem. Ber.* **1969**, *102*, 3460.
36. Pederes, G. D.; Jorgensen, W. L. *J. Org. Chem.* **1992**, *57*, 1904.
37. Snider, B. B. *Acc. Chem. Res.* **1980**, *13*, 426.
38. Woodward, R. B.; Hoffmann, R. *The Conservation of Orbital Symmetry*; Verlag Chemie: Weinheim, 1971. Academic Press: New York, 1970.
39. Kirby, A. J. *Stereoelectronic Effects*; New York: Oxford University Press, 1996 (chapter 8).
40. Fleming, I. *Molecular Orbitals and Organic Chemical Reactions*; Wiley: Chichester, 2009. Student Edition; Reference Edition, 2010 (chapter 7).
41. Giese, B.; Horler, H.; Leising, M. *Chem. Ber.* **1986**, *119*, 444.
42. Pasto, D. J. *J. Org. Chem.* **1992**, *57*, 1139.
43. Dewar, M. J. S.; Olivella, S. *J. Am. Chem. Soc.* **1978**, *100*, 5290.
44. Spellmeyer, D. C.; Houk, K. N. *J. Org. Chem.* **1987**, *52*, 959.
45. Beckwith, A. L. J.; Schiesser, C. H. *Tetrahedron* **1985**, *41*, 3925.
46. Beckwith, A. L. J.; Blair, I. A.; Phillipou, G. *Tetrahedron Lett.* **1974**, 2251.
47. Broeker, J. L.; Houk, K. N. *J. Org. Chem.* **1991**, *56*, 3651.
48. Fleming, I. *Molecular Orbitals and Organic Chemical Reactions*; Wiley: Chichester, 2009. Student Edition; Reference Edition, 2010 (chapter 8).
49. Turro, N. J. *Modern Molecular Photochemistry*; Benjamin/Cummings: Menlo Park, CA, 1978.
50. Schuster, D. I.; Lem, G.; Kaprinidis, N. A. *Chem. Rev.* **1993**, *93*, 3.

Chapter 5

# Introduction to stereospecificity, stereoselectivity and asymmetric synthesis

## 5.1 Conformation and reactivity[1]

In Chapters 1 and 3, we have discussed the conformational equilibria in acyclic and cyclic organic molecules. In this section, we will consider how conformational equilibria in general can affect chemical reactivity and product composition. The conformers of a reactant are rapidly interconvertible and the equilibration is usually much more rapid than the rate of the reaction. Under this situation, we consider that one conformer ($A_1$) of a reactant would give the product $P_1$ and another ($A_2$) would give the product $P_2$ as shown in Eq. (5.1).

$$P_1 \xleftarrow{k_1} A_1 \underset{k_{c'}}{\overset{k_c}{\rightleftharpoons}} A_2 \xrightarrow{k_2} P_2 \qquad (k_c, k_{c'} >> k_1, k_2) \tag{5.1}$$

where $k_1, k_2$ = rate constants for the conversion of the conformers into the products; $k_c, k_{c'}$ = rate constants for the conformational interconversion. Notably, the kinetic scheme shown above can be applied not only to conformational isomers but also to any pair of isomers that are rapidly interconverting.

We proceed now to describe the relative rates and product composition in reactions involving two competing reaction paths as shown in Eq. (5.1).

### 5.1.1 Relative rates: Winstein–Holness equation

Assuming the reaction (Eq. 5.1) to be first order, the observed rate is given by

$$\text{Rate} = k_{obs}\{[A_1] + [A_2]\} \tag{5.2}$$

where $k_{obs}$ = experimentally observed rate constant (also called Winstein–Holness rate constant); $[A_1]$, $[A_2]$ = concentrations of the conformers.

Stereochemistry and Organic Reactions. https://doi.org/10.1016/B978-0-12-824092-2.00005-8

Since the conformational equilibration is fast, the equilibrium is maintained throughout the reaction when the equilibrium constant ($K$) is given by

$$K = \frac{[A_2]}{[A_1]} = \frac{x_2}{x_1} \tag{5.3}$$

where $x_1$, $x_2$ = mole fractions of $A_1$ and $A_2$ at equilibrium. If no other components exist in the system, $x_1 + x_2 = 1$.

From Eq. (5.1), we also obtain

$$\text{Rate} = k_1[A_1] + k_2[A_2] \tag{5.4}$$

Equating Eqs (5.2) and (5.4), and substituting $[A_2]$ by $(x_2/x_1)[A_1]$, we get

$$k_{\text{obs}}\left(1 + \frac{x_2}{x_1}\right) = k_1 + k_2\frac{x_2}{x_1}$$

$$\text{or,}\quad k_{\text{obs}} = k_1x_1 + k_2x_2 \tag{5.5}$$

Eq. (5.5) is called the Winstein–Holness equation[2] which states that the observed rate constant is the weighted average of the rate constants at which the individual conformers react. For any number of contributing isomers, Eq. (5.5) can be generalized to

$$k_{\text{obs}} = \sum_i k_ix_i \tag{5.6}$$

where $k_i$ = rate constant for the conversion of the $i$th conformer to product; $x_i$ = mole fraction of the $i$th conformer.

To illustrate Winstein–Holness kinetics, consider E2 dehydrohalogenation of *meso* and chiral diastereomers of 2,3-dibromobutane (Fig. 5.1).[3] The *meso* isomer gives predominantly ($E$)-2-butene (96%), whereas the chiral diastereomer yields mainly ($Z$)-2-butene (91%). The stereoelectronic factor for E2 eliminations (see Section 4.3.3) is that the departing groups (Br, Br) must be antiperiplanar. Therefore, only one conformer (anti) out of three possible

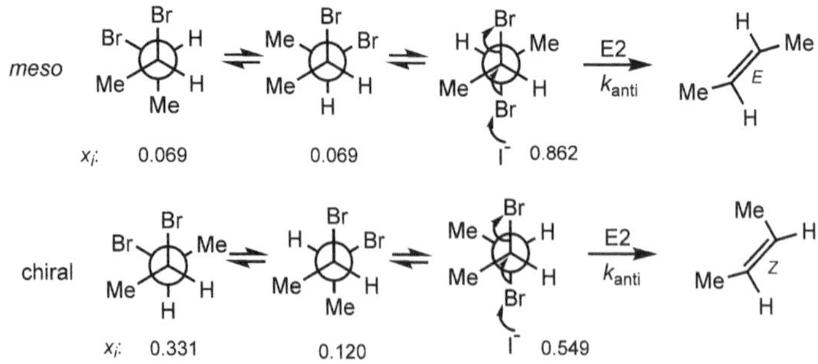


**FIG. 5.1** Application of the Winstein–Holness equation to E2 debromination of *meso* and chiral diastereomers of 2,3-dibromobutane with KI. $x_i$ denotes mole fraction of conformers.

conformers in each case (cf. Fig. 2.32B) can undergo E2 debromination with $I^-$. Using Eq. (5.5) or (5.6), we get

$$k_{meso} = k_{anti} x_{anti} = 0.862\, k_{anti}$$

$$k_{chiral} = k_{anti} x_{anti} = 0.549\, k_{anti}$$

where $k_{anti}$ = intrinsic rate constant for antiperiplanar conformation; $x_{anti}$ = mole fraction of anti conformer.[4]

The relative rates of reaction of *meso* and chiral diastereomers are then obtained as

$$k_{meso} : k_{chiral} = 1.57$$

This calculated value is found to be in reasonable agreement with the experimental value of 1.93 at 60°C.

### 5.1.2 Product composition: Curtin–Hammett principle

The Curtin–Hammett principle[5,6] for product composition is also derived for the situation where the reactant conformers are in equilibrium and are interconverted at a rate much faster than the reaction does (Eq. 5.1). The energy diagram for the Curtin–Hammett kinetics is shown in Fig. 5.2.

Of the two conformers, $A_1$ is assumed to be more stable (lower free energy). Thus, the equilibrium constant ($K$) for $A_2 \rightleftharpoons A_1$ is given by

$$K = \frac{[A_1]}{[A_2]} = e^{\frac{-\Delta G^0}{RT}} \tag{5.7}$$

As the product composition is determined by kinetic control, the product ratio is:

$$\frac{[P_1]}{[P_2]} = \frac{\text{Rate}_1}{\text{Rate}_2} = \frac{k_1[A_1]}{k_2[A_2]}$$

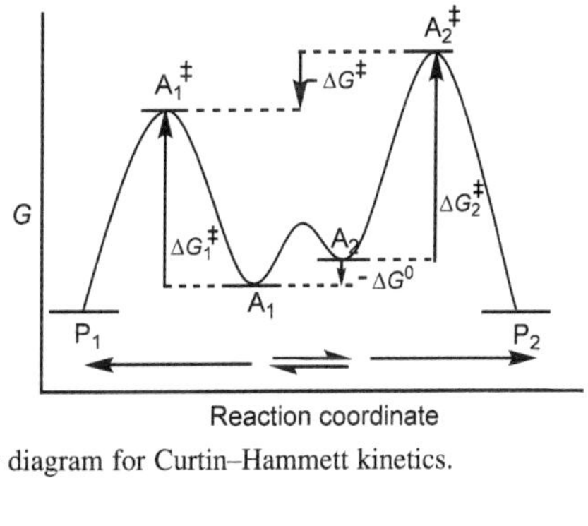


**FIG. 5.2** Free energy diagram for Curtin–Hammett kinetics.

$$\text{or,} \quad \frac{[P_1]}{[P_2]} = \frac{k_1}{k_2} e^{\frac{-\Delta G^0}{RT}} \tag{5.8}$$

Using the expression of rate constant ($k$) from transition state theory (see Eq. 1.6) into Eq. (5.8), we get

$$\frac{[P_1]}{[P_2]} = e^{\frac{-\Delta G_1^{\neq}}{RT}} e^{\frac{\Delta G_2^{\neq}}{RT}} e^{\frac{-\Delta G^0}{RT}} = e^{\frac{\left(-\Delta G^0 + \Delta G_2^{\neq} - \Delta G_1^{\neq}\right)}{RT}}$$

From Fig. 5.2, $-\Delta G^0 + \Delta G_2^{\neq} - \Delta G_1^{\neq} = -\Delta G^{\neq}$, where $\Delta G^{\neq}$ is the difference in free energy of the two TSs. Therefore,

$$\frac{[P_1]}{[P_2]} = e^{\frac{-\Delta G^{\neq}}{RT}} \tag{5.9}$$

Eq. (5.9) defines the Curtin–Hammett principle which states that the product composition from two pertinent conformers depends only on the relative free energy of the two transition structures and is independent of the relative populations of the conformers, provided the conformational interconversion is faster than the rates of the reaction. Thus, according to the principle, a reaction can proceed through a minor conformation if it is the one that leads to the lowest energy TS.

The Curtin–Hammett principle is illustrated with two examples in Fig. 5.3. The first example is E2 dehydrohalogenation of 2-bromobutane with a base (Fig. 5.3A). Here two antiperiplanar conformers are possible that give ($E$)-2-butene and ($Z$)-2-butene in a ratio of about 6:1. The major ($E$) isomer arises via the sterically less hindered TS of lower energy. Note that the more stable conformer leads to the more stable TS. Thus, in such cases it is difficult to ascertain whether the product ratio reflects the conformer population or the relative energies of the TSs. This problem is avoided in the $S_N2$ reaction shown in Fig. 5.3B. Here the major product results from the less stable conformer via

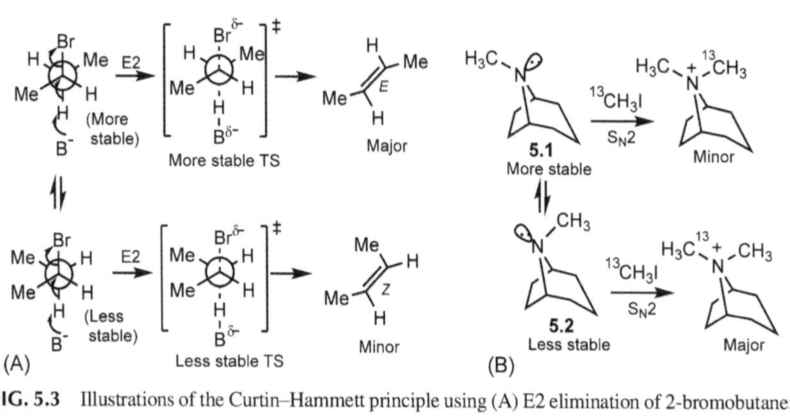


FIG. 5.3 Illustrations of the Curtin–Hammett principle using (A) E2 elimination of 2-bromobutane; (B) $S_N2$ reaction (quaternization) of a tropane.

the more stable TS. The reaction involves the quaternization of a tropane with $^{13}CH_3I$.[6] The use of isotopic label permits the product composition to be determined by $^{13}C$ NMR spectroscopy. Since the tropane is sterically more hindered on the side of the larger ring, the conformer **5.2** is less stable than the conformer **5.1**. But in **5.2**, the nitrogen lone pair resides in the sterically more accessible region on the side of the smaller ring, and therefore the quaternization of **5.2** will be faster via the more stable TS to give the major product.

*Remarks* The Curtin–Hammett principle doesn't apply to the situation where the rates of the reaction are faster than the rates of conformational interconversion, that is $k_1, k_2 \gg k_c, k_{c'}$ (cf. Eq. 5.1). In this situation, the conformer ratio will not change during the reaction since the interconversion is stopped (quenched) by the reaction in their equilibrium concentrations. The product ratio ($P_1$:$P_2$) will then reflect the conformer ratio ($A_1$:$A_2$). The product composition in this case is thus contrary to the Curtin–Hammett principle.

## 5.2 Stereospecific and stereoselective reactions[7,8]

The terms 'stereospecific' and 'stereoselective' are often confusing, as these are defined in a number of ways.[9] We will use the terms on mechanism basis as described below.

A reaction can take place by a mechanism which has a strict stereochemical requirement for an efficient, symmetry-allowed orbital overlap. If that mechanism leads to a single and specific product stereochemistry from a particular stereochemistry of the starting material, the reaction is stereospecific. For example, the E2 eliminations shown in Fig. 5.1 are stereospecific because a specific diastereomeric product is formed from a particular reactant stereochemistry by the antiperiplanar mechanism; the *meso* diastereomer gives an (*E*) alkene, whereas the chiral diastereomer produces a (*Z*) alkene. Another example with the Diels-Alder reaction is shown in Fig. 5.4. Here the supra/supra mechanism (suprafacial for both diene and dienophile) leads to the stereospecificity: the (*E*,*E*) diene gives a *cis* product, whereas the (*E*,*Z*) diene yields a *trans* product.

For a stereoselective reaction, the mechanism does not prevent the formation of two or more stereoisomeric products. This is possible when the mechanism allows for two or more possible conformations or geometries of the TS.

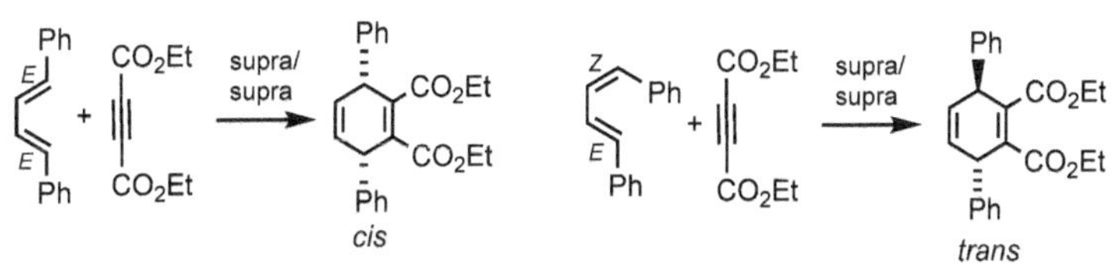


FIG. 5.4 Stereospecific Diels-Alder reactions.

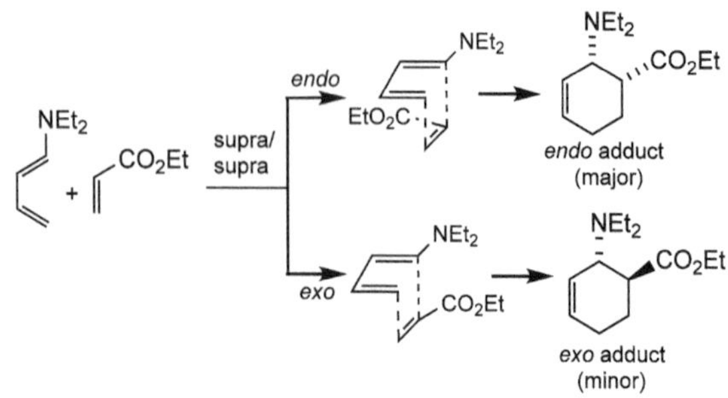


**FIG. 5.5** A stereoselective Diels-Alder reaction.

Usually one product predominates and the reaction becomes stereoselective. For example, the E2 elimination shown in Fig. 5.3A is stereoselective. Here two possible antiperiplanar conformations lead to two diastereomeric products of which one is the major product. The Diels-Alder reaction also shows stereoselectivity as the supra/supra mechanism can allow *endo* or *exo* geometry of the transition structure to give *endo* and *exo* adducts (Fig. 5.5). The major product is usually an *endo* diastereomer. Notably, the above definitions would permit many reactions to be described as both stereospecific and stereoselective, like the Diels-Alder reaction.

## 5.2.1 Stereoselectivity: Free energy diagram

If the two possible transition structures of a reaction are enantiomeric, they will be of equal energy and the two products will form at the same rate. Thus, under kinetic control, the products are formed in equal amounts, and there is no stereoselectivity. For a reaction to be stereoselective, it must proceed through diastereomeric transition structures ($TS_1$ and $TS_2$) of differing energies as shown in Fig. 5.6. The free energy of activation for the formation of product $P_1$ ($\Delta G_1^{\neq}$) is then less than that for the formation of product $P_2$ ($\Delta G_2^{\neq}$). Under kinetic control, the faster formed product ($P_1$) via the lower energy transition structure ($TS_1$) would be formed as a major product.

The ratio of the two products $[P_1]/[P_2]$ is given by

$$\frac{[P_1]}{[P_2]} = \frac{e^{\frac{-\Delta G_1^{\neq}}{RT}}}{e^{\frac{-\Delta G_2^{\neq}}{RT}}}$$

$$\text{or,} \quad \frac{[P_1]}{[P_2]} = e^{\Delta\Delta G^{\neq}/RT} \tag{5.10}$$

where $\Delta\Delta G^{\neq} = \Delta G_2^{\neq} - \Delta G_1^{\neq}$.

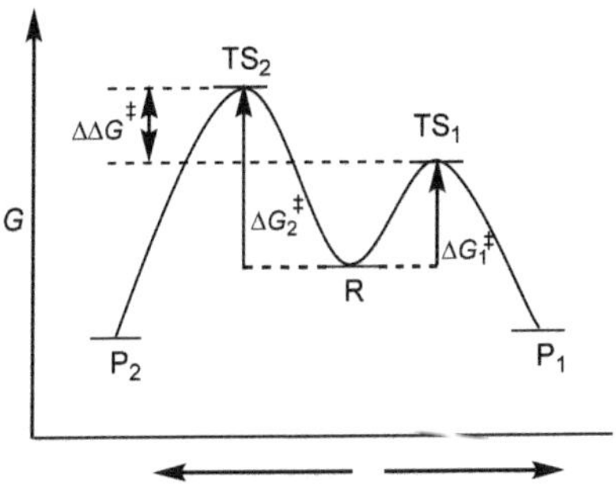


**FIG. 5.6** Free energy diagram of a stereoselective reaction involving diastereomeric transition structures. R stands for reactant(s), $P_1$ and $P_2$ denote two products, and $TS_1$ and $TS_2$ indicate two transition structures. $\Delta\Delta G^{\neq}$ is the difference between $\Delta G_2^{\neq}$ and $\Delta G_1^{\neq}$.

### 5.2.2 Diastereoselectivity or diastereotope-selectivity and enantioselectivity or enantiotope-selectivity

For stereoselective reactions, the terms 'diastereoselective' and 'enantioselective' refer to the *products* of the reaction, whereas the terms 'diastereotope-selective' and 'enantiotope-selective' relate to the *substrates*.

If the stereoisomeric products of a stereoselective reaction are diastereomeric, the reaction is said to be diastereoselective. A single diastereomeric product is the target molecule of a diastereoselective synthesis. Thus, a diastereoselective synthesis is the synthesis of a diastereomer. In a diastereoselective reaction, a chiral substrate possesses diastereotopic groups or faces. Selective reaction at only one of the diastereotopic groups or faces gives a single diastereomeric product. Such diastereotopos discrimination is called diastereotope-selective.

On the other hand, if the stereoisomeric products of a stereoselective reaction are enantiomeric, and one of the enantiomers is formed in excess, the reaction is termed enantioselective. In an enantioselective reaction, the substrate is prochiral and possesses two enantiotopic groups or faces. If the reaction occurs selectively at only one of the enantiotopic groups or faces, a single enantiomeric product will result. Such enantiotopos differentiation is called enantiotope-selective. An enantioselective synthesis is known as asymmetric synthesis.

In a diastereoselective synthesis, the extent of diastereoselectivity is expressed as diastereomer ratio (dr), or as diastereomeric excess (de) which denotes the excess of one diastereomer over the other. Similarly, in an enantioselective synthesis, the extent of enantioselectivity is expressed as enantiomeric excess (ee), that is the excess of one enantiomer over the other (see Section 2.4.3).

The extent of diastereoselectivity or enantioselectivity in terms of $\Delta\Delta G^{\neq}$ values at 25°C (298 K), as estimated using Eq. (5.10), is given in Table 5.1. It is seen that a free energy difference of only $11.4\,\text{kJ mol}^{-1}$ ($2.72\,\text{kcal mol}^{-1}$) favours the major product to the extent of 99: 1 (de or ee 98%).

**TABLE 5.1** Extent of diastereoselectivity or enantioselectivity as a function of $\Delta\Delta G^{\neq}$ at 298 K.

| $\Delta\Delta G^{\neq}$, kJ mol$^{-1}$ (kcal mol$^{-1}$)[a] | $P_1$:$P_2$ (dr) | de or ee (%) |
|---|---|---|
| 0 | 1:1 | 0 |
| 2.72 (0.65) | 3:1 | 50 |
| 5.44 (1.30) | 9:1 | 80 |
| 7.30 (1.74) | 19:1 | 90 |
| 11.4 (2.72) | 99:1 | 98 |
| 17.1 (4.09) | 999:1 | 99.8 |
| 22.8 (5.45) | 9999:1 | 99.98 |

[a] *1 cal = 4.184 J.*

**TABLE 5.2** Stereoselectivity as a function of temperature.

| Temperature (°C) | $\Delta\Delta G^{\neq}$ (kJ mol$^{-1}$) | $P_1$:$P_2$ (dr) |
|---|---|---|
| 25 | 8.0 | 25:1 |
| −78 | 8.0 | 139:1 |

According to Eq. (5.10), the product ratio also depends on temperature; lowering of temperature gives enhanced stereoselectivity (Table 5.2).

## 5.3 Asymmetric synthesis[10–12]

Asymmetric synthesis has been defined in several ways and the definitions have evolved with time. Fundamentally, an asymmetric synthesis is one that creates new stereogenic units (usually stereocentres) with control of stereochemistry. This process is called asymmetric induction. It may be diastereoselective or enantioselective or both, and may lead to products as racemates or as single enantiomers. Nowadays, asymmetric synthesis implies an enantioselective synthesis, that is an asymmetric synthesis is the synthesis of an enantiomer.

The importance of enantiomerically pure compounds relates primarily to the role of enantiomer recognition in biological activity[13] since the biological molecules themselves (nucleic acids, proteins and polysaccharides) are chiral. For example, in drugs tested for therapeutical use, the desired biological activity is often related to the absolute configuration.[14,15] Only one enantiomer may possess the desired activity while the other enantiomer is inactive or possesses a different activity or causes toxic side effects. For instance, the (*S*) enantiomer of propanolol (see Fig. 2.70) is a β-blocker, whereas the (*R*) enantiomer has contraceptive activity. The need for enantiomerically pure drugs and other chemicals such as agrochemicals[16] is immense to solve the pharmacodynamic, environmental and economic issues, and hence is the importance of asymmetric synthesis.

## 5.3.1 Evolution and designs of asymmetric synthesis

Approaches to designing an asymmetric synthesis are often classed by generations[12] as outlined briefly here. These approaches reflect increasing challenges and sophistications involved in the asymmetric synthesis design.

### *5.3.1.1 First-generation asymmetric synthesis: Substrate control*

The earliest design (first generation) of asymmetric synthesis is based on diastereoselective reactions in which a new chiral centre is created under the influence of an existing chiral centre in the same molecule. This is called substrate control. The chiral substrates are often racemic since the availability of chiral nonracemic or enantiopure substrates is limited (obtaining an enantioenriched or enantiopure substrate itself might require an asymmetric synthesis!). A diastereoselective reaction with a chiral racemic substrate gives diastereomers as racemates (Fig. 5.7). However, if the substrate can be used as a single enantiomer, the products are obtained as single enantiomers.

In the first case, the reaction is simply diastereoselective (de = 50%, ee = 0%) but in the second case, the reaction is both diastereoselective and enantioselective (de = 50%, ee = 100%). The second case denotes a substrate-controlled asymmetric synthesis in view of the present day definition. If the chiral nonracemic substrate is not enantiomerically pure, the diastereoselective product may be obtained with the corresponding ee.

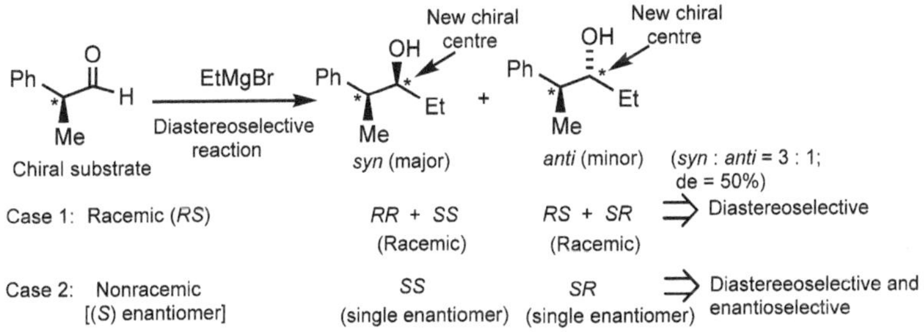


**FIG. 5.7** Nucleophilic addition to a chiral carbonyl substrate.

### 5.3.1.2 *Second-generation asymmetric synthesis: Chiral auxiliary control*

Chiral auxiliaries are enantiomerically pure compounds derived mostly from inexpensive chiral natural sources (i.e. chiral pool, see Section 2.13.1). The use of a chiral auxiliary in asymmetric synthesis is referred to as second-generation (2G) asymmetric synthesis. In this approach, a stoichiometric chiral auxiliary is covalently attached to the substrate followed by a diastereoselective reaction in which chirality in the auxiliary controls the asymmetric induction. After the new chiral centre is set up, the auxiliary is removed which may be recycled. A general scheme of the overall process is shown in Fig. 5.8.

The substrate-auxiliary is a new chiral compound which undergoes the diastereoselective reaction to give a major diastereomeric product-auxiliary which may be separated from the minor diastereomer and purified by conventional techniques such as column chromatography. The preferred diastereomer is then subjected to a suitable cleavage reaction that separates the chiral auxiliary from the final enantiomeric product without causing its racemisation. The recovered chiral auxiliary is recyclable; thus although stoichiometric quantities are needed, there is no waste. Notably, the attachment of the chiral auxiliary causes the enantiotopic ligands or faces of the prochiral substrate to become diastereotopic in the chiral substrate-auxiliary.

An arsenal of chiral auxiliaries[17] has been developed by several research groups and successfully applied in asymmetric synthesis. The most versatile chiral auxiliaries are available as both enantiomers. Two of the most widely used chiral auxiliary systems are oxazolidinones[18] (due to Evans[19]) derived from naturally occurring amino acids or amino alcohols (Fig. 5.9A), and sultams (due to Oppolzer[20]) derived from 10-camphorsulphonic acid (Fig. 5.9B).

### 5.3.1.3 *Third-generation asymmetric synthesis: Reagent control*

The use of an enantiomerically pure chiral reagent in asymmetric synthesis is referred to as third-generation (3G) asymmetric synthesis. Here the substrate is prochiral. The asymmetric induction is controlled by the chirality in the reagent. Consider the reduction of a prochiral ketone with a chiral borane (Fig. 5.10).

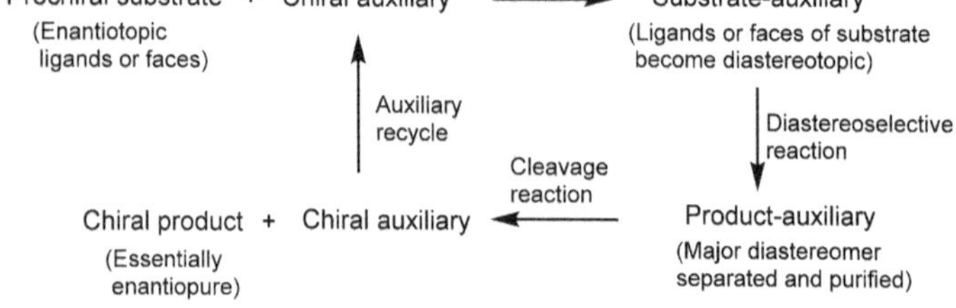


**FIG. 5.8** A general scheme for the use of chiral auxiliary in second-generation (2G) asymmetric synthesis.

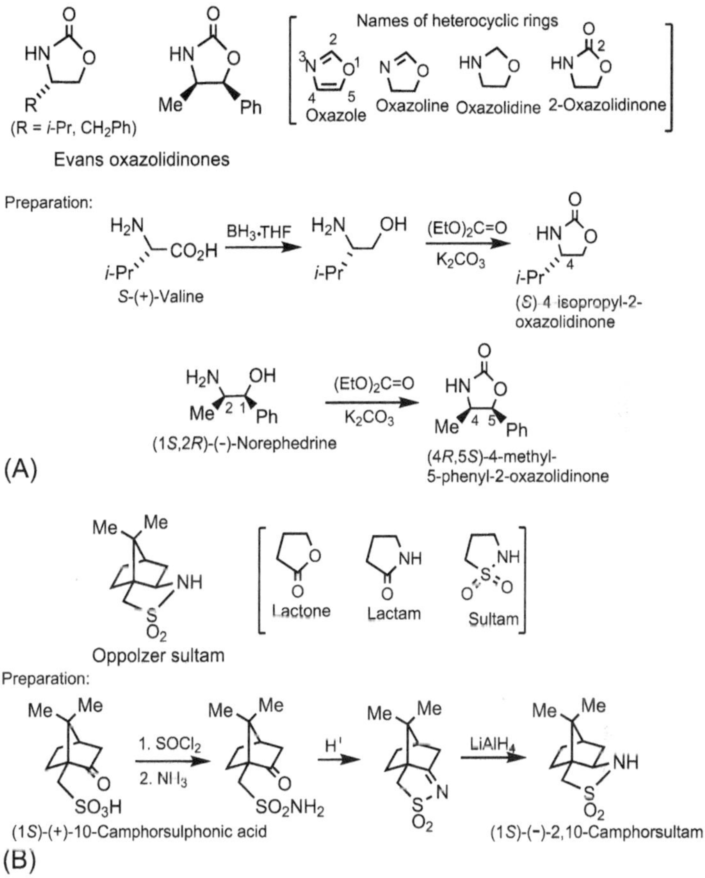


**FIG. 5.9** Two widely used chiral auxiliary systems and their methods of preparation: (A) Evans oxazolidinones; (B) Oppolzer sultams.

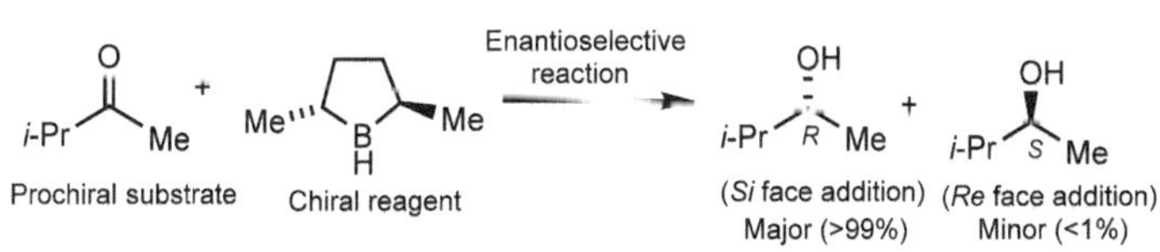


**FIG. 5.10** Use of a chiral reagent in third-generation (3G) asymmetric synthesis.

As shown, the enantioselective reaction gives almost exclusively the (*R*) enantiomer (ee > 99%). The TSs are diastereomeric which lead to the enantiomeric products in unequal amounts. No diastereomeric intermediates are involved. Since the chiral reagent is to be used in stoichiometric quantities, the synthesis may be quite expensive on a large scale.

### 5.3.1.4 Fourth-generation asymmetric synthesis: Asymmetric catalysis

In general, a catalyst speeds up a reaction and only a very small proportion of catalyst (i.e. catalytic amount) is needed for the reaction. Moreover, the catalyst could be recoverable after the reaction and is reusable. If the catalyst is chiral and nonracemic, the reaction is enantioselective. This is called asymmetric catalysis and is referred to as fourth-generation (4G) asymmetric synthesis. In practice, this is an extremely effective approach to asymmetric synthesis in comparison to second or third-generation approach that requires stoichiometric quantities of chiral auxiliary or chiral reagent. A general scheme of an asymmetric catalysis is shown in Fig. 5.11.

Initially, a catalyst–substrate complex is formed. Asymmetric induction then takes place under the influence of chiralty in the catalyst. This leads to a catalyst–product complex which on dissociation gives a product enantiomer with high ee. The TSs are diastereomeric and involve the chiral catalyst and the developing stereocentre, and lie between the catalyst–substrate and catalyst–product complexes.

### 5.3.1.5 Summary

The four generations of asymmetric synthesis are summarized in Fig. 5.12 with respect to chirality, stoichiometry and nature of the reaction and the product. For an asymmetric synthesis, the chiral participant (substrate, auxiliary, reagent or catalyst) must be a single enantiomer or an enantioenriched nonracemic sample.

Notably, first- and second-generation approaches involve diastereoselective (or diastereotope-selective) reactions in which the original substrate or the derived substrate (substrate-auxiliary) is chiral with diastereotopic groups or faces. The chiral auxiliary method however leads to an enantiomer as an end product by cleavage of the initial product diastereomer. In contrast, third- and fourth-generation approaches involve enantioselective (or enantiotope-selective) reactions in which the substrate is prochiral with enantiotopic groups or faces. Both chiral reagent and chiral catalyst controlled methods lead directly to a product enantiomer; however, the former approach is stoichiometric and the latter is most usefully catalytic.

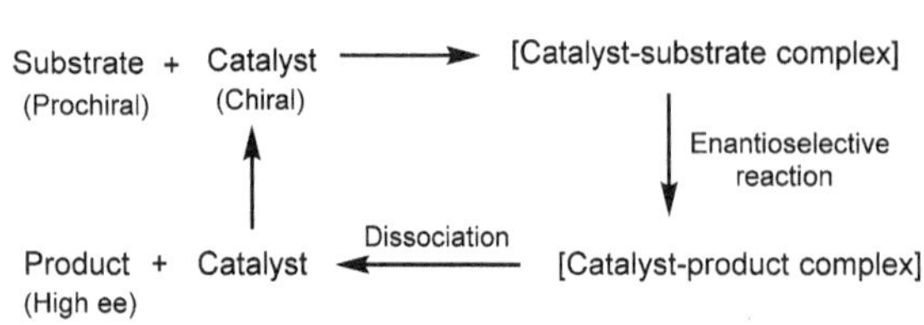


FIG. 5.11 A general scheme for asymmetric catalysis.

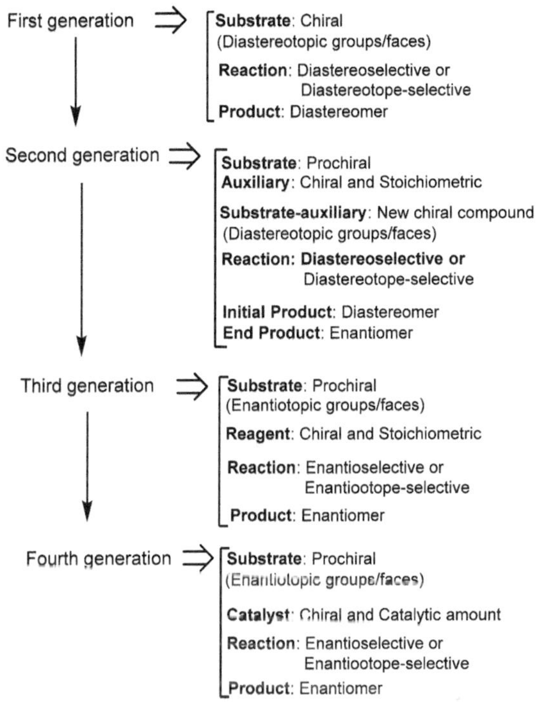


**FIG. 5.12** Summary of the four generations of asymmetric synthesis.

## 5.3.2 Nonlinear effects and asymmetric amplification[21]

It has been observed that a correlation between reactant ee and product ee could deviate from a linear relationship in certain cases of asymmetric synthesis. This nonlinear effects can be positive or negative (Fig. 5.13A). A positive deviation from linearity indicates that ee of the product is greater than ee of the chiral participant (chiral auxiliary, chiral reagent or chiral catalyst). This is called asymmetric amplification.[22] Its practical consequences are enormous because a low ee in the reactant or catalyst would be sufficient to achieve a high ee in the product. This will be particularly attractive when the chiral substance in question is not easy to obtain in enantiopure form. The troublesome separation of enantiomers as well as a strict stereo control in chiral ligand synthesis for asymmetric catalysis could then be avoided. Negative nonlinear effects have also been observed.[23]

An example of asymmetric amplification in a catalytic asymmetric synthesis is provided in Fig. 5.13B.[24,25] The enantioselective addition of diethylzinc to benzaldehyde, catalysed by the chiral catalyst (−)-DAIB, gives (*S*)-1-phenylpropanol. As shown, the catalyst with only 15% ee can lead to the product with 95% ee!

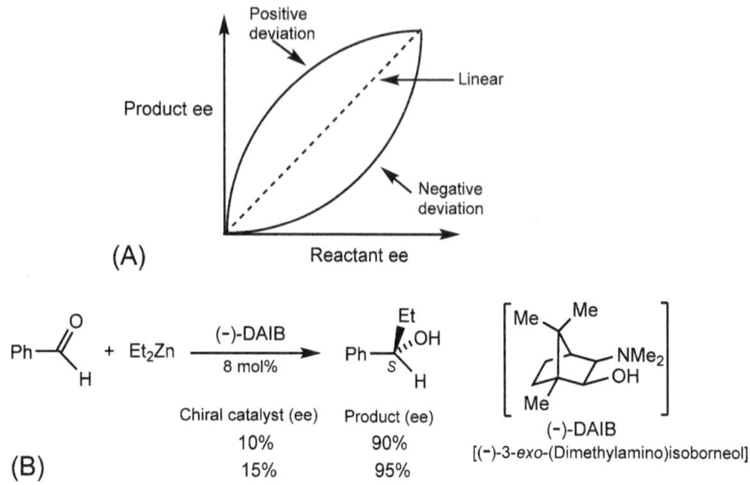


**FIG. 5.13** (A) Linear and nonlinear correlations between the product ee and the reactant ee in asymmetric synthesis; (B) An example of asymmetric amplification in an asymmetric synthesis involving a chiral catalyst (−)-DAIB.

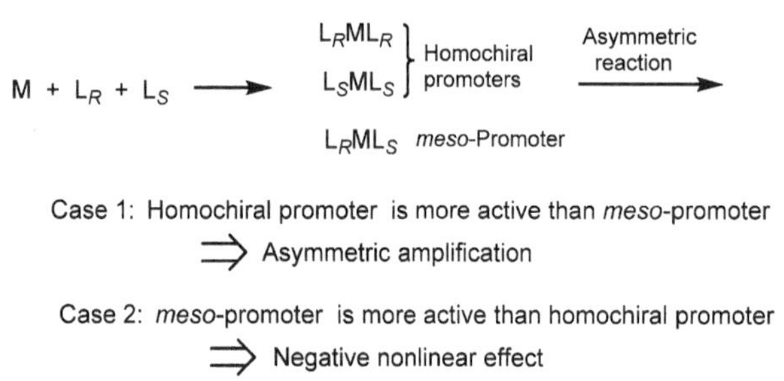


**FIG. 5.14** A general scheme for the rationalization of nonlinear effects in asymmetric synthesis. $L_R$ and $L_S$ denote (*R*) and (*S*) enantiomers of a nonracemic chiral ligand and M stands for the metal centre.

The positive nonlinear effect is attributed to the formation of aggregated complexes. Fig. 5.14 shows a general scheme to explain the nonlinear effects.[26,27]

If $L_R$ and $L_S$ denote (*R*) and (*S*) enantiomers of a nonracemic chiral ligand and M stands for the metal centre, three $ML_2$-type complexes are possible: $L_RML_R$ and $L_SML_S$, each bearing two ligands with the same chirality, and $L_RML_S$ having two ligands with the opposite absolute configuration. These aggregated complexes would then proceed to undergo the enantioselective reaction. The first two chiral complexes are called homochiral promoters while the third complex is called heterochiral or *meso*-promoter. Two cases can arise in the asymmetric transformation. If a homochiral promoter $L_RML_R$ or $L_SML_S$ is more active than the *meso*-promoter in the asymmetric reaction (Case 1), the product will have a higher ee than that expected from a linear correlation and the result will be asymmetric amplification. On the other hand, if the *meso*-promoter is more active (Case 2), the product will have a lower ee than that from a linear correlation and a negative nonlinear effect will occur.

Let us consider the asymmetric amplification shown in Fig. 5.13B. If the catalyst (−)-DAIB is enantiopure (ee 100%), only one dimer (−)-DAIBZn (−)-DAIB can form (Fig. 5.15A). But in this reaction, the dimer itself is not catalytically active. The monomer is the catalytically active species; the inactive dimer dissociates into the monomers which then enter into the catalytic cycle (see later Fig. 7.18B).[25,28]

If the catalyst (−)-DAIB is not enantiomerically pure and ee is, say, 50% with 75% (−)-DAIB and 25% (+)-DAIB, three dimers are possible. These are (−)-DAIBZn(−)-DAIB, (+)-DAIBZn(+)-DAIB and (−)-DAIBZn (+)-DAIB (Fig. 5.15B). The first two dimers are homochiral (optically active) promoters, and the third is a *meso* promoter. The *meso* dimer is most stable and its formation is essentially irreversible. Thus, almost the whole of (+)-DAIB (25%) is involved in the formation of the *meso* dimer in combination with 25% (−)-DAIB. As a result, the concentration of (+)-DAIBZn(+)-DAIB dimer is negligible. Effectively, it is the (−)-DAIBZn(−)-DAIB dimer that can dissociate into the monomers to enter into the catalytic cycle leading to high ee of the product. Note that (−)-DAIB available for the reaction is 50% (75% − 25%). Thus, $x$ mol% of nonracemic catalyst (ee 50%) in the reaction is equivalent to $0.5x$ mol% of the enantiopure catalyst.

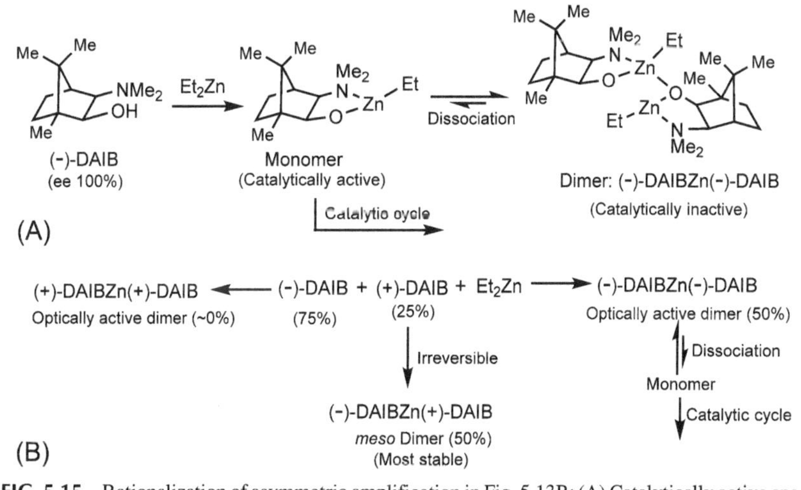


**FIG. 5.15** Rationalization of asymmetric amplification in Fig. 5.13B: (A) Catalytically active species in the presence of enantiopure (−)-DAIB; (B) Catalyst aggregation and mechanism of amplification using a nonracemic (−)-DAIB (ee 50%).

### 5.3.3 Asymmetric autocatalysis

Autocatalysis is the catalysis of a reaction by one of its products. Asymmetric autocatalysis with amplification of the ee was first achieved using the Soai autocatalytic reaction. Fig. 5.16A illustrates a Soai reaction that involves enantioselective addition of diisopropylzinc to pyrimidine-5-carbaldehyde.[29] Catalytic amounts of the (*S*) product (ee 5%) serves as asymmetric autocatalyst, and the newly formed product is obtained with an enhanced ee of 39%. The process can be repeated and the enantioselective product of each cycle can be used as the autocatalyst for the subsequent run to reach an ee of 85% after the third cycle.

Autocatalysis alone is not the answer for asymmetric amplification in the Soai reaction. This is because if each enantiomer of the catalyst catalyses its own production in the same way, the product at the end of the reaction will be obtained with the same ee as the catalyst used. Thus, an autocatalyst in order to generate asymmetric amplification needs to be able to not only reproduce itself but also act as an inhibitor for the production of its enantiomer. This idea is referred to as 'mutual antagonism'.[30,31] This concept can be related to a mechanism that allows dimer formation from two enantiomers. The dimeric $ML_2$ model proposed by Kagan is shown in Fig. 5.16B.[32] The model describes a system in which a metal centre (M) is in rapid coordination exchange with two enantiomeric ligands (*R* and *S*). Two homochiral complexes (*RR* and *SS*) and one heterochiral or *meso* complex (*RS*) are formed that can serve as active catalysts in the autocatalytic reaction (cf. Fig. 5.14; $L_SML_S \equiv SS$, and so on). If the *meso* catalyst is significantly less reactive than its homochiral partner,

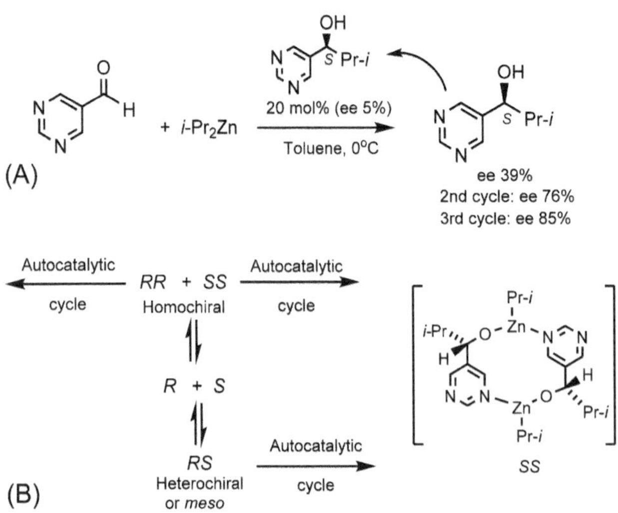


FIG. 5.16 (A) Soai asymmetric autocatalytic reaction with amplification of the enantiomeric excess (ee) in the product; (B) Kagan's $ML_2$ model for autocatalysis.

the autocatalytic reaction will show amplification in the product enantioselectivity. Theoretical analysis using density functional theory (DFT) also suggests that the enantioselective catalytic cycle of the homochiral dimer is the dominant mechanism.

## 5.3.4 Strategies to improve incomplete stereoselectivity

So far, we have considered situations in which there is a single chiral participant (substrate, chiral auxiliary, reagent or catalyst) in a diastereoselective or enantioselective reaction. But this strategy may not work perfectly in every situation. There are some additional strategies to improve incomplete stereoselectivity, such as double stereodifferentiation and kinetic resolution.

### *5.3.4.1 Double stereodifferentiation*

In double stereodifferentiation,[33] an additional chirality is introduced to bias stereoselectivity. Thus, if *two* chiral reactants are coupled, then they may act cooperatively ('matched') or antagonistically ('mismatched'). In the 'matched' situation, stereoselectivity is enhanced relative to a reference reaction in which a single reactant is chiral. In the 'mismatched' case, stereoselectivity is diminished. Fig. 5.17 illustrates double stereodifferentiation with aldol reactions. The aldol reactions are *syn*-selective with *cis* enolates (see Section 7.8). Fig. 5.17A provides the reference reaction when a single reactant (aldehyde) is chiral. The ratio of two *syn* products is 4.3:1. If both the aldehyde and the enolate are chiral,

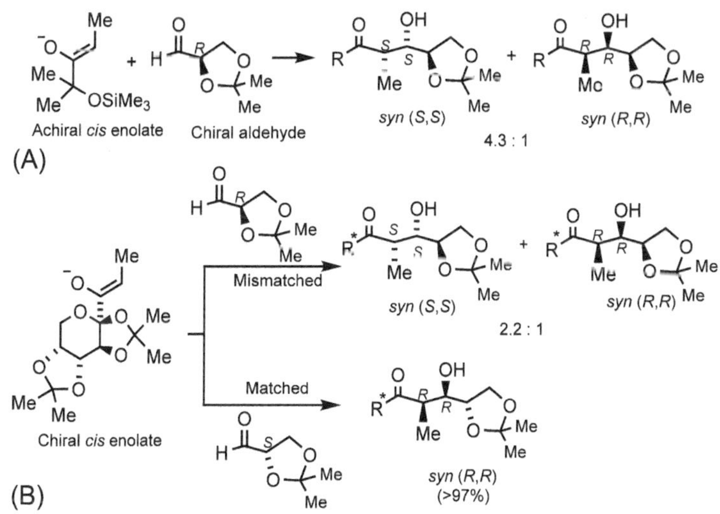


**FIG. 5.17** Double stereodifferentiation in aldol reactions: (A) Reference reaction; (B) Matched and mismatched reactions. In (A), R in products indicates the achiral substituent of achiral *cis* enolate; in (B), R* in products denotes the chiral moiety of chiral *cis* enolate.

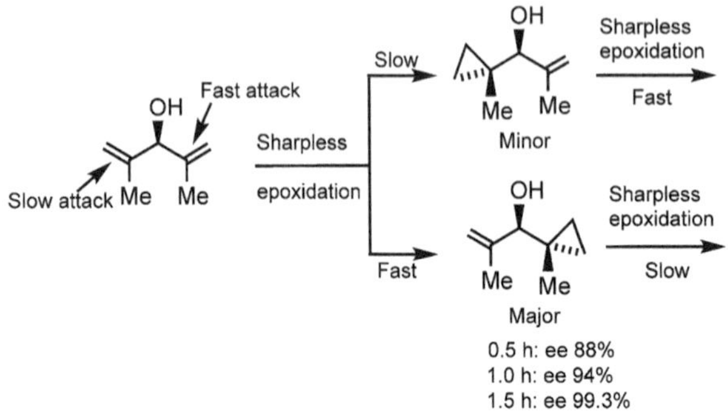


**FIG. 5.18** Kinetic resolution approach to improve enantioselectivity in an asymmetric synthesis (Sharpless asymmetric epoxidation).

their inherent diastereofacial preferences may reinforce (matched) or oppose (mismatched) as shown in Fig. 5.17B. The aldol reaction between the chiral enolate and the (*R*)-aldehyde gives a diastereomer ratio 2.2:1. This is a mismatched situation as the diastereoselectivity is diminished compared to the reference value of 4.3:1. But the reaction between the chiral enolate and the (*S*)-aldehyde gives the *syn* (*R*,*R*) product with a diastereoselectivity of >97% (matched).[34]

### 5.3.4.2 Kinetic resolution

Another approach to improve stereoselectivity is based on simultaneous asymmetric induction and kinetic resolution. This is illustrated using the Sharpless asymmetric epoxidation (see later Section 10.2.1) in Fig. 5.18.[35] Here the kinetic resolution involves a second Sharpless epoxidation of the products (major + minor) from the first Sharpless reaction. The key to this approach is to allow some of the products from the first reaction to proceed when the major product reacts slower but the minor product reacts faster, thereby achieving a very high ee of the remaining major product. Note that the minor product has a residual alkene moiety which is of the form that undergoes a fast reaction in the first step. Efficient kinetic resolution thus can enhance greatly the enantioselectivity of the initial asymmetric reaction.

## References

1. Eliel, E. L.; Wilen, S. H.; Mander, L. N. *Stereochemistry of Organic Compounds*; Wiley: New York, 1994; p. 647.
2. Winstein, S.; Holness, N. J. *J. Am. Chem. Soc.* **1955**, *77*, 5562.
3. Winstein, S.; Pressman, D.; Young, W. G. *J. Am. Chem. Soc.* **1939**, *61*, 1645.
4. Eliel, E. L.; Wilen, S. H.; Mander, L. N. *Stereochemistry of Organic Compounds*; Wiley: New York, 1994; p. 629.
5. Curtin, D. Y. *Rec. Chem. Prog.* **1954**, *15*, 111.

6. Seeman, J. I. *Chem. Rev.* **1983**, *83*, 83.
7. Robinson, M. J.; Organic, T. *Stereochemistry*; Oxford: Oxford University Press, 2000 (chapter 6).
8. Mandal, D. K. *Pericyclic Chemistry: Orbital Mechanisms and Stereochemistry*; Elsevier: Cambridge, MA, 2018; p. 58.
9. Eliel, E. L.; Wilen, S. H.; Mander, L. N. *Stereochemistry of Organic Compounds*; Wiley: New York, 1994; p. 837.
10. Procter, G. *Stereoselectivity in Organic Synthesis*; Oxford University Press: New York, 1998.
11. Procter, G. *Asymmetric Synthesis*; Oxford University Press: New York, 1996.
12. Stephenson, G. R., Ed. *Advanced Asymmetric Synthesis*; Chapman & Hall/Springer (India): London/New Delhi, 2010. 1996.
13. Bentley, R. In *Stereochemistry*; Tamm, C. H., Ed.; Elsevier: Amsterdam, 1982.
14. Stinson, S. C. *Chem. Eng. News* **1992**, *28*, 46.
15. Chan, A. S. C. *Chem. Tech.* **1993**, 46.
16. Tombo, R. G. M.; Bellus, D. *Angew. Chem. Int. Ed. Eng.* **1991**, *30*, 1193.
17. Rück-Braun, K.; Kunz, H. *Chiral Auxiliaries in Cycloadditions*; Wiley-VCH: Weinheim, 1999; p. 30.
18. Ager, D. J.; Prakash, I.; Schaad, D. R. *Chem. Rev.* **1996**, *96*, 835.
19. Evans, D. A. *Aldrichimica Acta* **1982**, *15*, 23.
20. Oppolzer, W.; Blagg, J.; Rodriguez, I.; Walther, E. *J. Am. Chem. Soc.* **1990**, *112*, 2767.
21. Bolm, C. In *Advanced Asymmetric Synthesis*; Stephenson, G. R., Ed.; Chapman & Hall: London, 1996.
22. Oguni, N.; Matsuda, Y.; Kaneko, T. *J. Am. Chem. Soc.* **1988**, *110*, 7877.
23. Agami, C. *Bull. Soc. Chim. Fr.* **1988**, 499.
24. Kitamura, M.; Suga, S.; Kawai, K.; Noyori, R. J. *Am. Chem. Soc.* **1986**, *108*, 6071.
25. Noyori, R.; Kitamura, M. *Angew. Chem. Int. Ed. Eng.* **1991**, *30*, 49.
26. Puchot, C.; Samuel, O.; Duñach, E.; Zhao, S.; Agami, C.; Kagan, H. B. *J. Am. Chem. Soc.* **1986**, *108*, 2353.
27. Guillaneux, D.; Zhao, S.-H.; Samuel, O.; Rainford, D.; Kagan, H. B. *J. Am. Chem. Soc.* **1994**, *116*, 9430.
28. Kitamura, M.; Okada, S.; Suga, S.; Noyori, R. *J. Am. Chem. Soc.* **1989**, *111*, 4028.
29. Soai, K.; Shibata, T.; Morioka, H.; Choji, K. *Nature* **1995**, *378*, 767.
30. Frank, F. C. *Biochim. Biophys. Acta* **1953**, *11*, 459.
31. Blackmond, D. G. *Proc. Natl. Acad. Sci. U. S. A.* **2004**, *101*, 5732.
32. Girard, C.; Kagan, H. B. *Angew. Chem. Int. Ed. Eng.* **1998**, *37*, 2922.
33. Masamune, S.; Choy, W.; Peterson, J. S.; Sita, L. R. *Angew. Chem. Int. Ed. Eng.* **1985**, *24*, 1.
34. Heathcock, C. H.; White, C. T.; Morrison, J. J.; Van Derveer, D. J. *Org. Chem.* **1981**, *46*, 1296.
35. Schreiber, S. L.; Schreiber, T. S.; Smith, D. B. *J. Am. Chem. Soc.* **1987**, *109*, 1525.

# Part III

# Stereochemistry of organic reactions

Part III consists of seven chapters (Chapters 6–12) which describe reaction stereochemistry with a major emphasis on diastereoselectivity and asymmetric synthesis, involving reactions of all mechanistic classes covering ionic, pericyclic, transition metal-catalysed, radical and photochemical reactions.

Chapter 6

# Ionic reactions 1: Fundamental stereochemistry

## 6.1 Substitution reactions

### 6.1.1 Nucleophilic substitution ($S_N2$, $S_N1$, $S_Ni$)

The stereoelectronic factor for an $S_N2$ reaction is the linear back-side approach of the nucleophile to the tetrahedral carbon bearing the leaving group (see Section 4.3.1). If the attacking carbon is a stereocentre, the stereochemical consequence is the inversion of configuration. On the other hand, an $S_N1$ reaction proceeds via a planar carbocation intermediate, and *if* the nucleophile can attack the two faces of the planar carbocation with equal probability, the result is 50% inversion and 50% retention, that is racemization.

Fig. 6.1A shows that the $S_N2$ reaction of 2-octyl tosylate with a moderately good nucleophile ($AcO^-$) takes place with complete inversion.[1] This $S_N2$ inversion can be used to convert an enantiomer of 2-octanol into its opposite enantiomer, as shown in Fig. 6.1B. The conversion of OH into a good leaving group OTs with TsCl (Ts = *p*-toluenesulphonyl) in the first step involves an $S_N2$ reaction on the tetrahedral S of the sulphonyl chloride and the hydrolysis of the acetate ester in the third step proceeds with attack at the carbonyl carbon. Thus, the first and third steps do not affect the stereochemistry of the chiral carbon. It is the $S_N2$ reaction in the second step that leads to the inversion at the stereocentre in the final product.

In $S_N1$ solvolysis of chiral substrates, the ionization step gives a carbocation that is planar and achiral. But only if the carbocation is sufficiently stable and long-lived under the reaction conditions so as to diffuse away from the leaving group, it can be solvated symmetrically and can lead to a racemic product. However, in practice, this condition is met rarely, and a partial racemization with *net inversion of configuration* is frequently observed.[2] If the carbocation is generated in super acid, and subsequently quenched with water, a racemic alcohol will result, as shown for the racemization of optically active 2-octanol in Fig. 6.2A.

Stereochemistry and Organic Reactions. https://doi.org/10.1016/B978-0-12-824092-2.00006-X

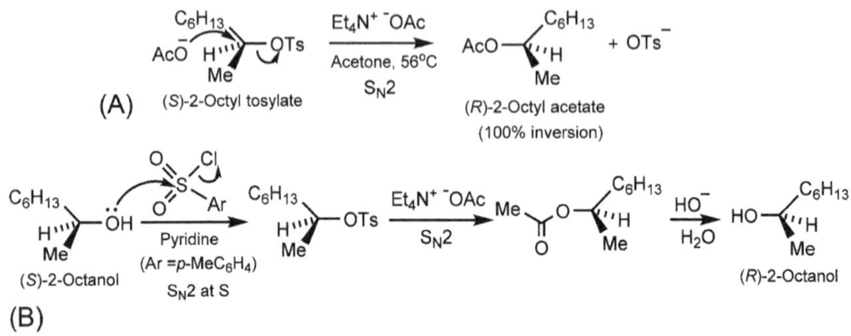


**FIG. 6.1** $S_N2$ reaction: (A) Inversion of configuration; (B) Conversion of (*S*)-2-octanol into (*R*)-2-octanol.

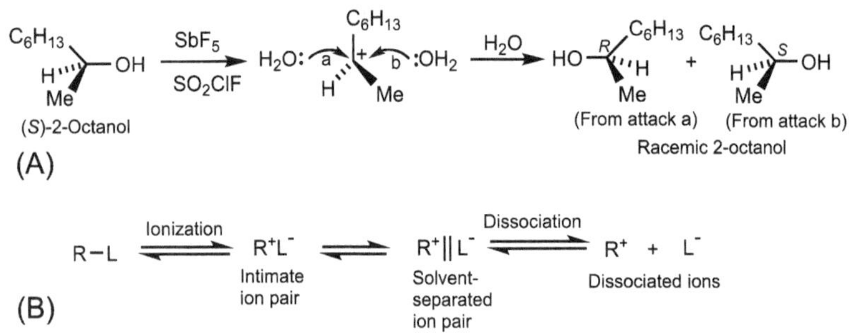


**FIG. 6.2** $S_N1$ reaction: (A) Racemization through a planar carbocation; (B) Different stages of the ionization process.

In general, the stereochemistry of $S_N1$ reactions can be rationalized in terms of three different stages of the ionization process (Fig. 6.2B).[3,4] The intimate or tight ion pair represents a very close association between the cation and the anion. This intimate ion pair can also recombine to give the original substrate, called internal return. The solvent-separated ion pair retains an association between the ions, but with intervening solvent molecules. In the dissociated ions, the carbocation is free and is symmetrically solvated. But in the first two stages, the solvation of the carbocation is asymmetric, and there occurs a tendency towards net inversion of configuration from the electrostatic shielding of one face of the carbocation by the anion in the ion pair.

A *part* of the leaving group in the intimate ion pair can sometimes attack the chiral centre, detaching itself from the rest of the leaving group in the process. This type of attack necessarily takes place from the front face (i.e. the original ionizing face) since it cannot get to the back side, and thereby results in the retention of configuration. This kind of mechanism is referred to as the $S_Ni$ mechanism (substitution nucleophilic internal).[5,6] The most important example is the reaction of a chiral alcohol with thionyl chloride (Fig. 6.3).

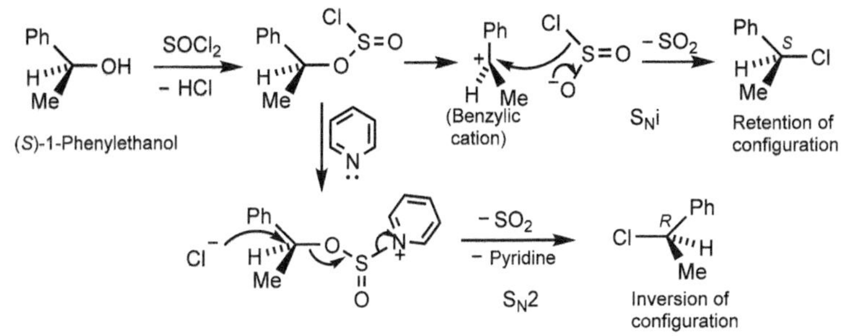


**FIG. 6.3** $S_Ni$ reaction between a chiral alcohol and thionyl chloride with retention of configuration, and $S_N2$ mechanism in the presence of pyridine.

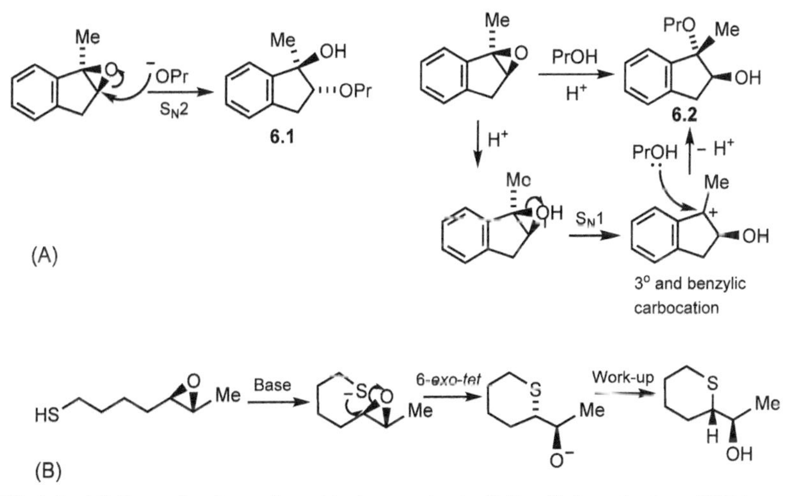


**FIG. 6.4** (A) Stereochemistry of epoxide ring opening by $S_N2$ or $S_N1$ mechanism; (B) Intramolecular $S_N2$ reaction in epoxide ring opening.

The intermediate alkyl chlorosulphite (which ionizes to give the intimate ion pair) can be isolated. Interestingly, addition of pyridine into the reaction mixture leads to the inversion of configuration (Fig. 6.3). This is because pyridine reacts with the chlorosulphite to give free $Cl^-$, which then attacks from the back side.

The ring opening of epoxides by the $S_N2$ or $S_N1$ mechanism is illustrated in Fig. 6.4A.[7] An $S_N2$ reaction is favoured with the more nucleophilic $PrO^-$, which involves attack at the less hindered carbon with inversion to give the product **6.1**. But the epoxide ring opening with the less nucleophilic PrOH in the presence of acid catalyst promotes an $S_N1$ reaction with formation of the stable carbocation (tertiary and benzylic). Thus, PrOH adds to the more substituted position of the epoxide to give the product **6.2**. Inversion occurs as the nucleophile prefers to add from the less hindered face opposite to the OH group.

(Alternatively, the reaction might be considered as an $S_N2$ attack to the benzylic centre having a partial positive charge when the carbocation is about to form but not formed yet.)

Fig. 6.4B shows an intramolecular $S_N2$ reaction (6-*exo-tet*) at one end of the epoxide with base. The reaction proceeds stereospecifically with inversion, and gives a single diastereomeric product.

**Problem 6.1** Predict the stereochemistry of the products in the following reactions:

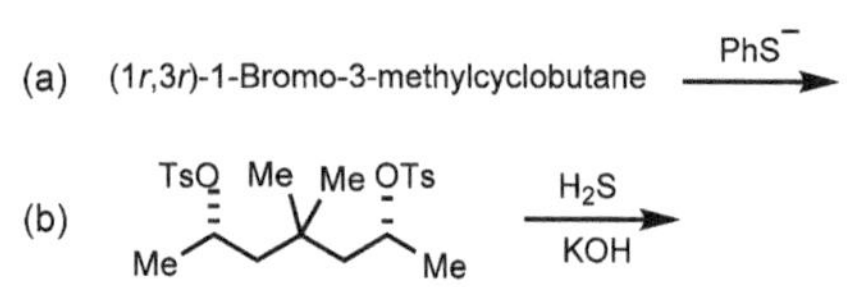


**Problem 6.2** Identify the compounds (**A–F**) in the Pfizer synthesis of the antidepressant reboxetine:

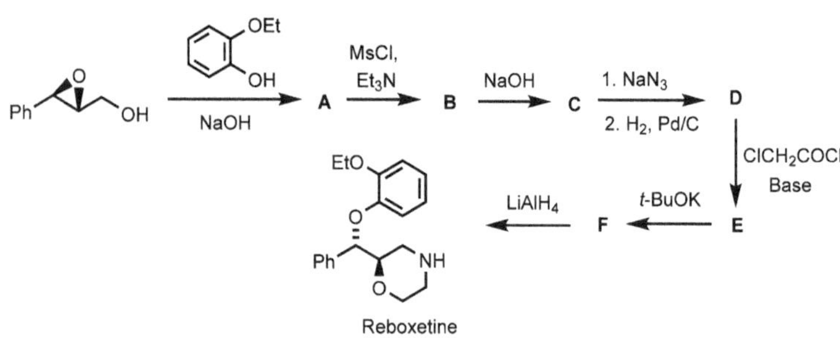


## 6.1.2 Neighbouring-group participation

The nucleophilic substitution of a substrate that possesses an internal nucleophile (a neighbouring group) can proceed through two sequential $S_N2$ steps and lead to *retention* of configuration (inversion + inversion = retention) at the substitution centre. This mechanism is known as the neighbouring-group participation (NGP).[8,9] A neighbouring group may be a lone pair, a π bond or a σ bond. Apart from stereochemistry, an important feature of NGP is the high acceleration of the reaction rate often by several orders of magnitude, called *anchimeric assistance*. The first $S_N2$ step gives a cyclic intermediate (usually three- or five-membered ring) favoured by the entropy factor, and the second $S_N2$ step involves attack by the external nucleophile leading to opening of the strained ring. Thus, each of these steps is much faster than the ordinary $S_N2$ reaction in the absence of NGP.

### 6.1.2.1 n (lone pair) participation

Consider the reactions of (*S*)-2-bromopropanoic acid with concentrated NaOH, and with dilute NaOH containing $Ag_2O$ (Fig. 6.5). The former is an $S_N2$ reaction that proceeds with inversion of configuration to give (*R*)-lactic acid. The second-order reaction is quite favourable in the presence of high concentration of nucleophile ($HO^-$) that attacks the carbon α to a carbonyl group. In contrast, the latter reaction gives (*S*)-lactic acid with retention of configuration. $Ag_2O$ as a halogen-selective Lewis acid assists the ionization of the C—Br bond and a neighbouring carboxylate (with lone pair on oxygen) readily participates by an intramolecular $S_N2$ mechanism to form a three-membered lactone intermediate. Note that formation of an unstable carbocation in an $S_N1$ pathway is also avoided by the participation. The strained lactone ring is then opened by hydroxide ion/water in a second intermolecular $S_N2$ step leading to overall retention of configuration.

Fig. 6.6 depicts a classic example of NGP in a cyclohexyl system with the acetolysis of *cis*- and *trans*-2-acetoxycyclohexyl tosylates.[10] Both substrates are chiral, and the acetolysis reactions are performed with one enantiomer of each diastereomer. The *cis* (1*R*,2*S*) enantiomer gives the (*S*,*S*) enantiomer of the *trans* product. Here the reaction proceeds by a simple $S_N2$ mechanism with inversion at the reaction centre (C1). On the other hand, the *trans* (1*S*,2*S*) enantiomer gives the *trans* diacetate which is racemic. Thus, the enantiomerically pure *trans* tosylate undergoes acetolysis with retention of relative stereochemistry but with loss of absolute stereochemical information. This can be rationalized by NGP.[11] In cyclohexyl system, the back-side attack ($S_N2$) requires that the neighbouring group (ÖAc) is anti to the leaving group (OTs). Therefore, participation occurs via the reactive but less populated diaxial conformation (a-OTs, a-OAc). Since the resulting five-membered ring intermediate (acetoxonium ion) is delocalized, symmetrical and achiral, attack by the nucleophile (AcÖH) on either carbon from the bottom side leads to equal amounts of

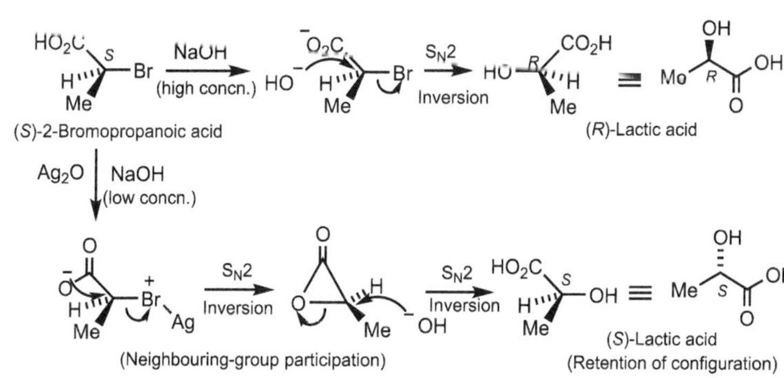


**FIG. 6.5** Neighbouring-group participation in the reaction of (*S*)-2-bromopropanoic acid with $Ag_2O$ and a low concentration of NaOH.

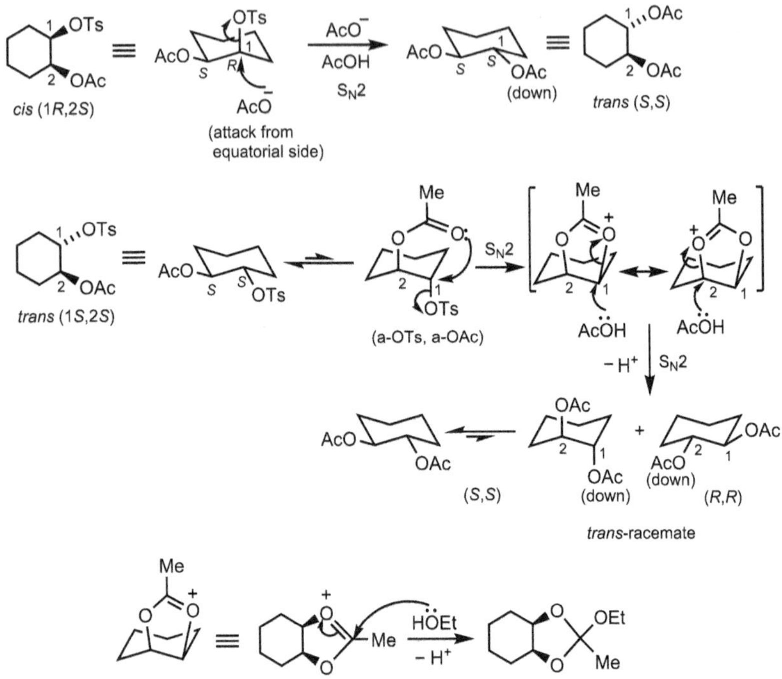


6.3

**FIG. 6.6** Acetolysis of *cis*- and *trans*-2-acetoxycyclohexyl tosylates.

(*R*,*R*) and (*S*,*S*) enantiomers, that is a racemic product. It should be noted that NGP is not possible in the axial-equatorial conformation of the *cis* isomer.

The intermediate oxonium ion can be trapped as a cyclic orthoester **6.3** when the solvolysis is carried out in ethanol, thereby providing an additional evidence for its formation.[12] The *trans* isomer is more reactive and undergoes acetolysis about 670 times faster than the *cis* isomer, thereby showing anchimeric assistance in NGP.[13]

**Problem 6.3** Predict the products of the following reactions:

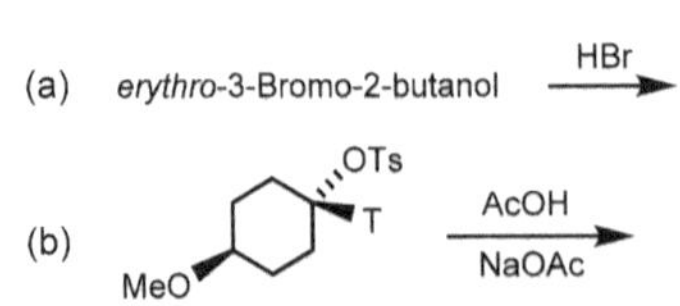


## 6.1.2.2 π bond participation

The participation of aromatic π bond has been well studied.[14] The intermediate species that results from participation by a β-phenyl group is termed a

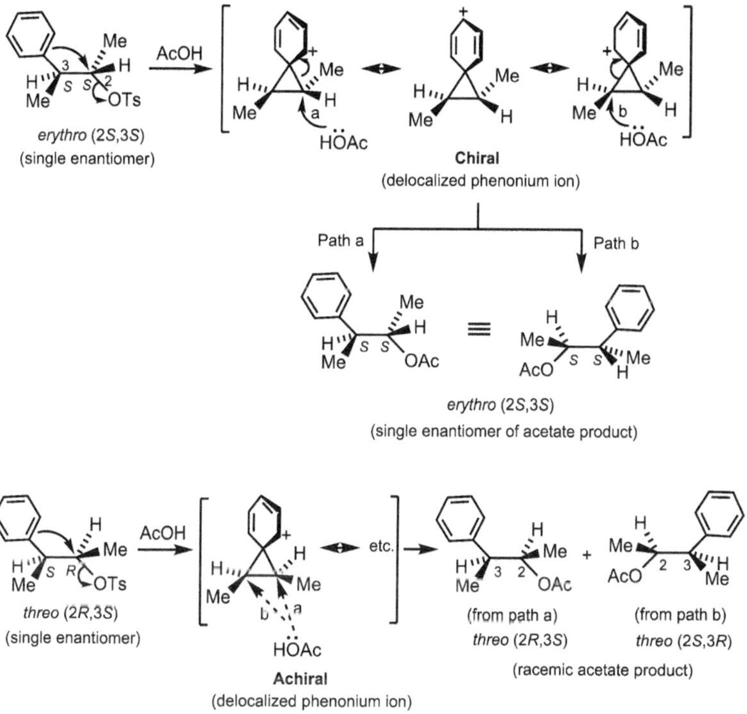


**FIG. 6.7** Acetolysis of *erythro*- and *threo*-3-phenyl-2-butyl tosylates.

phenonium ion, a bridged carbocation with a delocalized positive charge. Consider the solvolysis of *erythro*- and *threo*-3-phenyl-2-butyl tosylates in acetic acid (Fig. 6.7).[15] An enantiomerically pure *erythro* isomer (2*S*,3*S*) gives an enantiomerically pure *erythro* product (2*S*,3*S*), signifying retention of both relative and absolute configuration. Here the intermediate phenonium ion is chiral and attack by acetic acid on either end of the three-membered ring gives the same enantiomer of the *erythro* product.

In the case of enantiomerically pure *threo* tosylate (2*R*,3*S*), the intermediate phenonium ion is achiral. The nucleophilic attack on either carbon in the three-membered ring then leads to a racemic *threo* product, showing retention of relative stereochemistry but loss of absolute stereochemical information.

Fig. 6.8 illustrates an example of π bond participation in a bicyclic norbornenyl system. The acetolysis of *anti*-7-norbornenyl tosylate gives the *anti* acetate with retention of configuration.[16,17] The reaction is $10^{11}$ times faster than the saturated norbornyl analogue **6.4**. This indicates strong anchimeric assistance by the π bond for the departure of the tosyl group. The frontier orbital interaction is $HOMO_{\pi}/LUMO_{\sigma^*(C—O)}$ overlap. The intermediate is a nonclassical carbocation, a bridged cation with a pentacoordinate carbon and a

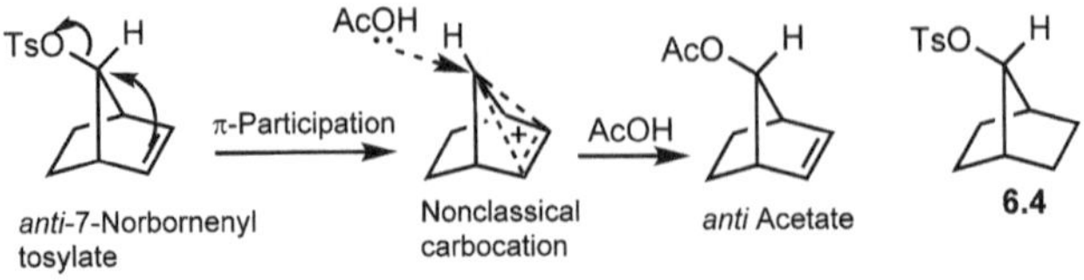


**FIG. 6.8** Acetolysis of *anti*-7-norbornenyl tosylate via π bond participation and consequent formation of a nonclassical carbocation.

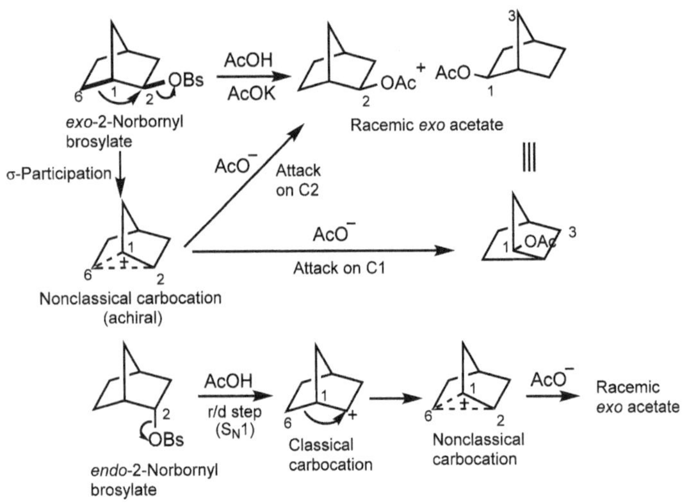


**FIG. 6.9** Acetolysis of *exo*-2-norbornyl brosylate with σ bond participation vis-a-vis acetolysis of *endo* brosylate without σ participation (Bs=*p*-$BrC_6H_4SO_2$).

three-centre two-electron bond. The acetic acid can attack only from the side which was previously occupied by the leaving group, and leads to retention of stereochemistry in the product.

### 6.1.2.3 σ bond participation

The electron pair in a σ bond can also participate in the assisted ionization of the leaving group in nucleophilic substitution. This is illustrated with reference to the solvolysis of *exo*- and *endo*-2-norbornyl brosylates in AcOH/KOAc (Fig. 6.9).[18,19] The acetolysis of the *exo* brosylate gives exclusively the *exo*-2-norbornyl acetate with retention of relative configuration. But if the *exo* brosylate is enantiomerically pure, the *exo* acetate product becomes completely racemic. These results are explained in terms of a nonclassical carbocation resulting from participation of 1–6 σ bond (highlighted in bold) that is anti to the leaving group (*exo*-OBs). Note that C6 is pentacoordinate in the bridged cation having a three-centre two-electron bond. The nonclassical cation has a plane of symmetry bisecting the bridging bonds, and is achiral. Thus the attack of the acetate ion on C2 and C1 takes place from the direction opposite the bridging interaction with equal probability, leading to a 1:1 mixture of the enantiomeric *exo* acetates.

The *exo* brosylate is more reactive and undergoes acetolysis 350 times faster than the *endo* brosylate does. The acetolysis of the *endo* brosylate does not involve participation by 1–6 σ bond (as it is not anti to *endo*-OBs) and proceeds by the unassisted formation of a classical carbocation in the rate-determining step. This explains the rate difference between the two isomers. However, the *endo* brosylate also produces the *exo* acetate that is racemic (>93%). This may arise from the formation of a nonclassical carbocation after the r/d step.

### 6.1.3 Relative rates of axial and equatorial isomers in nucleophilic substitution, chromic acid oxidation and ester hydrolysis

Relative rates of reactions of cyclohexyl axial and equatorial isomers refer to their free energy of activation $\Delta G_a^{\neq}$ and $\Delta G_e^{\neq}$. Two situations can arise in terms of steric effects: steric assistance (Fig. 6.10A) and steric hindrance (Fig. 6.10B).[20]

If the steric repulsion is more important in the ground state than in the transition state, the ground state free energy difference ($G_a - G_e$) between the axial and equatorial isomers will be greater than the transition state free energy difference ($G_a^{\neq} - G_e^{\neq}$). Therefore, $\Delta G_a^{\neq} < \Delta G_e^{\neq}$ as shown in Fig. 6.10A, and the axial isomer will react faster than the equatorial isomer. This is referred to as the case of steric assistance.

But if the steric repulsion is greater in the transition state than in the ground state, $\Delta G_a^{\neq} > \Delta G_e^{\neq}$ (Fig. 6.10B), and the axial isomer will react slower than the equatorial isomer. This situation is called the case of steric hindrance.

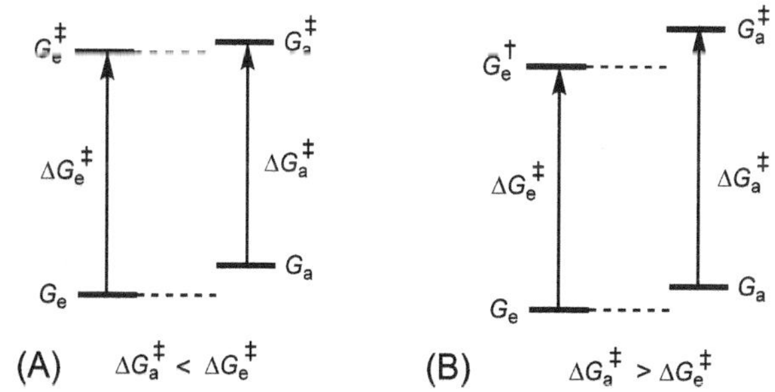


**FIG. 6.10** Free energy diagram of (A) steric assistance and (B) steric hindrance.

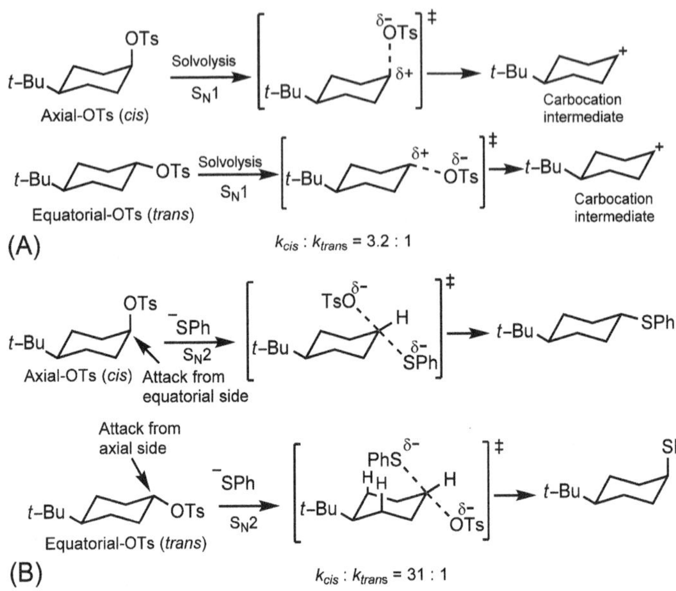


**FIG. 6.11** Steric assistance and relative rates in nucleophilic substitution of *cis*- and *trans*-4-*t*-butylcyclohexyl tosylates in (A) $S_N1$ solvolysis and (B) $S_N2$ reactions. The stereoelectronic back-side attack is an additional factor in the $S_N2$ reaction.

Fig. 6.11 illustrates the steric assistance in nucleophilic substitution of anancomeric *cis*- and *trans*-4-*t*-butylcyclohexyl tosylates in which the *t*-Bu group is equatorial. In $S_N1$ solvolysis such as ethanolysis or acetolysis, the rate-determining formation of a carbocation (hybridization changes from $sp^3$ to $sp^2$) relieves the steric strain mainly in the TS for the axial-OTs isomer (Fig. 6.11A). Since the ground state free energy $G_a > G_e$, and the TS free energy $G_a^{\neq} \sim G_e^{\neq}$ for the same carbocation involved, $\Delta G_a^{\neq} < \Delta G_e^{\neq}$. Thus the axial *cis* isomer reacts faster than the equatorial *trans* isomer ($k_{cis}/k_{trans} = 3.2$).

In $S_N2$ reactions with thiophenolate (Fig. 6.11B), the *cis* isomer reacts about 31 times faster than the *trans* isomer.[21] Here, besides steric assistance, an additional stereoelectronic factor is responsible for the larger rate difference. The back-side attack of the nucleophile from the axial side in the case of *trans* (equatorial-OTs) isomer gives rise to 1,3-diaxial steric strain, whereas no such strain is involved in the back-side attack from the equatorial side for the *cis* (axial-OTs) isomer.

The steric assistance also influences the relative rates of chromic acid oxidation of anancomeric cyclohexanols.[22] Fig. 6.12 shows that the axial-OH (*cis*) isomer of 4-*t*-butylcyclohexanol is oxidized at a faster rate than the equatorial-OH (*trans*) isomer does ($k_{cis}/k_{trans} = 3.23$). The chromic acid oxidation proceeds through the formation of a chromate ester intermediate in a fast step, followed

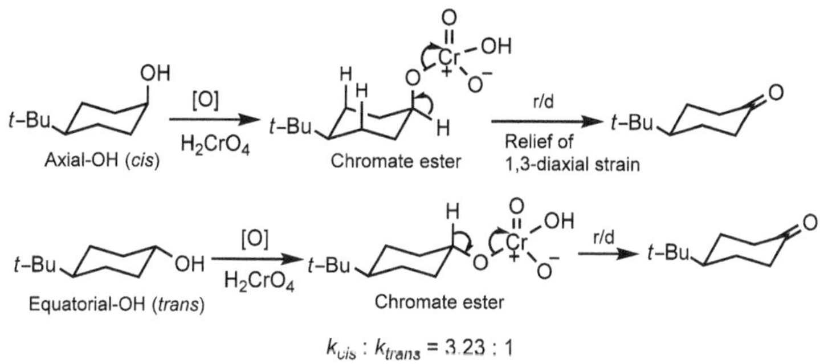


**FIG. 6.12** Steric assistance and relative rates in chromic acid oxidation of cyclohexanols.

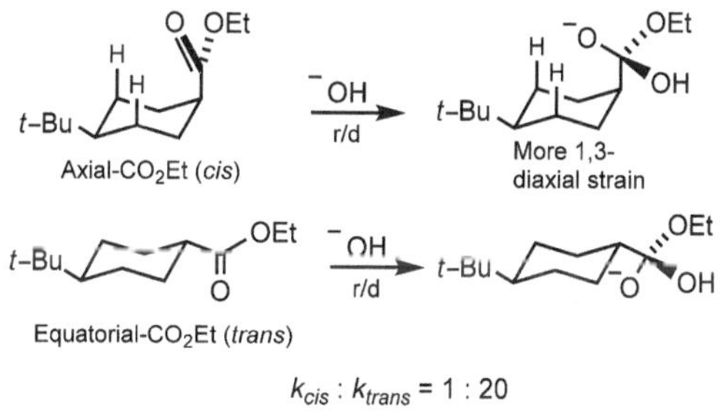


**FIG. 6.13** Steric hindrance and relative rates in saponification of *cis* and *trans* isomers of ethyl 4-*t*-butylcyclohexanecarboxylate.

by its conversion in the rate-determining step to the ketone with change of hybridization from $sp^3$ to $sp^2$. The *cis* isomer reacts faster because of the relief of two 1,3-diaxial interactions in the TS, whereas no such relief occurs in the case of the *trans* isomer.

The hydrolysis or saponification of anancomeric esters presents a case of steric hindrance (see Fig. 6.10B).[23,24] Fig. 6.13 shows that the *cis* (axial-$CO_2Et$) isomer of ethyl 4-*t*-butylcyclohexanecarboxylate undergoes hydrolysis about 20 times slower than the *trans* (equatorial-$CO_2Et$) isomer. The reaction proceeds through the formation of an intermediate in the rate-determining step with change of hybridization from $sp^2$ to $sp^3$. The steric strain in the case of *cis* (axial-$CO_2Et$) isomer is therefore increased in the TS relative to that in the ground state (cf. the conformational free energies of $sp^2$ and $sp^3$ substituents in Table 3.2; for the more solvated anionic $sp^3$ intermediate, the conformational free energy value will increase further). But no such increase in steric strain in the TS occurs for the *trans* isomer. The *trans* isomer therefore reacts about 20 times faster than the *cis* isomer. It may be mentioned that ethyl cyclohexanecarboxylate that exists as a mixture of equatorial (major) and axial (minor) conformers undergoes hydrolysis about 17 times faster than the *cis* isomer.

**Problem 6.4** Which cyclohexanol in each pair will react faster in chromic acid oxidation? Explain.

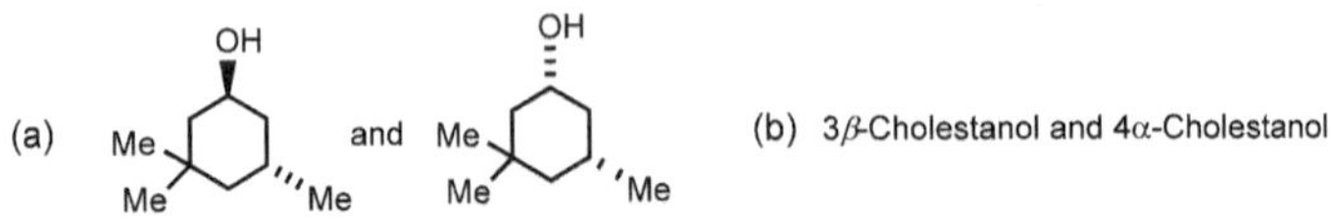


## 6.1.4 Electrophilic substitution ($S_E2$)

In contrast to $S_N2$ reactions, an electrophilic substitution ($S_E2$) on an organometallic substrate (usually an alkyllithium) can take place with retention or inversion of configuration, although the former is more common (see Section 4.3.2). In some cases, the two pathways are delicately balanced.[25] Fig. 6.14 illustrates $S_E2$ reactions using *exo*-2-norbornyllithium as a substrate.[26] Depending on the electrophile employed, retention or inversion has been observed. However, since organolithiums are not monomeric in solution, the stereochemical outcomes are often less clear-cut.

# 6.2 Elimination reactions

As regards stereochemistry, the most important ionic eliminations are E2 eliminations. The stereoelectronic factor for the E2 elimination is the *antiperiplanarity* of the two departing groups (see Section 4.3.3). In unconstrained systems, a bimolecular E2 elimination almost invariably proceeds in antiperiplanar fashion. Here we will describe briefly the stereochemistry of E2 eliminations in both acyclic and cyclic systems. Acyclic systems are generally flexible with many conformations available to the molecule that may satisfy the stereoelectronic requirements. In contrast, cyclic systems are relatively rigid with a single or a small number of conformations. Thus, the stereochemistry of E2 reactions in cyclic systems is well controlled.

The stereochemistry of carbon–carbon double bond formation by pyrolytic *syn* eliminations and the Wittig reaction which are pericyclic in nature will be described in Chapter 8.

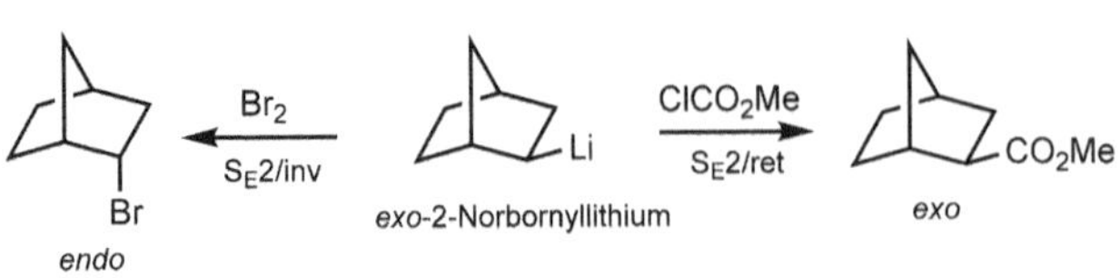


**FIG. 6.14** Examples of $S_E2$ reactions showing retention or inversion of configuration.

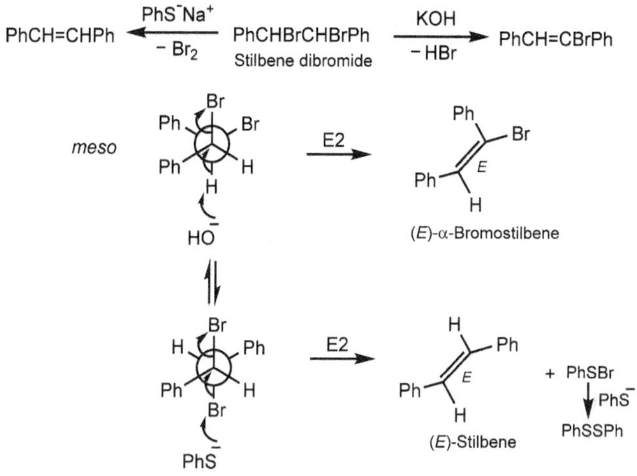


**FIG. 6.15** E2 eliminations of *meso*-stilbene dibromide in the presence of KOH and $PhS^-Na^+$ respectively.

## 6.2.1 E2 elimination in acyclic systems

Examples of stereospecific and stereoselective E2 eliminations in acyclic systems have been described previously with reference to the Winstein–Holness equation and the Curtin–Hammett principle (see Figs 5.1 and 5.3A). As further examples, E2 eliminations of *meso*-1,2-dibromo-1,2-diphenylethane (*meso*-stilbene dibromide) in the presence of KOH and $PhS^-Na^+$ respectively are shown in Fig. 6.15.[27] In KOH, the elimination is dehydrobromination but in $PhS^-Na^+$, debromination occurs. In each case, the *meso*-substrate has a single reactive conformer in which the two departing groups are anti. As shown in Fig. 6.15, the KOH induced elimination gives (*E*)-α-bromostilbene, and the elimination with $PhS^-Na^+$ produces (*E*)-stilbene. Similarly, the chiral diastereomer of stilbene dibromide would give (*Z*)-α-bromostilbene and (*Z*)-stilbene in KOH and $PhS^-Na^+$ induced eliminations, respectively (verify). These E2 eliminations are therefore stereospecific.

**Problem 6.5** Only one of the two diastereomers of stilbene dichloride undergoes dehydrohalogenation with pyridine at 200°C. Identify the diastereomer.

**Problem 6.6** Predict the stereochemistry of the products in the following Peterson reactions:

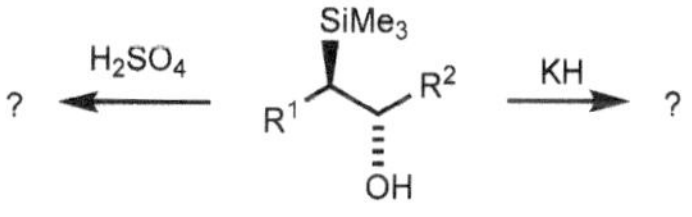

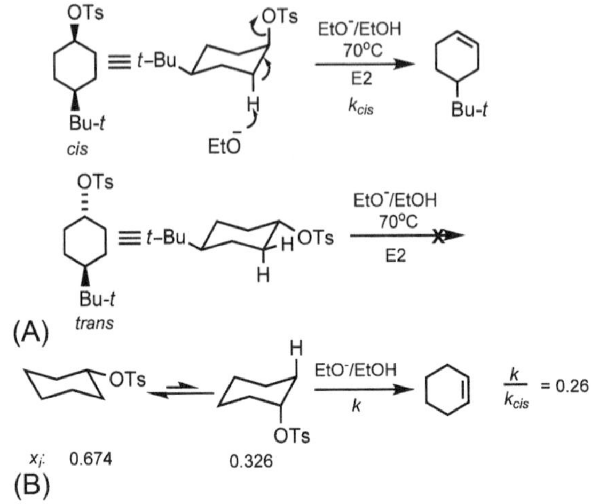


**FIG. 6.16** E2 eliminations with (A) *cis*- and *trans*-4-*t*-butylcyclohexyl tosylates and (B) cyclohexyl tosylate.

## 6.2.2 E2 elimination in cyclic systems[28,29]

In cyclohexyl systems, the required antiperiplanar conformation of the departing groups in E2 elimination is the chair diaxial conformation (see Fig. 4.9B). Fig. 6.16A shows that anancomeric *cis*-4-*t*-butylcyclohexyl tosylate undergoes E2 elimination readily with base but the *trans* isomer does not. Since in the *cis* isomer, OTs and H on adjacent carbons are diaxial (antiperiplanar), E2 elimination takes place readily to give 4-*t*-butylcyclohexene. Note that the conformation of the cyclohexene product will be half-chair. But in the *trans* isomer, OTs and H are diequatorial (e,e) or equatorial-axial (e,a), and E2 does not occur.

Cyclohexyl tosylate itself undergoes the E2 elimination slower than the *cis* isomer of 4-*t*-butylcyclohexyl tosylate ($k/k_{cis}=0.26$ where $k$ and $k_{cis}$ are the rate constants for cyclohexyl tosylate and the *cis* isomer respectively) (Fig. 6.16B). Using the Winstein–Holness equation (see Section 5.1.1),

$$k = x_e k_e + x_a k_a = x_a k_a \text{ (since } k_e = 0) = 0.326\, k_a$$

$$k_{cis} = x_a k_a = k_a \text{ (since } x_a \approx 1).$$

Thus, $k/k_{cis}=0.326$, which agrees reasonably well with the experimental value.

Fig. 6.17 shows that diastereomeric neomenthyl chloride and menthyl chloride (differing in the stereochemistry of Cl) undergo E2 dehydrohalogenation differently with respect to reactivity and product composition under the same conditions with base. Neomenthyl chloride reacts 250 times faster than menthyl chloride and gives a 3:1 mixture of 3-menthene and 2-menthene, whereas menthyl chloride produces only 2-menthene. Neomenthyl chloride can react

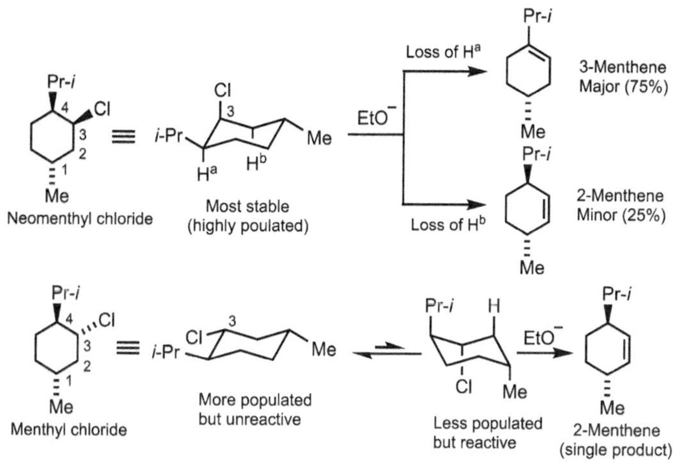


**FIG. 6.17** E2 eliminations of neomenthyl chloride and menthyl chloride.

via its most populated (lowest energy) conformer in which Cl and H are antiperiplanar (Fig. 6.17). In contrast, the all-equatorial lowest energy conformer of menthyl chloride in which Cl and H are not antiperiplanar, is unreactive. Ring flipping can produce the all-axial conformer which is much higher in energy but only in this conformer can E2 dehydrochlorination take place. Since the concentration of the reactive conformer of menthyl chloride is low, its rate is much lower than that of neomenthyl chloride. In the case of neomenthyl chloride, there are two antiperiplanar protons (labelled $H^a$, $H^b$) that can be eliminated producing two alkenes with a preference for the more substituted Saytzeff alkene (3-menthene). But the reactive conformer of menthyl chloride has only one antiperiplanar proton for removal by base giving a single alkene (2-menthene).

The different stereochemical course of E2 eliminations of two diastereomeric carboxylic acids is depicted in Fig. 6.18A. The *cis* acid undergoes dehydrobromination as Br and H are only antiperiplanar. But in the *trans* acid, since Br is also antiperiplanar with $CO_2^-$, decarboxylation takes place as it is expected to be faster than dehydrobromination.

Fig. 6.18B shows that both *cis*- and *trans*-1,2-dibromocyclohexanes on treatment with iodide give cyclohexene. For the *trans* isomer, the E2 debromination takes place readily from the antiperiplanar diaxial (a,a) conformation but the (e,a) conformer of the *cis* isomer cannot undergo E2 elimination directly. The kinetic studies show that the *cis* isomer reacts about 11.5 times slower than the *trans* isomer. Presumably, the *cis* dibromide undergoes an $S_N2$ reaction in a slow, rate-determining step followed by a fast diaxial (a-I,a-Br) elimination.

If the departing groups could not achieve antiperiplanarity, syn elimination can take place. For example, elimination from the deuterated norbornyl bromide (X=Br) or deuterated norbornyl trimethylammonium ion ($X=NMe_3^+$) with

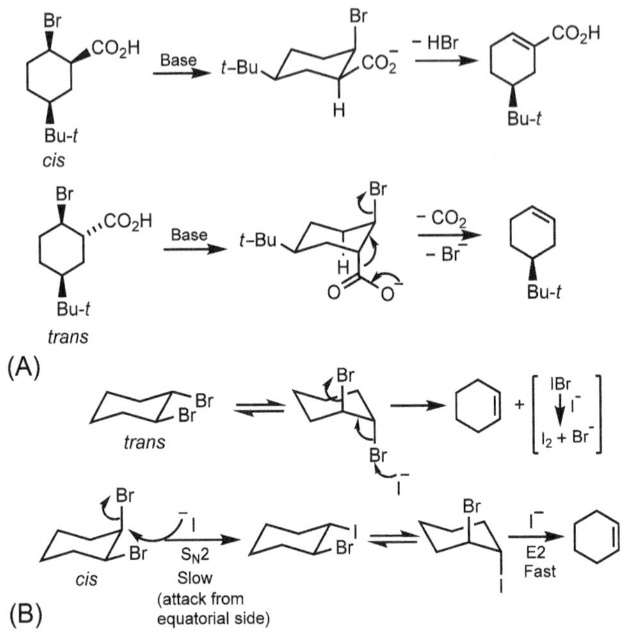


**FIG. 6.18** (A) Stereochemical course of E2 elimination of two diastereomers of 2-bromo-5-*t*-butylcyclohexanecarboxylic acid; (B) Iodide induced elimination of *cis*- and *trans*-1,2-dibromocyclohexanes.

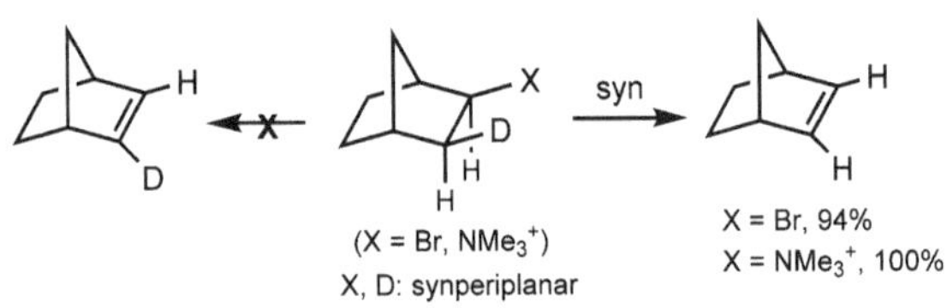


**FIG. 6.19** syn Elimination in norbornyl systems.

base is syn (X and D are synperiplanar), and gives norbornene containing no deuterium (Fig. 6.19).[30,31] Here the rigid ring system prohibits antiperiplanarity. Furthermore, there will be a steric constraint for the removal of the *endo* proton. In general, $NMe_3^+$ has a greater tendency for syn elimination than do other groups such as Br, Cl and OTs.

Finally, it should be noted that cyclohexane rings are the most important rings for antiperiplanar E2 eliminations, and other rings are not so selective. For example, in the decomposition of cycloalkyltrimethylammonium hydroxides, the extent of syn elimination with ring size is: four-membered (90%), five-membered (46%), six-membered (4%), seven-membered (31%–37%).[32]

**Problem 6.7** One diastereomer (β) of 1,2,3,4,5,6-hexachlorocyclohexane undergoes HCl elimination with base 7000 times slower than the slowest of the other seven diastereomers. Identify the β isomer.

**Problem 6.8** Which of the following bromides would eliminate HBr with base? Explain.

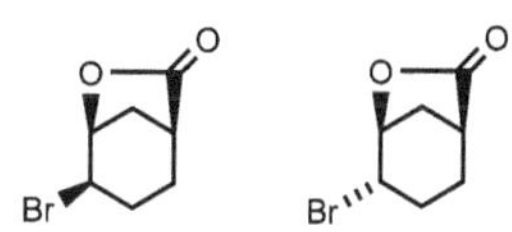


**Problem 6.9** Account for the following observations:

**(a)** The compound **A** undergoes HCl elimination much slower than the corresponding nonbridged compound.
**(b)** The compound **B** undergoes HCl elimination about 8 times faster than the compound **A**.

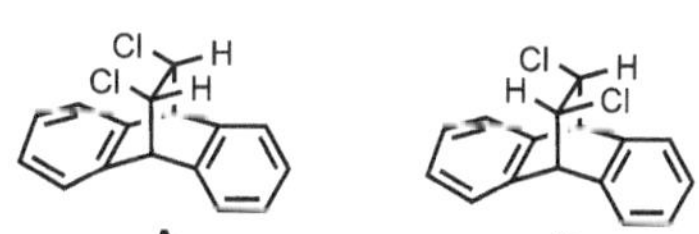


**A** **B**

## 6.3 Fragmentation reactions

The stereoelectronic requirement for fragmentation reactions is similar to E2 eliminations in the sense that the two breaking bonds are *antiperiplanar*. Fragmentations take place easily from 1,3-difunctionalized compounds, in which the breaking C—C and C—L bonds are aligned antiperiplanar (see Section 4.3.6). The fragmentation process however needs electron push by an electron pair (usually a lone pair) and electron pull by the departing group (L). To exert the electron push, the lone pair must also be antiperiplanar to the breaking C—C bond (see Section 4.3.6). The fragmentation reactions, called Grob fragmentations, are very useful for the synthesis of alkenals, alkenones and alkynones.[33–35]

Fig. 6.20A shows that *cis*-3-dimethylaminocyclohexyl tosylate **6.5** undergoes easy fragmentation on treatment with triethylamine in 80% ethanol to form an intermediate iminium ion which then hydrolyzes to give the product aldehyde.[36] Fragmentation occurs because the breaking C—C and C—O bonds

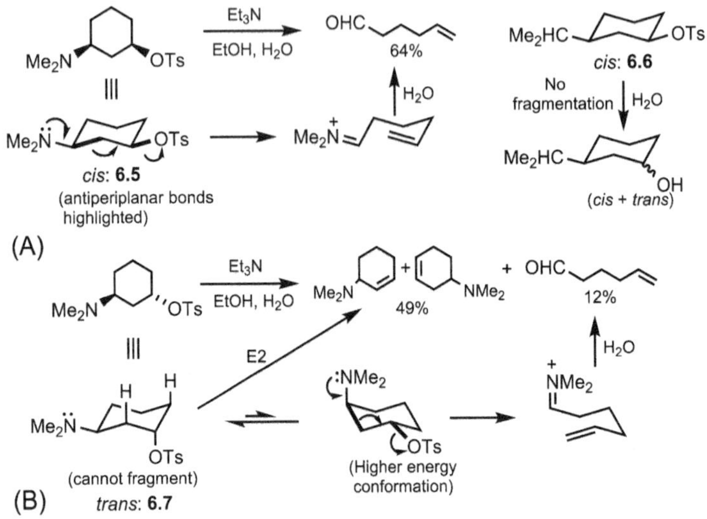


**FIG. 6.20** (A) Fragmentation reaction in the solvolysis of *cis*-3-dimethylaminocyclohexyl tosylate **6.5** vs simple solvolysis (without fragmentation) of *cis*-3-isopropylcyclohexyl tosylate **6.6**; (B) Solvolysis of *trans*-3-dimethylaminocyclohexyl tosylate **6.7**.

are antiperiplanar when the nitrogen lone pair pushes and the departing tosyloxy pulls. The reaction is about 40 times faster than the solvolysis of *cis*-3-isopropylcyclohexyl tosylate **6.6** when no fragmentation occurs in the absence of electron push by an electron pair. Here the products arise mainly from $S_N1$ solvolysis.

In contrast, *trans*-3-dimethylaminocyclohexyl tosylate **6.7** gives a low yield of fragmentation product since this can arise only from the higher energy conformation with axial $NMe_2$ group (Fig. 6.20B).[36] The major products arise from E2 eliminations from the lower energy conformation. The *trans* isomer **6.7** reacts about 8 times slower than the *cis* isomer **6.5** does.

Fragmentation reactions can be used to synthesize medium rings which are rather difficult to prepare. The stereochemistry of the double bond in the ring product is governed by the relative stereochemistry of the groups in the cyclic precursor. Fig. 6.21 shows that the decalin derivative **6.8** (in which the ring junction is *trans* and the two groups OH and OTs are also *trans*) fragments to give (*E*)-5-cyclodecenone, whereas the diastereomer **6.9** with *cis* ring junction gives (*Z*)-5-cyclodecenone. In these derivatives, the tosyloxy (OTs) group is antiperiplanar to the ring-fusion C—C bond, and the fragmentations are stereospecific. In contrast, the isomer **6.10** without an antiperiplanar arrangement of the breaking bonds does not practically fragment.

The control of double bond stereochemistry using fragmentation reaction has been exploited as a useful strategy in synthesis. The stereochemistry of the double bond can be set up by controlling the relative stereochemistry of

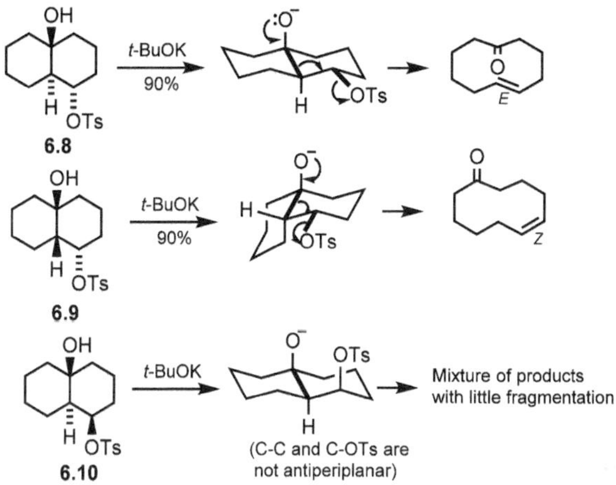


FIG. 6.21 Fragmentation reactions with diastereomeric decalin derivatives.

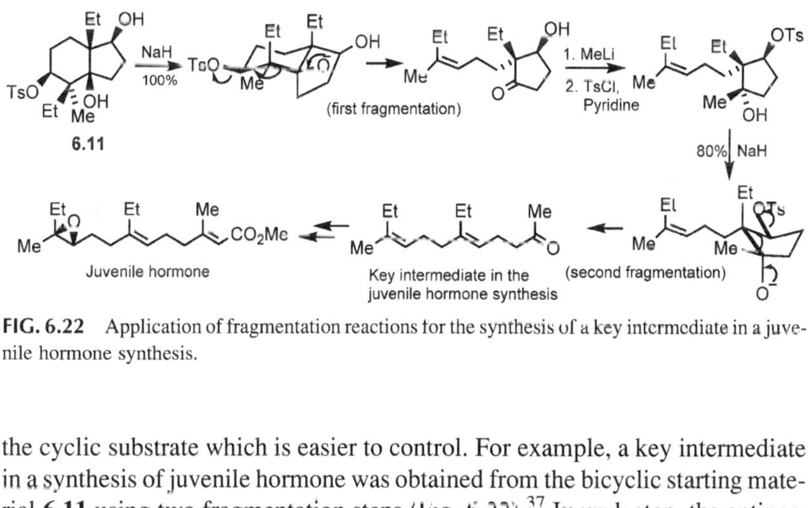


FIG. 6.22 Application of fragmentation reactions for the synthesis of a key intermediate in a juvenile hormone synthesis.

the cyclic substrate which is easier to control. For example, a key intermediate in a synthesis of juvenile hormone was obtained from the bicyclic starting material **6.11** using two fragmentation steps (Fig. 6.22).[37] In each step, the antiperiplanar geometry of the breaking bonds allows easy fragmentation.

An alkyne triple bond can be generated using the Eschenmoser fragmentation. This is the fragmentation of a hydrazone of the epoxide of α,β-unsaturated ketones. The use of the Eschenmoser fragmentation in a synthesis of an insect pheromone *exo*-brevicomin is shown in Fig. 6.23.[38] The starting epoxyketone **6.12** is converted into the corresponding tosylhydrazone which, upon deprotonation, undergoes ring opening of the epoxide. The resulting alkoxy compound then fragments to give a keto-alkyne **6.13** with elimination of nitrogen gas and the sulphinate species.

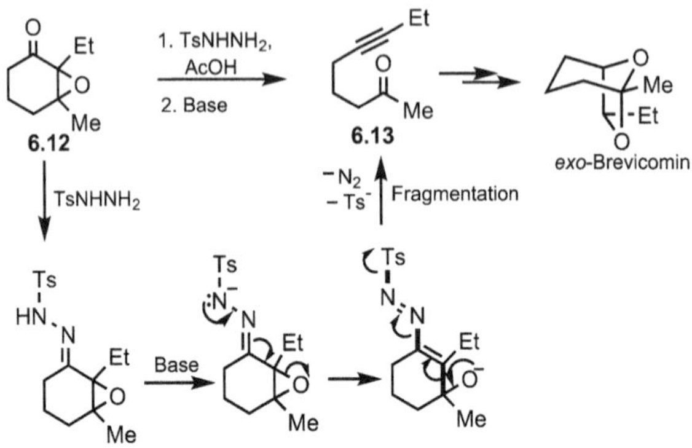


FIG. 6.23 Eschenmoser fragmentation for the preparation of an intermediate in a synthesis of an insect pheromone *exo*-brevicomin.

**Problem 6.10** Rationalize the formation of the product in the following fragmentation reaction:

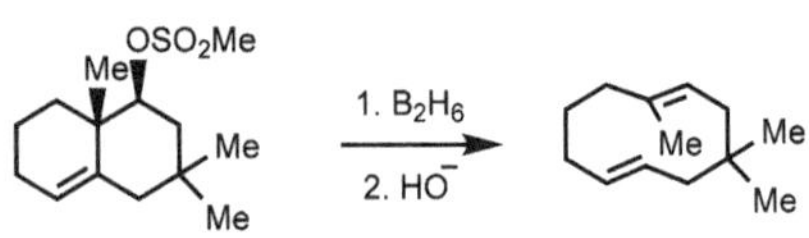


**Problem 6.11** Explain with mechanism the stereochemistry of the double bonds formed in the following fragmentation reaction:

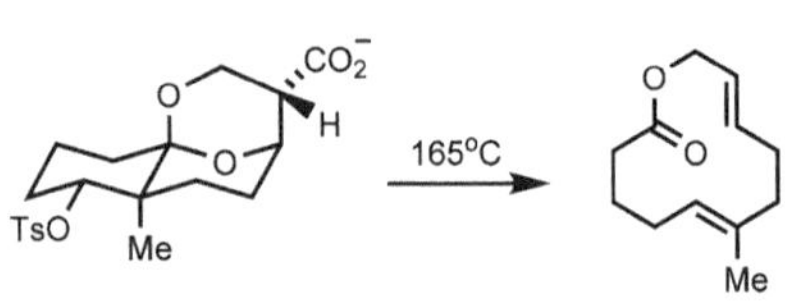


## 6.4 Addition reactions of alkenes and alkynes

### 6.4.1 Electrophilic addition to acyclic alkenes

The electrophilic addition to an alkene involves orthogonal and inside approach of the electrophile to form a cationic intermediate (see Section 4.3.5). With a bridging electrophile, the addition can give a bridged intermediate or product. The stereochemistry of electrophilic additions to alkenes has been extensively studied.[39] For example, consider the bromination of (*Z*)- and (*E*)-2-butenes (Fig. 6.24). The reaction proceeds via a cyclic bromonium ion intermediate

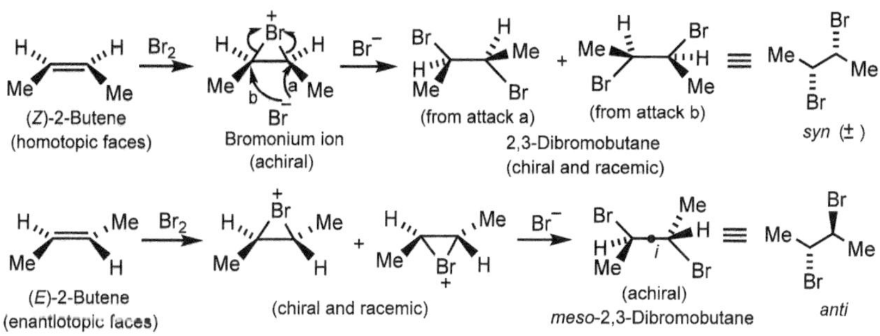


**FIG. 6.24** Stereospecific *anti* bromination of (*Z*)- and (*E*)-2-butenes.

and provides an overall *anti* addition. The reaction is stereospecific as the stereochemistry of the alkene determines which diastereomer of the product is formed. (*Z*)-2-Butene gives chiral 2,3-dibromobutane (as racemate), whereas the (*E*) isomer produces an achiral (*meso*) diastereomer.

As shown in Fig. 6.24, since the two faces of (*Z*)-2-butene are homotopic, electrophilic attack on either face leads to the same bromonium ion intermediate. As this intermediate is achiral, nucleophilic attack ($S_N2$) by bromide at either end of the bromonium ion gives equal amounts of the enantiomeric products (*syn*). On the other hand, (*E*)-2-butene faces are enantiotopic and hence electrophilic attack on the two faces provides a 1:1 mixture of enantiomeric bromonium ion intermediates. The $S_N2$ attack by bromide on either end of chiral bromonium ions then leads to a single *meso* (or *anti*) diastereomer. Note the presence of centre of inversion (*i*) in the achiral product.

Similarly, maleic acid on bromination gives chiral and racemic 2,3-dibromosuccinic acid and fumaric acid provides *meso*-2,3-dibromosuccinic acid (draw the scheme).

#### 6.4.1.1 Methods of dihydroxylation

Dihydroxylation of alkenes can be carried out in *syn* or *anti* fashion. Similar to *anti* bromination, *anti* dihydroxylation can be achieved via ring opening of isolable epoxide intermediates. For example, treatment of (*E*)-2-butene with *m*CPBA (*m*-chloroperoxybenzoic acid, a safe crystalline solid) gives the *trans*-epoxide which is then ring opened by nucleophilic water in aqueous acid ($H_2O/HClO_4$) to obtain *meso*-2,3-butanediol (draw the scheme).

The most popular method for dihydroxylation of alkenes involves reaction with osmium tetroxide ($OsO_4$).[40,41] The dihydroxylation is stereochemically *syn* (the two OH groups are added on the same face of the double bond). Fig. 6.25 illustrates *syn* dihydroxylation with $OsO_4$ using two diastereomeric alkenes: maleic acid gives *meso*-tartaric acid, whereas fumaric acid gives racemic tartaric acid. One plausible mechanism of dihydroxylation involves direct

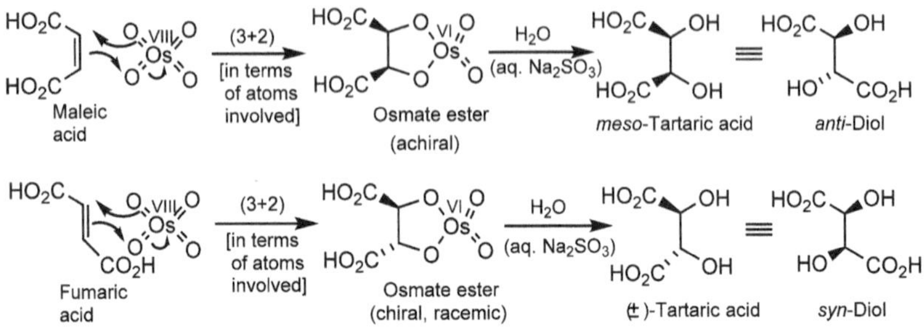


FIG. 6.25 *syn* Dihydroxylation of maleic acid and fumaric acid with osmium tetroxide.

(3+2) cycloaddition (in terms of atoms involved) of $OsO_4$ to the alkene to give the intermediate osmate ester (Fig. 6.25), whereas another mechanism proposes a [2+2] cycloaddition followed by a rearrangement to give the osmate ester (not shown).[42] The alkene stereochemistry is retained in the osmate ester as a result of suprafacial addition to the alkene. The osmate ester is then hydrolyzed to release the diol product. Note that *syn* or *anti* dihydroxylation refers to the stereochemical mode of addition, whereas the *syn*- or *anti*-diol product refers to the *syn*/*anti* descriptor in the staggered zigzag formula.

$OsO_4$ is toxic and expensive. For this reason, many co-oxidants have been used in conjunction with catalytic amounts of $OsO_4$. One popular co-oxidant is *N*-methylmorpholine-*N*-oxide (NMO). The use of NMO with $OsO_4$ (cat. amount) is often referred to as the Upjohn conditions.[43] The Os(VI) species generated upon hydrolysis of the osmate ester is reoxidized to Os (VIII) by NMO to continue the cycle.

Among other methods of dihydroxylation, Prévost and Woodward methods are of value because of their specificity and mild reaction conditions.[44–46] The Prévost reagent is a solution of $I_2$ in $CCl_4$ together with an equivalent of silver acetate (AgOAc) or silver benzoate (AgOBz). The Prévost reaction uses anhydrous conditions and leads to *anti* dihydroxylation, while the Woodward method under moist conditions gives *syn* dihydroxylation. Fig. 6.26 shows that the Prévost reaction with a *cis* alkene under anhydrous conditions in boiling $CCl_4$ gives the diester of *threo*-1,2-diol, whereas the monoester of *erythro*-1,2-diol is obtained in the Woodward reaction under moist conditions in $AcOH/H_2O$.

The reaction proceeds through the formation of an iodonium ion followed by consecutive ring-opening and ring-closing in the presence of carboxylate and silver ions respectively to form the resonance-stabilized oxonium ion intermediate **6.14** (cf. Fig. 6.6). An $S_N2$ attack by the carboxylate on the cation **6.14** under Prévost conditions gives the *threo* diester but in the presence of water (Woodward conditions), a hydroxyl acetal is formed which then undergoes the hemiacetal cleavage to give the *erythro* monoester.

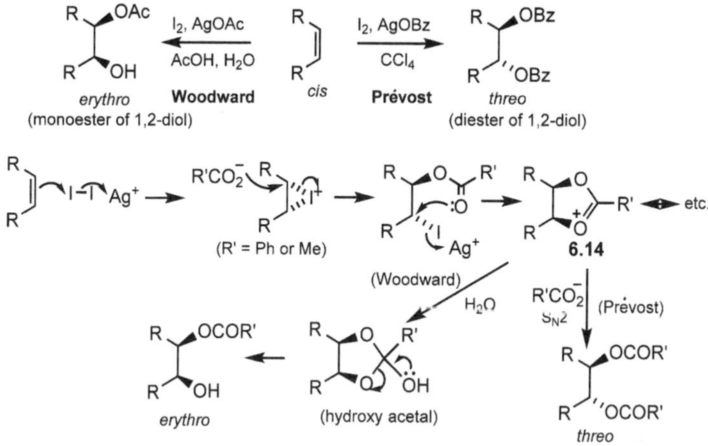


FIG. 6.26 Prévost and Woodward methods for dihydroxylation of alkenes.

**Problem 6.12** Explain the formation of products in the following reactions:

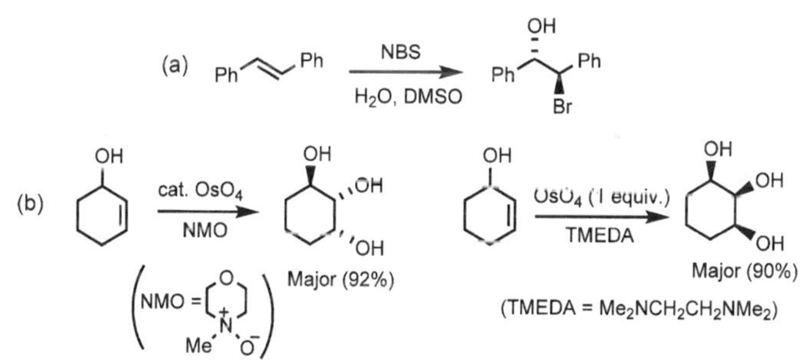


## 6.4.2 Electrophilic addition to cyclohexenes

Like *anti* bromine addition to acyclic alkenes, addition of bromine to a cyclohexene is diaxial and antiperiplanar. For example, consider the bromine addition to anancomeric 4-*t*-butylcyclohexene in which *t*-butyl is equatorial (Fig. 6.27).[47,48] Cyclohexene exists preferentially in the half-chair conformation, and there are two possible pathways for diaxial addition. As shown, bromide attack on the cyclic bromonium ion by path a leads directly to the diaxial chair conformation of the product dibromide, whereas path b gives rise to the diaxial twist-boat conformation which then goes to the diequatorial chair by ring flipping. Since the chair conformation is more stable than the twist-boat, the TS leading directly to the chair is favoured, and the diaxial dibromide is the predominant product (>94%).[49]

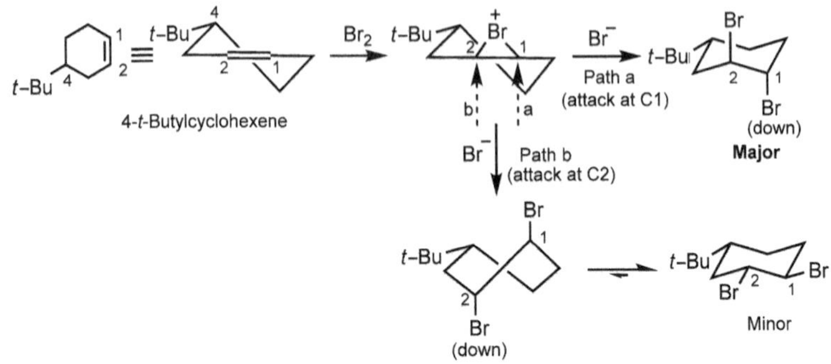


**FIG. 6.27** Electrophilic addition of bromine to anancomeric 4-*t*-butylcyclohexene.

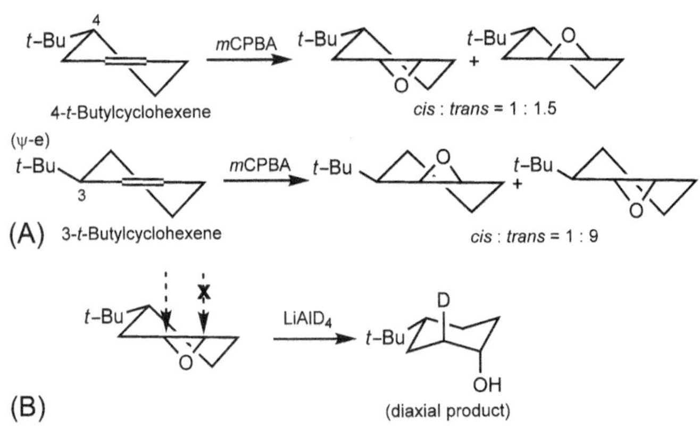


**FIG. 6.28** (A) Epoxidation of 4- and 3-*t*-butylcyclohexenes; (B) Regiochemical control in the ring opening of *cis*-4-*t*-butylcyclohexene oxide by reduction with $LiAlD_4$.

Epoxidation of cyclohexene with *m*CPBA gives cyclohexene oxide that also exists as a half-chair. With substituted cyclohexenes, attack on two faces of the double bond provides a mixture of *cis* and *trans* cyclohexene oxides (Fig. 6.28A).[50] In 4-*t*-butylcyclohexene, attack on either face is more or less likely, whereas in 3-*t*-butylcyclohexene, attack on the *trans* side is more favoured as the psudoequatorial *t*-Bu at 3-position sterically impedes the attack from the *cis* side.

The ring opening of cyclohexene epoxides generally gives diaxial products rather than diequatorial ones (cf. Fig. 6.27). This entails regiochemical control in cyclohexene epoxide ring opening, as shown for reduction of *cis*-4-*t*-butylcyclohexene oxide with $LiAlD_4$ in Fig. 6.28B.[51]

The enolates or enamines derived from cyclohexanones also exist in cyclohexene half-chair conformation. Thus, alkylation of the enolate or enamine will show a preference for axial attack that leads directly to a chair with the electrophile in the axial position. For example, consider an alkylation of the enamine derived from 4-*t*-butylcyclohexanone (Fig. 6.29). Electrophilic attack from the

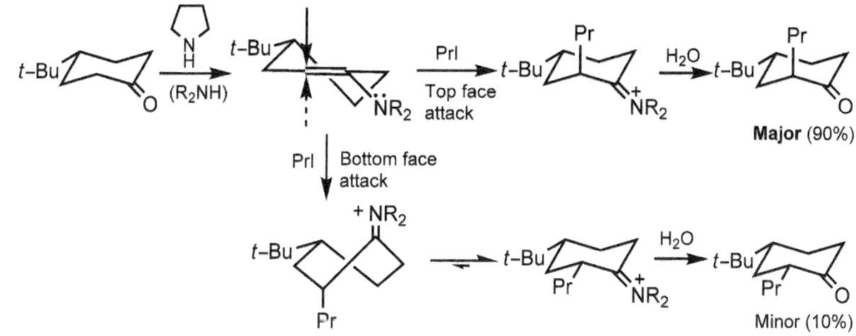


**FIG. 6.29** Alkylation of an enamine derived from anancomeric 4-*t*-butylcyclohexanone.

top face of the enamine double bond gives directly the chair form with axial Pr as a major product (90%), whereas the attack from the bottom face leads to a less stable twist-boat which by ring flipping gives the chair form with equatorial Pr as a minor product (10%).

**Problem 6.13** Predict the regio- and stereoselectivity of the following reactions:

**Problem 6.14** Suggest a plausible mechanism for the following transformation:

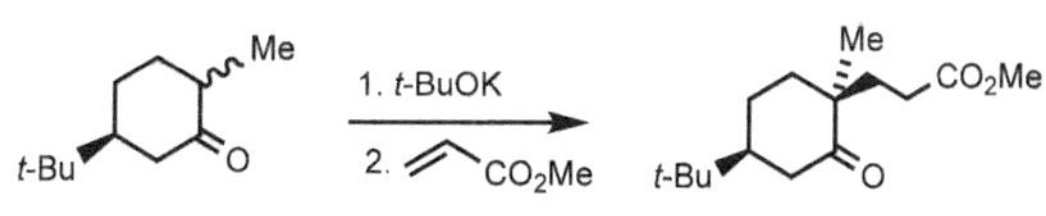


## 6.4.3 Partial reduction of alkynes

Partial reduction of alkynes can be performed stereoselectively to produce (*Z*)- or (*E*)-alkenes. The (*Z*)-selective reduction is effected conveniently in high yields using Lindlar's catalyst (palladium-calcium carbonate catalyst partially deactivated by addition of lead acetate or quinoline).[52] The (*Z*)-selectivity results because two H atoms bound to the catalyst surface are delivered simultaneously to the alkyne in a *syn* fashion (Fig. 6.30A). This stereoselective reduction has been exploited in the synthesis of carotenoids and other natural products. Fig. 6.30B illustrates the (*Z*)-selective reduction using Lindlar's catalyst with two examples.

The (*Z*)-selective reduction of alkynes is also possible by hydroboration and protonolysis. Monohydroboration using a dialkylborane such as disiamylborane

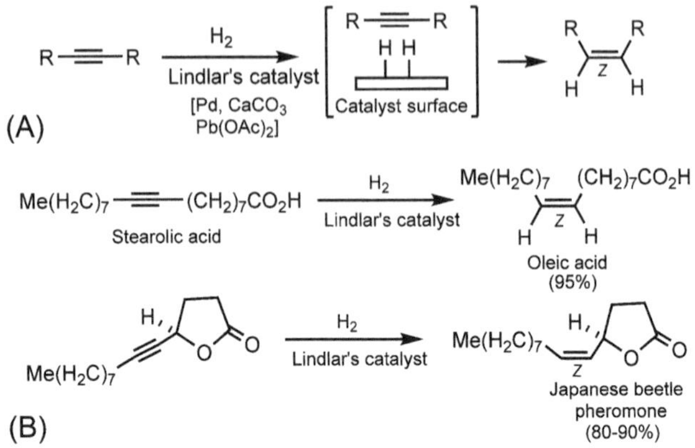


**FIG. 6.30** (A) (*Z*)-Selective reduction of alkynes with Lindlar's catalyst; (B) Two examples of the stereoselective reduction.

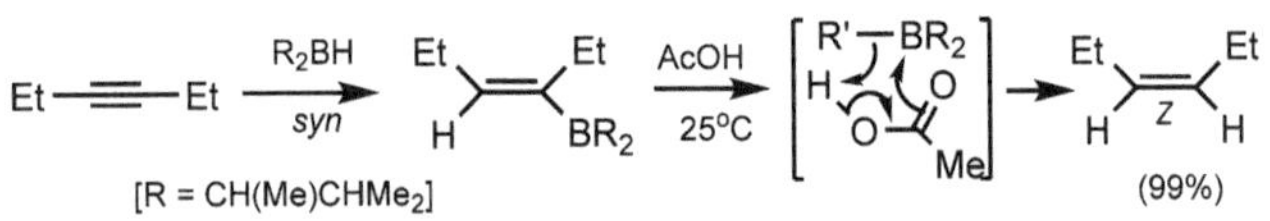


**FIG. 6.31** (*Z*)-Selective reduction of alkynes by hydroboration and protonolysis.

gives an alkenylborane which undergoes protonolysis with acetic acid with retention of alkene configuration (Fig. 6.31).[53]

For (*E*)-selective reduction of alkynes, see Section 11.4.

**Problem 6.15** Suggest a plausible mechanism for the following transformation:

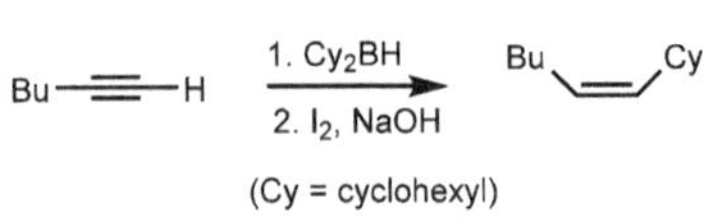


## 6.5 Nucleophilic addition to cyclohexanones

The stereoselectivity in nucleophilic addition to cyclohexanones is governed by the preference of the nucleophilic reagents to approach from the equatorial or the axial face of the carbonyl group. The equatorial attack gives the axial alcohol whereas the axial attack leads to the equatorial alcohol.

### 6.5.1 Steric approach control: Role of steric strain

Bulky hydrides (such as L-Selectride and lithium trisiamylborohydride) and large organometallics (e.g. Grignard reagents *i*-PrMgBr and *t*-BuMgCl) add

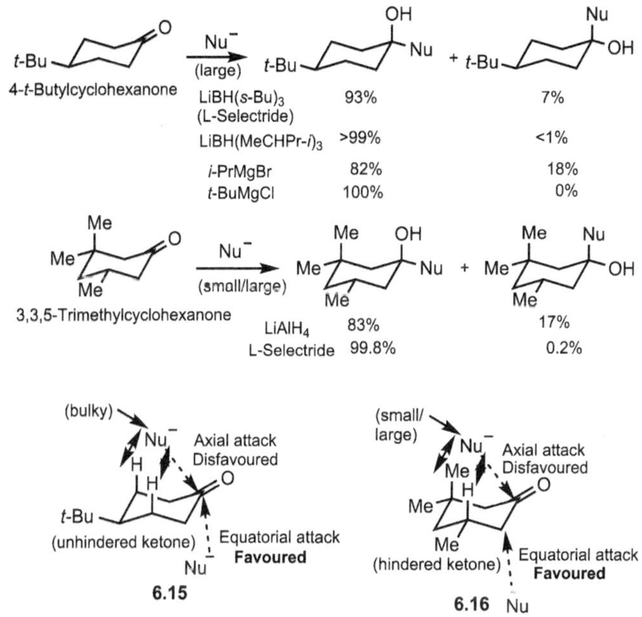


**FIG. 6.32** Steric approach control in stereoselective nucleophilic additions to cyclohexanones.

to anancomeric 4-*t*-butylcyclohexanone preferentially from the equatorial side to give the axial alcohols (Fig. 6.32).[54–56] Further, in cyclohexanones in which the axial side is hindered by synaxial substituents as in 3,3,5-trimethylcyclohexanone, predominant equatorial attack is observed even with a small nucleophile such as $LiAlH_4$.[57]

The nucleophile approaches along the Bürgi–Dunitz trajectory at an angle of 107 degrees with the C=O bond (see Section 4.3.4). The preferential addition of bulky nucleophiles to the unhindered cyclohexanone **6.15** or of small nucleophiles to the hindered ketone **6.16** is attributed to steric effects: the axial face is more hindered than the equatorial one because of 1,3-diaxial interactions with the incoming nucleophile (Fig. 6.32). This was termed *steric approach control* by Dauben,[58] as the consequence of steric repulsions with the 3,5-axial hydrogens or substituents

Bridged bicyclic systems such as norbornanones exist in the locked conformation, and the stereoselectivity of nucleophilic addition in these systems is also determined by steric approach control.[59] The *exo* attack gives the *endo* alcohol while the *endo* attack gives the *exo* alcohol. Fig. 6.33 shows that norbornanone **6.17** undergoes nucleophilic attack preferentially from the *exo* direction because the one-atom bridge offers less steric resistance than the two-carbon bridge on the *endo* side of the molecule.[54,57] Note that the *endo* hydrogen is in a relationship to the attacking site similar to the 1,3-diaxial interaction in a chair cyclohexane. This selectivity is completely reversed when the

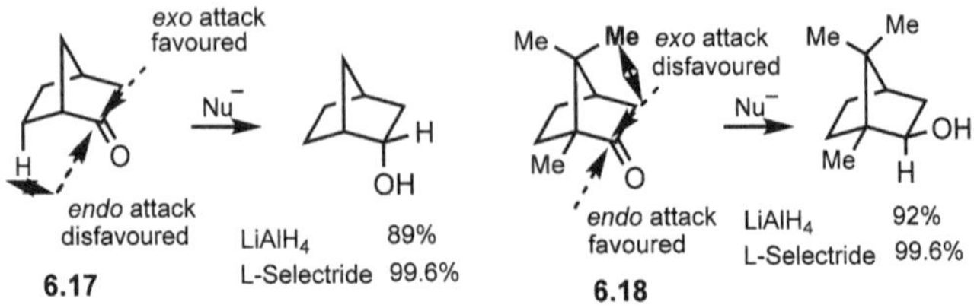


**FIG. 6.33** Steric approach control in stereoselective nucleophilic additions to bicyclic norbornanones.

one-atom bridge bears two Me groups, as in camphor **6.18**. The predominant attack then takes place from the relatively less hindered *endo* face because one Me on the *exo* side would interfere sterically with the approach of the nucleophile in that direction.[60]

## 6.5.2 The Felkin model: Role of torsional strain

It has been observed that small nucleophiles (such as $LiAlH_4$ and HC≡CNa) prefer to approach from the axial face of the anancomeric 4-*t*-butylcyclohexanone to give the equatorial alcohols, even though this is a more sterically congested approach (Fig. 6.34A).[56,57] The somewhat more bulky hydride, LiAlH(OBu-*t*)$_3$ still leads to the predominant axial attack.[61]

A number of theories have been put forward for the preference of axial attack. Historically, Dauben[58] attributed the axial preference to 'product development control', reflecting the greater stability of the resulting equatorial alcohol. However, this hypothesis seems unlikely since the addition involves presumably an early, reactant-like TS.

The most widely accepted model is the Felkin model[62] based on torsional effects (Fig. 6.34B). The C=O bond in cyclohexanone is not perfectly eclipsing the equatorial C—H bonds on C2 and C6 but is pointing 4–5 degrees lower. When viewed along C2—C1, the Newman projection clearly shows that C=O is pointing lower with respect to the equatorial H on C2. Thus the nucleophile along the Bürgi–Dunitz trajectory in the equatorial attack would be closer to the axial C—H bonds on C2 and C6, thereby inducing torsional strain. More importantly, torsional strain increases appreciably as the equatorial attack would push the oxygen atom past the two equatorial hydrogens through a fully eclipsed arrangement as the carbonyl carbon becomes tetrahedral. In contrast, the axial attack moves the oxygen atom down into an equatorial position without developing any eclipsing. The axial attack by small nucleophiles is also less affected by the 1,3-diaxial interactions with the 3,5-axial hydrogens. As a result, the energetic cost due to torsional or eclipsing strain in the equatorial approach is much greater than that due to steric strain in the axial approach.[63,64] The axial attack of small nucleophiles is therefore favoured.

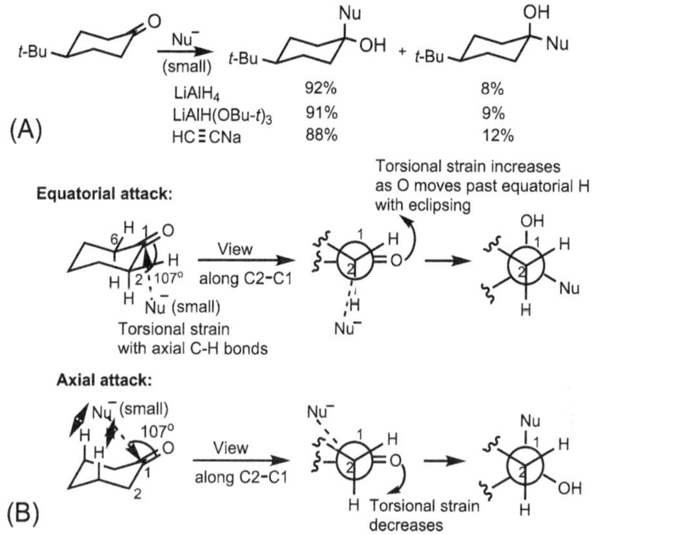


**FIG. 6.34** (A) Addition of small nucleophiles to 4-*t*-butylcyclohexanone via predominant axial attack; (B) The Felkin model based on torsional strain to rationalize the preferential axial approach.

## 6.5.3 The Cieplak model: Role of hyperconjugation

The stereoselectivity of nucleophilic addition to cyclohexanones has also been rationalized by a hyperconjugative model, called the Cieplak model.[65–67] The model assumes extensive pyramidalization of the carbonyl group in the TS, and is based on the interactions between the vacant σ* orbital of the incipient bond (between the nucleophile and the carbonyl carbon) and the filled σ orbitals of the vicinal C—H or C—C bonds (Fig. 6.35A). In the axial approach, a large lobe of the σ* orbital of the incipient bond interacts with the σ orbitals of the antiperiplanar axial C—H bonds on C2 and C6, whereas the 2–3 and 5–6 σ bonds are involved in the hyperconjugative interactions in the equatorial approach. Since the C—H bonds are better donors than the C—C bonds, the orbital interactions in the axial approach is more effective. Thus the TS for the axial approach is more stabilized leading to the preferential axial attack. There are however contrary opinions regarding electron donating ability of C—H and C—C bonds. The structural and spectroscopic evidences suggest that C—H bonds are intrinsically better donors, especially when the electron demand is not high as in the case of the attack of a nucleophile on a carbonyl group.[68,69]

The Cieplak model has been successfully applied to explain the stereoselectivity of nucleophilic addition in the norbornanone and adamantanone systems. For example, consider the $NaBH_4$ reduction of 5-substituted 2-adamantanones (Fig. 6.35B).[70] If the substituent is electron releasing (X = *p*-hydroxyphenyl or *p*-aminophenyl), the 1–9 and 3–4 bonds (highlighted) will be electron-rich (better donors), and hence the antiperiplanar equatorial attack occurs preferentially

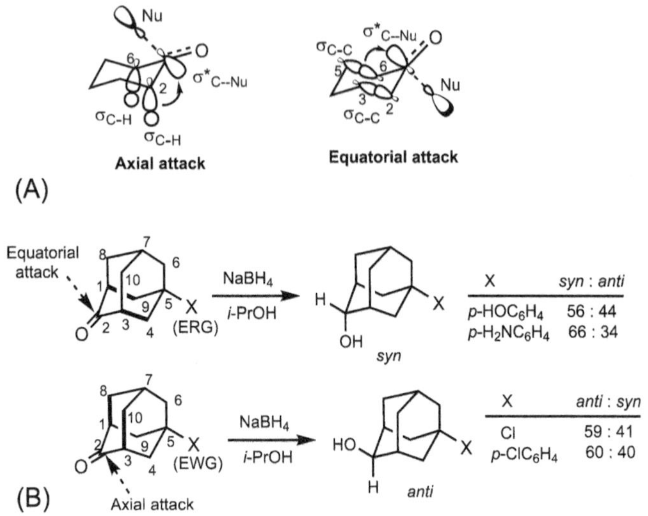


**FIG. 6.35** (A) Orbital interactions (hyperconjugation) in the Cieplak model; (B) Nucleophilic addition (hydride reduction) of 5-substituted 2-adamantanones. Participating σ bonds are highlighted. ERG = electron releasing group, EWG = electron withdrawing group.

to give the *syn* alcohols as major products. But with electron withdrawing substituents (X = Cl or *p*-chlorophenyl), the 1–9 and 3–4 bonds will be electron deficient. The 1–8 and 3–10 bonds (highlighted) will be then relatively more electron-rich (better donors) and the preferential antiperiplanar axial attack gives the predominant *anti* alcohols.

**Problem 6.16** Predict, with the help of the Cieplak model, the major product in the nucleophilic addition of methyllithium to 2,3-*endo*,*endo* disubstituted 7-norbornanone:

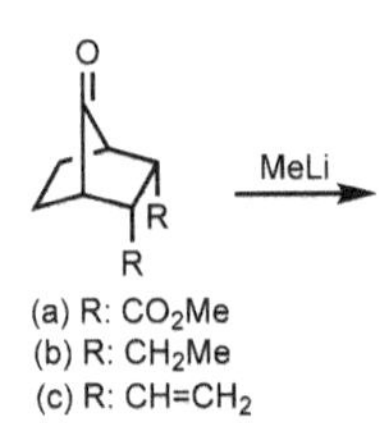


### 6.5.4 Conjugate addition to cyclohexenones

Like cyclohexenes, cyclohexenone rings assume half-chair conformations, but these are flatter than cyclohexenes. The conjugate addition to substituted cyclohexenones often involve organocopper reagents as nucleophiles.[71–73] For

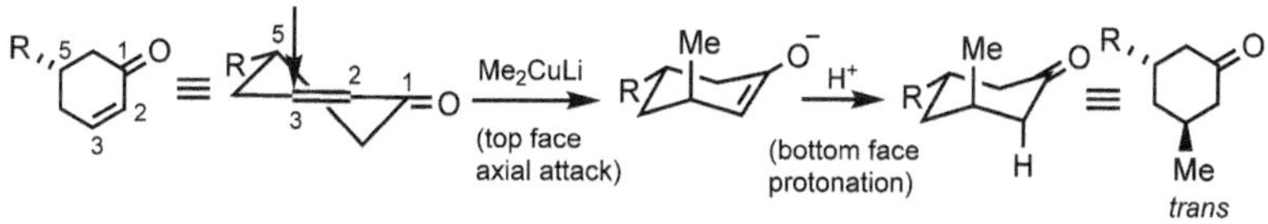


**FIG. 6.36** Conjugate addition to a 5-substituted cyclohexenone.

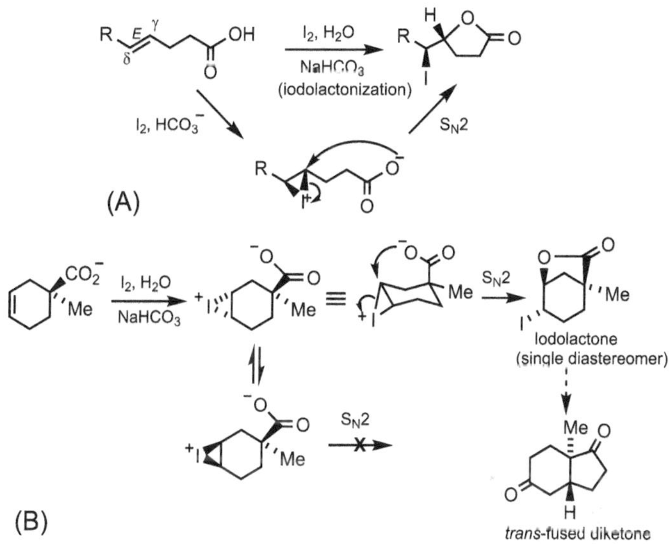


**FIG. 6.37** Diastereoselective iodolactonization using (A) an acyclic nonconjugated unsaturated acid and (B) a cyclic unsaturated acid in a synthesis of a diketone with steroid-like *trans*-fused 6–5 ring junction.

example, the conjugate addition of lithium dimethylcuprate to a 5-substituted cyclohexenone is shown in Fig. 6.36. The nucleophilic attack is an axial addition and takes place preferentially from the top face of the carbon–carbon double bond to give directly the chair, rather than a twist-boat from the bottom face attack (cf. Fig. 6.29). Subsequent protonation of the enolate intermediate then yields the *trans* cyclohexanone product. It may be noted that the protonation is also an axial addition, and the trapping of the enolate with a different electrophile (e.g. alkyl halide) will give the product via axial alkylation from the bottom face of the enolate.

## 6.6 Iodolactonization

Iodolactonization is the reaction of nonconjugated unsaturated acids with $I_2$ in aqueous $NaHCO_3$ to form an iodolactone.[74] The reaction can be diastereoselective, as illustrated using a $\gamma,\delta$-unsaturated acid in Fig. 6.37A. The reaction

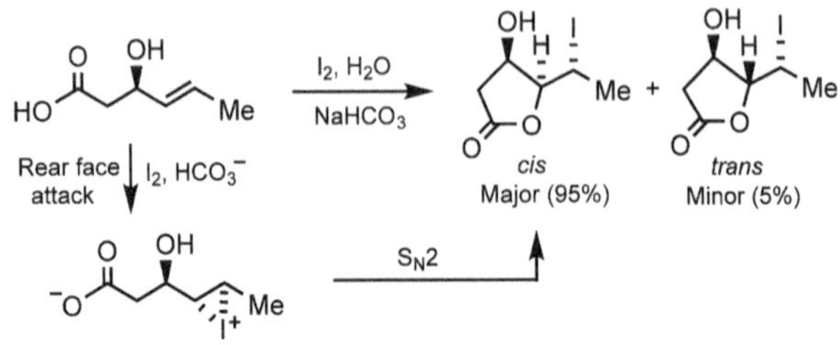


**FIG. 6.38** Diastereoselectivity of iodolactonization of an allylic alcohol.

proceeds through an intermediate iodonium ion followed by ring opening of the intermediate in an intramolecular $S_N2$ attack to give a single diastereomer of the iodolactone product. Note that (*E*) stereochemistry of the alkene is reproduced in the product.

Fig. 6.37B shows an iodolactonization with a cyclic substrate in a synthesis of a diketone with *trans*-fused 6–5 ring junction, as found in steroids. Here two diastereomeric iodonium ions can form but the $S_N2$ cyclization requires I and $CO_2^-$ to be *anti*, which leads to only one iodolactone diastereomer.

Iodolactonization of allylic alcohols usually shows high diastereoselectivity.[75] Fig. 6.38 illustrates the diastereoselectivity of iodolactonization of an (*E*) allylic alcohol that gives the *cis* isomer as a predominant product (*cis*: *trans* = 95:5).

## References


1. Streitwieser, A., Jr.; Walsh, T. D.; Wolfe, J. R. *J. Am. Chem. Soc.* **1965**, *87*, 3682.
2. Shiner, V. J., Jr.; Hartshorn, S. R.; Vogel, P. C. *J. Org. Chem.* **1973**, *38*, 3604.
3. Winstein, S.; Clippinger, E.; Fainberg, A. H.; Heck, R.; Robinson, G. C. *J. Am. Chem. Soc.* **1956**, *78*, 328.
4. Schleyer, P.v. R. In *Ions and Ion Pairs in Organic Reactions*; Raber, D. J., Harris, J. M., Szwarc, M., Eds.; Vol. 2; Wiley: New York, 1974; p. 247.
5. Lee, C. C.; Finlayson, A. J. *Can. J. Chem.* **1961**, *39*, 260.
6. Lee, C. C.; Clayton, J. W.; Lee, D. G.; Finlayson, A. J. *Tetrahedron* **1962**, *18*, 1395.
7. Parker, R. E.; Isaacs, N. S. *Chem. Rev.* **1959**, *59*, 737.
8. Capon, B. *Q. Rev. Chem. Soc.* **1964**, *18*, 45.
9. Capon, B.; McManus, S. *Neighboring Group Participation*, Vol. 1; Plenum Press: New York, 1976.
10. Winstein, S.; Hanson, C.; Grunwald, E. *J. Am. Chem. Soc.* **1948**, *70*, 812.
11. Winstein, S.; Hess, H. V.; Buckles, R. E. *J. Am. Chem. Soc.* **1942**, *64*, 2796.
12. Winstein, S.; Buckles, R. E. *J. Am. Chem. Soc.* **1943**, *65*, 613.
13. Winstein, S.; Grunwald, E.; Buckles, R. E.; Hanson, C. *J. Am. Chem. Soc.* **1948**, *70*, 816.
14. Schleyer, P.v. R. In *Carbonium Ions*; Lancelot, L. A., Cram, D. J., Olah, G. A., Schleyer, P.v. R., Eds.; Vol. 3; Wiley: New York, 1972; p. 1347.
15. Cram, D. J. *J. Am. Chem. Soc.* **1949**, *71*, 3863. **1952**, *74*, 2129.
16. Winstein, S.; Shavatsky, M.; Norton, C.; Woodward, R. B. *J. Am. Chem. Soc.* **1955**, *77*, 4183.

17. Winstein, S.; Lewin, A. H.; Pande, K. C. *J. Am. Chem. Soc.* **1963**, *85*, 2324.
18. Winstein, S.; Trifan, D. S. *J. Am. Chem. Soc.* **1949**, *71*, 2953. **1952**, *74*, 1147, 1154.
19. Winstein, S.; Clippinger, E.; Howe, R.; Vogelfanger, E. *J. Am. Chem. Soc.* **1965**, *87*, 376.
20. Eliel, E. L.; Wilen, S. H.; Mander, L. N. *Stereochemistry of Organic Compounds*; Wiley: New York, 1994; p. 720.
21. Eliel, E. L.; Ro, R. S. *J. Am. Chem. Soc.* **1957**, *79*, 5995.
22. Eliel, E. L.; Schroeter, S. H.; Brett, T. J.; Biros, F. J.; Richer, J.-C. *J. Am. Chem. Soc.* **1966**, *88*, 3327.
23. Eliel, E. L.; Haubenstock, H.; Acharya, R. V. *J. Am. Chem. Soc.* **1961**, *83*, 2351.
24. Allinger, N. L.; Freiberg, L. A.; Hu, S.-E. *J. Am. Chem. Soc.* **1962**, *84*, 2836.
25. McGahey, L. F.; Jensen, F. R. *J. Am. Chem. Soc.* **1979**, *101*, 4397.
26. Applequist, D. E.; Chmurny, G. N. *J. Am. Chem. Soc.* **1967**, *89*, 875.
27. Pfeiffer, P. *Z. Phys. Chem.* **1904**, *48*, 40. *Ber. Dtsch. Chem. Ges.* **1912**, *45*, 1819.
28. Eliel, E. L.; Allinger, N. L.; Angyal, S. J.; Morrison, G. A. *Conformational Analysis*; Wiley: New York, 1965.
29. Saunders, W. H.; Cockerill, A. F. *Mechanisms of Elimination Reactions*; Wiley: New York, 1973.
30. Kwart, H.; Takeshita, T.; Nyce, J. L. *J. Am. Chem. Soc.* **1964**, *86*, 2606.
31. Coke, J. L.; Cooke, M. P. *J. Am. Chem. Soc.* **1967**, *89*, 6701.
32. Cooke, M. P., Jr.; Coke, J. L. *J. Am. Chem. Soc.* **1968**, *90*, 5556.
33. Grob, C. A. *Angew. Chem. Int. Ed. Engl.* **1969**, *8*, 535.
34. Weyerstahl, P.; Marschall, H. In *Comprehensive Organic Synthesis*; Trost, B. M., Fleming, I., Eds.; Vol. 6; Pergamon Press: Oxford, 1991; p. 1041.
35. Ho, T.-L. *Heterolytic Fragmentation of Organic Molecules*; Wiley: New York, 1993.
36. Burckhardt, U.; Grob, C. A.; Kiefer, H. R. *Helv. Chim. Acta* **1967**, *50*, 231.
37. Zurflüh, R.; Wall, E. N.; Siddall, J. B.; Edwards, J. A. *J. Am. Chem. Soc.* **1968**, *90*, 6224.
38. Kocienski, P. J.; Ostrow, R. W. *J. Org. Chem.* **1976**, *41*, 398.
39. Fahey, R. C. *Top. Stereochem.* **1968**, *3*, 237.
40. Schröder, M. *Chem. Rev.* **1980**, *80*, 187.
41. Haines, A. H. In *Comprehensive Organic Synthesis*; Trost, B. M., Fleming, I., Eds.; Vol. 7; Pergamon Press: Oxford, 1991; p. 437.
42. Deubel, D. V.; Frenking, G. *Acc. Chem. Res.* **2003**, *36*, 645.
43. Van Rheenen, V.; Kelly, R. C.; Cha, D. Y. *Tetrahedron Lett.* **1976**, *17*, 1973.
44. Prévost, C. *C. R. Acad. Sci.* **1933**, *196*, 1129.
45. Wilson, C. V. *Org. React.* **1957**, *9*, 332.
46. Woodward, R. B.; Brutcher, F. B. *J. Am. Chem. Soc.* **1958**, *80*, 209.
47. Valls, J. *Bull. Soc. Chim. Fr.* **1961**, 432.
48. Valls, J.; Toromanoff, E. *Bull. Soc. Chim. Fr.* **1961**, 758.
49. Barili, P. L.; Bellucci, G.; Marioni, F.; Morelli, I.; Scartoni, V. *J. Org. Chem.* **1972**, *37*, 4353.
50. Berti, G. *Top. Stereochem.* **1973**, *7*, 93.
51. Lamaty, G.; Tapiero, C.; Wylde, R. *Bull. Soc. Chim. Fr.* **1966**, 4010.
52. Henrick, C. A. *Tetrahedron* **1977**, *33*, 1845.
53. Brown, H. C.; Molander, G. A. *J. Org. Chem.* **1986**, *51*, 4512.
54. Brown, H. C.; Krishnamurthy, S. *J. Am. Chem. Soc.* **1972**, *94*, 7159.
55. Krishnamurthy, S.; Brown, H. C. *J. Am. Chem. Soc.* **1976**, *98*, 3383.
56. Ashby, E. C.; Laemmle, J. T. *Chem. Rev.* **1975**, *75*, 521.
57. Brown, H. C.; Dickason, W. C. *J. Am. Chem. Soc.* **1970**, *92*, 709.
58. Dauben, W. G.; Fonken, G. J.; Noyce, D. S. *J. Am. Chem. Soc.* **1956**, *78*, 2579.

59. Brown, H. C.; Kawakami, J. H.; Liu, K.-T. *J. Am. Chem. Soc.* **1973**, *95*, 2209.
60. Brown, H. C.; Muzzio, J. *J. Am. Chem. Soc.* **1966**, *88*, 2811.
61. Klein, J.; Dunkelblum, E.; Eliel, E. L.; et al. *Tetrahedron Lett.* **1968**, *9*, 6127.
62. Chérest, M.; Felkin, H. *Tetrahedron Lett.* **1968**, *9*, 2205.
63. Wu, Y.-D.; Houk, K. N. *J. Am. Chem. Soc.* **1987**, *109*, 908.
64. Wu, Y.-D.; Tucker, J. A.; Houk, K. N. *J. Am. Chem. Soc.* **1991**, *113*, 5018.
65. Cieplak, A. S. *J. Am. Chem. Soc.* **1981**, *103*, 4540.
66. Cieplak, A. S. In *Structure Correlation*; Bürgi, H.-B., Dunitz, J. D., Eds.; Vol. 1; VCH: Weinheim, 1994; p. 205.
67. Cieplak, A. S. *J. Org. Chem.* **1998**, *63*, 521.
68. Edlund, U. *Org. Magn. Reson.* **1978**, *11*, 516.
69. Brown, H. C.; Periasamy, M.; Perumal, P. T. *J. Org. Chem.* **1984**, *49*, 2754.
70. Cheung, C. K.; Tseng, L. T.; Lin, M.-H.; Srivastava, S.; le Noble, W. J. *J. Am. Chem. Soc.* **1986**, *108*, 1598.
71. Posner, G. H. *Org. React.* **1972**, *19*, 1.
72. Kozlowski, J. A. In *Comprehensive Organic Synthesis*; Trost, B. M., Fleming, I., Eds.; Vol. 4; Pergamon Press: Oxford, 1991; p. 169.
73. Krause, N.; Gerold, A. *Angew. Chem. Int. Ed. Engl.* **1997**, *36*, 187.
74. Laya, M.; Banerjee, A. K.; Cabrera, E. V. *Curr. Org. Chem.* **2009**, *13*, 720.
75. Chamberlin, A. R.; Dezube, M.; Dussault, P.; McMills, M. C. *J. Am. Chem. Soc.* **1983**, *105*, 5819.

Chapter 7

# Ionic reactions 2: Diastereoselectivity and asymmetric synthesis

## 7.1 Diastereoselective carbonyl additions: Chiral substrates with α-stereocentres

If a carbonyl substrate is chiral with a stereocentre, its carbonyl faces become diastereotopic (see Section 2.12.5). The extent of diastereofaces differentiation by an attacking nucleophile depends on the influence of the stereocentre during asymmetric induction. If the stereocentre is situated far from the carbonyl group, the diastereoselection will be poor and the reaction will be largely unselective. For example, if the stereocentre is placed beyond β carbon, the diaster eotopic carbonyl faces, because of free rotation about single bonds, will average out to about same reactivity and practically no diastereoselectivity will be observed (Fig. 7.1).

An α-stereocentre to a carbonyl group has a large influence on the diastereofaces discrimination in asymmetric induction. These are called 1,2-aymmetric induction which has received a lot of attention, and several models have been proposed to rationalize the diastereoselectivity observed. Historically, formation of major diastereomer was predicted and explained using the Cram's rule (Fig. 7.2A).[1] This model is based on the conformation in which the carbonyl group is flanked by small (S) and medium (M) groups attached to the α-stereocentre when the large group (L) is eclipsed with the substituent (R) at the carbonyl group. The nucleophilic attack from the least hindered side of the small group (S) would lead to the major diastereomer. For example, (*S*)-2-phenylpropanal reacts with MeMgI to give *erythro* diastereomer as the major product (Fig. 7.2B). Although Cram's rule could predict successfully the major product in a number of cases, the model is empirical and not based on the stereoelectronic effects. Furthermore, the model encounters severe eclipsing strain in the TS leading to the product.

Stereochemistry and Organic Reactions. https://doi.org/10.1016/B978-0-12-824092-2.00007-1

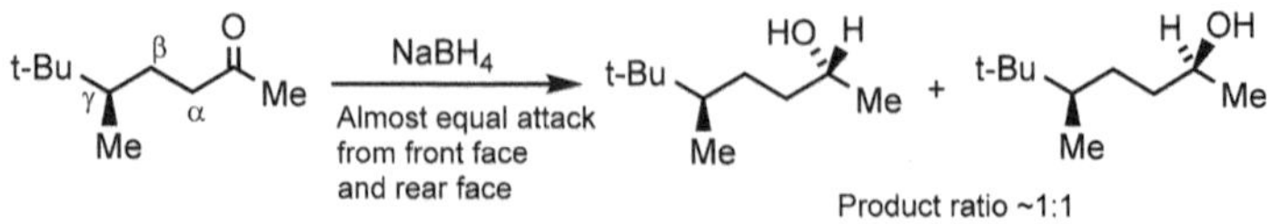


**FIG. 7.1** Lack of diastereoselectivity in the nucleophilic addition of a chiral carbonyl compound with a stereocentre located far away from the carbonyl group (beyond β-carbon).

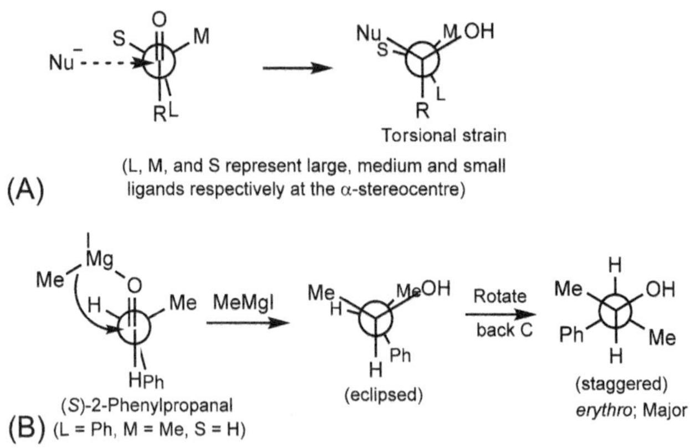


**FIG. 7.2** Nucleophilic addition to carbonyl compounds with an α-stereocentre: (A) Cram's acyclic model; (B) An example illustrating the Cram's rule.

## 7.1.1 Felkin–Anh model

The Felkin–Anh model[2,3] invokes a combination of stereoelectronic and steric effects in which the reactive conformation is different from that of the Cram's model. This is the most widely used and accepted model to rationalize the diastereoselectivity of the nucleophilic addition to carbonyl substrates with an α-stereocentre. Consider two nucleophilic additions as shown in Fig. 7.3. The products are same in the two cases but the diastereoselectivity is reversed. For the Grignard reaction, the diastereomer ratio (*syn*:*anti*) = 3:1, whereas in the case of $LiAlH_4$ reduction, *syn*:*anti* = 1:3. If the chiral carbonyl substrates

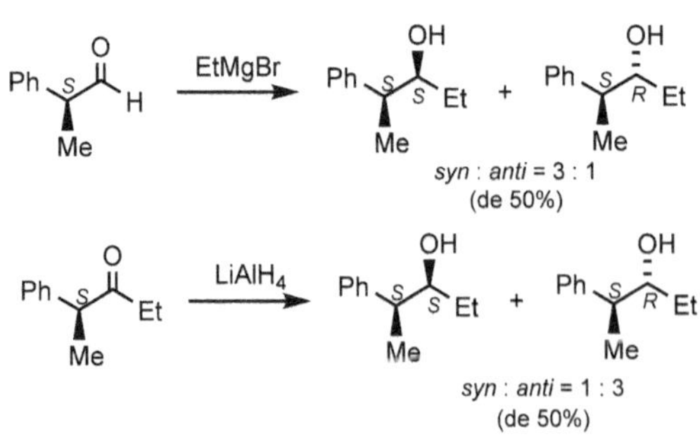


**FIG. 7.3** Diastereoselectivity in nucleophilic additions to chiral carbonyl substrates containing α-stereocentre.

are used as a single enantiomer (say, *S*), the diastereomers are also obtained as single enantiomers. The reactions are then both diastereoselective and enantioselective, and represent first-generation (1G) asymmetric synthesis (see Section 5.3.1.1). But if the chiral substrates are racemic, the diastereomeric products are also racemic and the reactions are then only diastereoselective.

The Felkin–Anh model is based on the following assumptions:

**(a)** The transition structure conformation is reactant-like (i.e. the reaction has an early TS).

**(b)** The reactive conformation is staggered and has the largest group perpendicular to C=O (i.e. the largest group is farthest from oxygen and the substituent at carbonyl group).

**(c)** The nucleophile approaches along the Bürgi–Dunitz trajectory[4] with an angle close to 107 degrees with the C=O bond vector (see Section 4.3.4).

Notably, the reactive conformation of the substrate is not necessarily the ground state conformation and the diastereoselectivity is controlled by the relative energies of the TSs in conformity with the Curtin–Hammett principle (see Section 5.1.2). The diastereoselectivity of the reactions in Fig. 7.3 can be rationalized using the Felkin–Anh model (Fig. 7.4). In each case, there are two reactive conformations for possible attacks by the nucleophile along the Bürgi–Dunitz trajectory. The major diastereomer results from the unhindered approach of the nucleophile from the side of the smallest group (H) in the most stable TS.

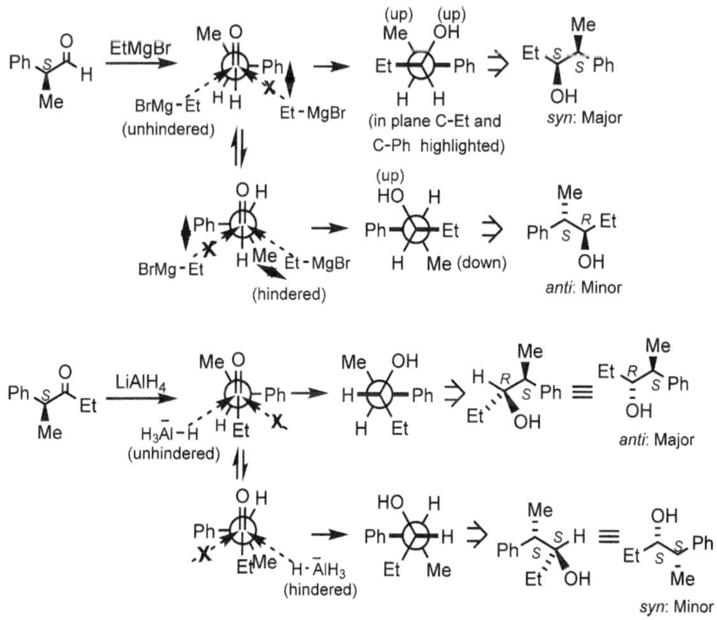


**FIG. 7.4** Rationalization of diastereoselectivity of the reactions in Fig. 7.3 using the Felkin–Anh model.

### 7.1.2 Polar Felkin–Anh model

If the α-stereocentre of the carbonyl substrate bears a polar (electronegative) substituent, the diastereoselectivity is often much greater. For example, the nucleophilic addition of an enolate to a chiral aldehyde with α-$NBn_2$ substituent exhibits high diastereoselectivity (dr 24:1, de 92%) in favour of the major product (Fig. 7.5). Here the chiral aldehyde can be derived as a single enantiomer from the natural amino acid (*S*,*S*)-isoleucine by a sequence of N-benzylation, $LiAlH_4$ reduction and oxidation.

The two ligands CH(Me)Et and $NBn_2$ at the α-stereocentre do not differ much in size. If CH(Me)Et is assumed to be larger, the simple Felkin–Anh model would predict a diastereoselectivity which is opposite to the observed selectivity. However, if the electronegative substituent ($NBn_2$) gets the preference, it would be perpendicular to C=O and the mechanism gives the observed selectivity (Fig. 7.6A). This is called the polar Felkin–Anh model.[5,6]

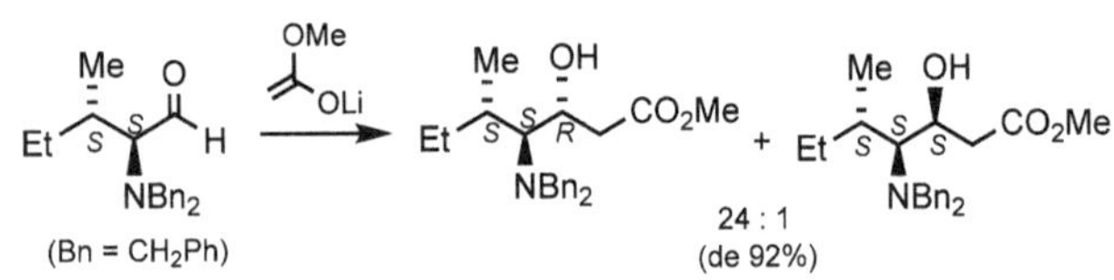


**FIG. 7.5** Diastereoselectivity in nucleophilic addition to a chiral aldehyde containing α-electronegative substituent.

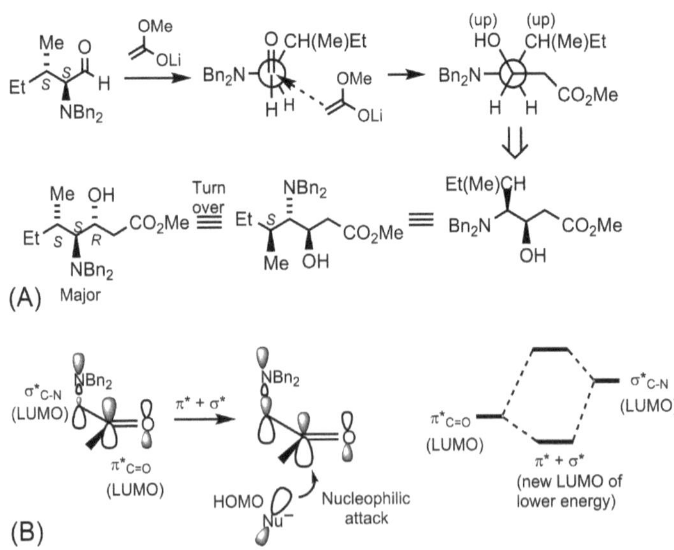


**FIG. 7.6** Polar Felkin–Anh model: (A) Rationalization of the diastereoselectivity of the reaction in Fig. 7.5; (B) The conformation in which electronegative substituent ($NBn_2$) is perpendicular to C=O has a lower energy LUMO and is more reactive.

Fig. 7.6B shows that when C—N bond is perpendicular to C=O, the two LUMOs ($\pi^*_{C=O}$ and $\sigma^*_{C-N}$) can combine to give a new LUMO of lower energy ($\pi^*+\sigma^*$) which would allow more effective interaction with the HOMO of the nucleophile. Thus, the conformation in which the polar substituent (irrespective of the size) is perpendicular to C=O will be more reactive and will give enhanced diastereoselectivity.

It may be mentioned that another model known as the Cornforth–Evans model[7] based on the electrostatic effects (minimization of dipole repulsion) was proposed to rationalize the effects of polar substituents on the stereoselectivity.

**Problem 7.1** Rationalize the diastereoselectivity of the following reaction:

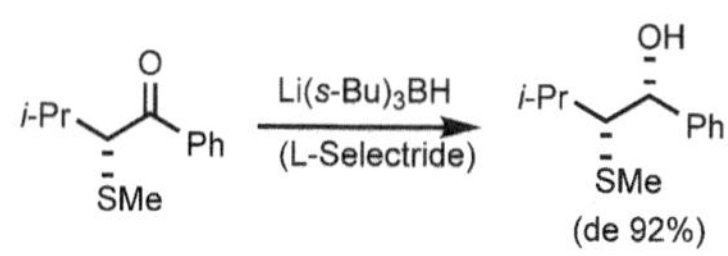


## 7.1.3 Cram chelate model

An important factor that controls the diastereoselectivity of carbonyl additions is chelation.[8,9] If the α-electronegative substituent can form a chelate with the metal ion (associated with the nucleophile) and the carbonyl oxygen, the diastereoselectivity is usually determined by the chelated conformation. This is known as the Cram chelate model. The most populated chelated conformation is usually most reactive because the metal ion as a Lewis acid in the complex enhances the reactivity of C=O by lowering its LUMO energy and raising its LUMO coefficient on carbonyl carbon. Of the metal ions, the highly charged ions such as $Mg^{2+}$, $Zn^{2+}$, $Al^{3+}$, $Ce^{3+}$, and $Ti^{4+}$ are generally very good at chelation; $Li^+$ is also involved in chelation but $Na^+$ or $K^+$ is generally poor at chelation.

Chelation can reverse the diastereoselectivity expected from the polar Felkin–Anh model. This is illustrated with an example in Fig. 7.7A. Here the metal ion ($Zn^{2+}$) from the reagent can form a strong chelate with SMe and carbonyl O, thereby changing the reactive conformation of the polar Felkin–Anh model. The chelated conformation leads to the reversal of stereoselectivity from *syn* to *anti*.

Fig. 7.7B shows that the diastereoselectivity is reversed when $NaBH_4$ is replaced by $Me_2Mg$. In the former, the reaction proceeds through a nonchelated conformation (polar Felkin–Anh model) to give a major *syn* diastereomer whereas in the latter, the reaction is under chelation control (Cram chelate model) to produce almost exclusively the *anti* product.[10]

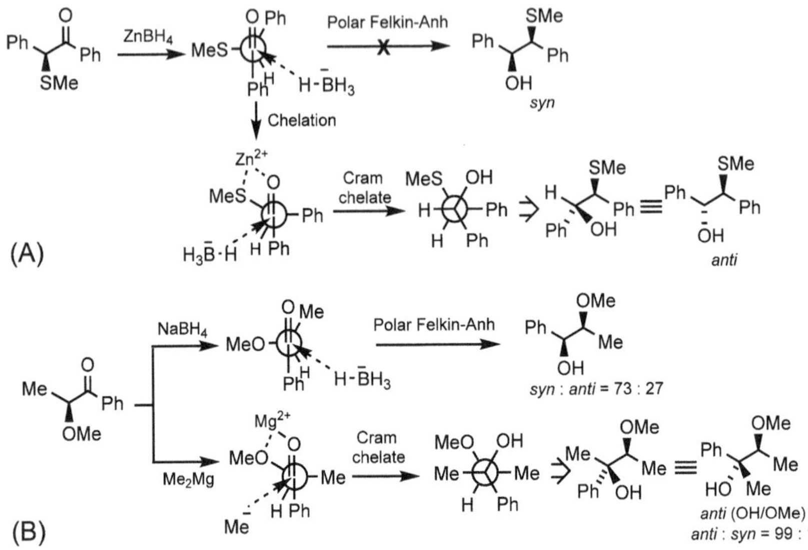


**FIG. 7.7** (A) Cram chelate model for diastereoselectivity in carbonyl additions; (B) Reversal of diastereoselectivity on changing the nucleophilic reagent from $NaBH_4$ to $Me_2Mg$.

**Problem 7.2** Predict the diastereoselectivity of the following reaction:

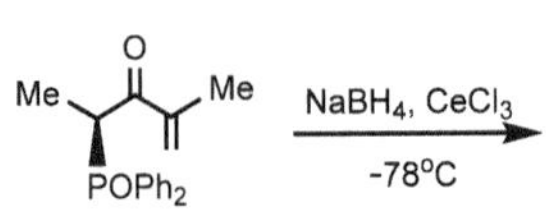


**Problem 7.3** Rationalize the stereochemical outcomes of the following reactions:

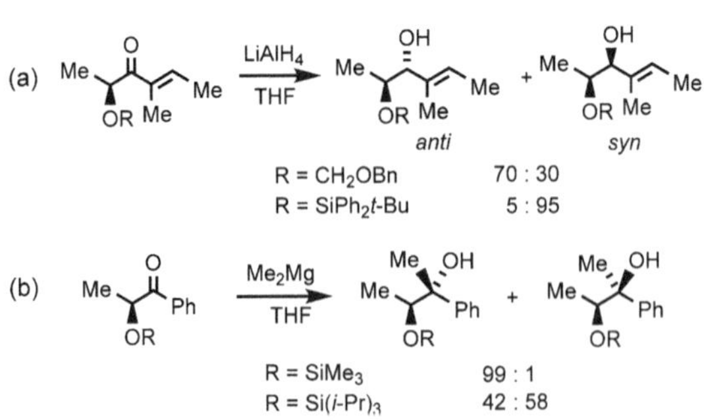

### 7.1.4 Prelog's rule

If a carbonyl substrate has an α-ester group containing a stereocentre ($CO_2C^*R_SR_MR_L$ where $R_S$, $R_M$ and $R_L$ indicate small, medium and large groups respectively), the stereochemical course of Grignard addition to the keto carbonyl can be predicted using the Prelog's rule (Fig. 7.8).[11] The substrate is a chiral α-ketoester which is obtained from the corresponding α-ketoacid by reaction with a chiral alcohol. The ligand priority order at the stereocentre in the ketoester substrate is assumed to be —O > $R_L$ > $R_M$ > $R_S$.

As shown in Fig. 7.8, the reactive conformation has two carbonyl groups being antiperiplanar to each other (to minimize the dipole repulsion) and the large group ($R_L$) in the same plane as the carbonyl groups so that the bulky complex of ester carbonyl with the reagent lies between two smaller groups ($R_S$ and $R_M$). The attack of the nucleophile ($R^2MgX$) will then take place preferentially from the side of the smallest group ($R_S$). This implies that if the alcohol has *S* configuration, the Grignard reagent will approach from the *Re* (front) face of the keto carbonyl in which the smallest group ($R_S$) is placed (Fig. 7.8). Thus the reaction is of *ul* (unlike) type since *S* and *Re* are opposite. Similarly, with an (*R*) alcohol, the attack will take place from the *Si* face of the keto carbonyl function.

If the α-ketoester is a pyruvate ($R^1$ = Me) and the Grignard reagent is PhMgBr ($R^2$ = Ph), the Grignard reaction followed by hydrolysis will produce atrolactic acid (Fig. 7.9A). The atrolactic acid can be identified polarimetrically as (*S*)-(+) or (*R*)-(−). As shown, the (*S*)-pyruvate gives *R*-(−)-atrolactic acid. This atrolactic acid synthesis and the application of Prelog's rule (*ul* reaction) have been used to assign the absolute configuration of many chiral secondary alcohols.[12] The correlation is:

$$(S)-(+)-\text{Atrolactic acid} \Rightarrow (R)-\text{Alcohol}$$
$$(R)-(-)-\text{Atrolactic acid} \Rightarrow (S)-\text{Alcohol}$$

(If the α-ketoester is a phenylglyoxylate ($R^1$ = Ph) and the Grignard reagent is MeMgBr ($R^2$ = Me), the correlation will be opposite, that is (*R*)-atrolactic acid ⇒ (*R*)-alcohol and (*S*)-atrolactic acid ⇒ (*S*)-alcohol.)

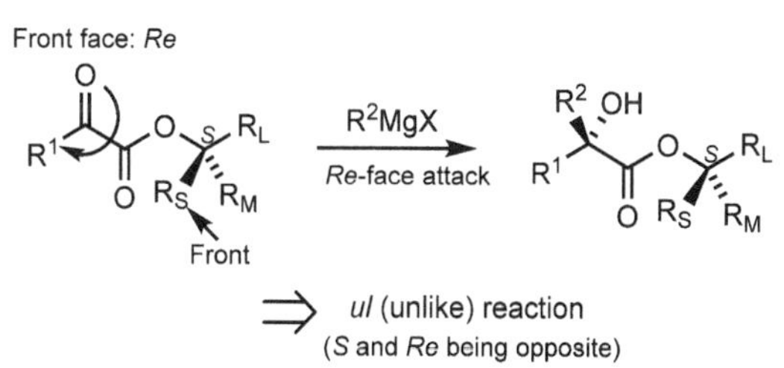


**FIG. 7.8** Prelog's rule for nucleophilic attack by Grignard reagent to a chiral α-ketoester. $R_S$, $R_M$ and $R_L$ represent small, medium and large groups, respectively, with the priority $R_L$ > $R_M$ > $R_S$.

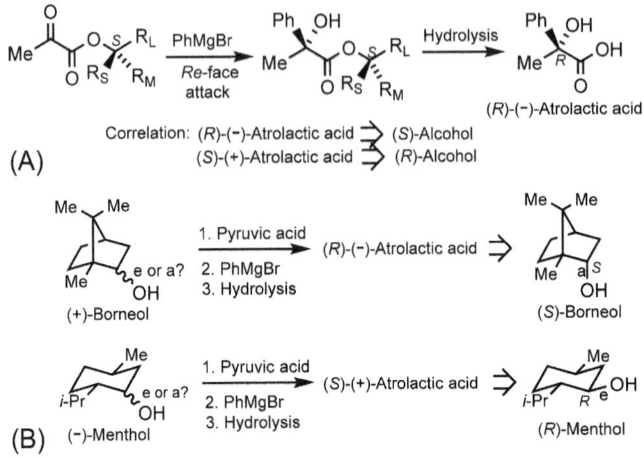


**FIG. 7.9** (A) Atrolactic acid synthesis in conformity with the Prelog's rule; (B) Applications of the Prelog's rule for the assignment of absolute configuration to chiral secondary alcohols. e = equatorial, a = axial

The Prelog's rule is empirical and not necessarily mechanistic, but the rule is useful for assigning absolute configuration of chiral secondary alcohols. For example, the assignment of *R*/*S* configuration to (+)-borneol and (−)-menthol is illustrated in Fig. 7.9B.

Note that the chiral centre in the α-ester group is not adjacent (α) to the keto carbonyl function (site of the reaction). The large distance between them often leads to a less efficient asymmetric induction. Consequently, the diastereoselectivity may not be high in all situations and the measured optical rotations of atrolactic acid may be quite small in some cases. Furthermore, hydrolysis needs to be complete so as to avoid the problems of kinetic resolution.

## 7.2 Diastereoselective carbonyl additions: Chiral substrates with β-stereocentres

Besides 1,2-asymmetric induction, stereoselective 1,3-asymmetric induction in chiral substrates with β-stereocentres is possible. For example, Fig. 7.10 shows the diastereoselective reductions of ketones having β-stereocentres with a bulky borohydride, lithium tri(*sec*-butyl)borohydride (known by the trade name L-Selectride).[13] The diastereoselectivity observed for the two chiral ketones **7.1** and **7.2** is opposite and can be rationalized by the Felkin–Anh type TS conformations **7.3** and **7.4**. For the ketone **7.1**, the TS **7.3** (use a molecular model) in which the steric interaction between *i*-Pr and R is minimized is favoured and gives the major *anti* diastereomer. In the case of ketone **7.2**, the favoured TS **7.4** which minimizes the dipole repulsion between C=O and C—O and also

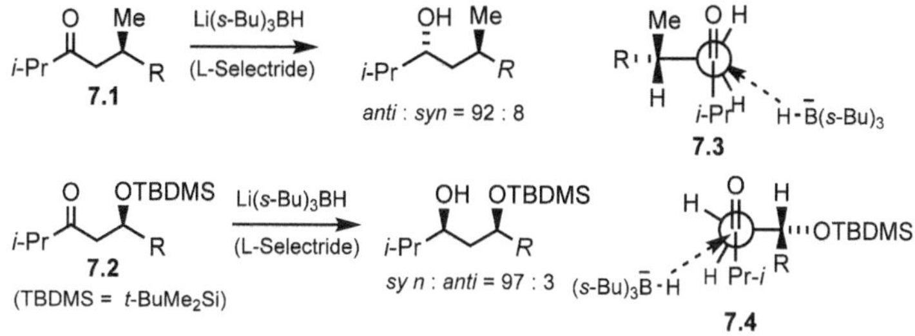


**FIG. 7.10** Diastereoselective reductions of chiral ketones having β-stereocentres with L-Selectride (lithium tri-*sec*-butylborohydride).

avoids steric crowding between *i*-Pr and OTBDMS, leads to *syn* diastereomer as the major product.

### 7.2.1 Narasaka–Prasad reduction: *syn*-Selectivity

An important strategy to improve asymmetric induction in the chiral ketone substrate with a β-stereocentre which is somewhat remote from the carbonyl group is to form a temporary ring involving the β-stereocentre. This is possible using a β-hydroxyketone in the presence of a boron chelating agent such as $R_2BOMe$. Narasaka–Prasad reduction[14,15] uses this strategy with a small borohydride $NaBH_4$ as a reducing agent. Fig. 7.11 shows a scheme for the Narasaka–Prasad reduction. The reaction is highly diastereoselective in favour of the *syn*-1,3-diols.

The reaction proceeds through a six-membered ring chelate in the half-chair conformation (Fig. 7.12). Nucleophilic addition of hydride takes place preferentially via an axial attack from the top face opposite to alcohol oxygen to give favourable chair-like TS (cf. Fig. 6.29). This TS leads directly to the chair conformation of the product, which on hydrolysis gives predominantly *syn*-1,3-diol. An axial attack from the bottom face on the same side of alcohol oxygen leads to a twist-boat TS and an initial twist-boat product, which is less favourable.

### 7.2.2 Evans–Saksena reduction: *anti*-Selectivity

The Evans–Saksena reduction[16,17] employs a borohydride, $Me_4NBH(OAc)_3$ as a reducing agent. The reduction of chiral β-hydroxyketones with this

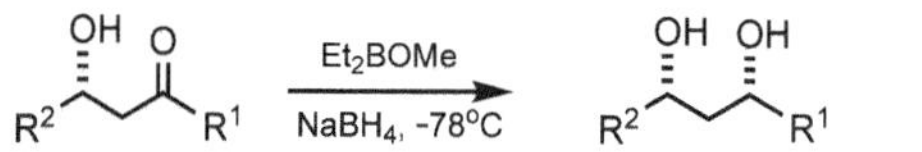


| | *syn* : *anti* | Yield (%) |
|---|---|---|
| $R^1 = R^2 = Bu$ | 99 : 1 | 99 |
| $R^1 = CH_2CO_2Et$, $R^2 = Ph$ | 98 : 2 | 85 |

**FIG. 7.11** *syn* Diastereoselectivity in the Narasaka–Prasad reduction of chiral β-hydroxyketones.

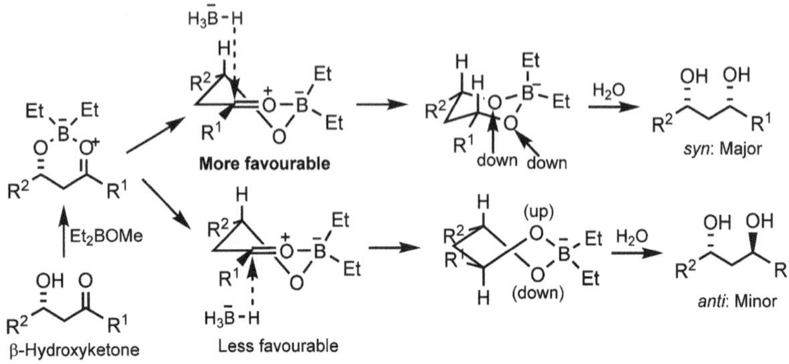


**FIG. 7.12** Rationalization of diastereoselectivity of the Narasaka–Prasad reduction in Fig. 7.11.

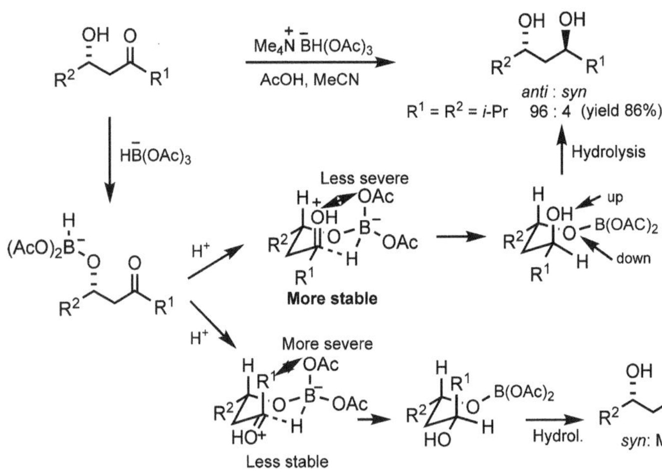


**FIG. 7.13** The Evans–Saksena reduction of chiral β-hydroxyketones and rationalization of *anti* diastereoselectivity.

borohydride exhibits high diastereoselectivity in favour of the *anti*-1,3-diols (Fig. 7.13). The reaction proceeds preferably through a more stable chair TS, which gives the *anti* diastereomer as the major product. The other chair TS suffers from a more severe 1,3-diaxial interaction and is less stable, which leads to the minor *syn* isomer.

**Problem 7.4** The following Evans–Tishchenko reduction is *anti*-selective. Explain.

## 7.3 Asymmetric carbonyl reductions: Chiral reagents

With prochiral carbonyl substrates, asymmetric reductions are possible using enantiopure chiral reagents in stoichiometric quantities. This approach refers to the third generation (3G) of asymmetric synthesis (see Section 5.3.1.3).

### 7.3.1 Chiral boranes

A variety of chiral borane reagents have been developed for asymmetric carbonyl reductions. The highly efficient chiral boranes include Alpine–Borane and $Ipc_2BCl$ or DIP-Cl (diisopinocamphenylboron chloride).[18] Fig. 7.14A shows the syntheses of (*R*)-Alpine–Borane and (−)-DIP-Cl from (1*R*)-(+)-α-pinene.

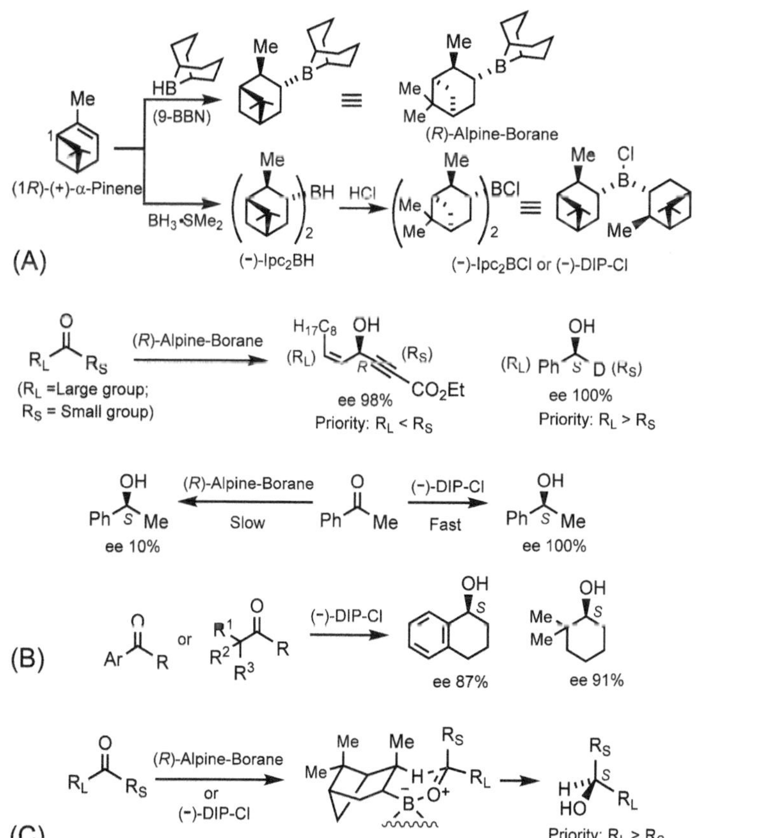


**FIG. 7.14** (A) Methods of preparation for Alpine–Borane and DIP-Cl (diisopinocamphenylboron chloride); (B) Asymmetric carbonyl reductions using (*R*)-Alpine–Borane and (−)-DIP-Cl; (C) Rationalization of enantioselectivity using a boat-like TS model.

The asymmetric reduction of ketones $R_LR_SC{=}O$ (where $R_L$ and $R_S$ stand for large and small substituents respectively) with (*R*)-Alpine–Borane is shown in Fig. 7.14B. As shown, the reaction is highly enantioselective with an acetylenic ketone[19] (ee 98%) and a deuterated aldehyde[20] (ee 100%). If the ligand priority is $R_L > R_S$, the configuration of the alcohol is *S*, and if $R_L < R_S$, the configuration is *R*. The reagent DIP-Cl is more reactive than Alpine–Borane because of increased Lewis acidity of the reagent. For example, the reaction of acetophenone with DIP-Cl is faster than with (*R*)-Alpine–Borane (Fig. 7.14B). In fact, DIP-Cl reacts rapidly with aromatic and *t*-alkyl ketones to give the corresponding alcohols with high ee, as shown also in Fig. 7.14B.[21,22]

The enantioselectivity of reductions with (*R*)-Alpine–Borane or (−)-DIP-Cl has been rationalized using a boat-like TS model (Fig. 7.14C).[23] The large substituent ($R_L$) of the ketone is away from the bulky reagent in the preferred TS which gives the major enantiomer (*S*) when $R_L > R_S$. Similarly, (*S*)-Alpine–Borane or (+)-DIP-Cl would give the opposite enantiomer of the alcohol.

### 7.3.2 Chiral hydrides

An achiral hydride such as $LiAlH_4$ can be converted into a new chiral hydride by reaction with an enantiopure diol as a chiral ligand. The most popular chiral diol is 1,1′-bi-2-naphthol (BINOL), which is available as both (*R*) and (*S*) enantiomers. The reaction of $LiAlH_4$ with (*R*)-BINOL or (*S*)-BINOL gives the chiral hydride (*R*)-BINAL-H or (*S*)-BINAL-H, as shown in Fig. 7.15A. These reagents are highly effective for reduction of ketones in which one of the groups attached to the carbonyl carbon is a π-substituent.[24,25] Fig. 7.15B illustrates the (*R*)-BINAL reduction of prochiral ketones **7.5** with three examples, which take place with good to high ee. The enantioselectivity is attributed to a chair TS model (Fig. 7.15C). The most basic oxygen in OEt acts as a bridging ligand in the chair TS. The preferred TS has the π-substituent in the equatorial position thereby avoiding π–π (filled/filled) repulsion from the naphthalene unit of (*R*)-BINAL-H, and gives the (*R*)-alcohol.

A simple mnemonic for the BINAL-H reduction is that (*R*)-BINAL-H gives (*R*)-alcohol and (*S*)-BINAL-H gives (*S*)-alcohol when the ligand priority is π-substituent > alkyl.

## 7.4 Asymmetric deprotonations: Chiral bases

Asymmetric deprotonations of weak carbon acids are possible with chiral bases. The most important chiral bases developed for asymmetric deprotonations include chiral lithium amides and chiral alkyllithiums.

### 7.4.1 Chiral lithium amides

An achiral lithium amide base such as LDA cannot differentiate between two enantiotopic protons in a prochiral carbonyl substrate. But a chiral lithium

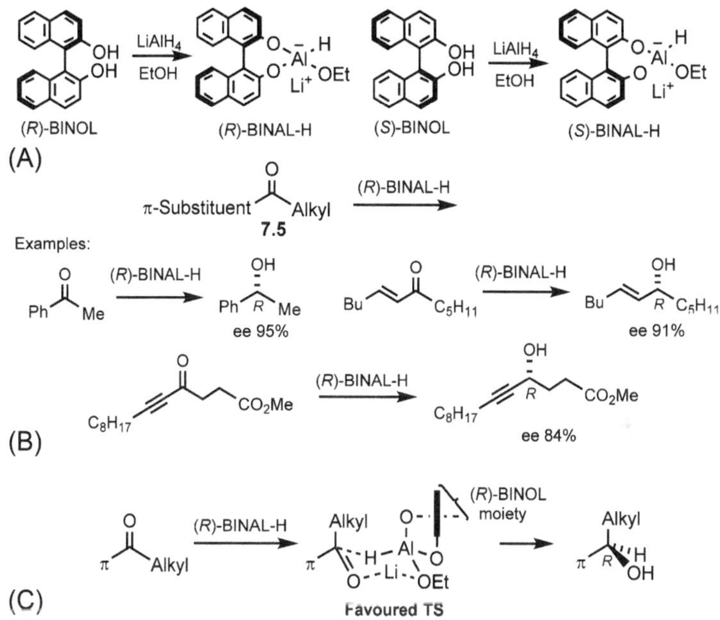


**FIG. 7.15** (A) Chiral hydride reagent (*R*)- or (*S*)-BINAL-H derived from (*R*)- or (*S*)-BINOL with $LiAlH_4$; (B) Asymmetric reduction of ketones **7.5** with (*R*)-BINAL-H; (C) A mechanistic model for reduction with (*R*)-BINAL-H.

amide base can selectively abstract one of the enantiotopic protons and these asymmetric deprotonation reactions are quite useful in asymmetric synthesis.[26] The enantiopure lithium amide bases are usually prepared in situ by treatment of the corresponding enantiopure secondary amines with butyllithium. A useful chiral lithium amide base is **7.6** (Fig. 7.16A), which can be prepared as either enantiomer. It is $C_2$-symmetric and is therefore advantageous in asymmetric induction. Fig. 7.16B shows that the base **7.6** can selectively remove the enantiotopic hydrogen $H_a$ (not $H_b$) of 4-*t*-butylcyclohexanone.[27,28] The selectivity is attributed to the least hindered approach of the ketone to the base away from the larger Ph substituent. The carbonyl oxygen is coordinated to Li for deprotonation of the cyclohexanone by basic nitrogen. This deprotonation (enolization) reaction has been used in an asymmetric synthesis of (*S*)-4-*t*-butyl-2-cyclohexenone (Fig. 7.16B).[29]

The deprotonation protocol has been successfully applied in the asymmetric synthesis of natural products. For example, a chiral α-hydroxy ketone **7.8** is obtained in moderate ee (65%) via deprotonation of *cis*-2,6-dimethylcyclohexanone with the camphor-derived lithium amide base **7.7**, as a key intermediate in a synthesis of the natural product (5*S*)-dihydroactinidiolide (Fig. 7.16C).[30]

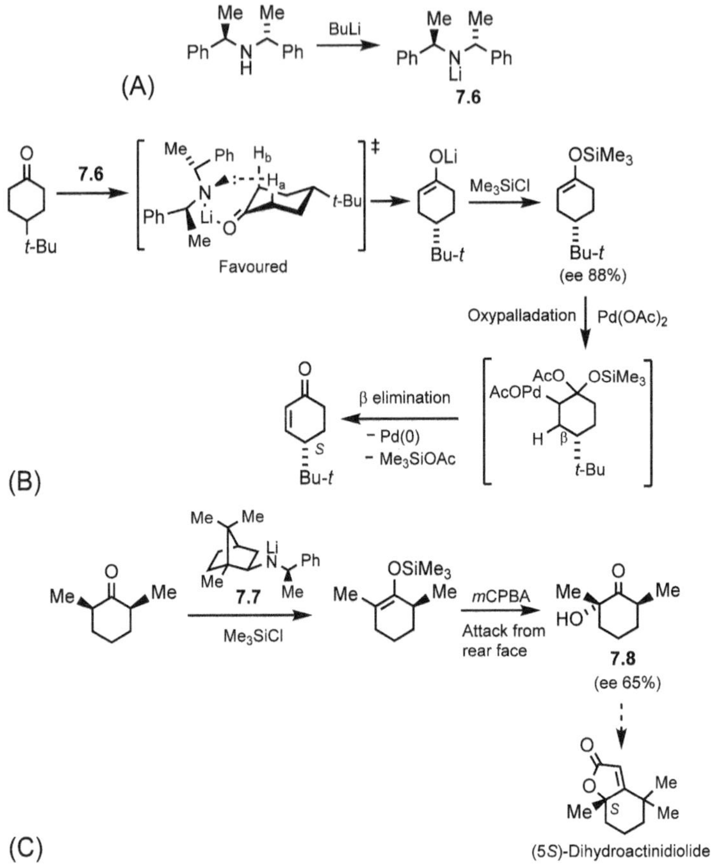


**FIG. 7.16** (A) A chiral lithium amide base **7.6** of $C_2$ symmetry; (B) Asymmetric deprotonation by selective abstraction of enantiotopic H atom from 4-*t*-butylcyclohexanone by **7.6**, applied to an asymmetric synthesis of (*S*)-4-*t*-butyl-2-cyclohexenone; (C) An application of asymmetric deprotonation in natural product synthesis.

## 7.4.2 Chiral alkyllithiums

Chiral alkyllithiums can also be used as bases for asymmetric deprotonations. These chiral bases are usually obtained by complexation of achiral alkyllithiums with a chiral ligand. A useful chiral ligand is the readily available natural product (−)-sparteine. The tetracyclic structure of (−)-sparteine folds around the Li atom of alkyllithium to form a complex which can act as a chiral base for asymmetric deprotonation.[31] Fig. 7.17 illustrates an asymmetric synthesis of a derivative of the unnatural amino acid (*R*)-proline via deprotonation of Boc-protected pyrrolidine by a chiral alkyllithium base which is a complex of *s*-BuLi with (−)-sparteine.

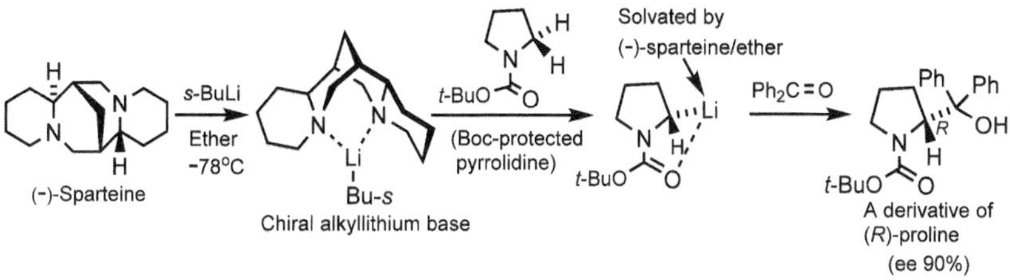


**FIG. 7.17** Asymmetric synthesis of a derivative of the unnatural (*R*)-proline via asymmetric deprotonation by a chiral alkyllithium base as a complex of (−)-sparteine and *s*-BuLi.

## 7.5 Asymmetric carbonyl additions/reductions: Chiral catalysts

In comparison to stoichiometric chiral reagents, only a catalytic (substoichiometric) amount of a chiral catalyst is needed for an enantioselective reaction. Furthermore, the reaction with chiral catalysts is much more efficient and asymmetric catalysis can be an extraordinarily effective approach to asymmetric synthesis. This catalytic approach refers to the fourth generation (4G) of asymmetric synthesis (see Section 5.3.1.4).

### 7.5.1 Addition of zinc organometallics

The asymmetric addition of dialkylzinc compounds to aldehydes can be catalysed by chiral 1,2-(or β-)aminoalcohols. This is illustrated with the addition of diethylzinc to aldehydes catalysed by (−)-DAIB [(−)-3-*exo*-(dimethylamino) isoborneol], as shown in Fig. 7.18A.[32] The reaction exhibits high to moderate enantioselectivity. If R = Ph, the reaction gives high yield (97%) with excellent enantioselectivity (ee 99%); however, aliphatic aldehydes generally give products with moderate ee (R = hexyl, ee 60%). The catalyst can be derived from camphor and both enantiomers of the catalyst are available. (+)-DAIB will give the opposite enantiomer of the product.

The enantioselectivity can be rationalized by a TS model shown in Fig. 7.18B.[33] The active catalyst is formed by the reaction of (−)-DAIB with $Et_2Zn$. The reaction then proceeds by coordination of the second molecule of $Et_2Zn$ and RCHO. Note that aldehyde oxygen is coordinated from the side opposite to the *gem*-dimethyl group. The nucleophilic attack by Et occurs to the *Si* (front) carbonyl face with the large substituent (R) avoiding the steric clash with Et (highlighted) in the favoured TS. The product is the (*S*) alcohol.

Besides DAIB, a range of chiral β-aminoalcohols and related systems can be used as catalysts.[33] Fig. 7.18C shows three examples of such catalysts employed in the reaction between $Et_2Zn$ and PhCHO.

The $Et_2Zn$/(−)-DAIB system has been extensively studied and a remarkable asymmetric amplification has been observed (catalyst with 15% ee can lead to a product with 95% ee) (see Section 5.3.2).

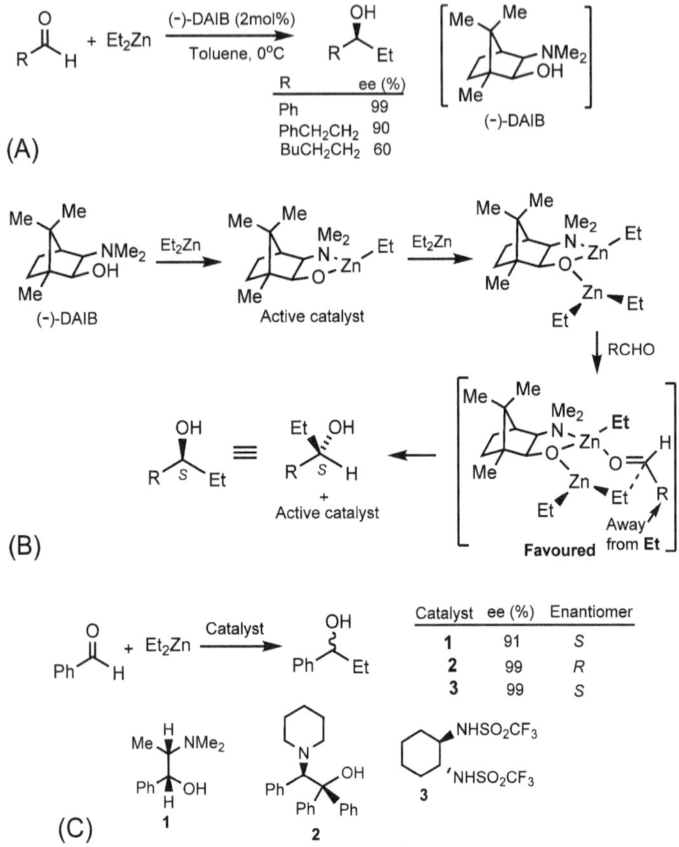


**FIG. 7.18** (A) Catalytic asymmetric addition of diethylzinc to aldehydes using (−)-DAIB as a catalyst; (B) A mechanistic TS model for enantioselectivity; (C) Examples of other related catalysts that give high enantioselectivity in the asymmetric addition.

## 7.5.2 CBS (Corey, Bakshi, Shibata) reduction

A highly effective catalytic asymmetric reduction of ketones is CBS (Corey, Bakshi, Shibata) reduction.[34,35] The CBS catalyst **7.9** is a chiral boron heterocycle (oxazaborilidine) derived from the amino acid (*S*)-proline (Fig. 7.19A). The unnatural (*R*) enantiomer of proline is also available. The active reducing agent is the chiral borohydride **7.10** formed by complexation of the CBS catalyst with $BH_3$. This active reductant is stable, and can be stored and used as a stoichiometric reagent, if so desired.[36] The reactivity and enantioselectivity of the catalyst towards a particular ketone can be optimized by modification in the substituent groups.[37] Nature's CBS catalyst is NADH.

The CBS reduction proceeds well when the two substituents at carbonyl carbon differ much in size, denoted by $R_L$ (large substituent) and $R_S$ (small substituent).

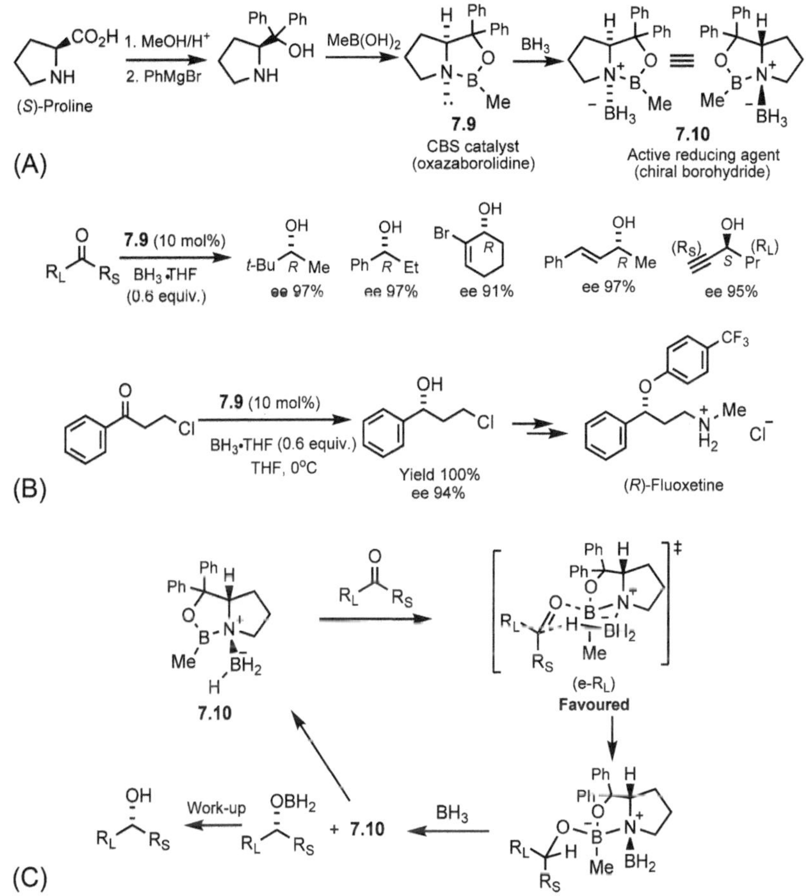


**FIG. 7.19** Catalytic asymmetric CBS reduction of ketones: (A) CBS catalyst and active reductant; (B) Examples and an application in a synthesis of the drug (*R*)-fluoxetine; (C) Mechanism based on a chair TS model.

Fig. 7.19B shows several examples of CBS reductions of prochiral ketones. An application of the CBS reduction in a synthesis of the antidepressant drug (*R*)-fluoxetine (Prozac) is also shown in Fig. 7.19B.[38]

The mechanism of the CBS reduction based on computational studies is delineated in Fig. 7.19C.[39,40] The enantioselectivity arises from a chair-like TS in which the controlling steric interaction is exerted by the alkyl substituent (Me) on boron. Thus the major enantiomer results from the preferred TS in which the large substituent ($R_L$) is equatorial, thus avoiding the 1,3-diaxial interaction with Me. It is of note that complexation of carbonyl oxygen to boron activates the carbonyl for hydride attack and allows the formation of a six-membered chair TS.

**Problem 7.5** Predict the enantioselectivity of the following CBS reduction:

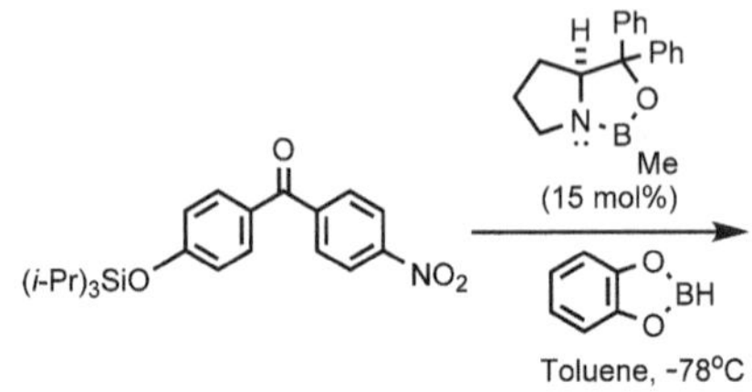


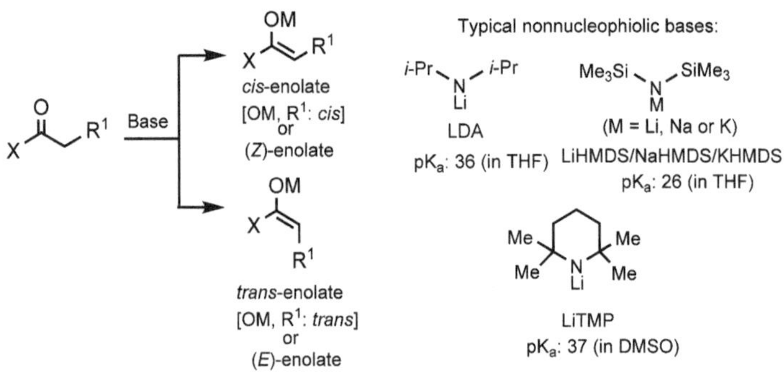


**FIG. 7.20** *cis* (or *Z*) and *trans* (or *E*) enolates, and some typical nonnucleophilic bases for enolization; LDA (lithium diisopropylamide), LiHMDS (lithium hexamethyldisilazide), NaHMDS, KHMDS and LiTMP (lithium tetramethylpiperidide). $pK_a$ of the base refers to its conjugate acid.

## 7.6 Diastereoselective enolate formation

For enolates, two diastereomers are possible: *cis* (or *Z*) and *trans* (or *E*) (Fig. 7.20). The *cis*/*trans* notation refers to the relationship between the substituent ($R^1$) and the oxygen associated with the metal ion of the base (OM). Depending on the ligand priority, *cis* and *trans* may not always correspond to *Z* and *E*, respectively.

The enolization requires a strong nonnucleophilic base for complete deprotonation of the carbonyl compound ($pK_{a\ (water)}$: ketones ~20, esters ~25, amides ~26). Some typical bases are lithium diisopropylamide (LDA), lithium hexamethyldisilazide (LiHMDS) (or NaHMDS, KHMDS) and lithium tetramethylpiperidide (LiTMP).

### 7.6.1 Lithium enolates: Control of enolate geometry

Asymmetric enolate reactions require that an enolate is formed preferentially in one of the two possible geometries. Thus, the control of enolate geometry is crucial. This can be achieved by the judicious choice of reagents, reaction conditions and the structure of carbonyl substrates.

### 7.6.1.1 Kinetic enolates: Ireland model

The lithium enolates of ketones, esters or amides are usually obtained under kinetic conditions in the presence of LDA in THF at low temperature (Fig. 7.21).[41–43] A ketones with a *t*-alkyl substituent (e.g. X=*t*-Bu) gives predominantly the *cis*-enolate but a less highly substituted ketone (e.g. X=Et) usually gives the mixture of *cis*- and *trans*-enolates (Fig. 7.21A). An amide (e.g. X=$NEt_2$) gives mainly the *cis*-enolate but in contrast, an ester (X=OMe) gives predominantly the *trans*-enolate (Fig. 7.21A).

Cyclic ketones such as cyclopentanone and cyclohexanone give the *trans*-enolates because the ring double bond can only be *cis* (*trans* double bond will be highly strained), as shown for cyclopentanone in Fig. 7.21B.

The diastereoselectivity in enolization with LDA in THF has been explained in terms of a six-centre cyclic TS model that is preferably chair-like (Ireland model).[41,44] The stereoelectronic requirement for the α-deprotonation of a carbonyl compound by a base to form the enolate is that the breaking C—H bond is orthogonal to the plane of the carbonyl group.[45] In this conformation, the p orbital on the α-carbon (evolving from the σ C—H orbital) would be parallel to the carbonyl π system, thereby stabilizing the TS for the enolate formation (Fig. 7.22A).

In THF, the coordination of $Li^+$ cation of LDA to the carbonyl oxygen allows the formation of closed (cyclic) transition structure. There are two possible chair TSs: *trans*-enolate TS (e-Me) and *cis*-enolate TS (a-Me) (Fig. 7.22B). The major strain in the a-Me conformation is 1,3-diaxial (*i*-Pr/Me) interaction while that in the e-Me conformation is allylic ($A^{1,2}$) strain. The diastereoselectivity in the enolate formation depends on which of these strains is dominant or a balance between them.

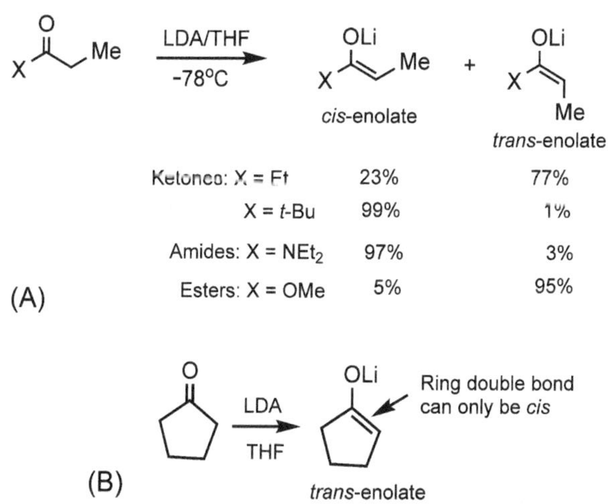


**FIG. 7.21** (A) Stereoselective formation of lithium enolates with ketones, amides and esters under kinetic conditions; (B) Formation of the *trans* enolate with a cyclic ketone cyclopentanone.

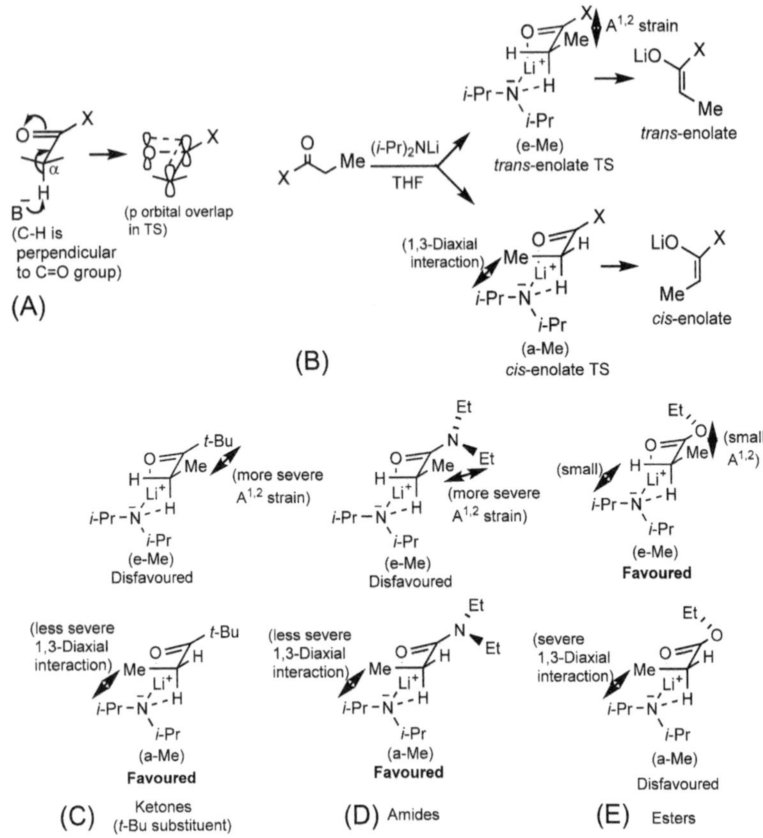


**FIG. 7.22** (A) Stereoelectronic requirement for enolization; (B) Ireland model for enolization through chair transition structure; (C) Rationalization of diastereoselectivity in enolization (see Fig. 7.21A) of (C) ketones, (D) amides and (E) esters based on the Ireland model.

For ketones with a bulky *t*-Bu substituent, the developing $A^{1,2}$ strain in the e-Me conformation is much more severe than 1,3-diaxial (*i*-Pr/Me) interaction in the a-Me conformation (Fig. 7.22C). As a result, the a-Me conformation is favoured which gives the *cis*-enolate. For the less hindered Et substituent, a balance between the strains gives a mixture of *cis*-and *trans*-enolates. For amides, the a-Me conformation is favoured for the similar reasoning as for ketones with a *t*-Bu substituent, and the formation of the *cis*-enolate is highly favoured (Fig. 7.22D). For esters, the developing $A^{1,2}$ strain in the e-Me conformation is quite small for the steric interaction of Me with the O atom of ester group (Fig. 7.22E). On the other hand, the a-Me conformation has 1,3-diaxial (*i*-Pr/Me) interaction which is much more severe than the diaxial (*i*-Pr/H) interaction present in the e-Me conformation. Consequently, the *trans*-enolate arising from the e-Me conformation is favoured.

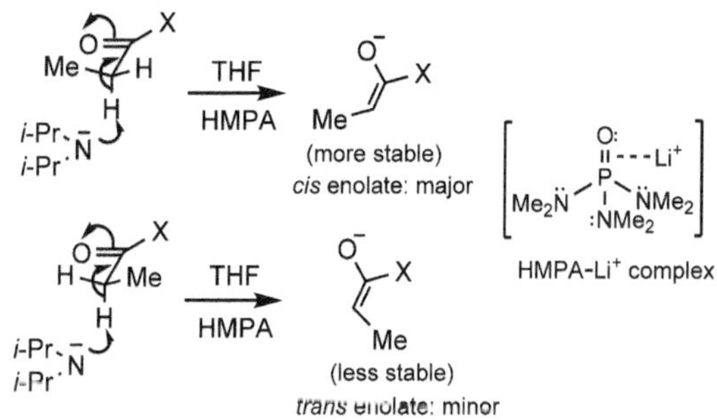


**FIG. 7.23** Formation of thermodynamic enolate through open (noncyclic) transition structure with LDA in THF/HMPA.

### *7.6.1.2 Thermodynamic enolates: Open model*

If enolization with LDA/THF is carried out in the presence of a co-solvent such as HMPA (hexamethylphosphoramide), the *cis*-enolates are favoured in all cases.[46] For example, enolization of pentan-3-one (X=Et in Fig. 7.21A) with LDA in THF/HMPA at −78°C gives *cis*:*trans*=95:5. Under these conditions, the *cis*-enolates are also favoured for esters (cf. Fig. 7.21A). HMPA can strongly coordinate with $Li^+$, and hence the association of $Li^+$ with carbonyl group is negligible. The enolization then proceeds through open (noncyclic) transition structure (Fig. 7.23). The more stable TS in which X and Me are far apart leads to the more stable *cis*-enolate as a major product. Thus, in THF/HMPA, a thermodynamic enolate is formed, whereas in THF alone, a kinetic enolate is produced.

## 7.6.2 Boron enolates: Control of enolate geometry

The boron enolates are normally prepared from the corresponding ketone and a dialkylboron triflate ($R_2BOTf$) or chloride ($R_2BCl$) in the presence of a tertiary amine base. It is possible to control the enolate geometry by the choice of the boron reagent. In general, dialkylboron reagents with bulky ligands (e.g. cyclohexyl) with a poor leaving group (such as Cl), and an unhindered base ($Et_3N$) give predominantly the *trans*-enolates (Fig. 7.24A).[47] Using smaller boron substituents (such as Bu) with a good leaving group (such as OTf) and a hindered base (*i*-$Pr_2NEt$), the *cis*-enolates are formed selectively (Fig. 7.24A).[48] Like $Bu_2OTf$, 9-BBNOTf (where 9-BBN stands for 9-borabicyclo[3.3.1]nonyl) also gives the *cis*-enolate with high selectivity. Although the bicyclic part of 9-BBN appears to be large, it is effectively tied back behind the boron. Thus, the steric demand of the bicyclic part is small and it resembles a smaller substituent on boron.

The diastereoselectivity in boron enolate formation can be explained in terms of the reactant conformation and the stereoelectronic requirement for deprotonation (Fig. 7.24B).[49,50] The model involves coordination of the boron

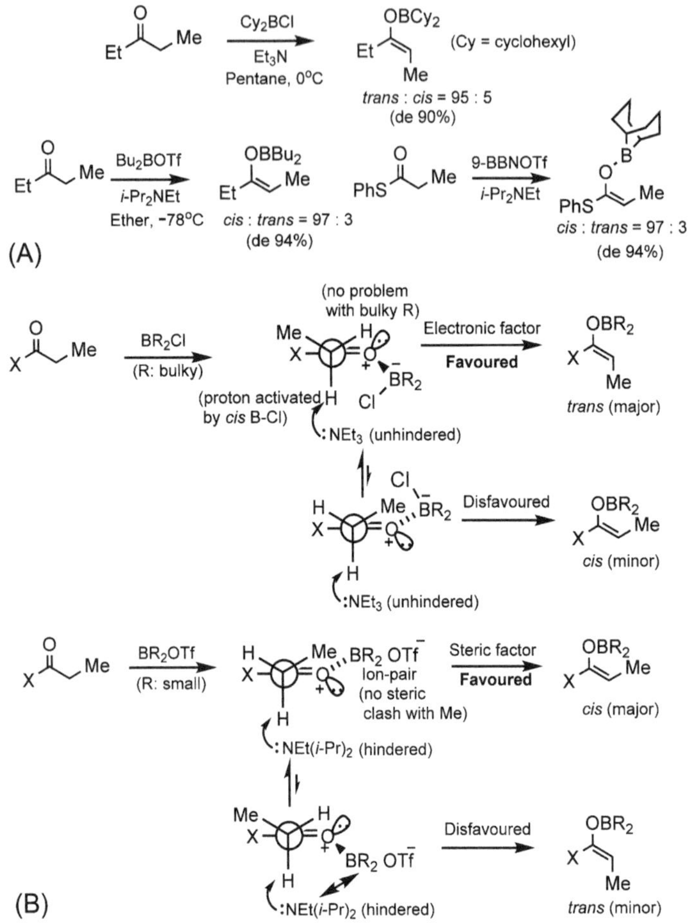


**FIG. 7.24** (A) Diastereoselective generation of *cis* or *trans* boron enolates; (B) Mechanistic model for rationalization of diastereoselectivity in boron enolate formation.

reagents to the ketone carbonyl and subsequent deprotonation by the tertiary amine. The difference exerted by the two types of boron reagents also lies in the extent of dissociation of the leaving group. With bulky boron substituents and unhindered base, the reactant conformation that provides the complex in which B—Cl bond activates the *cis* proton for deprotonation (favourable electronic factor) with no steric problem from bulky substituents is preferred, and leads to the *trans*-enolate. But with small substituents on boron and the triflate seems to exist as an ion pair, the sterically favoured complex in the presence of hindered base leads to the *cis*-enolate. Thus the diastereoselectivity is determined primarily by electronic control in the former and by steric control in the latter.

## 7.7 Asymmetric enolate alkylations

Asymmetric alkylations of enolates with high stereochemical control constitute an important area of asymmetric synthesis.[51] The methodologies for asymmetric enolate alkylations involve mainly chiral auxiliaries.

### 7.7.1 Chiral auxiliaries

The chiral auxiliary strategy of asymmetric enolate alkylations refers to second-generation (2G) asymmetric synthesis (see Section 5.3.1.2). However, new auxiliary-controlled reactions continue to evolve frequently.[52] Fig. 7.25 shows general schemes for chiral auxiliary-mediated asymmetric alkylations of carbonyl substrates using carboxylic acids or derivatives (Fig. 7.25A) and ketones (Fig. 7.25B).[53] For an efficient asymmetric synthesis, the chiral auxiliary ($X_c$), after its facile attachment to the carbonyl substrate, must provide a strong predisposition for highly stereoselective enolization, a strong bias for enolate diastereoface selection, and its easy removal under mild cleavage conditions without racemisation of the desired products. The frontier orbital picture for the electrophilic attack on an enolate has been described earlier (see Section 4.3.5, Fig. 4.11B). Since the trajectory of the electrophile is somewhat displaced from the perpendicular approach towards the chiral auxiliary, the steric effects from the auxiliary will be enhanced giving rise to more favourable diastereoselection.

#### *7.7.1.1 Evans oxazolidinone auxiliaries*

The oxazolidinone auxiliaries[54] (see Fig. 5.9A) introduced by Evans[55] have proved to be popular and efficient chiral auxiliaries for asymmetric alkylation reactions.[56,57] Either enantiomer of the desired product can be obtained by the choice of the oxazolidinone **7.11** or **7.12** (Fig. 7.26A). Fig. 7.26B illustrates the asymmetric alkylation reaction using the auxiliary **7.11**. The auxiliary is converted to its anion which then reacts readily with an acid chloride to give the acylated oxazolidinone.[58] Treatment of the *N*-acylated oxazolidinone with LDA in THF produces a chelated *cis*-enolate with >99:1 selectivity (cf. enolization of

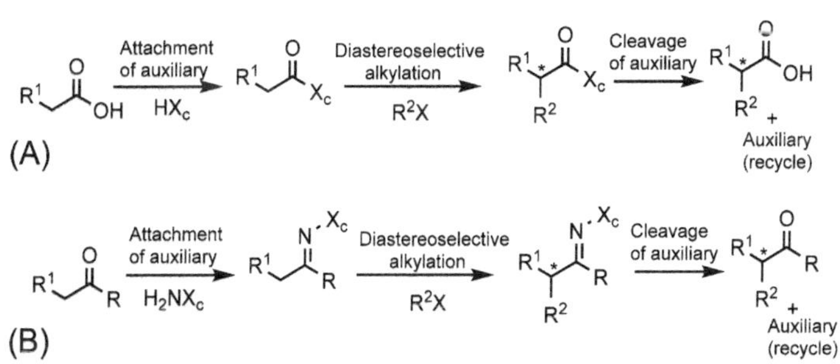


**FIG. 7.25** General schemes for chiral auxiliary-mediated alkylations of carbonyl substrates using (A) carboxylic acids and (B) ketones.

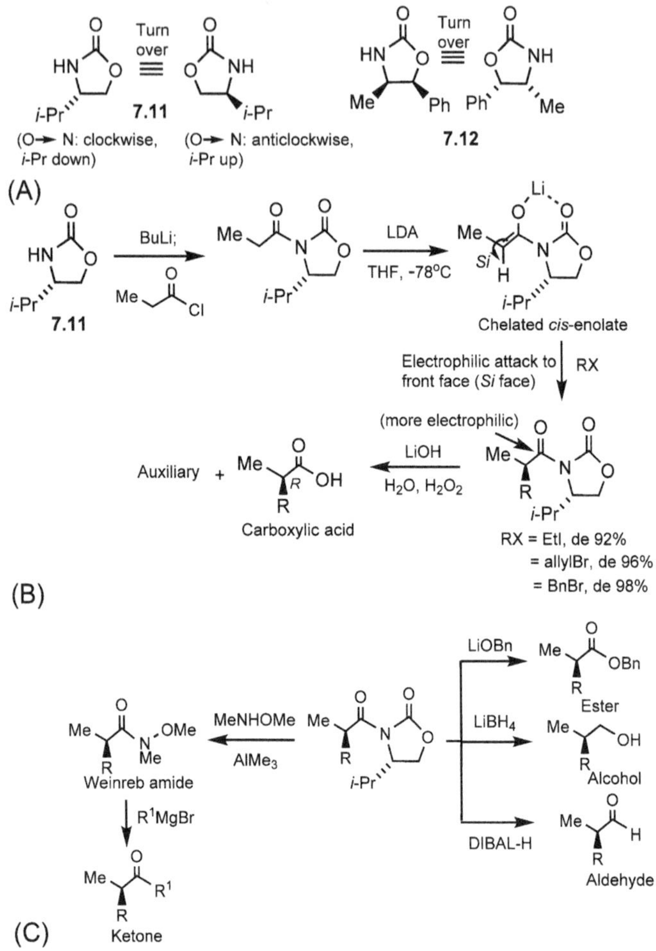


**FIG. 7.26** (A) Evans chiral oxazolidinone auxiliaries; (B) Asymmetric enolate alkylation using the oxazolidinone auxiliary **7.11**; (C) Methods of auxiliary cleavage besides hydrolysis.

amides, see Fig. 7.22D). Note that the enolate is held rigidly in one conformation by chelation of the Li to the carbonyl group of the oxazolidinone. In this conformation, the rear face (*Re* face) of the enolate is hindered by *i*-Pr substituent on the auxiliary, and hence alkylation occurs to the front face (*Si* face) with very high levels of diastereoselection. The major diastereomer can be easily purified by the conventional techniques, such as crystallization or column chromatography. The removal of the auxiliary by hydrolysis[59] with alkaline $H_2O_2$ gives the carboxylic acids essentially as a single enantiomer. If the auxiliary **7.12** is employed, the opposite enantiomer of the product will result.

(Note that the scheme can also be drawn with **7.11** (O→N: anticlockwise, *i*-Pr up); the alkylation will then be represented as occurring from the enolate rear face when this will be the *Si* face. Therefore, in whatever way the scheme is drawn, the attacking face of the enolate will remain the same in terms of the descriptor which characterizes its configuration.)

Besides hydrolysis, the removal of oxazolidinone auxiliary can also be performed in other useful ways to prepare essentially enantiopure esters, alcohols, aldehydes and ketones (Fig. 7.26C).

In asymmetric alkylations, usually an alkyl halide is used as an electrophile. However, other electrophilic reagents can also be employed to achieve asymmetric hydroxylation, bromination or amination. For example, an asymmetric enolate hydroxylation using the chiral auxiliary **7.12** is shown in Fig. 7.27.[60] Here the electrophilic reagent is an oxaziridine which acts, in effect, as a source of electrophilic OH ('HO$^{+}$'). The electrophilic attack takes place preferentially on the rear face (*Re* face) of the enolate as the front *Si* face is shielded by Me (reinforced by Ph) on the auxiliary. The method provides an effective asymmetric synthetic route to α-hydroxy acids or their derivatives.

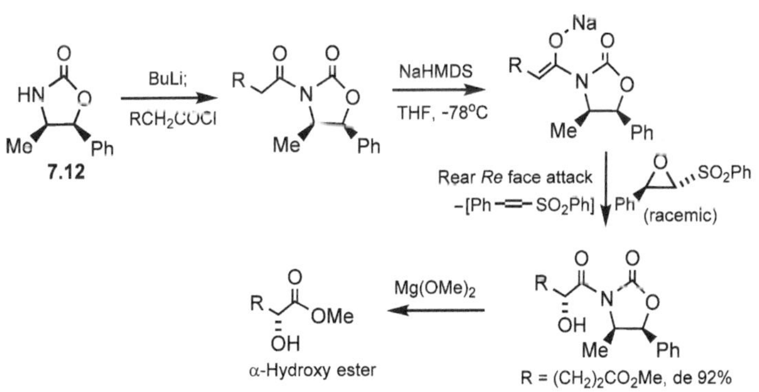


**FIG. 7.27** Asymmetric enolate hydroxylation using the oxazolidinone auxiliary **7.12**.

**Problem 7.6** Predict the stereochemistry of the chiral alcohol in the following scheme involved in a synthesis of the immunosuppressant (−)-sanglifehrin A.

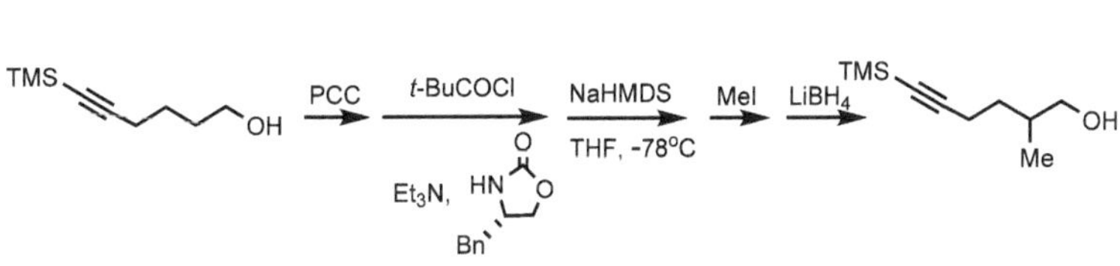

**Problem 7.7** Predict and explain the stereochemical outcomes of the following reactions:

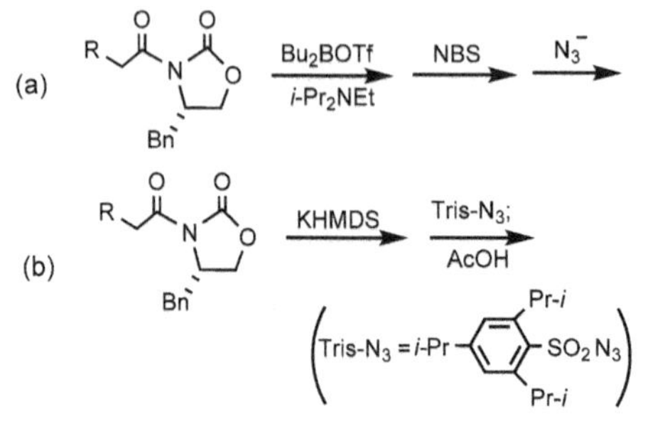


### *7.7.1.2 Oppolzer sultam auxiliaries*

The Oppolzer method for the asymmetric enolate alkylation is similar to Evans oxazolidinone approach but uses a camphor sultam[61] as the chiral auxiliary (see Fig. 5.9B). Fig. 7.28 illustrates an Oppolzer asymmetric alkylation using the sultam auxiliary **7.13**.[62] Both enantiomers of the sultam are available and hence the alkylation product can be obtained as either enantiomer. The sultam is converted to its anion which then reacts with an acid chloride to give the acylated sultam. Treatment of the N-acylated sultam with NaHMDS provides the chelated *cis*-enolate. Alkylation with benzyl bromide then occurs from the lower face (*Re*) of the enolate as the upper *Si* face is shielded by a Me group (highlighted) on the auxiliary. The desired diastereomer is produced with excellent selectivity (de 97%). After purification, the auxiliary is removed by hydrolysis with alkaline $H_2O_2$ to give essentially the enantiopure carboxylic acid.

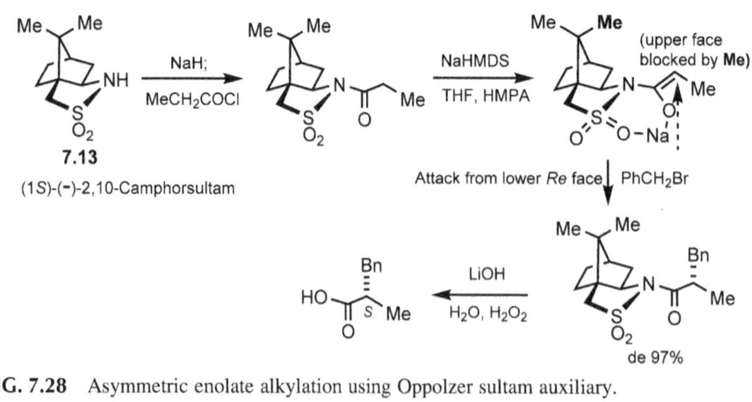


**FIG. 7.28** Asymmetric enolate alkylation using Oppolzer sultam auxiliary.

**Problem 7.8** Complete the following sequence of reactions used in a scheme for the synthesis of an alkaloid (−)-histrionicotoxin:

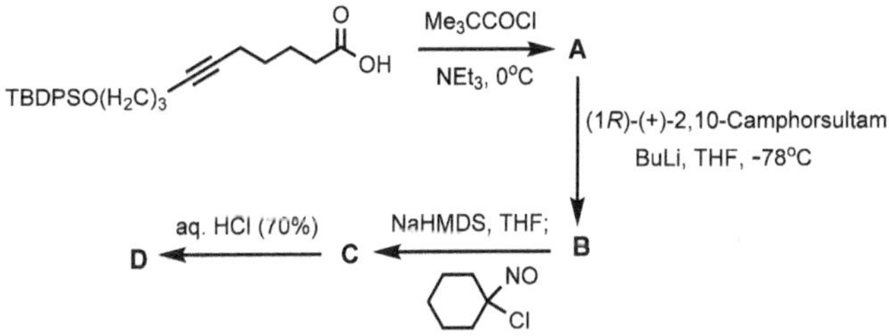


### 7.7.1.3 *Enders SAMP/RAMP auxiliaries*

The Enders chiral auxiliary (*S*)-1-amino-2-(methoxymethyl)pyrrolidine (SAMP) and its enantiomer (RAMP) are prolinol-derived chiral hydrazines (Fig. 7.29A).[63] Both SAMP and RAMP are available commercially. The SAMP/RAMP auxiliary strategy provides an efficient method for asymmetric alkylation of aldehydes and ketones, and the procedure has been exploited in the asymmetric synthesis of many natural products.[63,64] The auxiliary is introduced into ketones by condensation reaction to form chiral hydrazones (see Fig. 7.25B). An asymmetric synthesis of (*S*)-4-methyl-3-heptanone, the principal alarm pheromone of the

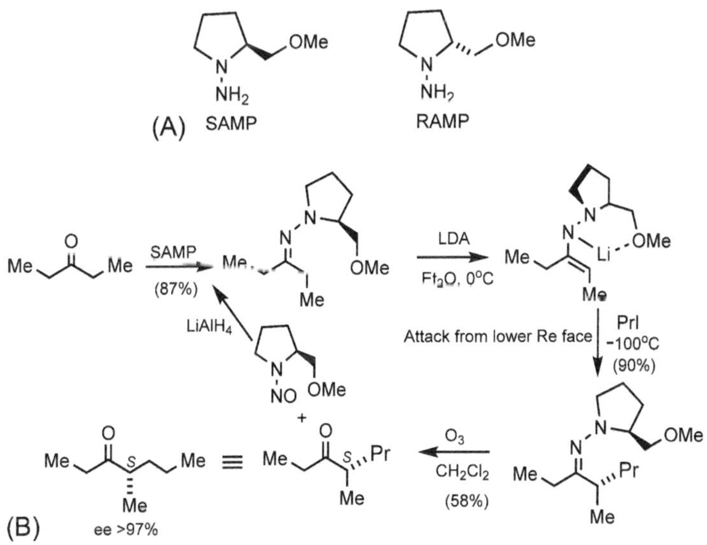


FIG. 7.29 (A) Enders SAMP and RAMP auxiliaries; (B) Asymmetric synthesis of a ketone (an alarm pheromone) using Enders SAMP hydrazone method. SAMP and RAMP stand for (*S*)- and (*R*)-1-amino-2-(methoxymethyl)pyrrolidine, respectively.

leaf-cutting ant *Atta texana*, from 3-pentanone using SAMP hydrazone method is shown in Fig. 7.29B. The alkylation of the lithiated SAMP hydrazone (formed by treatment of the hydrazone with LDA) occurs from the lower *Re* face since the upper *Si* face is hindered by the ring residues. Ozonolysis of the alkylated hydrazone gives the product ketone with excellent enantioselectivity (ee > 97%).[65] The generated (*S*)-nitrosamine can be reduced by $LiAlH_4$ for auxiliary (SAMP) recycling.

**Problem 7.9** Predict the stereochemistry of the final product in the following α-silyl-controlled asymmetric alkylation:

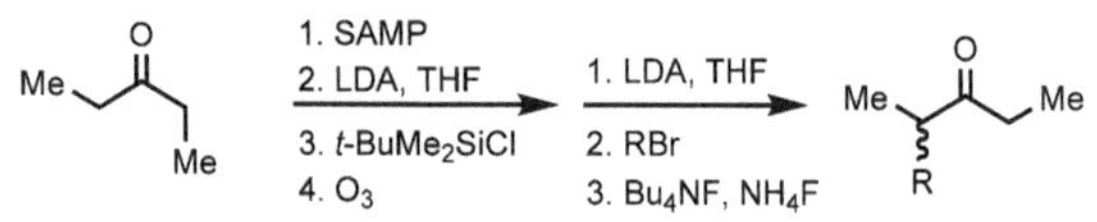


### 7.7.1.4 Schöllkopf bis-lactim ether auxiliaries

Cyclization of glycine with a chiral amino acid (e.g. valine) produces a bis-lactam, which on treatment with Meerwein's salt (trimethyloxonium tetrafluoroborate) is converted into a bis-lactim ether. The valine derived bis-lactim ethers are known as Schöllkopf auxiliaries,[66] and are commercially available as both enantiomers. The Schöllkopf method as an asymmetric alkylation of a masked glycine has been widely used for the synthesis of natural or unnatural (nonproteinogenic) α-amino acids, as illustrated in Fig. 7.30.[67,68] Enolization of the bis-lactim ether ((*S*)-2,5-dihydro-3,6-dimethoxy-2-isopropylpyrazine) with BuLi occurs by the abstraction of the less hindered methylene proton of the glycine residue. The enolate is essentially planar, and the steric bulk of the *i*-Pr group of valine imposes a strong facial bias by shielding the lower *Si* face of

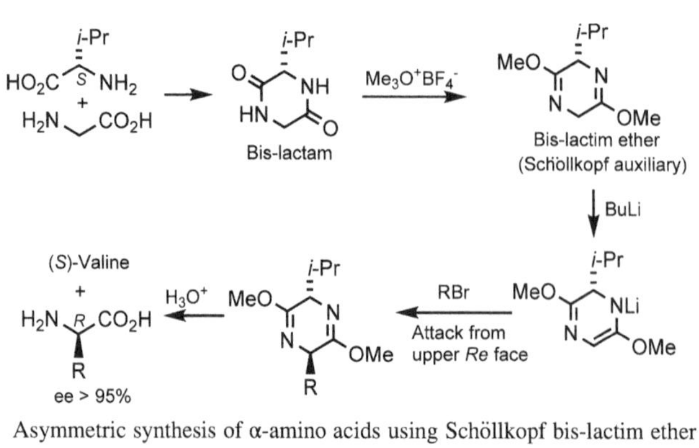


FIG. 7.30 Asymmetric synthesis of α-amino acids using Schöllkopf bis-lactim ether method.

the enolate, and hence electrophilic attack occurs preferentially from the upper *Re* face to give the predominant diastereomer. Subsequent hydrolysis provides the α-amino acid with high ee (typically >95%). A wide range of electrophiles including alkyl halides, aldehydes, ketones, enones and epoxides can be used.

The Schöllkopf method is also suitable for the synthesis of α-disubstituted amino acids. A bulky base (*t*-BuLi) is better for the desired deprotonation of the substituted bis-lactim ether.[69]

**Problem 7.10** Predict the stereochemistry of the product in the following scheme:

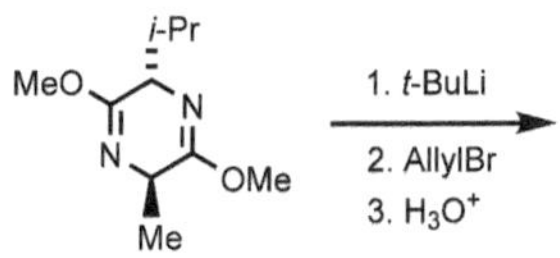


### 7.7.1.5 Meyers pseudoephedrine auxiliaries

Meyers asymmetric alkylation of enolates is based on pseudoephedrine auxiliaries.[70,71] These are cheap and available as both enantiomers (Fig. 7.31). The attachment of carbonyl substrates to the auxiliary produces amides (in contrast to imides in the case of Evans or Oppolzer auxiliaries). Notably, the amide enolates are more nucleophilic than imide enolates towards their reaction with electrophiles. An example of asymmetric alkylation using (*S*,*S*)-(+)-pseudoephedrine auxiliary is shown in Fig. 7.31. The enolate is generated with ~2 equivalents

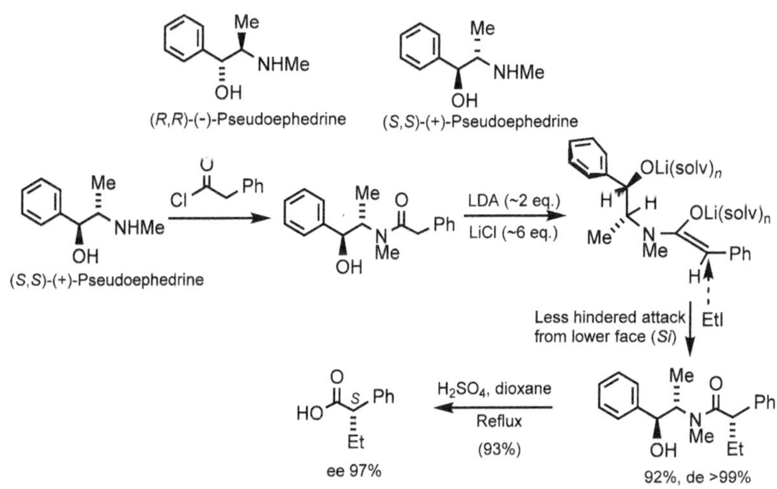


FIG. 7.31 Meyers method for asymmetric enolate alkylation using pseudoephedrine auxiliaries.

of LDA because of the presence of OH in the auxiliary. The asymmetric α-alkylation of the *cis*-enolate via the less hindered attack of the electrophile (EtI) from the lower face (*Si*) of the enolate (as the upper *Re* face is blocked by solvated lithium ions) provides excellent diastereoselectivity. Note that the addition of LiCl (~6 eq.) promotes a rapid and clean reaction. If the electrophile is an epoxide, selectivity will be reversed because of the coordination of lithium cation with the epoxide on the upper face of the enolate.[72,73]

The removal of the auxiliary can be performed in various ways[70,74]: (i) hydrolysis under acidic conditions ($H_2SO_4$, dioxane) or alkaline conditions ($Bu_4NOH$, *t*-BuOH, $H_2O$) gives a carboxylic acid, (ii) reduction with $LiH_2NBH_3$ [lithium amidotrihydroborate (LAB)] gives a primary alcohol, (iii) semireduction with $LiAlH(OEt)_3$ (lithium triethoxyaluminium hydride) produces an aldehyde, and (iv) addition of alkyllithium reagents such as BuLi (>2 eq.) gives a ketone.

**Problem 7.11** Asymmetric alkylation of glycine by the Meyers method is a useful strategy to prepare enantiopure α-amino acids. Predict the stereochemistry of the product in the following scheme:

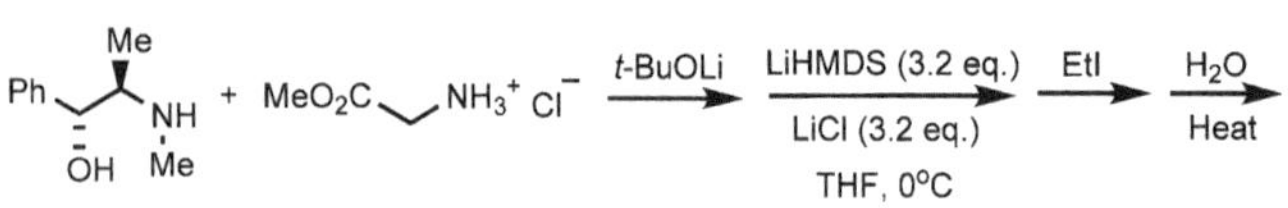


## 7.7.2 Seebach method: Self-regeneration of chirality

Seebach has developed a method for asymmetric alkylation involving substitution at a stereocentre in a chiral molecule without racemisation or loss of optical activity.[75] The method needs no chiral auxiliary and has been successfully applied for the α-alkylation of α-amino acids. The Seebach methodology involves four steps: (1) generation of a temporary chiral centre diastereoselectively, (2) destruction of the original tetrahedral chiral centre by trigonalization, for example by enolization, (3) creation of a new chiral centre at the original site diastereoselectively and (4) removal of the temporary chiral centre. This type of process in which the original chirality in the starting material is relayed to another position in order to functionalize the first position is called self-regeneration of chirality.[76]

To illustrate, an asymmetric α-alkylation of the amino acid (*S*)-proline is shown in Fig. 7.32. Reaction of proline with pivaldehyde (step 1) gives a single diastereomer which then produces the chiral enolate with LDA (step 2). Alkylation with MeI (step 3) occurs almost exclusively from the lower *Re* face of the enolate since the upper *Si* face is shielded by the saturated ring. Subsequent hydrolysis (step 4) provides α-methylproline essentially as a single enantiomer (*S*).

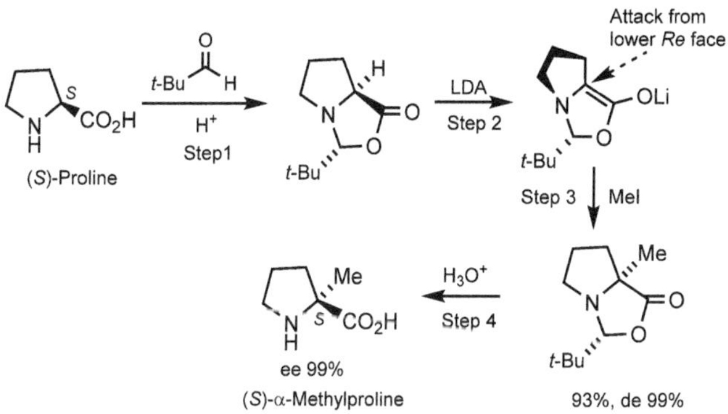


**FIG. 7.32** Asymmetric α-alkylation of the α-amino acid proline using Seebach method. No chiral auxiliary is needed.

### 7.7.3 Catalytic method: Phase transfer catalysis

Ionic reactants are usually soluble in aqueous phase but are insoluble in organic phase. A phase transfer catalyst is one that facilitates the migration of an ionic reactant from the aqueous phase to the organic phase where the reaction can take place. A phase transfer catalyst for anionic reactants is often a quaternary ammonium salt ($Q^+X^-$). The cation $Q^+$ has large nonpolar substituent groups which confer good solubility in organic solvents. As a consequence of electrostatic forces, the anion is also transferred to the organic phase.

Asymmetric enolate alkylation by phase transfer catalysis is a very useful method for the asymmetric synthesis of natural or unnatural α-amino acids.[77,78] This is illustrated with alkylation of a Schiff base protected glycine ester in Fig. 7.33. Reaction of the glycine ester with an alkyl halide (RBr) under phase transfer conditions in the presence of a chiral, enantiomerically pure quaternary ammonium salt $Q^{*+}Br^-$ **7.14** derived from (−)-cinchonidine (see Fig. 2.60A) as a phase transfer catalyst proceeds with good yields and very high enantioselectivity (Fig. 7.33A).[79] N- and C-deprotections provide the α-amino acid.

Mechanistic and kinetic studies indicate that the asymmetric alkylation involves three steps (Fig. 7.33B)[79]: (i) deprotonation of the glycine ester at the interface to form the *trans*-enolate, (ii) ion exchange to form catalyst/substrate ion pair ($Q^{*+}$/enolate anion) and its migration to the organic phase and (iii) enantioselective alkylation in the organic phase. The enantioselectivity has been rationalized with the help of a model shown in Fig. 7.33C.[80] The alkylation takes place from the front face (α-*Si*) of the enolate as the rear α-*Re* face is blocked by the quinoline ring of the catalyst and leads to the enantioselective product.

## 7.8 Diastereoselective aldol reactions

An aldol reaction is a nucleophilic addition of an enolate to an aldehyde or a ketone to form a β-hydroxycarbonyl compound (aldol). In the aldol reaction

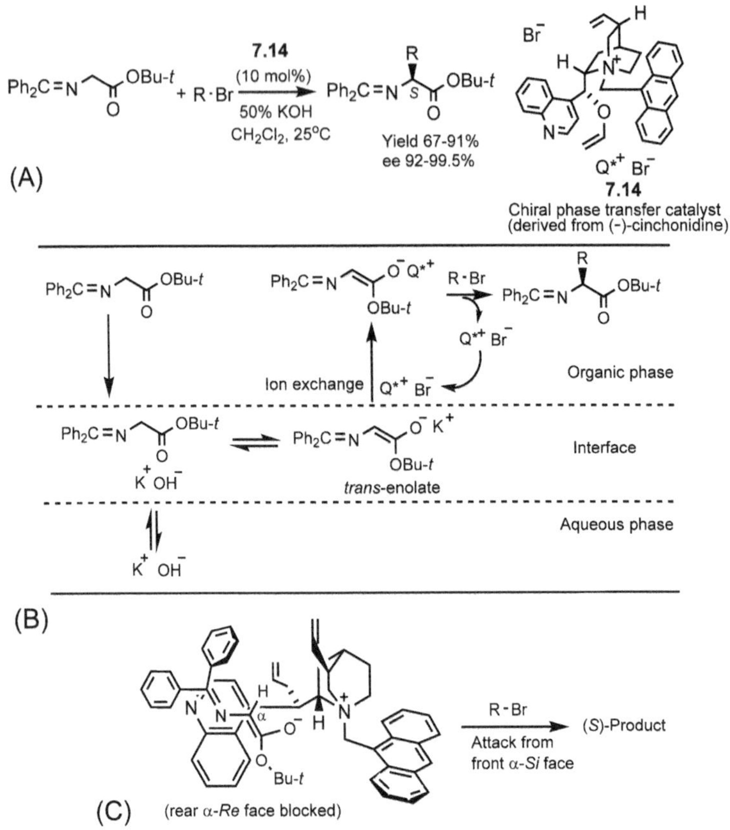


**FIG. 7.33** Catalytic asymmetric alkylation using (−)-cinchonidine based phase transfer catalyst: (A) Alkylation of a glycine Schiff base for the synthesis of α-amino acids; (B) Mechanism of the alkylation reaction under phase transfer conditions; (C) Model for the rationalization of enantioselectivity.

of a *prochiral* enolate with a *prochiral* aldehyde/ketone, two new stereocentres are created and the relative stereochemistry of the two stereocentres can lead to two aldol products, *syn* and *anti*. However, aldol reactions are often diastereoselective producing either the *syn* or the *anti* diastereomer as a major product. If the reacting partners are both achiral, the chiral *syn* or *anti* aldol product will be necessarily racemic. In this case, aldol reactions are simply diastereoselective, and not enantioselective.

The prochiral enolates can be *cis* or *trans* (see Section 7.6), and there often exists a relationship between the geometry of the enolate and the relative stereochemistry of the new stereocentres. Thus, which enolate is formed is an important factor for the diastereoselectivity of aldol reactions. Usually (but not always), *cis*-enolates give *syn* aldols preferentially and *trans*-enolates provide *anti* aldols preferentially (Fig. 7.34).

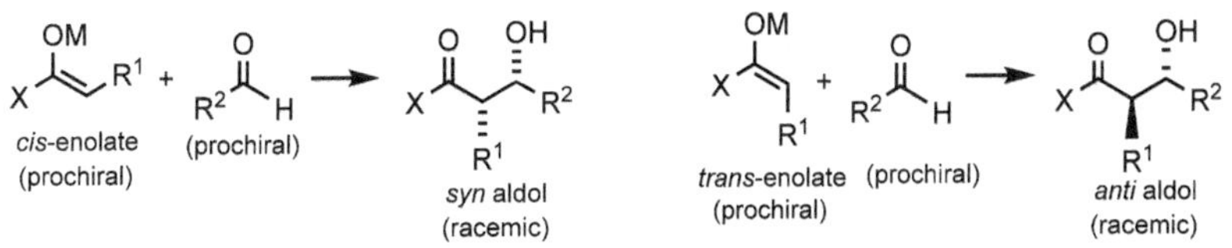


**FIG. 7.34** Enolate geometry and diastereoselectivity in aldol reactions.

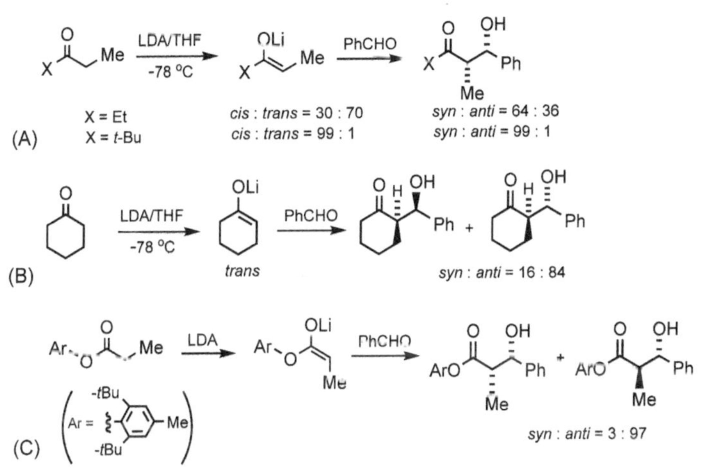


**FIG. 7.35** Diastereoselective aldol reactions using lithium enolates.

The diastereoselectivity in aldol reactions is illustrated using lithium enolates in Fig. 7.35. Usually, *cis*-lithium enolates are more stereoselective than *trans*-enolates in aldol reactions. For example, a mixture of *cis*- and *trans*-lithium enolates (30:70) derived from pentan-3-one (X=Et) reacts with benzaldehyde to give the *syn* aldol as a major product (*syn*:*anti*=64:36) (Fig. 7.35A).[41–43] With 2,2-dimethylpentan-3-one (X=*t*-Bu), the *cis*-enolate (99%) gives the *syn* aldol almost exclusively. However, the exclusive *trans*-enolate of cyclohexanone does not form the *anti* aldol exclusively but gives the *anti* isomer as a major product (84%) only under optimum conditions (Fig. 7.35B).[81] In general, the preferred *trans*-enolates derived from esters give poor selectivity for the *anti* aldol product. However, if the ester contains a bulky aromatic substituent, the *anti* aldol is formed predominantly (Fig. 7.35C).

The boron enolates give better stereoselectivity than lithium enolates in aldol reactions. The diastereoselectivity of aldol reactions using boron enolates is illustrated in Fig. 7.36. Both *cis*- and *trans*-boron enolates derived from pentan-3-one give excellent diastereoselectivity (Fig. 7.36A, cf. Fig. 7.35A, X=Et).[47,48] High diastereoselectivity has also been observed with boron enolates of thioesters (Fig. 7.36B). With S-*t*-butyl ester, the *trans*-enolate is favoured because of the severe steric interaction of the bulky *t*-Bu with boron ligands for *cis*-enolate formation.

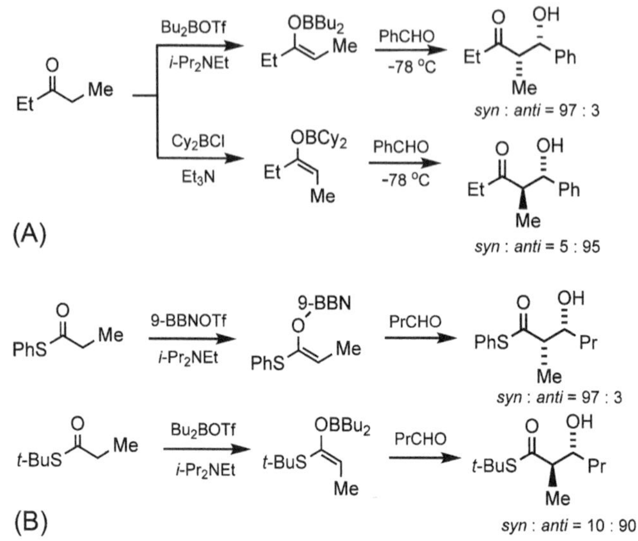


**FIG. 7.36** Diastereoselective aldol reactions using boron enolates.

It may be mentioned that titanium(IV) enolates (prepared from the ketone, $TiCl_4$ and a base such as $Bu_3N$) provide an alternative method for the formation of the *syn* aldol products in high yield.[82]

### 7.8.1 Zimmerman–Traxler model

Most aldol reactions are performed under conditions of kinetic control. The mechanistic model for the stereochemical control of aldol reactions under kinetic control is known as the Zimmerman–Traxler model.[83,84] The model is based on a chair-like six-membered cyclic transition structure in which both the carbonyl and enolate oxygens are coordinated to the ligated metal atom (Fig. 7.37). The coordination by the metal atom (Li or B) promotes the reaction by increasing the carbonyl reactivity and by bringing the reacting molecules together in the TS. The Zimmerman–Traxler models with *cis*- and *trans*-lithium enolates are shown in Fig. 7.37A. The major aldol product (*syn* for the *cis*-enolate, *anti* for the *trans*-enolate) arises via the more stable chair TS in which the substituent ($R^2$) at the aldehyde carbon is preferably equatorial. Note that *cis* stereochemistry of the enolate forces $R^1$ to be axial and *trans*-enolate stereochemistry forces $R^1$ to be equatorial.

The Zimmerman–Traxler models with boron enolates are similar to those for lithium enolates (Fig. 7.37B). As with lithium enolates, the stereochemistry of the boron enolates controls the diastereoselectivity. In general, the stereoselectivity with boron enolates is higher than with lithium enolates. This has been attributed to the shorter B—O bond length (B—O 1.36–1.47 Å vs Li—O

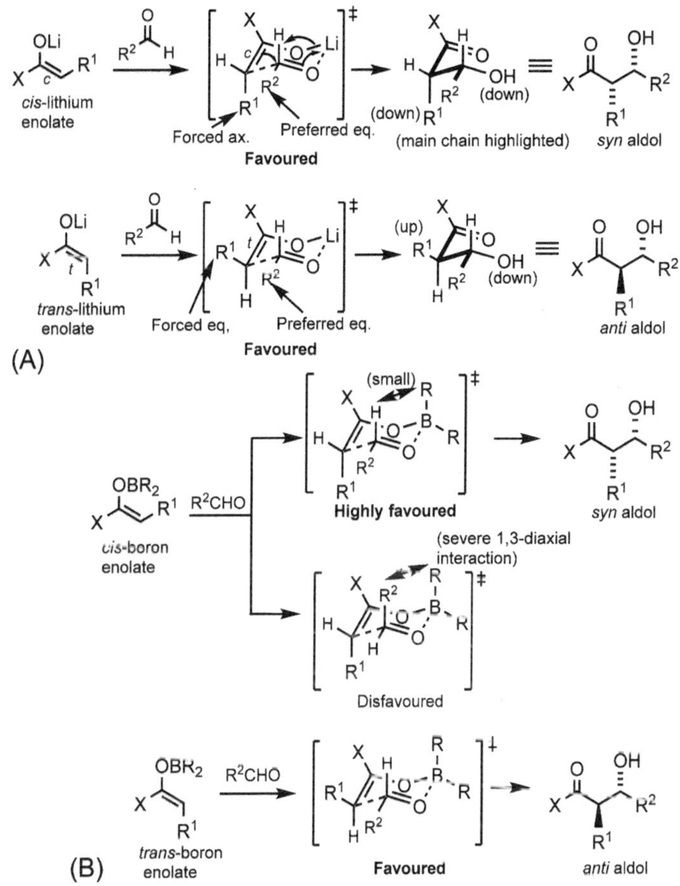


**FIG. 7.37** Zimmerman–Traxler models for diastereoselectivity in aldol reactions using (A) lithium enolates and (B) boron enolates.

1.92–2.0 Å), which leads to a more compact chair TS and thereby enhances the steric effects that control diastereoselectivity. As shown in Fig. 7.37B, the conformation in which the substituent ($R^2$) at the aldehyde carbon is equatorial is highly preferred as the axial $R^2$ conformation is destabilized by a severe 1,3-diaxial interaction involving a boron ligand (R) in the more tight chair TS. This explains the much greater diastereoselectivity with boron enolates than with lithium enolates.

## 7.9 Asymmetric aldol reactions

The asymmetric aldol reaction is a powerful method in organic synthesis for the construction of a new C—C bond with control over the absolute stereochemistry at the new chiral centres.[85–88] Since the aldol products are common features

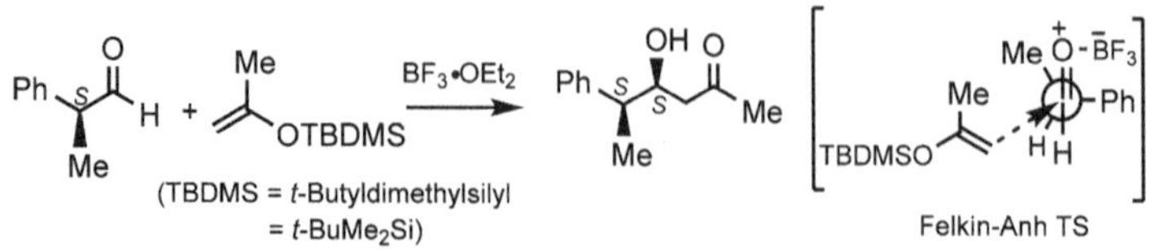


**FIG. 7.38** An asymmetric Mukaiyama aldol reaction using a chiral aldehyde and an achiral silyl enol ether.

in numerous biologically active natural products, the asymmetric aldol methodology is very useful for the synthesis of complex natural molecules, for example erythromycin A, a macrolide antibiotic.

A substrate controlled asymmetric aldol reaction (first generation) may involve chiral aldehydes and achiral enolates. This is illustrated with a Mukaiyama aldol reaction[89] in Fig. 7.38. The reaction involves the addition of achiral silyl enol ether (as the enolate component) to a chiral aldehyde in the presence of a Lewis acid. The silyl enol ether is unreactive towards the aldehyde alone, but can react when the aldehyde is activated by a Lewis acid. Since the silyl enol ether cannot coordinate with the carbonyl group, the reaction proceeds through an open transition structure and the Felkin–Anh control operates to provide the *syn* diastereoselection. Good levels of *syn* selectivity in the Mukaiyama aldol reaction can be achieved using $Ph_2BOH$ as a catalyst and water as a solvent.[90]

## 7.9.1 Chiral auxiliaries

The use of chiral auxiliaries in asymmetric aldol reactions refers to second-generation (2G) asymmetric synthesis (see Section 5.3.1.2). The most important asymmetric aldol methodology using chiral auxiliaries is due to Evans and this method is one of the most reliable and widely used methods in organic synthesis. The Evans oxazolidinone chiral auxiliaries are particularly effective for *syn*-selective asymmetric aldol reactions.[55,91]

Fig. 7.39 shows an example of highly efficient *syn*-selective asymmetric aldol reactions using a chiral oxazolidinone auxiliary. Formation of the *cis*-boron enolate followed by the addition of benzaldehyde leads to essentially complete selectivity in favour of one *syn* aldol product, called Evans-*syn*. The other minor *syn* aldol is called non-Evans-*syn*. The Evans-*syn* imide can be solvolysed in water or alcohols to give the corresponding enantiomer (*S*, *S*) of β-hydroxy acid or ester, or reduced to give an aldehyde or alcohol product (cf. Fig. 7.26C). An advantage of the chiral auxiliary strategy is the ease of purification of the product such that any undesired diastereomer can be removed usually by crystallization or chromatography. The removal of auxiliary then gives essentially the enantiopure product. It should be noted that the opposite enantiomer of the *syn* product can be obtained by the choice of the appropriate chiral auxiliary.

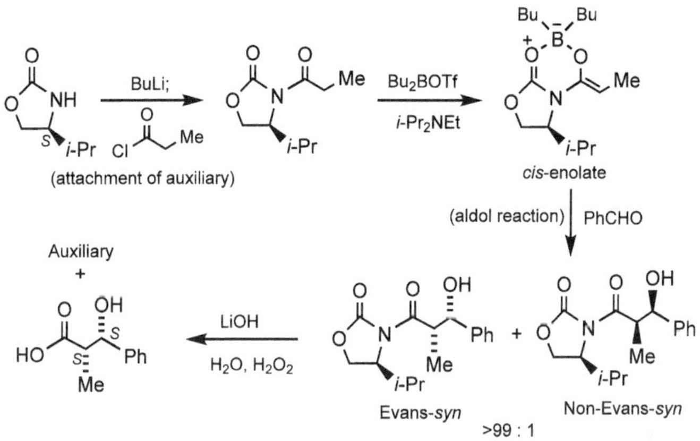


**FIG. 7.39** *syn*-selective asymmetric aldol reaction using Evans oxazolidinone chiral auxiliary.

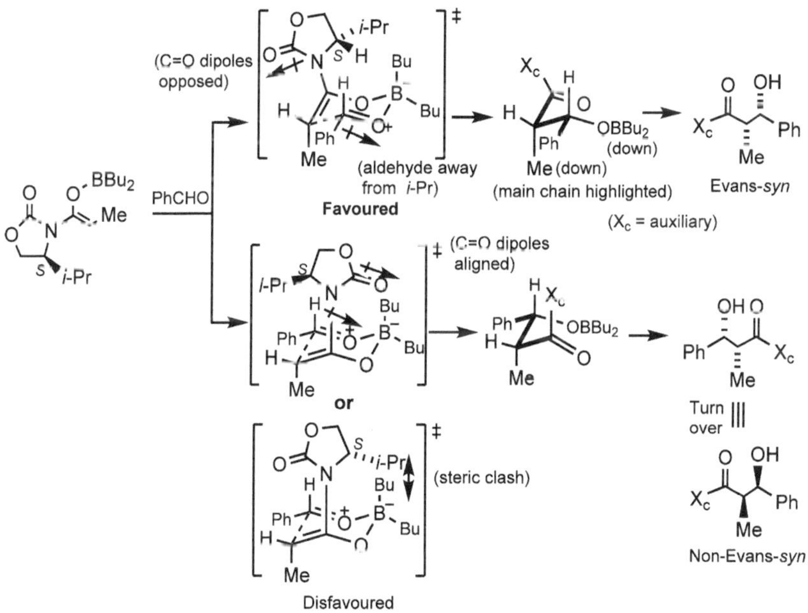


**FIG. 7.40** Rationalization of the asymmetric aldol reaction in Fig. 7.39 in terms of Zimmerman–Traxler TS.

The *syn*-selective asymmetric reaction can be rationalized by the Zimmerman–Traxler TS, in which the aldehyde approaches away from the large *i*-Pr group on the auxiliary when the two carbonyl dipoles on the auxiliary and the aldehyde are opposed minimizing dipole repulsion (Fig. 7.40). This transition structure is favoured and leads to the Evans-*syn* product. The other possible transition structure suffers from dipole repulsion or steric clash and is

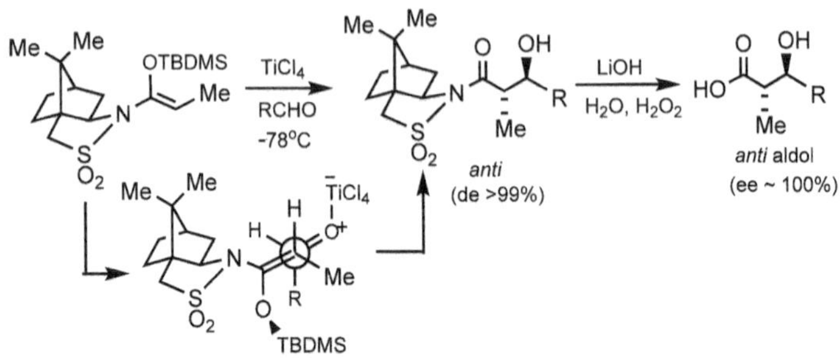


**FIG. 7.41** *anti*-selective asymmetric aldol reaction using Oppolzer sultam auxiliary.

disfavoured, which gives the non-Evans-*syn* aldol. Note that Ph is equatorial in both favoured and disfavoured TSs for *syn* (instead of *anti*) preference.

The *anti*-selective asymmetric aldol reactions with the Evans oxazolidinones are also possible.[92] However, an especially powerful method for the *anti*-selective asymmetric aldol reaction has been developed using Oppolzer chiral auxiliaries.[93] Fig. 7.41 illustrates an Oppolzer sultam auxiliary-mediated Mukaiyama aldol reaction. The diastereoselectivity appears to be maximal (de > 99%) for the *anti* aldol diastereomeric product when $TiCl_4$ is used as the Lewis acid. The reaction proceeds through open (acyclic) transition structure. The TS in which the aldehyde approaches from the less hindered side of the auxiliary and the substituent (R) is away from the bridged Me on the auxiliary is favoured and leads to the major *anti*-aldol diastereomeric product, which on hydrolysis gives the essentially enantiopure *anti* aldol β-hydroxy acid. Here the *cis* enolate gives the *anti* aldol product via open TS.

**Problem 7.12** Predict the stereochemistry of the product in the following asymmetric aldol reaction:

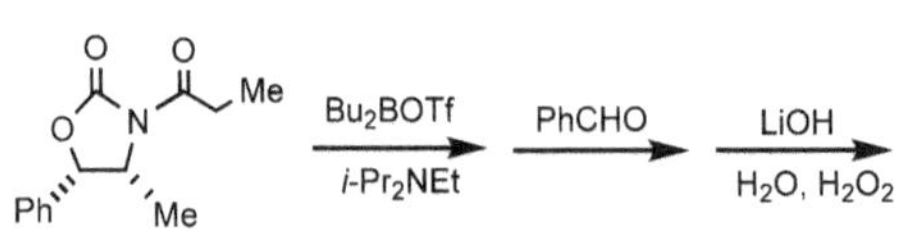


## 7.9.2 Chiral reagents

The asymmetric aldol reactions mediated by chiral reagents under stoichiometric conditions refer to third-generation (3G) asymmetric synthesis (see Section 5.3.1.3). Chiral boron reagents can provide very good levels of diastereofacial and enantiofacial selectivity with aldehydes, ketones or esters.

Fig. 7.42 illustrates *syn*-selective asymmetric aldol reactions using the chiral boron reagent (−)-diisopinocamphenylboron triflate [(−)-$Ipc_2BOTf$].[94,95] The boron triflate is easily prepared from the corresponding borane (−)-$Ipc_2BH$

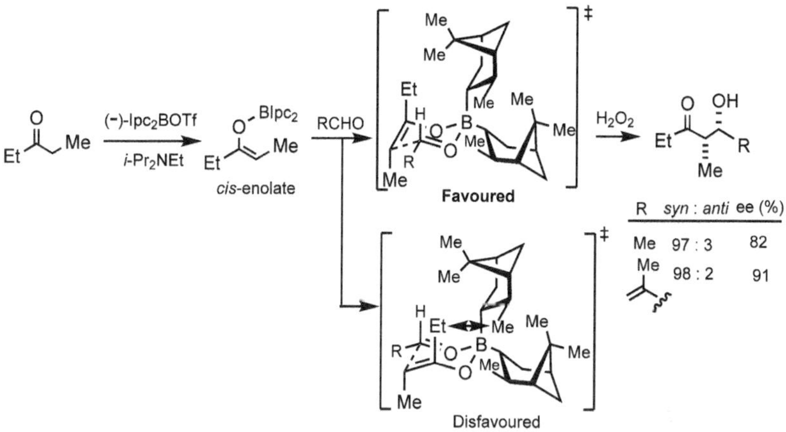


**FIG. 7.42** *syn*-selective asymmetric aldol reaction using chiral boron reagent (−)-$Ipc_2BOTf$.

with triflic acid (TfOH) (cf. (−)-$Ipc_2BCl$ in Fig. 7.14A). Either enantiomer of the boron triflate is available. As shown in Fig. 7.42, the *cis*-boron enolate is formed almost exclusively, which undergoes aldol reaction with aldehydes to give the *syn* aldol products with good enantioselectivity, after oxidative work-up. The major *syn* enantiomer results from the sterically favoured Zimmerman–Traxler TS.[96] Note that the other possible TS is disfavoured because of the destabilizing steric interaction between Et (enolate) and Me (reagent).

The use of $C_2$-symmetric dialkylboron reagent introduced by Masamune allows the formation of *trans*-boron enolate with bulky thioesters, and promotes the *anti*-selective asymmetric aldol reaction with excellent stereocontrol (Fig. 7.43).[97,98]

The $C_2$-symmetric boron reagent **7.11** in Fig. 7.44 introduced by Corey is effective for both *syn*- and *anti*-selective asymmetric aldol reactions, depending on the structure of carbonyl substrates and reaction conditions.[50,99,100] For example, Fig. 7.44A shows that phenylthiopropionate ester forms the

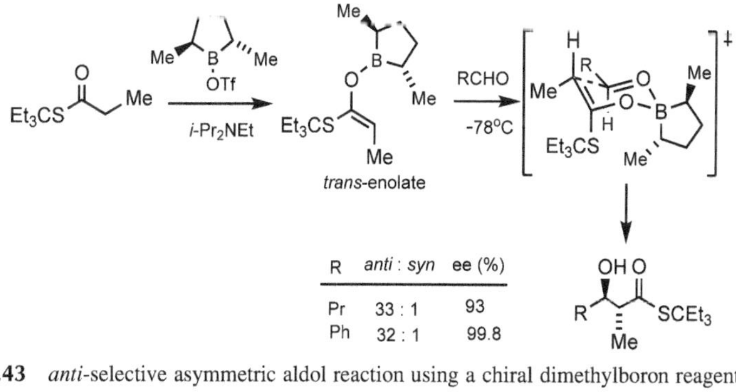


| R | *anti* : *syn* | ee (%) |
|---|---|---|
| Pr | 33 : 1 | 93 |
| Ph | 32 : 1 | 99.8 |

**FIG. 7.43** *anti*-selective asymmetric aldol reaction using a chiral dimethylboron reagent.

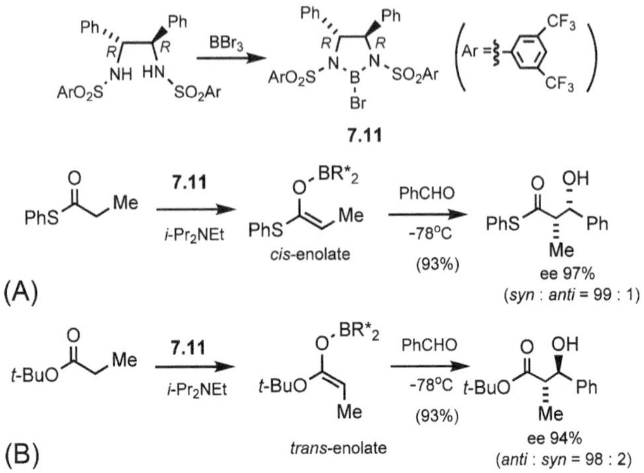


**FIG. 7.44** (A) *syn*-selective and (B) *anti*-selective asymmetric aldol reactions with the $C_2$-symmetric chiral borane reagent **7.11**.

*cis*-boron enolate which, on addition to benzaldehyde, produces the *syn* aldol product with high enantioselectivity. On the other hand, the aldol reaction with *t*-butylpropionate ester leads to the *anti* aldol product with high ee (Fig. 7.44B). The mechanistic models of these reactions are similar to the Zimmerman–Traxler TSs as described before. Since both enantiomers of the boron reagent are available, the asymmetric synthesis of either enantiomer of the aldol product is possible.

### 7.9.3 Chiral catalysts

The catalytic asymmetric aldol reactions refer to fourth-generation (4G) asymmetric synthesis (see Section 5.3.1.4). This is the most attractive approach requiring only a catalytic amount of a chiral promoter, and provides an atom economical alternative that avoids the formation of stoichiometric by-products while maintaining the high level of stereochemical control. A variety of Lewis acid systems with chiral ligands have been investigated as chiral catalysts for the asymmetric catalysis of aldol reactions which generally involve silyl enol ethers (silyl ketene acetals) as enolate equivalents.[101–103]

Fig. 7.45 illustrates asymmetric Mukaiyama aldol reaction with aldehydes using a chiral diamine ligand and tin(II) triflate as a Lewis acid.[104,105] The reaction with a *trans* enolate gives selectively the *syn* aldol products with high ee (cf. Fig. 7.41).

The chiral catalyst is the chiral amine–$Sn(OTf)_2$ complex which coordinates with the aldehyde through carbonyl oxygen **7.12**. The enantioselectivity arises from the preferential *Re* face (rear face) attack of the nucleophile

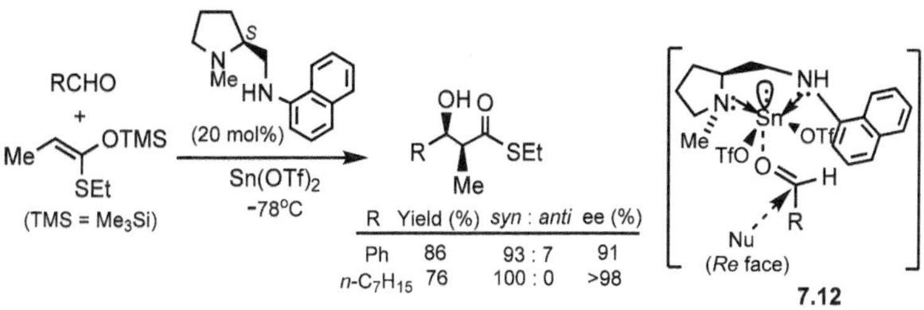


**FIG. 7.45** Catalytic asymmetric Mukaiyama aldol reaction and the TS model **7.12** for enantioselectivity.

(enolate) as the aldehyde *Si* face (front face) is sterically shielded by the naphthylamine unit (Fig. 7.45).

An effective catalyst system is box catalysts, the $C_2$-symmetric copper(II) complex of chiral **b**is-**ox**azolines, such as [Cu-(*S*,*S*)-*t*-Bu-box]$(OTf)_2$ **7.13**.[106] The metal ion ($Cu^{2+}$) is a Lewis acidic centre, and the carbonyl substrates that can engage in chelate-organized association with the metal centre are suitable for the catalytic aldol reactions. Fig. 7.46A depicts an asymmetric aldol reaction catalysed by **7.13** that exhibits excellent enantioselectivity.[107] The enolate equivalent is a silyl ketene acetal, and the carbonyl substrate is methyl pyruvate (an α-keto ester).

The enantioselectivity of the reaction has been rationalized as follows.[107] The pyruvate bears two carbonyl groups which could chelate with the copper (II) centre forming a well-defined catalyst substrate complex of square-planar

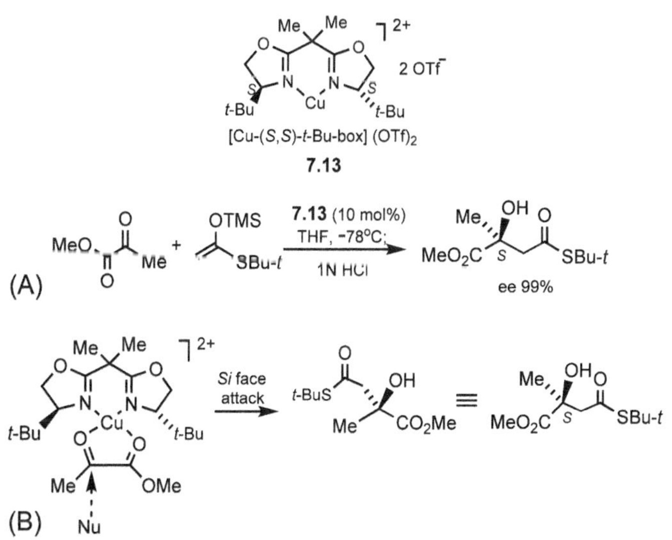


**FIG. 7.46** (A) Catalytic asymmetric aldol reaction using a chiral box catalyst; (B) Rationalization of enantioselectivity in terms of Cu(II) square planar model of catalyst–substrate complex.

geometry (Fig. 7.46B). Since the *Re* face (front face) of the pyruvate is sterically shielded by the bulky *t*-Bu group, the nucleophilic attack occurs to the rear *Si* face which leads to the (*S*) enantiomer of the product.

Another box-type catalyst system is the pybox catalysts, the Cu(II) complex of chiral **py**ridine-**b**is-**ox**azolines,[108] which can promote highly enantioselective asymmetric aldol reactions with α-alkoxy aldehydes.[109]

**Problem 7.13** Rationalize the stereochemical outcome of the following catalytic asymmetric aldol reaction:

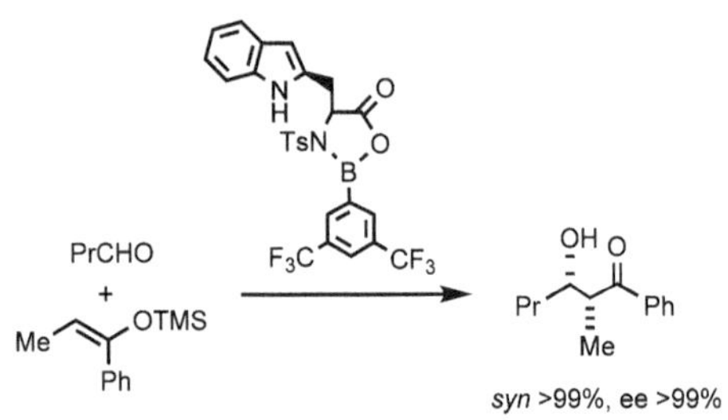


## 7.9.4 Organocatalysis

An organocatalyst, as distinct from a metal (or organometallic) catalyst, is a small organic molecule that does not contain a metal. As with other catalysis, these small molecules are required in catalytic (substoichiometric) amounts and are recovered at the end of the reaction. The early use of small molecules as catalysts in asymmetric synthesis was reported in mid-1970s but the field was not defined until the big potential of the field was discovered at the beginning of the 21st century (see Section 7.14.4). It was given the name 'organocatalysis'.[110]

### *7.9.4.1 Proline-catalysed aldol reactions*

The classic example of a small molecule catalyst or organocatalyst is the amino acid proline that can catalyse asymmetric aldol reactions between enolizable ketones (or aldehydes) and aldehydes that are nonenolizable or branched to disfavour enolization.[110–113] For example, the direct asymmetric aldol reaction between acetone and isobutyraldehyde catalysed by (*S*)-proline gives the (*R*)-aldol in high ee (Fig. 7.47A). The mechanism of the proline-catalysed aldol reaction is shown in Fig. 7.47B.[114] The catalytic cycle begins with the formation of a chiral enamine from proline and acetone, followed by aldehyde addition activated by the carboxyl group in a Zimmerman–Traxler chair-like TS. The resulting iminium ion on hydrolysis gives predominantly (*R*) enantiomer of the aldol product and regenerates the catalyst. It may be mentioned that this mechanism via an enamine intermediate is analogous to the function of Type I aldolases in

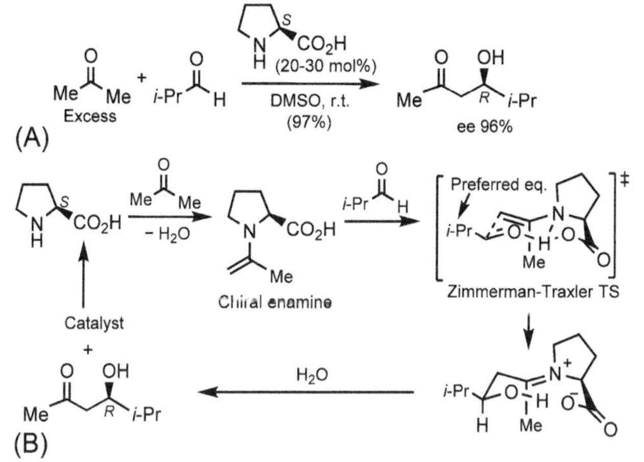


**FIG. 7.47** (A) Organocatalytic asymmetric aldol reaction using (*S*)-proline as an organocatalyst; (B) The catalytic cycle and mechanism via the Zimmerman–Traxler chair TS.

biological systems that catalyse the aldol reaction via an enamine mechanism, in which an enzyme lysine residue generates an enamine in the active site.[114]

**Problem 7.14** Rationalize the stereochemical outcomes of (a) intermolecular and (b) intramolecular organocatalytic asymmetric aldol reactions shown below:

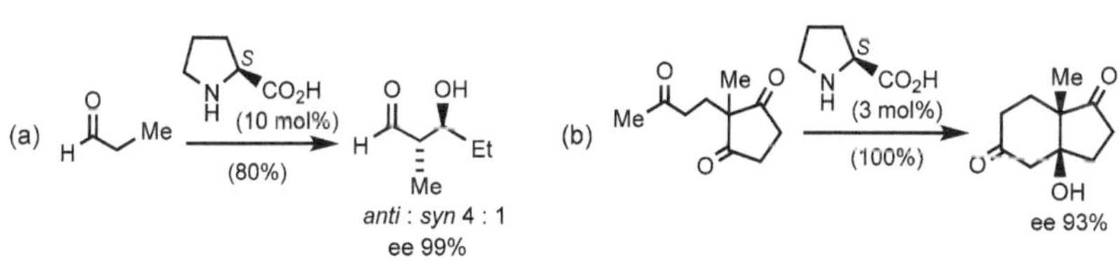


### 7.9.4.2 *Proline-catalysed Mannich reactions*

A Mannich reaction involves the nucleophilic attack of an enolate (or its equivalent) to imines (generated in situ from aldehydes and amines) to form the products β-amino ketones, so-called Mannich bases. Fig. 7.48 shows a *syn*-selective asymmetric Mannich reaction catalysed by (*S*)-proline.[115] The reaction between the more stable (*E*)-enamine intermediate (enol equivalent) and the (*E*)-imine proceeds via a chair-like TS to give selectively the *syn* product with excellent enantioselectivity.

The *anti*-selective approaches[116] to asymmetric Mannich reactions have also been developed with proper design of chiral organocatalysts such as a pyrrolidine derivative *trans*-5-methylpyrrolidine-3-carboxylic acid.[117]

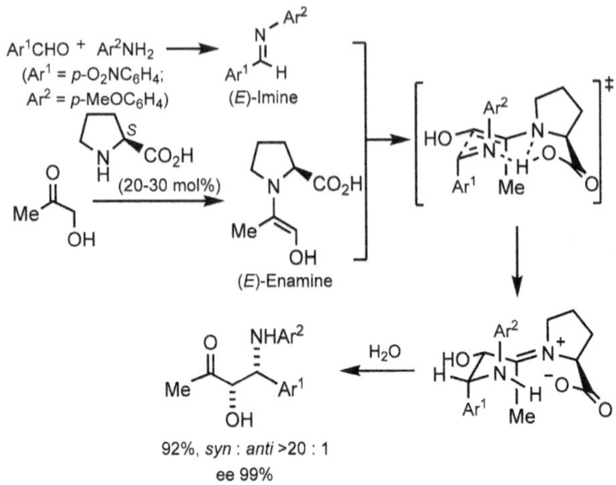


FIG. 7.48 Proline-catalysed *syn*-selective asymmetric Mannich reaction.

**Problem 7.15** Rationalize the stereochemical outcome in the following proline-catalysed asymmetric α-functionalization of aldehydes:

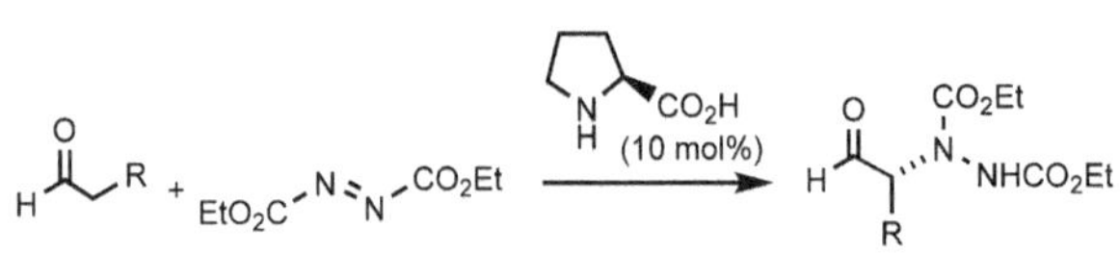


## 7.10 Asymmetric allylation reactions

Allylation reactions involve addition of allyl metal reagents to aldehydes which produce homoallyl alcohols. Asymmetric allylation reactions are usually performed under reagent control (third generation) and chiral allyl boron reagents are most widely used. As regards diastereoselectivity, allylation reactions are analogous to aldol reactions: (*Z*)-crotyl boranes usually give *syn* products, whereas (*E*)-crotyl boranes give *anti* products.[118]

### 7.10.1 Brown allylation reaction[119,120]

The chiral allyl borane reagent developed by Brown is *B*-allyldiisopinocamphenylborane ($Ipc_2B$-allyl) prepared from inexpensive α-pinene available commercially as either enantiomer (Fig. 7.49A).[121] The boron reagent can be prepared and used in situ after filtration of Mg salts. The asymmetric allylation reaction with aldehydes is quite general and low temperatures lead to enhanced enantioselectivity, as illustrated in Fig. 7.49B.[122,123]

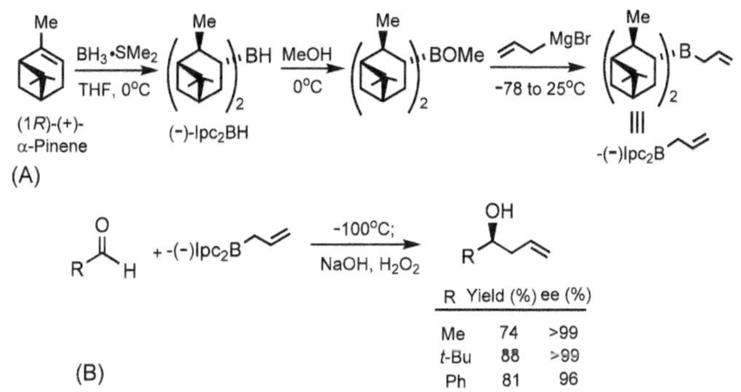


**FIG. 7.49** (A) Preparation of *B*-allyldiisopinocamphenylborane ($Ipc_2B$-allyl); (B) Asymmetric allylation of aldehydes with (−)-$Ipc_2B$-allyl at low temperatures.

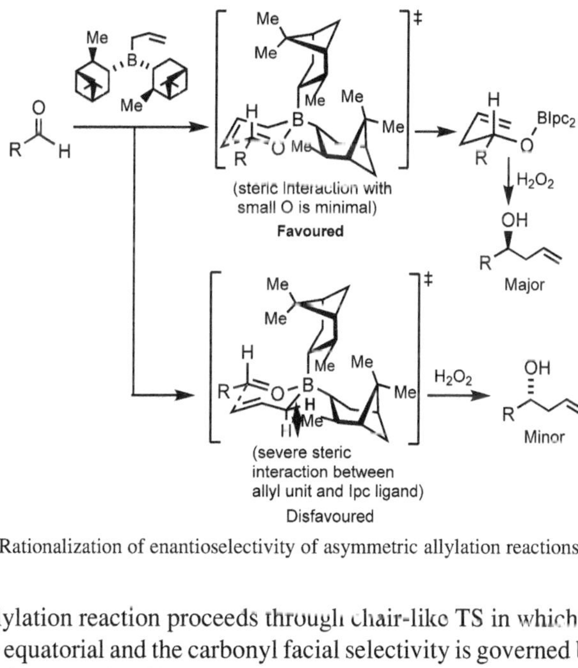


**FIG. 7.50** Rationalization of enantioselectivity of asymmetric allylation reactions in Fig. 7.49B.

The allylation reaction proceeds through chair-like TS in which the substituent (R) is equatorial and the carbonyl facial selectivity is governed by the steric interaction between the allyl group and the Ipc ligand (Fig. 7.50). The preferred transition structure with minimized steric interaction gives the major enantiomer of the product.

If the aldehyde is chiral containing α-stereocentre, double stereodifferentiation (see Section 5.3.4.1) comes into play. The diastereofacial selectivity of (−)-$Ipc_2B$-allyl or (+)-$Ipc_2B$-allyl then controls the product ratio (dr) under 'matched' or 'mismatched' situations (Fig. 7.51).[124]

The asymmetric allylation reactions using (*Z*)- and (*E*)-crotylboranes are shown in Fig. 7.52A.[125] The syntheses of crotylboranes, as shown in

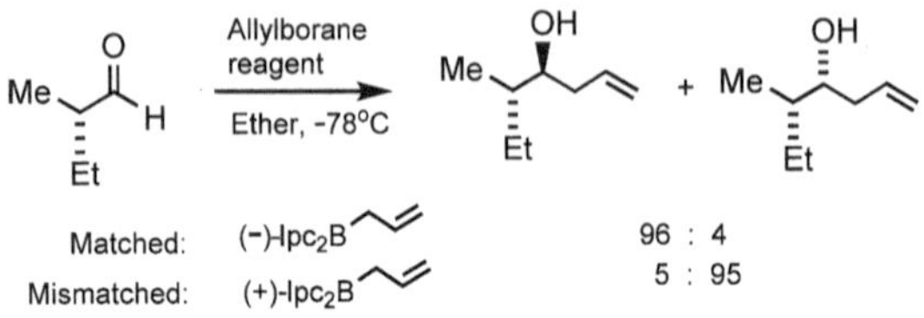


FIG. 7.51 Double stereodifferentiation in the diastereoselective allylboration of a chiral aldehyde.

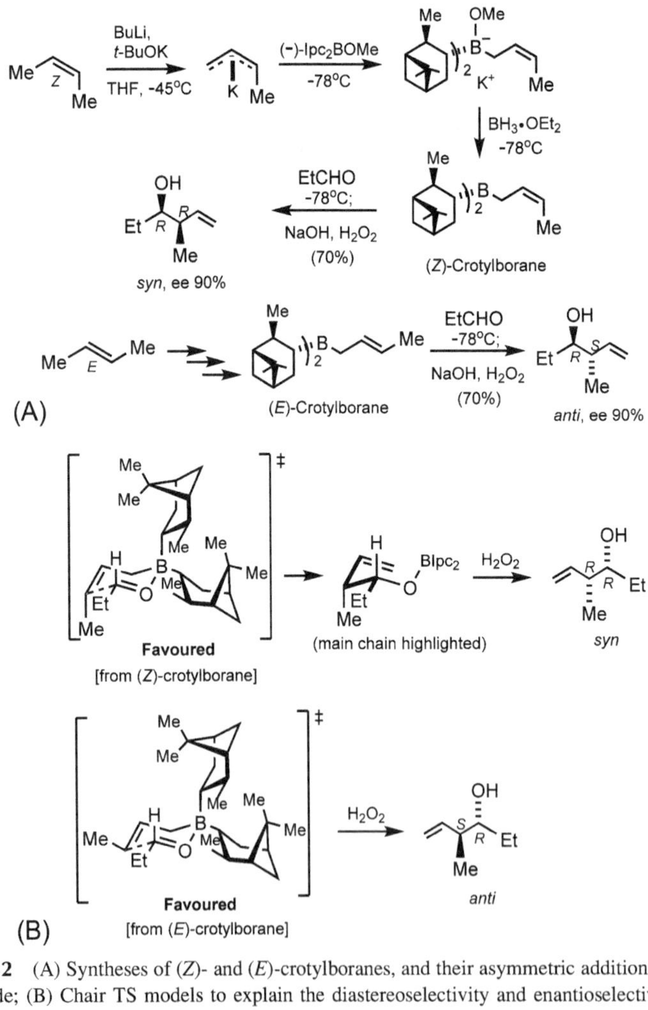


FIG. 7.52 (A) Syntheses of (Z)- and (E)-crotylboranes, and their asymmetric addition to propionaldehyde; (B) Chair TS models to explain the diastereoselectivity and enantioselectivity of the asymmetric crotylation reactions.

Fig. 7.52A, involve the metallation of weakly acidic (*Z*)- and (*E*)-2-butenes with the 'superbase' (1:1 mixture of BuLi and *t*-BuOK) when large atomic radius of potassium favours $\eta^3$–bonding involving the $\pi$-allyl unit.[126] The (*Z*)-crotylborane gives the *syn* product with high enantioselectivity (ee 90%), whereas the (*E*)-crotylborane gives the *anti* product with 90% ee. Fig. 7.52B rationalizes the diastereoselectivity and enantioselectivity of the asymmetric crotylation reactions with propionaldehyde based on the preferred chair TS (equatorial-Et) in each case. Note that the geometry of the double bond (*Z* or *E*) in the allyl unit forces the Mc to be axial or equatorial in the chair TS.

## 7.10.2 Roush allylation reaction[120,127]

Chiral allyl boron reagents developed by Roush are allyl- and crotylboronate reagents based on chiral tartrate esters. A synthesis of a chiral allylboronate reagent and its asymmetric addition to aldehydes is shown in Fig. 7.53A.[128] The reagent is derived by treating allylmagnesium bromide with triisopropyl borate, followed by exchange of the isopropyl groups on boron with chiral (*R*,*R*)-(+)- or (*S*,*S*)-(−)-diisopropyl tartrate (DIPT) ligand. The allylboronates

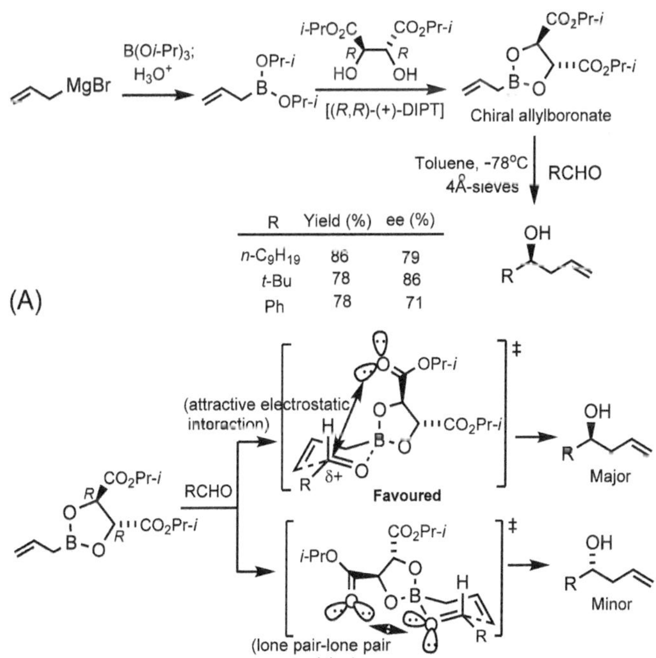


**FIG. 7.53** (A) Synthesis of a chiral allylboronate based on tartrate esters such as (*R*,*R*)-(+)-diisopropyl tartrate (DIPT), and its asymmetric addition to aldehydes; (B) Chair TS models to explain the enantioselectivity of the asymmetric allylation reactions.

are stable enough to permit their purification by distillation. The reactions typically exhibit moderate to good enantioselectivity, and 4 Å molecular sieves are necessary to obtain the highest enantioselectivity.

The enantioselectivity has been rationalized using chair TS models in which the substituent (R) at the aldehyde carbon is preferably equatorial (Fig. 7.53B).[128] The TS that minimizes the lone pair–lone pair repulsion and is believed to have attractive electrostatic interaction between a lone pair and the carbonyl carbon is favoured and leads to the major enantiomeric product. The other chair TS is destabilized by dipole repulsion between the two lone pairs and is disfavoured.

**Problem 7.16** Complete the following scheme for the asymmetric crotylation reaction. Explain the stereochemical outcome of the reaction.

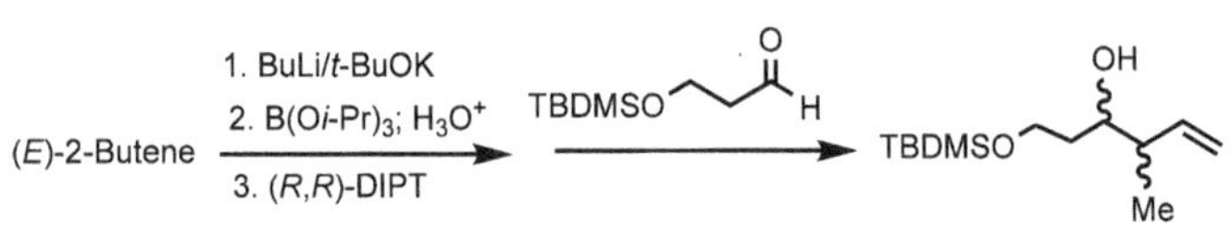


## 7.11 Asymmetric synthesis of amines

Chiral amines are prevalent in a vast majority of drugs, natural products and biological molecules. There are many valuable approaches to the enantioselective preparation of chiral amines based on chiral auxiliaries and chiral catalysts. Here we will describe briefly a popular and efficient approach based on Ellman chiral auxiliary.[129,130]

### 7.11.1 Ellman chiral auxiliary: *t*-Butylsulphinamide

Ellman chiral auxiliary is a chiral ammonia equivalent, (*R*)- or (*S*)-*t*-butylsulphinamide (or *t*-butanesulphinamide) **7.14** (Fig. 7.54A). The S atom in **7.14** is tetrahedral with a lone pair, and is the chiral centre. The Ellman auxiliary has been synthesized in excellent yield and enantiopurity,[131] and is commercially available as both enantiomers. As with chiral auxiliaries in general, three steps are involved in the asymmetric synthesis of amines.[129] (a) *Imine formation*: direct condensation of **7.14** with aldehydes or ketones proceeds in high yields under mild conditions to give *t*-butylsulphinimines (or *t*-butanesulphinyl imines), which are much less hydrolytically labile or prone to tautomerization than most *N*-alkyl or *N*-aryl imines. (b) *Nucleophilic addition*: addition of organometallic reagents to *t*-butanesulphinyl imines gives clean and high-yield adducts as the sulphinyl imines are more electrophilic than alkyl or aryl imines. These adducts are stable to strong bases, nucleophiles and many transition metal-catalysed reactions. (c) *Removal of sulphinyl*

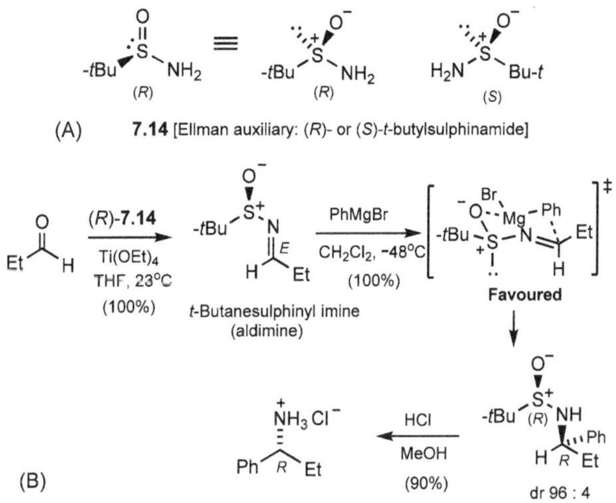


FIG. 7.54 (A) Ellman chiral auxiliary: *t*-butylsulphinamide or *t*-butanesulphinamide; (B) Asymmetric synthesis of an amine from an aldehyde using Ellman chiral auxiliary.

*group*: treatment of adducts with methanolic HCl causes easy removal of the sulphinyl group that resembles a Boc-protecting group.

Fig. 7.54B illustrates an asymmetric synthesis of a chiral amine using (*R*)-*t*-butylsulphinamide starting from an aldehyde.[132,133] The use of $Ti(OEt)_4$ (or $CuSO_4$) as a Lewis acid promoter gives a high yield of *t*-butanesulphinyl imine (aldimine). Addition of a Grignard reagent (PhMgBr) to the enantioenriched aldimine affords the corresponding sulphinamide with high diastereoselectivity (dr=96:4). Subsequent removal of the sulphinyl group with methanolic HCl gives the (*R*) primary amine (as a salt). The stereochemical outcome of the reaction can be rationalized by invoking chair-like TS involving coordination of oxygen to the metal (Mg) centre.[133] The chair TS in which the substituent (Et) at the carbonyl carbon is equatorial is favoured and gives the major diastereomer. It may be mentioned that a competitive or even dominant pathway via an open transition structure may take place depending on the nature of solvent, Lewis acid additive or organometallic reagent.[132]

Like aldimines, ketimines can also be prepared by condensation of ketones with *t*-butylsulphinamide **7.14** using $Ti(OEt)_4$ as a Lewis acid promoter, albeit at elevated temperatures.[132] Ketimines are however less reactive and their reactions with Grignard reagents give low yield and diastereoselectivity. The use of an organolithium reagent in the presence of Lewis acid additive such as $AlMe_3$ provides optimal results (Fig. 7.55A).[133,134] Removal of the sulphinyl group with methanolic HCl will give the (*R*) amine.

The allylation of ketimines (or aldimines) with allylmetal reagents such as allylmagnesium bromide gives high diastereoselectivity (Fig. 7.55B).[134,135]

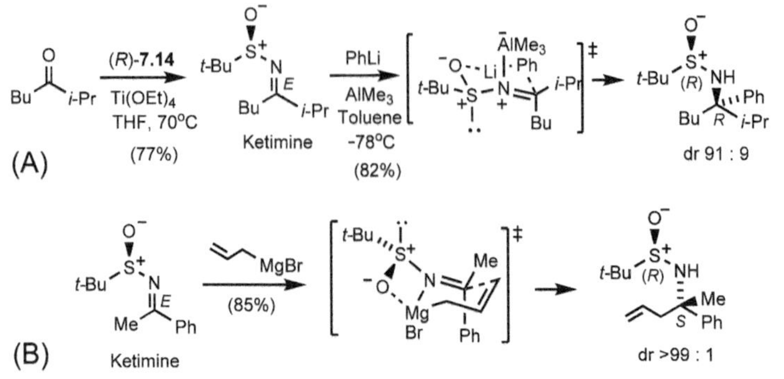


FIG. 7.55 Asymmetric syntheses of amines via *t*-butanesulphinyl ketimines using (A) organolithium reagent in the presence of a Lewis acid, and (B) allylation reaction.

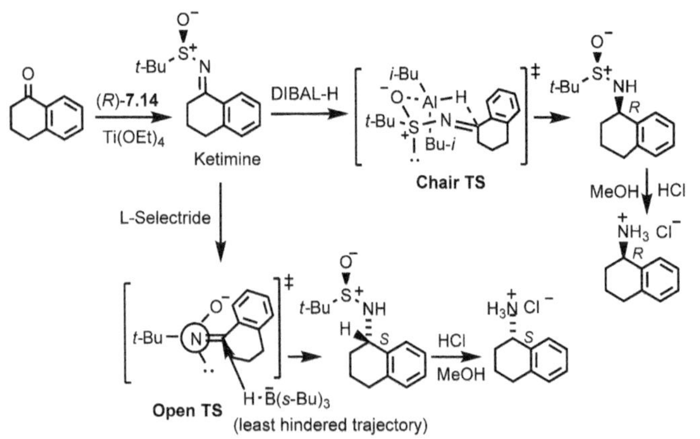


FIG. 7.56 Asymmetric synthesis of either enantiomer of an amine via hydride reduction of *t*-butanesulphinyl ketimine using a coordinating reductant (DIBAL-H) and a noncoordinating reductant (L-Selectride).

Here the TS involves coordination of both imino nitrogen and sulphinyl oxygen to the metal. Subsequent hydrolysis with methanolic HCl will yield the enantiopure amine.

Interestingly, hydride reduction of ketimines can lead to a different major diastereomer depending on the choice of the reductant and hence provides, after removal of the sulphinyl moiety, either enantiomer of the chiral amine (Fig. 7.56).[130,136,137] Coordinating reductants such as $NaBH_4$ or DIBAL-H ($i$-$Bu_2AlH$) give the product stereochemistry via chair TS, whereas noncoordinating and hindered reductants such as L-Selectride [$Li(s$-$Bu)_3BH$] provide the opposite stereochemistry via open TS.

**Problem 7.17** Explain the formation of the major diastereomer in the following asymmetric synthesis of an antihistamine drug cetirizine dihydrochloride.

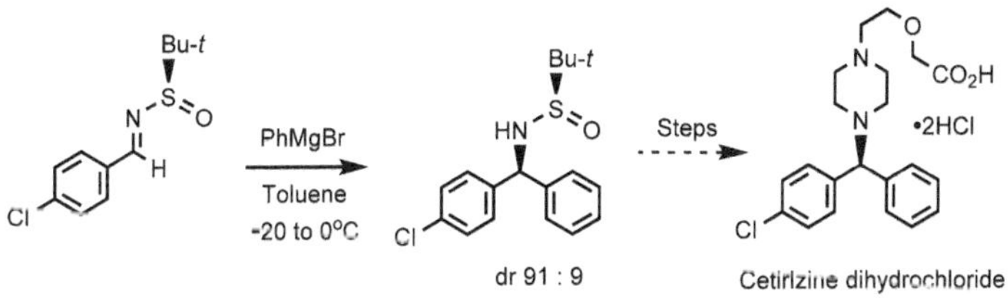


## 7.12 Diastereoselective addition to alkenes

The stereospecific additions to achiral alkenes to obtain single diastereomers have been described previously (see Section 6.4.1). If the alkene is chiral with a stereocentre, its two faces will be diastereotopic. The diastereoface differentiation by an attacking electrophile can then give one diastereomer as a major product, and the alkene addition will be diastereoselective. Here we will describe briefly diastereoselective epoxidation and hydroboration reactions of acyclic alkenes using Houk models.

### 7.12.1 Diastereoselective epoxidation: Houk model

Chiral alkene substrates with α-stereocentres exhibit low to high diastereoselectivity depending on the geometry (*E* or *Z*) of alkenes. Consider the epoxidation of chiral (*E*)- and (*Z*)-alkenes with *m*CPBA (*m*-chloroperbenzoic acid) shown in Fig. 7.57.[138] The (*E*)-alkene gives a low diastereoselectivity (dr = 61:39, de 22%), whereas the (*Z*)-alkene gives a high diastereoselectivity (dr = 95:5, de 90%). If the starting alkenes are racemic, the product diastereomers will also be racemic.

The diastereoselectivity has been rationalized on the basis of the preferred (low-energy) alkene conformation and the steric effect encountered in the electrophilic attack to the alkene diastereofaces. The transition structure model based on computational studies by Houk is known as the Houk model.[139]

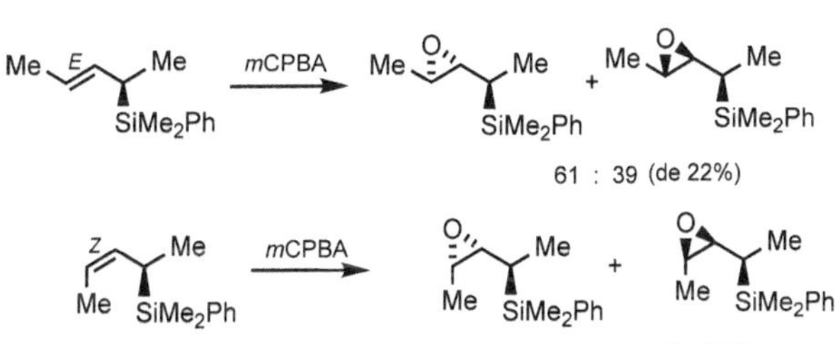


**FIG. 7.57** Diastereoselective epoxidation of chiral (*E*)- and (*Z*)-alkenes with *m*CPBA.

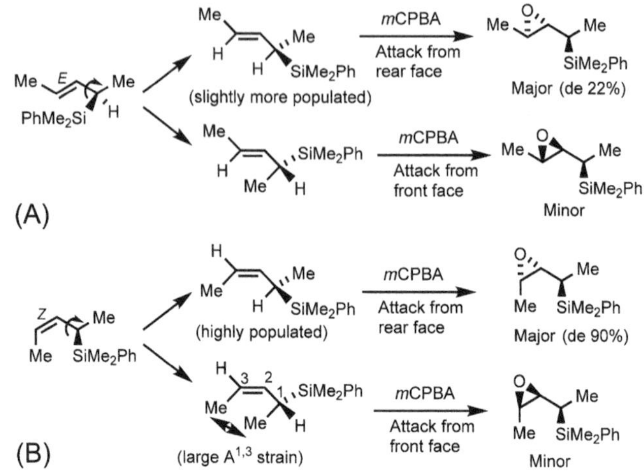


**FIG. 7.58** Rationalization of diastereoselectivity in the epoxidation of (A) chiral (*E*)-alkene and (B) chiral (*Z*)-alkene in Fig. 7.57 based on the Houk model.

The low-energy conformations of alkenes are those in which an allylic substituent *eclipses* the double bond (see Section 1.3.9). Eclipsing the smallest group with the double bond minimizes the allylic ($A^{1,3}$) strain. Fig. 7.58 depicts the alkene models for the diastereoselective epoxidation of (*E*)- and (*Z*)-alkenes. For the (*E*)-alkene, there are two low-energy conformations with H eclipsing and Me eclipsing the double bond, respectively (Fig. 7.58A); the conformation with H eclipsing being slightly more stable (cf. Fig. 1.15A). Note that the conformation with the bulky silyl group eclipsing the double bond is destabilized by a severe $A^{1,3}$ strain and is therefore neglected. Attack of *m*CPBA on the two conformations of the (*E*)-alkene occurs from the less hindered face of the double bond (opposite to the bulky silyl group) and gives different diastereomers. However, the selectivity is low presumably because the major diastereomer results from the slightly more populated conformer. In the case of (Z)-alkene, one conformer (H eclipsing) is much more populated than the other (Me eclipsing) that suffers from a large $A^{1,3}$ strain (Fig. 7.58B). Thus one diastereomer is formed predominantly from the heavily populated conformer via the more stable TS and the reaction becomes highly diastereoselective.

**Problem 7.18** Predict the major diastereomer formed in each of the following reactions:

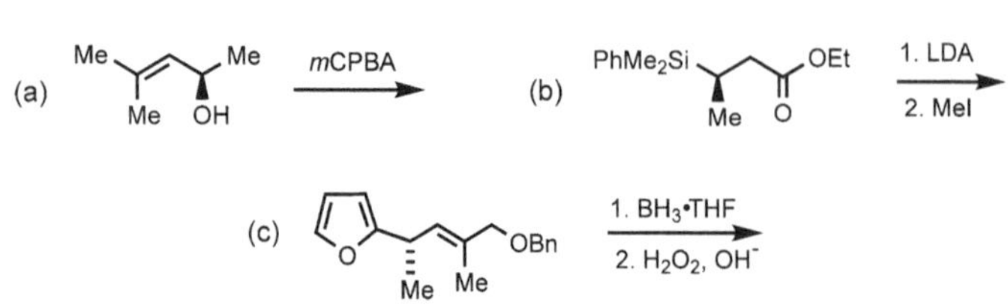

## 7.12.2 Diastereoselective hydroboration: Houk model

In general, the hydroboration reaction of a chiral alkene with a dialkylborane reagent ($R_2BH$) is diastereoselective, as shown in Fig. 7.59A. $R_M$ and $R_L$ are the medium and large substituents, respectively, at the stereocentre. The major diastereomeric product is predicted to be *anti*, using the Houk model.[140] Minimization of the steric interaction between $R_M$ and the boron ligand (R) provides the favoured transition structure which leads to the *anti* alcohol after oxidation with alkaline $H_2O_2$ involving retention of configuration. With borane itself ($BH_3$), a reversal of diastereoselectivity is predicted by the Houk model (Fig. 7.59B). Minimization of the $A^{1,2}$ strain gives the favoured TS that provides the *syn* diastereomer.

Fig. 7.60 illustrates examples of diastereoselective hydroborations,[141,142] consistent with Houk rules. With 9-BBN, the diastereoselectivity is in favour of the *anti* product, whereas the reaction is *syn*-selective with $BH_3$. However, in some cases, $BH_3$ gives the unexpected *anti*-selective hydroboration.[143]

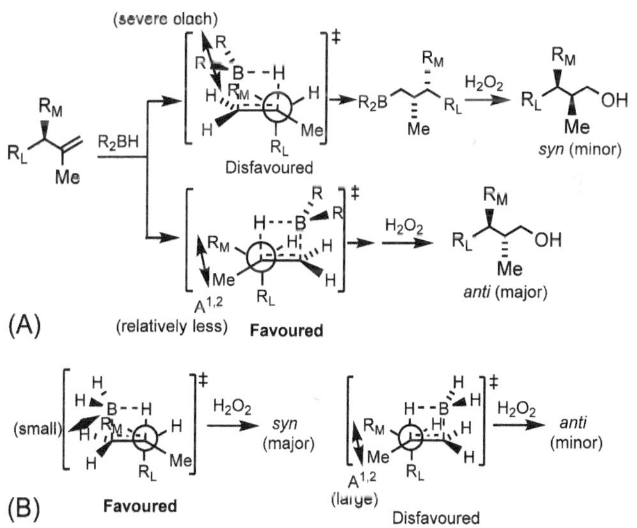


**FIG. 7.59** Diastereoselectivity in hydroboration of alkenes based on Houk models, using (A) dialkylborane ($R_2BH$) and (B) borane ($BH_3$).

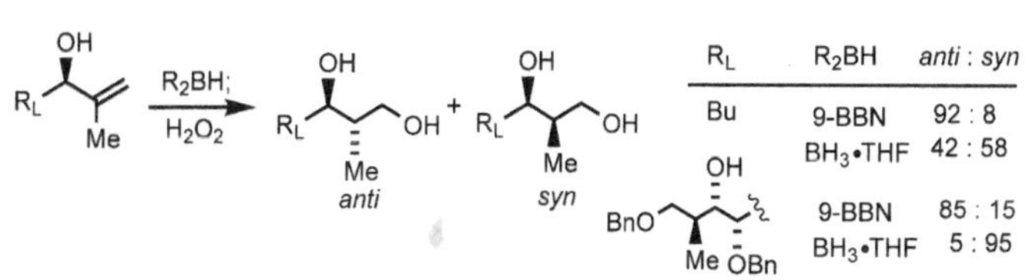


**FIG. 7.60** Examples of diastereoselective hydroboration of alkenes consistent with Houk models.

**Problem 7.19** Predict the diastereoselectivity of the following hydroborations involved in a synthesis of the antibiotic lonomycin A:

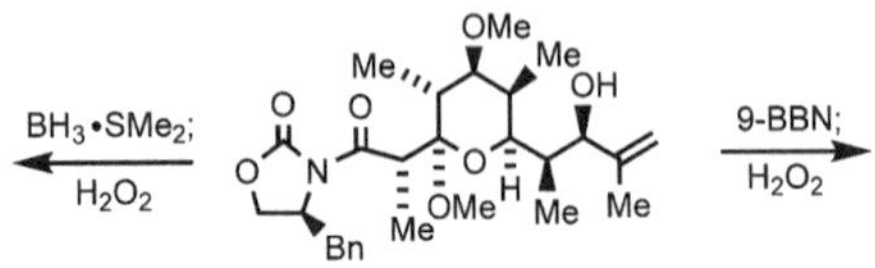


## 7.13 Asymmetric addition to alkenes

The most important transition metal-catalysed asymmetric oxidations and reductions of alkenes will be described in a separate chapter (Chapter 10). Here, we will describe briefly the asymmetric hydroboration under reagent control (Brown hydroboration), and an organocatalytic asymmetric epoxidation with dioxiranes (Shi epoxidation).

### 7.13.1 Asymmetric hydroboration: Brown's chiral Ipc reagents

Asymmetric hydroboration of alkenes can be performed with high enantioselectivity using chiral borane reagents.[144] The hydroboration product is then converted with retention of configuration into the corresponding enantiomeric alcohol with alkaline $H_2O_2$. The reagent diisopinocamphenylborane ($Ipc_2BH$) was developed by Brown from inexpensive α-pinene which is available commercially as both enantiomers. (+)-α-Pinene on hydroboration provides (−)-$Ipc_2BH$, whereas (−)-α-Pinene gives (+)-$Ipc_2BH$ (Fig. 7.61A). Notably, $Ipc_2BH$ is too hindered to add to a third molecule of α-pinene. The asymmetric hydroboration of *cis*-alkenes with $Ipc_2BH$ takes place with excellent enantioselectivity (Fig. 7.61A).[145] Note that boron is attached to the more nucleophilic end in the case of an unsymmetrical alkene.

The mechanism of the asymmetric hydroboration with (*Z*)-2-butene is depicted in Fig. 7.61B. The addition occurs preferably on the less hindered face (not obstructed by the proximate *gem*-dimethyl group) to give, after oxidation, the predominant (*R*)-alcohol.

Since $Ipc_2BH$ is a hindered reagent, *trans*-alkenes and trisubstituted alkenes give unsatisfactory results and poor enantioselectivity. For example, (−)-$Ipc_2BH$ reacts with (*E*)-2-butene to give the product with only 14% ee. To address this problem, Brown developed a less hindered reagent $IpcBH_2$ (monoisopinocamphenylborane)[146] from $Ipc_2BH$, and this reagent reacts with *trans*- and trisubstituted alkenes with good to high enantioselectivity (Fig. 7.62).[147]

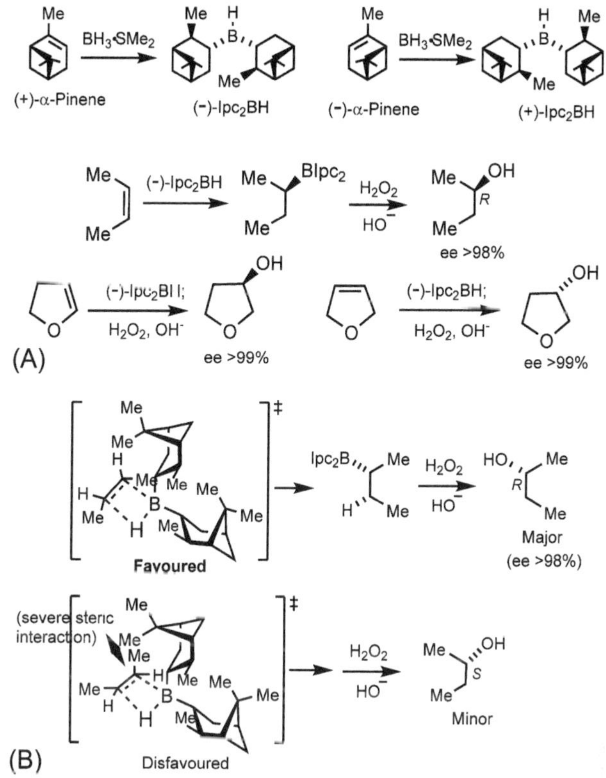


FIG. 7.61 (A) Brown's $Ipc_2BH$ reagent and its reaction with *cis*-alkenes; (B) Mechanism of the enantioselective hydroboration reaction with (*Z*)-2-butene.

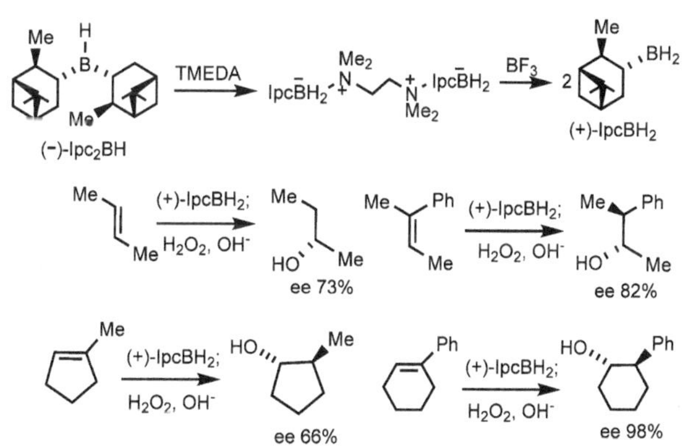


FIG. 7.62 Brown's $IpcBH_2$ reagent and its reaction with *trans*- and trisubstituted alkenes.

**Problem 7.20** Identify **A–G** in the following scheme:

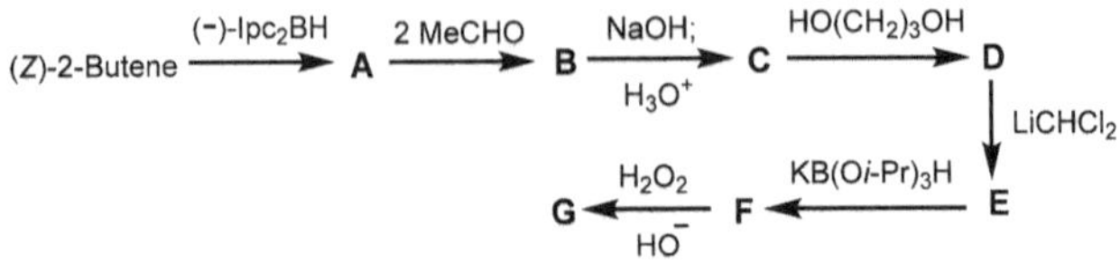


## 7.13.2 Asymmetric epoxidation: Shi epoxidation with organocatalysis[148–150]

Like peracids, dioxiranes can epoxidize an alkene efficiently. Dioxiranes are generated from the corresponding ketones and oxone (2:1:1 mixture of $KHSO_5$, $KHSO_4$, $K_2SO_4$) as shown in Fig. 7.63A.[151] The dioxirane epoxidation is mechanistically similar to the peracid epoxidation, and proceeds through spiro transition structure to give the epoxide and the starting ketone (Fig. 7.63B). The primary stereoelectronic interaction is $\pi_{C=C}/\sigma^*_{O-O}$ interaction.

Shi has developed an efficient asymmetric epoxidation of *trans*-alkenes with a chiral dioxirane generated in situ from the chiral ketone **7.15** as an organocatalyst (Fig. 7.64A).[152] The Shi organocatalyst **7.15** can be readily derived from the sugar D-fructose. L-Fructose (prepared from the readily available L-sorbose) provides the opposite enantiomer.

The Shi epoxidation method also works well with trisubstituted alkenes, dienes and enynes. These are illustrated with some examples in Fig. 7.64B.[152–154] *cis*-Alkenes and styrenes can also be epoxidized but with a modified catalyst.[155,156]

The enantioselectivity of the epoxidation reaction has been rationalized on the basis of a spiro TS model (Fig. 7.65).[152] As shown, the attack of the dioxirane moiety occurs preferentially from the lower face of the alkene to give the predominant epoxide enantiomer. Notably, small $R^1$ (as it projects over the pyranose ring) and large $R^2$ favour the TS more leading to enhanced enantioselectivity.

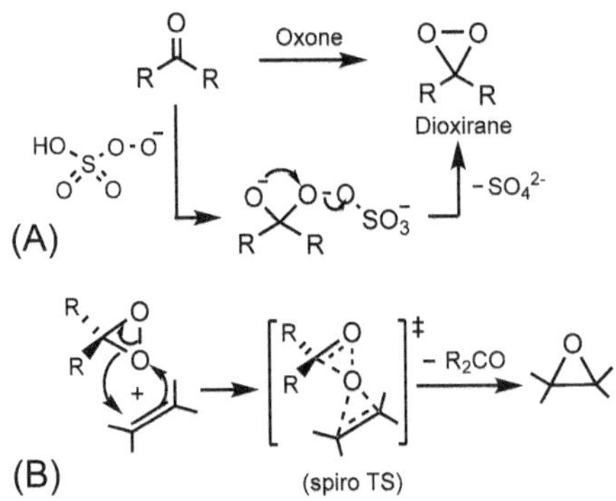


**FIG. 7.63** (A) Preparation of dioxiranes and (B) mechanism of dioxirane epoxidation.

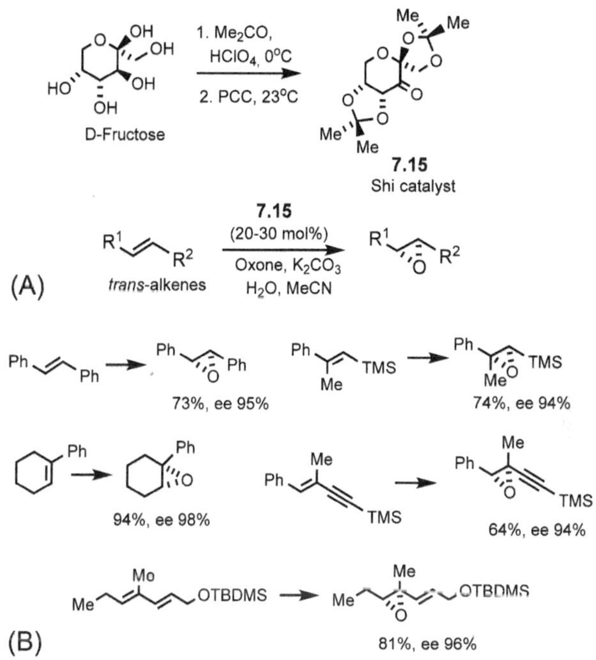


FIG. 7.64 (A) Shi asymmetric epoxidation with *trans*-alkenes; (B) Illustrative examples.

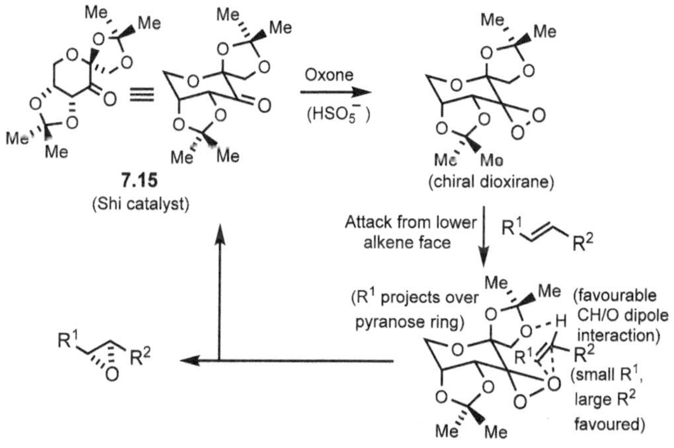


FIG. 7.65 Catalytic cycle and rationalization of enantioselectivity of Shi asymmetric epoxidation with *trans*-alkenes in Fig. 7.64A.

## 7.14 Asymmetric conjugate additions

A conjugate addition to α,β-unsaturated carbonyl systems involves two steps: nucleophilic attack on the β-carbon to generate an enolate anion, and trapping of the enolate with an electrophile, usually proton (Fig. 7.66A). The conjugate addition thus creates a new stereocentre, and an efficient asymmetric induction will lead to the formation of a single enantiomer. If the α-carbon bears a substituent or the electrophile is other than a proton, the conjugate addition will create two new stereocentres, and two possible diastereomeric products will be formed, of which one may predominate.

With a chiral enone substrate, the asymmetric induction is under substrate control (first generation). Fig. 7.66B shows that conjugate addition of the Grignard reagent MeMgBr to a chiral cyclic enone is diastereoselective in favour of the *cis* isomer. The diastereoselectivity arises from an axial addition of the Grignard reagent to the top face of C=C of the cyclohexenone to give directly a chair rather than a twist boat (cf. Fig. 6.36). A catalytic amount of copper(I) salt is added to promote the conjugate addition; the Cu(I) works by transmetalling the Grignard reagent to give an organocopper reagent ('Me-Cu') which is softer than Grignard and hence adds to the softer C=C. After the addition, the copper salt is available for transmetallation again.

Now we will describe briefly the asymmetric approaches to conjugate additions using chiral auxiliaries (second generation), chiral reagents (third generation) and chiral catalysts and organocatalysis (fourth generation) with a limited selection of examples.

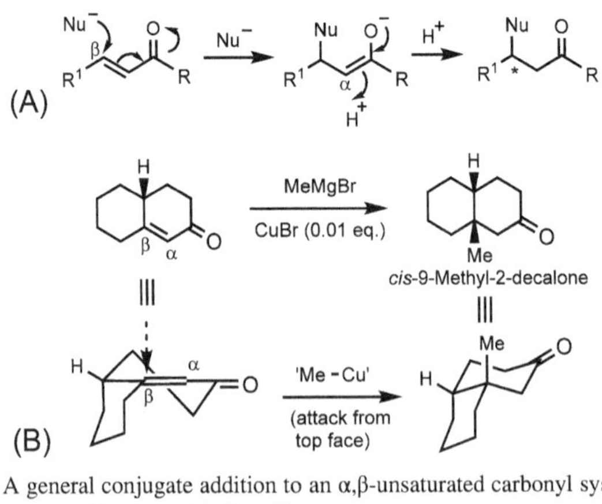


**FIG. 7.66** (A) A general conjugate addition to an α,β-unsaturated carbonyl system; (B) Diastereoselective conjugate addition to a chiral cyclic enone.

### 7.14.1 Chiral auxiliaries

Efficient asymmetric conjugate additions have been achieved using several chiral auxiliaries. Two effective methods involve Oppolzer sultam auxiliaries (see Fig. 5.9B) and chiral aminal auxiliaries.

#### *7.14.1.1 Oppolzer sultam auxiliaries*

An asymmetric conjugate addition using the Oppolzer sultam auxiliary in the synthesis of β-silylcarboxylic acids is illustrated in Fig. 7.67.[157] The reaction proceeds through a chelated transition structure involving the coordination of carbonyl and sulphonyl oxygens to the aluminium (Lewis acid). The attack of the organocopper reagent ($R_2CuLi$) occurs preferably from the less hindered face of C=C in the *s-cis* conformation of the enone to give the major diastereomer. Note that the *s-trans* conformation is sterically impeded by a methyl group on the auxiliary. Purification and subsequent hydrolysis provides the enantioselective β-silylcarboxylic acids, which can be converted into useful derivatives such as *anti*-aldol products by alkylation and conversion of the silyl group into a hydroxyl group.[157]

#### *7.14.1.2 Chiral aminal auxiliaries*

Aminals are the nitrogen analogues of acetals.[158] Chiral aminals can be easily prepared from the reaction of an enantiomerically pure chiral diamine with an aldehyde. The chiral aminals derived from $C_2$-symmetric diamines are very efficient, and provide good yields and high enantioselectivity. An asymmetric conjugate addition involving a chiral aminal prepared from a $C_2$-symmetric diamine[159] is illustrated in Fig. 7.68.[160] The chiral aminal moiety is introduced at the *ortho*-position of a cinnamate ester, and the chiral cinnamate then undergoes diastereoselective conjugate addition with $R_2CuLi$. After removal of the auxiliary by hydrolysis, the conjugate adduct is obtained with high yield

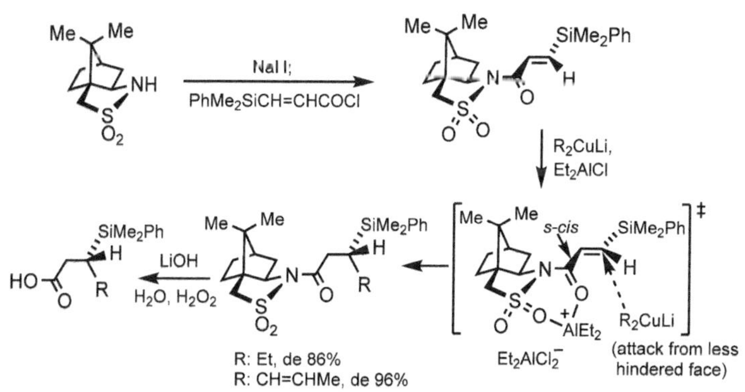


**FIG. 7.67** Asymmetric conjugate addition using the Oppolzer sultam auxiliary.

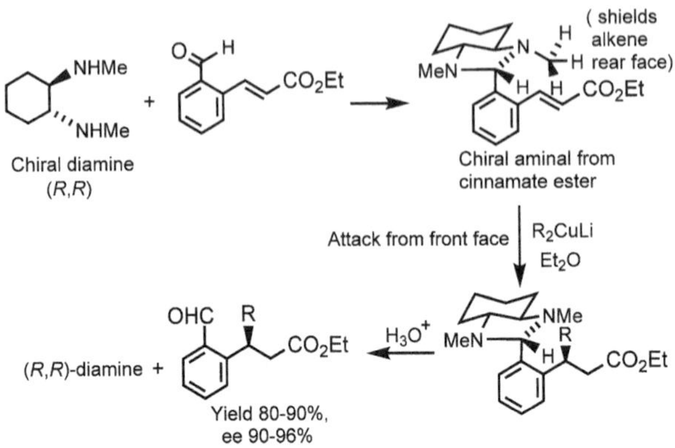


**FIG. 7.68** Asymmetric conjugate addition using a chiral aminal auxiliary.

and ee. The aminal has a rigid *trans*-fused bicyclic structure with little or no conformational flexibility, and thus allows one N-Me group to shield the rear face of the alkene double bond (Fig. 7.68). The attack of the organocopper reagent then occurs from the front face to give, after hydrolysis, the enantioselective product.

## 7.14.2 Chiral reagents: Davies chiral lithium amides

In asymmetric conjugate additions, the nucleophile can also act as a chiral partner. The most effective chiral nucleophilic reagents are chiral lithium amides (chiral ammonia equivalents). We have described earlier the use of chiral lithium amides as chiral bases in asymmetric deprotonation (see Section 7.4.1) but asymmetric conjugate additions rely upon chiral lithium amides to act as chiral nucleophiles.

Davies has developed a range of chiral secondary lithium amides from enantiopure α-methylbenzylamine for highly efficient conjugate additions to α,β-unsaturated esters and amides.[161,162] Some examples of enantiopure lithium amides are shown in Fig. 7.69A. Both enantiomers of a chiral lithium amide are available. The conjugate addition proceeds in high yield and with excellent diastereocontrol to give β-amino acids and their derivatives, as illustrated with a crotonate ester in Fig. 7.69B.[163,164] Hydrogenolysis of the conjugate adduct followed by ester deprotection provides the β-amino acid with high ee. Note that the N-benzyl unit in the chiral amide reagent permits easy hydrogenolysis, and the *t*-butyl ester moiety in the substrate allows simple deprotection under mild condition in TFA (trifluoroacetic acid) without racemisation.

The diastereoselectivity of the conjugate addition has been rationalized with a 'butterfly' TS model in which the reagent approaches with its large Ph group at the stereocentre away from the substrate. The transition structure that avoids Me/Me steric interaction is favoured and leads to the major diastereomer.

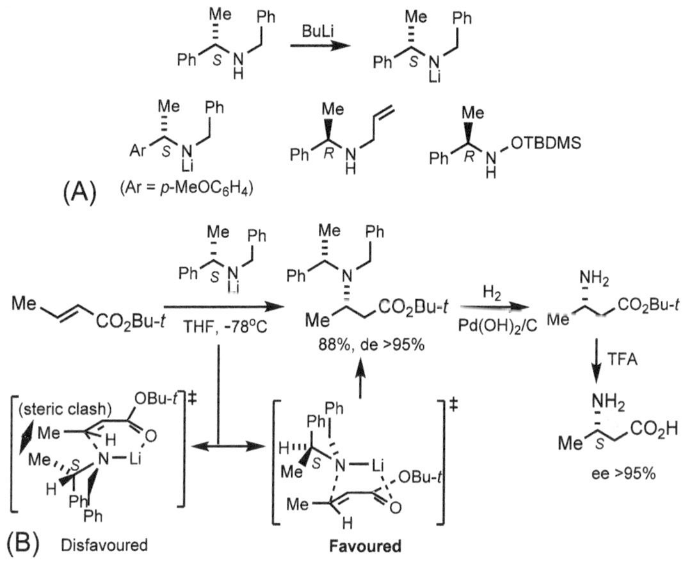


FIG. 7.69 (A) Davis chiral lithium amides; (B) Asymmetric conjugate addition of an enantiopure lithium amide to a crotonate ester, and rationalization of diastereoselectivity with a 'butterfly' TS model.

### 7.14.3 Chiral catalysts

A variety of chiral metal catalysts has been developed for asymmetric conjugate additions with moderate to high enantioselectivity. For example, a monodentate chiral phosphoramidite ligand **7.16** based on (*S*)-BINOL ligand provides good ee in the asymmetric conjugate addition as shown in Fig. 7.70A.[165] Here a Cu(I) species generated by in situ reduction with the zinc reagent is the active species. The enone is probably activated by zinc, and the copper coordinated by two molecules of the ligand **7.16** binds to the C=C and does the alkyl transfer in the asymmetric fashion.

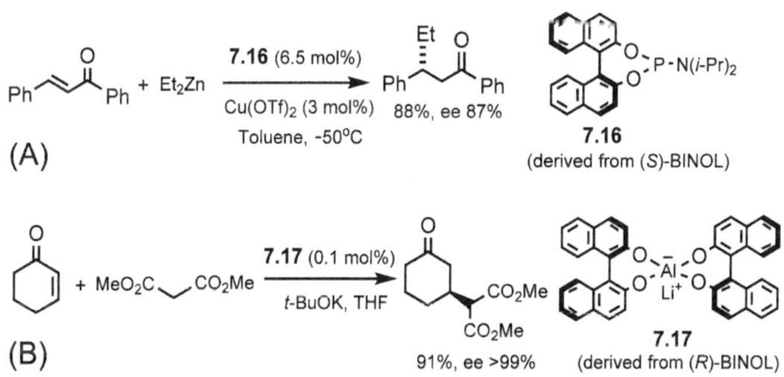


FIG. 7.70 Catalytic asymmetric conjugate additions using (A) copper catalyst with phosphoramidite ligand derived from (*S*)-BINOL, and (B) lithium-aluminium salt of (*R*)-BINOL as a chiral catalyst.

Fig. 7.70B shows the asymmetric conjugate addition of the enolate of dimethylmalonate to cyclohexenone in the presence of lithium-aluminium salt of (*R*)-BINOL **7.17** as a chiral catalyst. The reaction was employed in a synthesis of the alkaloid (−)-strychnine.[166]

### 7.14.4 Organocatalysis: MacMillan imidazolidinones

Organocatalysis as a general strategy was discovered by MacMillan in the early years of 21st century.[111,167] Here we will describe briefly the MacMillan imidazolidinone organocatalysts in asymmetric conjugate additions.

The 4-imidazolidinone catalyst **7.18** can be easily prepared from the ester of the amino acid (*S*)-phenylalanine (Fig. 7.71A). The catalyst is a cyclic secondary amine and can condense with α,β-unsaturated carbonyl substrate to give a chiral, enantiopure iminium ion. The conjugated iminium ion is more electrophilic than the parent enone towards nucleophilic attack in conjugate addition. In fact, the iminium ion catalysis by LUMO activation (lowering LUMO energy) is a general and useful strategy for asymmetric catalysis. An asymmetric conjugate addition using the imidazolidinone catalyst **7.18** is illustrated in Fig. 7.71B.[168] The geometry of the iminium C=N bond that avoids the severe steric clash with the *gem*-dimethyl group is favoured. The *gem*-dimethyl group also prevents the attack of the nucleophile to $C{=}N^+$. Since the front face (β-*Re*) of the alkene double bond in the favoured iminium ion is shielded by the benzyl

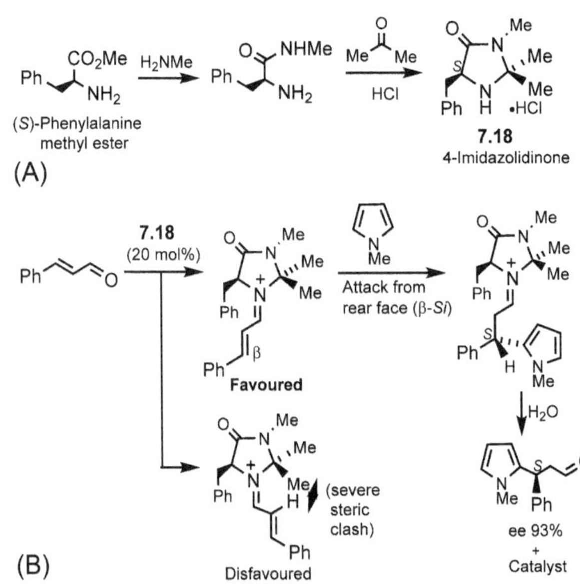


**FIG. 7.71** (A) Preparation of a MacMillan 4-imidazolidinone organocatalyst **7.18**; (B) An organocatalytic asymmetric conjugate addition using the imidazolidinone catalyst **7.18**.

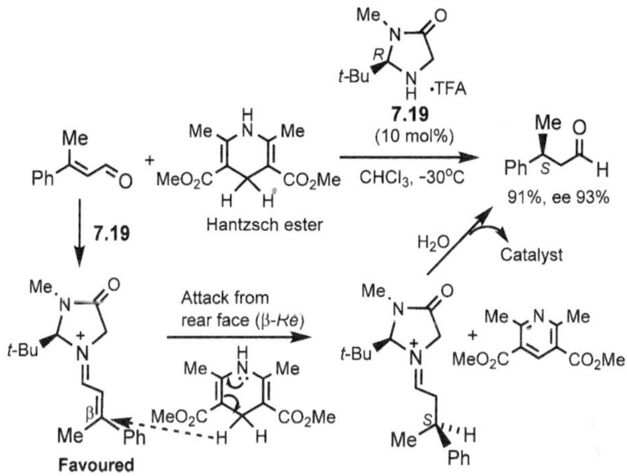


FIG. 7.72 Organocatalytic asymmetric conjugate reduction of an α,β-unsaturated aldehyde using the 4-imidazolidinone catalyst **7.19**.

group, the nucleophilic attack of *N*-methylpyrrole via its 2-position takes place from the exposed rear face (β-*Si*) of C=C to give the conjugate adduct with high enantioselectivity. Note that the pyrrole is rather a weak nucleophile but can be used successfully in the iminium ion catalysis of the conjugate addition.

Fig. 7.72 shows an organocatalytic asymmetric conjugate reduction using the 4-imidazolidinone catalyst **7.19**.[169,170] The reducing agent for the hydride transfer is a Hantzsch dihydropyridine ester that acts as an NADH analogue in the presence of the catalyst. The catalyst **7.19** condenses with the enone to form an iminium ion in a favourable geometry that avoids steric clash with the bulky *t*-butyl group. The hydride transfer then occurs from the rear face (β-*Re*) of C=C as the front face (β-*Si*) is shielded by the *t*-Bu substituent to give the reduction product with high yield and ee.

A mixture of the (*R*)- or (*S*)-catalyst (10 mol%) and the Hantzsch ester (120 mol%), called Mac-H-(*R*) or Mac-H(*S*), can be stored for several months and used later to perform asymmetric hydrogenation reaction.

**Problem 7.21** Predict the stereochemistry of the product (a COX-2 inhibitor) in the following asymmetric synthesis via an organocatalysed asymmetric conjugate addition using a MacMillan catalyst:

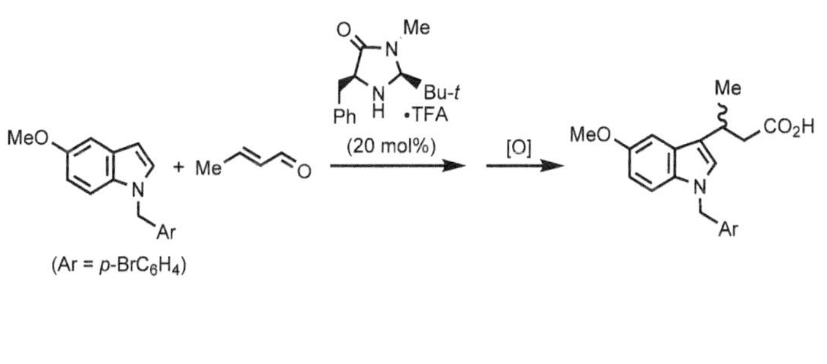

## 7.15 Enzyme-catalysed asymmetric reactions

Enzymes are proteins that act as highly efficient chiral biocatalysts in biological reactions. The use of enzymes (wild-type or mutants derived by site-directed mutagenesis) as chiral catalysts in organic chemical reactions is a valuable approach to asymmetric synthesis.[171] Here we will briefly describe the enzyme-catalysed asymmetric reactions using hydrolysis, esterification, reduction, oxidation and formation of C—C bonds.

### 7.15.1 Hydrolysis and esterification

The most widely used hydrolysis reactions are the hydrolysis of esters and a common enzyme is pig liver esterase.[172] For example, hydrolysis of a *meso*-diester with pig liver esterase is shown in Fig. 7.73A.[173] The two ester groups are enantiotopic, but since the enzyme is chiral and enatiomerically pure, it can make enantiotopos differentiation leading to the hydrolysis of one of the two ester groups. The esterase enzyme has a serine residue (side chain $CH_2OH$) at the active site. The reaction of the serine residue with the ester groups will give diastereomeric TSs, and the preferred transition structure gives predominantly one enantiomer of the half-ester product. This process of converting *meso* compounds into single enantiomers is often called desymmetrization and is a valuable tactic for asymmetric induction. The enantiomeric half-ester was used in a total synthesis of the natural product (−)-fortamine (Fig. 7.73A).[173]

Interestingly, the single enantiomer of the half-ester can be used to produce both enantiomers of a chiral lactone by selective reduction of the functional groups (Fig. 7.73B).[174] This is known as the '*meso*-trick'.

The enzyme-catalysed hydrolysis is reversible but since the hydrolysis is carried out in water, the reaction becomes effectively irreversible by the high concentration of water (55.5 M). However, under suitable conditions,

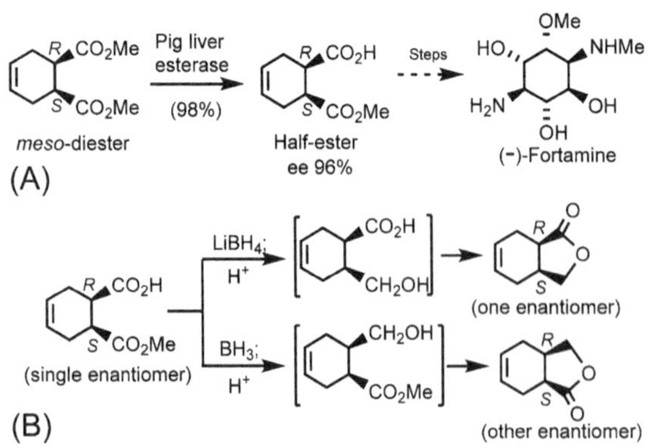


FIG. 7.73 (A) Desymmetrization of a *meso*-diester using an esterase-catalysed hydrolysis; and (B) the '*meso*-trick'.

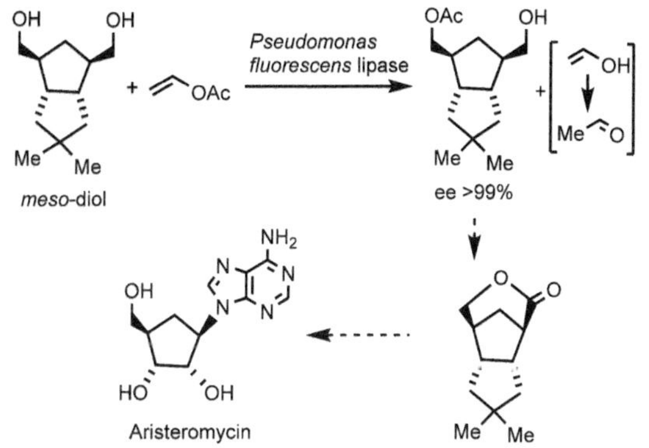


**FIG. 7.74** Desymmetrization of a *meso*-diol using a lipase-catalysed transesterification with vinyl acetate in a synthesis of aristeromycin.

the esterase enzymes can be used to catalyse the formation of esters.[175] The enantioselective ester formation is also possible by transesterification reaction between an ester and an alcohol, catalysed by lipases as acyl transferase enzymes. Fig. 7.74 illustrates a lipase-catalysed asymmetric transesterification between a *meso*-diol and vinyl acetate ester. The use of vinyl acetate makes the process irreversible as the vinyl alcohol formed immediately tautomerizes to acetaldehyde. The monoacetate product is produced essentially as a single enantiomer, which is then converted to a lactone as a precursor to the naturally occurring nucleoside aristeromycin.[176]

## 7.15.2 Reduction and oxidation

The enzyme-catalysed asymmetric reduction of prochiral carbonyl substrates can be performed using whole cells (such as yeast) or isolated enzymes. With isolated enzymes, the cofactor NADH or NAD(*P*)H [nicotinamide adenine dinucleotide (phosphate)] needs to be added in catalytic amounts whereas in whole cells, cofactor recycling is achieved automatically, though the work-up may be more complex. The reduction of carbonyl compounds is achieved with the enzyme alcohol dehydrogenase (ADH). Several ADHs are commercially available (such as yeast ADH and horse-liver ADH), and the choice of ADH is determined by the steric requirement of the substrate. The principles of enzyme-catalysed asymmetric reduction in the presence of yeast ADH and NADH were introduced earlier in the context of topicity, taking the conversion of acetaldehyde into ethanol (see Section 2.12.7).

Baker's yeast is cheap and readily available, and is probably the most popular reducing biocatalyst. In particular, it is highly efficient for the asymmetric reduction of β-ketoesters.[177,178] The facial selectivity in the reduction of simple ketones and β-ketoesters can be determined by the application of Prelog's rule, which predicts that hydride attack occurs to the *Re*-face producing (*S*)-alcohols as shown in Fig. 7.75A.[179] Exceptions to the rule, that is anti-Prelog specificity with some ADHs such as *Pseudomonas* sp. ADH are known. Fig. 7.75B illustrates the reduction of a β-ketoester involved in a synthesis of L-carnitine.[180] The reaction also works well with cyclic β-ketoesters.[181]

The enzyme-catalysed asymmetric oxidation of a variety of organic substrates has been investigated. A popular source for dihydroxylation of benzene and its derivatives is *Pseudomonas putida* strains that express the enzyme dioxygenases. The dihydroxylation reaction leads to the formation of cyclohexadienediols that are valuable synthetic precursors to natural and unnatural products.[182] Fig. 7.76A shows that benzene dioxygenase (BDO) from a strain of *P. putida* dihydroxylates benzene in the presence of NADH to give a

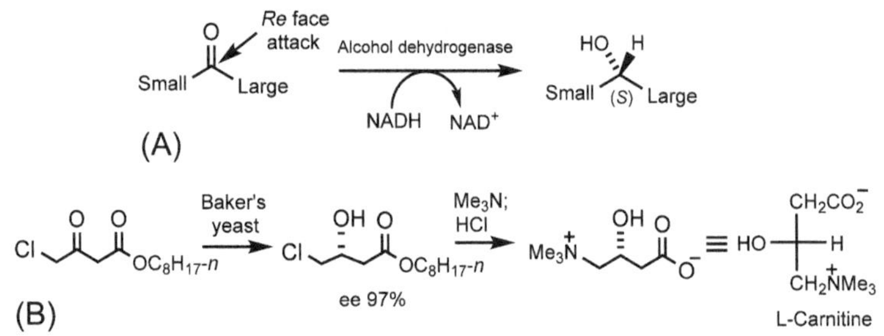


**FIG. 7.75** (A) Prelog's rule for asymmetric reduction with alcohol dehydrogenase; (B) Asymmetric synthesis of L-carnitine via reduction of a β-ketoester using Baker's yeast.

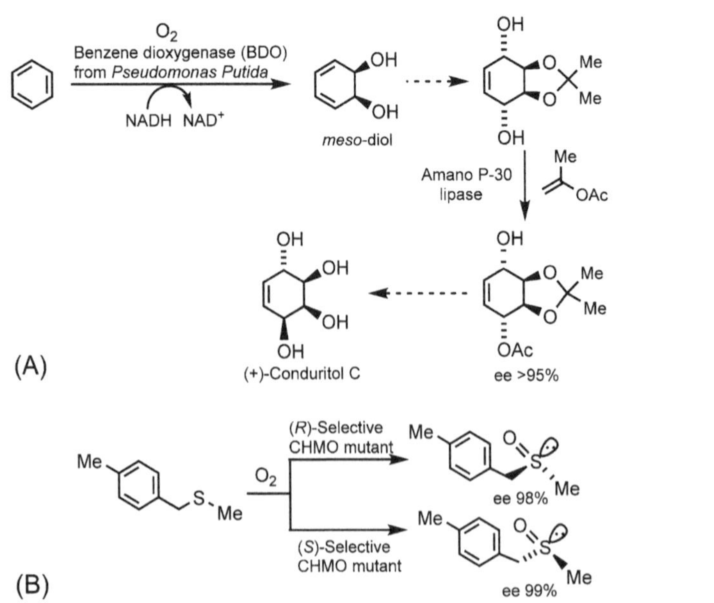


**FIG. 7.76** (A) Enzyme-catalysed dihydroxylation of benzene in a synthesis of (+)-conduritol C; (B) Enzyme-catalysed oxidation of a thioether to enantiopure sulphoxides.

*meso*-diol that was used in an elegant synthesis of (+)-conduritol C using desymmetrization strategy.[183]

The enzyme-catalysed asymmetric oxidation of sulphides serves as an efficient method for the asymmetric synthesis of chiral sulphoxides. For example, the mutants of cyclohexanone monooxygenase (CHMO), a bacterial flavoenzyme, catalyses the oxidation of prochiral thioethers to essentially enantiopure sulphoxides (Fig. 7.76B).[184]

### 7.15.3 C—C bond formation

A number of enzymatic methods have been developed for the asymmetric construction of C—C bonds. The enzyme oxynitrilase catalyses the enantioselective addition of HCN to a range of aldehydes to give enantiopure cyanohydrins[185] that are excellent starting materials for the asymmetric synthesis of drugs and other pharmacologically important compounds. For example, an asymmetric addition of HCN is used in a synthesis of (*R*)-salbutamol using (*R*)-specific oxynitrilase from almonds (Fig. 7.77A).[186]

Aldolases can catalyse C—C bond formation through aldol condensation. Fig. 7.77B shows the synthesis of enantiopure trihydroxyketones through aldol condensation of dihydroxyacetone phosphate with aldehydes using fructose-1,6-diphosphate aldolase (FDP-aldolase) from rabbit muscle, followed by dephosphorylation with acid phosphatase.[187] These polyhydroxyketones are efficient precursors to the synthesis of polyhydroxylated natural products, as illustrated with a synthesis of an α-glucosidase inhibitor deoxynojirimycin in Fig. 7.77B.[188,189] The asymmetric aldolase methods have been widely applied to the synthesis of carbohydrates and carbohydrate mimetics.[190–192]

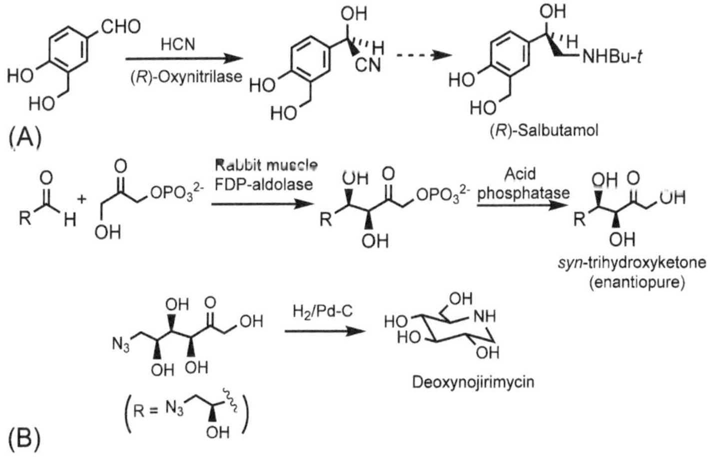


**FIG. 7.77** (A) Oxynitrilase-catalysed asymmetric cyanohydrin formation in a synthesis of (*R*)-salbutamol; (B) Aldolase-catalysed aldol condensation for the asymmetric synthesis of trihydroxyketones, and its application in a synthesis of deoxynojirimycin.

## References


1. Cram, D. J.; Elhafez, F. A. A. *J. Am. Chem. Soc.* **1952**, *74*, 5828.
2. Chérest, M.; Felkin, H.; Prudent, N. *Tetrahedron Lett.* **1968**, *9*, 2199.
3. Bürgi, H. B.; Dunitz, J. D.; Shefter, E. *J. Am. Chem. Soc.* **1973**, *95*, 5065.
4. Anh, N. T. *Top. Curr. Chem.* **1980**, *88*, 145.
5. Anh, N. T.; Eisenstein, O. *Nouv. J. Chim.* **1977**, *1*, 61.
6. Reiser, O. *Chem. Rev.* **1999**, *99*, 1191.
7. Evans, D. A.; Siska, S. J.; Cee, V. J. *Angew. Chem. Int. Ed. Engl.* **2003**, *42*, 1761.
8. Cram, D. J.; Kopecky, K. R. *J. Am. Chem. Soc.* **1959**, *81*, 2748.
9. Cram, D. J.; Wilson, D. R. *J. Am. Chem. Soc.* **1963**, *85*, 1245.
10. McGarvey, G. J.; Kimura, M. *J. Org. Chem.* **1982**, *47*, 5420.
11. Seebach, D.; Prelog, V. *Angew. Chem. Int. Ed. Engl.* **1982**, *21*, 654.
12. Fiaud, J. C. Prelog's methods. In *Stereochemistry, Fundamentals and Methods*; Kagan, H. B., Ed.; Vol. 3; Georg Thieme Publishers: Stuttgart, 1977; p. 19.
13. Evans, D. A.; Dart, M. J.; Duffy, J. L. *Tetrahedron Lett.* **1994**, *35*, 8541.
14. Narasaka, K.; Pai, F.-C. *Tetrahedron* **1984**, *40*, 2233.
15. Chen, K.-M.; Hardtmann, G. E.; Prasad, K.; Repič, O.; Shapiro, M. J. *Tetrahedron Lett.* **1987**, *28*, 155.
16. Saksena, A. K.; Mangiaracina, P. *Tetrahedron Lett.* **1983**, *24*, 273.
17. Evans, D. A.; Chapman, K. T.; Carreira, E. M. *J. Am. Chem. Soc.* **1988**, *110*, 3560.
18. Reviews: Brown, H. C.; Ramachandran, P. V. *Pure Appl. Chem.* **1991**, *63*, 307. *J. Organomet. Chem.* **1995**, *500*, 1.
19. Midland, M. M.; McDowell, D. C.; Hatch, R. L.; Tramontano, A. *J. Am. Chem. Soc.* **1980**, *102*, 867.
20. Midland, M. M.; Greers, S.; Tramontano, A.; Zderic, S. A. *J. Am. Chem. Soc.* **1979**, *101*, 2352.
21. Chandrasekharan, J.; Ramachandran, P. V.; Brown, H. C. *J. Org. Chem.* **1985**, *50*, 5446.
22. Brown, H. C.; Chandrasekharan, J.; Ramachandran, P. V. *J. Org. Chem.* **1986**, *51*, 3394.
23. Midland, M. M.; Zderic, S. A. *J. Am. Chem. Soc.* **1982**, *104*, 525.
24. Noyori, R.; Tomino, I.; Tanimoto, Y.; Nishizawa, M. *J. Am. Chem. Soc.* **1984**, *106*, 6709.
25. Noyori, R.; Tomino, I.; Yamada, M.; Nishizawa, M. *J. Am. Chem. Soc.* **1984**, *106*, 6717.
26. Review: Cox, P. J.; Simpkins, N. S. *Tetrahedron Asymmetry* **1991**, *2*, 1.
27. Simpkins, N. S. *J. Chem. Soc. Chem. Commun.* **1986**, 88.
28. Shirai, R.; Tanaka, M.; Koga, K. *J. Am. Chem. Soc.* **1986**, *108*, 543.
29. Majewski, M.; Gleave, D. M. *J. Org. Chem.* **1992**, *57*, 3599.
30. Cain, C. M.; Simpkins, N. S. *Tetrahedron Lett.* **1987**, *28*, 3723.
31. Hoppe, D.; Christoph, G. In *The Chemistry of Organolithium Compounds*; Rappoport, Z., Marek, I., Eds.; Book Series: PATAI's Chemistry of Functional Groups; Wiley: New York, 2004; p. 1055 (chapter 17).
32. Kitamura, M.; Suga, S.; Kawai, K.; Noyori, R. *J. Am. Chem. Soc.* **1986**, *108*, 6071.
33. Review: Noyori, R.; Kitamura, M. *Angew. Chem. Int. Ed. Engl.* **1991**, *30*, 49.
34. Corey, E. J.; Bakshi, R. K.; Shibata, S.; Chen, C.-P.; Singh, V. K. *J. Am. Chem. Soc.* **1987**, *109*, 7925.
35. Review: Corey, E. J.; Helal, C. J. *Angew. Chem. Int. Ed. Engl.* **1998**, *37*, 1986.
36. Mathre, D. J.; Thompson, A. S.; Douglas, A. W.; et al. *J. Org. Chem.* **1993**, *58*, 2880.
37. Douglas, A. W.; Tschaen, D. M.; Reamer, R. A.; Shi, Y.-J. *Tetrahedron Asymmetry* **1996**, *7*, 1303.
38. Corey, E. J.; Reichard, G. A. *Tetrahedron Lett.* **1989**, *30*, 5207.

39. Jones, D. K.; Liotta, D. C.; Shinkai, I.; Mathre, D. J. *J. Org. Chem.* **1993**, *58*, 799.
40. Linney, L. P.; Self, C. R.; Williams, T. H. *J. Chem. Soc. Chem. Commun.* **1994**, 1651.
41. Ireland, R. E.; Mueller, R. H.; Willard, A. K. *J. Am. Chem. Soc.* **1976**, *98*, 2868.
42. Kleschick, W. A.; Buse, C. T.; Heathcock, C. H. *J. Am. Chem. Soc.* **1977**, *99*, 247.
43. Fataftah, Z. A.; Kopka, I. E.; Rathke, M. W. *J. Am. Chem. Soc.* **1980**, *102*, 3959.
44. Ireland, R. E.; Wipf, P.; Armstrong, J. D., III. *J. Org. Chem.* **1991**, *56*, 650.
45. Corey, E. J. *J. Am. Chem. Soc.* **1954**, *76*, 175.
46. Ireland, R. E.; Willard, A. K. *Tetrahedron Lett.* **1975**, *16*, 3975.
47. Ganesan, K.; Brown, H. C. *J. Org. Chem.* **1993**, *58*, 7162.
48. Evans, D. A.; Nelson, J. V.; Vogel, E.; Taber, T. R. *J. Am. Chem. Soc.* **1981**, *103*, 3099.
49. Goodman, J. M.; Paterson, I. *Tetrahedron Lett.* **1992**, *33*, 7223.
50. Corey, E. J.; Kim, S. S. *J. Am. Chem. Soc.* **1990**, *112*, 4976.
51. Evans, D. A. In *Asymmetric Synthesis*; Morrison, J. D., Ed.; Vol. 3; Academic Press: New York, 1984; p. 1.
52. Roos, G., Ed. *Compendium of Chiral Auxiliary Applications*; Academic Press: New York, 2002.
53. Evans, D. A. In *Asymmetric Synthesis—The Essentials*; Christmann, M., Bräse, S., Eds.; Wiley-VCH: Weinheim, 2007; p. 3.
54. Ager, D. J.; Prakash, I.; Schaad, D. R. *Chem. Rev.* **1996**, *96*, 835.
55. Evans, D. A. *Aldrichimica Acta* **1982**, *15*, 23.
56. Evans, D. A. In *Asymmetric Synthesis*; Morrison, J. D., Ed.; Vol. 3; Academic Press: New York, 1984; p. 87.
57. Evans, D. A.; Britton, T. C.; Dorow, R. L.; Dellaria, J. F. *J. Am. Chem. Soc.* **1986**, *108*, 6395. *Tetrahedron* **1988**, *44*, 5525.
58. Evans, D. A.; Gage, D. A. *Organic Synth.* **1989**, *68*, 77.
59. Evans, D. A.; Britton, T. C.; Ellman, J. A. *Tetrahedron Lett.* **1987**, *28*, 6141.
60. Evans, D. A.; Morrisey, M. M.; Dorow, R. L. *J. Am. Chem. Soc.* **1985**, *107*, 4346.
61. Oppolzer, W.; Blagg, J.; Rodriguez, I.; Walther, E. *J. Am. Chem. Soc.* **1990**, *112*, 2767.
62. Oppolzer, W.; Moretti, R.; Thomi, S. *Tetrahedron Lett.* **1989**, *30*, 5603.
63. Enders, D. In *Asymmetric Synthesis*; Morrison, J. D., Ed.; Vol. 3; Academic Press: New York, 1984; p. 275.
64. Review: Job, A.; Janeck, C. F.; Bettray, W.; Peters, R.; Enders, D. *Tetrahedron* **2002**, *58*, 2253.
65. Enders, D.; Kipphardt, H.; Fey, P. *Organic Synth.* **1987**, *65*, 183.
66. Deng, C.; Groth, U.; Schöllkopf, U. *Angew. Chem. Int. Ed. Engl.* **1981**, *20*, 798.
67. Schöllkopf, U. *Tetrahedron* **1983**, *39*, 2085.
68. Schöllkopf, U. *Pure Appl. Chem.* **1983**, *55*, 1799.
69. Magriotis, P. A.; Vassiliou, S.; Dimitropoulos, C. *Synlett* **2003**, 2398.
70. Myers, A. G.; Yang, B. H.; Chen, H.; McKinstry, L.; Kopecky, D. J.; Gleason, J. L. *J. Am. Chem. Soc.* **1997**, *119*, 6496.
71. Morales, M. R.; Mellem, K. T.; Myers, A. G. *Angew. Chem. Int. Ed. Engl.* **2012**, *51*, 4568.
72. Myers, A. G.; McKinstry, L. *J. Org. Chem.* **1996**, *61*, 2428.
73. Askin, D.; Volante, R. P.; Ryan, K. M.; Reamer, R. A.; Shinkai, I. *Tetrahedron Lett.* **1988**, *29*, 4245.
74. Myers, A. G.; Yang, B. H.; Chen, H.; Kopecky, D. *J. Synlett* **1997**, *5*, 457.
75. Seebach, D.; Boes, M.; Naef, R.; Schweizer, W. B. *J. Am. Chem. Soc.* **1983**, *105*, 5390.
76. Review: Seebach, D.; Sting, A. R.; Hoffmann, M. *Angew. Chem. Int. Ed. Engl.* **1996**, *35*, 2708.
77. O'Donnell, M. J.; Bennett, W. D.; Wu, S. *J. Am. Chem. Soc.* **1989**, *111*, 2353.
78. Corey, E. J.; Xu, F.; Noe, M. C. *J. Am. Chem. Soc.* **1997**, *119*, 12414.
79. Ooi, T.; Maruoka, K. *Angew. Chem. Int. Ed. Engl.* **2007**, *46*, 4222.

80. Lygo, B.; Andrews, B. *Tetrahedron Lett.* **1999**, *40*, 1385.
81. Majewski, M.; Gleave, D. M. *Tetrahedron Lett.* **1989**, *30*, 5681.
82. Tanabe, Y.; Matsumoto, N.; Higashi, T.; et al. *Tetrahedron* **2002**, *58*, 8269.
83. Zimmerman, H. E.; Traxler, M. D. *J. Am. Chem. Soc.* **1957**, *79*, 1920.
84. Heathcock, C. H.; Buse, C. T.; Kleschick, W. A.; Pirrung, M. C.; Sohn, J. E.; Lampe, J. *J. Org. Chem.* **1980**, *45*, 1066.
85. Cowden, C. J.; Paterson, I. *Org. React.* **1997**, *51*, 1.
86. Arya, P.; Qin, H. *Tetrahedron* **2000**, *56*, 917.
87. Machajewski, T. D.; Wong, C.-H. *Angew. Chem. Int. Ed. Engl.* **2000**, *39*, 1352.
88. Palomo, C.; Oiarbide, M.; García, J. M. *Chem. A Eur. J.* **2002**, *8*, 36.
89. Mukaiyama, T.; Banno, K.; Narasaka, K. *J. Am. Chem. Soc.* **1974**, *96*, 7503.
90. Mori, Y.; Kobayashi, J.; Manabe, K.; Kobayashi, S. *Tetrahedron* **2002**, *58*, 8263.
91. Evans, D. A.; Bartrolli, J.; Shih, T. L. *J. Am. Chem. Soc.* **1981**, *103*, 2127.
92. Evans, D. A.; Tedrow, J. S.; Shaw, J. T.; Downey, C. W. *J. Am. Chem. Soc.* **2002**, *124*, 392.
93. Oppolzer, W.; Starkemann, C.; Rodriguez, I.; Bernadinelli, G. *Tetrahedron Lett.* **1991**, *32*, 61.
94. Meyers, A. I.; Yamamoyo, Y. *Tetrahedron* **1984**, *40*, 2309. *J. Am. Chem. Soc.* **1981**, *103*, 4278.
95. Paterson, I.; Goodman, J. M.; Lister, M. A.; Schumann, R. C.; McClure, C. K.; Norcross, R. D. *Tetrahedron* **1990**, *46*, 4663.
96. Bernardi, A.; Capelli, A. M.; Comotti, A.; et al. *Tetrahedron* **1991**, *47*, 3471.
97. Masamune, S.; Sato, T.; Kim, B. M.; Wollmann, T. A. *J. Am. Chem. Soc.* **1986**, *108*, 8279.
98. Tanimoto, N.; Gerritz, S. W.; Sawabe, A.; Noda, T.; Filla, S. A.; Masamune, S. *Angew. Chem. Int. Ed. Engl.* **1994**, *33*, 673.
99. Corey, E. J.; Imwinkelreid, R.; Pikul, S.; Xiang, Y. B. *J. Am. Chem. Soc.* **1989**, *111*, 5493.
100. Corey, E. J.; Lee, D.-H.; Choi, S. *Tetrahedron Lett.* **1992**, *33*, 6735.
101. Mahrwald, R. *Chem. Rev.* **1999**, *99*, 1095.
102. Nelson, S. G. *Tetrahedron Asymmetry* **1998**, *9*, 357.
103. Gröger, H.; Vogl, E. M.; Shibasaki, M. *Chem. A Eur. J.* **1998**, *4*, 1137.
104. Mukaiyama, T.; Kobayashi, S.; Uchiro, H.; et al. *Chem. Lett.* **1990**, *19*, 129.
105. Mukaiyama, T.; Furuya, M.; Ohtsubo, A.; et al. *Chem. Lett.* **1991**, *20*, 989.
106. Evans, D. A.; Murry, J. A.; Kozlowski, M. C. *J. Am. Chem. Soc.* **1996**, *118*, 5814.
107. Evans, D. A.; Kozlowski, M. C.; Burgey, C. S.; MacMillan, D. W. C. *J. Am. Chem. Soc.* **1997**, *119*, 7893.
108. Nishiyama, H.; Kondo, M.; Nakamura, T.; Itoh, K. *Organometallics* **1991**, *10*, 500.
109. Evans, D. A.; MacMillan, D. W. C.; Campos, K. R. *J. Am. Chem. Soc.* **1997**, *119*, 10859.
110. Sakthivel, K.; Notz, W.; Bui, T.; Barbas, C. F. *J. Am. Chem. Soc.* **2001**, *123*, 5260.
111. Ahrendt, K. A.; Borths, C. J.; MacMillan, D. W. C. *J. Am. Chem. Soc.* **2000**, *122*, 4243.
112. List, B. *Tetrahedron* **2002**, *58*, 5573.
113. Northrup, A. B.; MacMillan, D. W. C. *J. Am. Chem. Soc.* **2002**, *124*, 6798.
114. Trost, B. M.; Brindle, C. S. *Chem. Soc. Rev.* **2010**, *39*, 1600.
115. List, B. *J. Am. Chem. Soc.* **2000**, *122*, 9336.
116. Notz, W.; Tanaka, F.; Watanabe, S.-I.; et al. *J. Org. Chem.* **2003**, *68*, 9624.
117. Zhang, H.; Mifsud, M.; Tanaka, F.; Barbas, C. F., III. *J. Am. Chem. Soc.* **2006**, *128*, 9630.
118. Hoffmann, R. W.; Zeiβ, H.-J. *J. Org. Chem.* **1981**, *46*, 1309.
119. Srebnik, M.; Ramachandran, P. V. *Aldrichimica Acta* **1987**, *20*, 9.
120. Roush, W. R. In *Comprehensive Organic Synthesis*; Trost, B. M., Fleming, I., Eds.; Vol. 2; Pergamon Press: New York, 1991; p. 1.
121. Brown, H. C.; Desai, M. C.; Jadhav, P. K. *J. Org. Chem.* **1982**, *47*, 5065.
122. Brown, H. C.; Jadhav, P. K. *J. Am. Chem. Soc.* **1983**, *105*, 2092.

123. Racherla, U. S.; Brown, H. C. *J. Org. Chem.* **1991**, *56*, 401.
124. Brown, H. C.; Bhat, K. S.; Randad, R. S. *J. Org. Chem.* **1987**, *52*, 319. **1989**, *54*, 1570.
125. Brown, H. C.; Bhat, K. S. *J. Am. Chem. Soc.* **1986**, *108*, 293. **1986**, *108*, 5919.
126. Schlosser, M. *Pure Appl. Chem.* **1988**, *60*, 1627.
127. Roush, W. R.; Palkowitz, A. D.; Ando, K. *J. Am. Chem. Soc.* **1990**, *112*, 6348.
128. Roush, W. R.; Walts, A. E.; Hoong, L. K. *J. Am. Chem. Soc.* **1985**, *107*, 8186.
129. Xu, H.-C.; Chowdhury, S.; Ellman, J. A. *Nat. Protoc.* **2013**, *8*, 2271.
130. Robak, M. T.; Herbage, M. A.; Ellman, J. A. *Chem. Rev.* **2010**, *110*, 3600.
131. Weix, D. J.; Ellman, J. A. *Organic Synth.* **2005**, *82*, 157.
132. Liu, G.; Cogan, D. A.; Owens, T. D.; Tang, T. P.; Ellman, J. A. *J. Org. Chem.* **1999**, *64*, 1278.
133. Liu, G.; Cogan, D. A.; Ellman, J. A. *J. Am. Chem. Soc.* **1997**, *119*, 9913.
134. Cogan, D. A.; Liu, G.; Ellman, J. A. *Tetrahedron* **1999**, *55*, 8883.
135. Bertrand, M. B.; Wolfe, J. P. *Org. Lett.* **2006**, *8*, 2353.
136. Ellman, J. A.; Owens, T. D.; Tang, T. P. *Acc. Chem. Res.* **2002**, *35*, 984.
137. Tanuwidjaja, J.; Peltier, H. M.; Ellman, J. A. *J. Org. Chem.* **2007**, *72*, 626.
138. Fleming, I.; Lawrence, N. J.; Sarkar, A. K.; Thomas, A. P. *J. Chem. Soc., Perkin Trans 1* **1992**, 3303.
139. Paddon-Row, M. N.; Rondan, N. G.; Houk, K. N. *J. Am. Chem. Soc.* **1982**, *104*, 7162.
140. Houk, K. N.; Rondan, N. G.; Wu, Y.-D.; Metz, J. T.; Paddon-Row, M. N. *Tetrahedron* **1984**, *40*, 2257.
141. Still, W. C.; Barrish, J. C. *J. Am. Chem. Soc.* **1983**, *105*, 2487.
142. Paterson, I.; Channon, J. *Tetrahedron Lett.* **1992**, *33*, 797.
143. Mori, K. *Tetrahedron* **1976**, *32*, 1979.
144. Brown, H. C.; Jadhav, P. K. In *Asymmetric Synthesis*; Morrison, J. D., Ed.; Vol. 2; Academic Press: New York, 1983; p. 1.
145. Brown, H. C. *Chemtracts* **1988**, *1*, 77.
146. Brown, H. C.; Schwier, J. R.; Singaram, B. *J. Org. Chem.* **1978**, *43*, 4395.
147. Brown, H. C.; Jadhav, P. K.; Mandal, A. K. *Tetrahedron* **1981**, *37*, 3547.
148. Wong, O. A.; Shi, Y. *Chem. Rev.* **2008**, *108*, 3958.
149. Shi, Y. *Acc. Chem. Res.* **2004**, *37*, 488.
150. Frohn, M.; Shi, Y. *Synthesis* **2000**, *14*, 1979.
151. Murray, R. W.; Singh, M. *Organic Synth.* **1997**, *74*, 91.
152. Wang, Z.-X.; Tu, Y.; Frohn, M.; Zhang, J.-R.; Shi, Y. *J. Am. Chem. Soc.* **1997**, *119*, 11224.
153. Warren, J. D.; Shi, Y. *J. Org. Chem.* **1999**, *64*, 7675.
154. Cao, G.-A.; Wang, Z.-X.; Tu, Y.; Shi, Y. *Tetrahedron Lett.* **1998**, *39*, 4425.
155. Tian, H.; She, X.; Shu, L.; Yu, H.; Shi, Y. *J. Am. Chem. Soc.* **2000**, *122*, 11551.
156. Tian, H.; She, X.; Yu, H.; Shu, L.; Shi, Y. *J. Org. Chem.* **2002**, *67*, 2435.
157. Oppolzer, W.; Mills, R. J.; Pachinger, W.; Stevenson, T. *Helv. Chim. Acta* **1986**, *69*, 1542.
158. Duhamel, L. In *The Chemistry of Amino, Nitroso and Nitro Compounds and Their Derivatives: Supplement F*; Patai, S., Ed.; Wiley: New York, 1982; p. 849 (chapter 20).
159. Fiorini, M.; Giongo, G. M. *J. Mol. Catal.* **1979**, *5*, 303.
160. Alexakis, A.; Sedrani, R.; Mangeney, P.; Normant, J. F. *Tetrahedron Lett.* **1988**, *29*, 4411.
161. Davies, S. G.; Smith, A. D.; Price, P. D. *Tetrahedron Asymmetry* **2005**, *16*, 2833.
162. Davies, S. G.; Fletcher, A. M.; Roberts, P. M.; Thomson, J. E. *Tetrahedron Asymmetry* **2012**, *23*, 1111. **2017**, *28*, 1842.
163. Davies, S. G.; Fletcher, A. M.; Roberts, P. M. *Organic Synth.* **2010**, *87*, 143.
164. Davies, S. G.; Fletcher, A. M.; Roberts, P. M.; Thomson, J. E. *Organic Synth.* **2019**, *96*, 53.
165. de Vries, A. H. M.; Meetsma, A.; Feringa, B. L. *Angew. Chem. Int. Ed. Engl.* **1996**, *35*, 2374.

166. Ohshima, T.; Xu, Y.; Takita, R.; Shimizu, S.; Zhong, D.; Shibasaki, M. *J. Am. Chem. Soc.* **2002**, *124*, 14546.
167. MacMillan, D. W. C. *Nature* **2008**, *455*, 304.
168. Paras, N. A.; MacMillan, D. W. C. *J. Am. Chem. Soc.* **2001**, *123*, 4370.
169. Ouellet, S. G.; Tuttle, J. B.; MacMillan, D. W. C. *J. Am. Chem. Soc.* **2005**, *127*, 32.
170. Tuttle, J. B.; Ouellet, S. G.; MacMillan, D. W. C. *J. Am. Chem. Soc.* **2006**, *128*, 12662.
171. Davis, B. G.; Boyer, V. *Nat. Prod. Rep.* **2001**, *18*, 618.
172. Zhu, L.-M.; Tedford, M. C. *Tetrahedron* **1990**, *46*, 6587.
173. Kobayashi, S.; Kamiyama, K.; Ohno, M. *J. Org. Chem.* **1990**, *55*, 1169.
174. Schneider, M.; Engel, M.; Honicke, P.; Heinemann, G.; Gorisch, H. *Angew. Chem. Int. Ed. Engl.* **1984**, *23*, 67.
175. Faber, K.; Riva, S. *Synthesis* **1992**, 895.
176. Tanaka, M.; Yoshioka, M.; Sakai, K. *J. Chem. Soc. Chem. Commun.* **1992**, 1454.
177. North, M. *Tetrahedron Lett.* **1996**, *37*, 1699.
178. Medson, C.; Smallridge, A. J.; Trewhella, M. A. *Tetrahedron Asymmetry* **1997**, *8*, 1049.
179. Prelog, V. *Pure Appl. Chem.* **1964**, *9*, 119.
180. Gopalan, A. S.; Sih, C. *J. Tetrahedron Lett.* **1984**, *25*, 5235.
181. Cooper, J.; Gallagher, P. T.; Knight, D. W. *J. Chem. Soc. Chem. Commun.* **1988**, 509.
182. Hudlicky, T.; Olivo, H. F.; McKibben, B. *J. Am. Chem. Soc.* **1994**, *116*, 5108.
183. Johnson, C. R.; Ple, P. A.; Adams, J. P. *J. Chem. Soc. Chem. Commun.* **1991**, 1006.
184. Reetz, M. T.; Daligault, F.; Brunner, B.; Hinrichs, H.; Deege, A. *Angew. Chem. Int. Ed. Engl.* **2004**, *43*, 4078.
185. Effenberger, F. *Angew. Chem. Int. Ed. Engl.* **1994**, *33*, 1555.
186. Effenberger, F.; Jager, J. *J. Org. Chem.* **1997**, *62*, 3867.
187. Bednarski, M. D.; Simon, E. S.; Bishofberger, N.; et al. *J. Am. Chem. Soc.* **1989**, *111*, 62.
188. van der Osten, C. H.; Sinskey, A. J.; Barbas, C. F.; Pederson, R. L.; Wang, Y. F.; Wong, C.-H. *J. Am. Chem. Soc.* **1989**, *111*, 3924.
189. Turner, N. J.; Whitesides, G. M. *J. Am. Chem. Soc.* **1989**, *111*, 624.
190. Gijsen, H. J. M.; Qiao, L.; Fitz, W.; Wong, C.-H. *Chem. Rev.* **1996**, *96*, 443.
191. Wong, C.-H. *Pure Appl. Chem.* **1995**, *67*, 1609.
192. Wymer, N.; Toone, E. *J. Curr. Opin. Chem. Biol.* **2000**, *4*, 110.

Chapter 8

# Pericyclic reactions 1: Basic stereochemistry

## 8.1 The Diels–Alder reaction

The Diels–Alder reaction[1–3] between a diene and an alkene dienophile can create *up to* four new stereocentres in the cyclohexene adduct (Fig. 8.1). The new double bond produced is *Z* in the six-membered ring. The double bonds in the diene and dienophile may be *E* or *Z*.

### 8.1.1 Stereospecificity

Consider the Diels–Alder reaction between (*E*,*E*)-1,4-dimethylbutadiene and dimethyl acetylenedicarboxylate (DMAD) shown in Fig. 8.2. The diene has stereochemistry at both double bonds, whereas the dienophile has no stereochemistry. The Diels–Alder reaction is a $[_{\pi}4_s+_{\pi}2_s]$ cycloaddition, being suprafacial on both diene and dienophile (see Section 4.4.1). As shown, the suprafacial addition to the (*E*,*E*) diene gives *cis* stereochemistry of two Me groups in the cycloadduct. Similarly, the reaction of the (*E*,*Z*) diene with DMAD gives the *trans* adduct. Note that the *cis* diastereomer is achiral, whereas the *trans* diastereomer is chiral and racemic.

The supra/supra mechanism thus leads to a specific diastereomeric product from a particular stereochemistry of the reactant diene. The reactions are therefore stereospecific which implies that the relative stereochemistry of substituents at C1 and C4 of the diene is retained in the product. This is known as the '*cis* principle'.

Another case of '*cis* principle' operates in the Diels–Alder reaction with respect to the dienophile. As a result of suprafacial addition to the dienophile, the relative stereochemistry of substituents in the dienophile is retained in the product. Thus, an alkene dienophile with *cis* substituents will form a cycloadduct in which these substituents are also *cis*. Similarly, a *trans* alkene will give a *trans* adduct. These are illustrated with two examples in Fig. 8.3.

Stereochemistry and Organic Reactions. https://doi.org/10.1016/B978-0-12-824092-2.00008-3

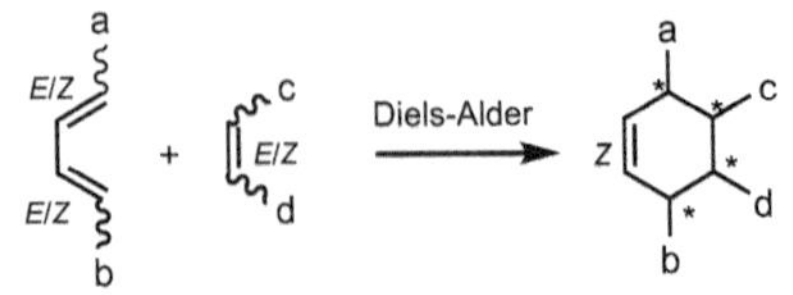


**FIG. 8.1** Schematic representation of the Diels–Alder reaction (* denotes a new stereocentre).

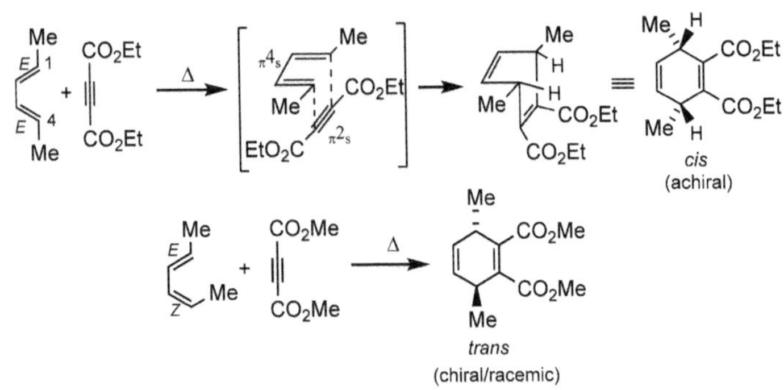


**FIG. 8.2** Stereospecificity of the Diels–Alder reaction with respect to the diene.

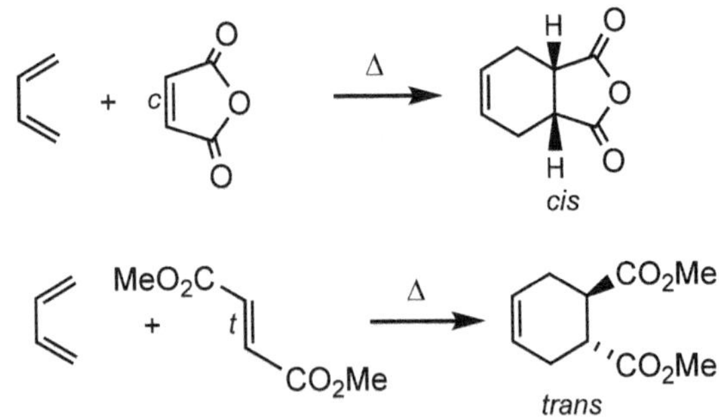


**FIG. 8.3** Stereospecificity of the Diels–Alder reaction with respect to the dienophile.

### 8.1.2 Stereoselectivity

If both diene and dienophile are prochiral, the supra/supra mechanism can lead to two diastereomeric products known as *endo* and *exo* adducts, as shown for a general case in Fig. 8.4. The dienophile can approach the diene in *endo* fashion (when the substituent z sits under the diene) or *exo* fashion (when the substituent z sits away from the diene). Note that 'ortho' regioselectivity is assumed for the reaction.

In general, the Diels–Alder reaction is highly diastereoselective, and the *endo* diastereomer though more hindered and thermodynamically less stable

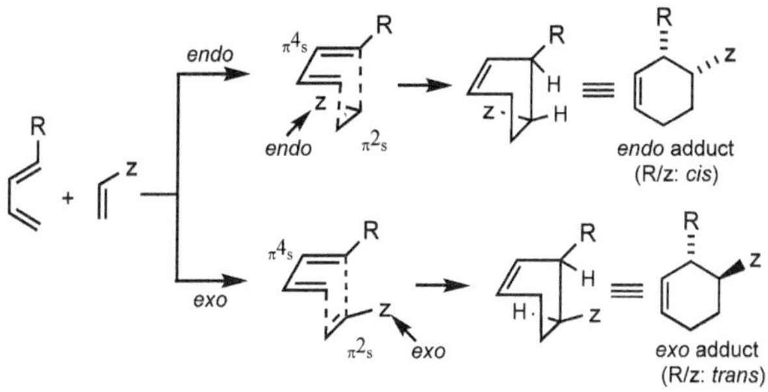


**FIG. 8.4** General scheme for *endo* and *exo* cycloadditions in the Diels–Alder reaction.

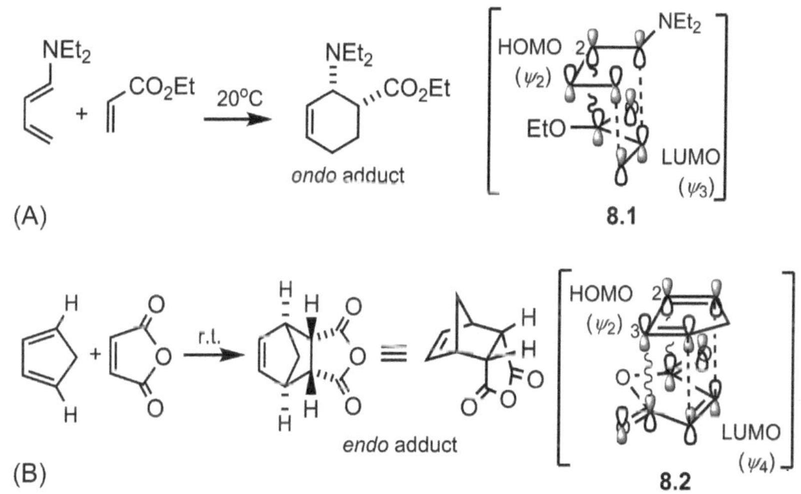


**FIG. 8.5** *endo* stereoselectivity of the Diels–Alder reaction using (A) acyclic and (B) cyclic substrates.

than the *exo* diastereomer, is often the major product under normal kinetic control of the reaction. This is known as the Alder *endo* rule, and is illustrated with acyclic and cyclic substrates in Fig. 8.5.

The preference for the *endo* addition has been rationalized by Woodward and Hoffmann in terms of secondary orbital interactions.[4] In the *endo* approach, the carbonyl π system (z-substituent) of the dienophile is oriented over the residual π orbitals of the diene in the TS **8.1** or **8.2**. Apart from primary bonding interactions (dashed lines), secondary overlap is possible between p orbital on C2/C3 of the diene and p orbital of the carbonyl carbon (wavy lines). The $HOMO_{diene}/LUMO_{dienophile}$ interaction shows that secondary interactions involving lobes of the same sign are bonding and hence stabilize the *endo* TS **8.1** or **8.2** relative to the *exo* TS where they are absent. Note that LUMO of the fully conjugated dienophile π system ($\psi_3$ for ethyl acrylate,

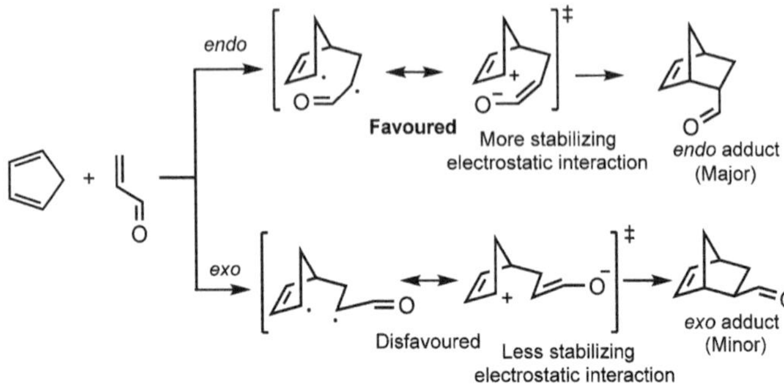


**FIG. 8.6** Diradicaloid TS model for the *endo* selectivity of the Diels–Alder reaction.

and $\psi_4$ for maleic anhydride) is considered to examine the secondary interaction. The Diels–Alder reaction involves a boat TS, and in the case of cyclic diene (Fig. 8.5B), the product is also locked in boat conformation.

It is pertinent to mention that Lewis acids can increase the rate and regioselectivity of the Diels–Alder reaction by lowering the energy of the LUMO of the dienophile. Lewis acids can also enhance the *endo* selectivity by a stronger secondary orbital interaction involving the increased LUMO coefficient on the carbonyl carbon of the dienophile–Lewis acid complex.[5]

The *endo*-selectivity of the Diels–Alder reaction can also be rationalized by the diradicaloid TS model[6] (see Section 4.4.1.2), as shown for the reaction of cyclopentadiene with acrolein in Fig. 8.6. Compared to the *exo* addition, the zwitterionic limiting structure of the *endo* TS is more stabilized by electrostatic interactions because of the shorter distance between the positively and negatively charged centres, thereby leading to the predominant *endo* adduct. The model also predicts that the rate of the reaction and the *endo* selectivity will increase with solvent polarity, as has been observed.[7]

### 8.1.3 Mnemonics for delineating product stereochemistry

The mechanistic basis for the observed stereochemistry of the Diels–Alder adduct has been described above. However, the product stereochemistry can be easily drawn with the help of two mnemonics (Fig. 8.7): mnemonic 1 for

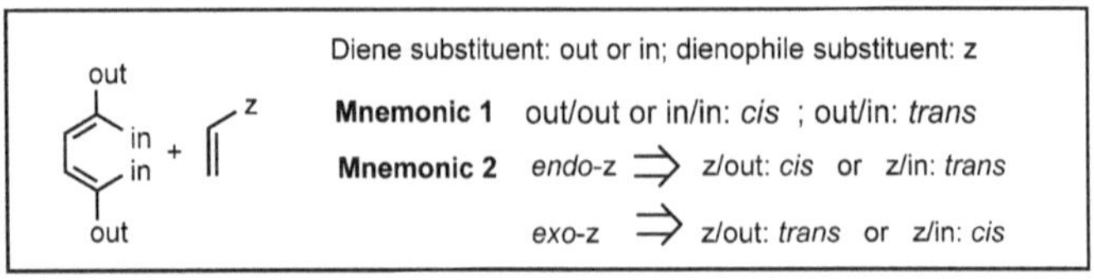


**FIG. 8.7** Mnemonic 1 for delineating stereospecificity with respect to the diene and mnemonic 2 for delineating *endo* or *exo* stereochemistry.

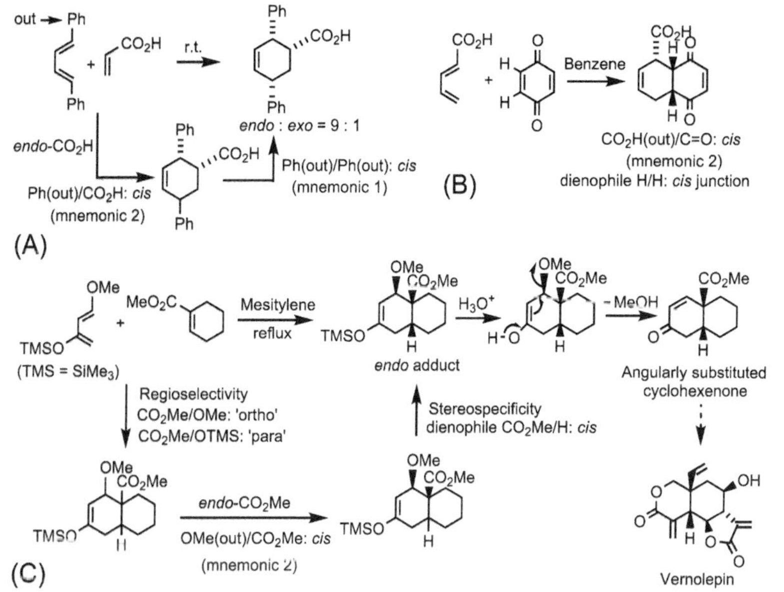


**FIG. 8.8** Three examples (A–C) of the Diels–Alder reaction for delineation of product stereochemistry with the help of mnemonics.

stereospecificity with respect to the diene, and mnemonic 2 for *endo* or *exo* selectivity. The delineation of stereospecificity with respect to the dienophile is straightforward (see Fig. 8.3).

In general, the product stereochemistry is delineated using the sequence: regioselectivity ('ortho/para') → *endo* or *exo* selectivity (mnemonic 2) → stereospecificity (mnemonic 1)/stereospecificity with respect to dienophile. The procedure is illustrated with three examples in Fig. 8.8.

Fig. 8.8A shows the Diels–Alder reaction between (*E*,*E*)-1,4-diphenylbutadiene and acrylic acid which gives predominantly an *endo* adduct at room temperature. As shown, the stereochemistry of the product can be delineated easily using mnemonic 2 and mnemonic 1 successively. The steps showing the use of mnemonics are given only as an aid to understanding; the product stereochemistry can be drawn straightway.

Woodward synthesis of reserpine[8] started with the Diels–Alder reaction of (*E*)-pentadienoic acid with *p*-benzoquinone (Fig. 8.8B). The reaction is stereospecific with respect to the quinone dienophile giving a *cis* ring junction and highly stereoselective forming an exclusive *endo* adduct.

Fig. 8.8C shows the use of Danishefsky diene (1-methoxy-3-trimethylsilyloxy-1,3-butadiene) in the Diels–Alder reaction for the formation of an angularly substituted cyclohexenone intermediate in a synthesis of the sesquiterpene lactone vernolepin.[9] Here the regioselectivity arises from the influence of both electron donating substituents, OMe and $OSiMe_3$: $CO_2Me$ is 'ortho' to OMe and 'para' to $OSiMe_3$. The stereochemistry of the adduct arises from the stereospecificity with respect to the dienophile (giving a *cis* ring junction) and stereoselectivity in favour of an *endo* adduct.

**Problem 8.1** Draw the stereochemistry of the products of the following Diels–Alder reactions:

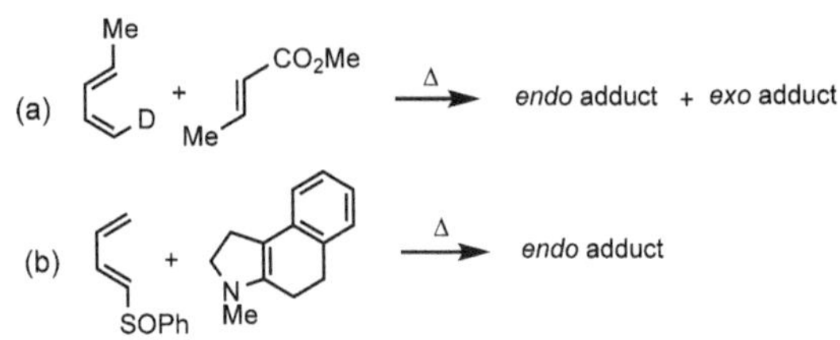


**Problem 8.2** Account for the following observations:

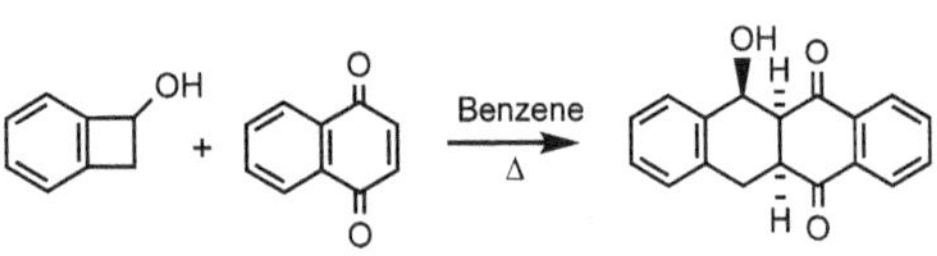


## 8.1.4 Intramolecular Diels–Alder reactions[10–14]

In intramolecular Diels–Alder reaction, the diene and dienophile units are held together in the same molecule by a tether and the intramolecular reaction can provide easy access to polycyclic compounds with high regio- and stereoselectivity. *Two new rings* are created, and the stereochemistry of the reaction refers to the stereochemistry of the ring junction as well as the relative stereochemistry of the substituents. Intramolecular reactions are more facile than intermolecular ones due to favourable entropy factor, and can occur with reasonable speed even in the absence of activating electron withdrawing substituent. Like intermolecular reactions, the formation of *endo* adduct is often favoured. However, the absence of activating electron withdrawing substituent or geometrical constraints can lead to the *exo* product.

Fig. 8.9 shows an intramolecular Diels–Alder reaction in which the dienophile moiety bears an activating carbonyl substituent. The reaction takes place readily and the major product is *endo* adduct with *cis* ring fusion. The *endo* TS **8.3** shows that the tether adopts a chair-like conformation when the ring junction is *cis*, one H is axial (*up*) and the other H is equatorial (*up*). See also the use of mnemonic 2 to delineate the *endo* adduct.

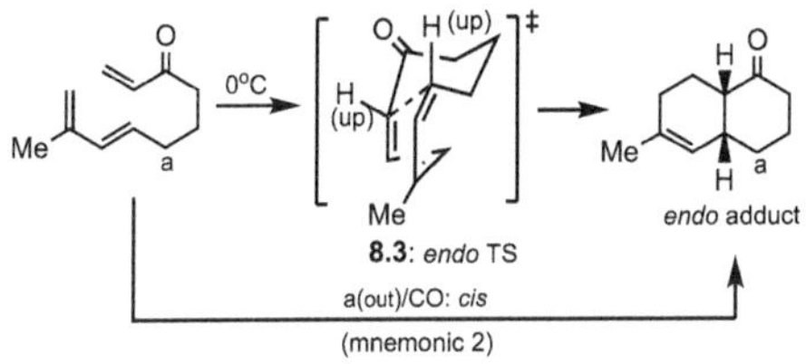


**FIG. 8.9** Formation of *endo* adduct in an intramolecular Diels–Alder reaction.

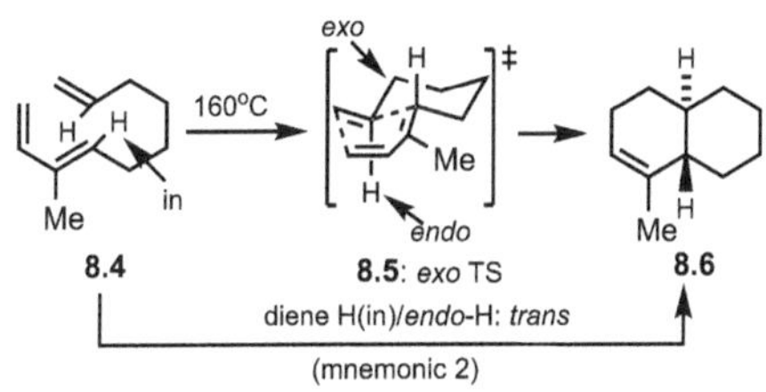


**FIG. 8.10** Formation of *exo* adduct in an intramolecular Diels–Alder reaction.

Intramolecular Diels–Alder reaction requires a much higher temperature if there is no activating carbonyl substituent on the dienophile unit of the substrate **8.4** (Fig. 8.10). The absence of secondary orbital interaction then favours the sterically favourable *exo* TS **8.5** (the connecting chain to the dienophile is *exo*) to give the *exo* adduct **8.6** with *trans* ring junction (H/H: *up/down* in the TS). See also the use of mnemonic 2 to draw the *exo* adduct.

**Problem 8.3** Predict the product of the following reaction:

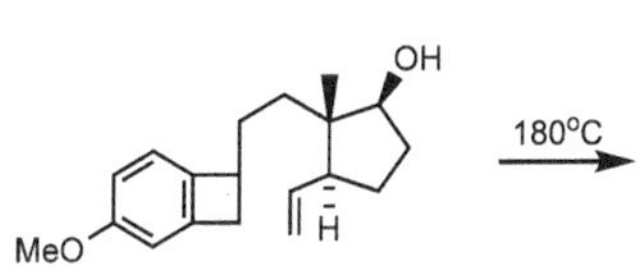


## 8.2 1,3-Dipolar cycloaddition reactions[15–18]

1,3-Dipolar cycloaddition (see Section 4.4.2) of a 1,3-dipole with an alkene or alkyne dipolarophile can produce a five-membered heterocyclic ring that may possess a *maximum* of four carbon stereocentres since the central atom of the dipole is always a heteroatom.

### 8.2.1 Stereospecificity

As a result of suprafacial addition, the relative stereochemistry of substituents in the alkene dipolarophile is retained in the product. Thus, an alkene with *cis*

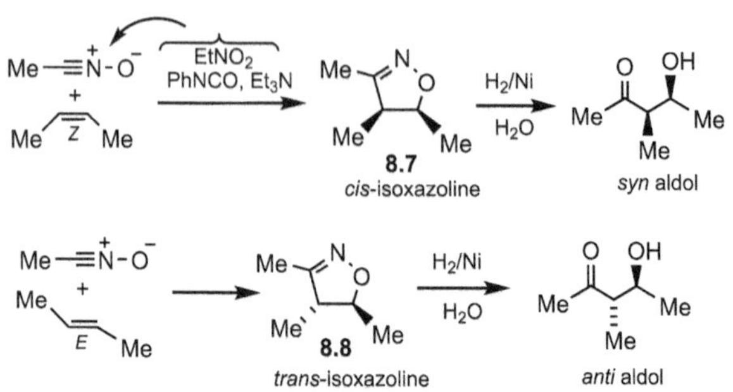

FIG. 8.11 Stereospecificity of nitrile oxide cycloadditions with (*Z*)- and (*E*)-2-butenes, and conversion of isoxazoline adducts into *syn* and *anti* aldols.

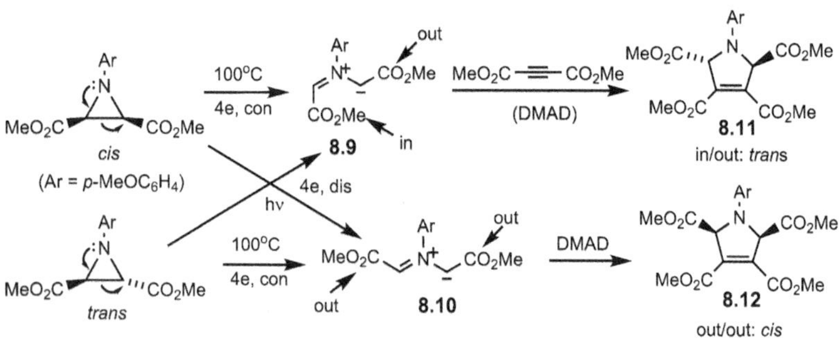

FIG. 8.12 Stereospecificity of 1,3-dipolar cycloadditions with respect to azomethine ylide dipoles generated from suitably substituted aziridines by electrocyclic ring opening.

substituents gives a *cis* adduct, and a *trans* alkene produces a *trans* product. Fig. 8.11 shows nitrile oxide cycloadditions with (*Z*)- and (*E*)-2-butenes.[19] The isoxazoline adducts **8.7** and **8.8** are sometimes not the final target and can be converted into diastereomeric *syn* and *anti* aldols, respectively, on hydrolytic reduction with Raney nickel in aqueous acid.

Stereospecificity is also observed with respect to 1,3-dipoles which can create two new stereocentres. Fig. 8.12 shows that electrocyclic ring opening of *cis*- and *trans*-disubstituted aziridines generates the diastereomeric azomethine ylides **8.9** and **8.10** in situ, which are trapped by an alkyne dipolarophile dimethyl acetylenedicarboxylate (DMAD).[20] The electron withdrawing $CO_2Me$ group stabilizes the negative charge in the azomethine ylide, thereby facilitates its formation. Under thermal condition, *cis* isomer of aziridine gives the ylide **8.9**, whereas the *trans* isomer gives **8.10**. On photolysis, *cis* aziridine gives **8.10**, and *trans* aziridine gives **8.9**. As a result of suprafacial addition to the ylide dipole, the relative stereochemistry of substituents in the ylide is retained in the cycloadduct. Thus **8.9** gives a *trans* $\Delta^3$-pyrroline product **8.11**, and **8.10** yields a *cis* pyrroline product **8.12**. See also the use of mnemonic 1 in Fig. 8.7 by replacing diene with dipole.

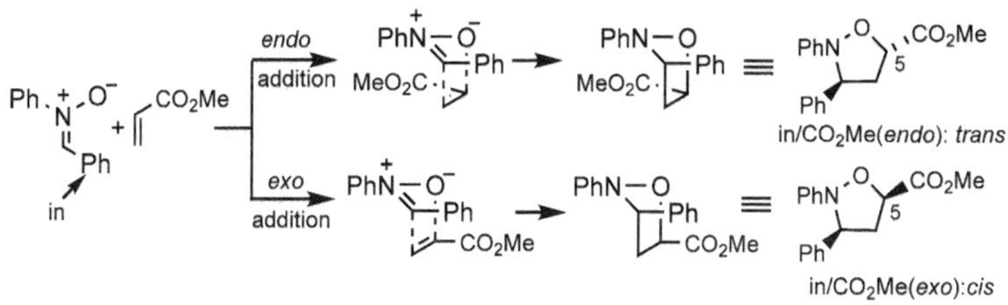


**FIG. 8.13** Formation of *endo* and *exo* adducts in a nitrone cycloaddition with methyl acrylate.

**Problem 8.4** Explain the following observation:

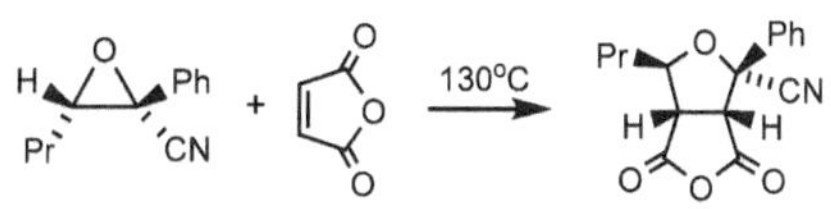


### 8.2.2 Stereoselectivity

The stereoselectivity of 1,3-dipolar cycloadditions is often low, although some reactions can give mainly the *endo* adduct while some others are *exo* selective. Both electronic and steric factors influence the selectivity. Fig. 8.13 shows a nitrone cycloaddition with methyl acrylate which gives a mixture of *endo* and *exo* adducts. The regioselectivity is in favour of 5-substituted isoxazolidine (see Fig. 4.17B). The *endo* addition gives a *trans* adduct while the *exo* addition leads to a *cis* adduct. The stereoselectivity is found to be low, the *exo* (*cis*) isomer is somewhat favoured (de 14%). See also the use of mnemonic 2 in Fig. 8.7 by replacing diene with dipole.

The cycloaddition of an azomethine ylide (generated from an imine by the addition of a metal salt and a base) to methyl acrylate is *endo* selective (Fig. 8.14A),[21,22] whereas the reaction between a cyclic nitrone and propene gives mainly the sterically favoured *exo* product (Fig. 8.14B).[23] Reductive N—O cleavage of the *exo* adduct gives the alkaloid sedridine.

### 8.2.3 Intramolecular 1,3-dipolar cycloadditions

Intramolecular cycloadditions of nitrones are quite useful in synthesis.[24,25] Fig. 8.15 shows that the alkenylnitrone **8.13** (generated by the condensation of (*E*)-5-heptenal with *N*-methylhydroxylamine) undergoes intramolecular cycloaddition to give a *cis* bicyclic isoxazolidine **8.15**. Here the intramolecularity of the reaction determines the stereochemical outcome. The alkene unit is tethered to the carbon atom of the nitrone dipole leading to an *endo*-Me TS **8.14**. Thus two H's at the ring junction and Me are all *cis* in the product **8.15**.

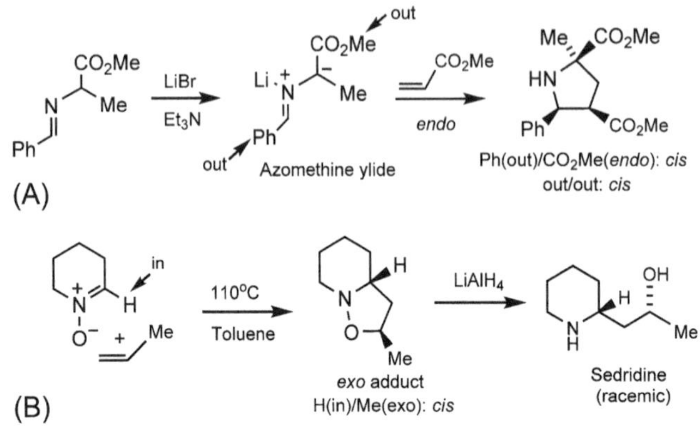


**FIG. 8.14** (A) *endo* selectivity in azomethine ylide cycloaddition; (B) *exo* selectivity of cycloaddition of a cyclic nitrone in a synthesis of sedridine.

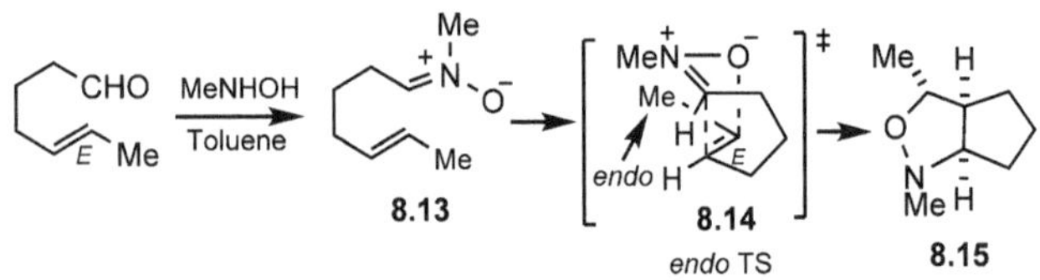


**FIG. 8.15** Intramolecular 1,3-dipolar cycloaddition of an unsaturated nitrone.

## 8.3 Ketene cycloadditions[26,27]

The [2+2] cycloaddition of a ketene with an alkene takes place under thermal condition and gives a cyclobutanone product. Many ketene cycloadditions are pericyclic and occur by supra/antara mechanism when the two reactants approach in orthogonal fashion (see Section 4.4.3).

### 8.3.1 Stereospecificity

The suprafacial addition of ketene to alkenes leads to the stereospecificity of ketene cycloaddition. The relative stereochemistry of alkene substituents is retained in the product. For example, Fig. 8.16A shows the cycloadditions of reactive dichloroketene (generated in situ by dehydrohalogenation of dichloroacetyl chloride with triethylamine) to (*Z*)- and (*E*)-cyclooctenes.[28] The (*Z*) alkene leads to a *cis* ring junction while the (*E*) alkene produces a *trans* ring fusion. Reductive dechlorination of the products with Zn/AcOH allows access to cyclobutanones, which are the products of addition of ketene itself.[29]

Fig. 8.16B shows another example of stereospecific cycloadditions when dimethylketene reacts with the (*Z*)- and (*E*)-enol ether to give predominantly the *cis*- and *trans*-cyclobutanone, respectively.[30] Note that the reaction is also regioselective (see Fig. 4.18B).

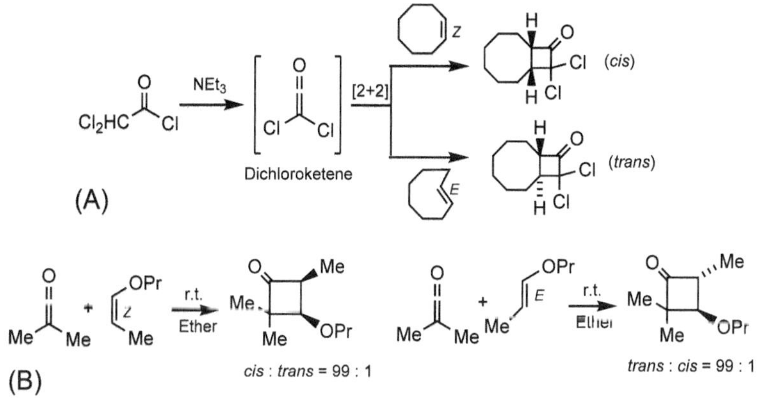


**FIG. 8.16** Stereospecificity of ketene cycloadditions with (A) (*Z*)- and (*E*)-cyclooctenes, and (B) (*Z*)- and (*E*)-enol ethers.

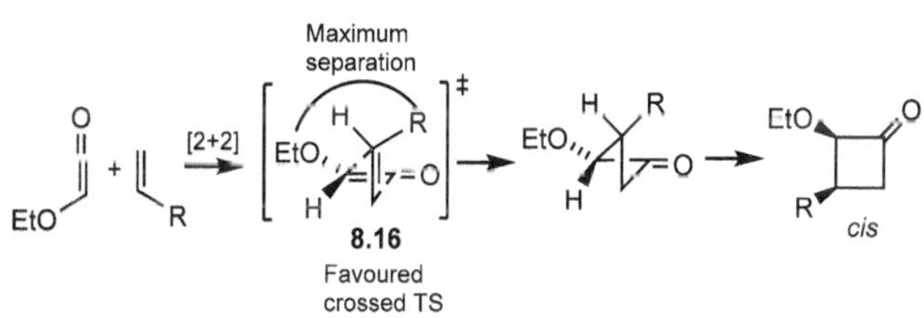


**FIG. 8.17** Stereoselectivity of the ketene cycloaddition.

## 8.3.2 Stereoselectivity

The cycloaddition of an unsymmetrical ketene with an unsymmetrical alkene can be stereoselective when the predominant product has bulky substituents *cis* to each other.[31] For example, the cycloaddition of ethoxyketene with a monosubstituted alkene produces the *cis* cyclobutanone (Fig. 8.17). The reaction is also regioselective (R being vicinal to OEt). The stereoselectivity results from the favoured crossed TS **8.16** in which OEt and R are maximally separated. As the twisted cyclobutanone ring moves to nearly planar form, R and OEt end up on the same face of the ring (use a model to see this). The reaction is under kinetic control and gives thermodynamically less stable *cis* product.

**Problem 8.5** Predict the stereochemistry of the product of the following ketene cycloaddition:

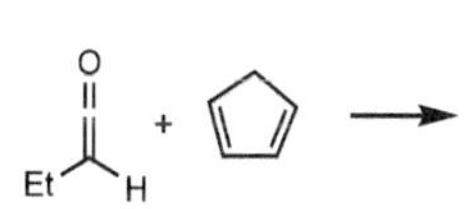

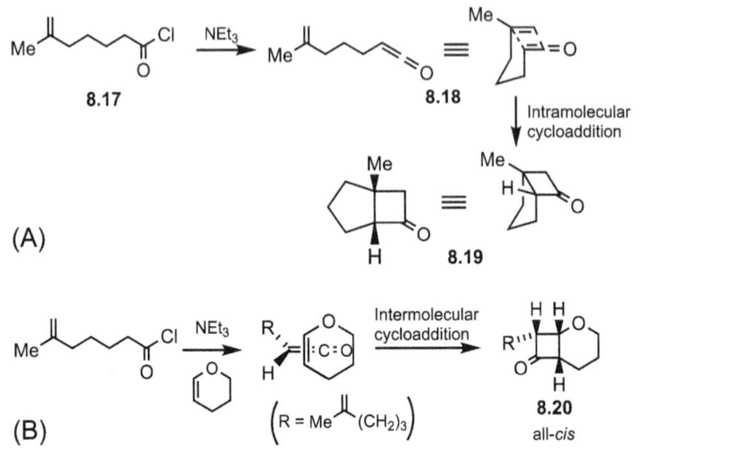


FIG. 8.18 (A) Intramolecular ketene cycloaddition versus (B) intermolecular ketene cycloaddition in the presence of external reactive alkene.

### 8.3.3 Intramolecular ketene cycloadditions

Intramolecular ketene cycloadditions[32] can occur when the ketene and alkene units are tethered in the same molecule. For example, treatment of the acid chloride **8.17** with base ($NEt_3$) leads to intramolecular cycloaddition of the resulting chloroketene **8.18** to produce a *cis* bicyclic ketone **8.19** (Fig. 8.18A).[33] However, in the presence of an external reactive alkene (an enol ether), the ketene cycloaddition takes the intermolecular route (Fig. 8.18B).[33] The intermolecular reaction is stereospecific with respect to the alkene and also stereoselective to give the all-*cis* product **8.20** (the large alkyl group and the six-membered ring end up on the same face of the cyclobutanone ring).

## 8.4 The Wittig reaction

The Wittig reaction creates a new carbon–carbon double bond from the reaction of aldehyde or ketone with a phosphonium ylide (alkylidene phosphorane).[34–36] The ylides are usually derived by action of a moderately strong base on phosphonium salts which are readily obtained from alkyl halides and triphenylphosphine.[37] The reaction proceeds through concerted [2+2] cycloaddition to form a four-membered oxaphosphetane intermediate which then fragments by retrocycloaddition to give an alkene and phosphine oxide. The driving force of the reaction is provided by the formation of the very strong P=O bond.

The Wittig reaction is stereoselective: *with nonstabilized ylides, the reaction is (Z)-selective, whereas with stabilized ylides, the reaction is (E)-selective.* Fig. 8.19A shows the Wittig reaction between a nonstabilized ylide and an aldehyde which gives predominantly the (Z)-alkene. The reaction proceeds via the favoured *cis* oxaphosphetane intermediate followed by stereospecific syn

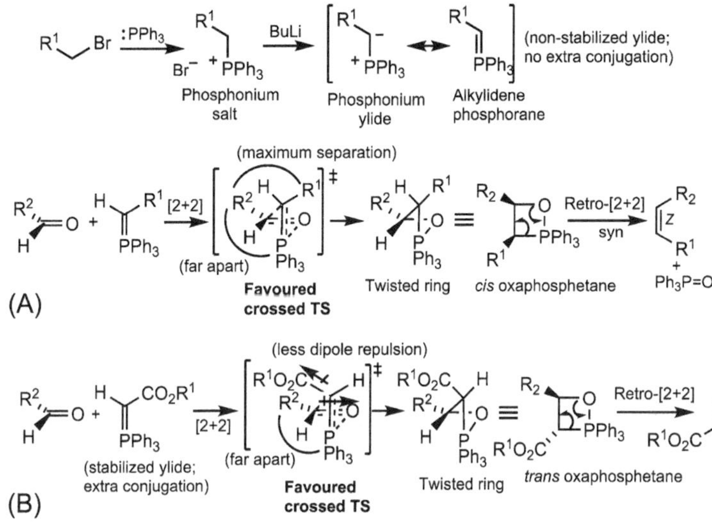


**FIG. 8.19** Stereoselectivity of the Wittig reaction: (A) with nonstabilized ylides; (B) with stabilized ylides.

elimination by retro-cycloaddition to give the (*Z*) alkene product. The preference for the formation of *cis* oxaphosphetane has been attributed to the minimized steric interactions in the crossed TS involving orthogonally aligned ylide and aldehyde (cf. also Fig. 8.17). As the twisted ring moves to nearly planar form, the substituents end up on the same face of the ring to give thermodynamically less stable *cis* diastereomer under kinetic control of the reaction.

The Wittig reaction with a resonance-stabilized ylide bearing an electron withdrawing carbonyl substituent is shown in Fig. 8.19B. The product is predominantly the (*E*)-alkene, which must arise from *trans* oxaphosphetane intermediate. There are several possible explanations for the (*E*) stereoselectivity. It used to be thought that with a stabilized ylide, oxaphosphetane formation is reversible and the equilibrium favours the more stable *trans* diastereomer under thermodynamic control. It now seems likely that (*E*)-selectivity also arises from kinetic control when *trans* oxaphosphetane is favoured via the more stable crossed TS in which the dipole–dipole repulsion is minimized (the carbonyl oxygen of the aldehyde and the electronegative substituent on the ylide are far apart) (Fig. 8.19B).[38]

**Problem 8.6** An insect pheromone bombykol has been synthesized using two successive Wittig reactions according to the following scheme. Give the structure of bombykol.

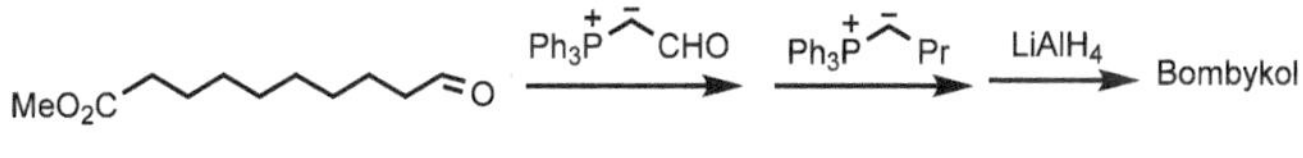

## 8.5 Carbene cycloadditions[39–42]

A singlet carbene allows concerted addition to an alkene to form a cyclopropane. The symmetry-allowed singlet carbene addition is a nonlinear cheletropic process (see Section 4.4.4). Simple carbenes such as dihalocarbenes are normally generated in situ as the singlet. Carbenes are also associated with a metal called carbenoids, for example Simmons–Smith reagent (a zinc carbenoid).

### 8.5.1 Stereospecificity

The cycloadditions of singlet dihalocarbenes to alkenes are stereospecific, arising from suprafacial addition to the alkene. For example, (*Z*)-2-butene gives the *cis* cyclopropane whereas (*E*)-2-butene yields the *trans* cyclopropane (Fig. 8.20).

A popular cyclopropanation reaction is the Simmons–Smith reaction of alkenes.[43] A variety of cyclopropane derivatives can be synthesized using this method since many functional groups remain unaffected during the reaction.[44–46] Fig. 8.21 shows the Simmons–Smith reaction of methyl oleate in a synthesis of dihydrosterculic acid. The Simmons–Smith reagent is a zinc carbenoid (derived from $CH_2I_2$ and Zn—Cu couple or $Et_2Zn$), which reacts with the alkene via a butterfly TS to give the cyclopropane product. The reaction occurs by *cis* addition to the alkene and is stereospecific.

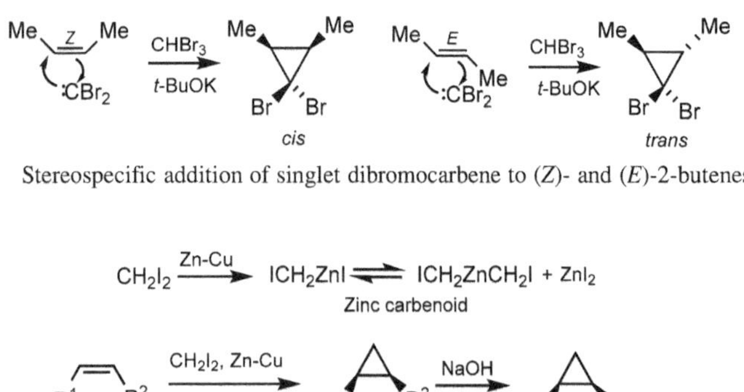


**FIG. 8.20** Stereospecific addition of singlet dibromocarbene to (*Z*)- and (*E*)-2-butenes.

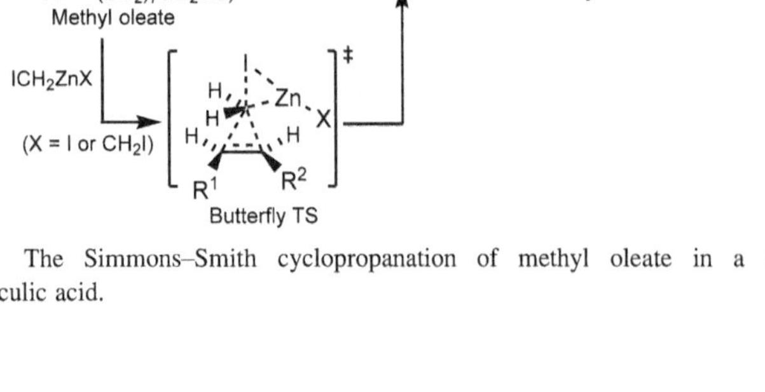


**FIG. 8.21** The Simmons–Smith cyclopropanation of methyl oleate in a synthesis of dihydrosterculic acid.

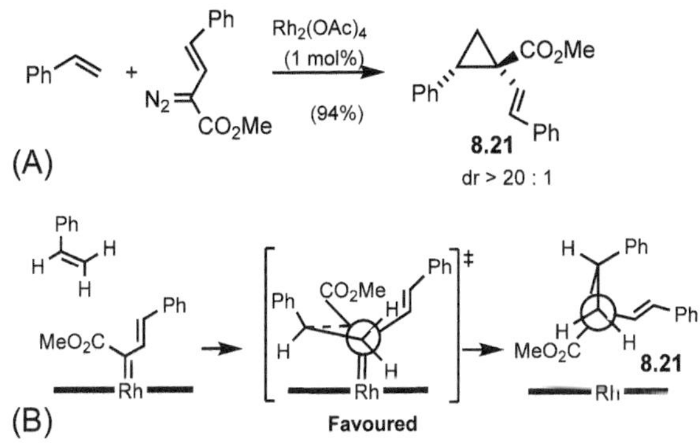


FIG. 8.22 (A) Stereoselective cyclopropanation using a diazo-mediated rhodium carbenoid; (B) Model of diastereoselectivity.

### 8.5.2 Stereoselectivity

Metal carbenoids [$R^1R^2C{=}ML_n$] can be successfully used to effect stereoselective cyclopropanation of alkenes.[47,48] Metal carbenoids are structurally related to singlet carbenes and have similar reactivity. These are generated by reacting salts of transition metals (e.g. Rh, Cu, Pd etc.) with the carbenoid precursors, typically diazo compounds. Fig. 8.22A illustrates a stereoselective cyclopropanation with a metal carbenoid (derived from rhodium-catalysed decomposition of an electron-deficient diazocarbonyl compound) to give the product **8.21**.[49] The reaction exhibits high diastereoselectivity (dr > 20:1), which has been rationalized in terms of a model shown in Fig. 8.22B.

### 8.5.3 Intramolecular carbene cycloadditions

The intramolecular carbene cycloaddition gives access to bicyclic systems in which the cyclopropane ring is commonly fused with five- or six-membered ring. Fig. 8.23A depicts intramolecular cyclopropanation of diastereomeric diazocarbonyl compounds in the presence of copper salt to give 5–3 bicyclic products. The (*E*)-isomer gives the *exo* product whereas the (*Z*)-isomer gives the *endo* product. The use of intramolecular cyclopropanation for the preparation of a 6–3 bicyclic intermediate in a synthesis of sirenin, the sperm attractant of a female water mould, is shown in Fig. 8.23B.[50]

## 8.6 The ene reaction[51–53]

The ene reaction involves the transfer of an allylic hydrogen atom from an alkene (the ene component) to an activated multiple bond (the enophile) (see Section 4.4.5). The all-carbon ene reactions require a high temperature; however, many ene reactions are catalysed by Lewis acids, and the catalysed reactions proceed under relatively mild conditions.[54]

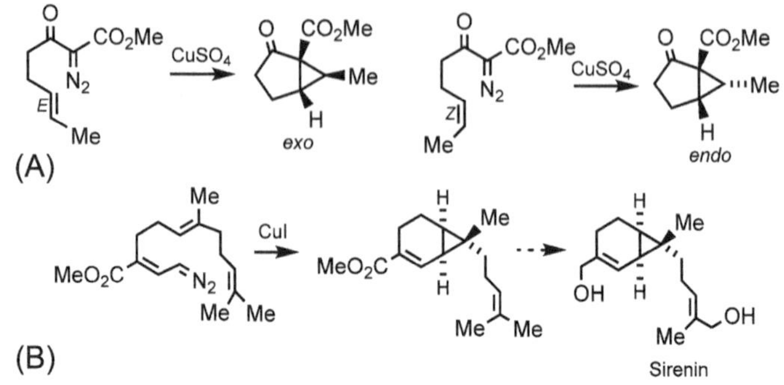


**FIG. 8.23** Formation of (A) 5–3 bicyclic products and (B) a 6–3 bicyclic intermediate in a synthesis of sirenin by intramolecular cyclopropanation reactions.

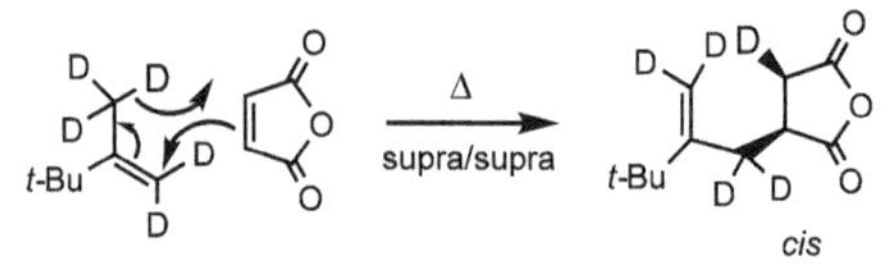


**FIG. 8.24** Stereospecificity of the ene reaction.

## 8.6.1 Stereospecificity

In the ene reaction, both the ene and enophile components are suprafacial (see Fig. 4.20A). The new C—H and C—C bonds are formed on the same face of the enophile which gives stereospecificity of the reaction.[55,56] This is illustrated with an example in Fig. 8.24. The new C—D and C—C bonds are formed on the same face of the enophile (maleic anhydride) leading to the *cis* product.

**Problem 8.7** Predict the product of the following reaction:

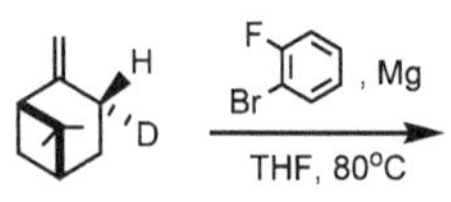


## 8.6.2 Stereoselectivity

The ene reaction proceeds through an envelope TS (see Fig. 4.20A), and can lead to a mixture of diastereomeric *endo* and *exo* adducts. With an electron withdrawing carbonyl substituent on the enophile, the ene reaction is highly *endo*-selective,[57] like the Diels–Alder reaction. Under Lewis acid catalysis, the ene reaction shows high regioselectivity and improved stereoselectivity.[54] Fig. 8.25 illustrates *endo*-selectivity of a Lewis acid-catalysed ene reaction. The reaction produces two new stereocentres and the predominant product arises from the preferred *endo*-$CO_2Me$ TS.

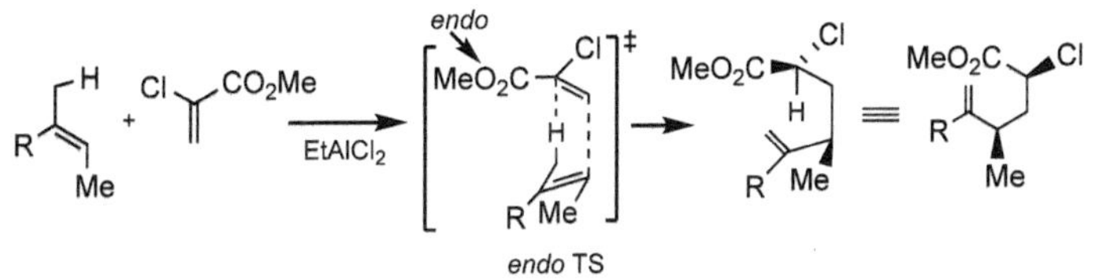


**FIG. 8.25** Stereoselectivity of a Lewis acid-catalysed ene reaction.

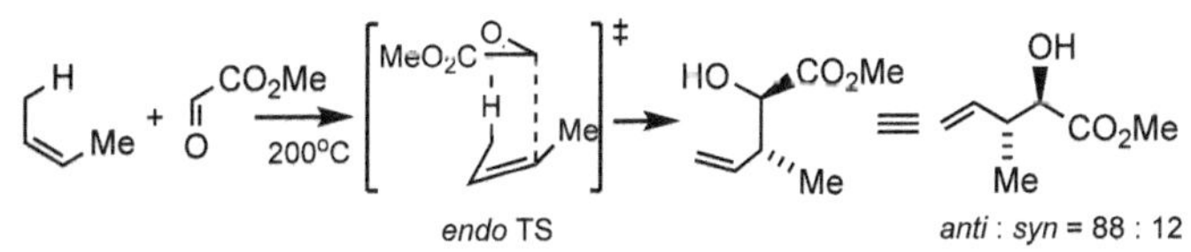


**FIG. 8.26** Diastereoselectivity of a carbonyl ene reaction.

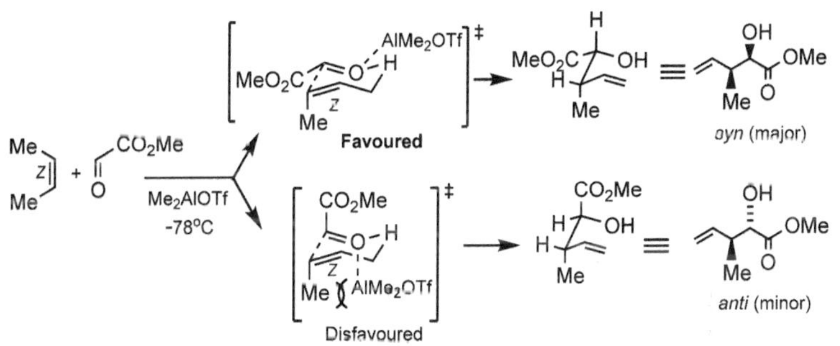


**FIG. 8.27** Diastereoselectivity in Lewis acid ($Me_2AlOTf$)-catalysed carbonyl ene reaction.

The carbonyl ene reaction[53,58] is more facile than all-carbon ene reaction (since the carbonyl enophile has a low energy LUMO) and gives high diastereoselectivity. For example, the reaction between (*Z*)-2-butene and methyl glyoxylate gives the *anti* diastereomer as the major product via the preferred *endo* TS (Fig. 8.26).[59]

If the same reaction is carried out in the presence of a Lewis acid ($Me_2AlOTf$), a reversal of stereoselectivity occurs producing the *syn* isomer as the major product (*syn*:*anti* = 91:9) (Fig. 8.27). The Lewis acid complexes with the carbonyl enophile as a monodentate ligand and the stereochemical outcome can be rationalized on the basis of a chair-like TS.[57,60] Note that Me is forced to be axial due to (*Z*) double bond. The favoured chair TS leads to the *syn* product.

### 8.6.3 Intramolecular ene reactions[52,61,62]

In intramolecular ene reactions, the geometrical constraints imposed by the tether often lead to high diastereoselectivity. An example is shown in

Fig. 8.28.[63] The reaction proceeds through *endo*-TS governed by the short tether to give a *cis* cyclopentane. The TS is called *endo* with respect to the alkyl chain attached to the enophile. See also the use of mnemonic 2 in Fig. 8.7 by replacing diene with ene component.

Fig. 8.29 depicts an intramolecular Lewis acid-catalysed carbonyl ene reaction as a key step in the synthesis of (−)-menthol from (*R*)-citronellal.[64] The stereochemistry of the adduct results from an all-chair bicyclic TS (cf. *trans*-decalin) with Me in the less hindered equatorial position.

In the Conia reaction, an enol is employed as a hetero-ene component. An intramolecular Conia reaction for the synthesis of camphor from dihydrocarvone is depicted in Fig. 8.30.[65] The dihydrocarvone tautomerizes to enol form which undergoes an intramolecular ene reaction via boat conformation to give camphor.

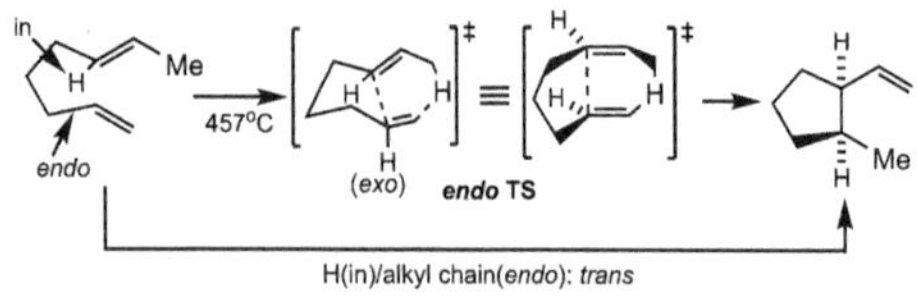


**FIG. 8.28** Diastereoselectivity of an intramolecular ene reaction.

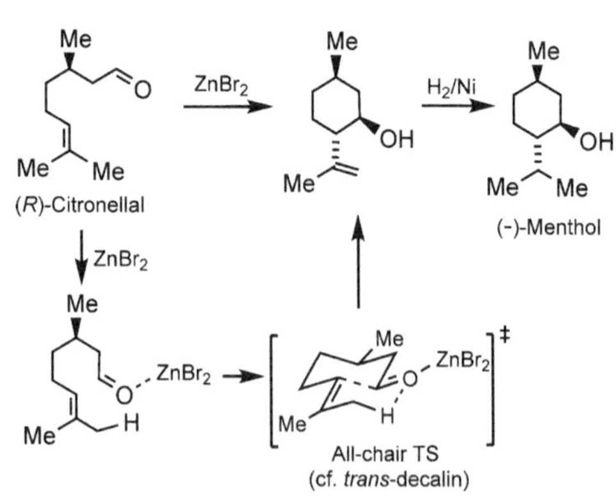


**FIG. 8.29** Intramolecular carbonyl ene reaction used in a synthesis of (−)-menthol.

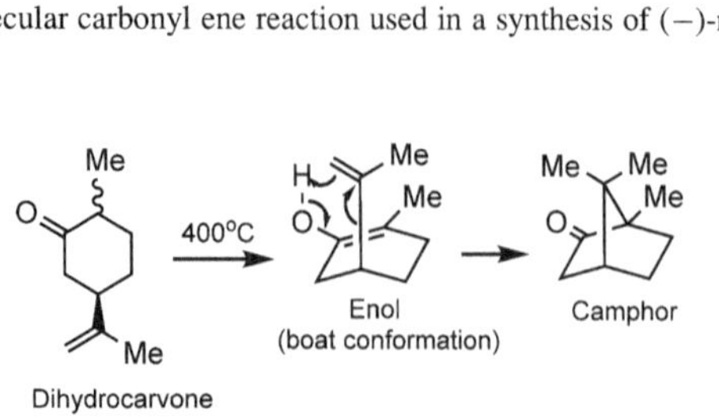


**FIG. 8.30** Synthesis of camphor from dihydrocarvone via an intramolecular Conia reaction.

**Problem 8.8** Give the structure of the intermediate **X** in a synthesis of the sesquiterpene capnellene:

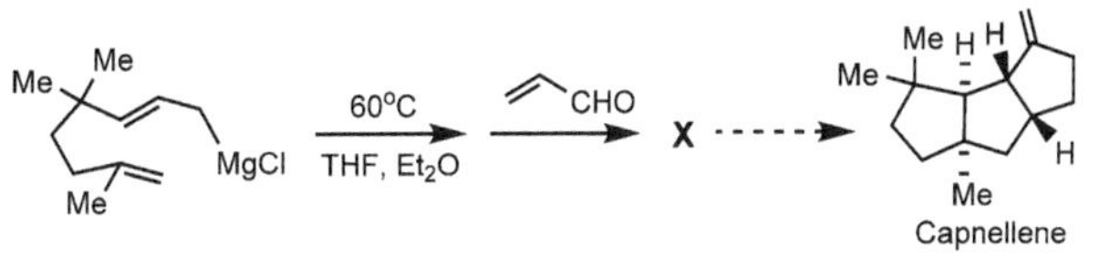


## 8.7 Retro-ene reactions: Pyrolytic *syn* eliminations

The most common retro-ene reactions (see Section 4.4.5.1) are pyrolytic eliminations[66] which proceed through cyclic TS and are stereospecifically *syn* (the H atom and the leaving group depart from the same side of the incipient double bond). These cycloeliminations are common when the substrate chain contains heteroatom(s), as shown in Fig. 8.31 for pyrolysis of esters and xanthates. The ester pyrolysis requires rather high temperatures (300–500°C), whereas the pyrolysis of xanthate (Chugaev reaction) can be performed under relatively mild conditions at temperatures of 150–250°C.[67] Fig. 8.31A shows that pyrolysis of *erythro*-3-deuterio-2-butyl acetate gives mainly a mixture of (*Z*)-butene (undeuterated) and (*E*)-butene (deuterated), each arising from the *syn* elimination.[68] The deuterated (*E*)-butene is the major product arising from the lower energy TS in which the two Me groups are far apart. Similarly, the *threo* diastereomer will give (*Z*)-butene containing deuterium and undeuterated (*E*)-butene (verify). An example of Chugaev reaction is shown in Fig. 8.31B. The xanthates are prepared from the corresponding alcohols with NaH, $CS_2$ and then MeI. Here the elimination of two alternative β hydrogens yields two products. The energy gained in the formation of stronger C=O bond at the expense of weaker C=S bond makes the Chugaev reaction more favourable requiring a lower temperature.

The pyrolytic eliminations using 1,2-dipolar substrates such as tertiary amine oxides, sulphoxides and selenoxides are more favourable (see Section 4.4.5.1) and proceed under milder conditions. Fig. 8.32A shows the Cope elimination[69]

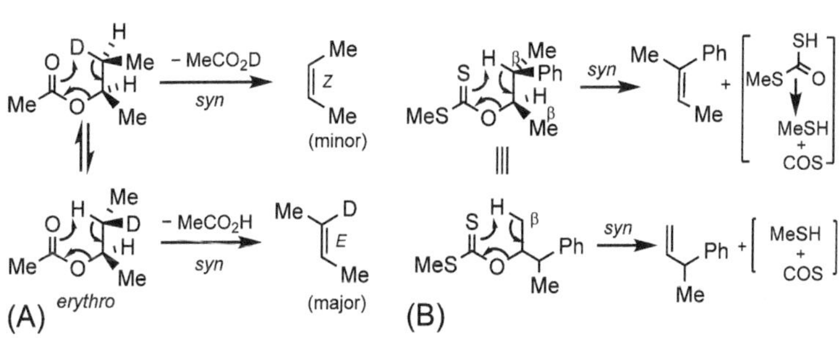


**FIG. 8.31** (A) Pyrolysis of an ester and (B) pyrolysis of a xanthate (Chugaev reaction).

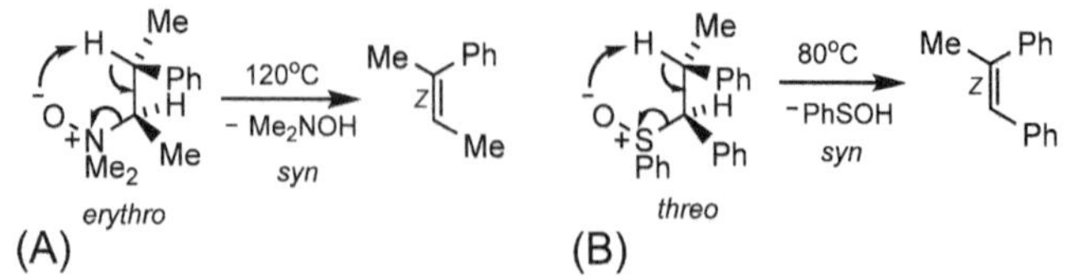


FIG. 8.32 (A) Cope elimination of an amine oxide and (B) 1,2-cycloelimination of a sulphoxide.

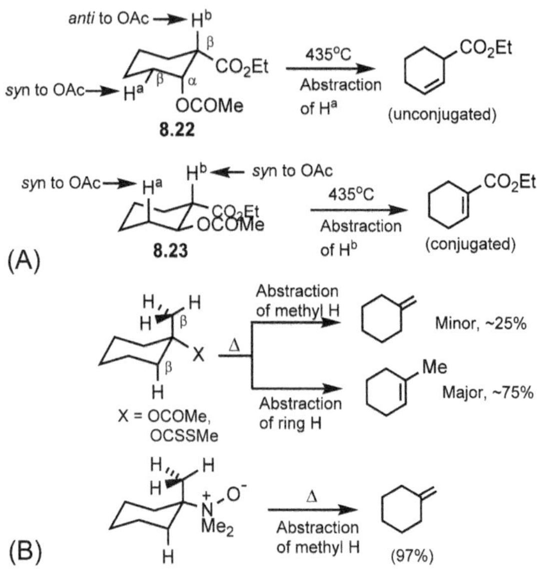


FIG. 8.33 (A) *syn* eliminations in the pyrolysis of diastereomeric *cis*- and *trans*-acetates; (B) Difference in *syn* eliminations in the pyrolysis of esters (or xanthates) and the Cope elimination.

of an *erythro* amine oxide (prepared by oxidation of the corresponding tertiary amine with $H_2O_2$ or *m*CPBA) at 120°C. The products are a (*Z*)-alkene and a hydroxylamine, arising from *syn* elimination. The pyrolysis of a sulphoxide can take place at still lower temperature (80°C) and gives the (*Z*) alkene from a *threo* isomer (Fig. 8.32B).

With cyclic substrates, there are some restrictions imposed by the ring conformation, conformation of the leaving groups and the necessity to form the cyclic TS. Fig. 8.33A shows that pyrolysis of the *cis* acetate **8.22** in which the leaving group (OCOMe) is axial, gives less stable unconjugated alkene instead of conjugated one because only abstraction of equatorial β-H (labelled $H^a$) *syn* to OAc can take place. The axial β-H (labelled $H^b$) in **8.22** is *anti* to OAc and not abstracted. In contrast, the *trans* acetate **8.23** gives the conjugated alkene. Here both axial H's ($H^a$, $H^b$) are *syn* to OAc but abstraction of $H^b$ is preferred since it gives the more stable alkene.

An important difference in the *syn* elimination pathway in the pyrolysis of esters (or xanthates) and the Cope elimination, using 1-methylcyclohexyl derivatives, is illustrated in Fig. 8.33B. Pyrolysis of the ester (or xanthate) leads to the preferential abstraction of ring hydrogen to give the more substituted 1-methylcyclohexene as a major product (~75%). The minor product (~25%) is the less substituted methylenecyclohexane resulting from the abstraction of a hydrogen atom of the methyl group. In contrast, the Cope elimination gives methylenecyclohexane almost exclusively. Since the five-membered cyclic TS is less flexible, the abstraction of ring hydrogen is more constrained; and the preferred option is to abstract a methyl hydrogen in the flexible geometry. But with more flexible six-membered cyclic TS, the abstraction of β-H from the ring as well as the methyl group is possible.

Fig. 8.34 shows that the Cope elimination of menthylamine oxide **8.24** gives a mixture of 2-menthene and 3-menthene, whereas heating neomenthylamine oxide **8.25** gives only 2-menthene. The results of these *syn* eliminations are in contrast with antiperiplanar E2 eliminations of menthyl and neomenthyl chlorides (cf. Fig. 6.17).

The *syn* eliminations of selenoxides can occur under milder conditions (room temperature or below) than sulphoxides because of the longer and weaker C—Se bond. The selenoxide elimination is a useful method for the synthesis of α,β-unsaturated carbonyl compounds, as shown in Fig. 8.35.[70]

**Problem 8.9** Predict **A** and **B** in the following scheme:

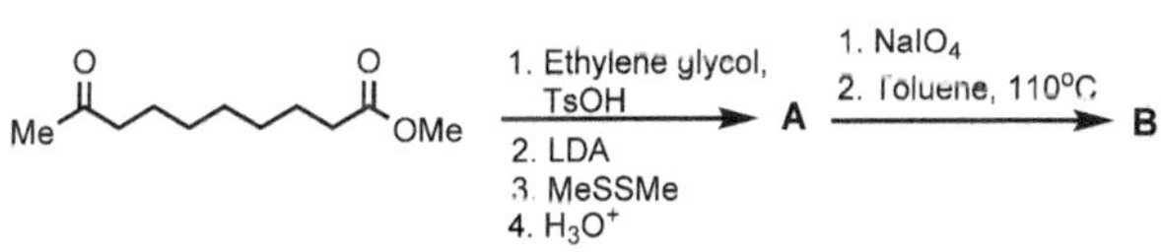


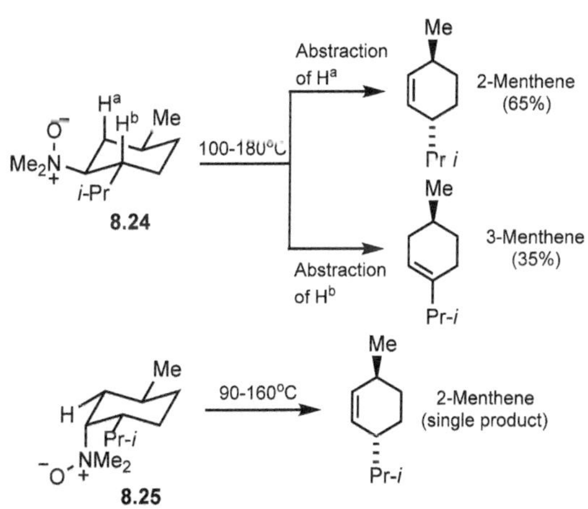


**FIG. 8.34** *syn* eliminations of menthylamine oxide **8.24** and neomenthylamine oxide **8.25**.

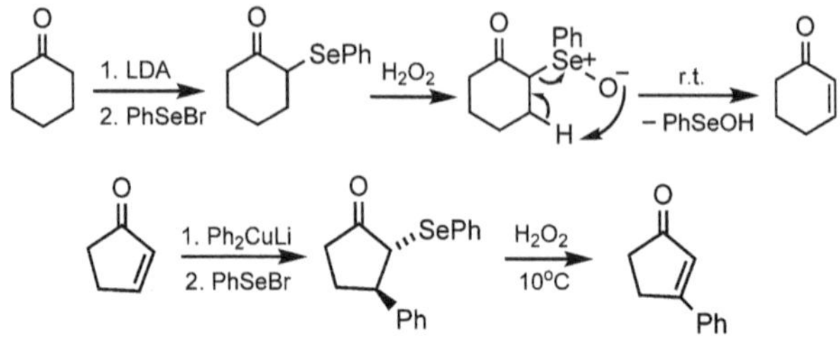


**FIG. 8.35** Synthesis of α,β-unsaturated ketones via selenoxide elimination.

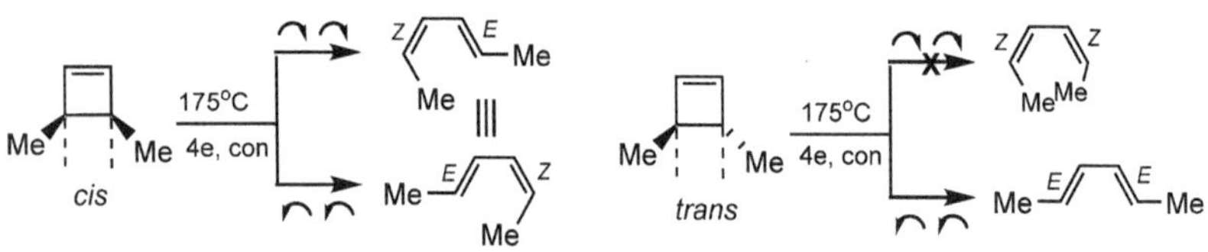


**FIG. 8.36** Thermal electrocyclic ring opening of *cis*- and *trans*-3,4-dimethylcyclobutenes.

## 8.8 Thermal electrocyclic reactions[71]

Thermal electrocyclic reactions proceed by conrotatory (con) or disrotatory (dis) pathway depending on the total number of electrons involved (see Section 4.4.6). There are two possible con and two possible dis modes. Two con modes may lead to the same or different diastereomeric products. The same is true for two dis modes. One of the two con (or dis) modes can get preference when it entails a more stable TS that leads to a major diastereomer. This kind of selectivity of one con (or dis) mode over an alternative one is called torquoselectivity.

The thermal 4-electron cyclobutene-butadiene interconversion[72] occurs by the conrotatory mode. Fig. 8.36 shows thermal electrocyclic ring opening (ERO) of *cis*- and *trans*-3,4-disubstituted cyclobutenes.

For the *cis* isomer, two con motions give the same diastereomeric (*E*,*Z*) diene, whereas in case of the *trans* isomer, (*E*,*E*) and (*Z*,*Z*) dienes could form; however, only the (*E*,*E*) diene is obtained since the TS for the formation of the (*Z*,*Z*) diene is severely crowded. Thus, the con motion leading to the (*E*,*E*) isomer is preferred over the other con motion, that is, a torquoselectivity operates due to a steric effect.

### 8.8.1 Mandal's stereochemical rule for electrocyclic reactions[73,74]

The product stereochemistry of electrocyclic reactions can be delineated easily using a simple and straightforward rule as described below. For the purpose of the rule, the stereochemical properties of the reactant and product are designated as syn (**S**) or anti (**A**) based on a directional sense as shown in Fig. 8.37. There are two possible syn (**S**) and two possible anti (**A**) relationships between the substituents X and Y at the two ends of a conjugated π system or at

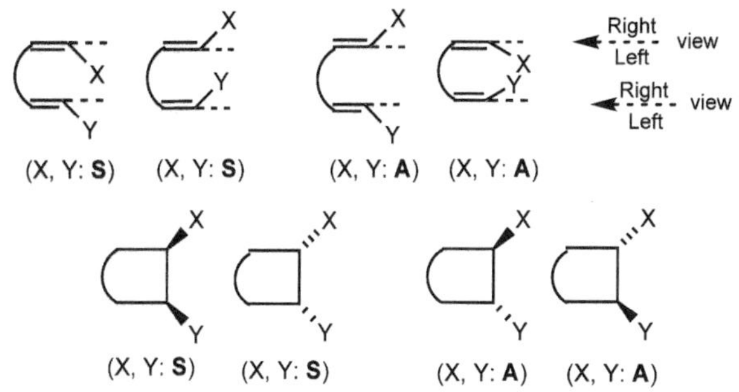


**FIG. 8.37** syn (**S**) and anti (**A**) relationships of the substituents X and Y at the ends of a conjugated π system or at the ends of a σ bond broken/formed in a cyclic system.

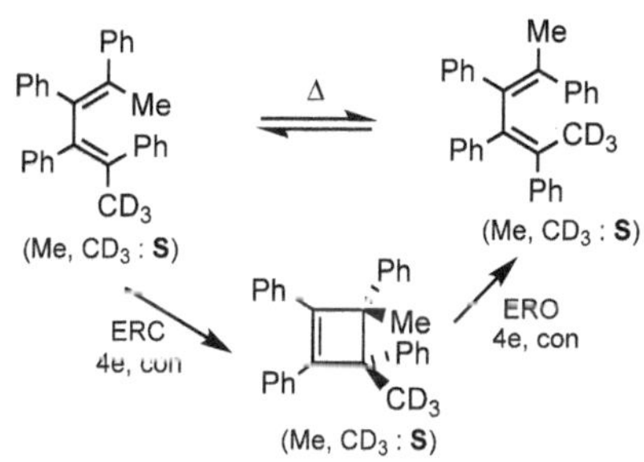


**FIG. 8.38** Thermal isomerization of two dienes.

the two ends of a σ bond broken or formed in a cyclic system. Note that, in the plane of the π system, a terminal substituent (X or Y) is either on the right side or left side when viewed along the respective terminal double bond.

The rule is

$$\mathbf{S} \times \text{con} = \mathbf{S}$$

$$\mathbf{A} \times \text{con} = \mathbf{A}$$

$$\mathbf{S} \times \text{dis} = \mathbf{A}$$

$$\mathbf{A} \times \text{dis} = \mathbf{S}$$

where the reactant stereochemistry undergoes the specified process to give the product stereochemistry.

The applications of the rule are illustrated below with several examples. Consider the thermal isomerization of two dienes (Fig. 8.38).[75] The isomerization occurs by two successive 4-electron electrocyclic ring closing (ERC) and electrocyclic ring opening (ERO) reactions. The product stereochemistry at each step is delineated using the rule: any two substituents, say Me and $CD_3$, which are syn (**S**), remain syn (**S**) throughout as **S** × con = **S**. It may be mentioned that no product by symmetry-forbidden dis mode was formed even after 51 days at 124°C.

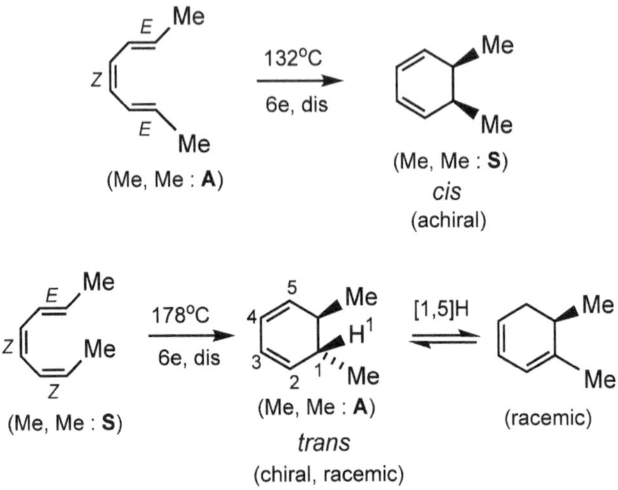


**FIG. 8.39** Thermal ring closing reactions of (2*E*,4*Z*,6*E*)- and (2*E*,4*Z*,6*Z*)-octatrienes.

The thermal 6-electron hexatriene-cyclohexadiene interconversion[76] proceeds through a disrotatory pathway. In ERC, the middle double bond of the triene must be *cis* (*Z*), and the molecule must adopt *s-cis* conformation so that the two ends can come closer to form a σ bond. Fig. 8.39 shows two thermal ring closing reactions: (2*E*,4*Z*,6*E*)-octatriene gives *cis*-5,6-dimethylcyclohexadiene, whereas (2*E*,4*Z*,6*Z*)-octatriene produces *trans*-5,6-dimethylcyclohexadiene. The latter reaction is slower than the former. In each case, the product stereochemistry is delineated using the stereochemical rule. The *trans* isomer is chiral, and the two possible dis modes lead to a racemic product. However, at a higher temperature (178°C), the *trans* isomer can undergo a sigmatropic [1,5]H shift (see Section 8.10) to give 1,6-dimethylcyclohexa-1,3-diene and a mixture of products is formed. Note that [1,5]H shift is also possible with the *cis* isomer at 178°C.

Fig. 8.40 shows that a *cis* fused 4–8 bicyclic system **8.26** is thermally transformed into a *trans* fused 6–6 bicyclic system **8.27**. The reaction proceeds through consecutive conrotatory 4-electron ring opening and disrotatory 6-electron ring closure. The intermediate triene with *cis* geometry of the middle double bond undergoes the ring closing to give the *trans* product **8.27**.

**Problem 8.10** Explain the following observation:

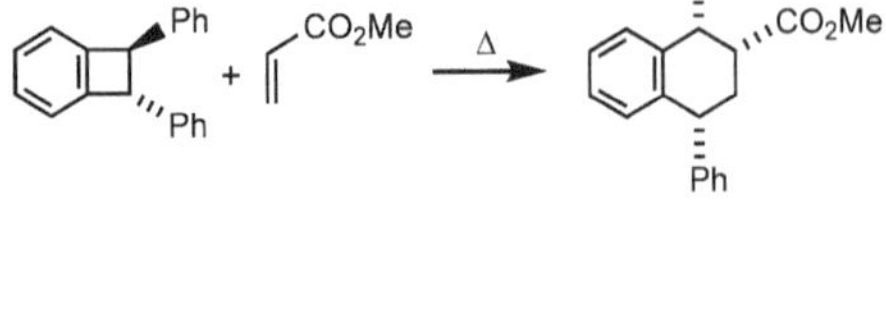

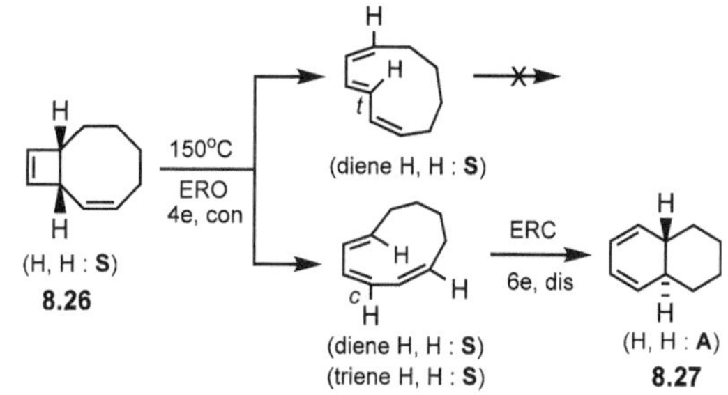


**FIG. 8.40** Thermal conversion of a *cis* fused 4–8 bicyclic system into a *trans* fused 6–6 bicyclic system.

**Problem 8.11** The optically active *trans*-2,3-di-*t*-butylcyclopropanone racemizes at 80°C. Explain.

## 8.8.2 The Nazarov cyclization[77,78]

The 4-electron electrocyclization of a pentadienyl cation to a cyclopentenyl cation provides a useful entry to five-membered ring compounds. An important reaction is the Nazarov cyclization of divinyl ketones which occurs thermally by a conrotatory process. Fig. 8.41 shows an acid-catalysed Nazarov cyclization of a divinyl ketone.[79] The stereochemistry of the cyclopentenone product results from the conrotatory ring closure. See also the use of the stereochemical rule.

In the above reaction, if R = H, the proton is lost preferentially from a ring junction of the cyclopentenyl cation and the product stereochemistry at the ring junctions could not be observed. However, with allylsilane **8.28**, the silyl group is lost in the final step ($Me_3Si^+$ is a better leaving group than $H^+$) and the stereochemistry at the ring junction remains intact in the cyclopentenone product (Fig. 8.42A).

If optically active divinyl ketone **8.29** is employed, torquoselectivity operates to give the optically active product **8.30** (Fig. 8.42B).[80] $Me_3Si$ is a donor substituent through C—Si hyperconjugation and directs the lobe (marked *) of a terminal p orbital anti to $SiMe_3$ for developing the σ bond in the TS **8.31**. Therefore, an anticlockwise con mode operates, resulting in the formation of enantiomer **8.30**. This is an asymmetric synthesis using Nazarov cyclization.

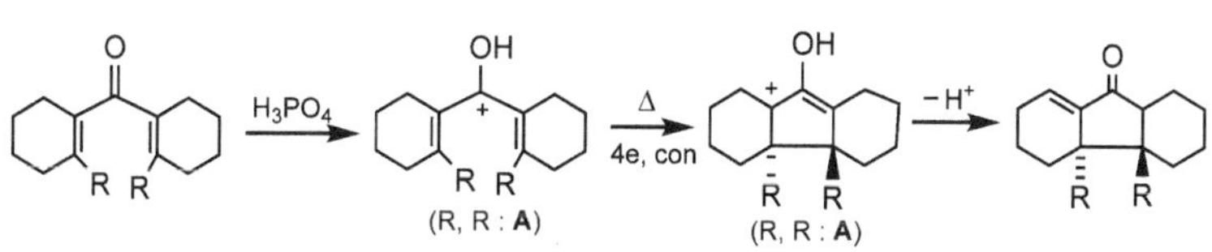


**FIG. 8.41** Nazarov cyclization of a divinyl ketone.

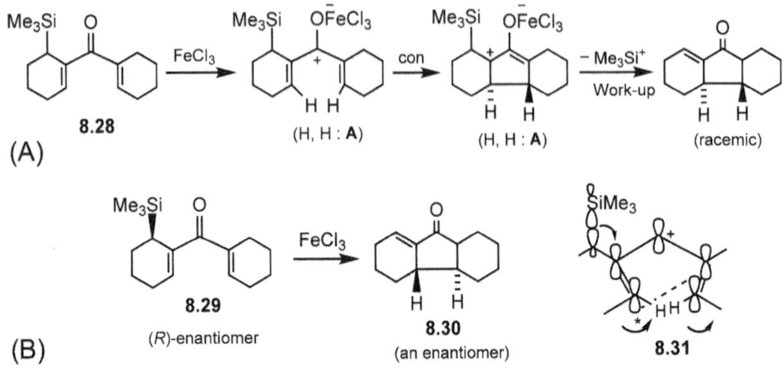


FIG. 8.42 (A) Stereochemistry of a silicon-directed Nazarov cyclization; (B) Asymmetric synthesis using Nazarov cyclization.

**Problem 8.12** Give the structures of the compounds **A–D** in the following scheme for the synthesis of the sesquiterpene capnellene which includes two silicon-directed Nazarov cyclizations (cf. Problem 8.8):

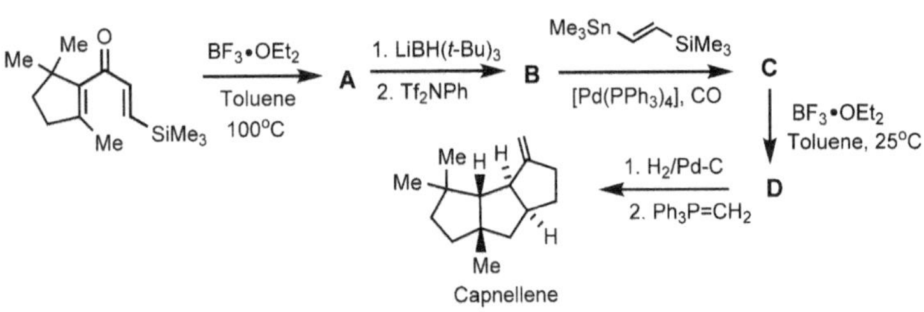


## 8.8.3 The Favorskii rearrangement

The Favorskii rearrangement[81] of two diastereomeric α-chloro ketones **8.32** and **8.33** with an alkoxide is shown in Fig. 8.43. The relationship between the product and the starting material shows an inversion of configuration at the migration terminus bearing Cl. The rearrangement proceeds via a reactive cyclopropanone intermediate which is opened by the alkoxide to give the ester product in each case. The ring closing of the diastereomeric α-chloro enolates **8.34** and **8.35** to cyclopropanones is a 2-electron disrotatory electrocyclic process. Here the stereoelectronic requirement of the backside approach of the π bond to provide assistance to the departure of the leaving group (Cl) is operative. The π bond approaches from the axial side in **8.34**, and from the equatorial side in **8.35** in the disrotatory ring closing (pericyclic mechanism). If the mechanism were not pericyclic, the ionic mechanism would entail a 3-*endo*-trig ring

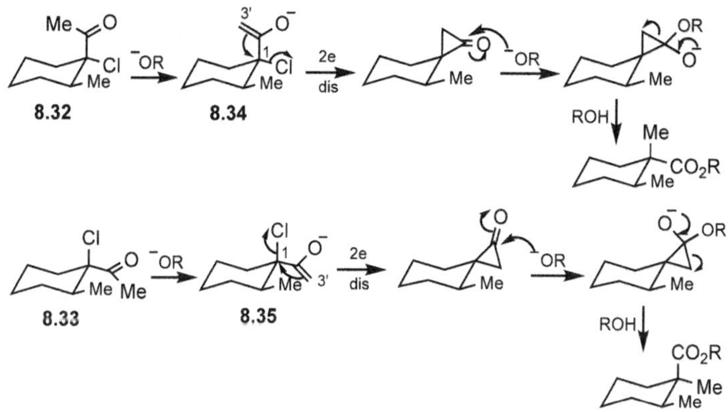


**FIG. 8.43** Favorskii rearrangements of diastereomeric α-chloro ketones.

closure at C3′ which is forbidden by Baldwin's rules (see Section 4.3.7). Pericyclic reactions are not subject to Baldwin's rules.

**Problem 8.13** Predict the product of the following reaction and comment on the stereochemistry:

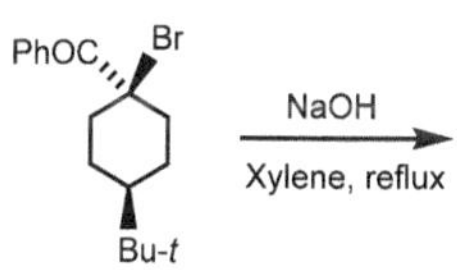


## 8.9 Rearrangement reactions: [1,2] sigmatropic shifts

The most common rearrangement reactions involve 1,2-shifts of a migrating group to an electron deficient centre generated by the loss of a leaving group (L). The migrating C—C bond and the breaking C—L bond ought to be antiperiplanar for the best overlap of $\sigma_{C-C}$ and $\sigma^*_{C-L}$ orbitals. These 1,2-shifts are termed [1,2] sigmatropic shifts and take place suprafacially with retention of configuration of the migrating group (see Fig. 4.24A).

The migrating centre and migration origin are usually carbons; however, the migration terminus can be carbon (Wagner–Meerwein/pinacol rearrangement), nitrogen (Beckmann/Hofmann rearrangement) or oxygen (Baeyer–Villiger rearrangement).[82]

Two examples of Wagner–Meerwein rearrangements from terpene chemistry are shown in Fig. 8.44.[83] The conversion of isoborneol into camphene is depicted in Fig. 8.44A. The driving force of the rearrangement seems to be the conversion of a 2° carbocation to a more stable 3° carbocation. However, there are three possibilities of C—C bond migration but it is the bond labelled

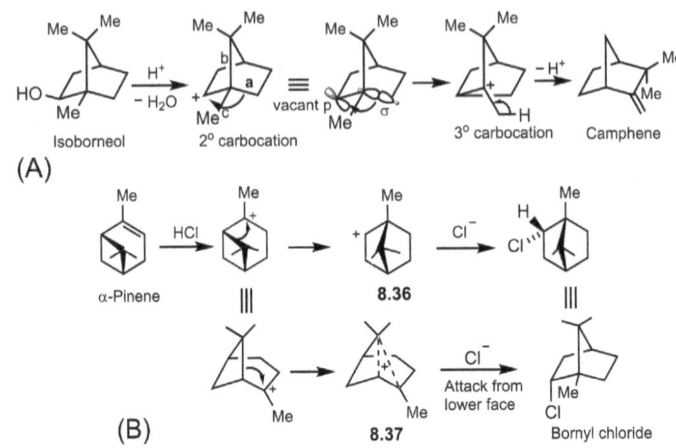


**FIG. 8.44** Wagner–Meerwein rearrangements: (A) Conversion of isoborneol into camphene and (B) conversion of α-pinene into bornyl chloride.

**a** which migrates. This is because this bond is nearly coplanar with the vacant p orbital on the carbocation that leads to an efficient $HOMO_\sigma/LUMO_p$ overlap. The other bonds (b, c) are incorrectly aligned to permit migration; moreover, migration of bond b would give a strained four-membered ring and methyl (bond c) has the lowest migratory aptitude. Loss of a proton from the rearranged cation gives camphene.

Fig. 8.44B shows the conversion of α-pinene into bornyl chloride. Here, although a stable 3° carbocation is transformed to a less stable 2° carbocation, the driving force of the rearrangement arises from the relief of strain in the expansion of a four-membered ring to a five-membered ring. However, the involvement of the classical carbocation **8.36** cannot explain the stereochemistry of the product. The observed stereochemistry of Cl in the product can be explained by invoking a nonclassical carbocation **8.37** with three-centre two-electron bond and a pentacoordinate carbon.

Fig. 8.45 shows two examples in which the configuration of the migrating group is retained. In the Baeyer–Villiger rearrangement (Fig. 8.45A), the migrating group has a stereocentre. The migration of the chiral group gives the major product (87%) with retention of configuration (98.5%). A minor product (13%) is also obtained from methyl migration. Note that product has the same absolute configuration (*S*) as reactant since the relative priority of ligands remains the same. Fig. 8.45B depicts a Hofmann rearrangement with an amide containing a chiral axis when the atropisomeric biphenyl ring migrates from carbon to nitrogen with retention of configuration.

As shown in Fig. 8.45, an antiperiplanar arrangement of the migrating C—C bond and the leaving group is easily achieved in acyclic molecules by rotation about single bonds. But in conformationally constrained cyclic systems, all possible migrating groups in the molecule cannot be antiperiplanar

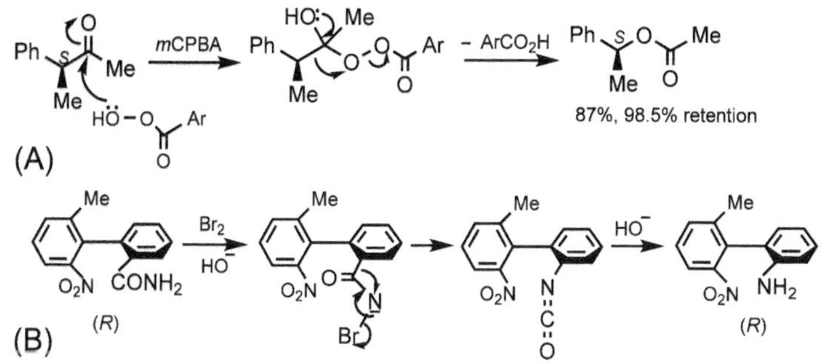

FIG. 8.45 Retention of configuration of the chiral migrating group in (A) Baeyer–Villiger rearrangement and (B) Hofmann rearrangement.

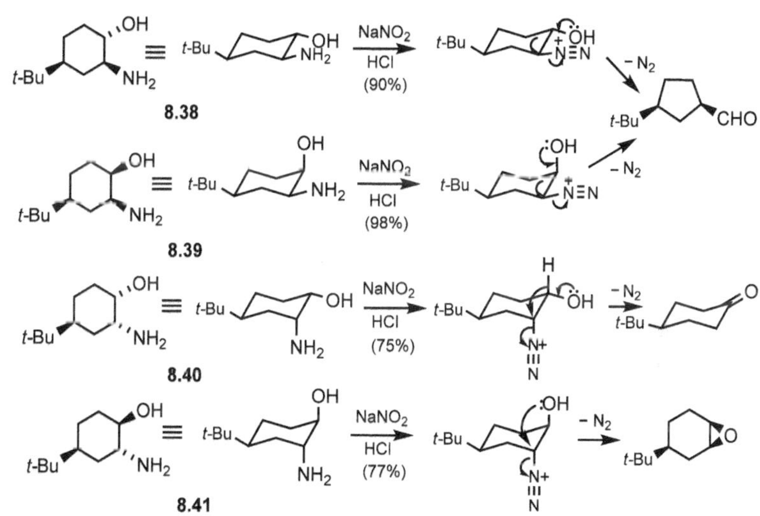

FIG. 8.46 Semipinacol rearrangements of diastereomeric 2-amino-4-*t*-butylcyclohexanols.

with the leaving group and conformation of the cyclic substrate then controls the migration. Consider the semipinacol rearrangement of four diastereomers (**8.38–8.41**) of anancomeric 2-amino-4-*t*-butylcyclohexanol in which *t*-Bu is equatorial (Fig. 8.46).[84] On diazotization, **8.38** and **8.39** undergo ring contraction with migration of the C—C bond antiperiplanar to the leaving group ($N_2^+$), and gives the same *cis* cyclopentane derivative. But **8.40** on diazotization leads to the migration of H atom antiperiplanar to $N_2^+$ and gives 4-*t*-butylcyclohexanone. An electronic 'push–pull mechanism' operates when the lone pair on OH pushes the migrating group and the loss of $N_2$ pulls it. In contrast, the diazonium cation derived from **8.41** lacks a group that can migrate; however, since OH is antiperiplanar to $N_2^+$, nucleophilic attack by OH (intramolecular $S_N2$) gives an epoxide.

It may be mentioned that *cis*-2-aminocyclohexanol which is conformationally mobile, gives a mixture of cyclopentanecarboxaldehyde and cyclohexanone (53 : 47) under the same conditions from (e-$NH_2$,a-OH) and (a-$NH_2$,e-OH) conformers, respectively.[44] Deamination of *trans*-2-aminocyclohexanol however gives the ring-contracted aldehyde via the preferred (e,e) conformation.

**Problem 8.14** Explain the formation of natural product α-caryophyllene alcohol in the following scheme:

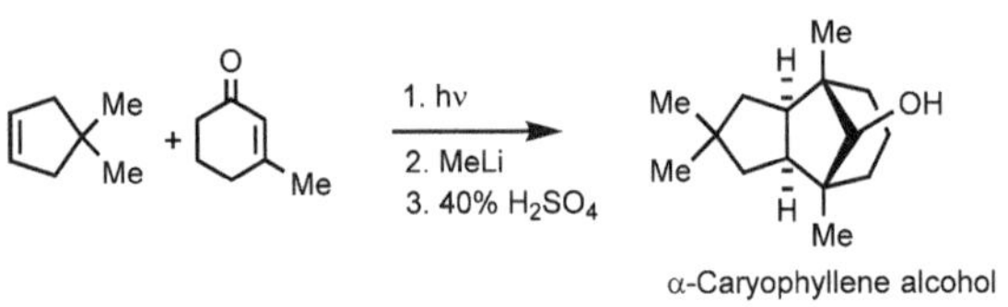


**Problem 8.15** Predict the products and explain the stereochemical outcomes of the following reactions:

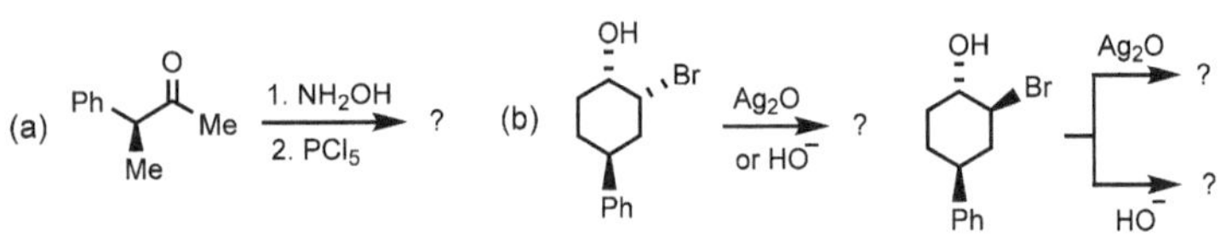


## 8.10 [1,3], [1,5] and [1,7] sigmatropic rearrangements

The suprafacial [1,5] sigmatropic shifts[85] are highly favourable (see Section 4.4.7) and are very common. Fig. 8.47 illustrates the thermal rearrangement of a diene via [1,5] suprafacial H shifts.[86] The (*E,R*) isomer of the diene on heating gives ~1:1 mixture of two stereoisomeric products, (*E,S*) and (*Z,R*). Since the diene is in equilibrium with two conformers by rotation about the C($sp^2$)-C($sp^3$) bond, each conformer undergoes a [1,5] suprafacial H shift to give the product mixture.

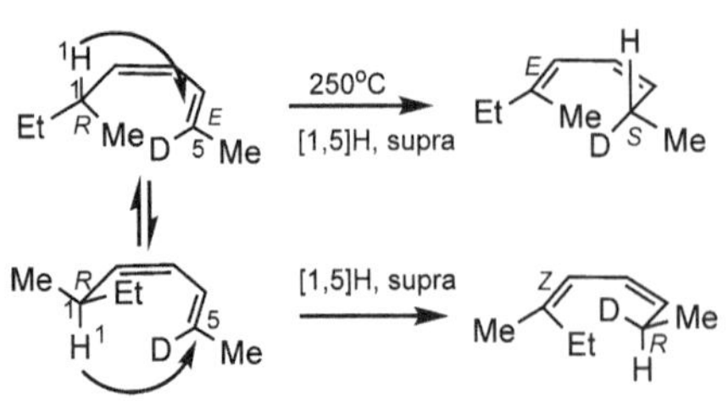


**FIG. 8.47** Thermal rearrangement of a diene via [1,5] suprafacial H shifts.

The [1,5]H shift is extremely facile in cyclopentadienes because the C1 and C5 ends of the pentadienyl system are held adjacent to each other.[87] Thus, the Diels–Alder reaction with 5-methylcyclopentadiene at room temperature involves the more populated 1-substituted diene generated by [1,5]H shift to give the adduct (Fig. 8.48A). In Corey's prostaglandin synthesis, an initial step required a Diels–Alder reaction with the 5-substituted isomer of a cyclopentadiene. The reaction was successful at 0°C in a $Cu(BF_4)_2$-catalysed process (Fig. 8.48B).[88] This is because the rate of isomerization of the 5-substituted isomer through [1,5]H shifts is unaffected by the Lewis acid $Cu^{2+}$, but the rate of Diels–Alder reaction is considerably increased by the coordination of the copper ion with the nitrile group making the dienophile more electrophilic and reactive.

The thermal [1,3]H shift requires antarafacial migration; however, this is geometrically impossible. This is why no thermal [1,3]H shift has been observed. In contrast, the antarafacial migration in the [1,7]H shift is geometrically favourable (see Fig. 4.23A). This has been clearly demonstrated in the thermal equilibrium of a complex triene **8.42** (Fig. 8.49A).[89] The antarafacial H shift leads to H (down) in the product **8.43** making H *trans* to OH, while the antarafacial D shift leads to D (up) in **8.44** making D *cis* to OH. We have previously seen that (*E,Z,Z*)-octatriene on electrocyclic ring closing is isomerized into *trans*-5,6-dimethylcyclohexadiene which can undergo a [1,5]H shift at 178°C (see Fig. 8.39). However, at 111°C, the (*E,Z,Z*) triene equilibrates into (Z,Z,Z) triene by two successive [1,7]H shifts (Fig. 8.49B).[90] The (*Z,Z, Z*) triene can undergo disrotatory electrocyclic isomerization into *cis*-5,6-dimethylcyclohexadiene at 100°C.

The thermal [1,3]C shift is possible suprafacially with inversion of configuration of the migrating group (see Fig. 4.23B). An example is shown in Fig. 8.50.[91] Notably, an *exo*-substituent remains *exo* as a result of inversion of configuration. The acetoxy group acts as a stereochemical marker; the *trans* relationship of D and OAc changes to *cis* in the product because of inversion at the stereocentre but *exo*-D remains *exo*.

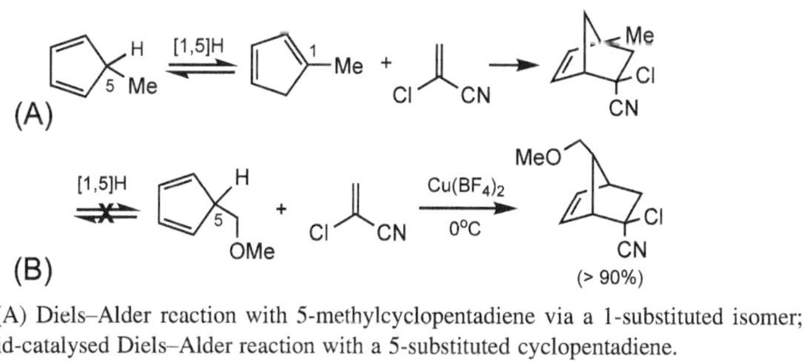


**FIG. 8.48** (A) Diels–Alder reaction with 5-methylcyclopentadiene via a 1-substituted isomer; (B) Lewis acid-catalysed Diels–Alder reaction with a 5-substituted cyclopentadiene.

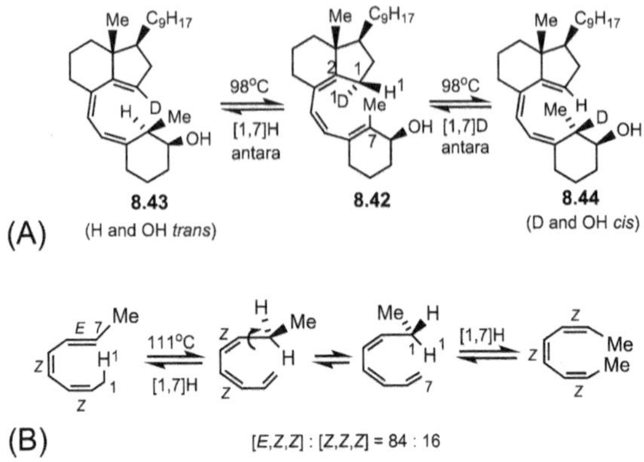


**FIG. 8.49** (A) Antarafacial [1,7]H/D shifts in the thermal equilibrium of a complex triene; (B) Equilibration of (*E,Z,Z*)- and (*Z,Z,Z*)-octatrienes by [1,7]H shifts.

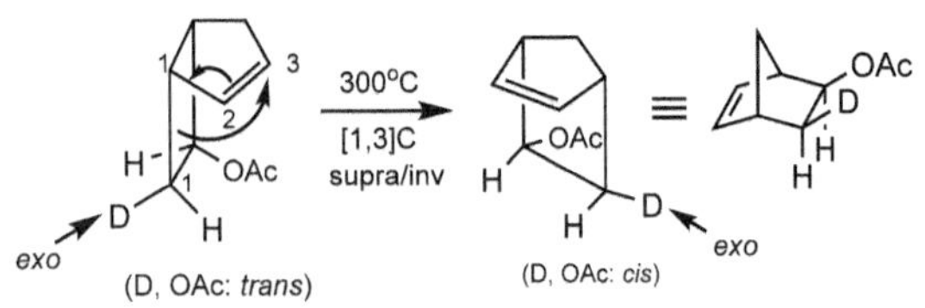


**FIG. 8.50** Thermal [1,3]C shift in a [3.2.0] bicyclic framework.

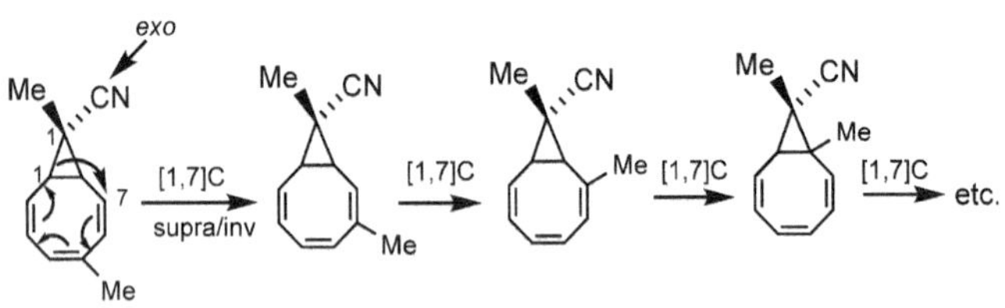


**FIG. 8.51** Walk rearrangement of a bicyclo[6.1.0]nonatrienyl system involving [1,7]C shifts.

In cyclopropane-fused bicyclic systems, the cyclopropane ring can walk around the periphery of the larger ring as a result of sequential [1,*j*]C shifts. This is known as 'walk' rearrangement.[92] Fig. 8.51 illustrates a walk rearrangement involving a sequential [1,7]C shifts.[93] The rearrangement proceeds suprafacially with inversion of configuration at the migrating carbon when *exo*-CN remains *exo*. Note that clockwise rotation of the cyclopropane ring is equivalently shown as the movement of Me on the eight-membered ring by one carbon forward anticlockwise at each step of the rearrangement.

**Problem 8.16** Explain the following observation:

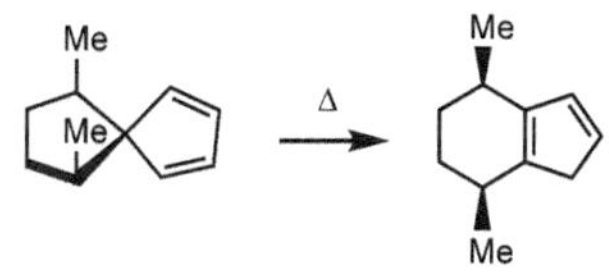


## 8.11 The Cope rearrangement[94,95]

The Cope rearrangement is a thermal [3,3] sigmatropic rearrangement of a 1,5-diene to another 1,5-diene and involves a chair or a boat TS (see Section 4.4.8). In general, there are two possible chair and two possible boat conformations, which may lead to four possible stereoisomeric products. For acyclic 1,5-dienes, when both chair and boat TSs are geometrically accessible, the Cope rearrangement proceeds through the more stable chair TS. If one chair conformation is preferred over the other, the preferred chair TS gives the major product. The Cope rearrangement is thus highly stereoselective. If steric factors make the chair TS prohibitively high in energy, the Cope rearrangement can then proceed through boat TS.

Consider the Cope rearrangements of *meso* and chiral diastereomers of 3,4-dimethyl-1,5-hexadiene (Fig. 8.52).[96] The *meso* diastereomer gives almost exclusively the (*E*,*Z*) diene, whereas the chiral diastereomer gives (*E*,*E*) diene as a major product with a small amount of (*Z*,*Z*) isomer. The stereochemical outcome of the reactions can be rationalized based on the chair TS, as shown in Fig. 8.52.

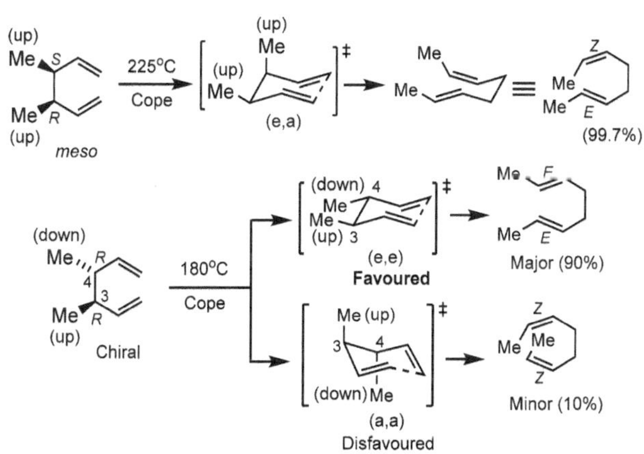


**FIG. 8.52** Cope rearrangements of *meso* and chiral (*R*,*R*) diastereomers of 3,4-dimethyl-1,5-hexadiene.

The Cope rearrangement of a chiral substrate with an (*E*) double bond is shown in Fig. 8.53.[97] The major product is (*E*) diene formed via more stable chair TS in which larger Ph is equatorial. The less stable chair TS with axial Ph leads to minor (*Z*) diene. Note that (*E*) double bond forces Me on it to be equatorial. If the substrate is used as (*R*) enantiomer, the major product is obtained with high enantioselectivity for (*S*) enantiomer (ee > 95%).

The oxy- or anionic oxy-Cope rearrangement with OH or $O^-$ at C3 of the 1,5-diene takes place at a much lower temperature. Fig. 8.54A shows an anionic oxy-Cope rearrangement of an (*E,E*) substrate, which gives a *threo* product.[98] Note that *E* stereochemistry in the starting diene forces Me on the double bond to be equatorial. Similarly, the diastereomeric (*E,Z*) substrate will give an *erythro* product (check). The high level of stereocontrol of the anionic oxy-Cope rearrangement makes this reaction a useful method in synthesis. For example, Fig. 8.54B shows an anionic oxy-Cope rearrangement used as a key step in the synthesis of the sesquiterpene juvabione.[99] The diastereoselectivity arises from the preferred chair TS with equatorial OMe.

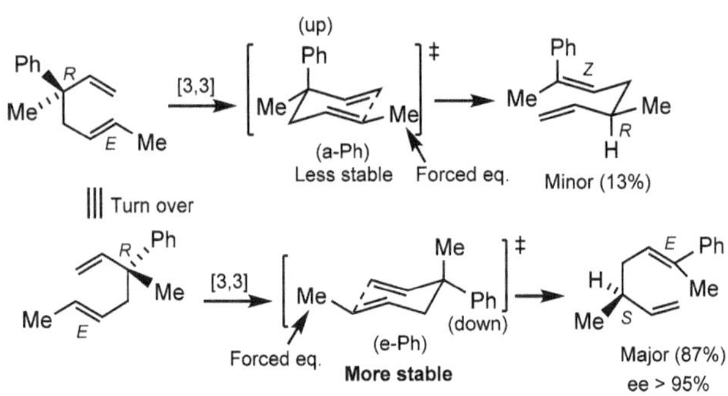


FIG. 8.53 Cope rearrangement of (*R,E*)-3-methyl-3-phenyl-1,5-heptadiene.

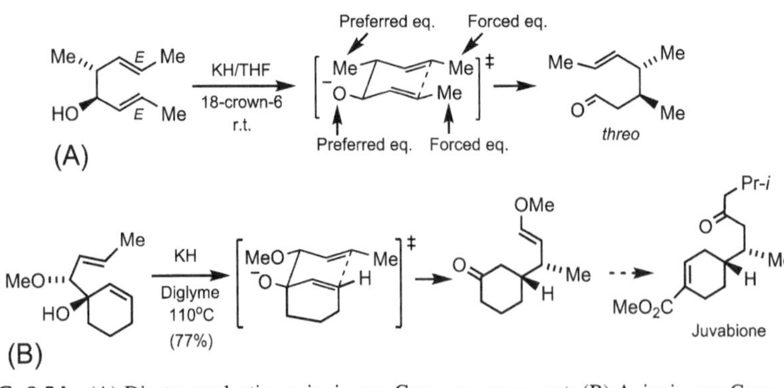


FIG. 8.54 (A) Diastereoselective anionic oxy-Cope rearrangement; (B) Anionic oxy-Cope rearrangement used as a key step in a synthesis of the sesquiterpene juvabione.

The condensation of an amino alcohol **8.45** with acetaldehyde produces an *N*-butenyl iminium ion **8.46** which undergoes an aza-Cope rearrangement followed by an intramolecular Mannich reaction to give the *cis* pyrrolidine **8.47** (Fig. 8.55). The aza-Cope-Mannich reaction sequence has been useful in synthesis, for example, as a key step in a synthesis of the alkaloid pancracine.[100]

The Cope rearrangement of *cis*-1,2-divinylcyclopropane[101,102] proceeds through boat TS via a higher energy *cisoid* conformation **8.48**, instead of a lower energy *transoid* conformation **8.49** (Fig. 8.56A). This is because the former leads to favourable *cis* double bonds in the cycloheptadiene **8.50**, while the latter leads to impossibly strained *trans*, *trans* double bonds in the seven-membered ring **8.51**. The rearrangement through chair TS is also not favourable because one double bond would then be unfavourably *trans* in the seven-membered ring (Fig. 8.56B).

The above reaction offers a useful entry to seven-membered rings in natural product synthesis. An example is shown in Fig. 8.57 which involves the construction of a cycloheptenone ring in a synthesis of the diterpene scopadulcic acid A.[103] Enolization of the ketone **8.52** in the presence of LDA/HMPA gives stereoselectively the *cis*-enolate (see Section 7.6.1.2), which is trapped as the (*Z*)-silyl enol ether **8.53**. The Cope rearrangement of **8.53** gives the desired diastereomer of cycloheptenone product **8.54** after desilylation.

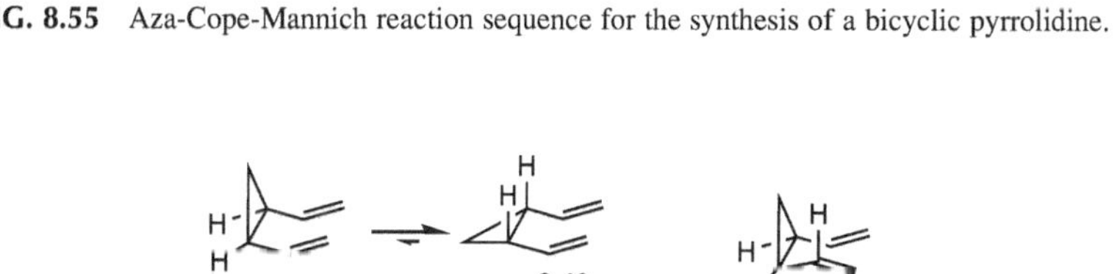


**FIG. 8.55** Aza-Cope-Mannich reaction sequence for the synthesis of a bicyclic pyrrolidine.

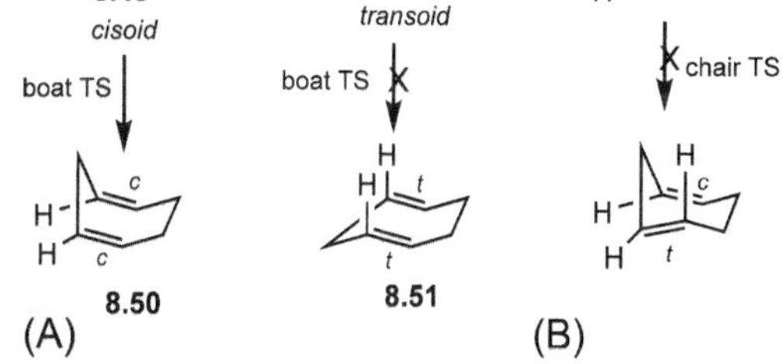


**FIG. 8.56** (A) Cope rearrangement of *cis*-1,2-divinylcyclopropane via a preferred boat TS. (B) A chair TS is unfavourable.

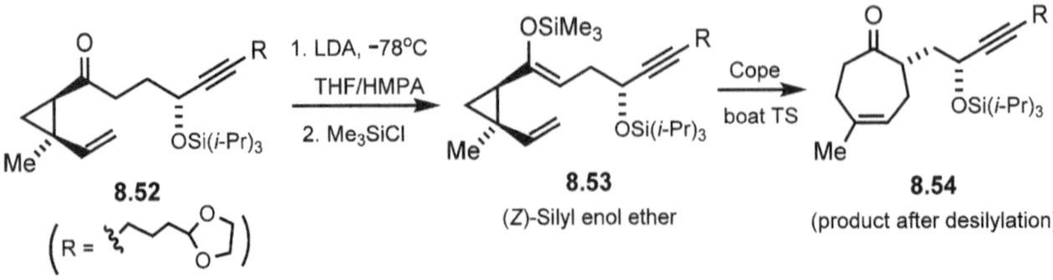


**FIG. 8.57** Cope rearrangement of *cis*-divinylcyclopropane for the construction of a cycloheptenone intermediate in a synthesis of the diterpene scopadulcic acid A.

**Problem 8.17** Suggest a mechanism for the following transformation:

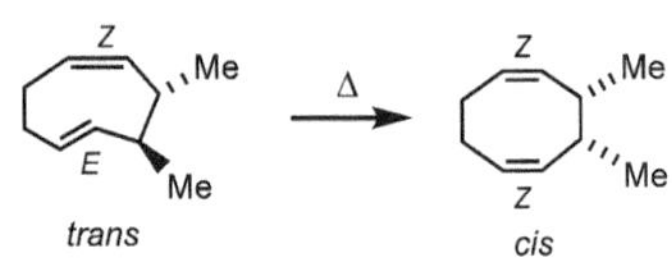


## 8.12 The Claisen rearrangement[104–106]

The Claisen rearrangement is a thermal [3,3] sigmatropic rearrangement of allyl vinyl ether to γ,δ-unsaturated carbonyl compound. The reaction is driven to the product side by the formation of a stable carbonyl π system. Like the Cope rearrangement, the Claisen rearrangement also involves a chair or a boat TS (see Section 4.4.8). The allyl vinyl ether can be prepared directly from the corresponding allyl alcohol by acid-catalysed ether exchange. There are some important variants, namely, Johnson–Claisen,[107] Eschenmoser–Claisen[108] and Ireland–Claisen[109] rearrangements (Fig. 8.58). These modifications require lower temperatures and are very useful for the stereoselective synthesis of γ,δ-unsaturated carbonyl compounds. The temperature of the Claisen rearrangement and its variants lies in the range from 200°C (Claisen) to room temperature (Ireland–Claisen). Another variant called Carroll rearrangement[110] (also called the Kimel-Cope rearrangement[111]) uses an allyl ester of a β-keto acid. The reaction proceeds through a [3] rearrangement of the enol form followed by loss of $CO_2$ to give γ,δ-unsaturated carbonyl compound (Fig. 8.58).

### 8.12.1 Stereospecificity

The stereospecificity of Claisen rearrangement arises from supra/supra interaction between allyl (C—C—C) and vinyl ether (C—C—O) components (cf. Fig. 4.25B). Fig. 8.59A illustrates the stereospecificity of a Johnson–Claisen rearrangement. A new C—C bond is formed at the expense of an

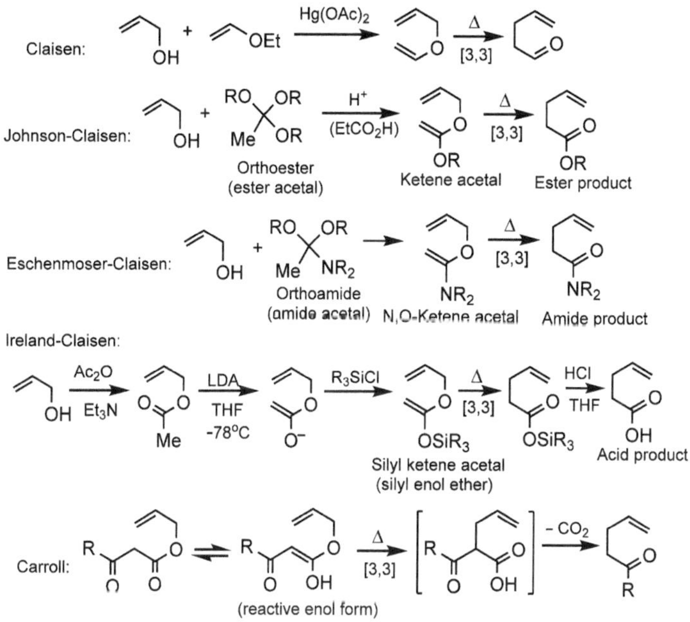


**FIG. 8.58** Claisen rearrangement and its variants: Johnson–Claisen, Eschenmoser–Claisen, Ireland–Claisen and Carroll rearrangements.

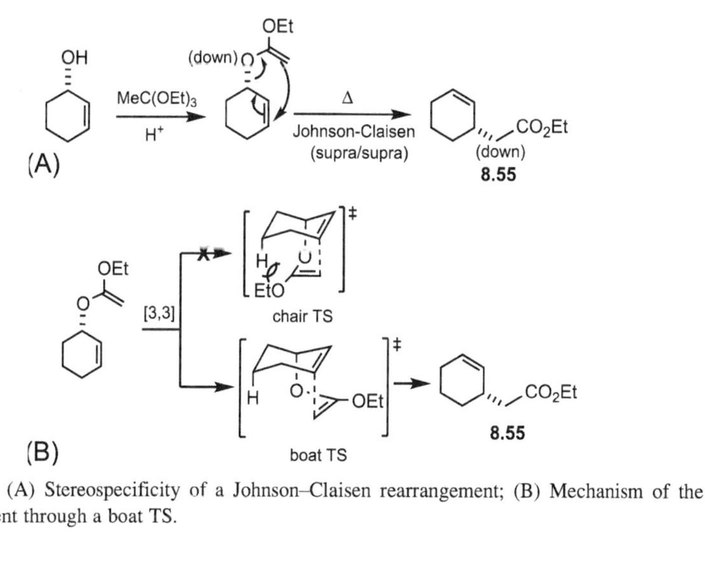


**FIG. 8.59** (A) Stereospecificity of a Johnson–Claisen rearrangement; (B) Mechanism of the rearrangement through a boat TS.

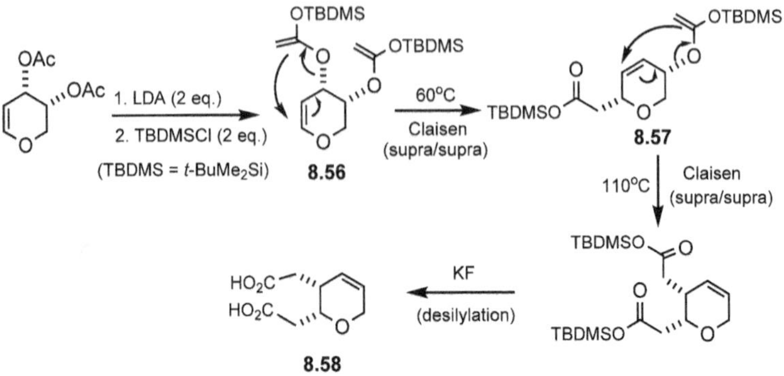


**FIG. 8.60** Stereospecificity of an Ireland–Claisen rearrangement.

old C—O bond. The supra/supra pathway leads to the formation of the C—C bond on the same lower face of the breaking C—O bond and thereby determines the stereochemistry of the product **8.55**.

In a conformationally constrained system in which three or more atoms of allyl vinyl ether is part of a ring, the Claisen rearrangement proceeds preferably through a boat TS.[112] Thus, the Johnson–Claisen rearrangement of the cyclic substrate, in which the three-atom allyl unit is part of the ring, proceeds through a boat TS to give the product **8.55** (Fig. 8.59B). The boat TS is favoured as the chair TS is destabilized by steric interaction.

The stereospecificity of an Ireland–Claisen rearrangement is shown in Fig. 8.60.[113] A [3] Claisen rearrangement of the bis silyl enol ether **8.56** generated from the corresponding diacetate produces **8.57** which can undergo a second Claisen rearrangement at a higher temperature. The stereospecificity at each rearrangement step is governed by supra/supra mode with migration occurring on the lower face of the allyl unit. Desilylation of the product after second Claisen rearrangement gives a *cis* dicarboxylic acid **8.58**.

**Problem 8.18** Predict the stereochemistry of the product in the following Eschenmoser–Claisen rearrangement:

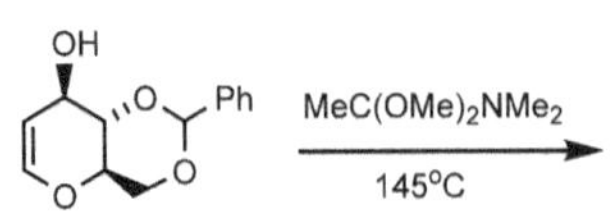


## 8.12.2 Stereoselectivity

Except in some sterically constrained systems (see Fig. 8.59B), the Claisen rearrangement takes place via the preferred chair TS. The more stable chair TS leads

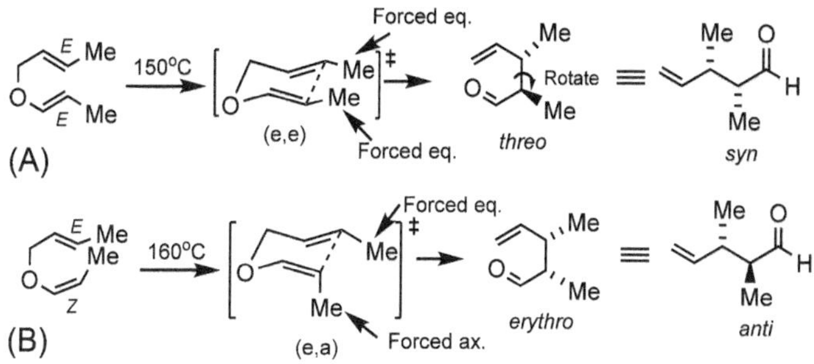


**FIG. 8.61** Stereoselectivity of Claisen rearrangement of (A) (*E*,*E*) and (B) (*E*,*Z*) allyl vinyl ethers.

to the predominant product. Fig. 8.61 illustrates the stereoselectivity of the Claisen rearrangement with the diastereomeric (*E*,*E*) and (*E*,*Z*) substrates.[114] During the reaction, both double bond stereochemistries are destroyed and two new stereocentres across the new C—C bond are created in the product. The creation of two stereocentres can give rise to two diastereomeric products (*threo* and *erythro*) in each case. However, the (*E*,*E*) substrate gives mainly a *threo* or *syn* diastereomer (Fig. 8.61A), whereas the (*E*,*Z*) isomer yields an *erythro* or *anti* diastereomer (Fig. 8.61B). The diastereoselectivity has been rationalized on the basis of the preferred chair TS of the rearrangement as shown in Fig. 8.61. It should be noted that there are two possible enantiomeric TSs which lead to a racemic product in each case. Also note that the (*E*,*E*) stereochemistry forces both Me's to be equatorial while (*E*,*Z*) stereochemistry forces one Me to be equatorial and the other Me to be axial.

As shown above, the geometry of the double bonds in the allyl unit and vinyl ether unit of the starting material determines the stereoselectivity. The stereochemistry of the double bond in the allyl unit is usually provided by an allyl alcohol used for the preparation of the allyl vinyl ether. The formation of a vinyl ether unit with stereochemical purity is however difficult to achieve. In this respect, the Ireland–Claisen procedure provides a suitable method in which the control of enolate geometry can be achieved by judicious choice of enolization conditions. Enolization of an ester with LDA/THF at low temperature favours the formation of a *trans*-enolate, whereas in the presence of the co-solvent HMPA, a *cis*-enolate is favoured (see Sections 7.6.1.1 and 7.6.1.2). In the Ireland–Claisen rearrangement, the *trans*- or *cis*-enolate is trapped as silyl enol ether (silyl ketene acetal) which then undergoes a [3] Claisen rearrangement to give γ,δ-unsaturated carboxylic acid after desilylation. The enolate geometry determines the geometry of the silyl ketene acetal, and therefore determines the stereochemistry of the final product.[115–117]

Fig. 8.62 illustrates an Ireland–Claisen procedure with an (*E*)-allyl ester **8.59** (derived from the corresponding (*E*)-allyl alcohol). Enolization in THF followed by silylation with *t*-butyldimethylsilyl chloride (TBDMSCl) produces an (*E*)-silyl ketene acetal which undergoes Claisen rearrangement via the

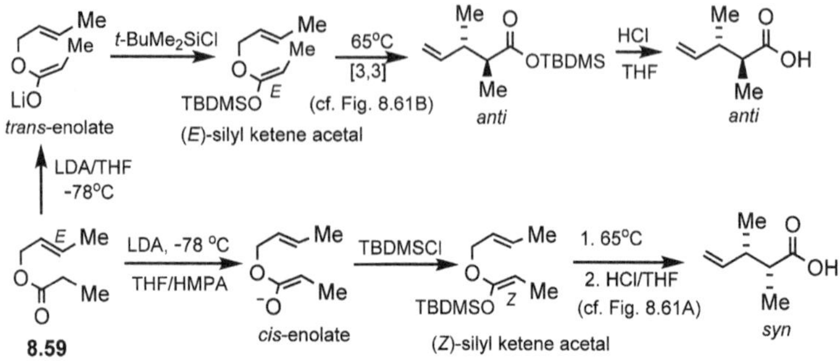


FIG. 8.62 Diastereoselectivity of Ireland–Claisen rearrangement.

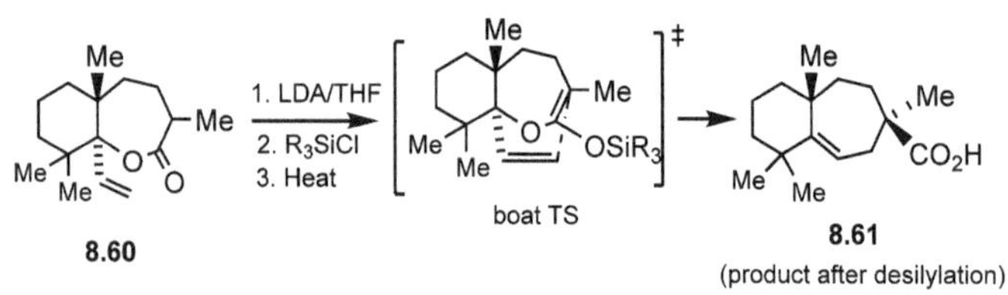


FIG. 8.63 Stereoselective Ireland–Claisen rearrangement via a boat TS.

preferred chair TS to give, after desilylation, an *anti* diastereomer of a γ,δ-unsaturated acid. In the presence of HMPA, a (*Z*)-silyl ketene acetal is obtained which leads to a *syn* diastereomer. Thus, either diastereomeric product can be obtained simply by choice of enolization conditions.

With a cyclic substrate **8.60**, four atoms become part of the seven-membered ring in the allyl vinyl ether formed. Here the Ireland–Claisen rearrangement proceeds via a boat TS to give the carboxylic acid **8.61** after desilylation (Fig. 8.63).[118] The reaction was employed in a synthesis of the sesquiterpene widdrol.

## 8.13 The Wittig rearrangement[119,120]

The [2,3] Wittig rearrangement of a carbanion of allyl ether proceeds suprafacially across the allyl unit via an envelope TS (see Section 4.4.9). Deprotonation of allyl ether can be made in a regioselective manner by incorporating an anion-stabilizing group in the substrate when the base removes the most acidic proton. The anion-stabilizing group is commonly a conjugating substituent (c-type) such as alkenyl, alkynyl or aryl group. This avoids the side reaction of carbanion condensation with an electron-withdrawing (z-type) carbonyl substituent. In the Wittig rearrangement, a C—C bond is formed at the expense of a C—O bond, and the product is a homoallylic alcohol.

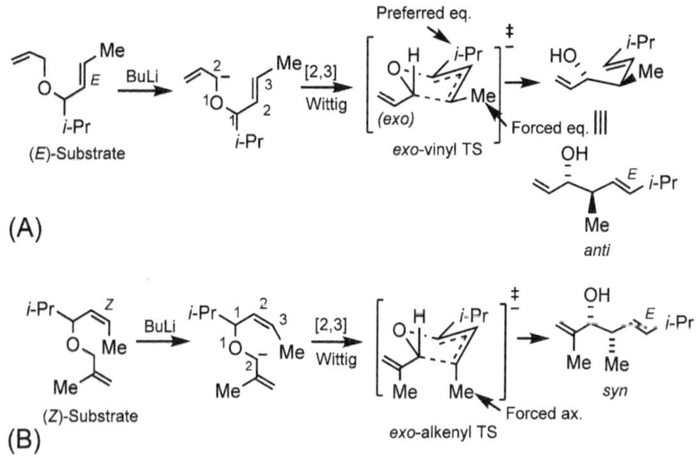


**FIG. 8.64** Diastereoselectivity of the Wittig rearrangement with (A) an (*E*)-substrate and (B) a (*Z*)-substrate containing an anion-stabilizing conjugating substituent (c-type).

The Wittig rearrangement is highly stereoselective.[120,121] Fig. 8.64 illustrates the diastereoselectivity of the Wittig rearrangement with (*E*)- and (*Z*)-substrates. In each case, the rearrangement proceeds through a preferred envelope TS in which the anion-stabilizing c-type substituent occupies the less hindered *exo* position, away from the fold of the envelope. The (*E*)-substrate gives predominantly the *anti* diastereomer (Fig. 8.64A), whereas the (*Z*)-substrate leads to the *syn* diastereomer (Fig. 8.64B).[122,123] Notably, the new double bond in the product is (*E*) as the *i*-Pr is preferably equatorial in the envelope TS.

If the anion-stabilizing group is an electron withdrawing z-type substituent, an *endo*-z TS is often preferred which leads to a reversal of stereoselectivity.[124] For example, the Wittig rearrangement with the (*E*)-allylic ether containing a z-type oxazoline substituent **8.62** gives predominantly the *syn* diastereomer **8.63** (Fig. 8.65). Similarly, the corresponding (*Z*) isomer would give mainly the *anti* diastereomer.

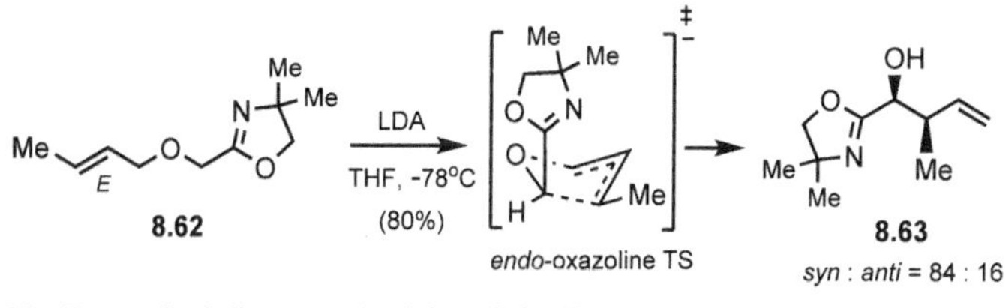


**FIG. 8.65** Reversal of diastereoselectivity of the Wittig rearrangement with an (*E*)-substrate containing an anion-stabilizing oxazoline substituent (z-type).

**Problem 8.19** Predict the product of the following reaction:

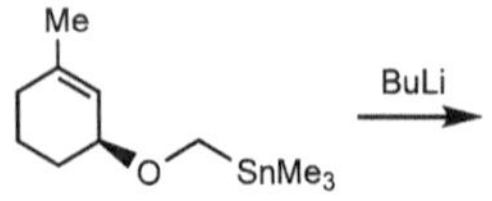


## References


1. Carruthers, W. *Cycloaddition Reactions in Organic Synthesis*; Pergamon Press: Oxford, 1990.
2. Oppolzer, W. In *Comprehensive Organic Synthesis*; Trost, B. M., Fleming, I., Eds.; Vol. 5; Pergamon Press: Oxford, 1991; p. 315.
3. Nicolaou, K. C.; Snyder, S. A.; Montagnon, T.; Vassilikogiannakis, G. *Angew. Chem. Int. Ed. Engl.* **2002**, *41*, 1668.
4. Woodward, R. B.; Hoffmann, R. *Angew. Chem. Int. Ed. Engl.* **1969**, *8*, 781.
5. Sauer, J. *Angew. Chem. Int. Ed. Engl.* **1967**, *6*, 16.
6. Vogel, P.; Houk, K. N. *Organic Chemistry: Theory, Reactivity and Mechanisms in Modern Synthesis*; Wiley-VCH: Weinheim, 2019; p. 406 (chapter 5).
7. Breslow, R.; Guo, T. *J. Am. Chem. Soc.* **1988**, *110*, 5613.
8. Woodward, R. B.; Bader, F. E.; Bickel, H.; Frey, A. J.; Kierstead, R. W. *Tetrahedron* **1958**, *2*, 1.
9. Danishefsky, S.; Kitahara, T.; Schuda, P. F.; Etheredge, S. J. *J. Am. Chem. Soc.* **1976**, *98*, 3028.
10. Ciganek, E. *Org. React.* **1984**, *32*, 1.
11. Roush, W. R. In *Comprehensive Organic Synthesis*; Trost, B. M., Fleming, I., Eds.; Vol. 5; Pergamon Press: Oxford, 1991; p. 513.
12. Winkler, J. D. *Chem. Rev.* **1996**, *96*, 167.
13. Fallis, A. G. *Acc. Chem. Res.* **1999**, *32*, 464.
14. Marsault, E.; Toró, A.; Nowak, P.; Deslongchamps, P. *Tetrahedron* **2001**, *57*, 4243.
15. Huisgen, R.; Padwa, A. *1,3-Dipolar Cycloaddition Chemistry*, Vol. 1; Wiley: New York, 1984; p. 1.
16. Padwa, A. In *Comprehensive Organic Synthesis*; Trost, B. M., Fleming, I., Eds.; Vol. 4; Pergamon Press: Oxford, 1991; p. 1069.
17. Wade, P. A. In *Comprehensive Organic Synthesis*; Trost, B. M., Fleming, I., Eds.; Vol. 4; Pergamon Press: Oxford, 1991; p. 1111.
18. Padwa, A.; Pearson, W. H., Eds. *Synthetic Applications of 1,3-Dipolar Cycloaddition Chemistry Toward Heterocycles and Natural Products*; New Jersey: Wiley, 2003.
19. Caramella, P.; Grünanger, P.; Padwa, A. *1,3-Dipolar Cycloaddition Chemistry*, Vol. 1; Wiley: New York, 1984; p. 291.
20. Huisgen, R.; Scheer, W.; Huber, H. *J. Am. Chem. Soc.* **1967**, *89*, 1753.
21. Tsuge, O.; Kanemasa, S.; Yoshioka, M. *J. Org. Chem* **1988**, *53*, 1384.
22. Barr, D. A.; Grigg, R.; Gunaratne, H. Q. N.; Kemp, J.; McMeekin, P.; Sridharan, V. *Tetrahedron* **1988**, *44*, 557.
23. Tufariello, J. J.; Ali, S. A. *Tetrahedron Lett.* **1978**, *19*, 4647.
24. LeBel, N. A.; Post, M. E.; Whang, J. J. *J. Am. Chem. Soc.* **1964**, *86*, 3759.
25. Mulzer, J.; Altenbach, H.-J. In *Organic Synthesis Highlights*; Mulzer, J., Braun, M., Krohn, K., Reissig, H.-U., Eds.; VCH: New York, 1991; p. 77.
26. Hyatt, J. A.; Raynolds, P. W. *Org. React.* **1994**, *45*, 159.

27. Tidwell, T. T. *Ketenes*; Wiley: New York, 1995.
28. Montaigne, R.; Ghosez, L. *Angew. Chem. Int. Ed. Engl.* **1968**, *7*, 221.
29. Brady, W. T. *Tetrahedron* **1981**, *37*, 2949.
30. Huisgen, R.; Feiler, L. A.; Binsch, G. *Chem. Ber.* **1969**, *102*, 3460.
31. Sustmann, R.; Ansmann, A.; Vahrenholt, F. *J. Am. Chem. Soc.* **1972**, *94*, 8099.
32. Snider, B. B. *Chem. Rev.* **1988**, *88*, 793.
33. Bisceglia, R. H.; Cheer, C. J. *J. Chem. Soc. Chem. Commun.* **1973**, 165.
34. Maercker, A. *Org. React.* **1965**, *14*, 270.
35. Maryanoff, B. E.; Reitz, A. B. *Chem. Rev.* **1989**, *89*, 863.
36. Vedejs, E.; Peterson, M. J. *Top. Stereochem.* **1994**, *21*, 1.
37. Bestmann, H. J.; Zimmerman, R. In *Comprehensive Organic Synthesis*; Trost, B. M., Fleming, I., Eds.; Vol. 6; Pergamon Press: Oxford, 1991; p. 171.
38. Robiette, R.; Richardson, J.; Aggarwal, V. K.; Harvey, J. N. *J. Am. Chem. Soc.* **2006**, *128*, 2394.
39. Padwa, A.; Krumpe, K. E. *Tetrahedron* **1992**, *48*, 5385.
40. Ye, T.; McKervey, M. A. *Chem. Rev.* **1994**, *94*, 1091.
41. Donaldson, W. A. *Tetrahedron* **2001**, *57*, 8589.
42. Lebel, H.; Marcoux, J.-F.; Molinaro, C.; Charette, A. B. *Chem. Rev.* **2003**, *103*, 977.
43. Simmons, H. E.; Cairns, T. L.; Vladuchick, S. A.; Hoiness, C. M. *Org. React.* **1973**, *20*, 1.
44. Helquist, P. In *Comprehensive Organic Synthesis*; Trost, B. M., Fleming, I., Eds.; Vol. 4; Pergamon Press: Oxford, 1991; p. 951.
45. Motherwell, W. B.; Nutley, C. J. *Contemp. Org. Synth.* **1994**, *1*, 219.
46. Charette, A. B.; Beauchemin, A. *Org. React.* **2001**, *58*, 1.
47. Davies, H. M. L.; Antoulinakis, E. G. *Org. React.* **2001**, *57*, 1.
48. Davies, H. M. L. In *Comprehensive Organic Synthesis*; Trost, B. M., Fleming, I., Eds.; Vol. 4; Pergamon Press: Oxford, 1991; p. 1031.
49. Davies, H. M. L.; Clark, T. J.; Church, L. A. *Tetrahedron Lett.* **1989**, *30*, 5057.
50. Corey, E. J.; Achiwa, K.; Katzenellenbogen, J. A. *J. Am. Chem. Soc.* **1969**, *91*, 4318.
51. Hoffmann, H. M. R. *Angew. Chem. Int. Ed. Engl.* **1969**, *8*, 556.
52. Oppolzer, W.; Snieckus, V. *Angew. Chem. Int. Ed. Engl.* **1978**, *17*, 476.
53. Snider, B. B. In *Comprehensive Organic Synthesis*; Trost, B. M., Fleming, I., Eds.; Vol. 5; Pergamon Press: Oxford, 1991; p. 1.
54. Snider, B. B. *Acc. Chem. Res.* **1980**, *13*, 426.
55. Friedrich, L. E.; Kampmeier, J. A.; Good, M. *Tetrahedron Lett.* **1971**, 2783.
56. Stephenson, L. M.; Mattern, D. L. *J. Org. Chem.* **1976**, *41*, 3614.
57. Mikami, K.; Shimizu, M. *Chem. Rev.* **1992**, *92*, 1021.
58. Adams, D. R.; Bhatnagar, S. P. *Synthesis* **1977**, 661.
59. Snider, B. B.; van Straten, J. W. *J. Org. Chem.* **1979**, *44*, 3567.
60. Mikami, K.; Loh, T.-P.; Nakai, T. *Tetrahedron Lett.* **1988**, *29*, 6305.
61. Fujita, Y.; Suzuki, S.; Kanehira, K. *J. Synth. Org. Chem. Jpn.* **1983**, *41*, 1152.
62. Taber, D. F. *Intramolecular Diels-Alder and Alder Ene Reactions*; Springer Verlag: Berlin, 1984.
63. Huntsman, W. D.; Curry, T. H. *J. Am. Chem. Soc.* **1958**, *80*, 2252.
64. Nakatani, Y.; Kawashima, K. *Synthesis* **1978**, 147.
65. Lange, G. L.; Conia, J. M. *Nouv. J. Chim.* **1977**, *1*, 189.
66. Astles, P. C.; Mortlock, S. V.; Thomas, E. J. In *Comprehensive Organic Synthesis*; Trost, B. M., Fleming, I., Eds.; Vol. 6; Pergamon Press: Oxford, 1991; p. 1011.
67. Nace, H. R. *Org. React.* **1962**, *12*, 57.

68. Skell, P. S.; Hall, W. L. *J. Am. Chem. Soc.* **1964**, *86*, 1557.
69. Cope, A. C.; Trumbull, E. R. *Org. React.* **1960**, *11*, 317.
70. Reich, H. J.; Wollowitz, S. *Org. React.* **1993**, *44*, 1.
71. Marvell, E. N. *Thermal Electrocyclic Reactions*; Academic Press: New York, 1980.
72. Durst, T.; Breau, L. In *Comprehensive Organic Synthesis*; Trost, B. M., Fleming, I., Eds.; Vol. 5; Pergamon Press: Oxford, 1991; p. 675.
73. Mandal, D. K. *J. Chem. Educ.* **2012**, *89*, 1041.
74. Mandal, D. K. *Pericyclic Chemistry: Orbital Mechanisms and Stereochemistry*; Elsevier: Cambridge, MA, 2018; p. 294.
75. Doorakian, G. A.; Freedman, H. H. *J. Am. Chem. Soc.* **1968**, *90*, 5310.
76. Okamura, W. H.; De Lera, A. R. In *Comprehensive Organic Synthesis*; Trost, B. M., Fleming, I., Eds.; Vol. 5; Pergamon Press: Oxford, 1991; p. 699.
77. Denmark, S. E. In *Comprehensive Organic Synthesis*; Trost, B. M., Fleming, I., Eds.; Vol. 5; Pergamon Press: Oxford, 1991; p. 751.
78. Habermas, K. L.; Denmark, S. E.; Jones, T. K. *Org. React.* **1994**, *45*, 1.
79. Woodward, R. B.; Lehr, R. *Aromaticity Chem. Soc. Spl. Publ.* **1967**, *21*, 237.
80. Denmark, S. E.; Wallace, M. A.; Walker, C. B., Jr. *J. Org. Chem.* **1990**, *55*, 5543.
81. Kende, A. S. *Org. React.* **1960**, *11*, 261.
82. Harwood, L. M. *Polar Rearrangements*; Oxford University Press: New York, 1992.
83. Harwood, L. M. *Polar Rearrangements*; Oxford University Press: New York, 1992; p. 18.
84. Chérest, M.; Felkin, H.; Sicher, J.; Šipoš, F.; Tichý, M. *J. Chem. Soc.* **1965**, 2513.
85. Spangler, C. W. *Chem. Rev.* **1976**, *76*, 187.
86. Roth, W. R.; König, J.; Stein, W. *Chem. Ber.* **1970**, *103*, 426.
87. Mironov, V. A.; Sobolev, E. V.; Elizarova, A. N. *Tetrahedron* **1963**, *19*, 1939.
88. Corey, E. J.; Weinshenker, N. M.; Schaaf, T. K.; Huber, W. *J. Am. Chem. Soc.* **1969**, *91*, 5675.
89. Hoeger, C. A.; Johnston, A. D.; Okamura, W. H. *J. Am. Chem. Soc.* **1987**, *109*, 4690.
90. Baldwin, J. E.; Reddy, V. P. *J. Org. Chem.* **1988**, *53*, 1129.
91. Berson, J. A.; Nelson, G. L. *J. Am. Chem. Soc.* **1967**, *89*, 5503.
92. Klärner, F.-G. *Top. Stereochem.* **1984**, *1*, 15.
93. Klärner, F.-G.; Wette, M. *Chem. Ber.* **1978**, *111*, 282.
94. Hill, R. K. In *Comprehensive Organic Synthesis*; Trost, B. M., Fleming, I., Eds.; Vol. 5; Pergamon Press: Oxford, 1991; p. 785.
95. Tantillo, D. J.; Hoffmann, R. *Angew. Chem. Int. Ed. Engl.* **2002**, *41*, 1033. *J. Org. Chem.* **2002**, *67*, 1419.
96. Doering, W.v. E.; Roth, W. R. *Tetrahedron* **1962**, *18*, 67.
97. Hill, R. K.; Gillman, N. W. *Chem. Commun.* **1967**, 619.
98. Wei, S.-Y.; Tomooka, K.; Nakai, T. *Tetrahedron* **1993**, *49*, 1025.
99. Evans, D. A.; Nelson, J. V. *J. Am. Chem. Soc.* **1980**, *102*, 774.
100. Overman, L. E.; Shim, J. *J. Org. Chem.* **1993**, *58*, 4662.
101. Piers, E. In *Comprehensive Organic Synthesis*; Trost, B. M., Fleming, I., Eds.; Vol. 5; Pergamon Press: Oxford, 1991; p. 971.
102. Hudlicky, T.; Fan, R.; Reed, J. W.; Gadamasetti, G. *Org. React.* **1992**, *41*, 1.
103. Fox, M. E.; Li, C.; Marino, J. P.; Overman, L. E. *J. Am. Chem. Soc.* **1999**, *121*, 5467.
104. Rhoads, S. J.; Raulins, N. R. *Org. React.* **1975**, *22*, 1.
105. Wipf, P. In *Comprehensive Organic Synthesis*; Trost, B. M., Fleming, I., Eds.; Vol. 5; Pergamon Press: Oxford, 1991; p. 827.
106. Ziegler, F. E. *Chem. Rev.* **1988**, *88*, 1423.
107. Johnson, W. S.; Wethermann, L.; Bartlett, W. R.; et al. *J. Am. Chem. Soc.* **1970**, *92*, 741.

108. Felix, D.; Gschwend-Steen, K.; Wick, A. E.; Eschenmoser, A. *Helv. Chim. Acta* **1969**, *52*, 1030.
109. Ireland, R. E.; Mueller, R. H. *J. Am. Chem. Soc.* **1972**, *94*, 5898.
110. Carroll, M. F. *J. Chem. Soc.* **1940**, *704*, 1266.
111. Kimel, W.; Cope, A. C. *J. Am. Chem. Soc.* **1943**, *65*, 1992.
112. Khaledy, M. M.; Kalani, M. Y. S.; Khuong, K. S.; et al. *J. Org. Chem.* **2003**, *68*, 572.
113. Curran, D. P.; Suh, Y. G. *J. Am. Chem. Soc.* **1984**, *106*, 5002. *Tetrahedron Lett.* **1984**, *25*, 4179.
114. Hansen, H.-J.; Schmid, H. *Tetrahedron* **1959**, *1974*, 30.
115. Ireland, R. E.; Willard, A. K. *Tetrahedron Lett.* **1975**, *16*, 3975.
116. Ireland, R. E.; Mueller, R. H.; Willard, A. K. *J. Am. Chem. Soc.* **1976**, *98*, 2868.
117. Chai, Y.; Hong, S.; Lindsay, H. A.; McFarland, C.; McIntosh, M. C. *Tetrahedron* **2002**, *58*, 2905.
118. Danishefsky, S.; Funk, R. L.; Kerwin, J. F. *J. Am. Chem. Soc.* **1980**, *102*, 6889.
119. Marshall, J. A. In *Comprehensive Organic Synthesis*; Trost, B. M., Fleming, I., Eds.; Vol. 3; Pergamon Press: Oxford, 1991; p. 975.
120. Nakai, T.; Mikami, K. *Org. React.* **1994**, *46*, 105.
121. Hoffmann, R. W. *Angew. Chem. Int. Ed. Engl.* **1979**, *18*, 563.
122. Tsai, D. J.-S.; Midland, M. M. *J. Org. Chem.* **1984**, *49*, 1842.
123. Tsai, D. J.-S.; Midland, M. M. *J. Am. Chem. Soc.* **1985**, *107*, 3915.
124. Brückner, R. In *Comprehensive Organic Synthesis*; Trost, B. M., Fleming, I., Eds.; Vol. 6; Pergamon Press: Oxford, 1991; p. 873.

Chapter 9

# Pericyclic reactions 2: Asymmetric synthesis

## 9.1 Asymmetric Diels–Alder reactions[1–5]

The ability to construct up to four new stereocentres via the highly ordered cyclic TS of the Diels–Alder reaction provides a great scope for its asymmetric designs. The advances made in the asymmetric strategies of Diels–Alder reactions have been the subject of several general reviews.[1–5] Here we will focus briefly on the asymmetric approaches involving mainly chiral auxiliaries (dia stereoselective approach) and chiral catalysts (enantioselective approach), including organocatalysis.

### 9.1.1 Chiral substrates

The chiral substrates employed in the asymmetric Diels–Alder reaction are usually enantiomerically pure chiral dienes. Fig. 9.1 illustrates an asymmetric Diels–Alder reaction of an (*S*)-diene with an achiral *N*-phenylmaleimide dienophile.[6] The major diastereomer (dr 82:18) arises from the diene conformation **9.1** which provides the less hindered approach of the dienophile from the top face of the diene. The hindered approach of the dienophile from the bottom face in an alternative diene conformation gives the minor product.

The most accessible chiral dienes are the sugar-substituted dienes or amino acid-derived dienes, which have been successfully employed in asymmetric Diels–Alder reactions.[7, 8]

### 9.1.2 Chiral auxiliaries

A popular method to achieve diastereoselectivity in the Diels–Alder reaction is to attach a chiral auxiliary[9] (see Section 5.3.1.2) to the dienophile. The reaction proceeds to give a major diastereomer which is separated and purified, and subjected to auxiliary cleavage to produce the enantiomerically pure product. Best diastereoselectivity is usually observed in the Lewis acid-catalysed Diels–Alder reaction at low temperatures. The Lewis acid lowers the LUMO energy of the

Stereochemistry and Organic Reactions. https://doi.org/10.1016/B978-0-12-824092-2.00009-5

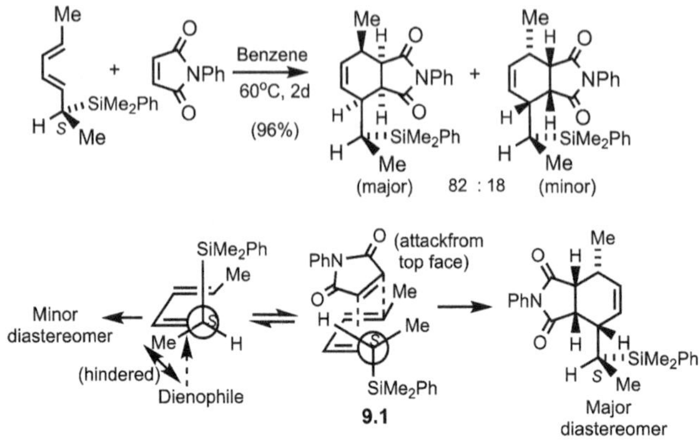


FIG. 9.1 Asymmetric Diels–Alder reaction with a chiral diene substrate.

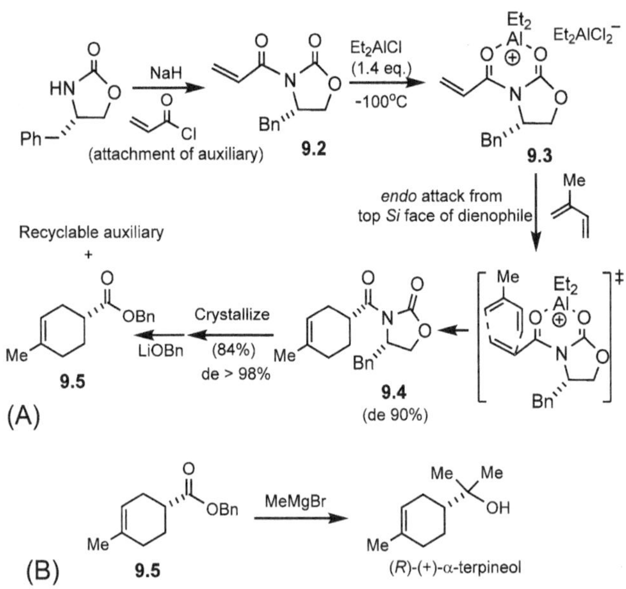


FIG. 9.2 (A) Asymmetric Diels–Alder reaction using Evans oxazolidinone auxiliary; (B) Application of the reaction in the synthesis of (*R*)-(+)-α-terpineol.

dienophile and increases its reactivity, and also restricts its conformational freedom by chelation, thereby reducing the number of possible TSs. Low temperatures also lead to high de (see Table 5.2).

Evans oxazolidinone chiral auxiliaries[10] are very useful for asymmetric Diels–Alder reactions. Fig. 9.2A illustrates an example.[11] Attachment of auxiliary gives the chiral dienophile **9.2** which undergoes diastereoselective reaction with 2-methylbutadiene via a chelated cationic complex **9.3** in the presence of Lewis acid $Et_2AlCl$ at low temperature (−100°C). The rigid aluminium

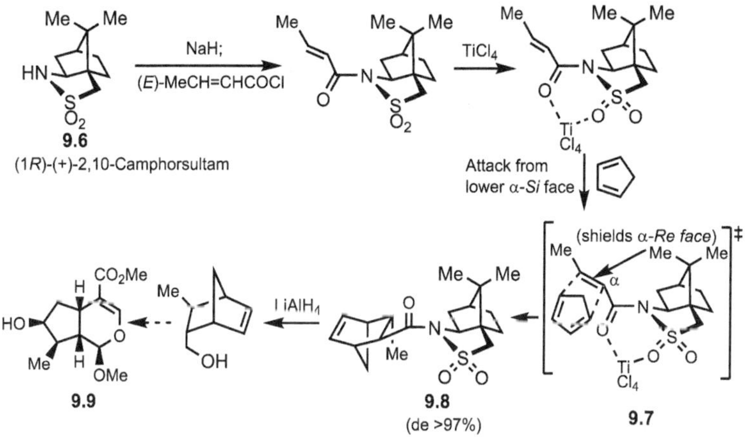


FIG. 9.3 Asymmetric Diels–Alder reaction using Oppolzer sultam auxiliary.

chelate **9.3**, because of restricted orientation of the dienophile, provides high levels of diastereofacial selectivity which is crucial to the success of the asymmetric Diels–Alder reaction. The cycloaddition is 'para' regioselective and highly *endo*-selective. Attack of the diene takes place preferentially on the less hindered *Si* face of the dienophile (the bottom *Re* face being shielded by the benzyl group on the auxiliary) to give the *endo* diastereomer **9.4** (de 90%). Subsequent purification (yield 84%, de >98%) and removal of auxiliary yields the enantiomeric product **9.5**. The chiral auxiliary can be recycled.

The synthetic utility of this reaction has been demonstrated by the conversion of **9.5** into (*R*)-(+)-α-terpineol by Grignard reaction (Fig. 9.2B).[11]

Fig. 9.3 depicts a Lewis acid-catalysed asymmetric Diels-–Alder reaction using an Oppolzer sultam auxiliary **9.6**.[12] Since both enantiomers of the sultam are available, the Diels–Alder adduct can be obtained as either enantiomer. The observed diastereoselectivity has been rationalized in terms of a chelated TS **9.7** involving *endo* attack of the diene from lower α-*Si* face of the dienophile (the upper α-*Re* face being shielded by a Me group on the auxiliary). The *endo*-adduct **9.8** (de >97%) is purified and subjected to auxiliary cleavage with $LiAlH_4$, and the resulting enantiopure alcohol is used in a synthesis of 1-*O*-methyl loganin aglucone **9.9**.[13]

**Problem 9.1** Explain the formation of the predominant product in the following intramolecular Diels–Alder reaction:

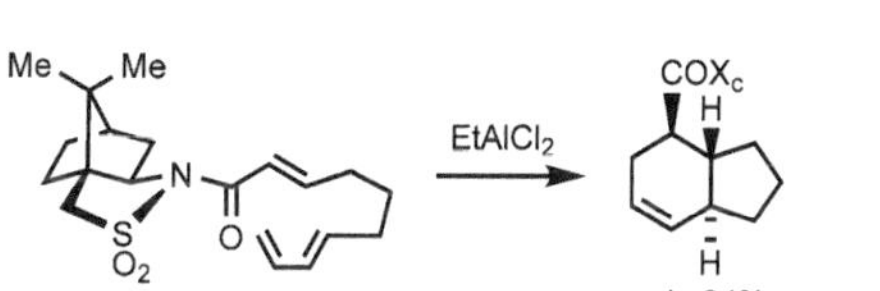


de 94%
($X_c$ = chiral auxiliary)

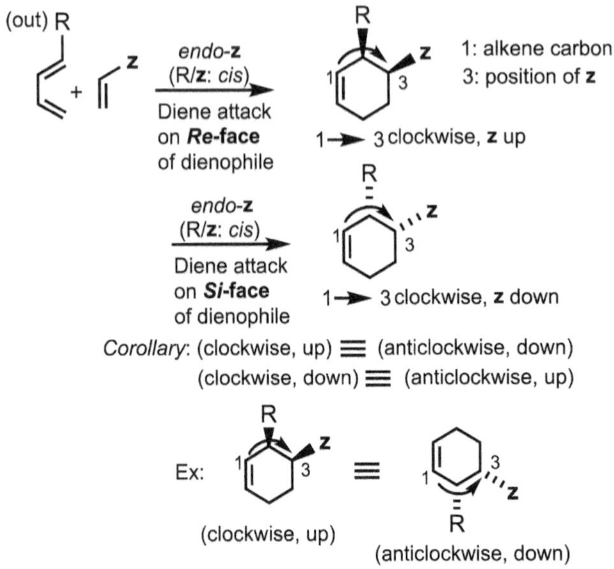


**FIG. 9.4** A mnemonic for delineating absolute stereochemistry of *endo* product in asymmetric Diels–Alder reactions. The clockwise or anticlockwise arc is drawn taking the centre of the cyclohexene ring as geometric centre.

### 9.1.3 A mnemonic for delineating asymmetric Diels–Alder reactions

The absolute stereochemistry of *endo* product in the asymmetric Diels–Alder reaction can be delineated easily using a mnemonic as shown in Fig. 9.4.[14]

To illustrate, consider the asymmetric Diels–Alder reaction in Fig. 9.5 to draw the absolute stereochemistry of the *endo* product. The diene will attack the top *Re* face of the dienophile since its bottom face is shielded by the benzyl group. Note that the preferred regioisomer is 'ortho' ($R^1/COX_c$). Using the mnemonic, the absolute stereochemistry of $COX_c$ is $1 \rightarrow 3$ clockwise, $COX_c$ up in the *endo* (*cis*) adduct. The complete stereochemistry of the product is then delineated considering stereospecificity with respect to dienophile. (Check that the product **9.4** drawn in Fig. 9.2 is consistent with the mnemonic.)

**Problem 9.2** Using retrosynthetic analysis, suggest a suitable oxazolidinone chiral auxiliary for the synthesis of enantiomer **1** and enantiomer **2** of a bicyclic compound, and outline the synthesis in each case.

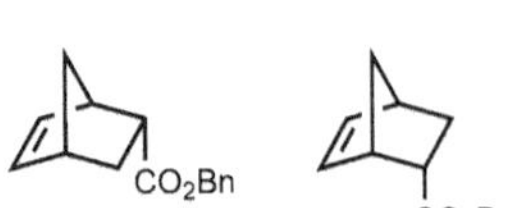


**1** **2**

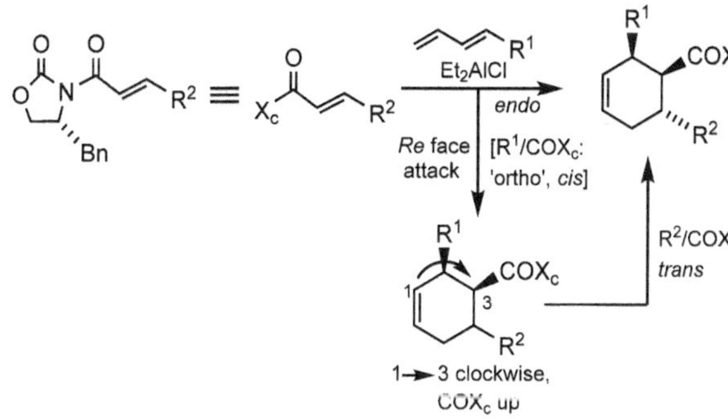


**FIG. 9.5** Delineation of absolute stereochemistry of an asymmetric Diels–Alder reaction with the help of mnemonic in Fig. 9.4.

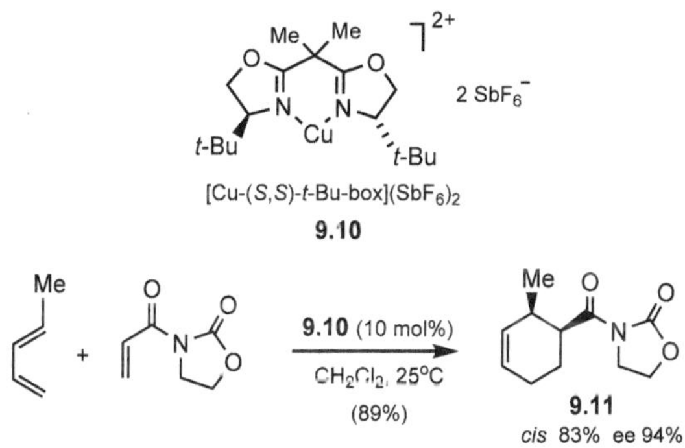


**FIG. 9.6** Asymmetric Diels–Alder reaction using a chiral box catalyst [Cu-(*S*,*S*)-*t*-Bu-box] $(SbF_6)_2$ **9.10**.

## 9.1.4 Chiral catalysts

A highly efficient general approach to asymmetric Diels–Alder reactions is to use a chiral catalyst, in the form of a Lewis acidic metal complex. Over the past decades, many chiral ligands and various metals have been tested as chiral catalysts, and some very effective catalysts have been developed that give excellent enantioselectivity with a range of dienes and dienophiles. The TS assembly of the cycloaddition is often based on the X-ray crystal structure of catalyst-dienophile complex and NMR studies.[15]

Particularly efficient catalysts are $C_2$-symmetric catalysts, derived from chiral ligands of $C_2$ symmetry and Lewis acidic metals. An advantage of a $C_2$-symmetric catalyst is that it reduces the number of competing diastereomeric TSs. One important $C_2$-symmetric catalyst system is known as box catalysts, the copper(II) complex of chiral **bis-ox**azolines[16, 17], such as [Cu(*S*,*S*)-*t*-Bu-box]$(SbF_6)_2$ **9.10** (Fig. 9.6). Best results are achieved when the dienophile

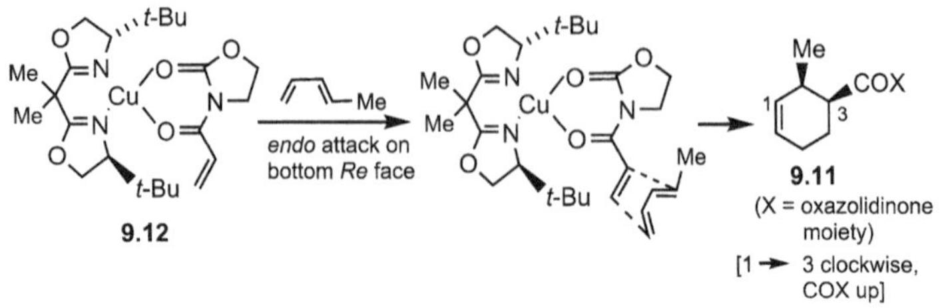


**FIG. 9.7** Mechanism of catalytic asymmetric Diels–Alder reaction in Fig. 9.6 (X = oxazolidinone ring).

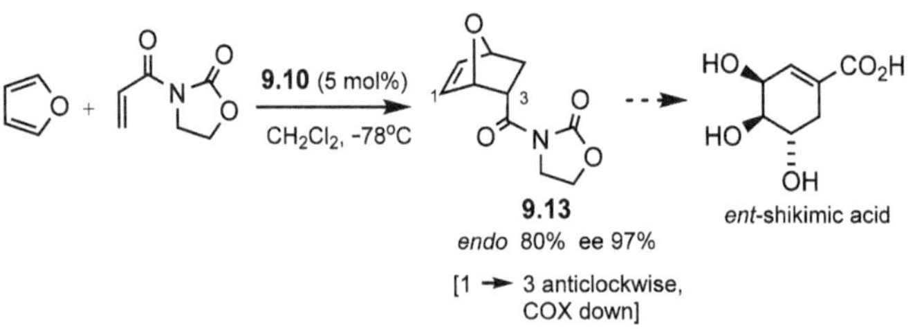


**FIG. 9.8** Asymmetric Diels–Alder reaction employed as a key step in the synthesis of *ent*-shikimic acid.

possesses two sites (e.g. two carbonyl groups) for coordination to the metal ion ($Cu^{2+}$) that acts as a Lewis acid. Fig. 9.6 shows an example of the asymmetric Diels–Alder reaction using the box catalyst **9.10**.[18, 19] The reaction gives the *endo* (*cis*) adduct **9.11** in high yield (83%) and with high ee (94%).

The formation of the observed product **9.11** has been rationalized as follows. The dienophile bears two carbonyl groups which could chelate with the copper forming a well-defined catalyst-substrate complex **9.12** of square-planar geometry (Fig. 9.7). Clearly, the diene would attack on the bottom *Re* face of the dienophile (top *Si* face being sterically shielded by bulky *t*-Bu group) to give the *endo* enatiomer **9.11** after dissociation of catalyst from the catalyst–product complex. (The use of mnemonic for the product stereochemistry is also shown.)

It is of note that, since the catalyst **9.10** has $C_2$ symmetry, the alternative orientation of the dienophile, in which the alkene double bond is on the side of the other oxazoline ring which has below the plane *t*-Bu, would give the same enantiomeric product **9.11**.

Fig. 9.8 shows an asymmetric Diels–Alder reaction with furan as a key step in the synthesis of an unnatural enantiomer of shikimic acid.[20] Attack of furan as a cyclic diene takes place from the bottom *Re* face of the dienophile of catalyst–dienophile complex (cf. **9.12**) to give the *endo* enantiomer **9.13** (1→3 anticlockwise, COX down; see *corollary* in Fig. 9.4). The *endo* adduct **9.13** of high

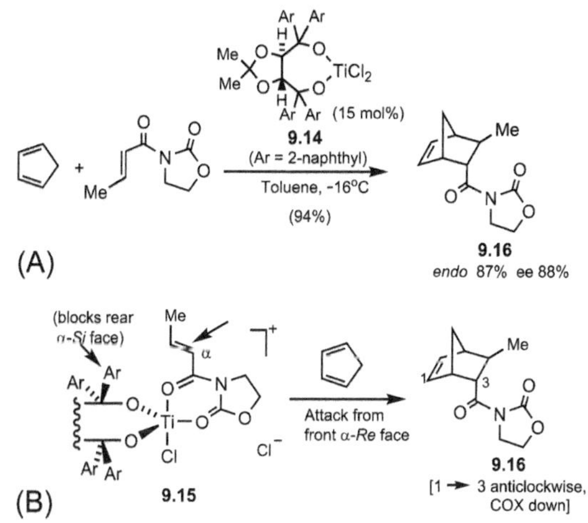


**FIG. 9.9** (A) Asymmetric Diels–Alder reaction using titanium(IV) TADDOL complex as a $C_2$-symmetric chiral catalyst and (B) model for enantioselectivity.

enantiomeric purity (ee 97%) is converted into the unnatural enantiomer of shikimic acid. The reaction is carried out at very low temperature (−78°C) to favour the formation of *endo* adduct.

Metal complexes of tartaric acid-derived diols (TADDOLs) and binaphthols (BINOLs) also serve as very effective $C_2$-symmetric catalysts in asymmetric Diels–Alder reactions. Fig. 9.9A illustrates an asymmetric Diels–Alder reaction between cyclopentadiene and a bidentate dienophile using the titanium(IV) TADDOL complex **9.14** as a chiral catalyst.[21–23] The predominant *endo* addition gives the product **9.16** with high enantioselectivity (ee 88%). The reaction is believed to occur via a cationic trigonal bipyramidal complex **9.15** in which titanium is coordinated with two carbonyl groups of the dienophile (Fig. 9.9B). The attack of cyclopentadiene then takes place preferentially from the exposed α-*Re* face of the dienophile (the rear α-*Si* face being blocked by an aryl group) to give the *endo* enantiomer **9.16** (see also the use of mnemonic).

Most successful asymmetric Diels–Alder reactions are achieved with dienophiles having two points of attachment to the metal as shown above. However, chiral catalysts are also developed which can be effective with dienophiles having single point of attachment to the metal. For example, Fig. 9.10 shows that the catalyst (*S*)-BINOL-Ti(IV) complex[24] **9.17** gives high enantioselectivity in the asymmetric Diels–Alder reaction with a simple dienophile benzoquinone, in a synthesis of the alkaloid (−)-ibogamine.[25]

Besides highly efficient $C_2$-symmetric catalysts, a variety of other chiral catalysts are known that promote highly enantioselective Diels–Alder reactions.[5, 26]

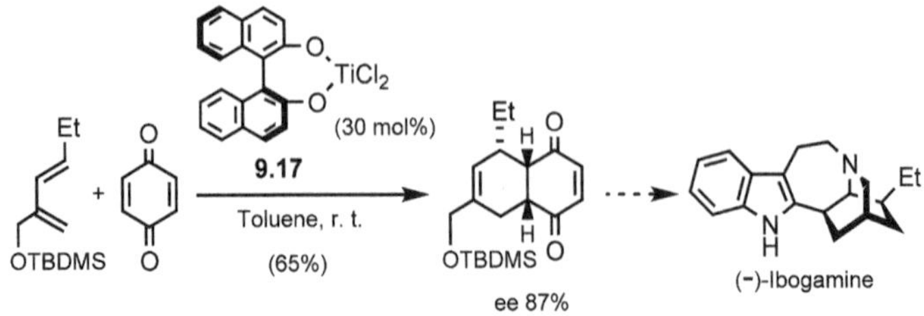


**FIG. 9.10** Asymmetric Diels–Alder reaction using (*S*)-BINOL-TiCl$_2$ catalyst in a synthesis of the alkaloid (−)-ibogamine.

**Problem 9.3** Explain the enantioselectivity observed in the following catalytic hetero-Diels–Alder reaction:

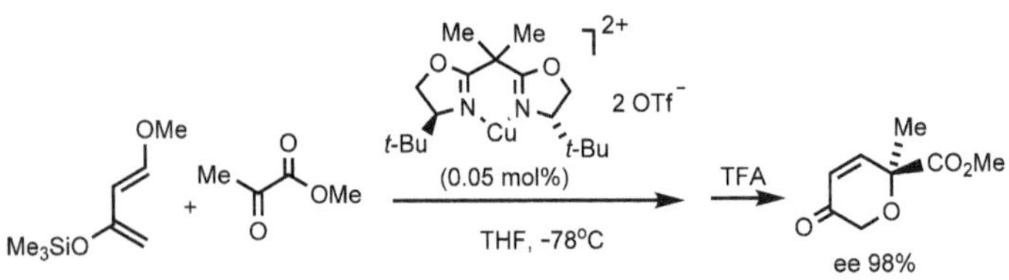


### 9.1.5 Organocatalysis

Organocatalysis as a general strategy for asymmetric synthesis was discovered by MacMillan [27], and the use of chiral imidazolidinones as organocatalysts in ionic conjugate additions was described earlier (see Section 7.14.4). Here we consider organocatalyzed asymmetric Diels–-Alder reactions with acyclic and cyclic dienes using the imidazolidinone catalyst **9.18** (Fig. 9.11A).[28, 29] The reactions give predominantly the *endo* products with high enantioselectivity.

The sense of asymmetric induction in these reactions is based on the iminium ion model (Fig. 9.11B). Condensation of the catalyst with acrolein dienophile gives selectively the (*E*)-iminium ion to avoid steric clash with the *gem*-dimethyl group. Since the front *Re* face of the alkene double bond in the favoured iminium ion is effectively shielded by the benzyl group on the catalyst framework, the diene attack takes place preferentially on the exposed *Si* face to give the predominant enantiomer of the *endo* aldehyde product after in situ hydrolysis. (See also the use of mnemonic.)

## 9.2 Asymmetric 1,3-dipolar cycloaddition reactions[30]

Like asymmetric Diels–Alder reactions, asymmetric 1,3-dipolar cycloaddition reactions have been performed using chiral substrates (dipole or dipolarophile),

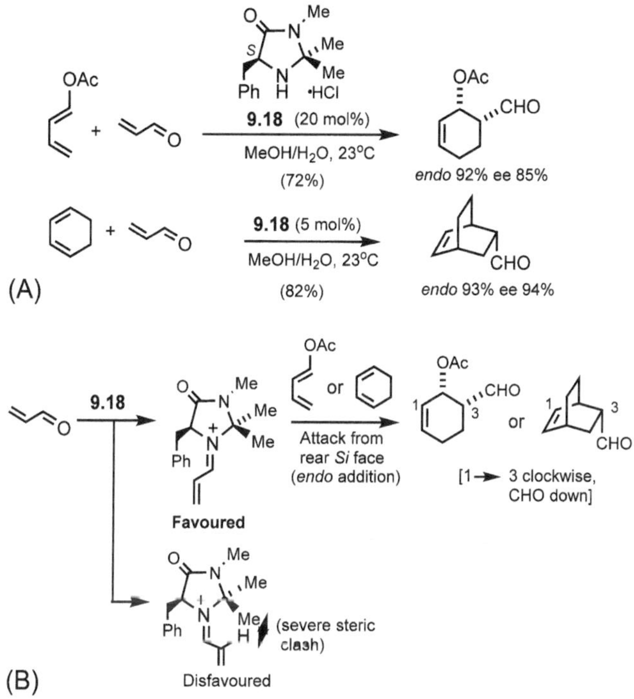


FIG. 9.11 (A) Organocatalyzed asymmetric Diels–Alder reactions between acyclic/cyclic diene and acrolein dienophile and (B) mechanism of enantioselective asymmetric induction based on the iminium ion model.

chiral auxiliaries and chiral catalysts, though relatively less progress has been made in this area.

## 9.2.1 Chiral substrates

Enantiopure 1,3-dipoles can be used as a chiral substrate in the asymmetric 1,3-dipolar cycloaddition. For example, a chiral azomethine ylide generated in situ by the condensation of an enantiopure (5*R*,6*S*) cyclic secondary amine with formaldehyde followed by deprotonation has been employed in the asymmetric synthesis of a proline derivative (Fig. 9.12).[31] The reaction is *endo*-selective producing three contiguous stereocentres in the newly formed five-membered ring with essentially complete stereochemical control. Hydrogenolysis gives the enantiopure proline derivative.

Vinyl sulphoxides are chiral and can act as a chiral dipolarophile in the asymmetric 1,3-dipolar cycloaddition. Fig. 9.13 shows the cycloaddition of a cyclic nitrone with an enantiopure sulphoxide. The regioselectivity is in favour of 4-SOAr-substituted isoxazolidine. The reaction is *exo*-selective with the approach of the nitrone preferentially to the face of the alkene anti to the bulky

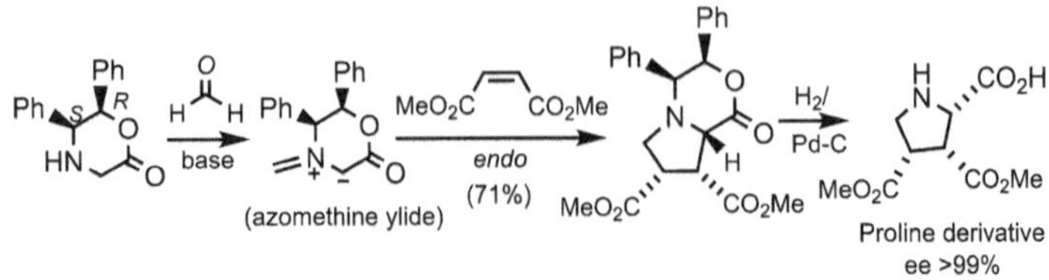


FIG. 9.12 Asymmetric 1,3-dipolar cycloaddition using a chiral dipole (azomethine ylide).

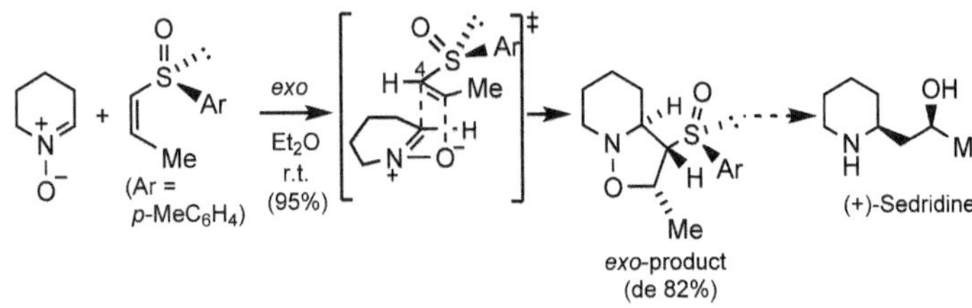


FIG. 9.13 Asymmetric 1,3-dipolar cycloaddition using a chiral dipolarophile (vinyl sulphoxide) in a synthesis of (+)-sedridine.

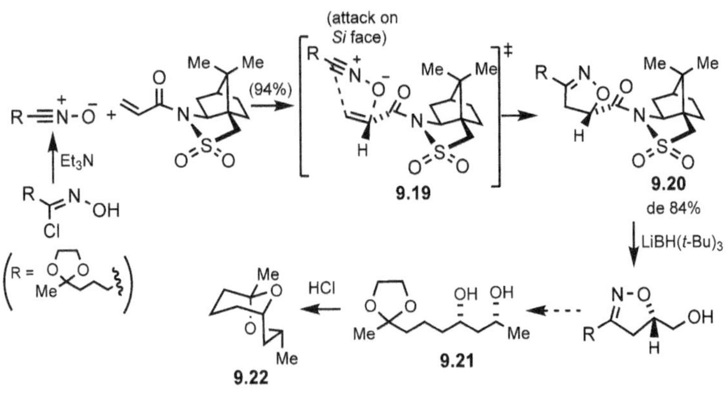


FIG. 9.14 Asymmetric nitrile oxide cycloaddition using Oppolzer sultam auxiliary in the synthesis of a natural product.

*p*-tolyl substituent. The *exo* product (de 82%) has been applied in the synthesis of the natural product (+)-sedridine.[32]

## 9.2.2 Chiral auxiliaries

The Oppolzer sultam chiral auxiliaries attached to a dipolarophile are found to be suitable in asymmetric 1,3-dipolar cycloadditions, as illustrated with the cycloaddition between a nitrile oxide dipole (generated in situ from an α-chlorooxime with a base) and an acrylic dipolarophile in Fig. 9.14.[33]

The reaction is highly diastereoselective (de 84%) and gives the adduct **9.20**. The diastereoselectivity can be rationalized in terms of a TS model **9.19**, in which the dipole approaches the *s-cis* conformation of the dipolarophile from the upper *Si* face opposite to the sulphonyl group on the auxiliary. Reductive removal of the auxiliary and further modifications produce the 1,3-diol **9.21** which is converted to the natural product **9.22** on treatment with acid.

### 9.2.3 Chiral catalysts

Many of the chiral catalysts used to achieve asymmetric 1,3-dipolar cycloaddition reactions are analogous to those employed in asymmetric Diels-Alder reactions.[30, 34] The catalysis of a normal electron demand 1,3-dipolar cycloaddition reaction ($HOMO_{dipole}$/$LUMO_{dipolarophile}$ control) involves coordination of a Lewis acidic metal to the dipolarophile, thereby lowering its LUMO energy. In order to favour coordination of the catalyst to the dipolarophile rather than to the dipole, a dipolarophile with two carbonyl groups is more suitable, which will chelate to the metal in a bidentate fashion. The organized TS model incorporating the metal and associated ligands then enforces a preferred orientation for the dipole addition. For example, the Ti(IV) TADDOL catalyst gives moderate to high enantioselectivity in the nitrone cycloaddition with *N*-acyloxazolidinone dipolarophiles, as shown in Fig. 9.15A.[35–37] Note that with strongly electron-deficient dipolarophile, the predominant regioisomer is 4-substituted isoxazolidine. The asymmetric nitrone cycloaddition with dipolarophiles with only one Lewis basic site is also possible by careful choice of Lewis acidic metals and chiral ligands.[38, 39]

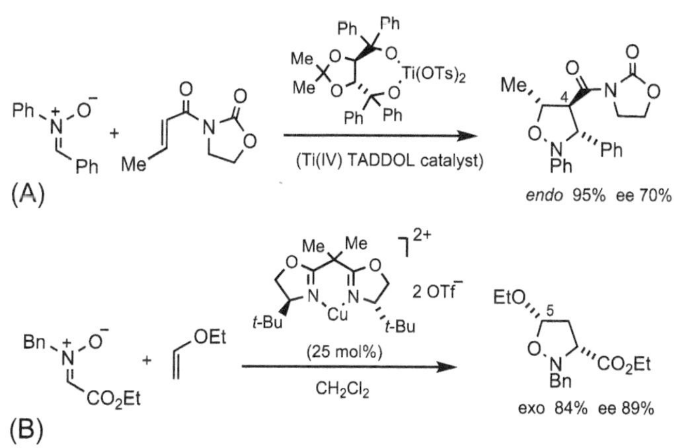


FIG. 9.15 Catalytic asymmetric nitrone cycloadditions: (A) normal electron demand reaction with an electron-deficient N-acyloxazolidinone dipolarophile and (B) inverse electron demand reaction with an electron-rich vinyl ether dipolarophile.

In an inverse electron demand nitrone cycloaddition with electron-rich dipolarophiles ($LUMO_{dipole}/HOMO_{dipolarophile}$ control), coordination of the catalyst to the dipole is generally responsible for the catalysis. For example, a nitrone cycloaddition with an electron-rich vinyl ether dipolarophile in the presence of the box catalyst [Cu-(*S*,*S*)-*t*-Bu-box]$(OTf)_2$ proceeds with high diastereo- and enantioselectivties (Fig. 9.15B).[40] Here the predominant regioisomer is 5-substituted isoxazolidine (see Fig. 4.17B).

### 9.2.4 Organocatalysis

Successful asymmetric 1,3-dipolar cycloadditions have been achieved using organocatalysis with MacMillan imidazolidinone catalysts (cf. Section 9.1.4). Fig. 9.16A illustrates a nitrone cycloaddition with crotonaldehyde using the organocatalyst **9.23**.[41] The reaction gives predominantly an *endo* adduct **9.24** with high enantioselectivity (ee 94%).

The cycloaddition proceeds through the reactive (*E*)-iminium ion **9.25** formed by condensation of the catalyst with the dipolarophile. The nitrone attack occurs preferentially from the top α-*Si* face of the alkene (bottom α-*Re* face being sterically shielded by benzyl group) in *endo* fashion to yield the enantioselective product **9.24** (Fig. 9.16B). As the dipolarophile bearing iminium ion moiety **9.25** is highly electron deficient, the nitrone addition is regioselective in favour of the 4-substituted isomer.

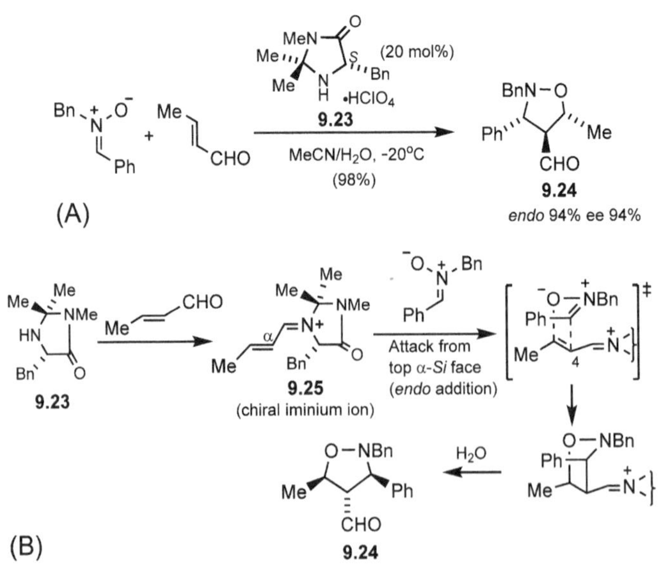


FIG. 9.16 (A) Organocatalytic asymmetric nitrone cycloaddition with crotonaldehyde and (B) mechanism of the asymmetric 1,3-dipolar cycloaddition.

## 9.3 Asymmetric cyclopropanation reactions[42–45]

Cyclopropanation is the reaction of alkenes with a carbene or a metal carbenoid to give cyclopropanes. The geometry of alkenes is often retained during cyclopropanation. A useful metal carbenoid is the Simmons–Smith reagent [$ICH_2ZnI$] or [$Zn(CH_2I)_2$] complex obtained by treating $CH_2I_2$ with Zn–Cu couple or $Et_2Zn$.[46, 47] The metal carbenoids can also be generated by the decomposition of diazo compounds in the presence of transition metal catalysts.[48, 49] Most common metal catalyst is rhodium acetate, although copper and other metal salts are also effective. The cyclopropane units are present in numerous biologically active natural products or their analogues, which have provided stimulus for investigations in asymmetric cyclopropanation reactions.

### 9.3.1 Chiral substrates

Diastereoselective cyclopropanation is possible using a chiral alkene in the Simmons–Smith reaction. The carbenoid usually reacts from the less hindered side of the alkene to provide the major diastereomer. However, directed cyclopropanation can be achieved using chiral allylic alcohols when the attack of the carbenoid is strongly directed by the hydroxyl group in the substrate, as shown with cyclic substrates in Fig. 9.17.[50] The reaction proceeds with coordination of the O atom to the Zn, followed by transfer of the $CH_2$ to the same face of the allylic double bond via a butterfly TS. Directed cyclopropanation is also possible in acyclic systems.[51]

### 9.3.2 Chiral auxiliaries

A chiral dioxaborolane auxiliary **9.26**, introduced by Charette[52, 53], has been shown to be highly effective in the asymmetric Simmons–Smith cyclopropanation of allylic alcohols. The auxiliary can be prepared from tetramethyltartramide and butylboronic acid. Fig. 9.18 illustrates an asymmetric cyclopropanation in the presence of the auxiliary **9.26** (a chiral Lewis acid)

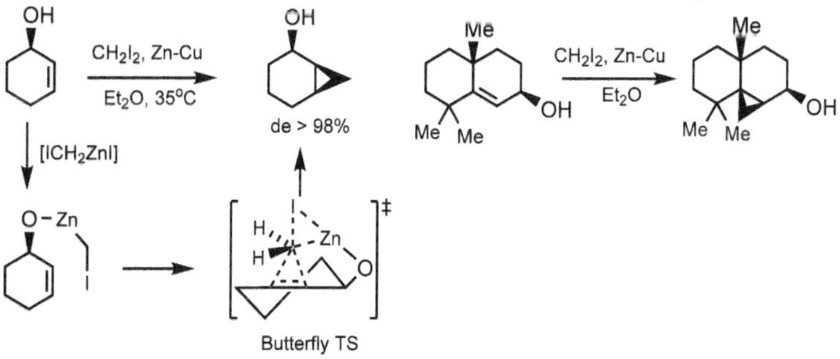


**FIG. 9.17** Diastereoselective cyclopropanation using chiral cyclic substrates.

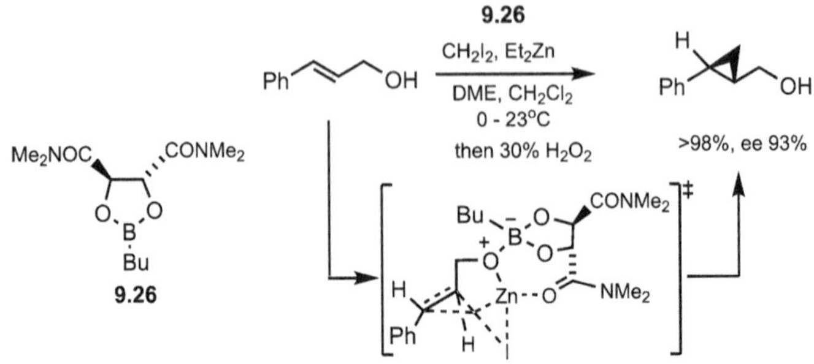


**FIG. 9.18** Asymmetric cyclopropanation using Charette chiral dioxaborolane auxiliary.

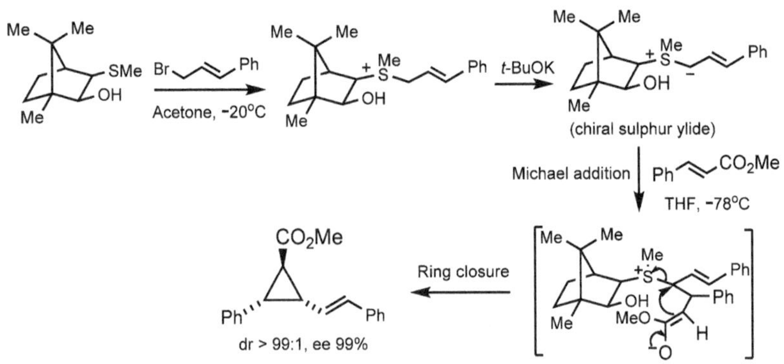


**FIG. 9.19** Asymmetric cyclopropanation through a camphor-based sulphur ylide.

which induces high levels of asymmetric induction in the resultant cyclopropane from an allylic alcohol.[53] A hydrogen peroxide work-up is used to eliminate boron side-products. This methodology has been employed in the synthesis of an antifungal agent FR-900848.[54]

Asymmetric cyclopropanation is also possible through chiral ylides, as shown in Fig. 9.19.[55] Camphor-based sulphur ylides are particularly effective for asymmetric cyclopropanation of electron-deficient alkenes. The cyclopropanation reaction involves Michael addition of the ylide followed by ring closure to give the product with excellent diastereo- and enantioselectivities (dr > 99:1, ee 99%).

### 9.3.3 Chiral catalysts

Transition metals can catalyse the cyclopropanation of alkenes in the presence of a chiral ligand in highly enantioselective fashion. The asymmetric cyclopropanation usually involves metal carbenoids from electron-deficient diazo compounds. A general catalytic cycle with Fischer type (electrophilic) metal carbenoid is shown in Fig. 9.20.

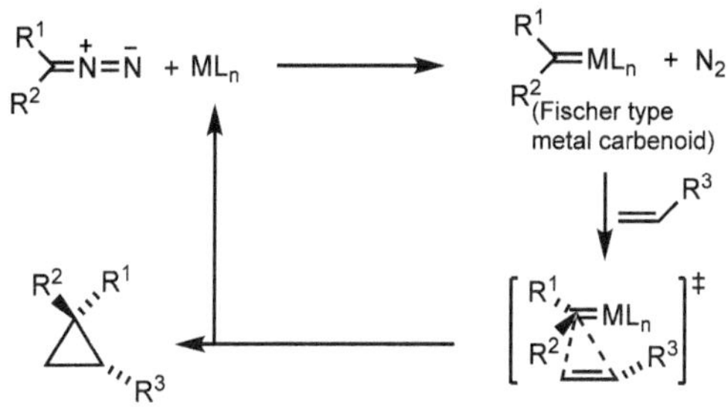


**FIG. 9.20** Catalytic cycle via Fischer-type metal carbenoid for asymmetric cyclopropanation. M = metal, L = chiral ligand.

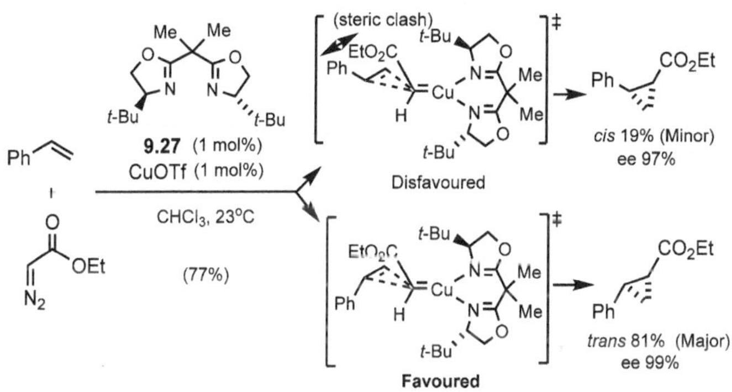


**FIG. 9.21** Catalytic asymmetric cyclopropanation in the presence of $C_2$ symmetric chiral bisoxazoline ligand **9.27** and copper(I) triflate.

Particularly effective catalysts are $C_2$ symmetric catalysts derived from enantiomerically pure bisoxazoline ligands and metal salts such as Cu(I) salt. The alkenes used are generally electron-rich, such as styrenes and enol ethers. Fig. 9.21 illustrates an asymmetric cyclopropanation reaction using catalytic amounts of the bisoxazoline ligand **9.27** and copper(I) triflate.[56, 57] The reaction gives predominantly the *trans*-cyclopropane product with excellent enantioselectivity.

The Cu(I) complex of chiral bisoxazolines is also effective in promoting asymmetric cyclopropanation of dienes. For example, the Cu(I) complex of the ligand **9.28** is an efficient catalyst for asymmetric cyclopropanation of 2,5-dimethyl-2,4-hexadiene, which gives the major product chrysanthemic acid *t*-butyl ester (a key intermediate to pyrethroid insecticides) with very high enantioselectivity (Fig. 9.22).[58] Notably, since the starting diene is more nucleophilic (has a high energy HOMO) than the single alkene in the ester product, the reaction can be stopped after one carbene addition.

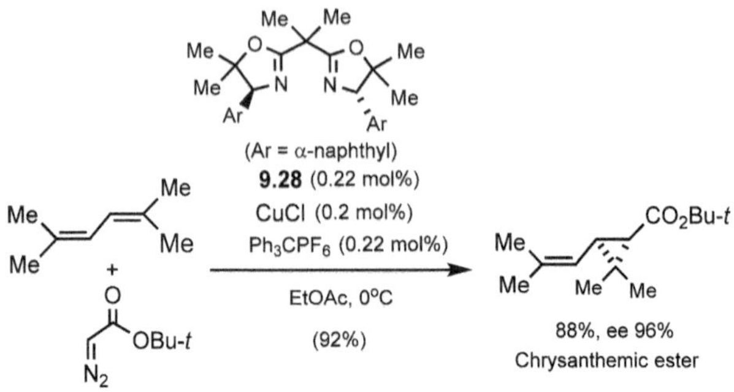


**FIG. 9.22** Synthesis of chrysanthemic acid *t*-butyl ester (a key intermediate in pyrethroid insecticides) using catalytic asymmetric cyclopropanation of a hexadiene in the presence of copper(I) complex of the bisoxazoline ligand **9.28**.

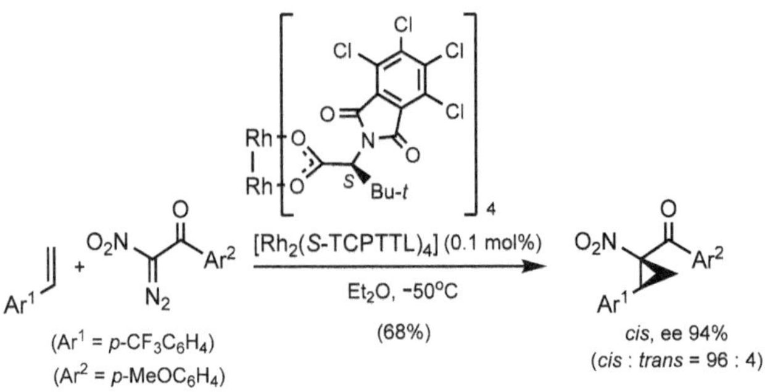


**FIG. 9.23** Catalytic asymmetric cyclopropanation using α-nitro-α-diazo carbonyl compounds in the presence of chiral rhodium complex $Rh_2$(*S*-TCPTTL)$_4$.

Highly enantioselective cyclopropanations have also been carried out using Rh(II) complexes of chiral ligands. Diazo compounds with two electron-withdrawing substituents serve as excellent substrates.[59] For example, Fig. 9.23 shows that α-nitro-α-diazo aryl ketones give selectively the *cis*-cyclopropanes with high enantioselectivity using the chiral rhodium complex, dirhodium(II) tetrakis(*N*-tetrachlorophthaloyl-*S*-*tert*-leucinate) [$Rh_2$(*S*-TCPTTL)$_4$].[60] The *cis*/*trans* diastereoselectivity has been shown to be sensitive to the nature of carbonyl substituent.[61]

Asymmetric intramolecular cyclopropanations can be performed with a variety of metal–ligand catalysts.[62, 63] Fig. 9.24 shows an intramolecular cyclopropanation using the chiral rhodium complex, dirhodium(II) tetrakis(methyl 2-pyrrolidone-5*S*-carboxylate) [$Rh_2$(5*S*-MEPY)$_4$]. The reaction gives the bicyclic lactone **9.29** (ee 95%), which was involved in a synthesis of the chiral cyclopropane unit in the antifungal drug ambruticin S.[64]

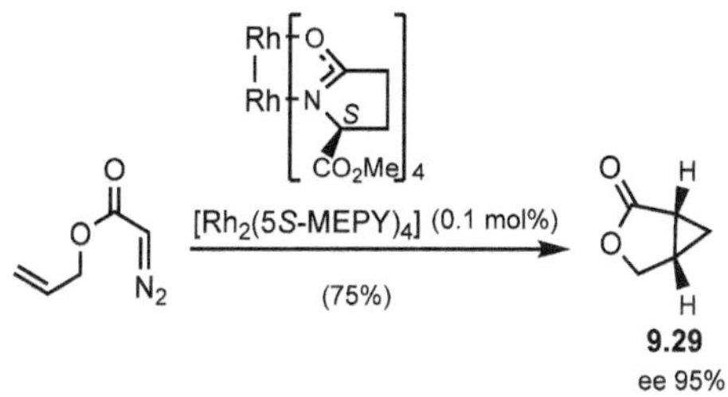


**FIG. 9.24** Catalytic asymmetric intramolecular cyclopropanation using the chiral rhodium complex $Rh_2(5S\text{-MEPY})_4$.

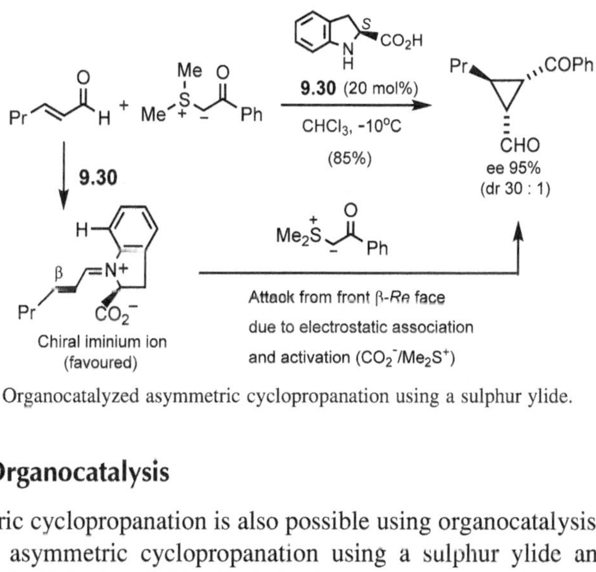


**FIG. 9.25** Organocatalyzed asymmetric cyclopropanation using a sulphur ylide.

## 9.3.4 Organocatalysis

Asymmetric cyclopropanation is also possible using organocatalysis. Fig. 9.25 shows an asymmetric cyclopropanation using a sulphur ylide and an α,β-unsaturated aldehyde (cf. Fig. 9.19) in the presence of a chiral indoline **9.30** as organocatalyst.[65] The reaction proceeds through a sterically favoured chiral iminium ion intermediate. The Michael addition of the ylide then occurs from the front β-*Re* face because of electrostatic association and activation by the $CO_2^-$ on the catalyst with the $Me_2S^+$ on the ylide. Subsequent ring closure leads to the cyclopropane product with high enantioselectivity.

**Problem 9.4** The following asymmetric aziridination reaction exhibits high enantioselectivity. Explain mechanistically.

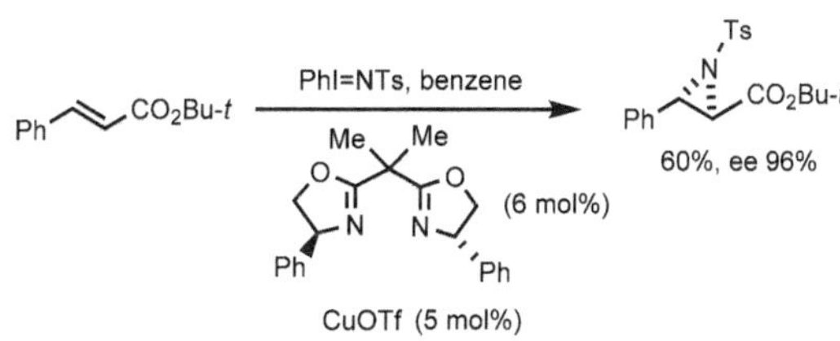

## 9.4 Asymmetric ene reactions[66, 67]

Asymmetric ene reactions are particularly successful with the carbonyl ene reaction of glyoxylate esters. Two main asymmetric ene approaches involve chiral auxiliaries and chiral catalysts.

### 9.4.1 Chiral auxiliaries

Various chiral auxiliaries can be used for asymmetric carbonyl ene reactions of glyoxylate esters in conjunction with Lewis acid catalysts. (−)-8-Phenylmenthol is a suitable chiral auxiliary for the purpose. Fig. 9.26A illustrates an asymmetric carbonyl ene reaction of 8-phenylmenthyl glyoxylate ester **9.31** with 1-substituted alkene in the presence of Lewis acid $SnCl_4$.[68] The reaction exhibits excellent diastereoselectivity (de > 97%). The geometry of the new double bond is preferably (*E*). Removal of the auxiliary with $LiAlH_4$ will give the (*S*) enantiomer of the glycol product.

The Lewis acid $SnCl_4$ chelates with two carbonyl groups of the glyoxylate ester (enophile) and the reaction proceeds through a preferred chair TS (e-R) involving attack of the alkene (ene component) on the front *Si* face of the aldehyde carbonyl (the rear face being shielded by the phenyl group on the auxiliary) to give the predominant diastereomer (Fig. 9.26B).[69]

Oppolzer chiral sultam auxiliary can also be used to effect asymmetric glyoxylate-ene reaction, as shown in Fig. 9.27.[70] The diastereoselectivity has

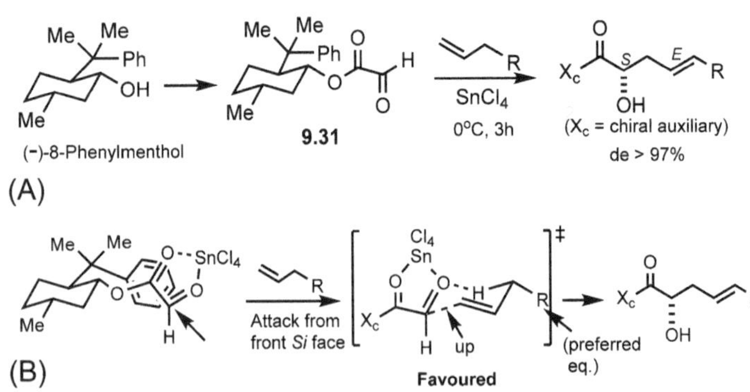


**FIG. 9.26** (A) Asymmetric carbonyl ene reaction using the chiral auxiliary (−)-8-phenylmenthol and (B) model for diastereoselectivity.

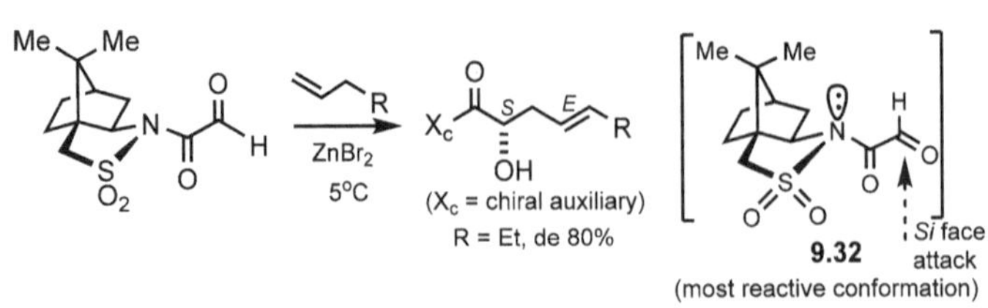


**FIG. 9.27** Asymmetric carbonyl ene reaction using Oppolzer sultam auxiliary.

been ascribed to the alkene attack on the lower *Si* face of the most reactive conformation **9.32** of the *N*-glyoxylyl-sultam.

**Problem 9.5** Explain the formation of the major product in each of the following chiral auxiliary-mediated asymmetric ene reactions:

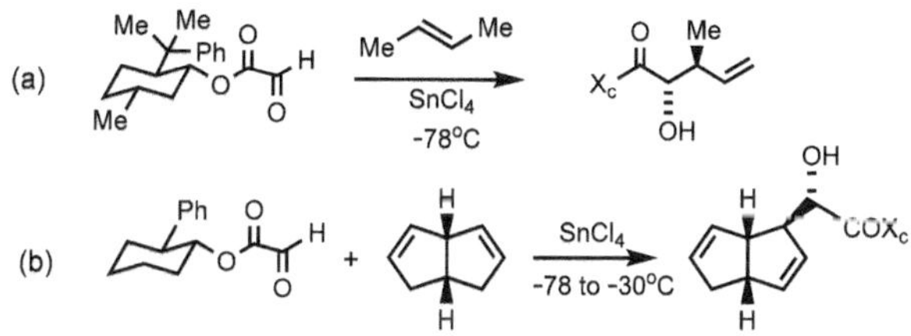


## 9.4.2 Chiral catalysts

A variety of metal–chiral ligand complexes have been employed successfully in asymmetric ene reactions. For example, 1,1′-bi-2-naphthol (BINOL) complex of titanium(IV) salt exhibits excellent enantioselectivity in the carbonyl ene reaction.[71, 72] This is illustrated with two examples in Fig. 9.28. The catalyst is formed from enantiomerically pure (*R*)- or (*S*)-BINOL and $(i\text{-PrO})_2TiCl_2$ in the presence of an additive, 4 Å molecular sieves. The use of molecular sieves facilitates the alkoxy–ligand exchange for the formation of metal–BINOL catalyst complex. The product is a homoallylic alcohol with very high enantiomeric purity. The second reaction has been applied to the synthesis of the terpene (*R*)-(−)-ipsdienol.[73, 74]

The enantioselectivity of the reaction has been explained in terms of a mechanistic model shown in Fig. 9.29.[75] The catalyst–enophile complex **9.33** is formed by the coordination of an aldehyde oxygen lone pair to titanium and H···O hydrogen bonding between aldehyde H and the closer and more

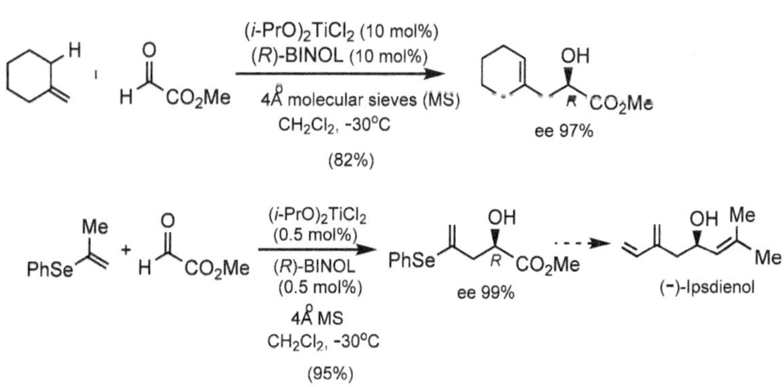


FIG. 9.28 Asymmetric carbonyl ene reactions using $(i\text{-PrO})_2TiCl_2$-(*R*)-BINOL catalyst.

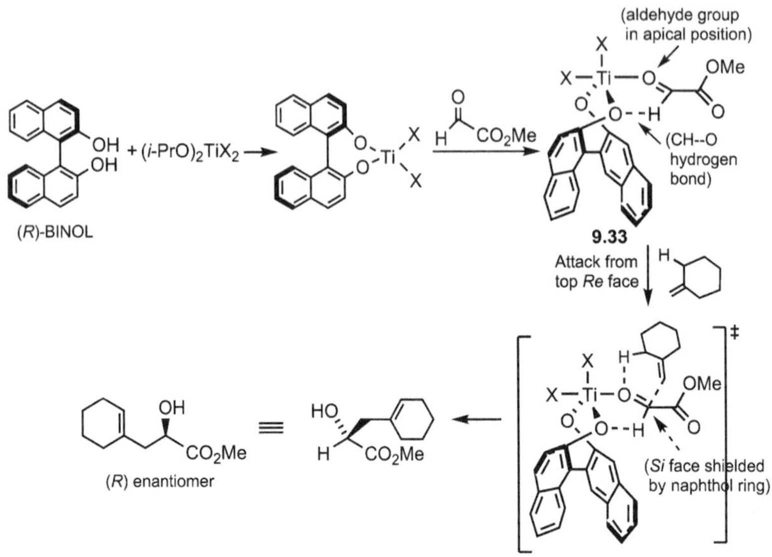


**FIG. 9.29** Model for enantioselectivity of the asymmetric carbonyl ene reaction with $(i\text{-PrO})_2$-$TiX_2$-(*R*)-BINOL catalyst in Fig. 9.28.

accessible O of the BINOL ligand. The complex **9.33** has a pentacordinate trigonal bipyramidal titanium structure with aldehyde group and one electronegative ligand (X) in apical positions. The attack of alkene (ene component) takes place preferentially on the more sterically accessible top *Re* face of aldehyde enophile (the bottom face being sterically shielded by neighbouring naphthol moiety) to give the (*R*) enantiomer of the product. As both (*R*)- and (*S*)-BINOL are available in optically pure form, the asymmetric methodology can be used to synthesize each enantiomer of the product.

Remarkably, this asymmetric glyoxylate–ene reaction exhibits a positive nonlinear effect (see Section 5.3.2).[76–78] It has been observed that enantioselectivity of the reaction far exceeds the optical purity of the (*R*)-BINOL reagent; for example the use of BINOL (ee 33%) leads to the product with ee 91.4%. This nonlinear asymmetric amplification is attributed to the aggregation of the catalyst in which (*R*)(*R*)-(BINOL-$TiX_2$)$_2$ promoter is more active than the *meso* (*R*)(*S*)-(BINOL-$TiX_2$)$_2$ promoter.

The box catalysts, the copper(II) complex of chiral bis-oxazolines, are also highly effective in the asymmetric carbonyl ene reaction. Fig. 9.30A illustrates an asymmetric glyoxylate-ene reaction using a box catalyst [Cu-(*S*,*S*)-*t*-Bu-box] $(SbF_6)_2$ (**9.10** in Fig. 9.6).[79]

The reaction proceeds through chelation of the two carbonyl groups of the glyoxylate ester with the copper forming a well-defined catalyst–enophile complex of square-planar geometry **9.34** (Fig. 9.30B).[16] The alkene attack then takes place from the bottom *Si* face of the aldehyde enophile (the top face being

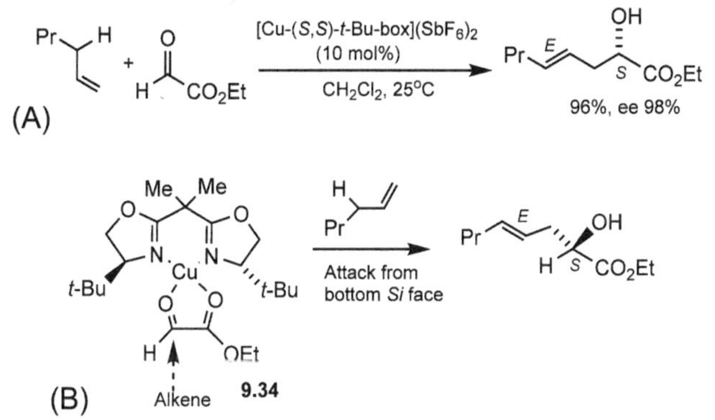


FIG. 9.30 (A) Asymmetric carbonyl ene reaction using the box catalyst [Cu-(*S*,*S*)-*t*-Bu-box] $(SbF_6)_2$ and (B) mechanism of the reaction.

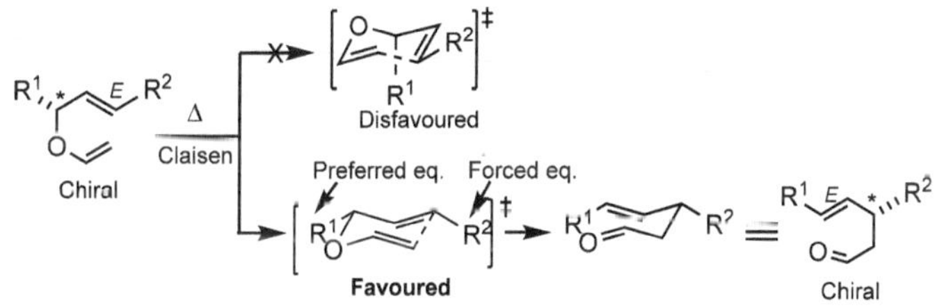


FIG. 9.31 Chirality transfer in the Claisen rearrangement.

sterically shielded by bulky *t*-Bu group) to give the (*S*) enantiomer of the (*E*) product (*E*:*Z*=96:4).

## 9.5 Asymmetric Claisen rearrangements[80–83]

The development of asymmetric methods for the Claisen rearrangement remains an important goal in organic synthesis. The methodology includes all four types of control for asymmetric induction using chiral substrates, chiral auxiliaries, chiral reagents and chiral catalysts. The diastereo- and enantioselectivitiy of the rearrangement usually results from the highly ordered chair transition structure.

### 9.5.1 Chiral substrates

A general Claisen rearrangement with a chiral substrate having (*E*) stereochemistry of the allylic double bond is shown in Fig. 9.31.

In the reaction, the original stereocentre is destroyed and a new stereocentre is created giving a chiral product. This is called 'chirality transfer' (transfer of chirality from one site in the molecule to another). The reaction proceeds

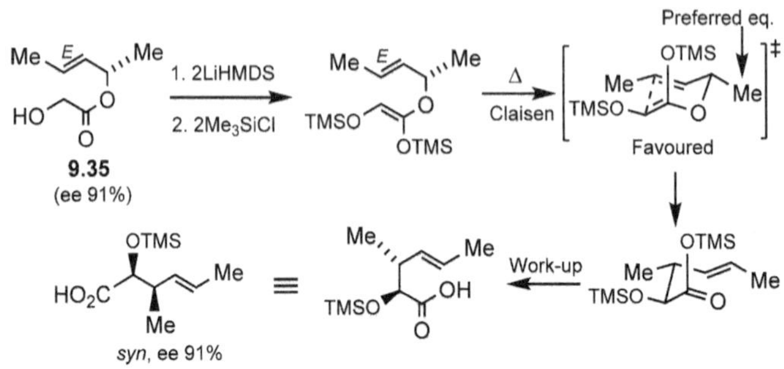


FIG. 9.32 Asymmetric Claisen rearrangement using a chiral substrate.

preferably through a chair TS in which $R^1$ at the old stereocentre becomes equatorial, which leads to (*E*) configuration of the new double bond. The (*E*) stereochemistry of the old double bond forces $R^2$ to be equatorial. If the chiral substrate is used as a single enantiomer, the reaction will be enantioselective.

Fig. 9.32 shows an asymmetric Claisen rearrangement using an enantiomerically enriched ester **9.35** (ee 91%) which gives a *syn* product (ee 91%). The reaction proceeds through a more stable chair TS (in which Me at the old stereocentre becomes equatorial) to give the enantioselective product.

Usually an enantiomer of a substrate gives a particular enantiomer of the product. However, it is possible to convert both enantiomers of a substrate into the same product enantiomer. This is referred to as 'enantioconvergent' strategy. Fig. 9.33 illustrates an enantioconvergent synthesis using a chiral acetylenic alcohol.[84] Catalytic hydrogenation of (*R*) enantiomer **9.36** gives a (*Z*) alkene by *syn* addition, whereas SET reduction of (*S*) enantiomer **9.37** in Na/liquid $NH_3$ provides a more stable (*E*) alkene. Both (*R*,*Z*) and (*S*,*E*) isomers of an allyl alcohol then undergo Eschenmosher–Claisen rearrangement to give the same (*S*) enantiomer of the (*E*) product. The rearrangements are shown to proceed with almost complete chirality transfer. The sequence was used in a synthesis of the side chain of α-tocopherol.[84]

Asymmetric Claisen rearrangements with chiral substrates can also be performed at a lower temperature using Lewis acid catalysis. Various metal salts have been investigated as Lewis acid catalysts. However, a major problem encountered is the dissociation and racemisation of starting allyl vinyl ethers. Lanthanide cations such as holmium(III) in combination with 2,2-dimethyl-6,6,7,7,8,8,8-heptafluoro-3,5-octanedionate (fod) ligand are found to be particularly effective.[85] Fig. 9.34 shows an asymmetric Claisen rearrangement using $Ho(fod)_3$ catalyst which proceeds with complete chirality transfer and produces a single (*E*) diastereomer (cf. Fig. 9.31).

A popular method for the asymmetric synthesis of allylic amines from enantioenriched allylic alcohol precursors makes use of the [3,3] aza-Claisen-type rearrangement of their imidates, called the Overman reaction.[86] This is illustrated with an example in Fig. 9.35. The imidates are usually prepared from

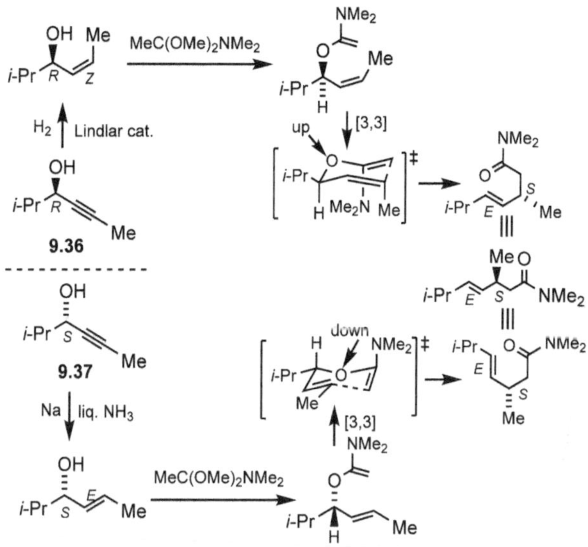


FIG. 9.33 An enantioconvergent synthesis via Eschenmoser–Claisen rearrangement.

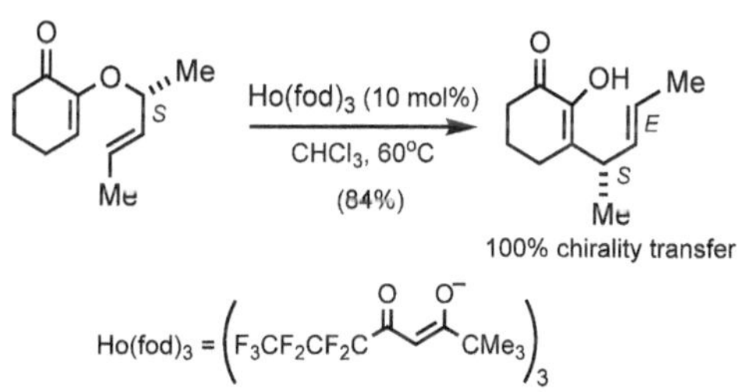


FIG. 9.34 Lewis acid-catalysed asymmetric Claisen rearrangement with a chiral substrate.

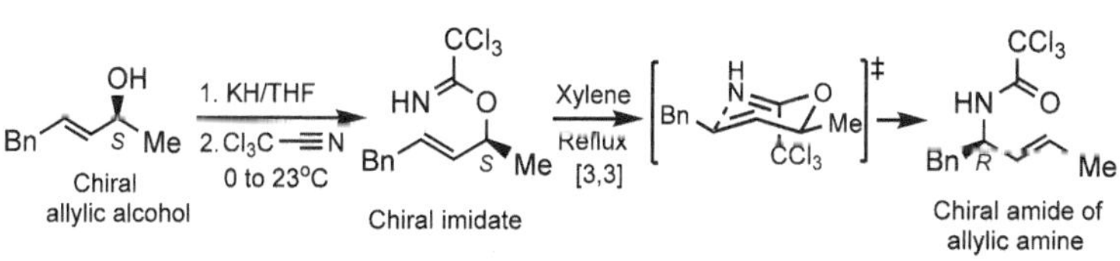


FIG. 9.35 Asymmetric Overman reaction via a chiral imidate of an allylic alcohol.

the allylic alkoxides and trichloroacetonitrile. The reaction proceeds with efficient chirality transfer and high stereoselectivity for the (*E*) double bond in the product. Hydrolysis of the amide product will give the allylic amine. The catalytic asymmetric version of this rearrangement has been described later (see Section 9.5.4, Fig. 9.40).

**Problem 9.6** Explain the formation of the product **X** in the following asymmetric aza-Claisen rearrangement involved in a synthesis of (−)-verrucarinolactone:

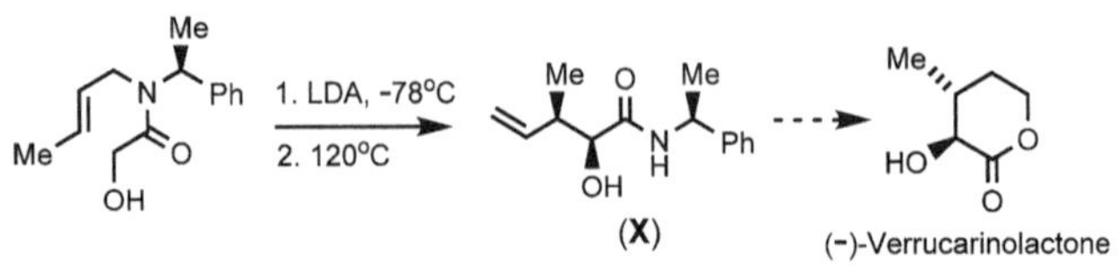


**Problem 9.7** The following asymmetric Claisen rearrangement involves chirality transfer from a substrate with a chiral centre to the product with a chiral axis. Explain the correlation between chiral centre and chiral axis in the reaction.

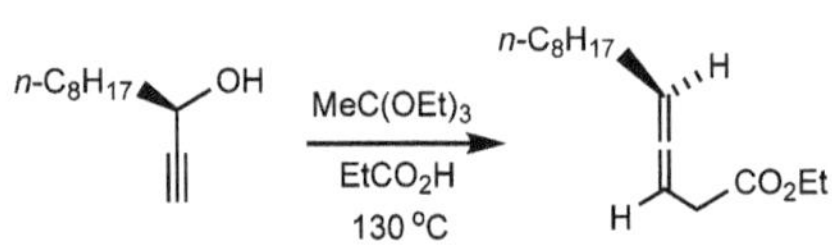


## 9.5.2 Chiral auxiliaries

The chiral auxiliary control provides an important strategy for the asymmetric Claisen rearrangement. This is illustrated using Enders RAMP/SAMP auxiliary (see Fig. 7.29) in the Carroll–Claisen rearrangement in Fig. 9.36.[87] An allyl ester of a β-keto acid **9.38** is covalently linked with RAMP. The resulting chiral intermediate then forms a lithium chelate with LDA, which undergoes a [3,3]

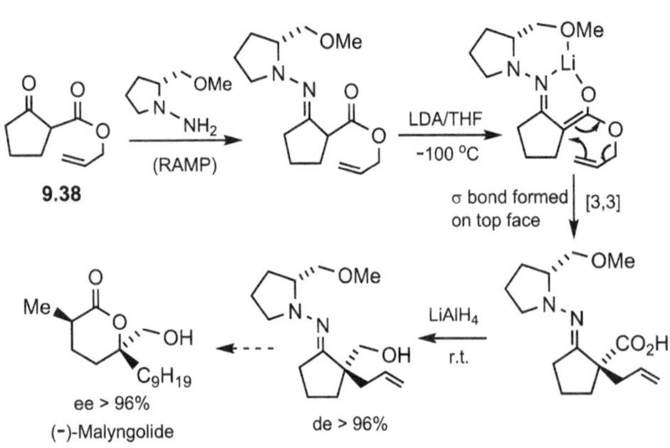


FIG. 9.36 Asymmetric Claisen rearrangement using RAMP chiral auxiliary in a synthesis of an antibiotic (−)-malyngolide.

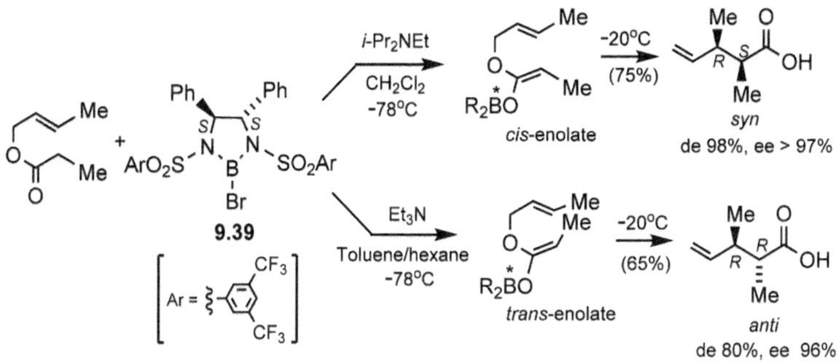


**FIG. 9.37** Asymmetric Claisen rearrangement using a chiral boron reagent.

rearrangement. The new σ bond is formed on the top face of the cyclopentane ring (the bottom face being sterically shielded) to give rise to a new quaternary chiral centre. The asymmetric rearrangement was used as a key step in the synthesis of an antibiotic (−)-malyngolide.[87]

## 9.5.3 Chiral reagents

Highly enantioselective Claisen rearrangements can be performed under reagent control using an enantiopure boron reagent **9.39** (Fig. 9.37).[88] In the presence of the base $Et_3N$, the *trans*-boron enolate is formed preferentially whereas with a more hindered Hünig's base (*i*-$Pr_2NEt$), the *cis*-boron enolate is favoured (see Section 7.6.2). The boron enolates undergo Claisen rearrangement to give the products with a very high diastereo- and enantioselectivity. The *cis*-enolate leads almost exclusively to a *syn* product (ee > 97%) whereas the *trans*-enolate gives predominantly an *anti* product (ee 96%) (cf. Fig. 8.62).

The enantioselectivity of the reaction can be rationalized in terms of a chair TS model, as shown for the *cis* boron enolate in Fig. 9.38. The (*S*,*S*) stereocentres of the $C_2$-symmetric boron reagent control the rotational preference of the sulphonyl groups, and the sterically favoured TS (that avoids the interaction between the bulky Ar group and Me substituent) leads to the major enantiomer of the *syn* product.

## 9.5.4 Chiral catalysts

The development of catalytic asymmetric Claisen rearrangements with acyclic substrates and chiral catalysts remains a challenging task. However, successful enantioselective rearrangements have been achieved with Lewis acid catalysts[89, 90], hydrogen bond donor catalysts[91] and transition metal catalysis[92]. But the substrate scopes for these methods are generally limited. Chiral Cu (II) bisoxazoline complexes (box catalysts) are found to be efficient Lewis acid catalysts for the asymmetric Claisen rearrangement.[89] An example with

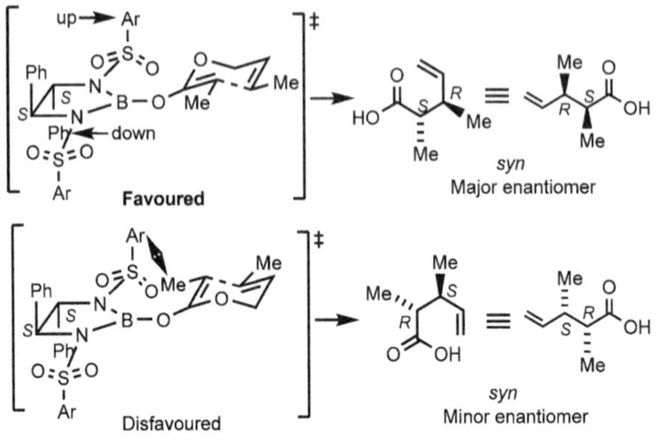


**FIG. 9.38** The chair TS model for enantioselectivity of the reaction using *cis* boron enolate in Fig. 9.37.

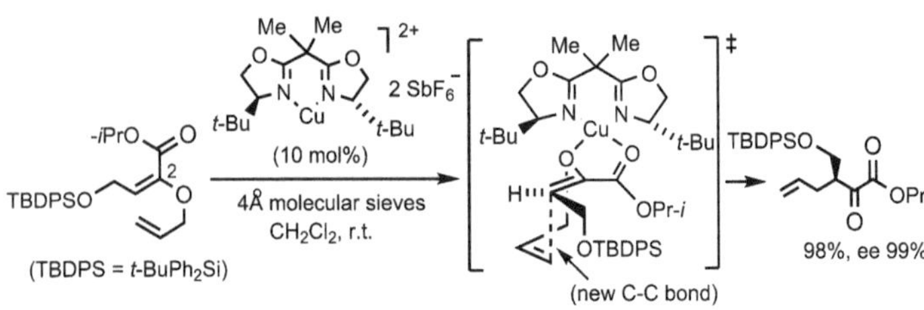


**FIG. 9.39** Catalytic asymmetric Claisen rearrangement using chiral Cu(II)-box catalyst.

2-alkoxycarbonyl-substituted allyl vinyl ether is shown in Fig. 9.39. The ester carbonyl and the vinyl ether oxygen could chelate with the copper forming a well-defined catalyst-substrate complex which allows C—C bond formation preferably from the less hindered side (opposite to the up *t*-Bu) to give the enantioselective product.

Although excellent enantioselectivities were observed in these reactions, there were severe substrate restrictions. The alkoxycarbonyl moiety on the vinyl ether is essential for the bidentate activation by the copper(II)-box catalyst. Furthermore, the diastereoselectivity of the reaction is dependent on the allyl ether geometry with the (*Z*) isomer providing better results.

Highly efficient catalytic asymmetric Claisen rearrangements have been developed using transition metal catalysts without rigid substrate requirements. The late transition metals such as Pd are better able to coordinate with the soft C=C bond than the lone pair on the oxygen atom in the allyl vinyl ether, and thus lead to the metal-alkene complex. A very popular enantioselective Claisen rearrangement is the Overman imidate–amide rearrangement (see Fig. 9.35) using Pd(II) catalysis.[93, 94] Fig. 9.40 illustrates an example using dimeric cobalt oxazoline palladacycle complex ([COP-Cl]$_2$) as chiral Pd(II) catalyst.[95]

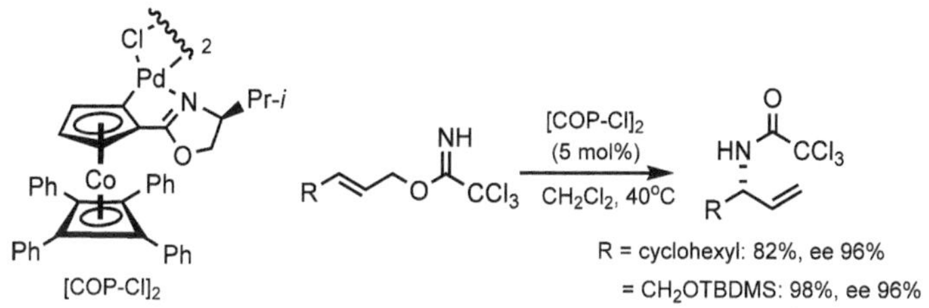


**FIG. 9.40** Catalytic asymmetric Overman imidate-amide rearrangement using Pd(II) catalyst [COP-Cl]$_2$ (COP = cobalt oxazoline palladacycle).

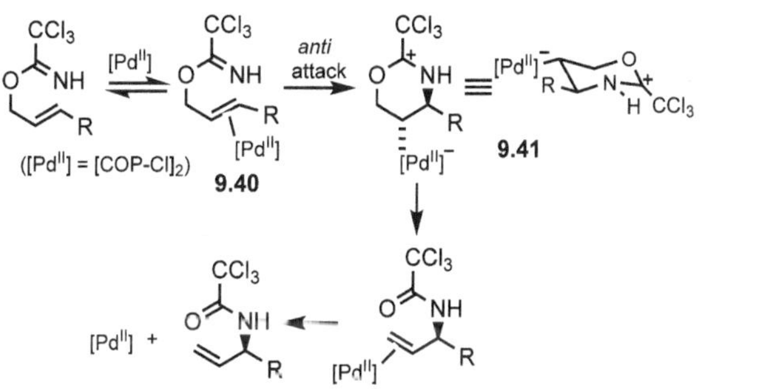


**FIG. 9.41** Cyclization-induced rearrangement mechanism for the enantioselective Overman reaction in Fig. 9.40.

The reaction involves a 'cyclization-induced' rearrangement mechanism (Fig. 9.41).[93, 95] Pd(II) coordinates to the allyl alkene moiety of the allyl vinyl ether, then *anti* attack by the imidate nitrogen on the alkene complex **9.40** generates the cyclopalladate intermediate **9.41**, which rearranges to the observed product and regenerates the catalyst. The *anti* pathway is preferred over the *syn* pathway as the *anti* attack leads to the most favoured complex **9.41** in which all substituents are equatorial.

## 9.6 Asymmetric Wittig and related rearrangements[96–100]

Asymmetric Wittig rearrangements provide a general tool for asymmetric synthesis of homoallylic alcohols from allyl ether derivatives. In this section, we will describe mainly asymmetric [2,3]-Wittig rearrangement, and some examples of other [2,3] rearrangements involving ylides, and sulphoxides (Mislow rearrangement).

### 9.6.1 Chiral substrates

With a chiral substrate, asymmetric Wittig rearrangement can take place with essentially complete chirality transfer. For example, the chiral (*Z*)-substrate

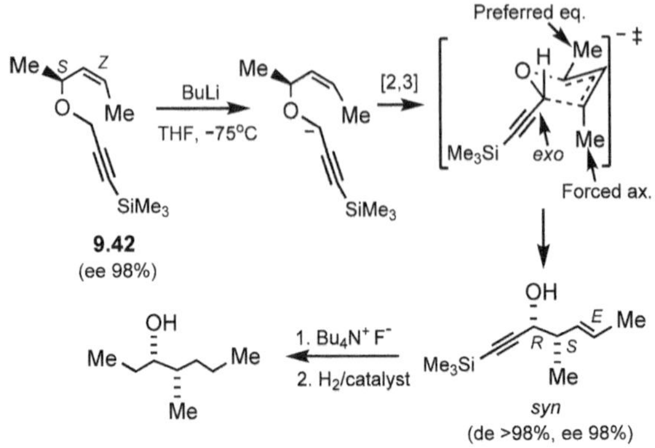


**FIG. 9.42** Asymmetric Wittig rearrangement with a chiral (*Z*)-substrate.

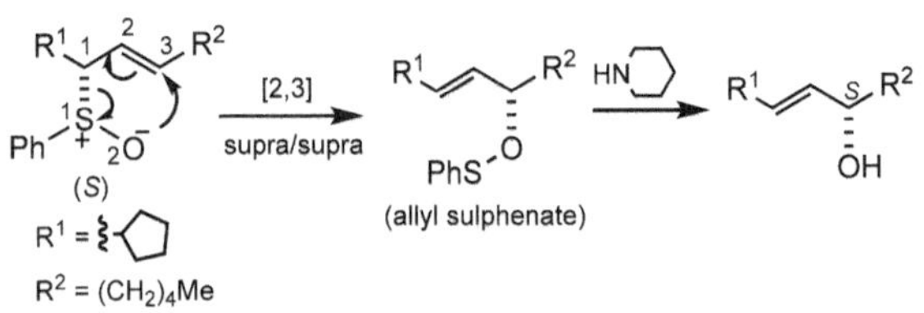


**FIG. 9.43** Asymmetric Mislow rearrangement with a chiral allyl sulphoxide.

**9.42** (ee 98%) is converted into the *syn* homoallylic alcohol product (de > 98%, ee 98%) with (*E*) stereochemistry of the new double bond (Fig. 9.42).[101] The diastereo- and enantioselectivity of the rearrangement arises from the favoured *exo*-alkynyl TS in which Me at the old stereocentre becomes preferably equatorial. Note that (*Z*) allylic double bond forces Me on it to be axial. Desilylation and catalytic hydrogenation afford the saturated *syn* alcohol in essentially pure enantiomeric form.

Asymmetric synthesis of allylic alcohols can be performed using the [2,3] sigmatropic rearrangement of chiral allyl sulphoxides (Mislow rearrangement)[102] or chiral allyl selenoxides[103]. An asymmetric Mislow rearrangement of an (*S*)-sulphoxide is shown in Fig. 9.43. The rearrangement proceeds suprafacially across the allyl unit to give the allyl sulphenate, which on treatment with a thiophile (piperidine) yields the (*S*) allylic alcohol.

## 9.6.2 Chiral auxiliaries

A number of chiral auxiliaries have been used to effect asymmetric [2,3]-Wittig rearrangements. Very high diastereoselectivity has been observed with the auxiliary 8-phenylmenthol (see Fig. 9.26) using an alkoxycarbonyl substrate **9.43** (Fig. 9.44).[104] The chiral auxiliary is attached to the substrate through an ester linkage. Here the anion-stabilizing substituent is an electron-withdrawing

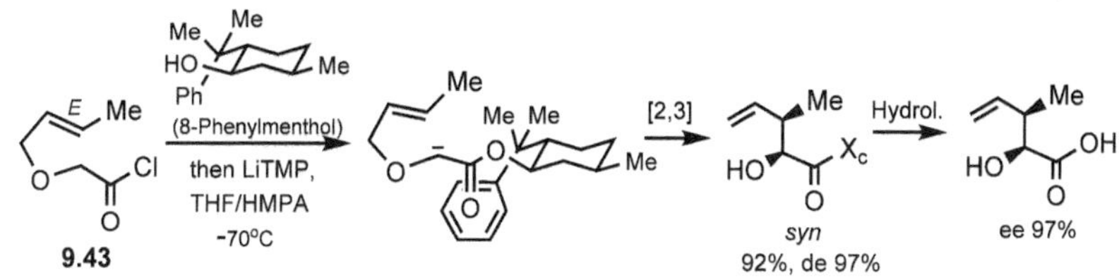

**FIG. 9.44** Asymmetric Wittig rearrangement using the chiral auxiliary 8-phenylmenthol.

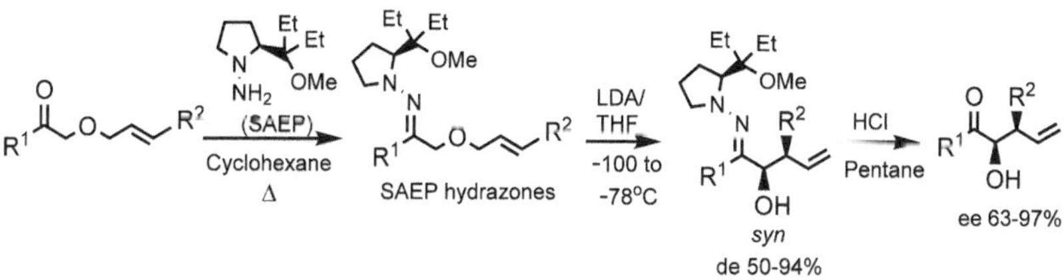

**FIG. 9.45** Asymmetric Wittig rearrangement of SAEP hydrazones with base. (SAEP=[(*S*)-1-amino-2-(1-ethyl-1-methoxypropyl)pyrrolidine]).

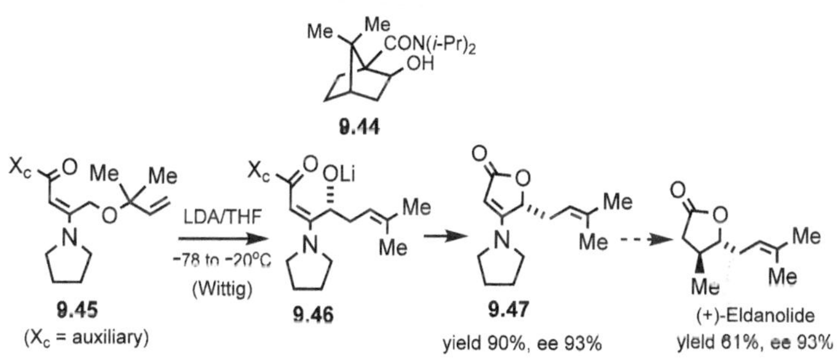

**FIG. 9.46** Application of asymmetric Wittig rearrangement in a synthesis of (+)-eldanolide.

carbonyl group instead of the simple conjugating substituent, and the (*E*)-substrate leads to the *syn* product (cf. Fig. 8.65) with excellent de. Removal of the auxiliary by hydrolysis gives the enantioselective α-hydroxyacid product.

The chiral auxiliary SAEP [(*S*)-1-amino-2-(1-ethyl-1-methoxypropyl)pyrrolidine] is also used for asymmetric induction in the Wittig rearrangement (Fig. 9.45).[105–107] The SAEP hydrazones of alkoxycarbonyl compounds undergo diastereoselective Wittig rearrangement with base and gives, after removal of the auxiliary, α-hydroxyketones with moderate to very high enantioselectivity.

An application of chiral auxiliary **9.44**-mediated asymmetric Wittig rearrangement in a synthesis of (+)-eldanolide, the pheromone of the male African sugar stem borer is shown in Fig. 9.46.[108] The Wittig rearrangement of the chiral unsaturated ester **9.45** gives the lithium alkoxide **9.46**, which undergoes

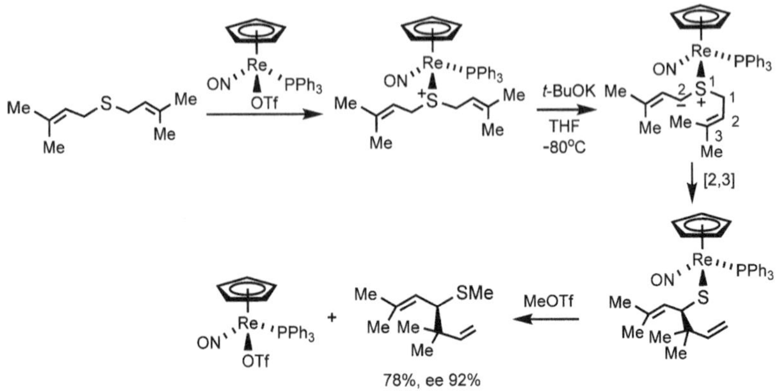


**FIG. 9.47** Asymmetric [2,3] rearrangement of a sulphur ylide using a chiral rhenium complex as an auxiliary.

spontaneous lactonization to form the unsaturated lactone **9.47** in very good yield and with high enantioselectivity. Interestingly, no additional step is needed to remove the auxiliary. The lactone **9.47** is then converted into (+)-eldanolide with conserved ee.

The chiral auxiliary strategy has also been employed in the asymmetric [2,3] rearrangement of ylides. Fig. 9.47 depicts an asymmetric synthesis of an organic sulphide via a [2,3] rearrangement of a sulphur ylide using a rhenium complex as a chiral auxiliary.[109] The substrate forms an adduct with the chiral complex, which gives a sulphur ylide with base. The [2,3] rearrangement of the ylide followed by release of the auxiliary through *S*-alkylation gives the sulphide with high enantioselectivity.

### 9.6.3 Chiral reagents

Chiral nonracemic bases can be used as chiral reagents in the asymmetric version of [2,3]-Wittig rearrangements. The first success was achieved in the ring-contracting rearrangement of the 13-membered cyclic ether with the chiral lithium amide base (Fig. 9.48A).[110] However, no appreciable enantioselectivity has been observed with the acyclic substrates.[111]

The asymmetric Wittig rearrangement has also been effected with a chiral boron reagent (Fig. 9.48B).[112, 113] The reaction proceeds through a chiral boron enolate and gives the product with high diastereo- and enantioselectivity (*anti*:*syn* = 83:17, ee 96%).

### 9.6.4 Organocatalysis

Although catalytic asymmetric [2,3]-Wittig rearrangements seem to be very attractive, they have remained largely insensitive to enantioselective catalysis

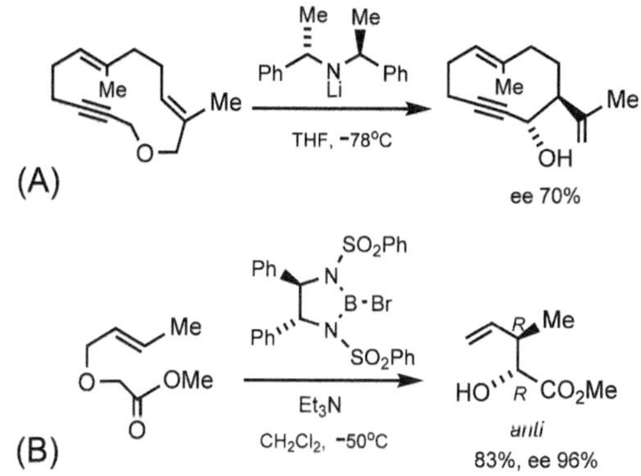


**FIG. 9.48** Asymmetric Wittig rearrangements using (A) a chiral lithium amide base and (B) a chiral boron reagent.

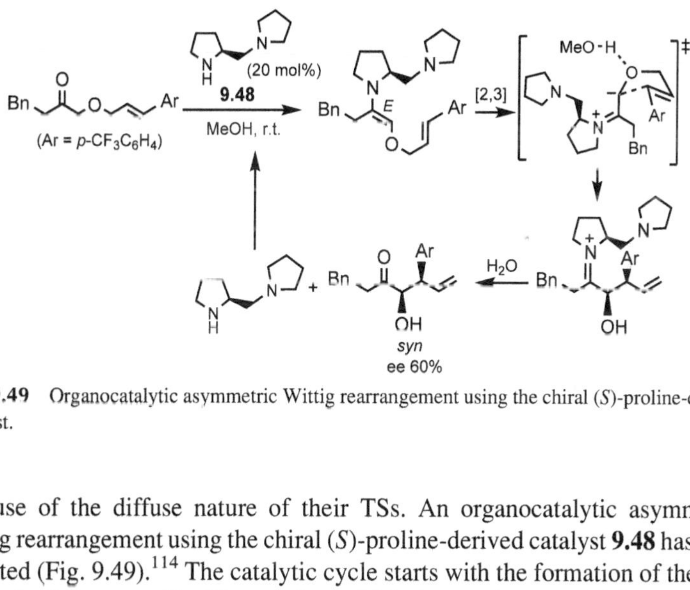


**FIG. 9.49** Organocatalytic asymmetric Wittig rearrangement using the chiral (*S*)-proline-derived catalyst.

because of the diffuse nature of their TSs. An organocatalytic asymmetric Wittig rearrangement using the chiral (*S*)-proline-derived catalyst **9.48** has been reported (Fig. 9.49).[114] The catalytic cycle starts with the formation of thermodynamically more stable (*E*)-enamine, which then undergoes a [2,3] rearrangement to give the *syn* α-hydroxyketone with modest enantioselectivity (ee 60%) after hydrolytic release of the catalyst.

## 9.6.5 Phase transfer catalysis

Recently, chiral phase transfer catalysts such as **9.49** are employed to facilitate compound transfer between different phases and for asymmetric induction in the [2,3]-Wittig rearrangement (Fig. 9.50).[115] The α-hydroxyketone products are isolated in excellent yields, but only modest enantioselectivities are

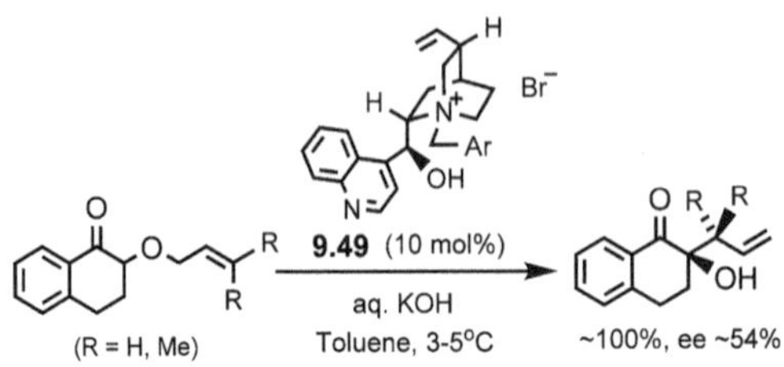


FIG. 9.50 Asymmetric [2,3]-Wittig rearrangement using phase transfer catalysis.

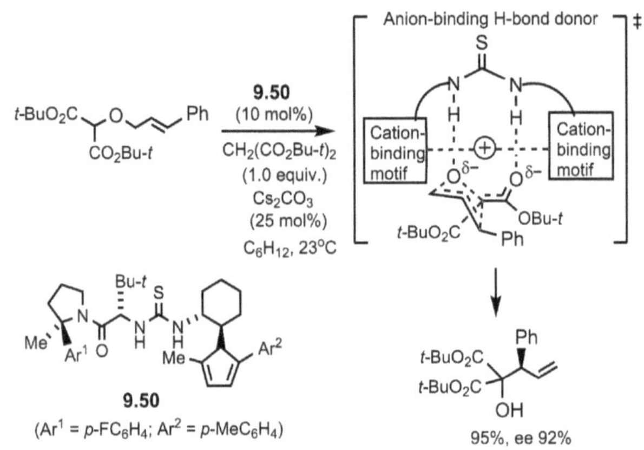


FIG. 9.51 Synergistic ion-binding strategy for asymmetric catalysis of the [2,3]-Wittig rearrangement.

observed despite extensive catalyst screening. Nevertheless, the approach clearly holds promise for the future with a new strategy for asymmetric sigmatropic rearrangement using phase transfer catalysis.

## 9.6.6 Synergistic ion-binding catalysis

Very recently, a novel synergistic ion-binding catalysis has been demonstrated with the catalytic asymmetric [2,3]-Wittig rearrangement of allyloxymalonates using the chiral thiourea catalysts such as **9.50** (Fig. 9.51).[116] The reaction proceeds through a preferred transition structure which is ion-binding stabilized by the catalyst. The thiourea moiety of the catalyst forms H-bonds with the deprotonated substrate, while aryl and amide groups encapsulate the $Cs^+$ ion through cooperative cation-π and -Lewis base interactions. The homoallylic alcohol product is obtained in excellent yield (95%) and with high enantioselectivity (ee 92%), showing the great utility of this approach.

## References


1. Oppolzer, W. *Angew. Chem. Int. Ed. Engl.* **1984**, *23*, 876.
2. Kagan, H. B.; Riant, O. *Chem. Rev.* **1992**, *92*, 1007.
3. Noyori, R. *Asymmetric Catalysis in Organic Synthesis*; Wiley: New York, 1994; p. 212 (chapter 4).
4. Evans, D. A.; Johnson, J. S. In *Comprehensive Asymmetric Catalysis*; Jacobsen, E. N., Pfaltz, A., Yamamoto, H., Eds.; Springer: Berlin, 1999; p. 1177.
5. Corey, E. J. *Angew. Chem. Int. Ed. Engl.* **2002**, *41*, 1650.
6. Fleming, I.; Sarkar, A. K.; Doyle, M. J.; Raithby, P. R. *J. Chem. Soc., Perkin Trans. 1* **1989**, 2023.
7. Gupta, R. C.; Raynor, C. M.; Stoodley, R. J.; Slawin, A. M. Z.; Williams, D. J. *J. Chem. Soc., Perkin Trans. 1* **1988**, 1773.
8. Gupta, R. C.; Larsen, D. S.; Stoodley, R. J.; Slawin, A. M. Z.; Williams, D. J. *J. Chem. Soc., Perkin Trans. 1* **1989**, 739.
9. Rück-Braun, K.; Kunz, H. *Chiral Auxiliaries in Cycloadditions*; Wiley-VCH: Weinheim, 1999; p. 30.
10. Evans, D. A. *Aldrichimica Acta* **1982**, *15*, 23.
11. Evans, D. A.; Chapman, K. T.; Bisaha, J. *J. Am. Chem. Soc.* **1984**, *106*, 4261. *J. Am. Chem. Soc.* **1988**, *110*, 1238.
12. Oppolzer, W.; Chapuis, C.; Bernardinelli, G. *Helv. Chim. Acta* **1984**, *67*, 1397.
13. Vandewalle, M.; Van der Eycken, J.; Oppolzer, W., Vullioud, C. *Tetrahedron* **1986**, *42*, 4035.
14. Mandal, D. K. *Pericyclic Chemistry: Orbital Mechanisms and Stereochemistry*; Elsevier: Cambridge, MA, 2018; p. 217.
15. Corey, E. J.; Sarshar, S. *J. Am. Chem. Soc.* **1992**, *114*, 7938.
16. Johnson, J. S.; Evans, D. A. *Acc. Chem. Res.* **2000**, *33*, 325.
17. Ghosh, A. K.; Mathivanam, P.; Cappiello, J. *Tetrahedron Asymmetry* **1998**, *9*, 1.
18. Evans, D. A.; Miller, S. J.; Lectka, T.; von Matt, P. *J. Am. Chem. Soc.* **1999**, *121*, 7559.
19. Evans, D. A.; Barnes, D. M.; Johnson, J. S.; et al. *J. Am. Chem. Soc.* **1999**, *121*, 7582.
20. Evans, D. A.; Barnes, D. M. *Tetrahedron Lett.* **1997**, *38*, 57.
21. Seebach, D.; Beck, A. K.; Heckel, A. *Angew. Chem. Int. Ed. Engl.* **2001**, *40*, 92.
22. Seebach, D.; Dahinden, R.; Marti, R. E.; Beck, A. K.; Plattner, D. A.; Kühnle, F. N. M. *J. Org. Chem.* **1995**, *60*, 1788.
23. Narasaka, K.; Iwasawa, N.; Inoue, M.; Yamada, T.; Nakashima, M.; Sugimori, J. *J. Am. Chem. Soc.* **1989**, *111*, 5340.
24. Mikami, K.; Motoyama, Y.; Terada, M. *J. Am. Chem. Soc.* **1994**, *116*, 2812.
25. White, J. D.; Choi, Y. *Org. Lett.* **2000**, *2*, 2373.
26. Ryu, D. H.; Lee, T. W.; Corey, E. J. *J. Am. Chem. Soc.* **2002**, *124*, 9992.
27. MacMillan, D. W. C. *Nature* **2008**, *455*, 304.
28. Northrup, A. B.; MacMillan, D. W. C. *J. Am. Chem. Soc.* **2002**, *124*, 2458.
29. Ahrendt, K. A.; Borths, C. J.; MacMillan, D. W. C. *J. Am. Chem. Soc.* **2000**, *122*, 4243.
30. Gothelf, K. V.; Jørgensen, K. A. *Chem. Rev.* **1998**, *98*, 863.
31. Williams, R. M.; Zhai, W.; Aldous, D. J.; Aldous, S. C. *J. Org. Chem.* **1992**, *57*, 6527.
32. Louis, C.; Hootelé, C. *Terahedron: Asymmetry* **1997**, *8*, 109. **1995**, *6*, 2149.
33. Curran, D. P.; Heffner, T. A. *J. Org. Chem.* **1990**, *55*, 4585.
34. Gothelf, K. V.; Jørgensen, K. A. *Chem. Commun.* **2000**, 1449.
35. Gothelf, K. V.; Jørgensen, K. A. *Acta Chem. Scand.* **1996**, *50*, 652.

36. Jensen, K. B.; Gothelf, K. V.; Hazell, R. G.; Jørgensen, K. A. *J. Org. Chem.* **1997**, *62*, 2471.
37. Jensen, K. B.; Gothelf, K. V.; Jørgensen, K. A. *Helv. Chim. Acta* **1997**, *80*, 2039.
38. Viton, F.; Bernardinelli, G.; Kündig, E. P. *J. Am. Chem. Soc.* **2002**, *124*, 4968.
39. Mita, T.; Ohtsuki, N.; Ikeno, T.; Yamada, T. *Org. Lett.* **2002**, *4*, 2457.
40. Jensen, K. B.; Hazell, R. G.; Jørgensen, K. A. *J. Org. Chem.* **1999**, *64*, 2353.
41. Jen, W. S.; Wiener, J. J. M.; MacMillan, D. W. C. *J. Am. Chem. Soc.* **2000**, *122*, 9874.
42. Li, A.-H.; Dai, L. X.; Aggarwal, V. K. *Chem. Rev.* **1997**, *97*, 2341.
43. Lebel, H.; Marcoux, J.-F.; Molinaro, C.; Charette, A. B. *Chem. Rev.* **2003**, *103*, 977.
44. Davies, H. M. L.; Beckwith, R. E. J. *Chem. Rev.* **2003**, *103*, 2861.
45. Roy, M.-N.; Lindsay, V. N. G.; Charette, A. B. In *Stereoselective Synthesis: Reactions of Carbon–Carbon Double Bonds (Science of Synthesis)*; de Vries, J. G., Ed.; Vol. 1; Thieme: Stuttgart, 2011; p. 731.
46. Simmons, H. E.; Smith, R. D. *J. Am. Chem. Soc.* **1958**, *80*, 5323.
47. Simmons, H. E. *Org. React.* **1973**, *20*, 1.
48. Davies, H. M. L.; Antoulinakis, E. G. *Org. React.* **2001**, *57*, 1.
49. Davies, H. M. L. In *Comprehensive Organic Synthesis*; Trost, B. M., Fleming, I., Eds.; Vol. 4; Pergamon Press: Oxford, 1991; p. 1031.
50. Dauben, W. G.; Berezin, G. H. *J. Am. Chem. Soc.* **1963**, *85*, 468.
51. Charette, A. B.; Lebel, H. *J. Org. Chem.* **1995**, *60*, 2966.
52. Charette, A. B.; Marcoux, J.-F. *Synlett* **1995**, 1197.
53. Charette, A. B.; Juteau, H.; Lebel, H.; Molinaro, C. *J. Am. Chem. Soc.* **1998**, *120*, 11943.
54. Barrett, A. G. M.; Kasdorf, K. *J. Am. Chem. Soc.* **1996**, *118*, 11030.
55. Deng, X.-M.; Cai, P.; Ye, S.; et al. *J. Am. Chem. Soc.* **2006**, *128*, 9730.
56. Evans, D. A.; Woerpel, K. A.; Hinman, M. M.; Faul, M. M. *J. Am. Chem. Soc.* **1991**, *113*, 726.
57. Haufe, G.; Rosen, T. C.; Meyer, O. G. J.; Fröhlich, R.; Rissanen, K. *J. Fluor. Chem.* **2002**, *114*, 189.
58. Itagaki, M.; Masumoto, K.; Suenobu, K.; Yamamoto, Y. *Org. Process Res. Dev.* **2006**, *10*, 245.
59. Marcoux, D.; Charette, A. B. *Angew. Chem. Int. Ed. Engl.* **2008**, *47*, 10155.
60. Lindsay, V. N. G.; Lin, W.; Charette, A. B. *J. Am. Chem. Soc.* **2009**, *131*, 16383.
61. Charette, A. B.; Wurz, R. P.; Ollevier, T. *Helv. Chim. Acta* **2002**, *85*, 4468.
62. Doyle, M. P.; Protopopova, M. N. *Tetrahedron* **1998**, *54*, 7919.
63. Singh, V. K.; DattaGupta, A.; Sekar, G. *Synthesis* **1997**, 137.
64. Kirkland, T. A.; Colucci, J.; Geraci, L. S.; et al. *J. Am. Chem. Soc.* **2001**, *123*, 12432.
65. Kunz, R. K.; MacMillan, D. W. C. *J. Am. Chem. Soc.* **2005**, *127*, 3240.
66. Mikami, K.; Shimizu, M. *Chem. Rev.* **1992**, *92*, 1021.
67. Dias, L. C. *Curr. Org. Chem.* **2000**, *4*, 305.
68. Whitesell, J. K.; Bhattacharya, A.; Aguilar, D. A.; Henke, K. *J. Chem. Soc. Chem. Commun.* **1982**, 989.
69. Whitesell, J. K.; Bhattacharya, A.; Buchanan, C. M.; et al. *Tetrahedron* **1986**, *42*, 2993.
70. Jeżewski, A.; Chajewska, K.; Wielogórski, Z.; Jurczak, J. *Terahedron: Asymmetry* **1997**, *8*, 1741.
71. Mikami, K.; Terada, M.; Nakai, T. *J. Am. Chem. Soc.* **1990**, *112*, 3949.
72. Yuan, Y.; Zhang, X.; Ding, K. *Angew. Chem. Int. Ed. Engl.* **2003**, *42*, 5478.
73. Terada, M.; Matsukawa, S.; Mikami, K. *J. Chem. Soc. Chem. Commun.* **1993**, 327.
74. Terada, M.; Mikami, K. *J. Chem. Soc. Chem. Commun.* **1995**, 2391.
75. Corey, E. J.; Barnes-Seeman, D.; Lee, T. W.; Goodman, S. N. *Tetrahedron Lett.* **1997**, *38*, 6513.
76. Terada, M.; Mikami, K.; Nakai, T. *J. Chem. Soc. Chem. Commun.* **1990**, 1623.

77. Mikami, K.; Terada, M. *Tetrahedron* **1992**, *48*, 5671.
78. Girard, C.; Kagan, H. B. *Angew. Chem. Int. Ed. Engl.* **1998**, *37*, 2922.
79. Evans, D. A.; Tregay, S. W.; Burgey, C. S.; Paras, N. A.; Vojkovsky, T. *J. Am. Chem. Soc.* **2000**, *122*, 7936.
80. Nubbemeyer, U. *Synthesis* **2003**, 961.
81. Ito, H.; Taguchi, T. *Chem. Soc. Rev.* **1999**, *28*, 43.
82. Enders, D.; Knopp, M.; Schiffers, R. *Terahedron: Asymmetry* **1996**, *7*, 1847.
83. Kazmaier, U.; Krebs, A. *Angew. Chem. Int. Ed. Engl.* **1995**, *34*, 2012.
84. Chan, K. K.; Cohen, N.; DeNoble, J. P.; Specian, A. C., Jr.; Saucy, G. *J. Org. Chem.* **1976**, *41*, 3497.
85. Trost, B. M.; Schroeder, G. M. *J. Am. Chem. Soc.* **2000**, *122*, 3785.
86. Overman, L. E. *Acc. Chem. Res.* **1980**, *13*, 218.
87. Enders, D.; Knopp, M. *Tetrahedron* **1996**, *52*, 5805.
88. Corey, E. J.; Lee, D. H. *J. Am. Chem. Soc.* **1991**, *113*, 4026.
89. Abraham, L.; Czerwonka, R.; Hiersmann, M. *Angew. Chem. Int. Ed. Engl.* **2001**, *40*, 4700.
90. Abraham, L.; Körner, M.; Schwab, P.; Hiersemann, M. *Adv. Synth. Catal.* **2004**, *346*, 1281.
91. Uyeda, C.; Jacobsen, E. N. *J. Am. Chem. Soc.* **2008**, *130*, 9228. **2011**, *133*, 5062.
92. Akiyama, K.; Mikami, K. *Tetrahedron Lett.* **2004**, *45*, 7217.
93. Overman, L. E. *Angew. Chem. Int. Ed. Engl.* **1984**, *23*, 579.
94. Overman, L. E.; Carpenter, N. E. In *Organic Reactions*; Overman, L. E., Ed.; Vol. 66; Wiley: Hoboken, NJ, 2005; p. 1.
95. Cannon, J. S.; Overman, L. E. *Acc. Chem. Res.* **2016**, *49*, 2220.
96. Marshall, J. A. In *Comprehensive Organic Synthesis*; Trost, B. M., Fleming, I., Eds.; Vol. 3; Pergamon Press: Oxford, 1991; p. 975.
97. Nakai, T.; Mikami, K. *Org. React.* **1994**, *46*, 105.
98. Hiersemann, M.; Jaschinski, T. In *Comprehensive Chirality*; Carreira, E. M., Yamamoto, H., Eds.; Vol. 2; Elsevier: Amsterdam, 2012; p. 625.
99. Seashore-Ludlow, B.; Somfai, P. In *Stereoselective Synthesis of Drugs and Natural Products*; Andrushko, V., Andrushko, N., Eds.; Vol. 1; Wiley: Hoboken, NJ, 2013; p. 475.
100. West, T. H.; Spoehrle, S. S. M.; Kasten, K.; Taylor, J. E.; Smith, A. D. *ACS Catal.* **2015**, *5*, 7446.
101. Mikami, K.; Azuma, K.; Nakai, T. *Tetrahedron* **1984**, *40*, 2303.
102. Evans, D. A.; Andrews, G. C. *Acc. Chem. Res.* **1974**, *7*, 147.
103. Nishibayashi, Y.; Uemura, S. *Top. Curr. Chem.* **2000**, *208*, 201.
104. Takahashi, O.; Mikami, K.; Nakai, T. *Chem. Lett.* **1987**, 69.
105. Enders, D.; Plant, A.; Backhaus, D.; Reinhold, U. *Tetrahedron* **1995**, *51*, 10699.
106. Enders, D.; Backhaus, D.; Runsink, J. *Tetrahedron* **1996**, *52*, 1503.
107. Enders, D.; Bartsch, M.; Backhaus, D.; Runsink, J.; Raabe, G. *Synthesis* **1996**, 1438.
108. Li, Y.-. J.; Ho, G.-. M.; Chen, P.-. Z. *Tetrahedron Asymmetry* **2009**, *20*, 1854.
109. Meyer, O.; Cagle, P. C.; Weickhardt, K.; Vichard, D.; Gladysz, J. A. *Pure Appl. Chem.* **1996**, *68*, 79.
110. Marshall, J. A.; Lebreton, J. *J. Am. Chem. Soc.* **1988**, *110*, 2747.
111. Marshall, J. A.; Wang, X. J. *J. Org. Chem.* **1992**, *57*, 2747.
112. Fujimoto, K.; Nakai, T. *Tetrahedron Lett.* **1994**, *35*, 5019.
113. Fujimoto, K.; Matsuhashi, C.; Nakai, T. *Heterocycles* **1996**, *42*, 423.
114. McNally, A.; Evans, B.; Gaunt, M. J. *Angew. Chem. Int. Ed. Engl.* **2006**, *45*, 2116.
115. Denmark, S. E.; Cullen, L. R. *J. Org. Chem.* **2015**, *80*, 11818.
116. Kennedy, C. R.; Guidera, J. A.; Jacobsen, E. N. *ACS Cent. Sci.* **2016**, *2*, 416.

Chapter 10

# Transition metal-catalysed reactions: Diastereoselectivity and asymmetric synthesis

## 10.1 Diastereoselective epoxidation

Epoxides are extremely valuable as building blocks in organic synthesis since they can be readily transformed into a wide range of products because of their spring-loaded nature.[1] An excellent method for epoxidation involves the reaction of alkenes with *t*-butyl hydroperoxide (*t*-BuOOH) in the presence of a transition metal catalyst.[2] The epoxidation of allylic alcohols using the vanadium (V) or titanium(IV) complex is of particular importance, which shows high diastereoselectivity.[3] Fig. 10.1A illustrates a diastereoselective epoxidation of acyclic allylic alcohol using *t*-BuOOH and vanadium acetylacetonate [$VO(acac)_2$] as a catalyst, which gives predominantly the *syn* product (*syn*:*anti* = 90:10). In comparison, practically no diastereoselectivity is observed using *m*CPBA, the common epoxidizing agent; however, epoxidation of cyclic allylic alcohol with *m*CPBA is usually *syn*-selective. The transition metal-catalysed epoxidation thus provides a more effective approach for the stereoselective epoxidation.

The high diastereoselectivity in the vanadium-catalysed epoxidation presumably arises via the formation of an intermediate in which the allylic OH group and the peroxide are coordinated to the metal **10.1**, as shown in Fig. 10.1B.

## 10.2 Asymmetric epoxidation

Asymmetric epoxidation is the most versatile and efficient asymmetric approach towards the synthesis of single enantiomer products. The highly reliable and widely used method for asymmetric epoxidation is due to K.B. Sharpless (Nobel Prize 2001), called the Sharpless asymmetric epoxidation. The Sharpless epoxidation is general for allylic alcohols with a range of substitution patterns and is highly predictable for the absolute configuration of the epoxide formed. However, the reaction is limited by the requirement for allylic alcohol

Stereochemistry and Organic Reactions. https://doi.org/10.1016/B978-0-12-824092-2.00010-1

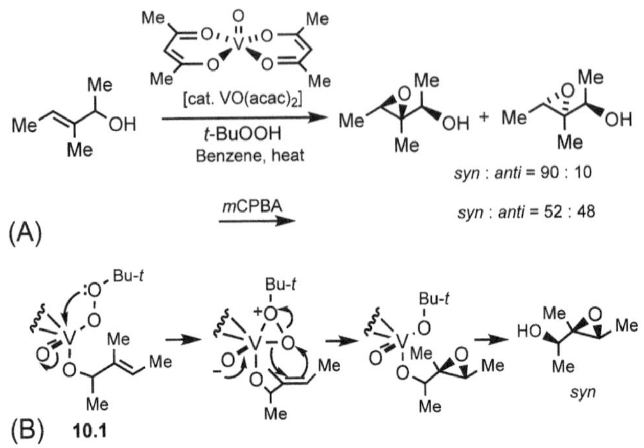


**FIG. 10.1** (A) Transition metal-catalysed diastereoselective epoxidation of acyclic allylic alcohol using vanadium acetylacetonate $[VO(acac)_2]$ as a catalyst; (B) Mechanism of epoxidation.

to be the substrate. The asymmetric epoxidation of an unfunctionalized alkene is also possible but is considerably less advanced than the Sharpless reaction. One such asymmetric approach is known as the Jacobsen–Katsuki epoxidation.

### 10.2.1 Sharpless asymmetric epoxidation[4–6]

A general scheme for the Sharpless asymmetric epoxidation (SAE) is shown in Fig. 10.2A. The allylic alcohol substrate **10.2** can have a wide range of substitution patterns; however, the *cis*-disubstituted alkenes with bulky $R^3$ appear to be somewhat problematic. The reagents are *t*-butyl hydroperoxide (*t*-BuOOH) as the oxidant, titanium tetraisopropoxide [Ti(O*i*-Pr)$_4$] as the Lewis acidic transition metal reagent and an enantiomer of diethyl tartrate (DET) (or diisopropyl tartrate, DIPT) as the chiral ligand. The asymmetric epoxidation is normally carried out using 5–10mol% of the chiral titanium catalyst. To be truly catalytic in titanium, scrupulously dry conditions (activated molecular sieves) are required; the solvent of choice is $CH_2Cl_2$. The reaction gives the epoxide **10.3** with high enantioselectivity.

A striking feature of the SAE is that the absolute configuration of the enantioselective product can be reliably predicted using the Sharpless rule or mnemonic (Fig. 10.2B). A particular enantiomer of the tartrate ligand directs the delivery of the O atom selectively to one face of the double bond, irrespective of the substitution pattern. If the allylic alcohol is drawn with the OH group at the lower right as shown in Fig. 10.2B, the O atom is delivered to the bottom face in the presence of L-(+)-DET (the natural isomer), and to the top face in the presence of D-(−)-DET (the unnatural isomer). The rule is reversed if the OH group occupies the lower left.

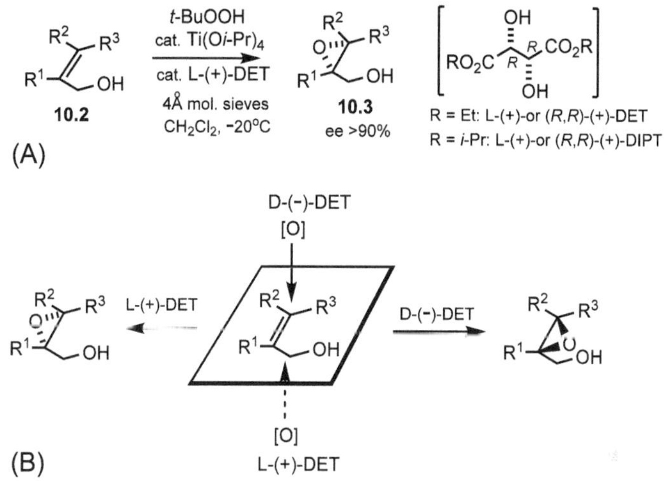


**FIG. 10.2** (A) A general scheme for the Sharpless asymmetric epoxidation; (B) Sharpless rule or mnemonic.

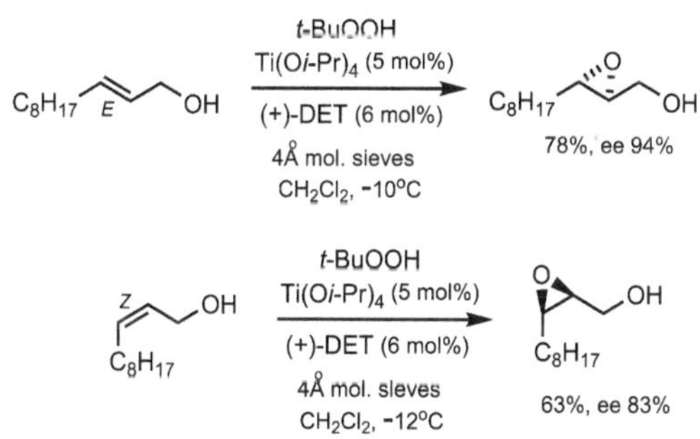


**FIG. 10.3** Sharpless asymmetric epoxidations with (*E*)- and unhindered (*Z*)-allylic alcohols.

Fig. 10.3 shows two examples of Sharpless asymmetric epoxidation. (*E*)-Allylic alcohols are found to be excellent substrates, whereas (*Z*)-allylic alcohols are less reactive. However, the hindered (*Z*)-allylic alcohols are poor substrates.

The Sharpless epoxidation is a powerful reagent-controlled reaction that can override the influence of substrate chirality (Fig. 10.4).[7] The sense of asymmetric induction is dominated by the chiral catalyst system rather than by the chiral substrate. In the absence of a chiral ligand, the diastereoselectivity is controlled by the chiral substrate but the selectivity is low. In contrast, both enantiomers of the chiral ligand (DIPT) override the substrate control. As shown, the selectivity is 'matched' with (−)-DIPT and 'mismatched' with (+)-DIPT.

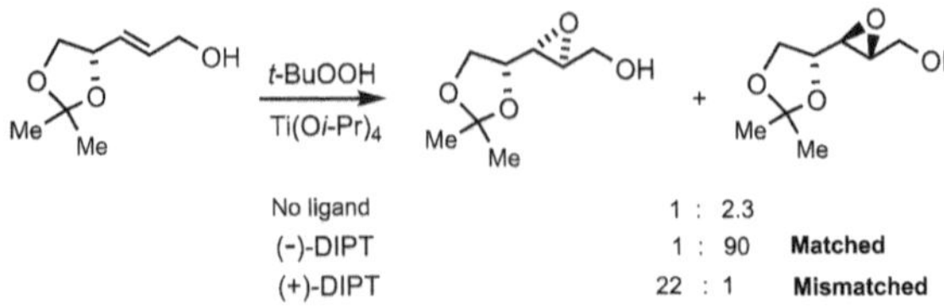


**FIG. 10.4** Reagent control in the Sharpless asymmetric epoxidation.

### 10.2.1.1 Mechanism

The mechanism of the Sharpless asymmetric epoxidation in the presence of L- or (*R*,*R*)-(+)-DET is depicted in Fig. 10.5. The active catalyst **10.4** is believed to be a dimeric complex formed from two Ti atoms bridged by two DET ligands. The coordination of each DET ligand to $Ti(Oi\text{-}Pr)_4$ involves ligand exchange (displacement of two isopropoxide groups). Each Ti atom is also coordinated to a C=O group of the DET ligand. The equilibrium lies in favour of the tartrate complex because of the higher binding constant of the chelating tartrate ligand.[8] The titanium and tartrate are normally left to stir for a while to allow the clean formation of the dimer.

To understand the molecular details of the catalytic cycle, it is sufficient to consider the events at one of the Ti centres, although the cooperative involvement of both Ti atoms cannot be ruled out.[9,10] It is believed that both *t*-BuOOH and the allylic alcohol are bound to the same Ti for the oxygen delivery step. The *t*-BuOOH (the source of oxygen) displaces the equatorial isopropoxide group and bind to Ti in a bidentate fashion, forming a triangular dioxametallacycle **10.5**. This ability to coordinate with both oxygens activates the

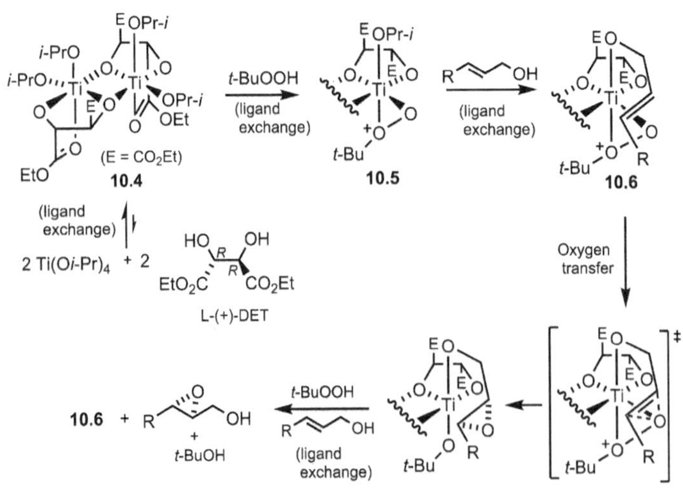


**FIG. 10.5** Mechanism of the Sharpless asymmetric epoxidation.

hydroperoxide for oxygen transfer to the allylic alcohol which is bound axial by the displacement of the remaining isopropoxide group. Owing to the shape of the complex **10.6**, the alkene attacks the equatorial oxygen of the peroxide from its lower face (as drawn). Ligand exchange then gives the epoxide and continues the catalytic cycle.

### *10.2.1.2 Applications*

The Sharpless epoxidation method is a reliable choice, wherever possible, to induce asymmetry in a synthetic design for chiral target molecules. Asymmetric synthesis of the β-blocker drug propranolol is shown in Fig. 10.6A.[11] Herein, the use of silicon-substituted allylic alcohol is beneficial since the epoxide from the allylic alcohol itself turns out to be difficult to handle. The epoxide is subjected to simple transformations followed by removal of the silyl group. In the final step, the epoxide ring is opened with isopropylamine to give (*S*)-propranolol.

Fig. 10.6B depicts an asymmetric synthesis of an antidepressant drug fluoxetine.[12] The epoxide enantiomer is ring-opened by a hydride nucleophile Red-Al [$NaAlH_2(OCH_2CH_2OMe)_2$] followed by functional group interconversions to give the drug (*R*)-fluoxetine. Note that the less hindered primary OH group is mesylated and displaced with $MeNH_2$.

The synthesis of unnatural L-aldohexoses has been achieved involving the Sharpless procedure.[13] For example, a scheme for the synthesis of L-glucose and L-mannose is outlined in Fig. 10.7.

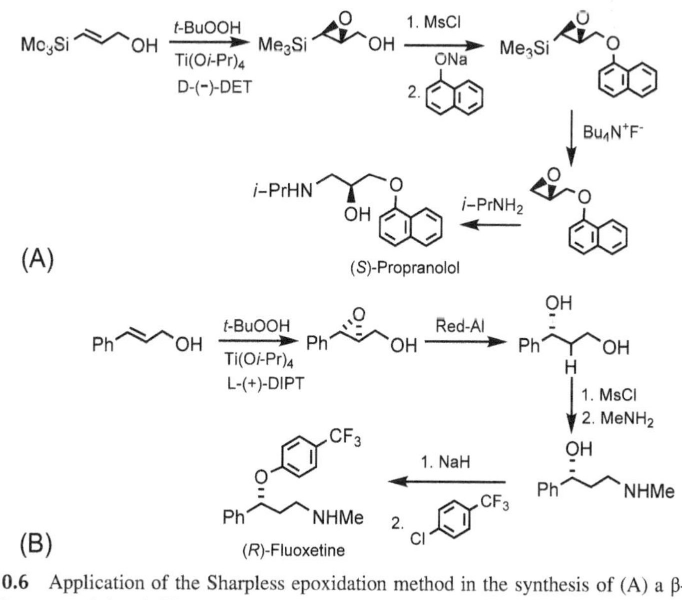


**FIG. 10.6** Application of the Sharpless epoxidation method in the synthesis of (A) a β-blocker drug (*S*)-propranolol and (B) an antidepressant drug (*R*)-fluoxetine.

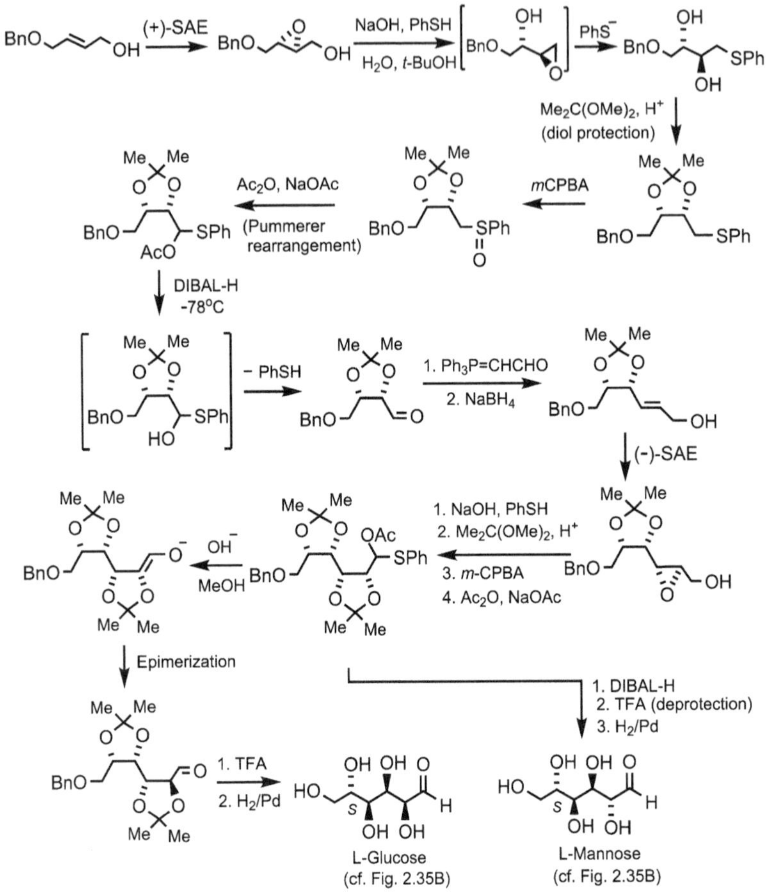


**FIG. 10.7** Application of the Sharpless procedure for the synthesis of L-sugars (L-glucose and L-mannose). (+)-SAE = *t*-BuOOH, Ti(O*i*-Pr)$_4$, (+)-DIPT; (−)-SAE = *t*-BuOOH, Ti(O*i*-Pr)$_4$, (−)-DIPT.

**Problem 10.1** Using the Sharpless asymmetric epoxidation, suggest a synthesis of the compound shown in the following:

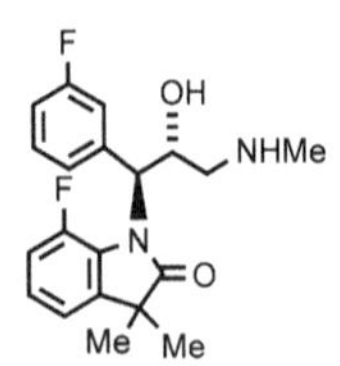

### 10.2.1.3 *Kinetic resolution*[14]

If a racemic secondary allylic alcohol is subjected to Sharpless asymmetric epoxidation, one of the two enantiomers of the allylic alcohol usually reacts much faster than the other. As a result, the ee of the recovered allylic alcohol substrate increases in favour of the slower reacting enantiomer. This is a kinetic resolution (see Section 5.3.4.2). Fig. 10.8A shows TS analysis of the Sharpless epoxidation of racemic allylic alcohols with (+)-DET ligand and predicts which enantiomer will react faster or slower. The extended Sharpless rule about the kinetic resolution of secondary allylic alcohols is shown in Fig. 10.8B. If the enantiomer of the substrate matches with that of the ligand, the reaction is fast; otherwise, the reaction is slow for a mismatch.

The kinetic resolution methodology is quite useful for the asymmetric synthesis of chiral secondary allylic alcohols or their corresponding epoxides. Fig. 10.9 illustrates three examples, which include the use of the faster-formed

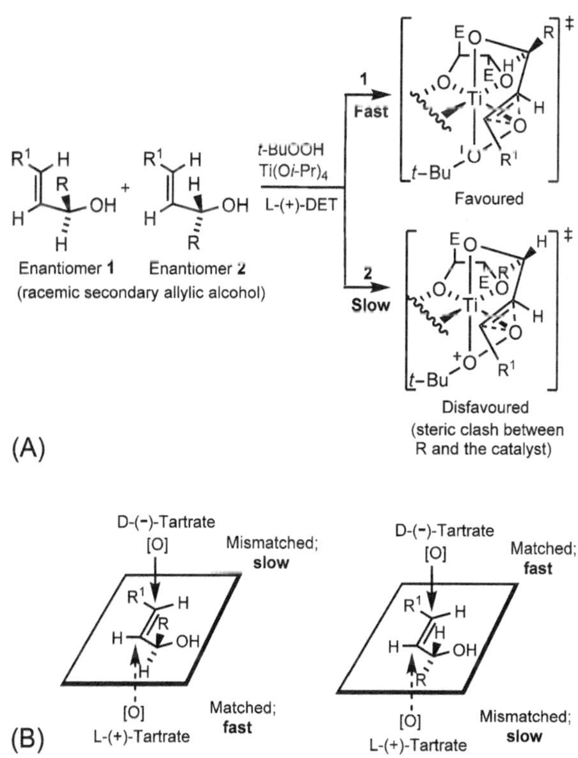


**FIG. 10.8** (A) Analysis of the TSs for the relative rates of the two enantiomers of a racemic allylic alcohol in the Sharpless epoxidation with the L-(+)-DET ligand; (B) Extended Sharpless rule for the kinetic resolution of racemic secondary allylic alcohols.

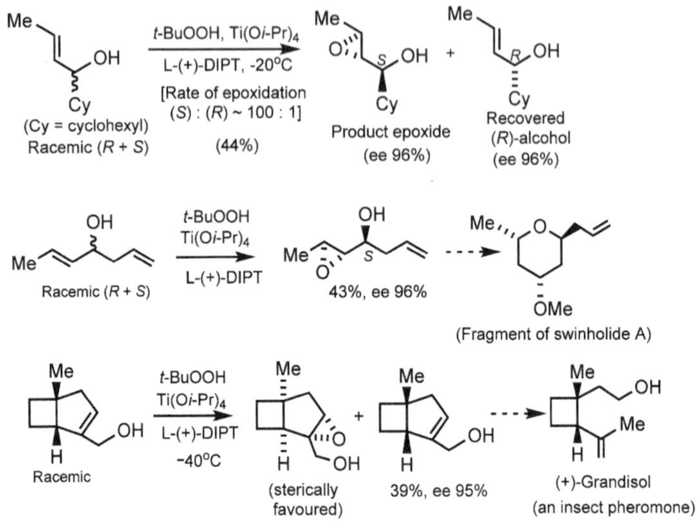


**FIG. 10.9** Kinetic resolution methodology for the synthesis of chiral allylic alcohols and their epoxides.

epoxide in the synthesis of a fragment of the marine toxin swinholide A,[15] and the use of the recovered (slower reacting) allylic alcohol in a synthesis of the sexual attracting insect pheromone (+)-grandisol.[16] The observed kinetic resolution is consistent with the extended Sharpless rule. However, the yield of the allylic alcohol or the epoxide is necessarily limited to 50%.

## 10.2.2 Jacobsen–Katsuki asymmetric epoxidation[17–20]

The asymmetric epoxidation of unfunctionalized alkenes is possible using the Jacobsen–Katsuki approach involving chiral manganese catalysts. The chiral manganese complexes are based on chiral salen ligands. The parent salen ligand is a tetradentate ligand obtained from salicylaldehyde (sal) and ethylenediamine (en). *cis*-Di- and trisubstituted alkenes with conjugating substituents are suitable substrates but *trans*-disubstituted alkenes are not (cf. Shi epoxidation, Section 7.13.2). Fig. 10.10 illustrates a Jacobsen–Katsuki asymmetric epoxidation using the chiral (*S*,*S*)-(salen)Mn(III)Cl catalyst **10.7** and sodium hypochlorite as the oxidant.[17] The reaction leads to *cis*-epoxide **10.8** with high enantioselectivity (ee 92%). The epoxidation reaction of *cis*-alkenes is not always stereospecific and some *trans*-epoxide can be formed via a radical intermediate. In the epoxidation of the stytrene derivative, *cis*:*trans* = 92:8.

The use of additives such as aromatic N-oxides increases the rate of Jacobsen–Katsuki epoxidation. The additive stabilizes the catalyst and transports the bulk oxidant hypochlorite into the organic phase.[21] The epoxidation reaction in the presence of additives has been applied in many useful syntheses, for

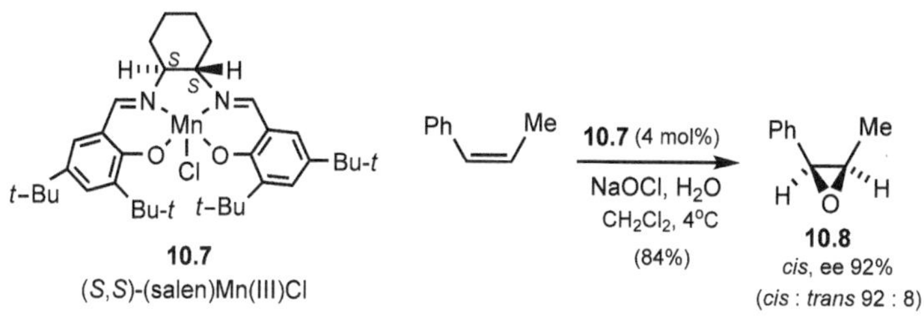


**FIG. 10.10** Jacobsen–Katsuki asymmetric epoxidation using chiral (salen)Mn catalyst.

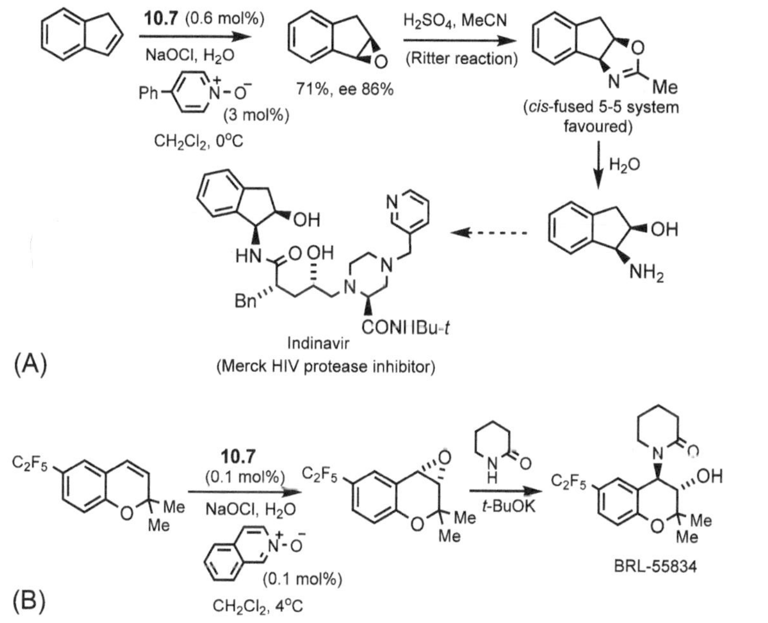


**FIG. 10.11** Application of the Jacobsen–Katsuki epoxidation in the presence of aromatic N-oxide as an additive in a synthesis of (A) Merck HIV protease inhibitor and (B) potassium channel activator BRL-55834.

example, in a synthesis of Merck HIV protease inhibitor (Fig. 10.11A),[22] and in a synthesis of the airways-selective potassium channel activator BRL-55834 (Fig. 10.11B).[23]

#### *10.2.2.1 Mechanism*

The active species in the Jacobsen–Katsuki epoxidation is thought to be a nearly planar Mn(IV)-oxo complex **10.9**; however, the exact mechanism of the reaction has remained controversial. Different kinds of models have been proposed regarding the alkene trajectory to the metal centre in the Mn-oxo complex.[24,25]

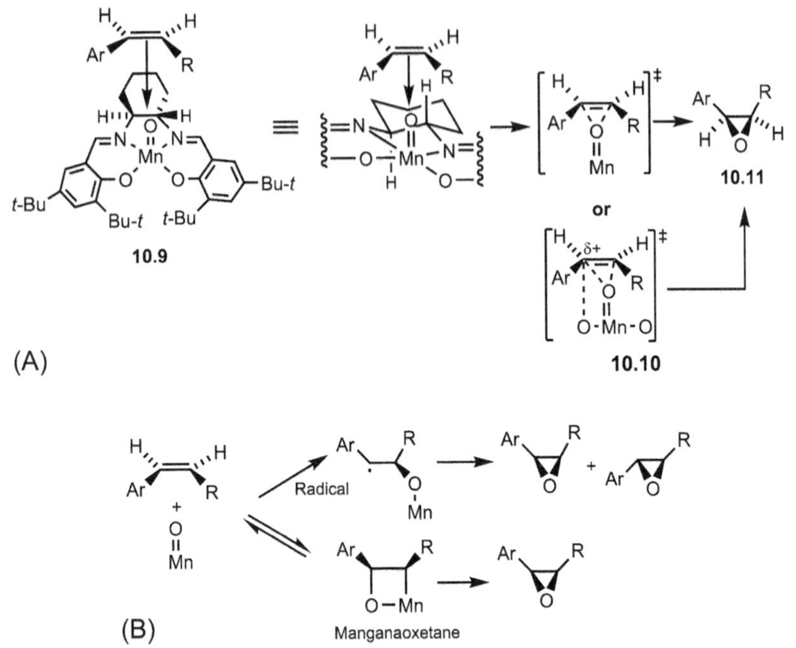


**FIG. 10.12** (A) A mechanistic model for the Jacobsen–Katsuki asymmetric epoxidation; (B) Alternative radical and manganaoxetane pathways for oxygen transfer.

The Jacobsen model based on steric effects, assuming a planar oxo species, is shown in Fig. 10.12A. The alkene approaches over the cyclohexane ring to avoid steric interactions with the bulky *t*-Bu groups. In the preferred pathway, the larger substituent (Ar) is away from the axial H atom, which leads to the epoxide enantiomer **10.11**. Note that the two nitrogens are placed in the equatorial positions. There may be alternative TS **10.10** for the concerted oxygen transfer which is quite asynchronous with the development of positive charge at the benzylic carbon, stabilized by donation of electron density from phenolic oxygen of the salen ligand.

Besides the concerted mechanism of oxygen transfer from the Mn(IV)-oxo species to the alkene double bond, two other mechanisms, viz., a radical pathway and a pathway involving a manganaoxetane have been proposed (Fig. 10.12B). The lack of stereospecificity can be explained by assuming the formation of a radical intermediate (stabilized by the conjugating substituent), which allows C—C bond rotation to give both *cis*- and *trans*-epoxides. Based on theoretical calculations, the mechanistic pathway via a reversibly formed manganaoxetane intermediate has also been suggested.[26]

### *10.2.2.2 Hydrolytic kinetic resolution*

Enantiopure or enantioenriched terminal epoxides are not commonly accessible by Sharpless or Jacobsen–Katsuki epoxidation methods. But these epoxides can

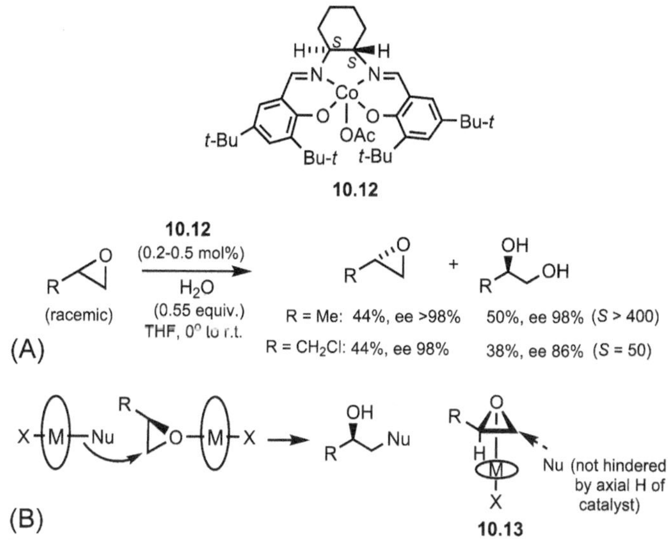


**FIG. 10.13** (A) Hydrolytic kinetic resolution method for the asymmetric synthesis of terminal epoxides; (B) Schematic representation of the mechanism of asymmetric ring-opening of epoxides.

be obtained with excellent ee from racemic epoxides using a hydrolytic kinetic resolution method.[27] In this method, a racemic epoxide is treated with a chiral (salen)Co(III) complex and water, which gives both recovered epoxide and diol product with very high enantioselectivity. The reaction is normally performed with slightly greater than 0.5 equivalents of water to obtain the unreacted epoxide enantiomer with excellent optical purity. The product diol is also obtained with a reasonable yield (maximum 50%) and high enantiomeric purity if the selectivity factor, $S$ ($k_{fast}/k_{slow}$) > 50. Besides water, other nucleophiles such as azides, carboxylates and phenoxides can be used, which lead to a wide scope of asymmetric ring-opening processes.

Fig. 10.13A illustrates the hydrolytic kinetic resolution procedure using the (*S*,*S*)-(salen)Co(III) catalyst **10.12**.[28] Note that the (*R*,*R*)-catalyst will give the opposite enantiomer of the recovered epoxide.

The hydrolytic reaction exhibits a second-order rate dependence, and is believed to proceed through a cooperative bimetallic mechanism, as shown schematically in Fig. 10.13B.[29] **10.13** represents that nucleophilic attack occurs at the unhindered carbon of that epoxide enantiomer which avoids unfavourable steric interaction with the axial H atom of the (*S*,*S*)-catalyst (cf. Fig. 10.12A).

## 10.3 Asymmetric dihydroxylation and aminohydroxylation

Asymmetric dihydroxylation has been developed as an asymmetric version of *syn* dihydroxylation of alkenes by $OsO_4$ (see Fig. 6.25). Like the dihydroxylation, the direct asymmetric conversion of alkenes into 1,2-amino alcohols, called aminohydroxylation has also been achieved. Herein, we will describe briefly the Sharpless asymmetric dihydroxylation reaction and the aminohydroxylation reaction.

## 10.3.1 Sharpless asymmetric dihydroxylation[30–32]

Chiral amine ligands can accelerate the dihydroxylation of alkenes with $OsO_4$, and provide asymmetric induction by coordinating to the Os atom. This ligand-accelerated catalysis overrides the influence of the slow background racemic reaction without the involvement of ligand and is responsible for the success of the asymmetric dihydroxylation reaction. In the Sharpless method, the chiral ligands are based on the alkaloids dihydroquinine (DHQ) and dihydroquinidine (DHQD). The alkaloids are connected to an aromatic ring, the choice of which depends on the alkene substrate. The most common chiral ligands are $(DHQ)_2PHAL$ and $(DHQD)_2PHAL$, in which two alkaloid molecules are attached to a phthalazine ring (Fig. 10.14). These chiral ligands are $C_2$-symmetric. Fig. 10.14 also depicts the Sharpless asymmetric dihydroxylation of *trans*-alkenes using the reagent combination AD-mix-α containing $(DHQ)_2PHAL$ and AD-mix-β containing $(DHQD)_2PHAL$, respectively. These AD-mix reagents are available commercially. $OsO_4$ is used in catalytic amounts

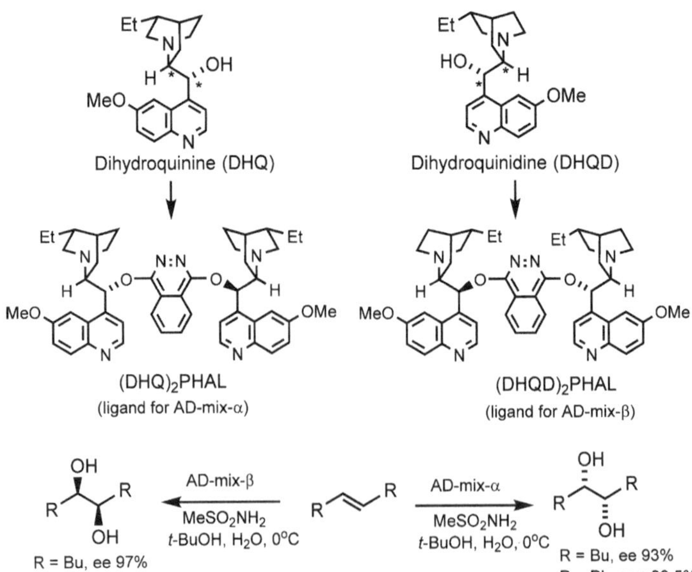


[AD-mix-α: $(DHQ)_2PHAL$ (1 mol%), $K_2OsO_2(OH)_4$ (0.2 mol%), $K_3Fe(CN)_6$ (3 equiv.), $K_2CO_3$ (3 equiv.)]

[AD-mix-β: $(DHQD)_2PHAL$ (1 mol%), $K_2OsO_2(OH)_4$ (0.2 mol%), $K_3Fe(CN)_6$ (3 equiv.), $K_2CO_3$ (3 equiv.)]

**FIG. 10.14** The Sharpless asymmetric dihydroxylation reaction.

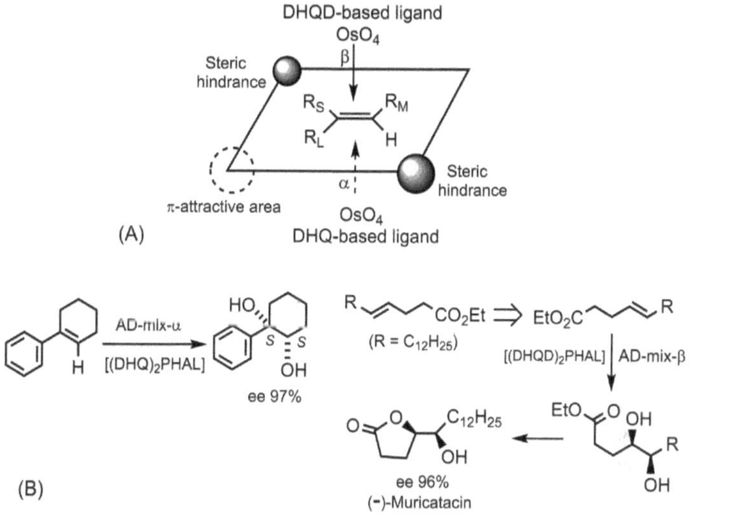


**FIG. 10.15** (A) Sharpless model or mnemonic for asymmetric dihydroxylation; (B) Examples illustrating the use of the mnemonic.

(typically, the osmium is added in the lower oxidation state (VI) as $K_2OsO_2(OH)_4$, with the co-oxidant $K_3Fe(CN)_6$). The additive $MeSO_2NH_2$ increases the rate of hydrolysis of the osmate ester producing the diol and Os (VI) for reoxidation. The chiral ligands $(DHQ)_2$PHAL and $(DHQD)_2$PHAL give the opposite enantiomers of the product. Notably, DHQ and DHQD are not enantiomeric but the chiral centres (marked by asterisks) in the two alkaloids are enantiomeric, and thus provide the mirror image environments around the Os atom which coordinates to the bridge-head N atom.

Sharpless asymmetric dihydroxylation is highly predictive, and a simple mnemonic or model is provided for face selectivity of the asymmetric dihydroxylation reaction.[33,34] The model is depicted in Fig. 10.15A for a trisubstituted alkene with large ($R_L$), medium ($R_M$) and small ($R_S$) substituents. Steric hindrance from the catalyst at two opposite corners of the plane drawn around the alkene leaves a cleft in which the alkene lies with $R_L$ to the lower left. $R_L$ is preferably a π-substituent such as aryl group. With this arrangement, the DHQ-based ligand will direct dihydroxylation from the lower (α) face of the alkene, whereas the DHQD-based ligand will promote dihydroxylation from the top (β) face.

The model thus predicts that *trans*-disubstituted alkenes in Fig. 10.14 with the larger substituent (Ph being more effective) to the lower left are excellent substrates in the asymmetric dihydroxylation reaction. The products observed are consistent with the model. The application of the mnemonic is further illustrated with two examples in Fig. 10.15B. The second example involves the asymmetric dihydroxylation of a γ,δ-unsaturated ester in a synthesis of the natural product (−)-muricatacin (a γ-lactone).[35]

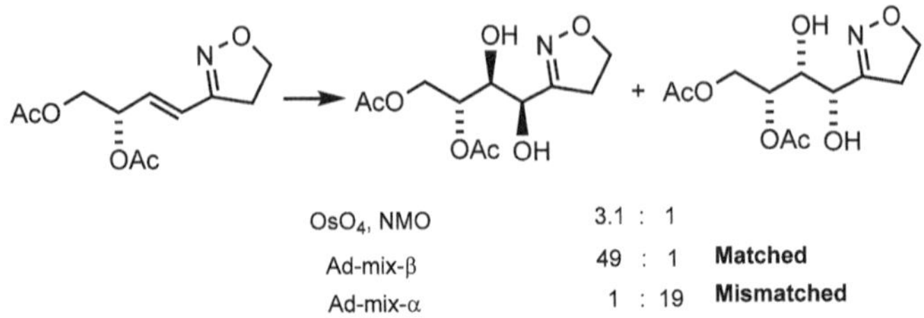


**FIG. 10.16** Reagent control in the Sharpless asymmetric dihydroxylation.

The selectivity with *cis*-disubstituted alkenes with a substituent in the more hindered position is often poor. However, the enantioselectivity of *cis*-disubstituted and even some tetrasubstituted alkenes can be improved using a modified ligand in which DHQ or DHQD is connected to a different aromatic unit such as *N*-carboxy-indoline (IND)[36] or better, anthraquinone (AQN).[37] Several polymer-supported catalysts have also been developed, which are advantageous because the products are easier to purify and the ligands can be isolated by simple filtration.[38]

Similar to the asymmetric epoxidation (see Fig. 10.4), reagent control often dominates over substrate control in the asymmetric dihydroxylation reaction with a chiral alkene substrate (Fig. 10.16).[30] As shown, the selectivity is matched with AD-mix-β, whereas the selectivity is mismatched with AD-mix-α.

**Problem 10.2** Give the structures of the compounds **A–F** in the following scheme, used in a synthesis of an anti-HIV drug:

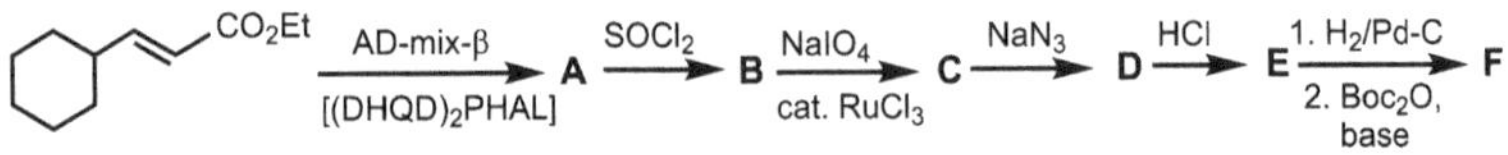


### *10.3.1.1 Mechanism*

The catalytic cycle of the Sharpless asymmetric dihydroxylation is shown in Fig. 10.17. The balance of evidence for osmium(VIII) addition is in favour of a (3+2) cycloaddition (see Fig. 6.25) rather than a [2+2] cycloaddition followed by a rearrangement.[39,40]

The detailed mechanism of the reaction is not fully understood. Corey has proposed a model with a U-shaped binding pocket **10.14** in which osmium (bonded to a bridge-head N atom of the chiral ligand) sits at the bottom of the pocket (Fig. 10.18).[41] A part of the chiral pocket is attractive to aromatic or strongly hydrophobic groups which accommodate $R_L$ substituent of the alkene (see Fig. 10.15A). The steric effects then lead to the optimum alignment

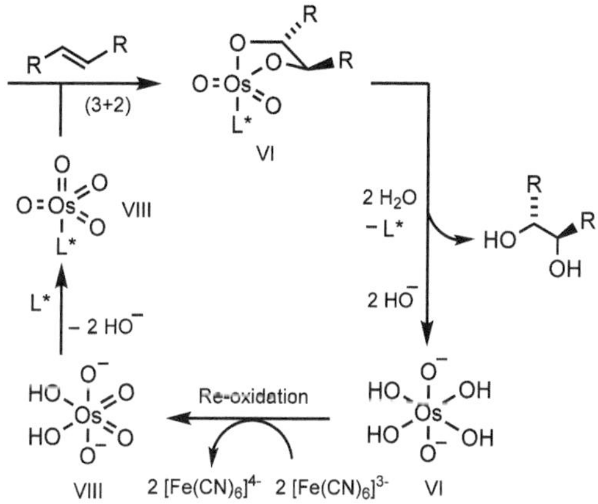


**FIG. 10.17** Catalytic cycle of the Sharpless asymmetric dihydroxylation (L* = chiral ligand).

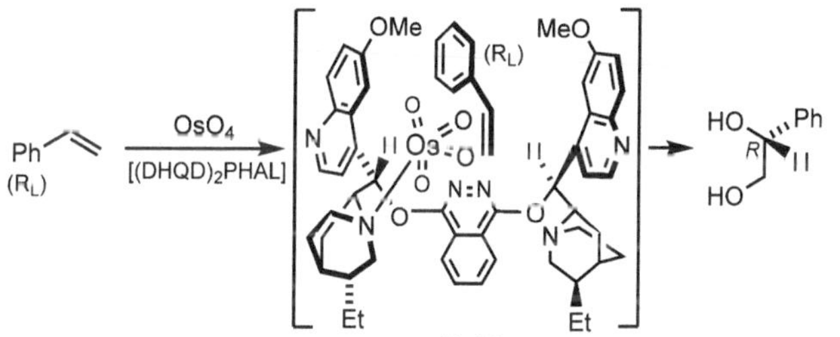


**10.14**

**FIG. 10.18** Corey's model with a U-shaped binding pocket for the Sharpless asymmetric dihydroxylation.

(best fit) of the alkene in the chiral pocket when only one face of the alkene can approach the osmium. This model resembles an enzyme, active site model.

### *10.3.1.2 Applications*

The enantioselective diol products from asymmetric dihydroxylation reactions can be easily converted into epoxides with retention of configuration. This transformation provides a valuable method of asymmetric synthesis of epoxides particularly from unfunctionalized alkenes when the Jacobsen–Katsuki method performs rather poorly. Fig. 10.19 shows the transformation of a chiral diol (obtained from asymmetric dihydroxylation) into the corresponding epoxide in a general case.[42] Treatment of the diol with trimethyl orthoacetate in the presence of pyridinium *p*-toluene sulphonate (PPTS) as a mild acid catalyst gives a cyclic orthoester, which then proceeds via ring-opening by bromide to give a mixture of regioisomeric bromoacetates. Acetyl bromide releases bromide ion on reacting with methanol generated in the reaction. Both bromoacetates,

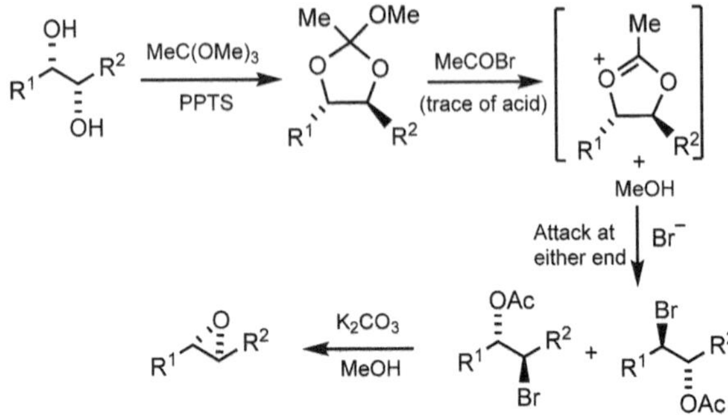


FIG. 10.19 Transformation of a chiral diol (obtained from asymmetric dihydroxylation) into an epoxide with retention of configuration.

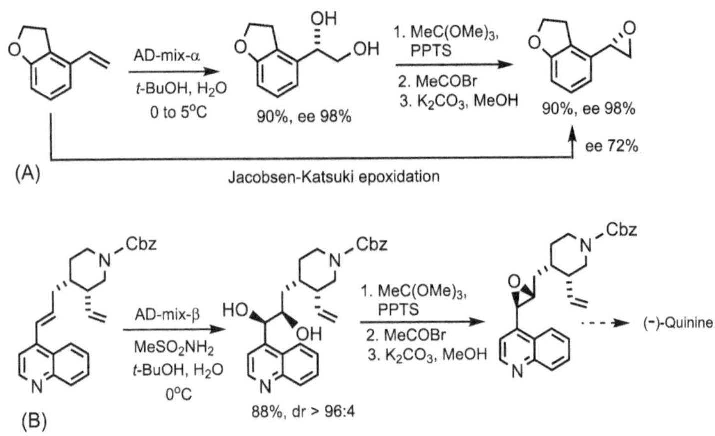


FIG. 10.20 Applications of the Sharpless asymmetric dihydroxylation in the asymmetric synthesis of (A) an epoxide of very high enantiopurity, and (B) a key epoxide intermediate in a synthesis of quinine.

on treatment with a base, give the same epoxide by hydrolysis of the ester group followed by ring-closing.

The usefulness of the previously mentioned procedure is illustrated with two examples in Fig. 10.20. Fig. 10.20A shows that the desired epoxide is formed with 98% ee, which is much greater than that obtained by direct Jacobsen–Katsuki epoxidation (ee 72%). An application of the method in a synthesis of the natural product (−)-quinine is shown in Fig. 10.20B.[43]

## 10.3.2 Sharpless asymmetric aminohydroxylation[44,45]

Like the dihydroxylation with $OsO_4$, the aminohydroxylation reaction of alkenes with $OsO_4$ in the presence of a nitrogen source is also diastereoselectively *syn*

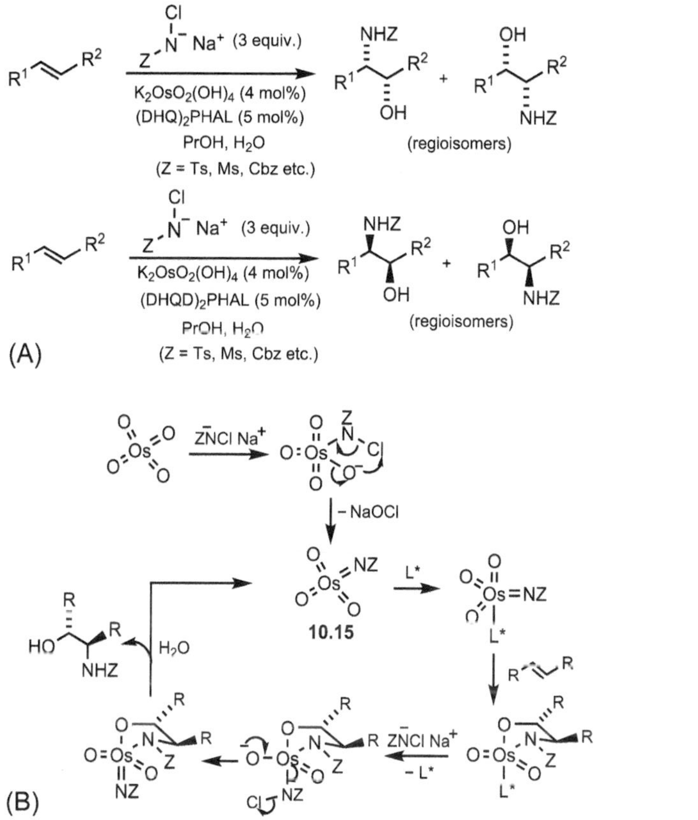


**FIG. 10.21** (A) The Sharpless asymmetric aminohydroxylation reaction; (B) Catalytic cycle of the reaction (L* = chiral ligand).

($NH_2$/OH). However, a major problem is poor regioselectivity with unsymmetrical alkenes. The asymmetric version of the aminohydroxylation reaction has also been developed by Sharpless and co-workers. Fig. 10.21A depicts the Sharpless asymmetric aminohydroxylation with *trans*-alkenes. The design of the reaction is similar to that of the dihydroxylation, with $(DHQ)_2PHAL$ or $(DHQD)_2PHAL$ as the chiral ligand.[46] Various nitrogen sources are possible, such as *N*-chloramine-T (TsNClNa),[46] *N*-chloramine-M (MsNClNa),[47] *N*-chloro-*N*-sodio-carbamates (e.g. CbzNClNa)[47] and *N*-bromo-*N*-lithio-carbamides (e.g. RCONBrLi).[48] The active osmium reagent in the aminohydroxylation is the imido osmium(VIII) species **10.15** that enters into the catalytic cycle in the presence of the chiral ligand (Fig. 10.21B).

Fig. 10.22 illustrates the Sharpless asymmetric aminohydroxylation using *N*-chloramine-T (Fig. 10.22A)[46] and sodium salt of *N*-chloro-benzylcarbamate (CbzNClNa) (Fig. 10.22B)[49] as nitrogen-atom sources. With the cinnamate, major regioisomer is a β-amino ester, and with the styrene, a benzylamine is

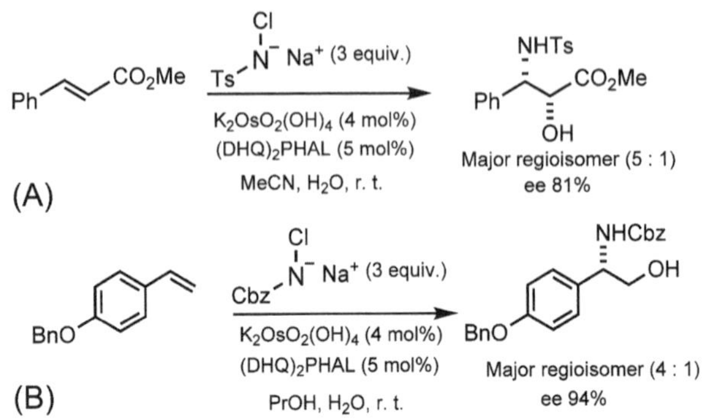


**FIG. 10.22** Examples of the Sharpless asymmetric aminohydroxylation using (A) *N*-chloramine-T and (B) *N*-chloro-benzylcarbamate as the nitrogen sources (Ts = *p*-$MeC_6H_4SO_2$; Cbz = BnOCO).

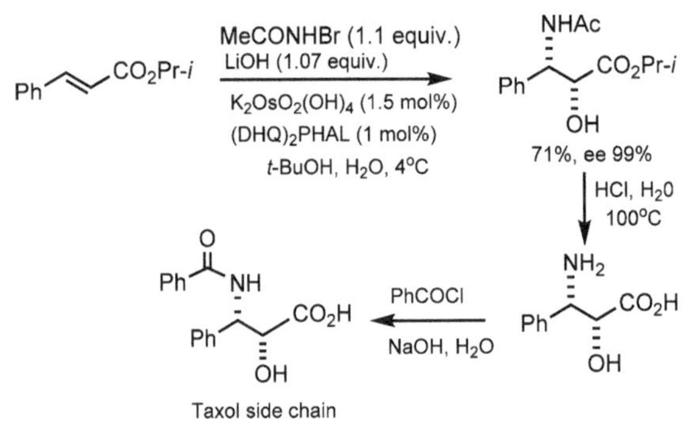


**FIG. 10.23** Application of the Sharpless asymmetric aminohydroxylation in a synthesis of the side chain of Taxol.

a major regioisomer. The reactions show good to high enantioselectivity of the major regioisomeric product. N-deprotection under mild conditions will give the corresponding amino alcohols (Ts can be removed with Na in Liquid $NH_3$ and Cbz by hydrogenolysis). The structure of the ligand can influence the regiochemical preference of the reaction. For example, in the reaction shown in Fig. 10.22B, the ligand $(DHQ)_2$AQN (AQN = anthraquinone) causes a reversal of regioselectivity forming predominantly a secondary alcohol.

The asymmetric aminohydroxylation reaction has been applied successfully in the asymmetric synthesis of natural products and drugs. For example, Fig. 10.23 depicts a synthesis of the side chain of the anticancer drug Taxol.[50] The reaction with the chiral ligand $(DHQ)_2$PHAL and *N*-bromo-*N*-lithio-acetamide as the nitrogen source provides the required N-acylated amino alcohol essentially as a single enantiomer after recrystallization. Hydrolysis and subsequent N-benzoylation then gives the taxol side chain.

## 10.4 Asymmetric hydrogenation[51,52]

A large number of chiral compounds occurring in nature or synthesized in the laboratory contain a hydrogen atom at the stereocentre. Catalytic hydrogenation of alkenes can create one or two new stereocentres, whereas a single stereocentre is formed in the catalytic hydrogenation of ketones. The goal of the asymmetric hydrogenation is to obtain the hydrogenation product as a single enantiomer. In general, asymmetric hydrogenation requires homogeneous (one phase) catalysis with transition metals. The impetus for the studies of asymmetric hydrogenations came from the work of Wilkinson on homogeneous hydrogenation of alkenes using the rhodium catalyst [$(Ph_3P)_3RhCl$] (known as Wilkinson's catalyst[53]). The catalytic cycle of the homogeneous hydrogenation using Wilkinson's catalyst is shown in Fig. 10.24. Addition of $H_2$ to the catalyst occurs by *cis* oxidative addition. Dissociation of a bulky $PPh_3$ ligand *trans* to a hydride (*trans* effect) and coordination of the alkene to the metal is followed by migratory insertion (migration of H from metal to alkene as the alkene inserts into the Rh—H bond) to form an intermediate with a carbon-metal bond. A *cis* arrangement of the two ligands (alkene and H) is required for the migratory insertion when the metal effectively loses a ligand. The insertion is driven by the addition of an external ligand ($PPh_3$) to form a coordinatively saturated complex. Finally, reductive elimination (reverse of oxidative addition) gives the product and continues the catalytic cycle.

The homogeneous hydrogenation reaction is amenable to asymmetric induction with a chiral modification of the achiral catalyst, most commonly by using chiral ligands on the rhodium(I) or ruthenium(II) metal centre. Thus the development of efficient chiral phosphine ligands has been crucial to the success of asymmetric hydrogenation reactions. W. S. Knowles and R. Noyori

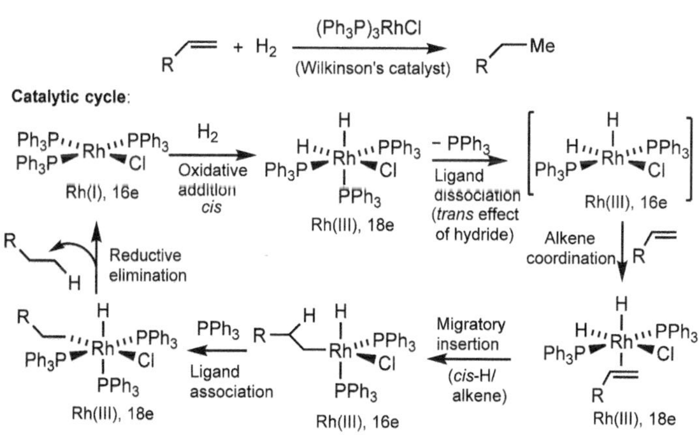


FIG. 10.24 Homogeneous hydrogenation of alkenes using Wilkinson's catalyst.

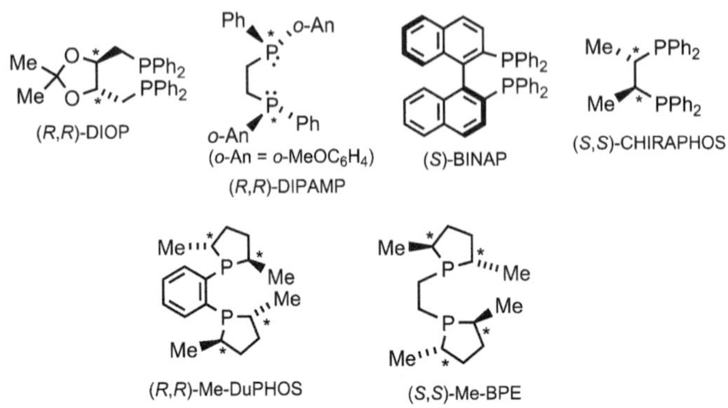


**FIG. 10.25** Chiral diphosphines as ligands for transition metal catalysts in asymmetric hydrogenation.

shared the 2001 Nobel Prize with K. B. Sharpless for their work on catalytic asymmetric hydrogenation. Fig. 10.25 shows some popular diphosphine ligands along with their acronyms, such as DIOP, DIPAMP, BINAP, CHIRAPHOS, DuPHOS and BPE.[54] Note that chiral centre is either at the P atom or at the carbon backbone of the ligands except BINAP which possesses axial chirality. Some monodentate phosphine ligands have also been found to be effective for asymmetric hydrogenation.[55]

### 10.4.1 Rhodium-catalysed asymmetric hydrogenation

Much of the early work on catalytic asymmetric hydrogenation was focussed on the development of chiral rhodium(I) catalysts. Asymmetric hydrogenation requires that the alkene substrate contains a chelating group that can coordinate to the metal. With Rh-catalysts, the suitable substrates are α-acylaminoacrylic acid derivatives (also called α-enamides or dehydroamino acid derivatives). Kagan[56] and Knowles[57–59] reported the $C_2$-symmetric chiral diphosphine ligands DIOP and DIPAMP, respectively, for Rh-catalysed asymmetric hydrogenations. Fig. 10.26A illustrates an asymmetric hydrogenation of (*Z*)-2-(acetamido) cinnamic acid ester with the rhodium catalyst [Rh(cod)(*R*,*R*-DIPAMP)]$BF_4$ (cod = 1,5-cyclooctadiene) **10.16**. The reaction gives the (*S*) enantiomer of the amino acid derivative with high ee (96%).

Rhodium catalysis is of practical significance; Knowles aimed to develop an industrial synthesis, while working at Monsanto, of a chiral drug L-DOPA [(*S*)-3′,4′-dihydroxyphenylalanine] used in the treatment of Parkinson's disease. The synthesis was performed using the Rh-(*R*,*R*)-DIPAMP catalyst **10.16** (Fig. 10.26B), which represented the first successful industrial application of a catalytic asymmetric synthesis.[51,60]

In general, the mechanism of asymmetric hydrogenation with Rh-catalysts involves a sequence of coordination of the substrate, oxidative addition of $H_2$,

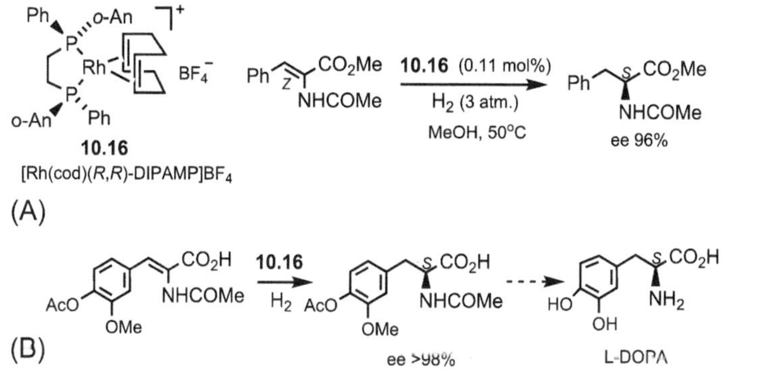

**FIG. 10.26** (A) Rhodium-catalysed asymmetric hydrogenation using (*R*,*R*)-DIPAMP as a chiral ligand; (B) An industrial synthesis of L-DOPA.

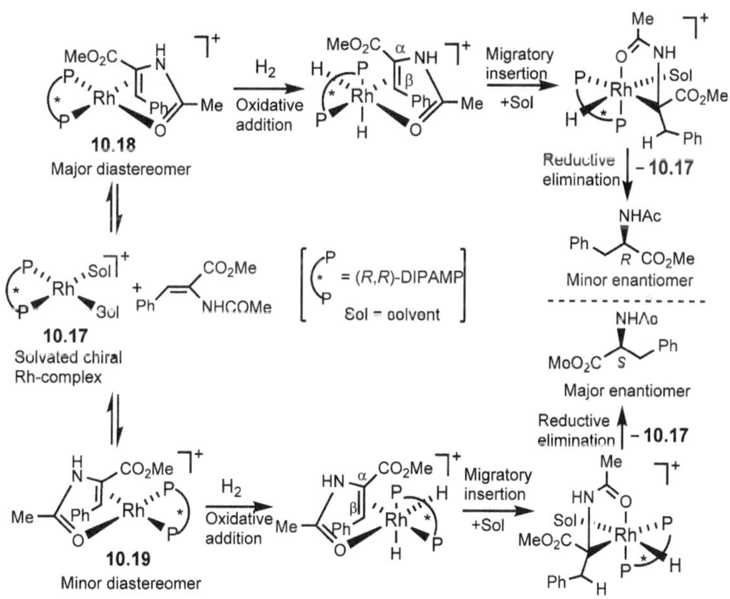

**FIG. 10.27** Halpern mechanism for the rhodium-catalysed asymmetric hydrogenation.

migratory insertion and reductive elimination to give the product. The catalytic cycle of the asymmetric hydrogenation with a chiral diphosphine ligand (CHIRAPHOS or DIPAMP) was investigated in detail by Halpern, and is shown in Fig. 10.27.[61] The chiral Rh-catalysts are typically diene adducts such as **10.16**, which reacts rapidly with $H_2$ in methanol to form a solvated complex $[RhP_2{}^*(Sol)_2]^+$ **10.17** ($P_2{}^*$ = chiral diphosphine ligand, Sol = solvent such as methanol). The complex **10.17** is the starting point of the catalytic cycle.

The alkene substrate can bind to the complex **10.17** via two different faces of the alkene by displacing two solvent (Sol) molecules, leading to the formation of two diastereomeric catalyst-substrate complexes **10.18** and **10.19**. The formation of the diastereomeric complex is rapid and reversible, and the two complexes remain in equilibrium. Each diastereomeric complex then follows the sequential pathway of oxidative addition of $H_2$, migratory insertion and reductive elimination. During migratory insertion, H-transfer from Rh occurs to the β-carbon of the alkene, while α-carbon is bonded to Rh. The binding mode of the substrate determines the sense of asymmetric induction. The diastereomeric complex **10.18** leads to the (*R*)-enantiomer of the product, whereas **10.19** gives the (*S*)-enantiomer.

Based on the combined kinetic and NMR studies, it was shown that the predominant enantiomer of the product arises from the *minor* diastereomer of the catalyst-substrate complex because of its much higher reactivity towards oxidative addition of $H_2$, compared with that of the major diastereomer. Thus the enantioselectivity of the asymmetric hydrogenation is governed not by the favoured initial binding of the substrate to the chiral catalyst but by the higher reactivity of the minor diastereomer of the catalyst-substrate complex formed via less-favoured binding. This mechanism thus conforms to the Curtin-Hammett principle.

It has been observed that with (*R,R*)-DIPAMP as the diphosphine ligand, the minor diastereomer is**10.19**, and the reaction is highly enantioselective in favour of the (*S*) enantiomer of the product (Fig. 10.27; see Fig. 10.26). If (*S,S*)-CHIRAPHOS is used, the minor diastereomer is **10.18**, and the reaction gives predominantly the (*R*) product.

(An alternative dihydride mechanism can also operate in some cases, particularly with bulky diphosphines, in which oxidative addition of $H_2$ occurs before substrate binding to the chiral catalyst.)[62]

Several chiral ligands have proved to be very efficient and selective for the Rh-catalysed asymmetric hydrogenation. The Rh-catalysed hydrogenation of the (*Z*)- and (*E*)-isomers, using BINAP as a ligand, affords opposite enantiomers of the product.[63] Interestingly, chiral bis(phopholane) ligands such as DuPHOS and BPE (see Fig. 10.25), offer general Rh-catalysed hydrogenation to give the same enantiomeric product, irrespective of the *E*/*Z* geometry of the substrate. Further, the substrates can have a wide selection of β-substituents that are tolerated by these catalysts to produce novel amino acids.[64] Fig. 10.28A illustrates an asymmetric hydrogenation using the catalyst Rh-(*R,R*)-Me-DuPHOS or Rh-(*R,R*)-Me-BPE. The reaction is highly enantioselective in favour of (*R*)-enantiomer of the amino acid derivative. Typically, the cationic rhodium complexes of the type [Rh(cod)(DuPHOS or BPE)]$^+$SbF$_6^-$ serve as catalyst precursors.

A stereochemical model with a quadrant diagram has been proposed to understand the origin of enantioselection (Fig. 10.28B).[65] The substrate can chelate to Rh through *Re* or *Si* face of the alkene unit and the N-acylcarbonyl

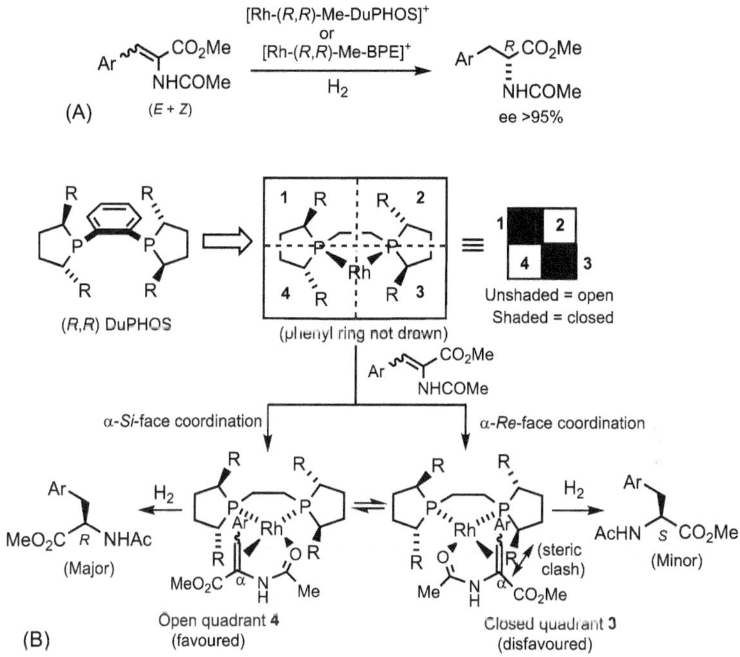


**FIG. 10.28** (A) General Rh-catalysed asymmetric hydrogenation with a bis(phopholane) ligand DuPHOS or BPE; (B) Stereochemical model in terms of quadrant diagram.

oxygen, and form two diastereomeric complexes analogous to those described earlier (cf. Fig. 10.27). The chiral environment around the rhodium atom is represented in terms of four quadrants, two of which have severe steric hindrance from the ligand R-substituents (closed quadrants), while the other two are devoid of such unfavourable steric repulsions (open quadrants). The α-*Re*-face addition involves a closed quadrant (marked 3) and is, therefore, disfavoured. In contrast, the α-*Si*-face addition via an open quadrant (marked 4) is favoured. The β-aryl substituent on the enamide will have little effect on the reaction because of minimal steric interactions between the ligand R-substituent and the (*E*)- or (*Z*)-aryl group. Thus, the preferred or major diastereomeric complex via the α-*Si*-face binding leads to the major enantiomer (*R*) of the product.

## 10.4.2 Noyori asymmetric hydrogenation with ruthenium catalysts

Noyori observed that the substrate scope of asymmetric hydrogenation can be greatly broadened by the use of ruthenium instead of rhodium.[66,67] The asymmetric hydrogenation of alkenes and ketones using Ru(II)-BINAP and related catalysts is called Noyori asymmetric hydrogenation. The catalytic cycle starts with the reaction of [$RuX_2$(BINAP)] complex with $H_2$ which does not involve oxidative addition but forms a Ru monohydride intermediate [RuHX(BINAP)] by heterolysis of $H_2$ by the Ru complex. Thus the Ru centre remains in the +2 oxidation state throughout the catalytic cycle,[68,69] unlike the Rh complex in which the oxidation state changes back and forth between +1 and +3.

#### *10.4.2.1 Asymmetric hydrogenation of alkenes*

The asymmetric hydrogenation of alkenes catalysed by Ru(II)-BINAP dicarboxylate complexes[67] has a wide scope but still the substrates need a heteroatom in the functional group that can allow coordination to the Ru centre. Both enantiomers of BINAP are commercially available; they can also be prepared from inexpensive racemic BINOL (1,1′-bi-2-naphthol) using the method of resolution.[70] Fig. 10.29 illustrates asymmetric hydrogenation of some functionalized alkenes such as α,β-unsaturated carboxylic acids, allyl alcohols and α-acylaminoacrylic acids using [Ru(OAc)$_2$(*S*)-BINAP] catalyst.[67,71,72] The reactions occur with high enantioselectivity up to 99% ee.

The application of the Ru-catalysed asymmetric hydrogenation in the synthesis of an antiinflammatory drug (*S*)-naproxen, and a monoterpenoid perfumery compound (*R*)-or (*S*)-citronellol are shown in Fig. 10.30.[71–73] The hydrogenation of geraniol or nerol is highly chemoselective, reducing only 2–3 double bonds. The substrate to catalyst ratio is generally quite high, often requiring <1 mol% Ru. The hydrogenation reaction has also been applied in the asymmetric synthesis of isoquinoline alkaloids including morphine.[74]

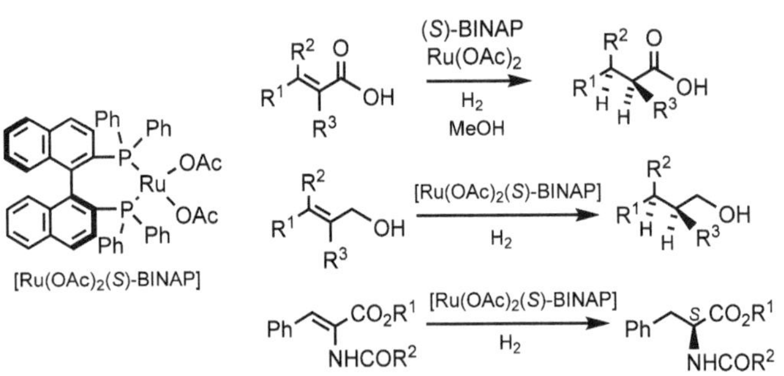


**FIG. 10.29** Asymmetric hydrogenation of some functionalized alkenes with Ru(II)-(*S*)-BINAP catalyst.

**Problem 10.3** Suggest a strategy for the synthesis of the artificial sweetener NC-00637:

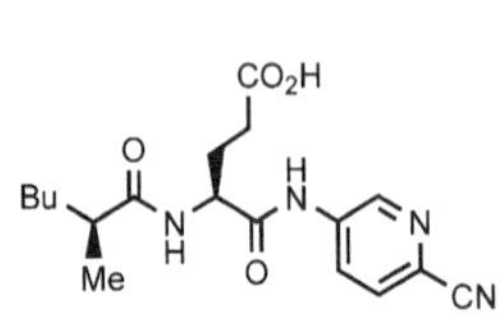


NC-00637

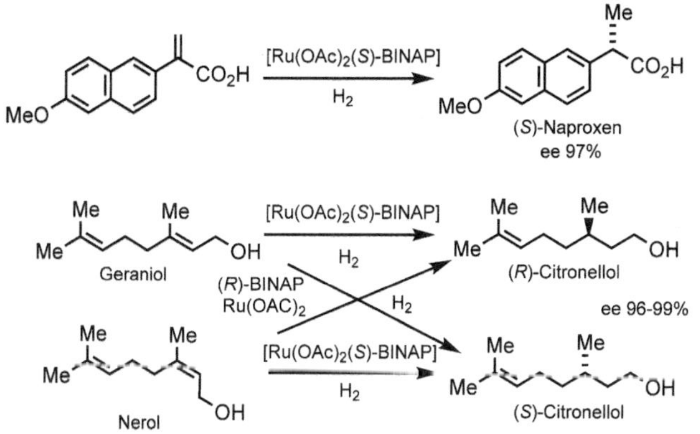


**FIG. 10.30** Application of the Ru-catalysed asymmetric hydrogenation in the synthesis of an antiinflammatory drug (*S*)-naproxen and a perfumery compound (*R*)- or (*S*)-citronellol.

## 10.4.2.2 Asymmetric hydrogenation of ketones[75–78]

The BINAP-Ru(II)-catalysed hydrogenation exhibits a wider scope of substrates and can be extended to a range of functionalized ketones (Fig. 10.31).[79] β-Keto esters are typically the best substrates and lead to the β-hydroxy esters with excellent enantioselectivity.[80] High levels of enantioselection have also been observed with other oxygen- and nitrogen-containing functional groups in the ketone substrates. The halogen-containing BINAP-Ru(II) complexes (oligomers) are found to be efficient catalysts for these asymmetric hydrogenations.[81]

Fig. 10.32 illustrates the synthetic applications in the industrial synthesis of antibiotic levofloxacin, and a synthesis of the antiviral drug (+)-brefeldin A.[82] The dynamic kinetic resolution based on the BINAP-Ru-catalysed hydrogenation when one enantiomer of the racemic substrate reacts much faster than the other, has also been used for the asymmetric synthesis of various biologically important compounds.[83,84]

The catalytic cycle of the BINAP-Ru-catalysed hydrogenation of β-keto esters is shown in Fig. 10.33A.[83] The halide ligand in the Ru complex generates

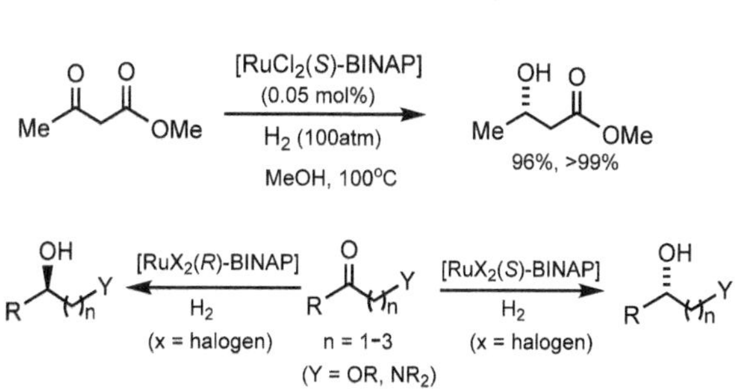


**FIG. 10.31** Asymmetric hydrogenation of a β-keto ester and other functionalized ketones using BINAP-Ru-dihalide complexes.

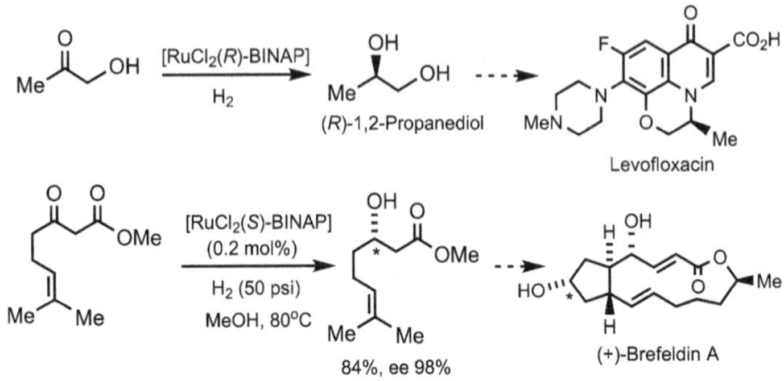


**FIG. 10.32** Synthetic applications of BINAP-Ru-catalysed asymmetric hydrogenation of ketones.

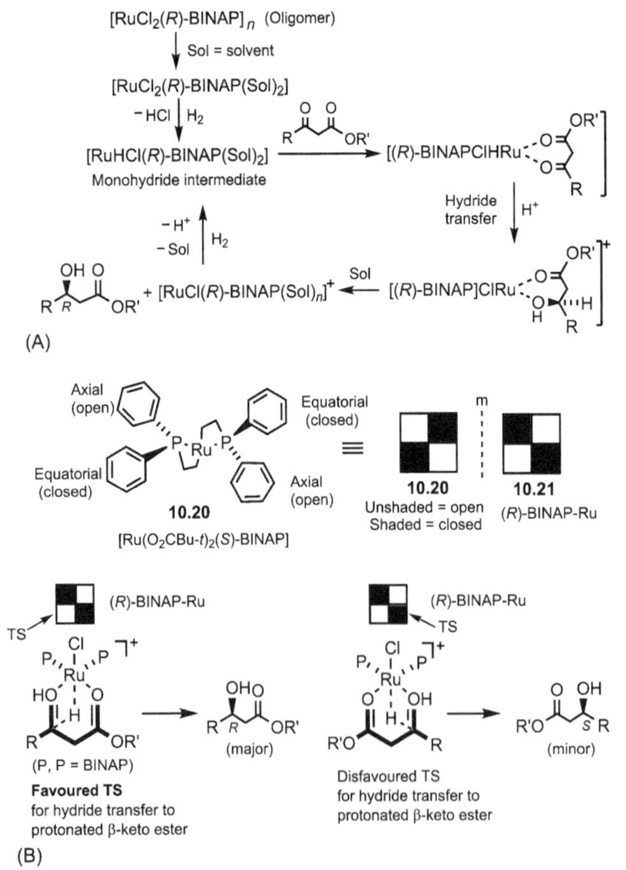


**FIG. 10.33** (A) Catalytic cycle of (*R*)-BINAP-Ru-catalysed asymmetric hydrogenation of β-keto esters; (B) Model for enantioselectivity.

a strong acid (HCl) and a Ru monohydride species [RuHCl(*R*)-BINAP(Sol)$_2$] (Sol = solvent) by the action of $H_2$. This intermediate then enters into the cycle with the formation of a monohydride-substrate complex, followed by the hydride transfer from the Ru centre to the carbonyl group.[85,86] The resulting complex leads to the product, and the monohydride species is regenerated by the action of $H_2$.

The crystal structure of [Ru(O$_2$CBu-*t*)$_2$(*S*)-BINAP] complex,[87] as determined by X-ray diffraction, reveals that the complex has a distorted octahedral geometry with the rigid BINAP backbone forcing the *P*-phenyl rings to adopt the conformation **10.20** (the naphthyl rings of BINAP are not shown for clarity) (Fig. 10.33B). The accessible face of Ru (the other face being blocked by naphthyl rings) can be divided into four quadrants, of which only two quadrants on the side of axial *P*-phenyl rings can allow the substrate coordination. These two quadrants are called open quadrants, while the other two on the side of equatorial *P*-phenyl rings are called closed or blocked quadrants where no coordination can take place. The situation for the (*R*)-BINAP-Ru complex will be a mirror image representation **10.21**.

The two possible diastereomeric TSs for the hydride transfer from the Ru to carbonyl group for the (*R*)-BINAP-Ru complex is shown in Fig. 10.33B. The TS leading to the (*R*)-product lies in the open quadrant and is highly favoured over the other (*S*)-product-directing TS in the closed quadrant with severe steric repulsion between the R substituent of the substrate and an equatorial *P*-phenyl ring of the catalyst.[88]

**Problem 10.4** Explain the following observations:

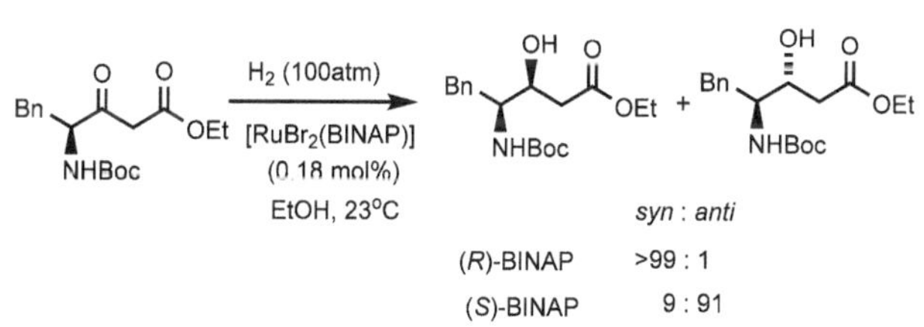


### 10.4.2.3 *Asymmetric transfer hydrogenation*

Asymmetric hydrogenation of ketones is also possible by transfer hydrogenation using a hydrogen transfer reagent. The chiral ruthenium complex **10.22** can catalyse the asymmetric transfer hydrogenation of aromatic ketones by using formic acid/triethylamine mixture as a reducing agent, which give the corresponding chiral alcohols with high enantioselectivity (Fig. 10.34A).[89] A *i*-PrOH/KOH system was originally used for the transfer hydrogenation but it is relatively less efficient due to the reversibility of isopropanol-acetone redox process. The synthetic utility of the transfer hydrogenation method is illustrated with two examples in Fig. 10.34B. The method can be extended to the

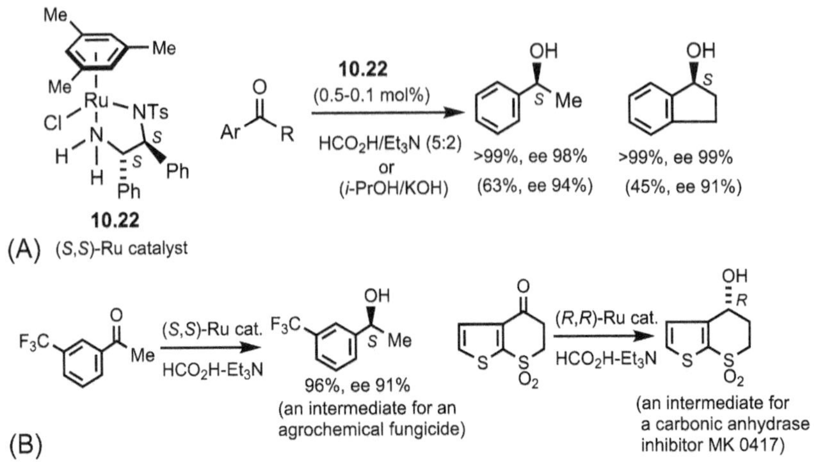


**FIG. 10.34** (A) Asymmetric transfer hydrogenation of aromatic ketones using $HCO_2H/Et_3N$ or (i-PrOH/KOH) mixture as a hydrogen transfer reagent in the presence of a chiral Ru catalyst; (B) Examples of synthetic applications.

asymmetric reduction of imines.[90] In general, the reaction is chemoselective for C=O or C=NR functions and is tolerant of many unsaturated systems including C=C, nitro, sulphone, aromatic and heteroaromatic groups.

A metal–ligand bifunctional mechanism for the asymmetric transfer hydrogenation, without involving substrate–metal complexation, is shown in Fig. 10.35.[91–93] The active Ru catalyst formed from the precatalyst **10.22** by the action of the base and formic acid (a formal adduct of $H_2$ and $CO_2$), is a coordinatively saturated 18-electron complex, [RuH{(*S*,*S*)-NTsCHPhCHPhNH$_2$}($\eta^6$-mesitylene)] **10.24**. The reaction proceeds by a metal–ligand bifunctional mechanism that allows for simultaneous delivery of the Ru—H and N—H to the C=O function via a six-membered pericyclic TS, to give an *S* alcohol. The released 16-electron Ru-amide complex **10.23** then regenerates the active catalyst **10.24** by the action of formic acid to continue the catalytic cycle. The recognition of the preferred carbonyl enantioface in the hydrogen transfer is governed mainly by the attractive CH/π interaction between the $\eta^6$-arene ligand and the aromatic substituent in the carbonyl substrate. The CH/π attractive force arises from both electrostatic and charge-transfer interactions. This mechanism shows that the substrate–metal complexation is not essential for hydrogenation of unsaturated functions.

## 10.5 Asymmetric palladium-catalysed coupling reactions

Palladium-catalysed coupling reactions are widely used in organic synthesis, and the 2010 Nobel Prize has been awarded to R.F. Heck, E. Negishi and A. Suzuki for their work in this area. The key to the attraction of this palladium

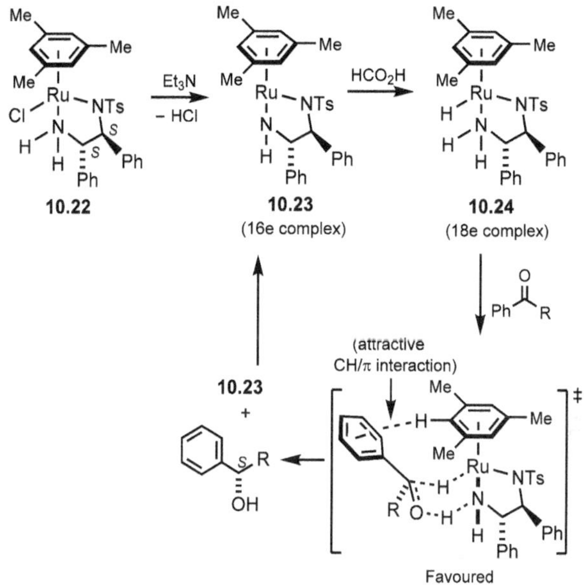


**FIG. 10.35** Metal–ligand bifunctional mechanism of the asymmetric transfer hydrogenation of aromatic ketones using (*S*,*S*)-Ru complex **10.22**.

chemistry is that two unsaturated components can be linked directly through a new C—C bond in the presence of a Pd(0) catalyst in a highly selective and efficient fashion. There are two main classes of coupling reactions: (1) the Heck coupling reactions[94,95] and (2) the cross-coupling reactions,[96,97] for example, the Suzuki cross-coupling.[98] A standard Heck reaction and its mechanism (catalytic cycle) are shown in Fig. 10.36A. The Pd(0)-catalysed Heck reaction is the vinylation of aryl (or vinyl) halides/triflates in the presence of a base. The mechanism involves a sequence of oxidative addition of ArX, migratory insertion (or carbopalladation) of the alkene via the formation of a π-complex, and *syn* β-hydride elimination (via a π-complex) to give the product. Reductive elimination in the presence of a base regenerates the catalyst. Fig. 10.36B shows a cross-coupling reaction and its mechanism. The cross-coupling reaction between a halide or triflate (ArX) and an organometallic reagent (RM) catalysed by a Pd (0) complex results in the formation of a carbon–carbon bond and gives the product (Ar-R). The catalytic cycle involves oxidative addition, transmetallation and reductive elimination. The transmetallation is possible with a wide variety of organometallic reagents of varying nucleophilicity, such as tin reagents $R_4Sn$ (Stille coupling),[99] boron reagents $RB(OR')_2$ (Suzuki coupling) and zinc reagents RZnX (Negishi coupling).[100] The mechanism of transmetallation appears to involve a concerted transfer of R to Pd through a four-centre TS.

The active catalyst is Pd(0), which can be added as a stable complex such as Pd $(PPh_3)_4$ or $Pd_2(dba)_3$ (dba = *trans*, *trans*-dibenzylideneacetone). Another option is

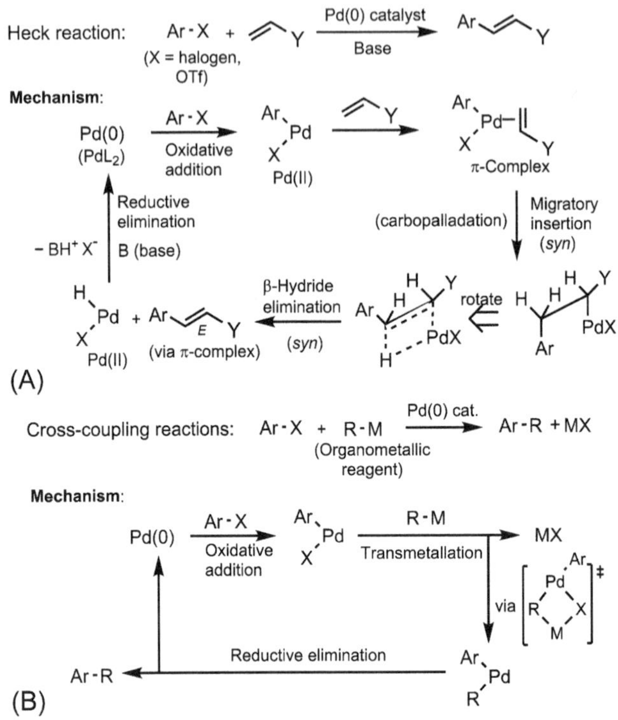


**FIG. 10.36** Palladium-catalysed coupling reactions: (A) Heck reaction and its mechanism; (B) Cross-coupling reactions and the catalytic cycle.

to use a Pd(II) precatalyst such as $Pd(OAc)_2$ together with a ligand. The Pd(II) is reduced to Pd(0) before the catalytic cycle can start. The reduction is usually brought about by a component of the reaction such as $PPh_3$, $Et_3N$ or RM.

At first glance, the asymmetric modification of palladium-catalysed coupling reactions looks unpromising, since no new chiral centre is formed. To induce asymmetry, several strategies have been developed to direct the course of the reaction so that a chiral product is obtained in the presence of a chiral palladium catalyst.

## 10.5.1 Asymmetric Heck reaction[101–103]

In the Heck reaction, the migratory *syn*-insertion and *syn*-β-hydride elimination steps are reversible, which can lead to isomerization of the alkene product. If the $sp^3$ centre formed in the insertion step lacks an H *syn* to Pd, the $sp^3$ centre cannot convert back to an $sp^2$ centre. Under this situation, the migrating group (Ar) is incorporated onto an $sp^3$ centre. The asymmetric design of the Heck reaction is thus possible. Fig. 10.37A shows two consecutive Heck reactions to give a diastereoselective *trans*-product **10.26**. Each Heck reaction leads to double-bond

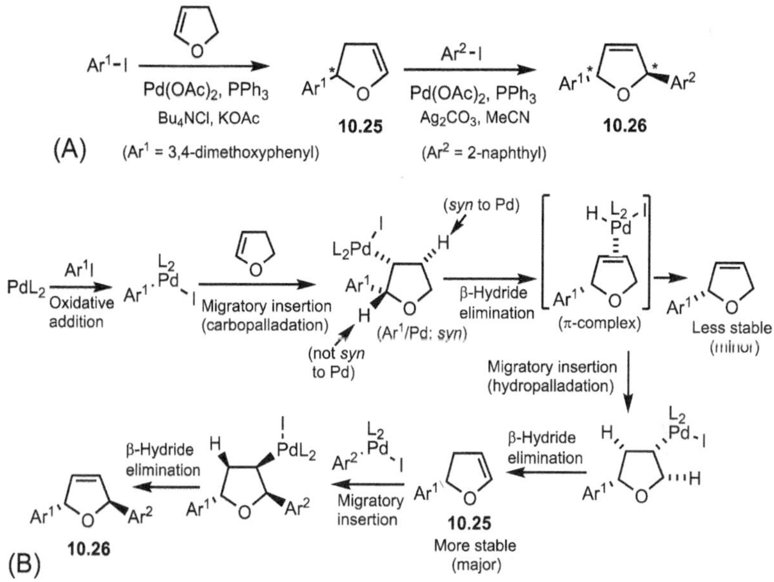


**FIG. 10.37** (A) Consecutive Heck reactions: double-bond isomerization, stereocentre creation and diastereoselectivity; (B) Mechanism and rationalization of diastereoselectivity.

isomerization with the migrating Ar group placed onto an $sp^3$ centre. The active Pd(0) complex is formed from $Pd(OAc)_2$ by reduction with $PPh_3$. Fig. 10.37B shows the mechanism of product formation. In the first Heck reaction, after oxidative addition of $Ar^1I$, the migratory insertion (carbopalladation) leads to the aryl addition to the less electron-rich end of the alkene in a *syn* fashion (Pd being attached to the more electron-rich end as a result of the interaction of the filled π orbital of the alkene with the vacant d orbital on Pd). Since the $sp^3$ centre generated lacks a *syn* H, β-hydride elimination involves the *syn* H on the other β carbon. The π-complex of the β-hydride elimination can give either a less stable alkene (minor product) in which the double bond moves one position around the ring, or a more stable alkene **10.25** (major product) by way of a second migratory insertion (hydropalladation) followed by a second β-hydride elimination. The second Heck reaction then leads to the final *trans* product **10.26**. Here the migratory insertion places $Ar^2$ *anti* to $Ar^1$ via a less hindered π-complex and the subsequent β-hydride elimination gives **10.26**. Further migration of the double bond by hydropalladation does not occur because the silver carbonate rapidly removes iodide from the π-complex of the β-hydride elimination, thereby preventing the addition of Pd—H to the alkene.

The asymmetric Heck reaction has been designed successfully using the chiral (P,P) ligands such as BINAP or the chiral (P,N) ligands such as PHOX (phosphinooxazoline). Fig. 10.38A illustrates an asymmetric arylation of 2,3-dihydrofuran with aryl triflates using the Pd/(*R*)-BINAP catalytic system.

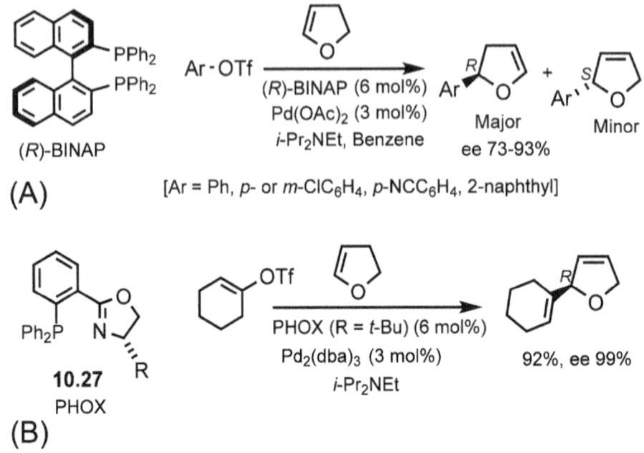


**FIG. 10.38** Asymmetric intermolecular Heck reactions using (A) Pd/BINAP and (B) Pd/PHOX catalyst systems.

This is the first example of the asymmetric intermolecular Heck reaction reported by Hayashi et al.[104,105] The major and minor products differ in double bond isomerization and absolute configuration at the stereocentre. Subsequently, the Pd/BINAP system has been extended to the asymmetric synthesis of various oxygen- and nitrogen-containing heterocycles.[106,107]

Chiral oxazoline-containing ligands such as PHOX[108] **10.27** can exhibit dramatic enhancement in enantioselectivity. For example, the cyclohexenylation of 2,3-dihydrofuran using the Pd/PHOX (R = *t*-Bu) catalyst proceeds with high yield and excellent enantioselectivity (ee 99%) (Fig. 10.38B).[109] Here, in contrast to the Pd/BINAP system, the product with no double-bond isomerization forms almost exclusively; only trace amounts of the isomerized product are observed. In general, PHOX ligands are known to effect minimal double bond migration.[110]

BINAP and PHOX ligands also promote good to excellent levels of enantioselectivity in the asymmetric intramolecular Heck reaction. Fig. 10.39A illustrates an intramolecular Heck reaction using the Pd/BINAP catalyst.[111] The reaction gives the desilylated indolinone **10.28** with high ee, used in a synthesis of an acetylcholinesterase inhibitor (−)-physostigmine. The enantioselectivity is enhanced by using the highly basic tertiary amine (PMP) and the catalyst system prepared in a polar solvent ($MeCONMe_2$).

The use of Pd/PHOX (R = *t*-Bu) catalyst in the asymmetric intramolecular Heck reaction, in a total synthesis of an indole alkaloid (+)-minfiensine, is shown in Fig. 10.39B.[112] The Heck reaction gives the desired product (no double bond isomerization) with excellent enantioselectivity. It may be mentioned that the same reaction with the Pd/BINAP catalyst gives the product with double bond migration.

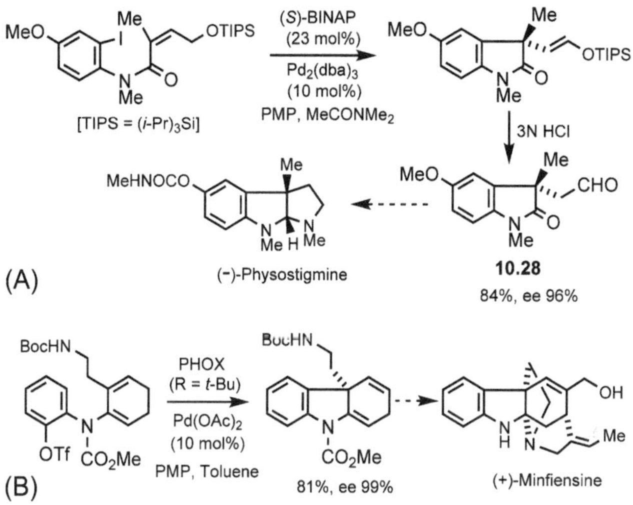


**FIG. 10.39** Asymmetric intramolecular Heck reactions using (A) Pd/BINAP and (B) Pd/PHOX catalyst systems.

## 10.5.2 Asymmetric Suzuki cross-coupling reaction[113]

Asymmetric palladium-catalysed cross-coupling reactions are much less explored than the asymmetric Heck reaction. The Suzuki cross-coupling reaction is particularly advantageous in synthesis because of its tolerance of a broad range of functional groups and its nontoxic by-products. The organometallic partner in the Suzuki coupling is an organoboronic acid derivative which is thermally stable, and inert to air and water. The presence of a base in the Suzuki coupling is crucial; coordination of the boron with a nucleophile such as $HO^-$, $F^-$ or an amine giving a tetrahedral boronate anion, is required to drive transmetallation.

Axially chiral biaryls are prevalent in bioactive natural products[114] and are widely useful in asymmetric homogeneous catalysis.[115] The synthesis of axially chiral biaryls has been reported using palladium-catalysed asymmetric Suzuki cross-coupling reactions. The asymmetric coupling of aryl boronic acid with aryl halide takes place in the presence of a suitable chiral ligand. The catalytic cycle of the asymmetric Suzuki reaction is shown in Fig. 10.40, which involves a sequence of oxidative addition, transmetallation and reductive elimination (cf. Fig. 10.36B). The base accelerates the transmetallation step, via a more nucleophilic 'ate' complex.

Fig. 10.41 illustrates asymmetric Suzuki cross-coupling reactions using chiral ligands (*R*)-BINAP and a ferrocene-based monophosphine (*S*)-(*R*)-PFNMe **10.29**, respectively.[116] (*R*)-BINAP leads to an axially chiral binaphthalene product with a modest ee of 25%. However, the ferrocene-based tertiary amine monophosphine ligand **10.29** gives the product with high enantioselectivity

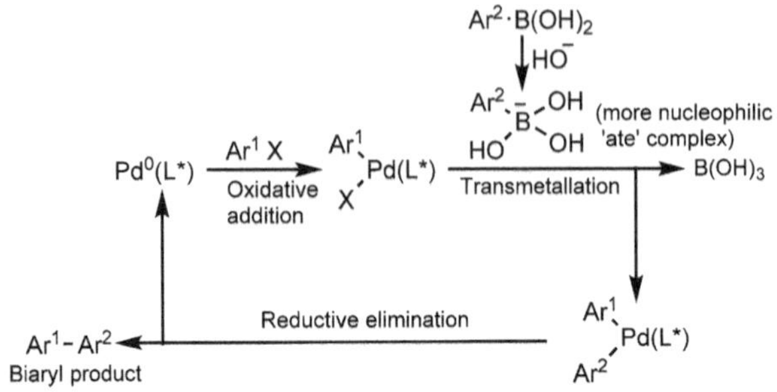


**FIG. 10.40** Catalytic cycle of the asymmetric Suzuki cross-coupling reaction to form chiral biaryl derivatives (L* is a chiral ligand).

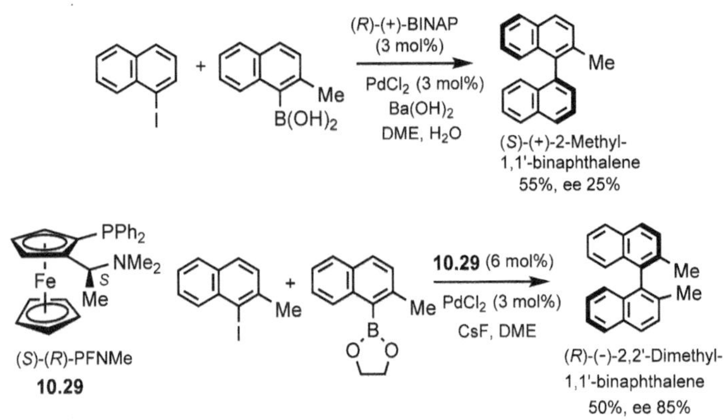


**FIG. 10.41** Asymmetric Suzuki cross-coupling reactions using ($R$)-BINAP and a monophosphine ligand ($S$)-($R$)-PFNMe, respectively ($R$ denotes planar chirality of the ferrocene).

(ee 85%) using ethylene glycol ester of an aryl boronic acid as the organometallic component. The high enantioselectivity arises presumably due to precomplexation between nitrogen and boron, before transmetallation.

Notably, some novel chiral ligands such as biaryl backbone-containing monophosphines,[117] phosphorous-free bis-hydrazones[118] and hybrid phosphine-hydrazones[119] have been developed and applied to asymmetric Suzuki cross-coupling reactions using a broad range of aryl electrophiles.

## References

1. Rao, A. S. In *Comprehensive Organic Synthesis*; Trost, B. M., Fleming, I., Eds.; Vol. 7; Pergamon Press: Oxford, 1991; p. 357.
2. Jørgensen, K. A. *Chem. Rev.* **1989**, *89*, 431.
3. Adam, W.; Wirth, T. *Acc. Chem. Res.* **1999**, *32*, 703.
4. Sharpless, K. B. In *Comprehensive Organic Synthesis*; Trost, B. M., Fleming, I., Eds.; Vol. 7; Pergamon Press: Oxford, 1991; p. 389.

5. Katsuki, T.; Martin, V. S. *Org. React.* **1996**, *48*, 1.
6. Johnson, R. A.; Sharpless, K. B. In *Catalytic Asymmetric Synthesis*; Ojima, I., Ed.; 2nd ed.; Wiley-VCH: New York, 2000; p. 231 (chapter 6A).
7. Ko, S. Y.; Lee, A. W. M.; Masamune, S.; Reed, L. A., III; Sharpless, K. B.; Walker, F. J. *Tetrahedron* **1990**, *46*, 245.
8. Finn, M. G.; Sharpless, K. B. *J. Am. Chem. Soc.* **1991**, *113*, 106.
9. Finn, M. G.; Sharpless, K. B. *J. Am. Chem. Soc.* **1991**, *113*, 113.
10. Carlier, P. R.; Sharpless, K. B. *J. Org. Chem.* **1989**, *54*, 4016.
11. Klunder, J. M.; Ko, S. Y.; Sharpless, K. B. *J. Org. Chem.* **1986**, *51*, 3710.
12. Gao, Y.; Sharpless, K. B. *J. Org. Chem.* **1988**, *53*, 4081.
13. Ko, S. Y.; Lee, A. W. M.; Masamune, S.; Reed, L. A., III; Sharpless, K. B.; Walker, F. J. *Science* **1983**, *220*, 949.
14. Keith, J. M.; Larrow, J. F.; Jacobsen, E. N. *Adv. Synth. Catal.* **2001**, *343*, 5.
15. Paterson, I.; Cumming, J. G.; Ward, R. A.; Lamboley, S. *Tetrahedron* **1995**, *51*, 9393.
16. Hamon, D. P. G.; Tuck, K. L. *J. Org. Chem.* **2000**, *65*, 7839.
17. Jacobsen, E. N. In *Comprehensive Organometallic Chemistry II*; Abel, E. W., Stone, F. G. A., Wilkinson, G., Eds.; Vol. 12; Elsevier: Oxford, 1995; p. 1097.
18. Katsuki, T. In *Catalytic Asymmetric Synthesis*; Ojima, I., Ed.; 2nd ed.; Wiley-VCH: New York, 2000; p. 287 (chapter 6B).
19. Katsuki, T. *Adv. Synth. Catal.* **2002**, *344*, 131.
20. Mcgarrigle, E. M.; Gilheany, D. G. *Chem. Rev.* **2005**, *105*, 1563.
21. Hughes, D. L.; Smith, G. B.; Liu, J.; et al. *J. Org. Chem.* **1997**, *62*, 2222.
22. Senanayake, C. H.; Roberts, F. E.; DiMichele, L. M.; et al. *Tetrahedron Lett.* **1995**, *36*, 3993.
23. Bell, D.; Davies, M. R.; Finney, F. J. L.; Geen, G. R.; Kincey, P. M.; Mann, I. S. *Tetrahedron Lett.* **1996**, *37*, 3895.
24. Linker, T. *Angew. Chem. Int. Ed. Engl.* **1997**, *36*, 2060.
25. Kürti, L.; Blewett, M. M.; Corey, E. J. *Org. Lett.* **2009**, *11*, 4592.
26. Norrby, P.-O.; Linde, C.; Aakermark, B. *J. Am. Chem. Soc.* **1995**, *117*, 11035.
27. Schaus, S. E.; Brandes, B. D.; Larrow, J. F.; et al. *J. Am. Chem. Soc.* **2002**, *124*, 1307.
28. Jacobsen, E. N. *Acc. Chem. Res.* **2000**, *33*, 421.
29. Nielsen, L. P. C.; Stevenson, C. P.; Blackmond, D. G.; Jacobsen, E. N. *J. Am. Chem. Soc.* **2004**, *126*, 1360.
30. Kolb, H. C.; Van Nieuwenhze, M. S.; Sharpless, K. B. *Chem. Rev.* **1994**, *94*, 2483.
31. Johnson, R. A.; Sharpless, K. B. In *Catalytic Asymmetric Synthesis*; Ojima, I., Ed.; 2nd ed.; Wiley-VCH: New York, 2000; p. 357 (chapter 6D).
32. Noe, M. C.; Letavic, M. A.; Snow, S. L.; McCombie, S. W. *Org. React.* **2005**, *66*, 109.
33. Amberg, W.; Bennani, Y. L.; Chada, R. K.; et al. *J. Org. Chem.* **1993**, *58*, 844.
34. Kolb, H. C.; Andersson, P. G.; Sharpless, K. B. *J. Am. Chem. Soc.* **1994**, *116*, 1278.
35. Wang, Z.-M.; Zhang, X.-L.; Sharpless, K. B. *Tetrahedron Lett.* **1992**, *33*, 6407.
36. Wang, L.; Sharpless, K. B. *J. Am. Chem. Soc.* **1992**, *114*, 7568.
37. Becker, H.; Sharpless, K. B. *Angew. Chem. Int. Ed. Engl.* **1996**, *35*, 448.
38. Toy, P. H.; Janda, K. D. *Acc. Chem. Res.* **2000**, *33*, 546.
39. Corey, E. J.; Noe, M. C.; Grogan, M. J. *Tetrahedron Lett.* **1996**, *37*, 4899.
40. DelMonte, A. J.; Haller, J.; Houk, K. N.; et al. *J. Am. Chem. Soc.* **1997**, *119*, 9907.
41. Corey, E. J.; Guzman-Perez, A.; Noe, M. C. *Tetrahedron Lett.* **1995**, *36*, 3481.
42. Kolb, H. C.; Sharpless, K. B. *Tetrahedron* **1992**, *48*, 10515.
43. Raheem, I. T.; Goodman, S. N.; Jacobsen, E. N. *J. Am. Chem. Soc.* **2004**, *126*, 706.
44. Bodkin, J. A.; McLeod, M. D. *J. Chem. Soc. Perkin Trans. 1* **2002**, 2733.

45. Nilov, D.; Reiser, O. *Adv. Synth. Catal.* **2002**, *344*, 1169.
46. Li, G.; Chang, H.; Sharpless, K. B. *Angew. Chem. Int. Ed. Engl.* **1996**, *35*, 451.
47. Li, G.; Chang, H.; Sharpless, K. B. *Angew. Chem. Int. Ed. Engl.* **1996**, *35*, 2810.
48. Demko, P. Z.; Bartsch, M.; Sharpless, K. B. *Org. Lett.* **2000**, *2*, 2221.
49. Reddy, K. L.; Sharpless, K. B. *J. Am. Chem. Soc.* **1998**, *120*, 1207.
50. Bruncko, M.; Schlingloff, G.; Sharpless, K. B. *Angew. Chem. Int. Ed. Engl.* **1997**, *36*, 1483.
51. Knowles, W. S. *Angew. Chem. Int. Ed. Engl.* **1998**, *2002*, 41.
52. Noyori, R. *Angew. Chem. Int. Ed. Engl.* **2002**, *41*, 2008.
53. Osborn, J. A.; Jardine, F. H.; Young, J. F.; Wilkinson, G. *J. Chem. Soc. A* **1966**, 1711.
54. Ojima, I.; Clos, N.; Bastos, C. *Tetrahedron* **1989**, *45*, 6901.
55. van den Berg, M.; Minnaard, A. J.; Haak, R. M.; et al. *Adv. Synth. Catal.* **2003**, *345*, 308.
56. Dang, T. P.; Kagan, H. B. *J. Chem. Soc. Chem. Commun.* **1971**, 481.
57. Knowles, W. S.; Sabacky, M. J.; Vineyard, B. D. *J. Chem. Soc. Chem. Commun.* **1972**, 10.
58. Vineyard, B. D.; Knowles, W. S.; Sabacky, M. J.; Bachman, G. L.; Weinkauff, D. J. *J. Am. Chem. Soc.* **1977**, *99*, 5946.
59. Knowles, W. S. *Acc. Chem. Res.* **1983**, *16*, 106.
60. Knowles, W. S. *Adv. Synth. Catal.* **2003**, *345*, 3.
61. Halpern, J. *Science* **1982**, *217*, 401.
62. Gridnev, I. D.; Imamoto, T. *Acc. Chem. Res.* **2004**, *37*, 633.
63. Miyashita, A.; Takaya, H.; Souchi, T.; Noyori, R. *Tetrahedron* **1984**, *40*, 1245.
64. Burk, M. J.; Feaster, J. E.; Nugent, W. A.; Harlow, R. L. *J. Am. Chem. Soc.* **1993**, *115*, 10125.
65. Burk, M. J. *Acc. Chem. Res.* **2000**, *33*, 363.
66. Noyori, R.; Ohta, M.; Hsiao, Y.; Kitamura, M.; Ohta, T.; Takaya, H. *J. Am. Chem. Soc.* **1986**, *108*, 7117.
67. Kitamura, M.; Tokunaga, M.; Noyori, R. *J. Org. Chem.* **1992**, *57*, 4053.
68. Ohta, T.; Takaya, H.; Noyori, R. *Tetrahedron Lett.* **1990**, *31*, 7189.
69. Ashby, M. T.; Halpern, J. *J. Am. Chem. Soc.* **1991**, *113*, 589.
70. Takaya, H.; Akutagawa, S.; Noyori, R. *Org. Synth.* **1989**, *67*, 20.
71. Ohta, T.; Takaya, H.; Kitamura, M.; Nagai, K.; Noyori, R. *J. Org. Chem.* **1987**, *52*, 3174.
72. Takaya, H.; Ohta, T.; Sayo, N.; et al. *J. Am. Chem. Soc.* **1987**, *109*, 1596.
73. Kitamura, M.; Yoshimura, M.; Tsukamoto, M.; Noyori, R. *Enantiomer* **1996**, *1*, 281.
74. Noyori, R. In *Catalytic Asymmetric Synthesis*; Ohkuma, T., Kitamura, M., Ojima, I., Eds.; 2nd ed.; Wiley-VCH: New York, 2000; p. 1.
75. Noyori, R.; Ohkuma, T. *Angew. Chem. Int. Ed.* **2001**, *40*, 40.
76. Tang, W.; Zhang, X. *Chem. Rev.* **2003**, *103*, 3029.
77. Kitamura, M.; Nakatsuka, H. *Chem. Commun.* **2011**, *47*, 842.
78. Noyori, R. *Angew. Chem. Int. Ed.* **2013**, *52*, 79.
79. Kitamura, M.; Ohkuma, T.; Inoue, S.; et al. *J. Am. Chem. Soc.* **1988**, *110*, 629.
80. Noyori, R.; Ohkuma, T.; Kitamura, M.; et al. *J. Am. Chem. Soc.* **1987**, *109*, 5856.
81. Mashima, K.; Kusano, K.; Sato, N.; et al. *J. Org. Chem.* **1994**, *59*, 3064.
82. Taber, D. F.; Silverberg, L. J.; Robinson, E. D. *J. Am. Chem. Soc.* **1991**, *113*, 6639.
83. Noyori, R. *Asymmetric Catalysis in Organic Synthesis*; Wiley: New York, 1993; p. 56.
84. Noyori, R.; Ikeda, T.; Okhuma, T.; et al. *J. Am. Chem. Soc.* **1989**, *111*, 9134.
85. Kitamura, M.; Tokunaga, M.; Noyori, R. *J. Am. Chem. Soc.* **1993**, *115*, 144.
86. Kitamura, M.; Tokunaga, M.; Noyori. *Tetrahedron* **1853**, *1993*, 49.
87. Ohta, T.; Takaya, H.; Noyori, R. *Inorg. Chem.* **1988**, *27*, 566.
88. Noyori, R.; Tokunaga, M.; Kitamura, M. *Bull. Chem. Soc. Jpn.* **1995**, *68*, 36.
89. Noyori, R.; Yamakawa, M.; Hashiguchi, S. *J. Org. Chem.* **2001**, *66*, 7932.

90. Uematsu, N.; Fujii, A.; Hashiguchi, S.; Ikariya, T.; Noyori, R. *J. Am. Chem. Soc.* **1996**, *118*, 4916.
91. Haack, K.-J.; Hashiguchi, S.; Fujii, A.; Ikariya, T.; Noyori, R. *Angew. Chem. Int. Ed. Engl.* **1997**, *36*, 285.
92. Yamakawa, M.; Ito, H.; Noyori, R. *J. Am. Chem. Soc.* **2000**, *122*, 1466.
93. Yamakawa, M.; Yamada, I.; Noyori, R. *Angew. Chem. Int. Ed. Engl.* **2001**, *40*, 2818.
94. Heck, R. F. *Org. React.* **1982**, *27*, 345.
95. Whitcombe, N. J.; Hii, K. K.; Gibson, S. E. *Tetrahedron* **2001**, *57*, 7449.
96. de Meijere, A.; Diederich, F., Eds. *Metal-Catalyzed Cross-Coupling Reactions*; 2nd ed.; Wiley-VCH: Weinheim, 2004.
97. Buchwald, S. L., Ed. Cross Coupling. *Acc. Chem. Res.* **2008**, *41* (Special Issue (11)), 1439–1564.
98. Hall, D. G., Ed. *Boronic Acids*; Wiley-VCH: Weinheim, 2005.
99. Espinet, P.; Echavarren, A. M. *Angew. Chem. Int. Ed. Engl.* **2004**, *43*, 4704.
100. Negishi, E.; Hu, Q.; Huang, Z.; Qian, M.; Wang, G. *Aldrichimica Acta* **2005**, *38*, 71.
101. Dounay, A. B.; Overman, L. E. *Chem. Rev.* **2003**, *103*, 2945.
102. Shibasaki, M.; Vogl, E. M. *J. Organomet. Chem.* **1999**, *576*, 1.
103. McCartney, D.; Guiry, P. J. *Chem. Soc. Rev.* **2011**, *40*, 5122.
104. Ozawa, F.; Kubo, A.; Hayashi, T. *J. Am. Chem. Soc.* **1991**, *113*, 1417.
105. Ozawa, F.; Hayashi, T. *J. Organomet. Chem.* **1992**, *428*, 267.
106. Ozawa, F.; Kobatake, Y.; Hayashi, T. *Tetrahedron Lett.* **1993**, *34*, 2505.
107. Sonesson, C.; Larhed, M.; Nyqvist, C.; Hallberg, A. *J. Org. Chem.* **1996**, *61*, 4756.
108. von Matt, P.; Pfaltz, A. *Angew. Chem. Int. Ed. Engl.* **1993**, *32*, 566.
109. Loiseleur, O.; Hayashi, M.; Schmees, N.; Pfaltz, A. *Synthesis* **1997**, 1338.
110. Loiseleur, O.; Meier, P.; Pfaltz, A. *Angew. Chem. Int. Ed. Engl.* **1996**, *35*, 200.
111. Matsuura, T.; Overman, L. E.; Poon, D. J. *J. Am. Chem. Soc.* **1998**, *120*, 6500.
112. Dounay, A. B.; Humphreys, P. G.; Overman, L. E.; Wrobleski, A. D. *J. Am. Chem. Soc.* **2008**, *130*, 5368.
113. Zhang, D.; Wang, Q. *Coord. Chem. Rev.* **2015**, *286*, 1.
114. Bringmann, G.; Gulder, T.; Gulder, T. A. M.; Breuning, M. *Chem. Rev.* **2011**, *111*, 563.
115. Genet, J.-P.; Ayad, T.; Ratovelomanana-Vidal, V. *Chem. Rev.* **2014**, *114*, 2824.
116. Cammidge, A. N.; Crépy, K. V. L. *Chem. Commun.* **2000**, 1723.
117. Yin, J.; Buchwald, S. L. *J. Am. Chem. Soc.* **2000**, *122*, 12051.
118. Bermejo, A.; Ros, A.; Fernández, R.; Lassaletta, J. M. *J. Am. Chem. Soc.* **2008**, *130*, 15798.
119. Ros, A.; Estepa, B.; Bermejo, A.; Álavarez, R.; Fernández, R.; Lassaletta, J. M. *J. Org. Chem.* **2012**, *77*, 4740.

Chapter 11

# Radical reactions: Stereochemistry and asymmetric synthesis

## 11.1 Radical chain mechanisms

Radical reactions are classified broadly into two types: (a) radical abstraction reactions—abstraction of hydrogen or another atom by radicals; (b) radical addition reactions—addition of radicals to π bonds. The reverse of radical addition is called a β-fragmentation or β-elimination. Most of the radical reactions, used in organic synthesis, involve a chain mechanism.[1–6]

Radicals can be generated from spin-paired molecules by homolysis of weak σ bonds or by electron transfer (addition of an electron). Radicals are also formed from other radicals by abstraction, addition or elimination. A common method to initiate a radical chain reaction is to heat azobisisobutyronitrile (AIBN) which generates the radical $^{\bullet}C(CN)Me_2$. This radical is rather unreactive but is capable of abstracting a hydrogen atom from the tin hydride $Bu_3SnH$ containing a weak Sn—H bond.[7] The resulting tributyltin radical ($^{\bullet}SnBu_3$) can then react readily with alkyl halides, selenides or other substrates to form a carbon-centred radical. Fig. 11.1 illustrates a radical chain mechanism for the radical addition reaction. The relative rates of the possible reactions of the intermediate radicals are critical to the success of the addition reaction. The initial alkyl radical ($^{\bullet}R$) is nucleophilic and undergoes addition faster to the electron-deficient (electrophilic) alkene than it abstracts H atom from $Bu_3SnH$, whereas the adduct radical **11.1** is electrophilic and prefers to abstract H atom rather than addition to another molecule of electrophilic alkene (see Sections 4.5.1 and 4.5.2). The addition of alkyl radical to the alkene is also regioselective and takes place predominantly at the unsubstituted end of the alkene (see Section 4.5.2).

The reagent allyltributylstannane is commonly used in a radical fragmentation method to prepare allylated products, and the corresponding radical chain mechanism is shown in Fig. 11.2. The utility of this allylstannane

Stereochemistry and Organic Reactions. https://doi.org/10.1016/B978-0-12-824092-2.00011-3

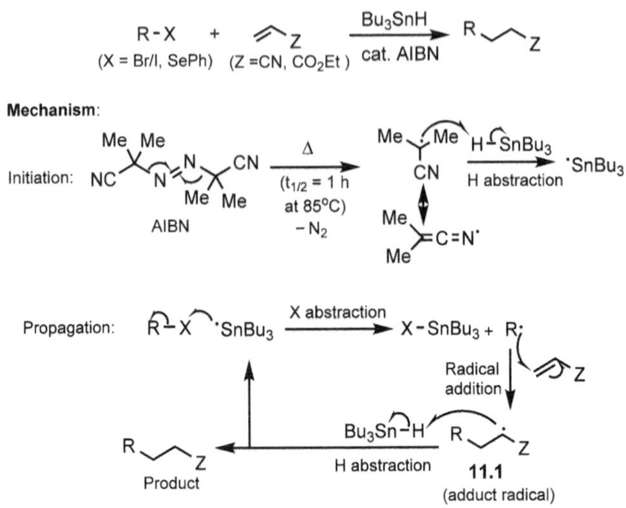


FIG. 11.1 Chain mechanism for radical addition reaction.

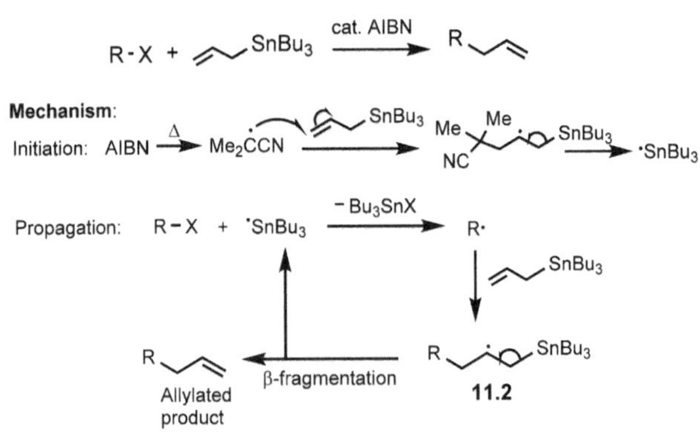


FIG. 11.2 Chain mechanism for radical allylation reaction.

method lies in the fact that the β-fragmentation is rapid and unimolecular, which differentiates the adduct radical **11.2** from other radicals.[8, 9]

## 11.2 Diastereoselective intermolecular radical reactions[10]

Like ionic and pericyclic reactions, radical reactions can also be stereoselective. It is thought that radicals as reactive intermediates will be subject to the same types of stereoelectronic and steric effects as all other organic molecules are.

## 11.2.1 Cyclic radicals

### 11.2.1.1 Steric effects on diastereoselectivity

Diastereoselective intermolecular radical reactions can occur when a suitable steric bias is present. For example, a substituent at the β- or γ-position of a cyclopentenyl or a related radical would shield its *syn* face, and, therefore, *anti*-attack predominates (Fig. 11.3).[11, 12]

An application of this *anti*-selectivity of five-membered cyclic radical is shown in Fig. 11.4, in a synthesis of the β-lactam antibiotic monobactam starting from the Barton thiohydroxamic ester **11.3**.[13] The ester can be prepared from the corresponding activated carboxylic acid and the sodium salt of *N*-hydroxypyridine-2-thione. Homolysis of the N—O bond followed by decarboxylation generates a five-membered ring radical, which then adds

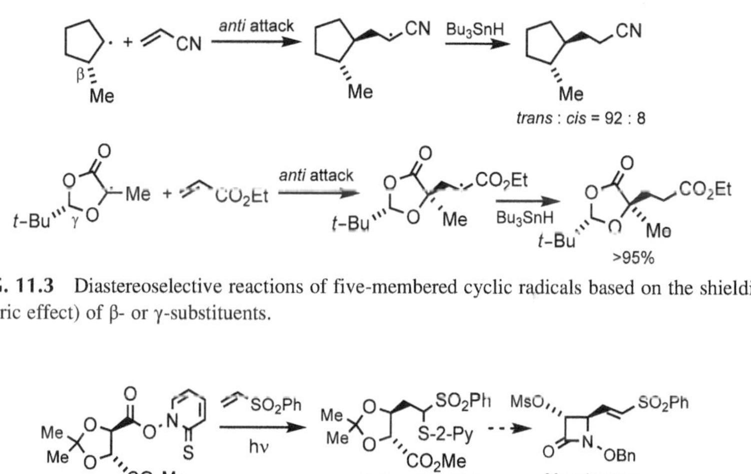


FIG. 11.3 Diastereoselective reactions of five-membered cyclic radicals based on the shielding (steric effect) of β- or γ-substituents.

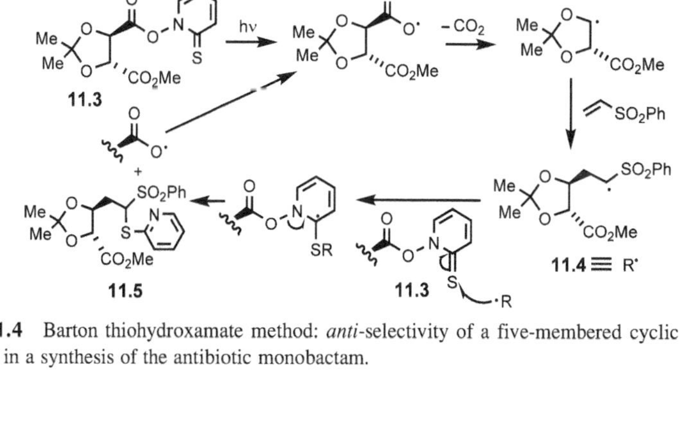


FIG. 11.4 Barton thiohydroxamate method: *anti*-selectivity of a five-membered cyclic radical applied in a synthesis of the antibiotic monobactam.

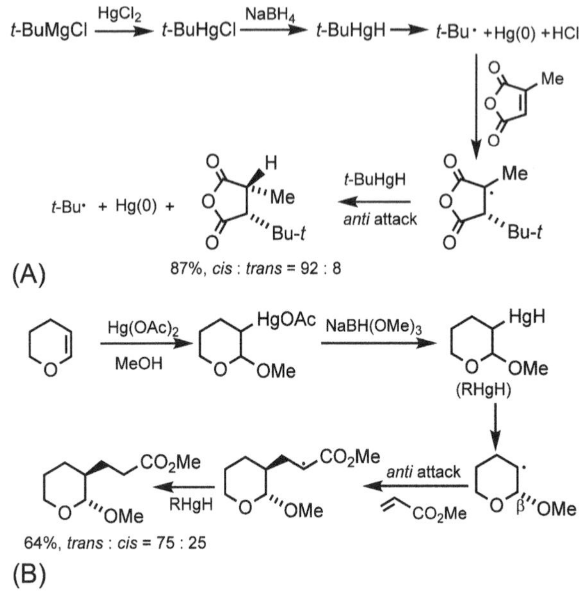


**FIG. 11.5** Alkylmercury hydride methods: (A) *anti*-selectivity of a five-membered cyclic radical in hydrogen abstraction and (B) *anti*-selectivity of a six-membered cyclic radical in alkene addition.

to an electron-deficient alkene to give a new *trans* radical **11.4**. This radical reacts with the thiohydroxamic ester **11.3** by attacking the S atom of the thiocarbonyl group to give the product **11.5** and propagate the cycle.

An alkylmercury hydride (RHgH), generated in situ by reduction of an alkylmercury halide (or acetate) with a borohydride, can give access to a carbon radical species.[14, 15] RHgH is unstable and fragments at room temperature to give the alkyl radical. The alkylmercury halide or acetate can be prepared from the corresponding Grignard reagent or from the addition of $Hg(OAc)_2$ to an alkene. Fig. 11.5A shows that the alkyl radical *t*-Bu· generated from *t*-BuHgCl/$NaBH_4$ reacts with methyl maleic anhydride to give predominantly the *cis* product, resulting from *anti* abstraction of H by the five-membered ring radical.[16, 17] Oxygen functions (e.g. OMe) at the β-position of a six-membered cyclic radical can also act as a shielding substituent, as shown in Fig. 11.5B.[18] Here, the alkylmercury hydride is generated by the reduction of alkylmercury acetate obtained by addition of $Hg(OAc)_2$ to dihydropyran.

Diastereoselective radical allylation reaction (see Fig. 11.2) can occur in the presence of a suitable steric bias.[19] This is illustrated with an example in Fig. 11.6.[9, 20] The thiocarbonyl substrate gives exclusively the *exo* allylated product since allyltributylstannane would approach favourably from the less hindered face of the cyclic radical.

Alkyl radical additions to heteroatoms are possible and can provide a method for the formation of carbon-heteroatom bond. The use of a sulphonyl

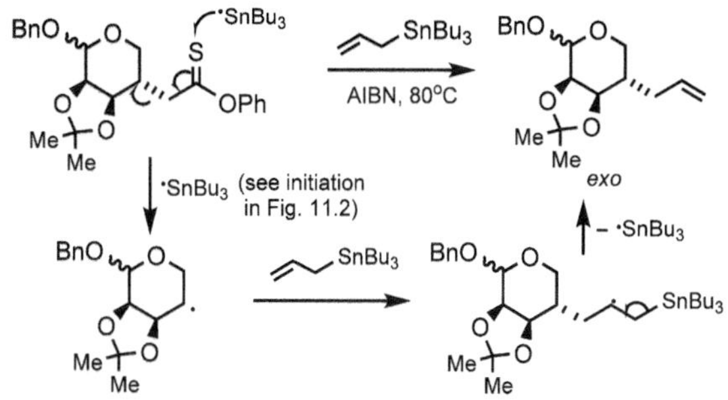


**FIG. 11.6** Allyltributylstannane fragmentation method: diastereoselective radical allylation.

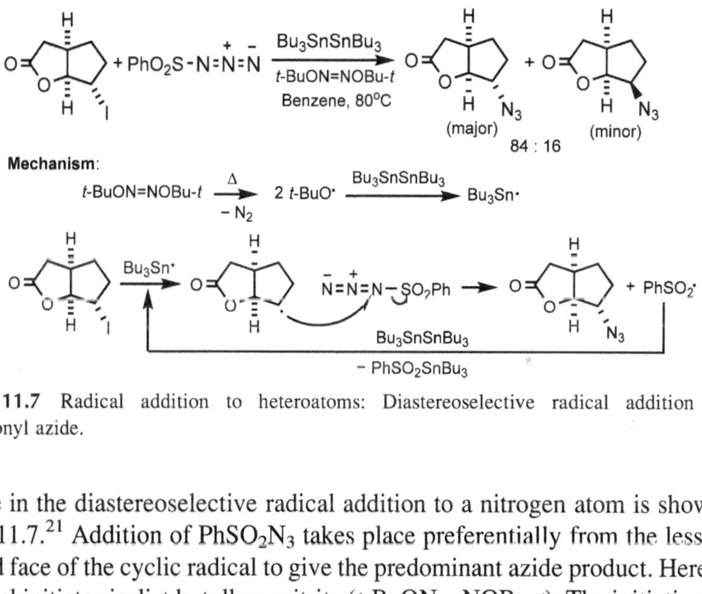


**FIG. 11.7** Radical addition to heteroatoms: Diastereoselective radical addition to a sulphonyl azide.

azide in the diastereoselective radical addition to a nitrogen atom is shown in Fig. 11.7.[21] Addition of $PhSO_2N_3$ takes place preferentially from the less hindered face of the cyclic radical to give the predominant azide product. Here, the radical initiator is di-*t*-butylhyponitrite (*t*-BuON=NOBu-*t*). The initiation step involves homolysis of the hyponitrite to form the *t*-butoxy radical, which reacts with hexabutyltin ($Bu_3SnSnBu_3$) to give $Bu_3Sn\cdot$ radical that enters into the chain mechanism. Notably, $Bu_3SnH$ is not used in this reaction to avoid the reduction of the cyclic radical, since this might compete with the slower addition to the sulphonyl azide.

### *11.2.1.2 Stereoelectronic effects on diastereoselectivity*

Stereoelectronic effects play important role in determining the diastereoselectivity of carbohydrate radicals. Because of the anomeric effect, the radicals are attacked predominantly from the axial direction, as shown in Fig. 11.8A.[22]

Amino acids can be attached to the anomeric centre of the carbohydrate using this radical approach (Fig. 11.8B).[23] The high diastereoselectivity of

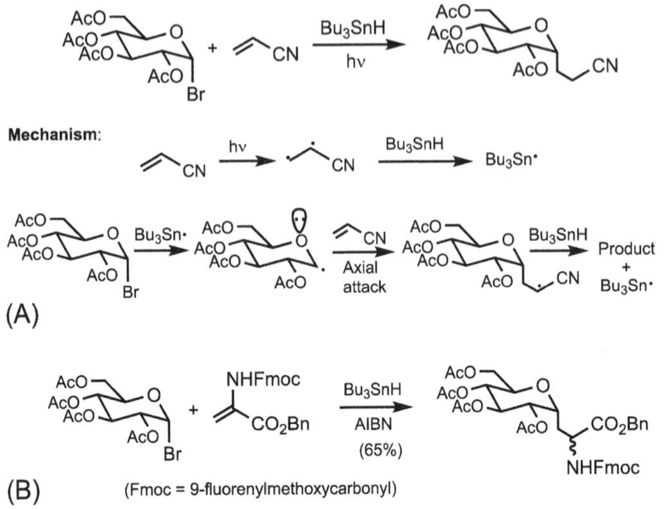


**FIG. 11.8** (A) Diastereoselectivity of a carbohydrate radical based on stereoelectronic (anomeric) effects and (B) radical reaction of carbohydrate with a dehydro amino acid derivative.

the radical addition arises via an axial attack; however, the stereoselectivity of H abstraction by the acyclic radical is found to be low.

**Problem 11.1** Account for the formation of major product in the following radical addition reaction:

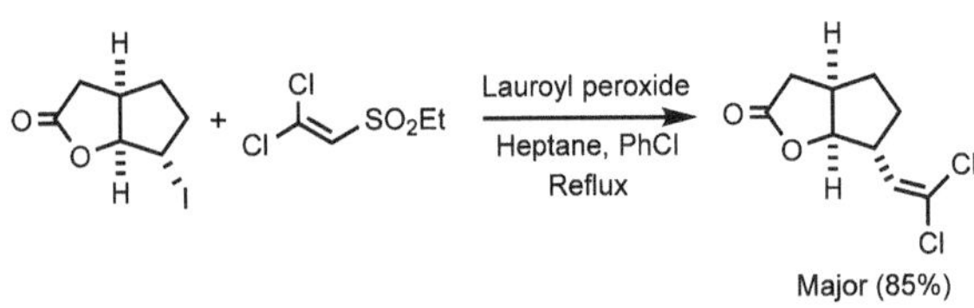


**Problem 11.2** Predict the products of the following radical reactions:

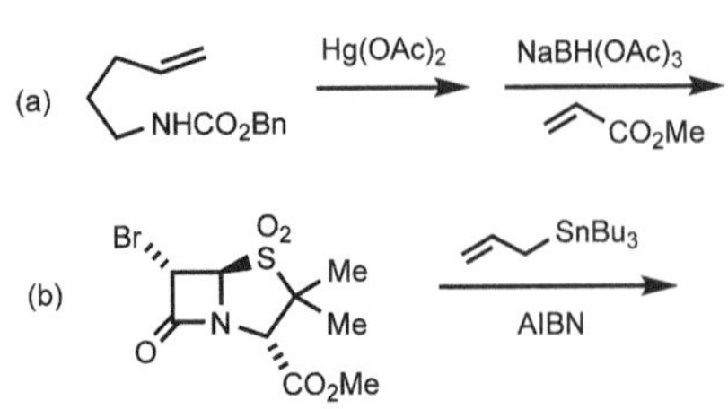

### 11.2.2 Acyclic radicals

Acyclic radicals are flexible and can assume a number of conformations, which will make their reactions largely unselective. The stereoselective reactions of acyclic radicals thus require the existence of a preferred conformation in which the two faces of the prochiral radical centre will be shielded to a different extent by the substituent(s).[24]

Consider a radical allylation reaction with an acyclic substrate, as shown in Fig. 11.9.[25]

The reaction proceeds via the preferred radical conformation **11.5** based on allylic strain ($A^{1,3}$) effects,[26] ESR and theoretical calculations[27] and the attack of allyltributylstannane occurs *anti* to the large *t*-Bu group to give the major product via the lowest energy TS. The less preferred conformation **11.6** will lead to a minor product.

High diastereoselectivity of acyclic radicals can be achieved by chelation with a Lewis acid.[28] Fig. 11.10 depicts a radical allylation reaction in the presence of Lewis acid $AlMe_3$. As shown, the allylation of the selenide proceeds through a chelated complex when the approach of the allylstannane is directed to the less hindered upper face (the lower face being shielded by Me group) to give the predominant product.

If a radical is formed adjacent to a three-membered ring, rapid fragmentation of the strained ring can take place (Fig. 11.11).[29, 30] The ring-opening gives the more stable benzylic radical intermediate. The stereochemical integrity of the stereocentre not affected by the ring-opening is retained in the product.

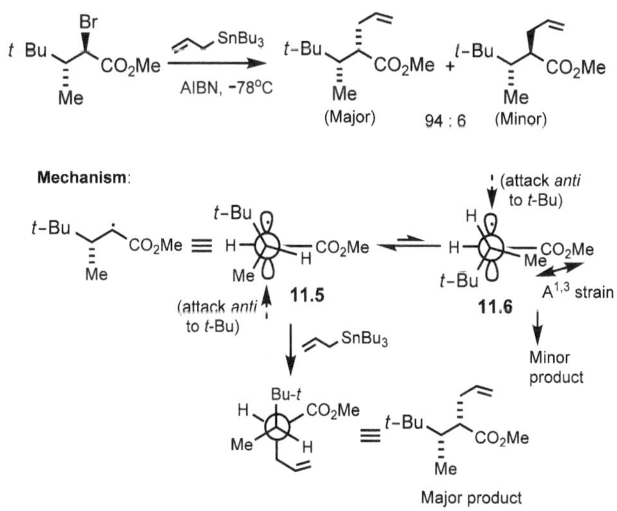


**FIG. 11.9** Diastereoselectivity of radical allylation reaction with an acyclic substrate.

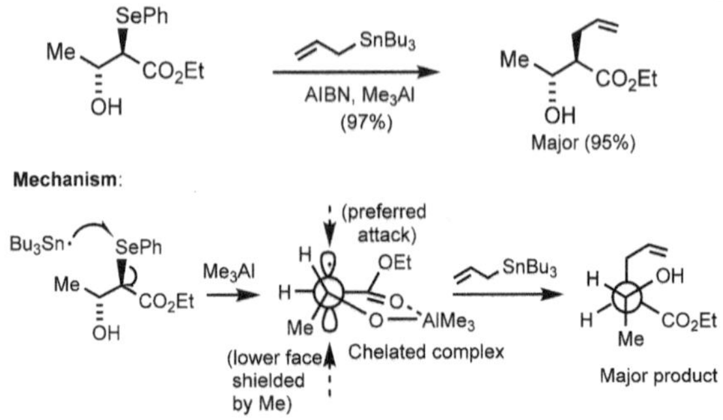


FIG. 11.10 Diastereoselectivity of radical allylation reaction via Lewis acid complexation.

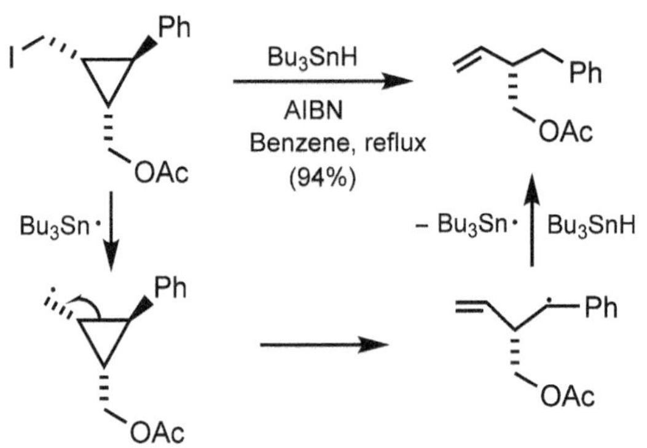


FIG. 11.11 Cyclopropyl fragmentation in radical reaction.

**Problem 11.3** Explain the following observations in radical reductions:

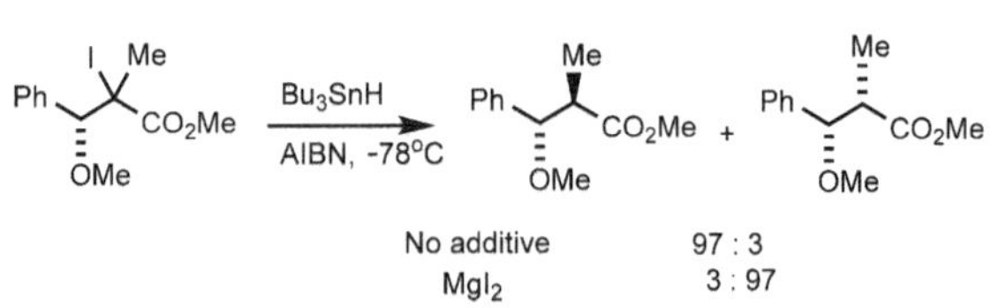


## 11.3 Diastereoselective radical cyclization reactions[31, 32]

Intramolecular radical reactions, called radical cyclizations, provide a valuable strategy in organic synthesis.[1–6, 33] Intramolecular reactions are fast processes and are more favourable than intermolecular reactions due to entropic factors. Most diastereoselective radical cyclizations form a five-membered ring by a 5-*exo* cyclization. Hexenyl radical cyclizations are particularly common, and occur mostly by a stereoelectronically favourable 5-*exo-trig* process (see Section 4.5.3).

### 11.3.1 Hexenyl radical cyclizations

Substituents within the hexenyl chain tend to increase the rate of cyclization (Thorpe-Ingold effect), and lead to a modest to high diastereoselection. 1-Substituted hexenyl radicals cyclize to give a modest *cis/trans* selectivity with a range of different substituents.[34] However, a 2-substituted hexenyl radical can cyclize to give predominantly *trans*-1,3-disubstituted cyclopentane, whereas a 3-substituted radical gives the *cis* isomer as a major product (Fig. 11.12).[34–36] Chair-like TS is generally favoured with the substituent adopting the less hindered pseudoequatorial position. The alternative boat-like TS leads to the minor product.

Commonly, $Bu_3SnH$ is used to initiate carbon radical formation. However, an efficient radical cyclization initiator makes use of trialkylboranes with $O_2$ at a low temperature (−78°C).[37] $O_2$ is a triplet diradical and adds to the vacant p orbital of B to release the alkyl radical that starts the chain reaction. Fig. 11.13 shows a radical cyclization of an acyl selenide via the formation of an acyl radical[38] in the presence of $Et_3B$ and $O_2$, as a key step in a synthesis of the nonisoprenoid sesquiterpene (−)-kamausallene.[39] Tris(trimethylsilyl)silane[40] is used as the hindered terminal hydride donor instead of toxic tin hydride. The hindered silicon hydride is a poorer hydride donor than the tin hydride and hence reduces the problem of competing reduction of the initial carbonyl radical. The cyclization reaction of the 3-substituted radical is highly selective in favour of the *cis* isomer of the product (*cis*:*trans* = 32:1).

Among simple hexenyl radicals, 4-substituted radicals usually cyclize with the highest levels of diastereoselectivity. This is interpreted in terms of the chair-equatorial TS that minimizes allylic ($A^{1,3}$) strain, which gives the diastereoselective *trans* product (Fig. 11.14A).[41] Fig. 11.14B illustrates a dominant role of the 4-substituent in the stereoselectivity of hexenyl radical cyclization,

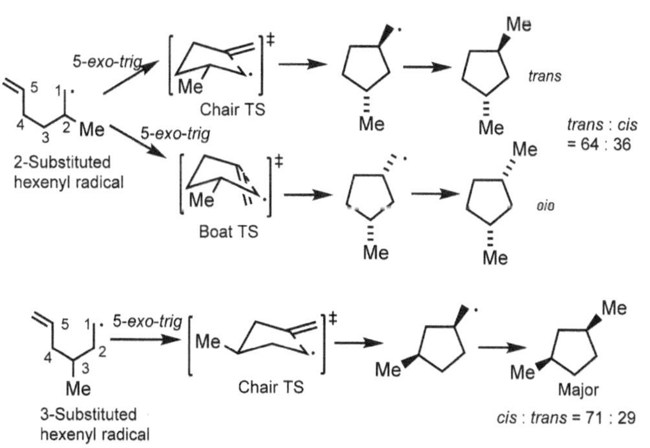


**FIG. 11.12** Diastereoselectivity of radical cyclizations of 2- and 3-substituted hexenyl radicals.

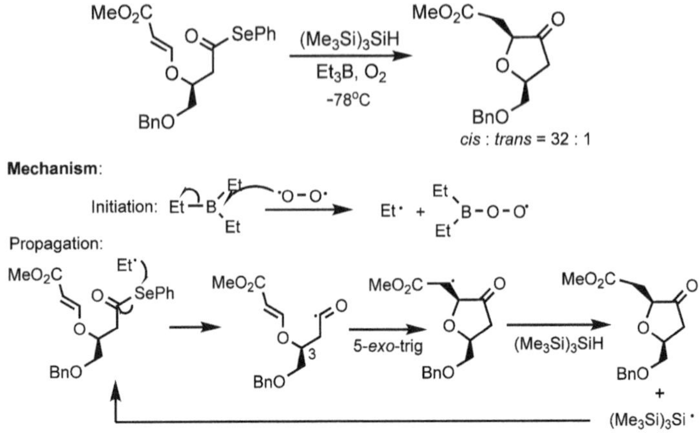


**FIG. 11.13** Radical cyclization of an acyl selenide using $Et_3B$ and $O_2$ as a radical initiator at low temperature.

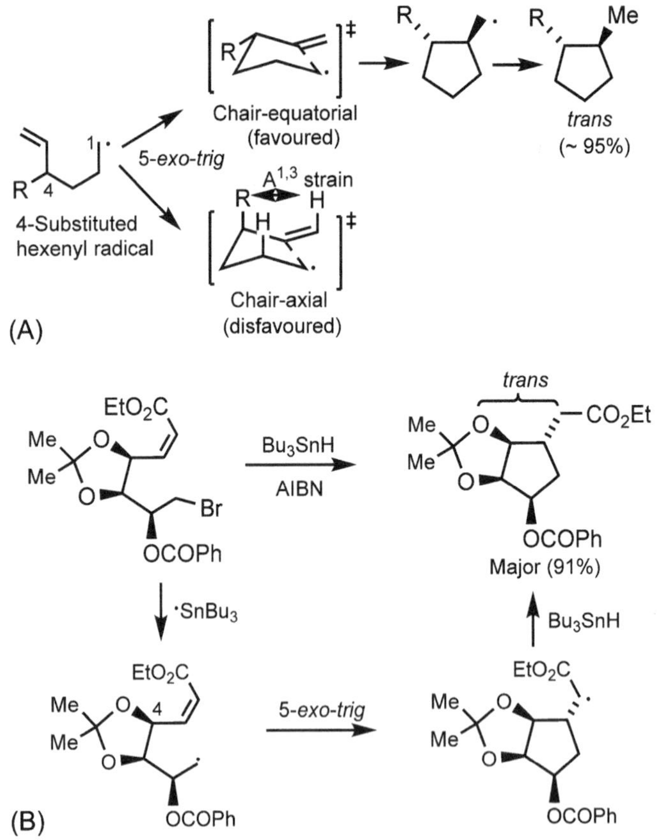


**FIG. 11.14** (A) Radical cyclization of 4-substituted hexenyl radicals with high levels of diastereoselectivity; (B) Diastereoselective hexenyl radical cyclization with a dominant influence of the 4-substituent.

forming predominantly the product with *trans* relationship between the 4-substituent and the new chiral centre.[42]

**Problem 11.4** Predict the predominant diastereomer formed in the following radical cyclization:

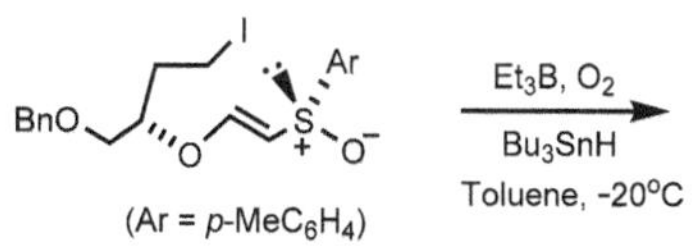


**Problem 11.5** Suggest a mechanism for the following radical cyclization reaction:

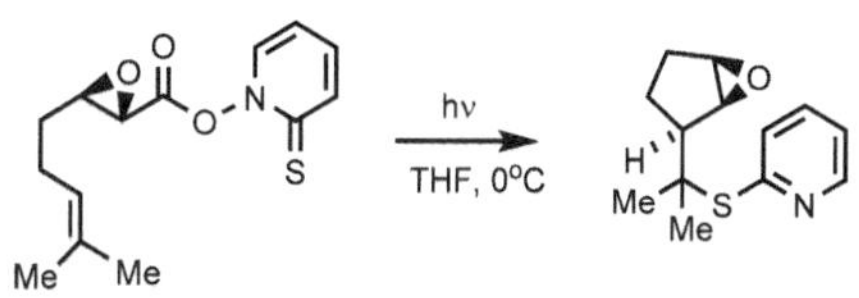


## 11.3.2 *cis* ring fusion in radical cyclizations

If a ring preexists in a hexenyl radical, ring fusion is possible by cyclization. The existing ring may contain either the radical centre or the acceptor double bond. The presence of a short tether between the preexisting cyclic radical (or acceptor) and its partner readily permits a favoured geometry for the *cis* ring fusion. The *cis* selectivity for 5-5 or smaller ring fusion is almost complete.[33, 43] The 6-5 or even 6-6 ring fusion also proceeds with good to high *cis* selectivity.[44, 45] The adduct radical formed by a radical cyclization usually abstracts H atom from the tin hydride to give the *cis*-fused product.

Fig. 11.15A illustrates a radical cyclization to a *cis*-fused hydrindane using an alkenyl radical.[46] The favoured geometry of the TS is *cis*-chair, whereas ring strain must be introduced for *trans*-chair geometry. An alternative *cis*-boat TS is also possible. Note that the protection of OH is not necessary since the bond dissociation energy of C—Br bond (67 kcal mol$^{-1}$) is much less than that of the O—H bond (~110 kcal mol$^{-1}$).

The formation of a *cis*-hydrindane derivative using an acyl radical from acyl selenide is shown in Fig. 11.15B.[47]

The *cis*-ring fusion using an alkenyl radical is highly efficient, and an application of this procedure is shown in Fig. 11.16, in a synthesis of the sesquiterpene (±)-merrilactone A.[48]

Alkenyl radicals can also be generated by radical addition to alkynes. An example of this method, developed by Stork, is shown in Fig. 11.17.[49]

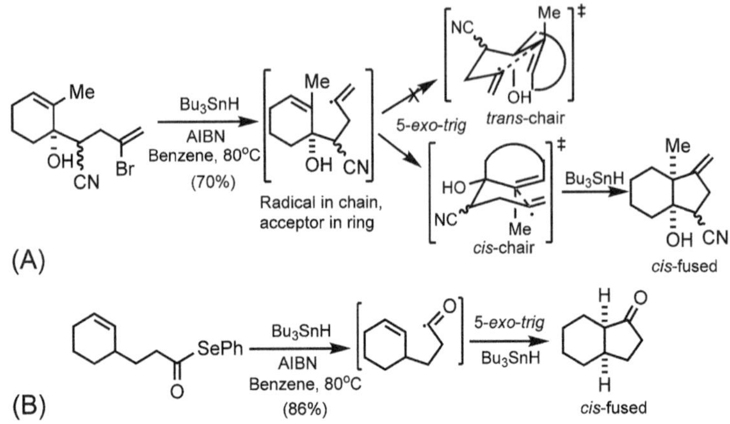


**FIG. 11.15** Radical cyclization using (A) an alkenyl radical and (B) an acyl radical.

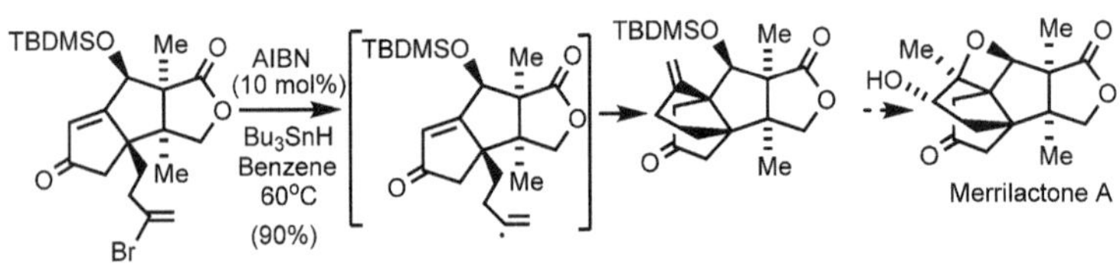


**FIG. 11.16** Application of radical-mediated *cis* ring fusion in a synthesis of the sesquiterpene (±)-merrilactone A.

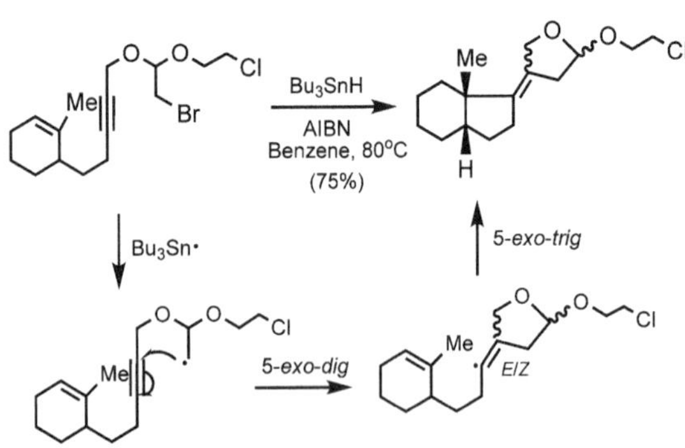


**FIG. 11.17** Radical cyclization via an alkenyl radical generated from an alkyne.

The initial alkyl radical generated from the bromide adds regioselectively to the alkyne to form an intermediate alkenyl radical, which then undergoes radical cyclization with the cyclohexene double bond to give the *cis*-fused product. Because of the rapid *E*/*Z* isomerization of the alkenyl radical, the product has a mixture of alkene stereoisomers.

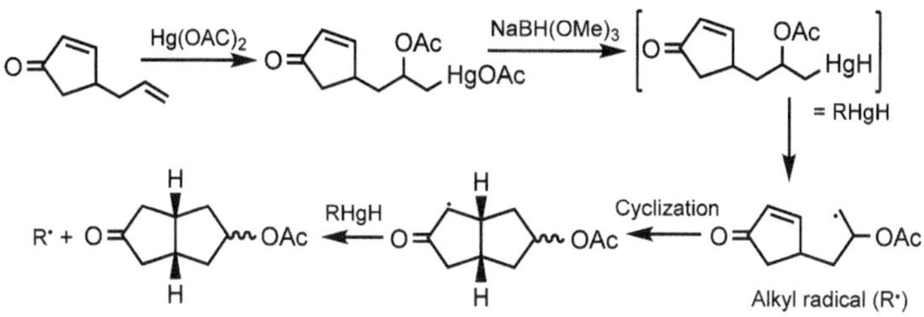


**FIG. 11.18** Radical cyclization using an alkylmercury hydride method.

Fig. 11.18 depicts an alkylmercury hydride method (see Fig. 11.5B) for the synthesis of a bicyclic ketone, which would be difficult to prepare by other methods. Here the alkyl radical does not undergo β-elimination of acetoxy radical (in contrast to the corresponding anion which would eliminate acetoxy anion to generate the starting alkene) but undergoes cyclization by forming the C—C bond.

In radical cyclization, the final carbon radical usually abstracts H atom from the hydride; however, it is possible to intercept the carbon radical with an external trap (alkene). An elegant example of this tandem radical cyclization-intermolecular radical addition procedure in a synthesis of (+)-prostaglandin $F_{2\alpha}$ ($PGF_{2\alpha}$) is shown in Fig. 11.19.[50] The iodide **11.7** is treated with a catalytic amount of tributyltin hydride (generated in situ by reduction of tributyltin chloride with $NaBH_3CN$) and an excess of an α-silyl-substituted vinyl ketone. Rad ical cyclization provides a cyclic radical which is trapped by intermolecular addition to the vinyl ketone to form a stabilized α-keto-α-silyl radical. This radical then abstracts H atom from the tin hydride to give the product **11.8**, which is then converted into $PGF_{2\alpha}$ via the formation of the Stork intermediate **11.9**. The presence of only a catalytic amount of $Bu_3SnH$ is necessary to promote intermolecular alkene addition instead of H abstraction by the cyclic radical. After each catalytic cycle, the $Bu_3SnH$ is converted to $Bu_3SnI$, and the stoichiometric $NaBH_3CN$ reduces it back to $Bu_3SnH$. Thus, as much $Bu_3SnH$ is produced as is needed. Further, this catalytic method obviates the formation of stoichiometric amounts of toxic tin by-products.

In an alternative method (Fig. 11.20),[51] the iodide **11.7** is treated with a β-stannyl vinyl ketone (as an external trap) in the presence of AIBN. After radical cyclization, the cyclic radical is trapped by the vinyl ketone to produce directly the Stork intermediate **11.9** (via a β-fragmentation), towards the synthesis of the prostaglandin $F_{2\alpha}$.

**Problem 11.6** Predict the products of the following radical cyclizations:

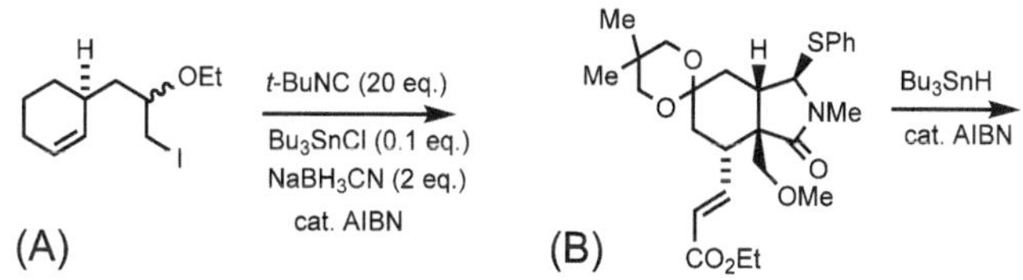

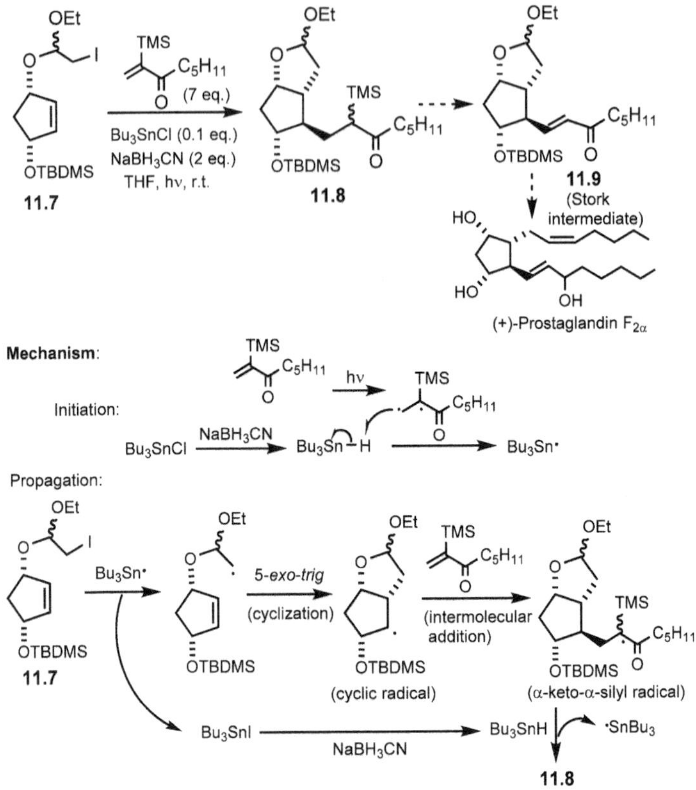

**FIG. 11.19** Tandem radical cyclization-intermolecular radical addition, used in a synthesis of (+)-prostaglandin $F_{2\alpha}$.

**Problem 11.7** Rationalize the formation of product in the following radical reaction:

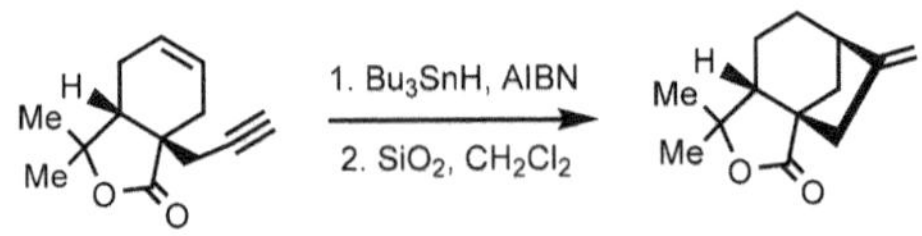

## 11.3.3 Tandem radical cyclizations[52, 53]

Many natural products possess polycyclic ring systems, and radical-based tandem cyclizations provide efficient approaches towards their synthesis. A classic example is Curran's synthesis of the triquinane hirsutene (Fig. 11.21).[54] The natural product is obtained directly by two consecutive radical cyclizations. The initial alkyl radical leads to a *cis*-fused bicyclic radical via a 5-*exo-trig*

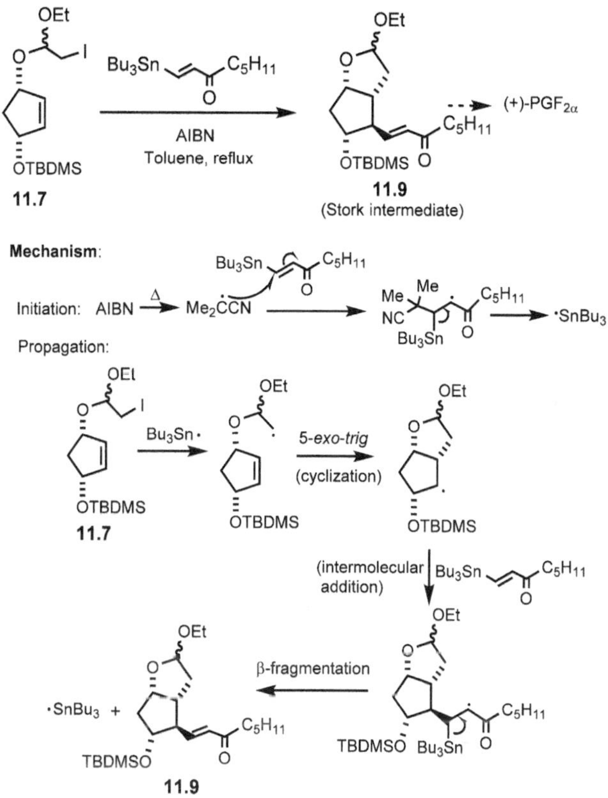


**FIG. 11.20** An alternative tandem radical cyclization-intermolecular radical addition method for the synthesis of (+)-prostaglandin $F_{2\alpha}$

cyclization, which then undergoes a 5-*exo-dig* cyclization to give a further *cis*-fused ring system. The resulting vinyl radical abstracts H atom from the tin hydride to provide hirsutene and tributyltin radical to propagate the chain reaction.

Fig. 11.22 depicts a tandem radical cyclization initiated by an alkenyl radical. The reaction proceeds through a rapid *E*/*Z* isomerization of the alkenyl radical followed by two consecutive 5-*exo-trig* cyclizations to give the product.

An excellent demonstration of a radical cascade, in a synthetic approach to linear triquinanes, is shown in Fig. 11.23.[55] A series of 5-*exo* cyclizations, an intermolecular radical addition and a 1,5-hydrogen transfer are involved. In the final step, β-elimination of phenylsulphonyl radical gives the triquinane and carries on the radical chain.

Fig. 11.24 depicts a radical cyclization cascade in a synthesis of (±)-estrone.[56] The cascade involves a sequence of radical macrocyclization, cyclopropane fragmentation, transannular addition and 5-*exo-trig* closure.

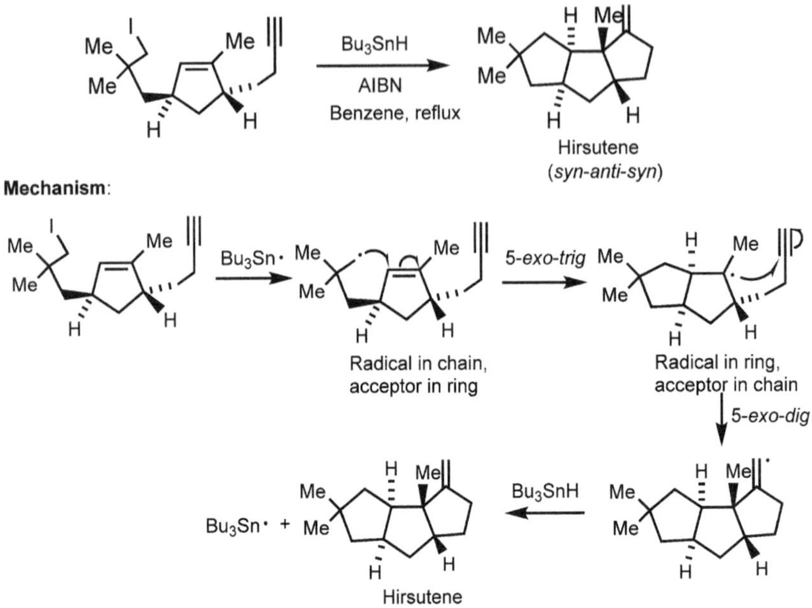


FIG. 11.21 A tandem radical cyclization for the synthesis of the triquinane hirsutene.

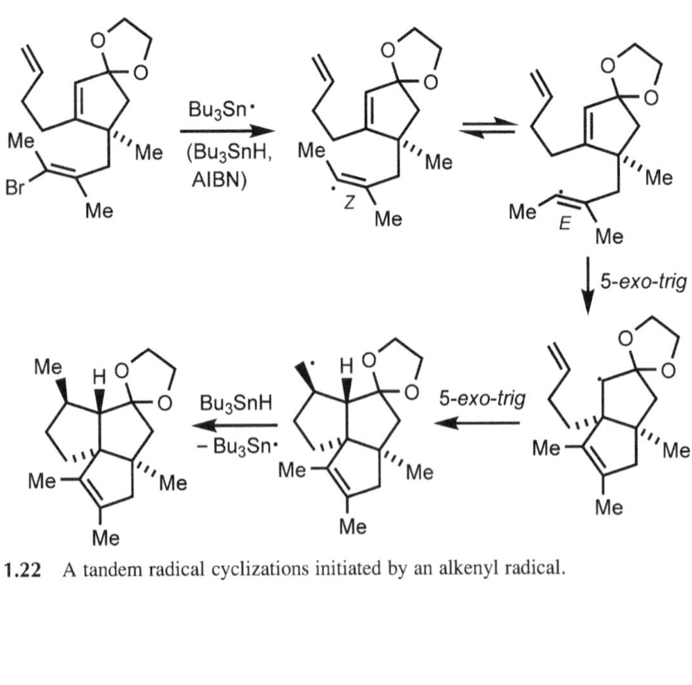


FIG. 11.22 A tandem radical cyclizations initiated by an alkenyl radical.

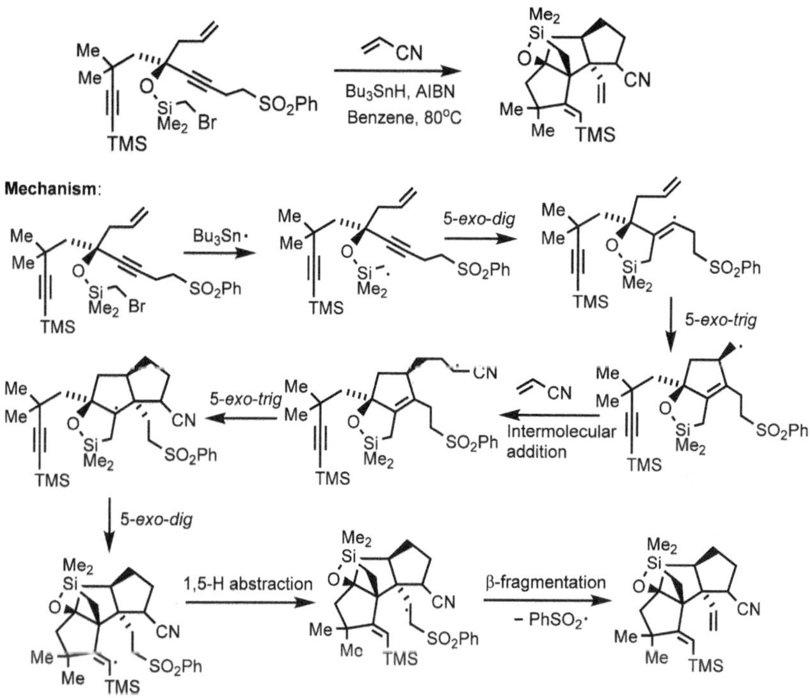


**FIG. 11.23** A radical cascade approach for the construction of a triquinane from an acyclic precursor.

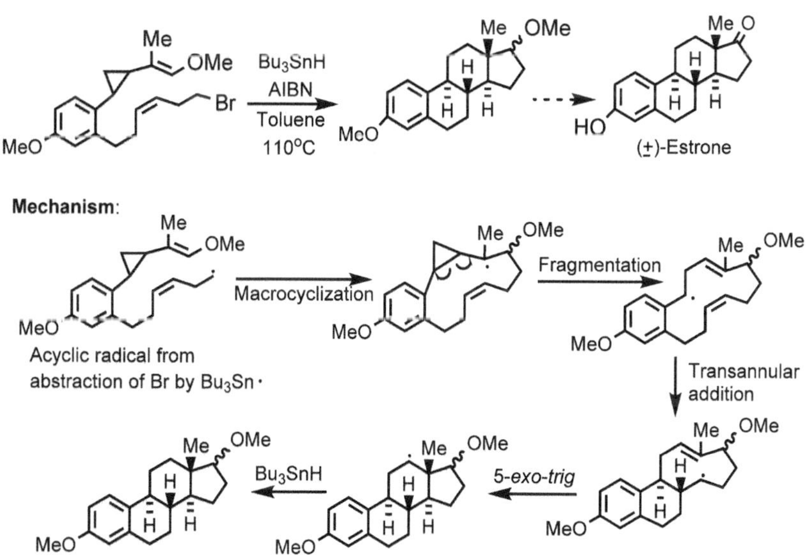


**FIG. 11.24** A radical cyclization cascade in a synthesis of (±)-estrone.

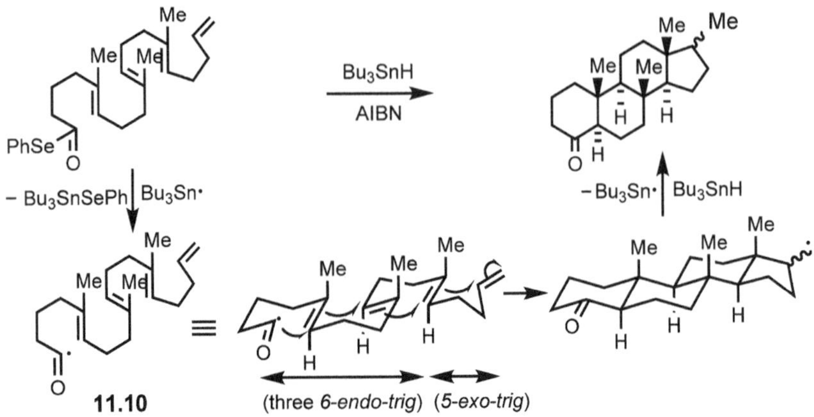


**FIG. 11.25** A tandem radical cyclization of a tetraene acyl radical for the construction of a steroid ring.

For 5-hexenyl radicals, 5-*exo-trig* cyclizations are generally preferred over 6-*endo-trig*; however, 5-hexen-1-oyl (acyl) radicals can cyclize via 6-*endo-trig* processes to form six-membered rings.[57, 58] A range of polyalkene acyl radicals generated from the corresponding phenylselenyl esters undergo tandem radical cyclizations via 6-*endo-trig* modes to give six-ring fused polycycles. Fig. 11.25 shows an example of a tandem radical cyclization for a steroid ring construction.[59] The tetraene acyl radical **11.10** obtained from the corresponding selenyl ester undergoes three consecutive 6-*endo-trig* cyclizations followed by a 5-*exo-trig* closure to give the steroid ketone. The *trans* ring fusions can be rationalized based on concerted mechanisms via chair-like conformations of the polyalkene acyl radical **11.10**.

**Problem 11.8** Suggest mechanisms for the following radical cyclization reactions:

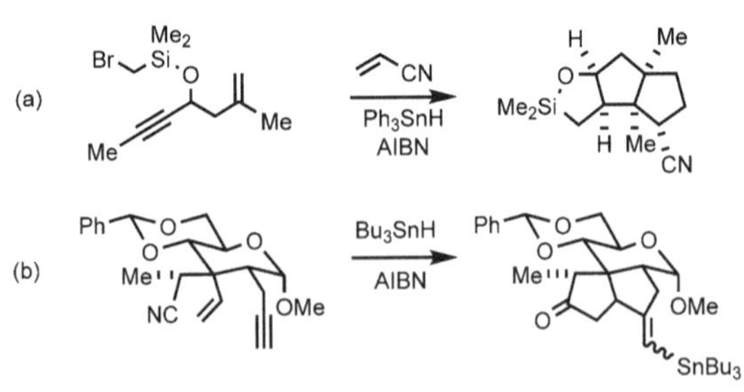

### 11.3.4 Radical cyclizations onto carbon-nitrogen double bonds

Radical additions are not restricted to alkenes or alkynes. Imines or imine derivatives, such as oximes or hydrazones are popular acceptors in radical cyclizations.[60] Fig. 11.26 illustrates a radical cyclization onto an oxime. The *trans* stereoselectivity of the reaction arises from the preferred chair-like TS (cf. Fig. 11.12).

A radical cyclization onto an *N*-aziridinyl hydrazone **11.11**, in a synthesis of the sesquiterpene α-cedrene, is shown in Fig. 11.27.[61] The initial alkyl radical cyclizes onto C=N of the hydrazone by the 5-*exo-trig* process to form nitrogen radical. Fragmentation of the aziridine ring then gives a carbon radical with the elimination of $N_2$ and stilbene. Notably, the new carbon radical is set up on the carbon atom of the original C=N bond, which undergoes a second 5-*exo-trig* cyclization to give, after H atom abstraction from the tin hydride, the product **11.12**.

### 11.3.5 Ketyl radical cyclizations

Aldehydes or ketones can accept a single electron from a suitable donor to form a ketyl radical anion. Metal naphthalenides (obtained by reduction of naphthalene with alkali metals) are useful one-electron donors.[62] Naphthalenides are soluble in ethereal solvents and form an intensely coloured solution.

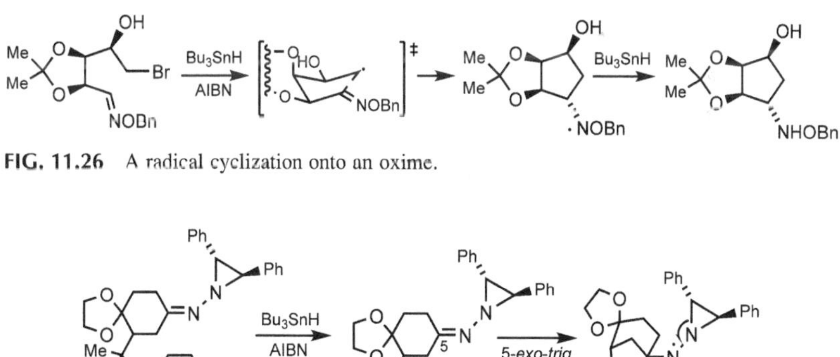


**FIG. 11.26** A radical cyclization onto an oxime.

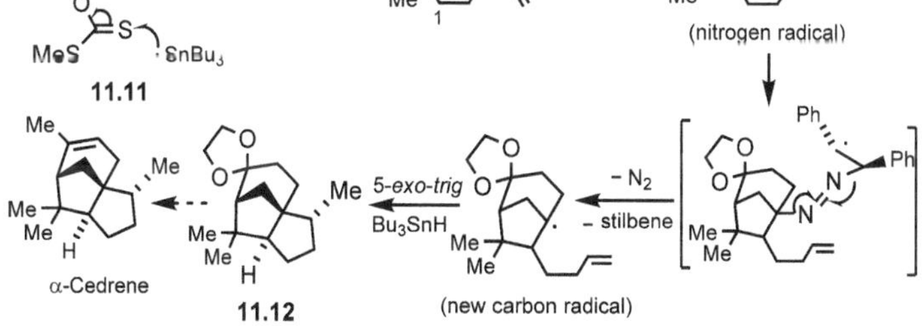


**FIG. 11.27** A radical cyclization onto an *N*-aziridinyl hydrazone in a synthesis of the sesquiterpene α-cedrene.

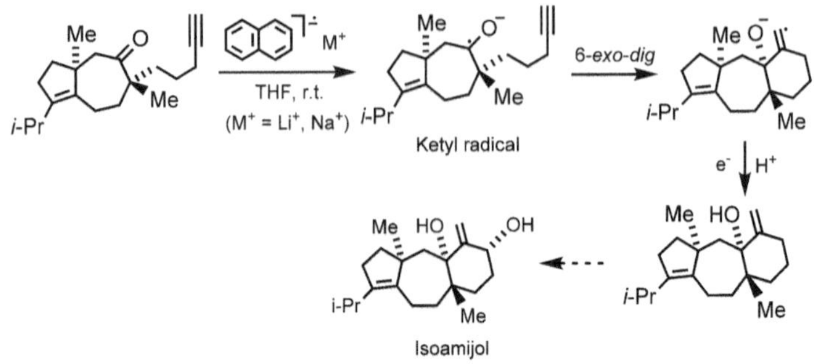


**FIG. 11.28** Ketyl-alkyne radical cyclization using lithium or sodium naphthalenide.

Thus radical cyclizations using a metal naphthalenide can be conducted titrimetrically. Fig. 11.28 illustrates a ketyl-alkyne radical cyclization, in a synthesis of isoamijol.[63] Treatment of the keto alkyne with lithium or sodium naphthalenide leads to a 6-*exo-dig* cyclization and gives tertiary allylic alcohol with a *trans* 6-7 ring fusion.

### *11.3.5.1 Samarium diiodide-mediated cyclizations*[64–66]

Samarium(II) diiodide ($SmI_2$) is the most popular and powerful single electron reducing agent for the generation of ketyl radicals in radical cyclizations. Very high diastereoselectivities have been achieved in the synthesis of cyclopentanols from a keto alkene precursor, as shown in Fig. 11.29.[67] The addition of HMPA increases the reduction potential of $SmI_2$.

Fig. 11.30 depicts a tandem radical cyclization beginning with $SmI_2$-mediated ketyl-alkene cyclization, as a key step in a synthesis of the sesquiterpene hypnophilin.[68] The $SmI_2$ concentration is kept low, so the reduction of the intermediate alkyl radical does not compete with the second cyclization.

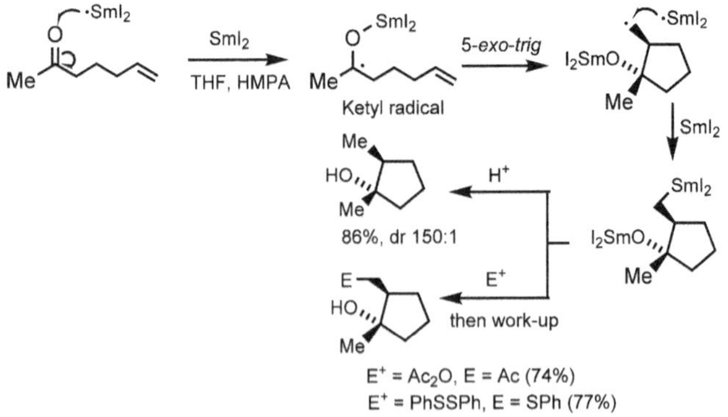


**FIG. 11.29** Samarium diiodide ($SmI_2$)-mediated ketyl radical cyclization for the synthesis of cyclopentanols.

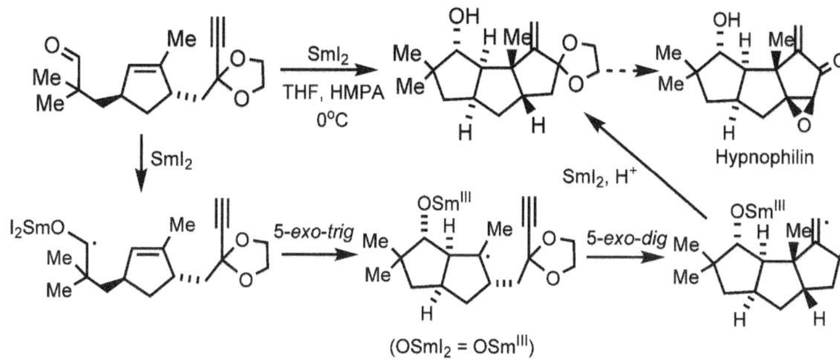

**FIG. 11.30** $SmI_2$-mediated tandem radical cyclization in a synthesis of the sesquiterpene hypnophilin.

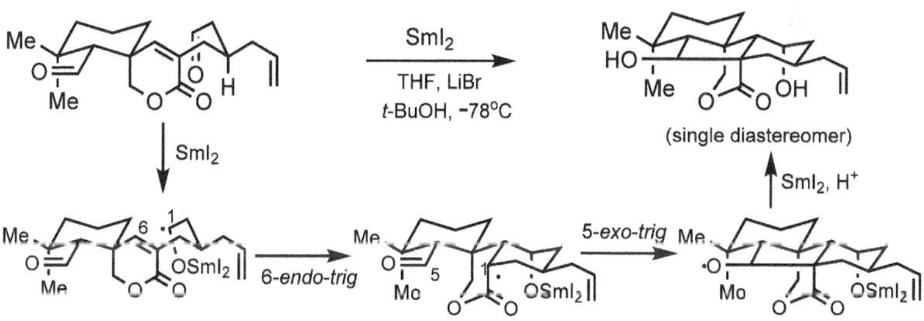

**FIG. 11.31** $SmI_2$-mediated radical cyclization cascade in the synthesis of the core structure of the diterpene maoecrystal Z.

A cyclization cascade, used in the synthesis of the core structure of the diterpene maoecrystal Z, is illustrated in Fig. 11.31.[69] The cascade involves a 6-*endo-trig* ketyl-alkene cyclization, followed by a 5-*exo-trig* radical-carbonyl cyclization.

**Problem 11.9** Draw mechanisms for the following $SmI_2$-mediated radical cyclization reactions:

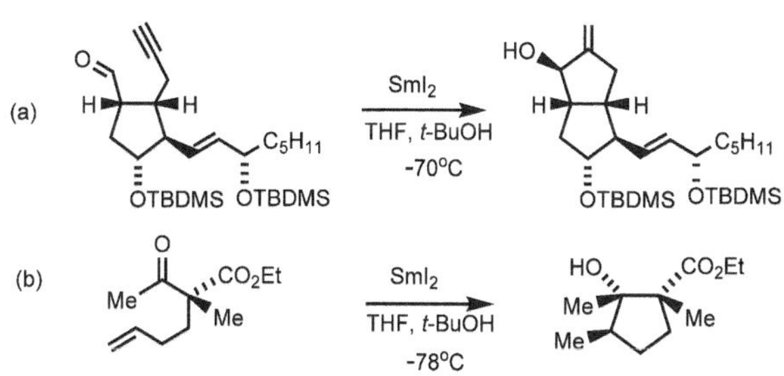

## 11.4 Dissolving metal reductions

In dissolving metal reductions, the solvent is usually liquid ammonia, in which the electrons from the metal are solvated. The proton source is generally ammonia itself or alcohol. As a complementary method to *syn* hydrogenation of alkynes (see Section 6.4.3), *anti* reduction of alkynes can be carried out using alkali metal (Na or Li) in liquid $NH_3$. The alkenes are not normally reduced; thus the reduction of the triple bond is selective. The reaction proceeds by stepwise addition of two electrons (Fig. 11.32). The addition of the first electron to the triple bond forms a radical anion that takes up a proton to give a vinyl radical. The second electron transfer then occurs to produce a vinyl anion which adopts the more stable (*E*) configuration. Protonation of the vinyl anion gives the thermodynamically more stable (*E*)-alkene.

Dissolving metals can also be used to perform the conjugate reduction of α,β-unsaturated ketones.[70] The conjugate addition also occurs by stepwise addition of two electrons but in contrast to alkyne addition, the two electrons are added in succession (Fig. 11.33). The first addition provides a radical anion, and a consecutive addition of the second electron gives a closed-shell dianion. Protonation of the dianion gives the more stable *trans* product. This is attributed

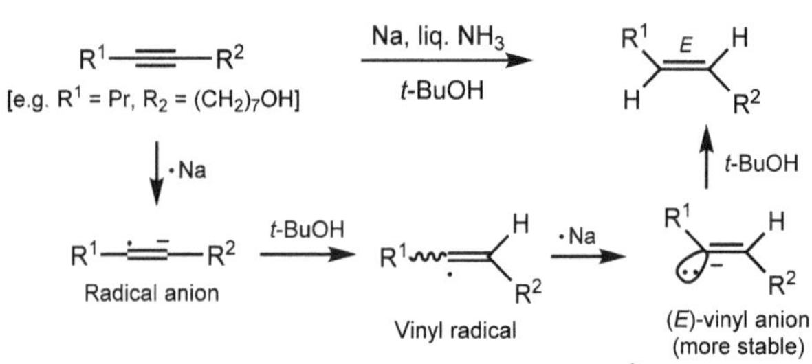


**FIG. 11.32** Stereoselective dissolving metal reduction of alkynes.

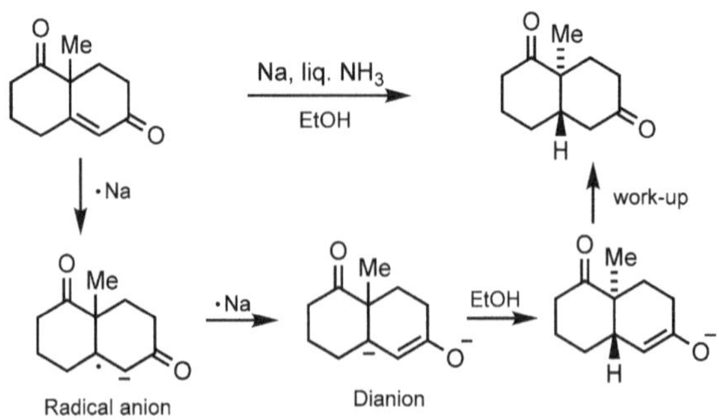


**FIG. 11.33** Stereoselective conjugate reduction of an α,β-unsaturated ketone with sodium and liquid ammonia.

to axial protonation (orthogonal to the enolate double bond), leading to a chair diaxial (Me/H) conformation for a *trans*-decalin system.

**Problem 11.10** Accomplish the following transformation:

## 11.5 Asymmetric radical reactions

Asymmetric control in radical reactions is rather difficult to achieve because of the intrinsic fast reactivity of the radicals. In recent years, several strategies have been developed to control asymmetry in radical synthesis.[4–6, 10, 71–75] Here, we will describe briefly three methods of asymmetric control based on chiral auxiliaries, chiral Lewis/Brønsted acids and organocatalysis.

### 11.5.1 Chiral auxiliaries

Oppolzer camphorsultams (see Fig. 5.9B) serve as an effective chiral auxiliary in asymmetric radical synthesis. Fig. 11.34A shows that a high degree of asymmetric control is achieved in the radical addition to the glyoxylic oxime ether using (1*R*)-(+)-2,10-camphorsultam as a chiral auxiliary.[76] The reductive removal of the OBn group by treatment with $Mo(CO)_5$[77] and the subsequent hydrolysis lead to the enantiomerically pure α-amino acids. This is the first reported asymmetric synthesis of α-amino acids based on the asymmetric radical addition.

The preferred site of radical addition to the camphorsultam derivative of glyoxylic oxime ether, leading to the observed stereoselection, is shown in Fig. 11.34B.[76] It has been suggested that in imides derived from camphorsultam, the dipole–dipole repulsion control[78] would tend to favour the conformers **11.13** and **11.14** with the C=O *anti* to an S=O (O marked in bold) of the sulphonyl group. Between these, the conformer **11.13** (*anti*, *s-cis*) is sterically more favoured than the conformer **11.14** (*anti*, *s-trans*) Thus based on the combined dipole repulsion and steric factors, the preferred conformer of the sultam derivative is **11.13** (*anti*, *s-cis*) The radical addition then takes place preferentially to the front *Re* face (the rear *Si* face being shielded by the axial oxygen of the sulphonyl group), to give the predominant diastereomer.

Oxazolidinone chiral auxiliaries have also been employed to achieve asymmetric radical synthesis. Lewis acid chelation to the chiral auxiliary gives high regio- and diastereoselectivity in conjugate radical addition reactions.[28, 79] A lanthanide Lewis acid such as $Sm(OTf)_3$ is particularly effective. This strategy is illustrated with an example in Fig. 11.35.[80] Chelation of the Lewis acid Sm $(OTf)_3$ with the auxiliary carbonyl and imide carbonyl groups activates the

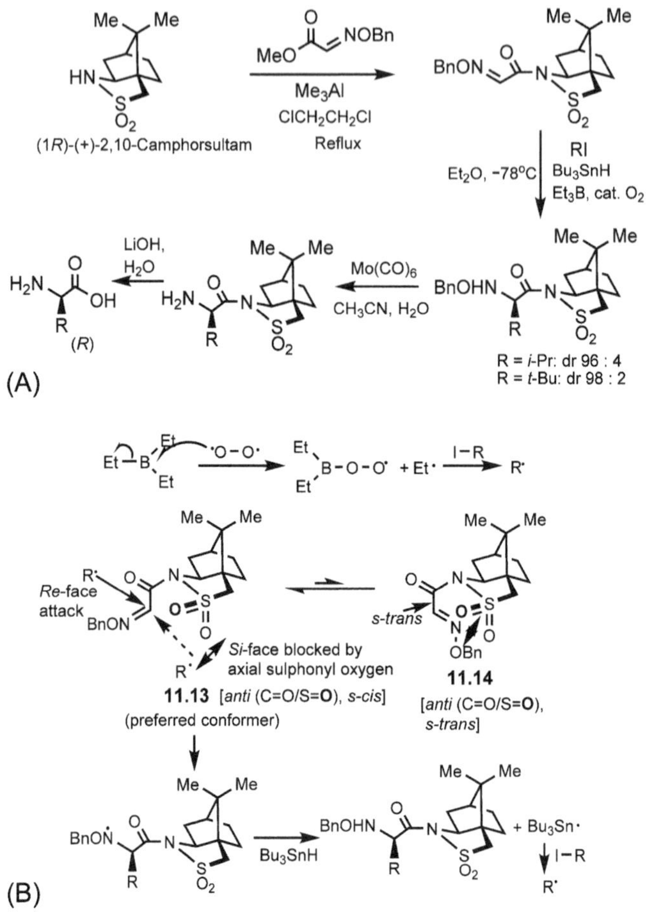


**FIG. 11.34** (A) Asymmertric radical synthesis of α-amino acids using Oppolzer camphorsultam auxiliary and (B) mechanism of the asymmetric radical addition.

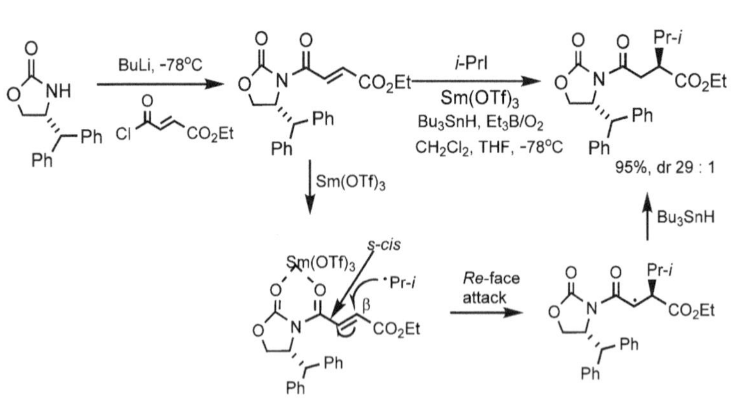


**FIG. 11.35** Asymmetric conjugate radical addition with oxazolidinone chiral auxiliary using Lewis acid chelation.

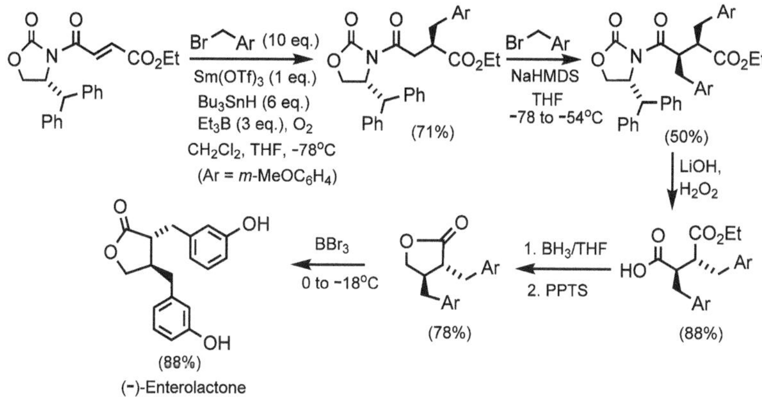


**FIG. 11.36** Application of an asymmetric radical strategy (shown in Fig. 11.35) for the synthesis of the lignan natural product (−)-enterolactone.

substrate for conjugate addition in which radical attack occurs preferably at a β position to the imide carbonyl, resulting in high regioselectivity. The conjugated substrate is in *s-cis* conformation, and a radical addition to the upper *Re* face (opposite to the bulky diphenylmethyl substituent on the auxiliary) gives the product with high diastereoselectivity (dr = 29:1).

Fig. 11.36 shows an application of the previously mentioned strategy in a highly efficient synthesis of the liganan natural product (−)-enterolactone.[80] The target enterolactone[81] has two stereocentres. One stereocentre is established using the asymmetric radical strategy as shown in Fig. 11.35, and the other is introduced via standard ionic asymmetric alkylation reaction (see Section 7.7.1.1). Thus, a single stereocentre in the chiral auxiliary allows for the sequential introduction of two stereocentres in the product with high regio- and diastereocontrol. After removal of the auxiliary, selective reduction of the carboxyl group using borane gives the ester alcohol, which is then cyclized to a lactone with pyridinium *p*-toluenesulphonate (PPTS). Finally, $BBr_3$-mediated demethylation[82] provides (−)-enterolactone in excellent yield.

## 11.5.2 Chiral acids

A combination of chiral ligands and Lewis acids has been used in the asymmetric radical synthesis with moderate enantioselectivity. This is illustrated with a carbon radical addition to a glyoxylic oxime ether in Fig. 11.37.[76] The isopropyl radical addition is run in $CH_2Cl_2$ in the presence of a chiral bisoxazoline (box) ligand **11.15** and a Lewis acid $MgBr_2$. The reaction exhibits excellent yield (97%) with moderate enantioselectivity (52%) in favour of (*R*) enantiomer of the product. The enantioselectivity results from the preferential *Re* face radical addition to the chiral Lewis acid complex **11.16**.

Asymmetric conjugate radical addition to α′-hydroxy α,β-unsaturated ketones (containing bidentate donors) with various radical precursors has

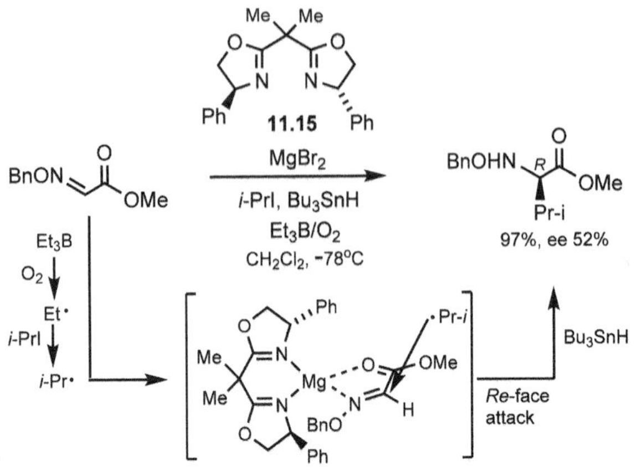


**FIG. 11.37** Asymmetric radical addition to a glyoxylic oxime ether using a chiral Lewis acid.

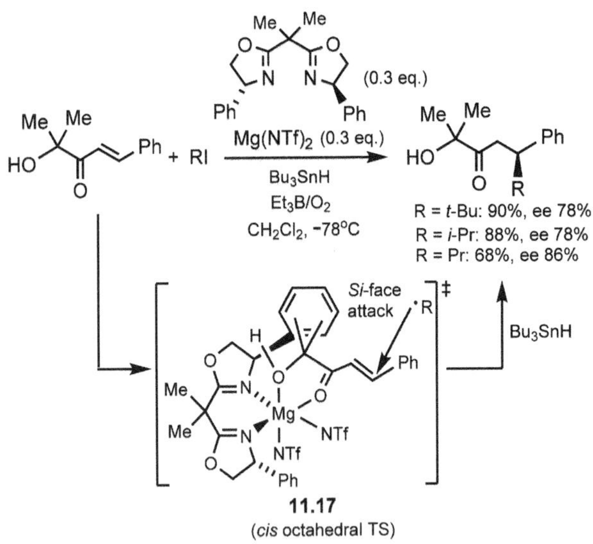


**FIG. 11.38** Asymmetric conjugate radical addition to an α′-hydroxy enone using a chiral Lewis acid.

been performed in the presence of a chiral bisoxazoline (box) ligand and the Lewis acid $Mg(NTf)_2$ with high yields and good ee's, as shown in Fig. 11.38.[83] The observed enantioselectivity has been rationalized with a *cis* octahedral TS model **11.17**. In this model, the box ligand occupies two equatorial sites with the substrate carbonyl at an equatorial position and the bulky alcohol moiety at an axial position. Radical addition from the unhindered *Si* face (the lower *Re* face being shielded by the rear Ph group of the ligand) leads to the major enantiomer of the product.

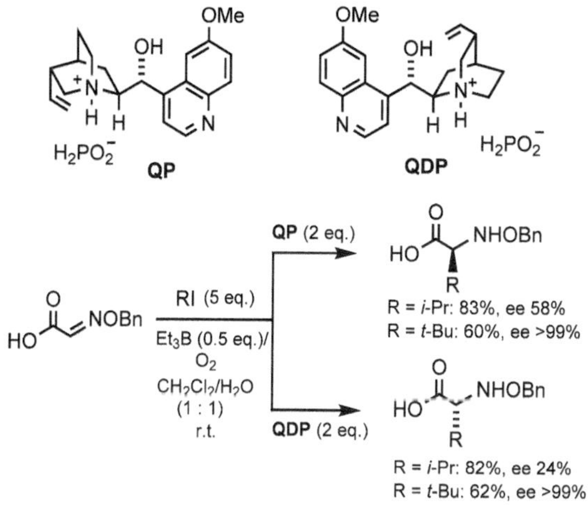


**FIG. 11.39** Asymmetric radical synthesis of α-amino acids using a chiral Brønsted acid, quinine hypophosphite (QP) or quinidine hypophosphite (QDP).

Asymmetric radical addition to glyoxylic oxime ethers for the synthesis of α-amino acids has also been carried out using chiral Brønsted acids in the form of chiral ammonium salts of quinine or quinidine with hypophosphorous acid ($H_3PO_2$). Fig. 11.39 illustrates the asymmetric radical addition using quinine hypophosphite (QP) and quinidine hypophosphite (QDP), respectively.[84] The reaction is run in a biphasic dichloromethane-water system using $Et_3B/O_2$ as an initiator at room temperature. The addition product is converted into its methyl ester derivative to determine ee. The reaction also yields a side product via the addition of ethyl radical formed during initiation. For a secondary alkyl halide, the desired product is obtained in high yield but with moderate ee. In the case of a tertiary alkyl halide, the yield is moderate but the ee is very high.

The asymmetric radical reactions with QP and QDP give opposite enantiomers of the product. The TS models for rationalizing the enantioselectivity is shown in Fig. 11.40.[84] The substrate is bound to QP or QDP by H-bonding between N—H and C=O/C=N, and also through π-stacking. In this arrangement, the *Re*-face of the C=N bond is shielded by the quinoline ring of QP, and radical addition occurs on the exposed *Si*-face to afford the product. On the other hand, the quinoline ring of QDP blocks the *Si*-face, and radical attack takes place preferentially on the *Re*-face of C=N bond to give the product with opposite configuration.

## 11.5.3 Organocatalysis

Organocatalysis via iminium activation (LUMO catalysis) was described previously for ionic and pericyclic reactions. For asymmetric radical synthesis, another mode of organocatalytic activation termed singly occupied molecular orbital (SOMO) catalysis[85] has been introduced.

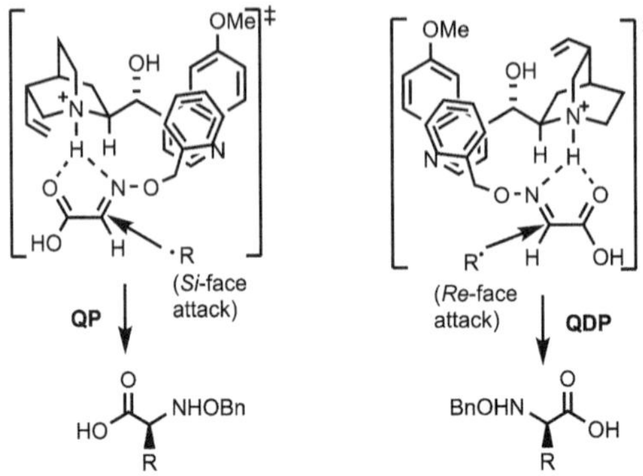


**FIG. 11.40** TS models for rationalization of enantioselectivity of the asymmetric radical reactions in Fig. 11.39 with QP and QDP, respectively.

Fig. 11.41A shows an asymmetric α-allylation of an aldehyde using the imidazolidinone catalyst **11.18** in the presence of cerium(IV) ammonium nitrate [CAN, $Ce(NH_4)_2(NO_3)_6$] as a suitable oxidant.[85] The reaction gives the product in good yield and with high enantioselectivity.

The mechanistic hypothesis is based on SOMO catalysis, which involves one-electron oxidation (SET) of the enamine intermediate (derived from the aldehyde and the chiral amine catalyst) to give a 3π-electron SOMO-activated species (Fig. 11.41B). The density functional theory (DFT) model of the SOMO-activated radical cation suggests that the 3π-electron system is projected away from the bulky *t*-Bu group, and the carbon-centred radical selectively assumes (*E*) geometry to minimize steric interactions. The benzyl group on the catalyst effectively shields the *Re*-face of the radical centre, thereby allowing the exposed *Si*-face for addition reaction. The adduct radical then undergoes β-fragmentation to give the enantioselective product.

Organo-SOMO catalysis has also been successful in achieving asymmetric α-vinylation of aldehydes. Fig. 11.42 illustrates an asymmetric α-vinylation of an aldehyde using a vinyl potassium trifluoroborate salt[86] and the imidazolidinone catalyst **11.18**.[87] Notably, the method provides a β,γ-unsaturated aldehyde without alkene transposition. The reaction proceeds through the formation of a β-borato-stabilized radical (via the *Si*-face attack of the SOMO-activated carbon radical), which readily forms the corresponding cation by one-electron oxidation. Subsequent *trans*-selective Peterson elimination followed by iminium hydrolysis affords the α-vinyl aldehyde product with very high enantioselectivity.

Asymmetric α-enolation of aldehydes is also possible using SOMO catalysis. This is illustrated with an example in Fig. 11.43.[88] The method allows direct access to enantioenriched γ-keto aldehydes from simple aldehydes and enolsilanes in the presence of the imidazolidinone catalyst **11.18**. It is of note that the reaction forms a umpolung polarity by a radical process.

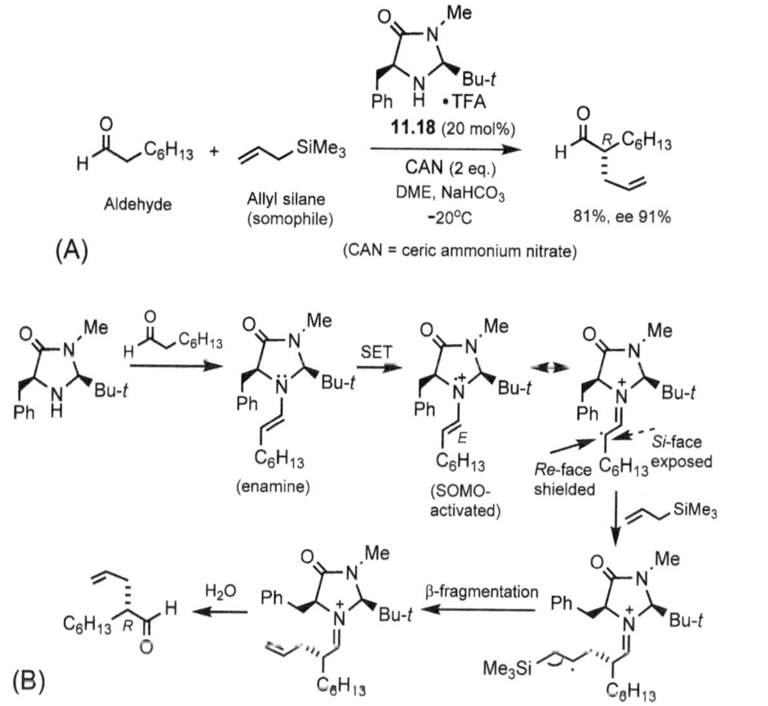


**FIG. 11.41** (A) Organocatalytic asymmetric α-allylation of an aldehyde and (B) mechanism of the reaction based on SOMO catalysis.

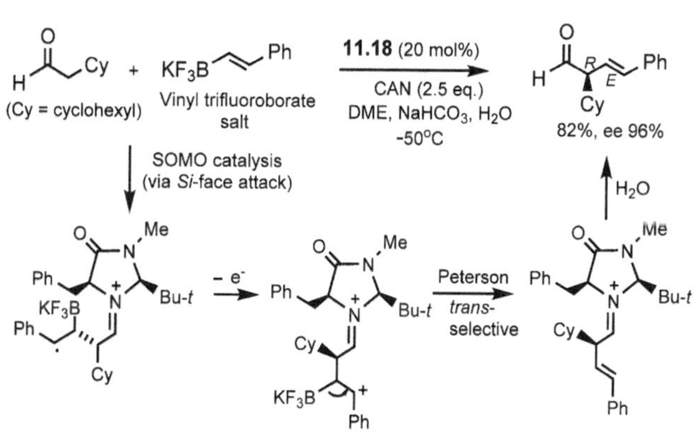


**FIG. 11.42** Organocatalytic asymmetric α-vinylation of an aldehyde.

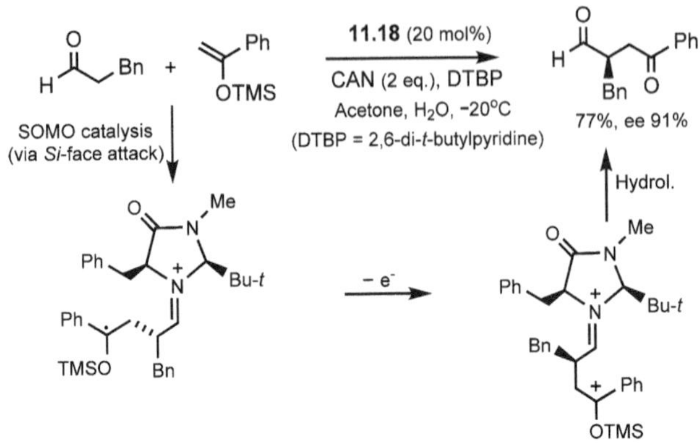


FIG. 11.43 Organocatalytic asymmetric α-enolation of an aldehyde.

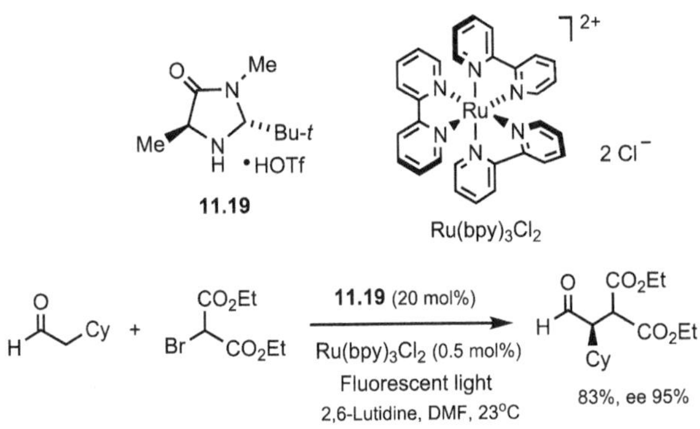


FIG. 11.44 Asymmetric α-alkylation of an aldehyde using a combination of photoredox catalysis and organocatalysis.

Finally, we consider asymmetric α-alkylation of aldehydes that has been accomplished by merging photoredox catalysis with organocatalysis. Fig. 11.44 shows an example of asymmetric α-alkylation using a combination of the photoredox catalyst $Ru(bpy)_3Cl_2$ (bpy = 2,2′-bipyridine)[89] and the imidazolidinone organocatalyst **11.19**.[90] The reaction gives the alkylated product with high enantioselectivity.

The dual-catalysis mechanism is depicted in Fig. 11.45.[90] Photoexcitation of the photoredox catalyst $Ru(bpy)_3^{2+}$ creates the excited state oxidant that removes a single electron from a sacrificial quantity of the enamine to *initiate* the photoredox catalytic cycle and provide the reductant $Ru(bpy)_3^{+}$. Then a single electron transfer (SET) from the reductant $Ru(bpy)_3^{+}$ to the bromomalonate substrate gives the electron-deficient alkyl radical and continues the photoredox catalytic cycle. Concurrent with this photoredox pathway, the organocatalytic cycle begins with the condensation of the imidazolidinone organocatalyst

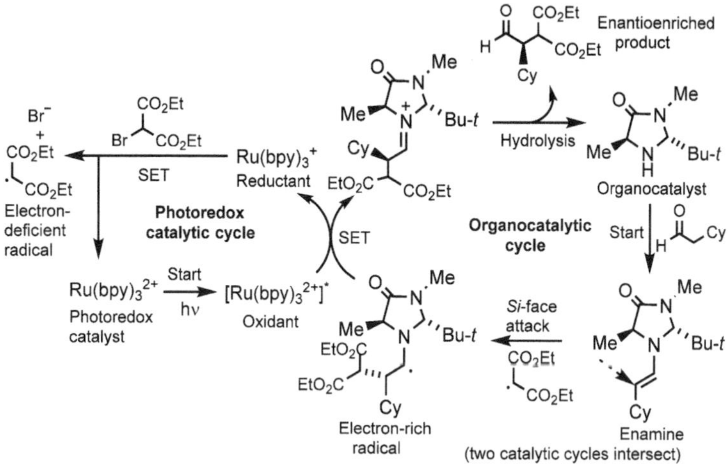


**FIG. 11.45** Dual-catalysis mechanism of asymmetric α-alkylation of an aldehyde in Fig. 11.44.

and the aldehyde substrate to form an enamine. At this stage, the two catalytic cycles intersect via the addition of the electron-deficient alkyl radical to the exposed *Si*-face of the enamine to produce an electron-rich α-amino radical. The two catalytic cycles then converge with SET from the electron-rich radical to the excited state oxidant $[Ru(bpy)_3^{2+}]^*$ to give the iminium ion and regenerate the active reductant $Ru(bpy)_3^+$. The iminium hydrolysis then produces the enantioenriched α-alkylated aldehyde and reconstitutes the imidazolidinone catalyst.

**Problem 11.11** Predict the major product in the following asymmetric α-oxyamination reaction:

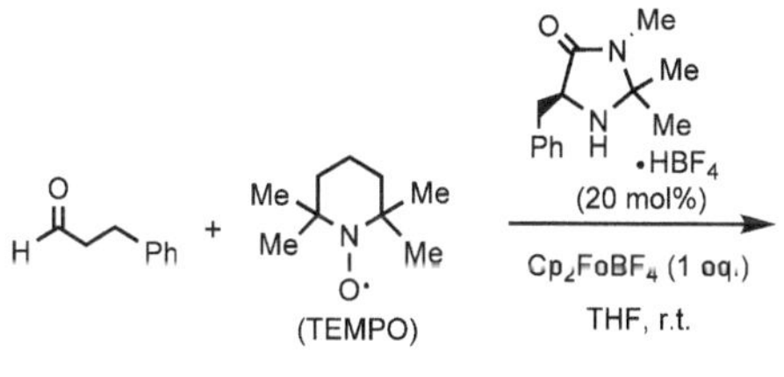


## References


1. Giese, B. *Radicals in Organic Synthesis: Formation of Carbon-Carbon Bonds*; Pergamon Press: Oxford, 1986.
2. Curran, D. P. In *Comprehensive Organic Synthesis*; Trost, B. M., Fleming, I., Eds.; Vol. 4; Pergamon Press: Oxford, 1991; p. 715. 779.

3. Motherwell, W. B.; Crich, D. *Free Radical Chain Reactions in Organic Synthesis*; Academic Press: London, 1992.
4. Renaud, P.; Sibi, M. P. *Radicals in Organic Synthesis*; Wiley: New York, 2001.
5. Zard, S. Z. *Radical Reaction in Organic Synthesis*; Oxford University Press: New York, 2003.
6. Romero, K. J.; Galliher, M. S.; Pratt, D. A.; Stephenson, C. R. *J. Chem. Soc. Rev.* **2018**, *47*, 7851.
7. Neuman, W. P. *Synthesis* **1987**, 665.
8. Curran, D. P.; van Elburg, P. A.; Giese, B.; Gilges, S. *Tetrahedron Lett.* **1990**, *31*, 2861.
9. Keck, G. E.; Enholm, E. J.; Yates, J. B.; Wiley, M. R. *Tetrahedron* **1985**, *41*, 4079.
10. Curran, D. P.; Porter, N. A.; Giese, B. *Stereochemistry of Radical Reactions: Concepts, Guidelines and Synthetic Applications*; VCH: Weinheim, 1996.
11. Giese, B. *Angew. Chem. Int. Ed. Engl.* **1989**, *28*, 969.
12. Kneer, G.; Mattay, J. *Tetrahedron Lett.* **1992**, *33*, 8051.
13. Barton, D. H. R.; Cléophax, J.; Gateau-Olesker, A.; Géro, S. D.; Tachdjian, C. *Tetrahedron* **1993**, *49*, 8381.
14. Giese, B. *Angew. Chem. Int. Ed. Engl.* **1985**, *24*, 553.
15. Barluenga, J.; Yus, M. *Chem. Rev.* **1988**, *88*, 487.
16. Giese, B.; Meixner, J. *Tetrahedron Lett.* **1977**, 2783.
17. Giese, B.; Farshchi, H.; Hartmann, J.; Metzger, J. O. *Angew. Chem. Int. Ed. Engl.* **1991**, *30*, 600.
18. Giese, B.; Heuck, K.; Lenhardt, H.; Lüning, U. *Chem. Ber.* **1984**, *117*, 2132.
19. Sibi, M. P.; Rheault, T. R. *J. Am. Chem. Soc.* **2000**, *122*, 8873.
20. Keck, G. E.; Yates, J. B. *J. Am. Chem. Soc.* **1982**, *104*, 5829.
21. Ollivier, C.; Renaud, P. *J. Am. Chem. Soc.* **2001**, *123*, 4717.
22. Kahne, D.; Yang, D.; Lim, J. J.; Miller, R.; Paguaga, E. *J. Am. Chem. Soc.* **1988**, *110*, 8716.
23. Kessler, H.; Wittmann, V.; Köck, M.; Kottenhahn, M. *Angew. Chem. Int. Ed. Engl.* **1992**, *31*, 902.
24. Durkin, K. A.; Liotta, D. C.; Rancourt, J.; Lavallée, J.–. F.; Boisvert, L.; Guindon, Y. *J. Am. Chem. Soc.* **1992**, *114*, 4912.
25. Hart, D. J.; Krishnamurthy, R. *J. Org. Chem.* **1992**, *57*, 4457.
26. Hart, D. J.; Huang, H.-C.; Krishnamurthy, R.; Schwartz, T. *J. Am. Chem. Soc.* **1989**, *111*, 7507.
27. Giese, B.; Damm, W.; Wetterich, F.; Zeitz, H.-G. *Tetrahedron Lett.* **1992**, *33*, 1863.
28. Renaud, P.; Gerster, M. *Angew. Chem. Int. Ed. Engl.* **1998**, *37*, 2562.
29. Takekawa, Y.; Shishido, K. *J. Org. Chem.* **2001**, *66*, 8490.
30. Gansäuer, A.; Lauterbach, T.; Narayan, S. *Angew. Chem. Int. Ed. Engl.* **2003**, *42*, 5556.
31. Jasperse, C. P.; Curran, D. P.; Fevig, T. L. *Chem. Rev.* **1991**, *91*, 1237.
32. Gilmore, K.; Alabugin, I. V. *Chem. Rev.* **2011**, *111*, 6513.
33. Giese, B.; Kopping, B.; Göbel, T.; et al. *Org. React.* **1996**, *48*, 301.
34. Spellmeyer, D. C.; Houk, K. N. *J. Org. Chem.* **1987**, *52*, 959.
35. Nakamura, E.; Machii, D.; Inubushi, T. *J. Am. Chem. Soc.* **1989**, *111*, 6849.
36. Beckwith, A. L. J.; Zimmermann, J. *J. Org. Chem.* **1991**, *56*, 5791.
37. Ollivier, C.; Renaud, P. *Chem. Rev.* **2001**, *101*, 3415.
38. Chatgilialoglu, C.; Crich, D.; Komatsu, M.; Ryu, I. *Chem. Rev.* **1999**, *99*, 1991.
39. Evans, P. A.; Murthy, V. S.; Roseman, J. D.; Rheingold, A. L. *Angew. Chem. Int. Ed. Engl.* **1999**, *38*, 3175.
40. Chatgilialoglu, C. *Acc. Chem. Res.* **1992**, *25*, 188.
41. Hoffmann, R. W. *Chem. Rev.* **1989**, *89*, 1841.
42. Zhang, W. *Tetrahedron* **2001**, *57*, 7237.
43. Ziegler, F. E.; Petersen, A. K. *J. Org. Chem.* **1994**, *59*, 2707.
44. Beckwith, A. L. J.; Phillipou, G.; Serelis, A. K. *Tetrahedron Lett.* **1981**, *22*, 2811.
45. Danishefsky, S.; Chackalamannil, S.; Uang, B.–. J. *J. Org. Chem.* **1982**, *47*, 2231.

46. Stork, G.; Baine, N. H. *J. Am. Chem. Soc.* **1982**, *104*, 2321.
47. Boger, D. L.; Mathvink, R. J. *J. Org. Chem.* **1988**, *53*, 3377.
48. Birman, V. B.; Danishefsky, S. J. *J. Am. Chem. Soc.* **2002**, *124*, 2080.
49. Stork, G.; Mook, R. *J. Am. Chem. Soc.* **1983**, *105*, 3720.
50. Stork, G.; Sher, P. M. *J. Am. Chem. Soc.* **1986**, *108*, 303.
51. Keck, G. E.; Burnett, D. A. *J. Org. Chem.* **1987**, *52*, 2958.
52. McCarroll, A. J.; Walton, J. C. *Angew. Chem. Int. Ed. Engl.* **2001**, *40*, 2225.
53. Albert, M.; Fensterbank, L.; Lacôte, E.; Malacria, M. *Top. Curr. Chem.* **2006**, *264*, 1.
54. Curran, D. P.; Rakiewicz, D. M. *J. Am. Chem. Soc.* **1985**, *107*, 1448.
55. Devin, P.; Fensterbank, L.; Malacria, M. *J. Org. Chem.* **1998**, *63*, 6764.
56. Pattenden, G.; Gonzalez, M. A.; McCulloch, S.; Walter, A.; Woodhead, S. J. *Proc. Nat. Acad. Sci.* **2004**, *101*, 12024.
57. Boger, D. L.; Mathvink, R. J. *J. Org. Chem.* **1988**, *53*, 3377. **1992**, *57*, 1429.
58. Crich, D.; Fortt, S. M. *Tetrahedron Lett.* **1987**, *28*, 2895. *Tetrahedron* **1989**, *45*, 6581.
59. Chen, L.; Gill, G. B.; Pattenden, G. *Tetrahedron Lett.* **1994**, *35*, 2593.
60. Friestad, G. K. *Tetrahedron* **2001**, *57*, 5461.
61. Lee, H.–. Y.; Lee, S.; Kim, D.; Kim, B. K.; Bahn, J. S.; Kim, S. *Tetrahedron Lett.* **1998**, *39*, 7713.
62. Pradhan, S. K.; Radhakrishnan, T. V.; Subramanian, R. *J. Org. Chem.* **1976**, *41*, 1943.
63. Begley, M. J.; Pattenden, G.; Robertson, G. M. *J. Chem. Soc. Perkin Trans.* **1988**, *1*, 1085.
64. Molander, G. A.; Harris, C. R. *Chem. Rev.* **1996**, *96*, 307.
65. Molander, G. A.; Harris, C. R. *Tetrahedron* **1998**, *54*, 3321.
66. Edmonds, D. J.; Johnston, D.; Procter, D. J. *Chem. Rev.* **2004**, *104*, 3371.
67. Molander, G. A.; McKie, J. A. *J. Org. Chem.* **1992**, *57*, 3132.
68. Fevig, T. L.; Elliott, R. L.; Curran, D. P. *J. Am. Chem. Soc.* **1988**, *110*, 5064.
69. Cha, J. Y.; Yeoman, J. T. S.; Reisman, S. E. *J. Am. Chem. Soc.* **2011**, *133*, 14964.
70. Caine, D. *Org. React.* **1976**, *23*, 1.
71. Sibi, M. P.; Manyem, S.; Zimmerman, J. *Chem. Rev.* **2003**, *103*, 3263.
72. Hashimoto, T.; Kawamata, Y.; Maruoka, K. *Nat. Chem.* **2014**, *6*, 702.
73. Wu, X.; Zhu, C. *Chem. Rec.* **2018**, *18*, 587.
74. Li, W.; Xu, W.; Xie, J.; Yu, S.; Zhu, C. *Chem. Soc. Rev.* **2018**, *47*, 654.
75. Wu, X.; Zhu, C. *Chin. J. Chem.* **2019**, *37*, 171.
76. Miyabe, H.; Ushiro, C.; Ueda, M.; Yamakawa, K.; Naito, T. *J. Org. Chem.* **2000**, *65*, 176.
77. Cicchi, S.; Goti, A.; Brandi, A.; Guarna, A.; De Sarlo, F. *Tetrahedron Lett.* **1990**, *31*, 3351.
78. Nanni, D.; Curran, D. P. *Tetrahedron Asymmetry* **1996**, *7*, 2417.
79. Sibi, M. P.; Porter, N. A. *Acc. Chem. Res.* **1999**, *32*, 163.
80. Sibi, M. P.; Liu, P.; Ji, J.; Hajra, S.; Chen, J.-X. *J. Org. Chem.* **2002**, *67*, 1738.
81. Stitch, S. R.; Funke, C. W.; Groen, M. B.; et al. *Nature* **1980**, *287*, 738.
82. Doyle, M. P.; Bode, J. W.; Lynch, V. J.; Protopopova, M. N.; Simonsen, S. H.; Zhou, Q.-L. *J. Org. Chem.* **1995**, *60*, 6654.
83. Lee, S.; Lim, C. J.; Kim, S.; Subramaniam, R.; Zimmerman, J.; Sibi, M. P. *Org. Lett.* **2006**, *8*, 4311.
84. Cho, D. H.; Jang, D. O. *Chem. Commun.* **2006**, 5045.
85. Beeson, T. D.; Mastracchio, A.; Hong, J. B.; Ashton, K.; MacMillan, D. W. C. *Science* **2007**, *316*, 582.
86. Molander, G. A.; Figueroa, R. *Aldrichimica Acta* **2005**, *38*, 49.
87. Kim, H.; MacMillan, D. W. C. *J. Am. Chem. Soc.* **2008**, *130*, 398.
88. Jang, H.–. Y.; Hong, J.–. B.; MacMillan, D. W. C. *J. Am. Chem. Soc.* **2007**, *129*, 7004.
89. Kalyanasundaram, K. *Coord. Chem. Rev.* **1982**, *46*, 159.
90. Nicewicz, D. A.; MacMillan, D. W. C. *Science* **2008**, *322*, 77.

Chapter 12

# Photochemical reactions: Stereochemistry and asymmetric synthesis

## 12.1 Photochemical [2+2] cycloadditions[1–4]

Photochemical [2+2] cycloadditions involve alkenes, α,β-unsaturated carbonyl compounds (enones), and aldehydes or ketones. Both intermolecular and intramolecular [2+2] cycloadditions are useful in synthesis.

### 12.1.1 Stereospecificity of alkene photocycloadditions

The [2+2] cycloaddition of two alkenes provides a cyclobutane ring with a maximum of four stereocentres. The photochemical reaction is allowed by orbital symmetry under supra/supra mode (see Section 1.6.1). Fig. 12.1 shows the dimerization of (*Z*)- and (*E*)-2-butenes.[5] (*Z*)-2-butene gives the cyclobutane derivatives **12.1** and **12.2**, while (*E*)-2-butene yields **12.2** and **12.3**. These simple alkenes absorb light in the far-UV, and the stereospecificity of the cycloaddition from the concerted supra/supra mechanism is observed only at low conversions (short reaction times). At longer reaction times, mainly fragmentation and *E*/*Z* isomerization occur, and cycloaddition reaction then leads to a mixture of all stereoisomeric cyclobutanes.[5]

Photochemical retrocycloaddition is also possible, as shown in Fig. 12.2. The *cis* and *trans* stereochemistry at the ring junction of the substrates are retained in the 12-membered ring products. Note that the large 12-membered ring can accommodate a triple bond and *E* or *Z* double bond. It is likely that the stereospecificity of the reaction arises from a concerted $[_{\sigma}2_s + _{\sigma}2_s]$ mechanism.[6]

Intramolecular photocycloadditions are quite useful synthetically.[2, 7] These can form highly strained polycyclic and cage systems, which are not easily obtainable by other routes.[8, 9] For example, the synthesis of a cubane is shown in Fig. 12.3.[10] The Diels–Alder reaction between dibromoquinone and cyclobutadiene (generated in situ by the oxidative decomposition of its iron complex

Stereochemistry and Organic Reactions. https://doi.org/10.1016/B978-0-12-824092-2.00012-5

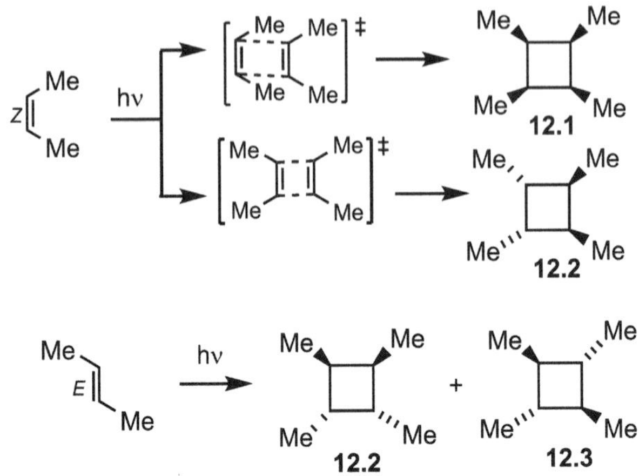


**FIG. 12.1** [2+2] Dimerization of (*Z*)- and (*E*)-2-butenes.

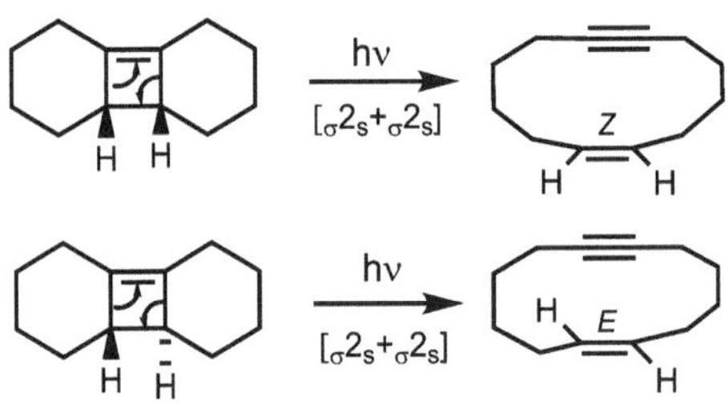


**FIG. 12.2** Stereospecificity of a photochemical [2+2] retrocycloaddition.

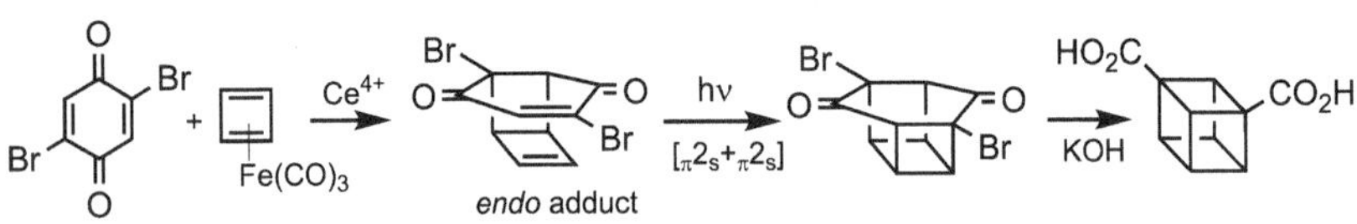


**FIG. 12.3** Application of intramolecular [2+2] photocycloaddition in the synthesis of a cubane.

with ceric ion) gives an *endo* adduct in which the two alkene units are suitably oriented one above the other for an efficient [2+2] photocycloaddition. The photoadduct is converted to a cubanedicarboxylic acid by the Favorskii rearrangement.

**Problem 12.1** Predict the products of the following reactions:

(a) 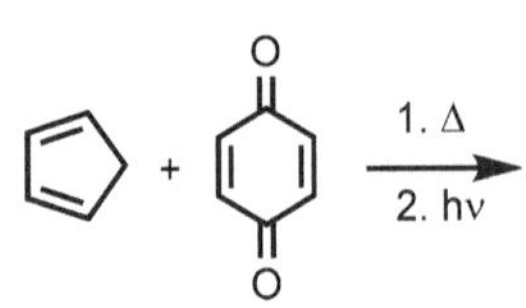


(b) 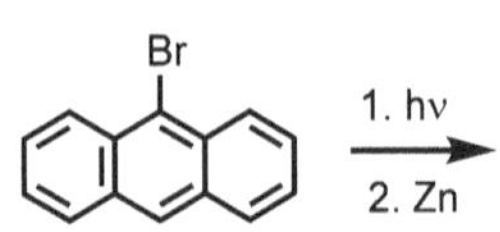

**Problem 12.2** Give the structures (**A–C**) in the following scheme for the synthesis of basketene:

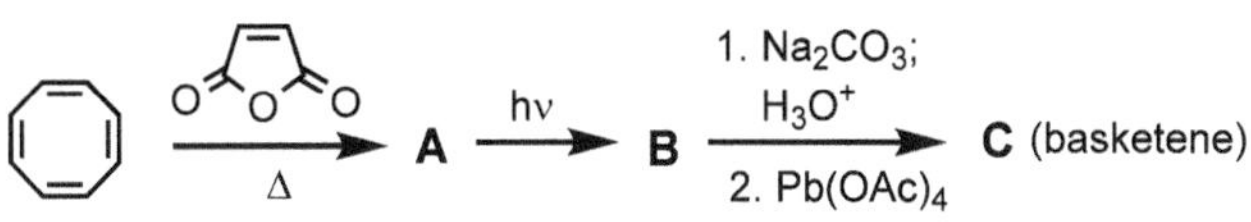


## 12.1.2 Diastereoselectivity of enone photocycloadditions

The most common and useful [2+2] photocycloadditions involve enones (α,β-unsaturated carbonyl compounds).[11, 12] These compounds absorb light at longer wavelengths (greater than 300nm); thus the destructive effect of short wavelength irradiation is avoided. In general, enone cycloadditions are stepwise and involve a triplet excited enone formed by intersystem crossing of the initial (n,π*) singlet[3] (see Section 4.6.2). The cycloadditions are often regioselective. An electron-rich alkene generally produces a head-to-tail regioisomer as a major product, whereas an electron-deficient alkene gives predominantly a head-to-head isomer (see Section 4.6.2).

The enone cycloadditions can exhibit good diastereoselectivity. For example, [2+2] photocycloaddition of cyclohexenone with the electron-rich isobutene gives predominantly a head-to-tail adduct with *trans* ring fusion, as shown in Fig. 12.4A.[13, 14] This reaction is the first step in Corey's synthesis of caryophyllene. The excited state of cyclohexenone seems to be a (π,π*) triplet with a twisted geometry. It is possible that an exciplex with a highly twisted character is involved, which leads to the preferential formation of the *trans*-fused product.

Fig. 12.4B shows the photocycloaddition of a cyclohexenone with an electron-deficient cyclobutene.[15] The major product is *exo* diastereomer of a head-to-head adduct. Note that the terminal six-membered and four-membered rings are *anti* to the middle cyclobutane ring in the *exo* isomer. The excited state of the cyclohexenone behaves like a diradical, and the preferred interaction involves the nucleophilic radical centre (tail) of the excited cyclohexenone with the electrophilic β-end (tail) of the cyclobutene. Since the cycloaddition is stepwise through a diradical intermediate, bond rotation can occur before spin inversion to produce a mixture of diradical species for *exo* and *endo* ring closures. The diastereoselectivity is in favour of thermodynamically more stable *exo* (*anti*) cyclobutane derivative.

On direct irradiation, coumarin dimerizes to give only *endo* (*syn*) head-to-head dimer, whereas the sensitized reaction gives a mixture of *endo* (*syn*) and *exo* (*anti*) head-to-head dimers (Fig. 12.5).[16] Coumarin behaves as an enone. A single diastereomeric product in the direct photochemical reaction could result if a singlet excited state is involved. The reaction may take a concerted pathway leading to the cycloadduct directly or proceed through a singlet exciplex giving a diradical when the cyclization is much faster than bond rotation. The formation of

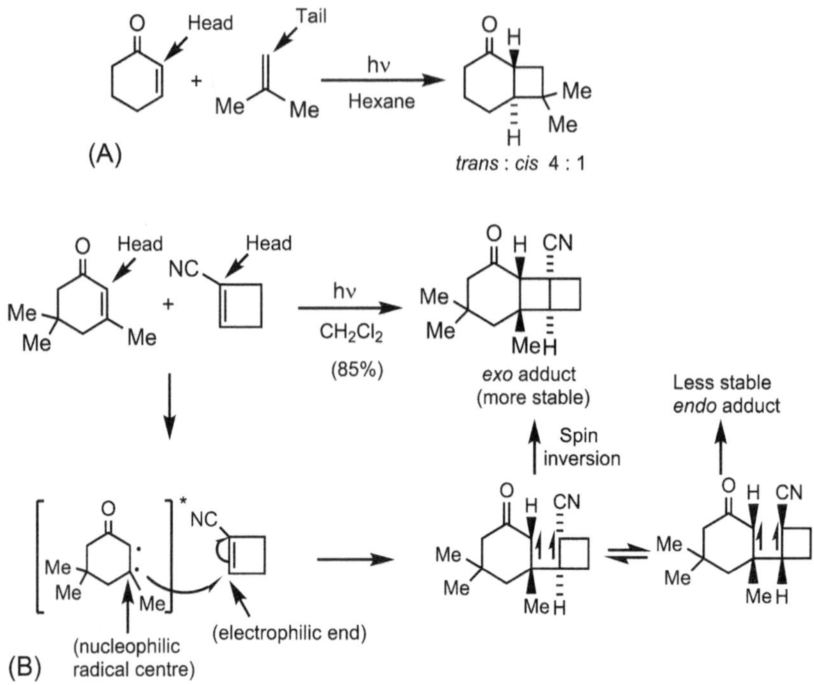


**FIG. 12.4** Diastereoselectivity of enone photocycloadditions: (A) reaction of cyclohexenone with electron-rich alkene (isobutene) and (B) reaction of a cyclohexenone with electron-deficient alkene (1-cyanocyclobutene).

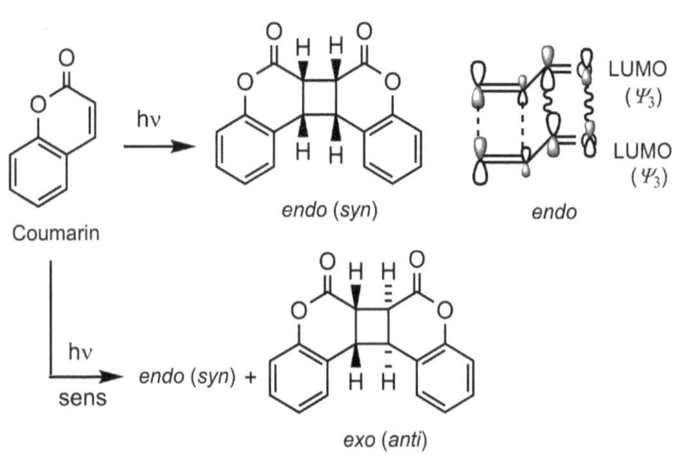


**FIG. 12.5** Direct and sensitized photochemical dimerization of coumarin.

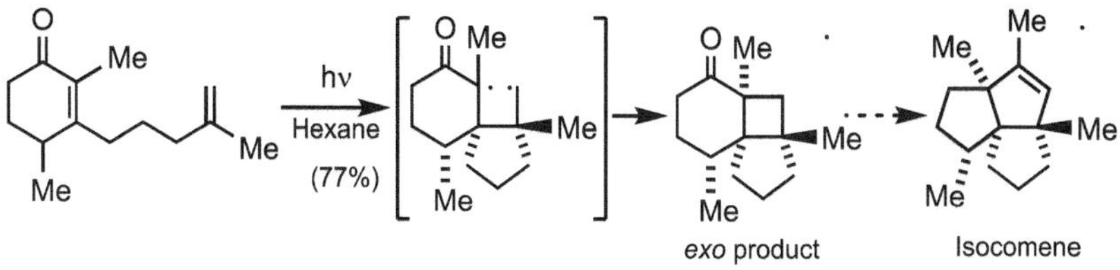


**FIG. 12.6** Stereoselective intramolecular enone photocycloaddition used in a synthesis of the sesquiterpene isocomene.

*endo* adduct can be rationalized in terms of bonding secondary orbital interactions (wavy lines) in the dominant LUMO/LUMO interaction. On the other hand, the sensitized cycloaddition is a stepwise triplet state reaction, when the cyclization step involves a spin inversion barrier. Consequently, bond rotation would lead to a mixture of both *endo* and *exo* isomers. The head-to-head regiochemical preference in both direct and sensitized reactions arises from the large/large coefficient interaction.

Intramolecular enone photocycloadditions provide a valuable strategy for the synthesis of bicyclic and polycyclic compounds. Fig. 12.6 illustrates an intramolecular enone cycloaddition forming two new rings and three contiguous quaternary stereocentres, used in a synthesis of the sesquiterpene isocomene.[17] The conformational constraints of the intramolecular reaction determine the regioselectivity of the cycloaddition with the initial C—C bond formed between the nearer ends of the two components, thereby leading to the formation of a five-membered ring (not a six-membered). The stereoselectivity of the reaction is in favour of the thermodynamically more stable *exo* product.

The synthetic utility of enone photocycloadditions can be extended via ring-opening of the strained cyclobutane ring to form complex ring systems. This is illustrated with an example in Fig. 12.7.[18] Photocycloaddition of the enol benzoate **12.4** gives almost exclusively the tricyclic compound **12.5**, which is dimethylated and then subjected to ester hydrolysis. Under the hydrolysis conditions, the alkoxide undergoes spontaneous retroaldol reaction with cyclobutane ring opening to produce an eight-membered ring in the resulting bicyclic diketone **12.6**, in a synthesis of the sesquiterpene *epi*-precapnelladiene.

**Problem 12.3** Account for the following observation:

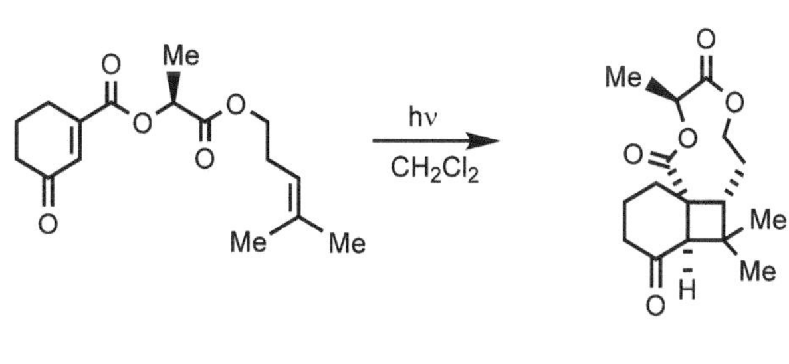

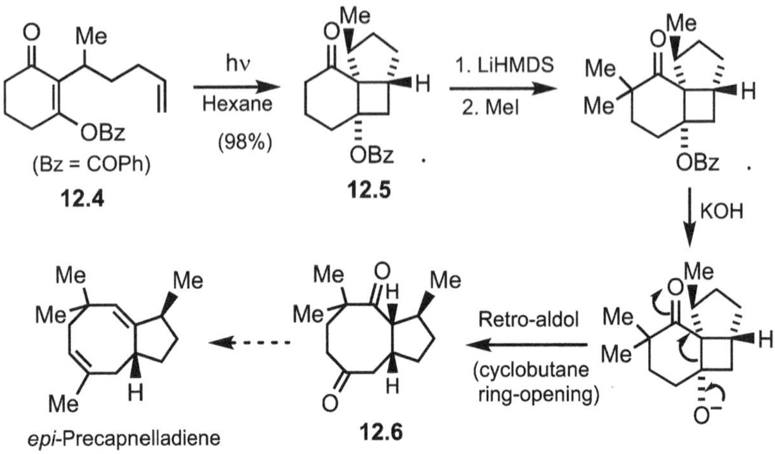


**FIG. 12.7** Application of intramolecular enone photocycloaddition via cyclobutane ring-opening to a synthesis of the sesquiterpene *epi*-precapnelladiene.

### 12.1.3 Visible light photocatalysis of enone cycloadditions

Irradiation of the photoredox catalyst $Ru(bpy)_3^{2+}$ (see Fig. 11.44) with visible light ($\lambda_{max}$ 452 nm) produces a photoexcited state with a relatively long lifetime (~600 ns).[19, 20] This photoexcited state $[Ru(bpy)_3^{2+}]^*$ is a powerful oxidant capable of single electron transfer (SET). The visible light photocatalysis of an intramolecular [2+2] cycloaddition of a bis(enone) **12.7** using the photoredox catalyst $Ru(bpy)_3Cl_2$ with *i*-$Pr_2NEt$ and $LiBF_4$ as additives is shown in Fig. 12.8A.[21] The reaction gives the *meso* diastereomer of bicyclic dione **12.8** with high diastereoselectivity.

The mechanism of the intramolecular photocycloaddition is shown in Fig. 12.8B. Photoexcitation of the catalyst results in a photoexcited state $[Ru(bpy)_3^{2+}]^*$ that removes a single electron from the amine (*i*-$Pr_2NEt$) to form the active reductant $Ru(bpy)_3^{+}$. Then a single electron transfer to the lithium-activated enone initiates the [2+2] cycloaddition and regenerates the photocatalyst.

Visible light photocatalysis of intermolecular enone cycloaddition is also possible, as shown in Fig. 12.9.[22] The aryl enone acts as an electron acceptor because of its greater ease of generating the key radical anion intermediate, and its reacting partner could be any suitable Michael acceptor. The reaction gives the crossed [2+2] cycloadduct in good yield and with high diastereoselectivity.

### 12.1.4 Diastereoselectivity of the Paternò–Büchi reaction

The [2+2] photocycloaddition of an aldehyde or ketone with an alkene produces an oxetane, which is known as the Paternò–Büchi reaction.[23, 24]

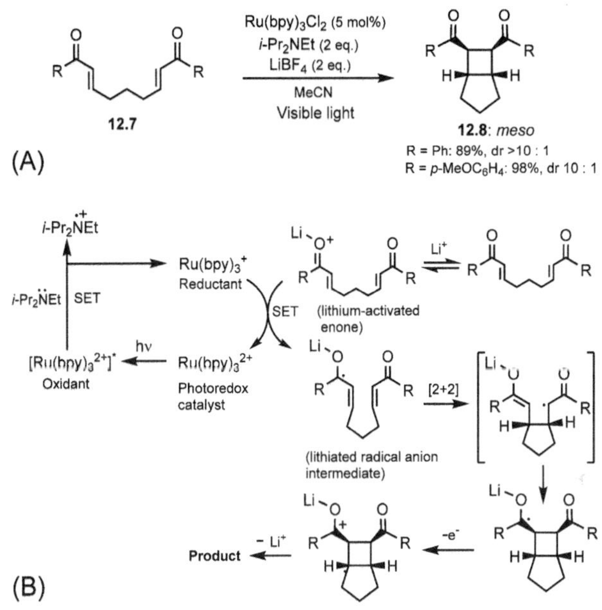


**FIG. 12.8** (A) Visible light photocatalysis of intramolecular [2+2] cycloaddition of a bis(enone) and (B) mechanism of the reaction.

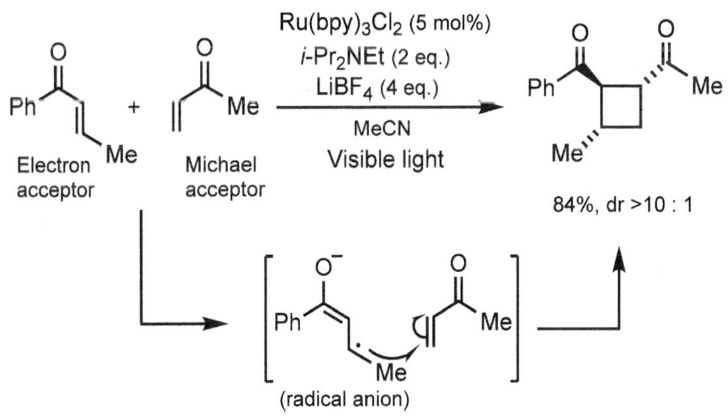


**FIG. 12.9** Visible light photocatalysis of crossed intermolecular [2+2] enone cycloaddition.

The mechanism of the reaction is shown in Fig. 12.10.[25, 26] The light-absorbing component responsible for the reaction is usually carbonyl addend. The carbonyl compound is excited to ($n$,$\pi^*$) singlet state which by intersystem crossing (ISC) goes to ($n$,$\pi^*$) triplet state. Addition of the triplet excited carbonyl to the alkene forms a triplet 1,4-diradical via an exciplex. The triplet 1,4-diradical must undergo spin inversion (ISC by spin-orbit coupling) to form a singlet

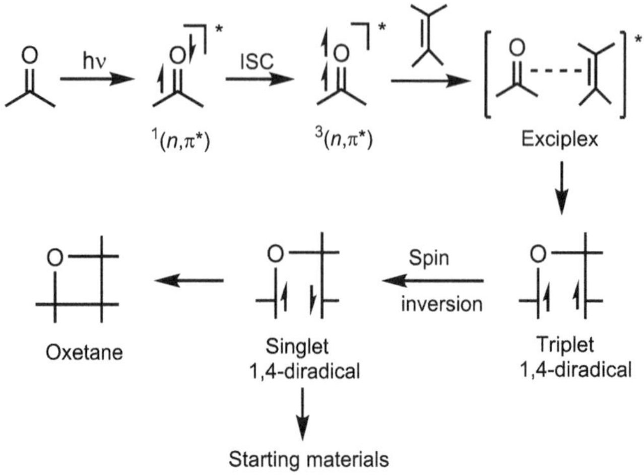


**FIG. 12.10** Mechanism of the Paternò–Büchi reaction. ISC = intersystem crossing; $^1(n,\pi^*)$ and $^3(n,\pi^*)$ indicate singlet and triplet $n$–$\pi^*$ excited state, respectively.

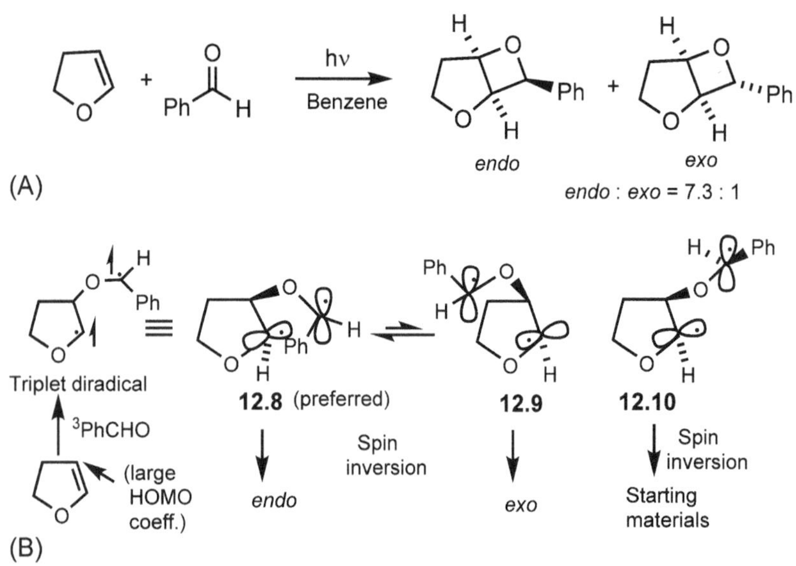


**FIG. 12.11** (A) Diastereoselectivity in the Paternò–Büchi reaction of 2,3-dihydrofuran with benzaldehyde and (B) model for *endo*-selectivity.

1,4-diradical which can lead to C—C bond formation to give the oxetane or cleave to give the starting materials.

The spin inversion geometry of the triplet 1,4-diradical is crucial for determining the diastereoselectivity of Paternò–Büchi reaction. Fig. 12.11A shows a Paternò–Büchi reaction of 2,3-dihydrofuran with benzaldehyde.[27] The reaction is highly diastereoselective and gives predominantly the thermodynamically less stable *endo* diastereomer (*endo*-Ph).

The *endo*-selectivity has been rationalized based on the possible spin inversion geometries of the triplet 1,4-diradical.[28] With electron-rich alkenes, it is the singly occupied *n* orbital of excited carbonyl that interacts with the π orbital (HOMO) of alkene (*n*–π interaction) in a perpendicular approach to form the first O—C bond; thus the radical orbitals of the resulting triplet 1,4-diradical would be orthogonal. The regioselectivity of the reaction results from the bonding of the O atom to the alkene carbon with larger HOMO coefficient, which also produces a more stable 1,4-diradical. As shown in Fig. 12.11B, spin inversion from the triplet 1,4-diradical conformers **12.8** and **12.9** results in product formation. Spin inversion is coupled with a torque, which leads to the formation of new C—C bond. The torque induced in the preferred conformer **12.8** rotates the large Ph over the plane to give the *endo* product. The less preferred conformer **12.9** will lead to the *exo* adduct. The spin inversion from the conformer **12.10** leads to cleavage of the resulting singlet diradical to revert to the starting materials.

The photocycloaddition of benzaldehyde with the dihydropyrrole **12.11** also gives predominantly the *endo*-Ph product (Fig. 12.12). The ring opening of the *endo* adduct by hydrogenolysis of the benzylic C—O bond, followed by reduction of the carbamate affords the antifungal agent (+)-preussin.[29, 30]

In contrast to *endo*-selective cycloaddition of 2,3-dihydrofuran, the photocycloaddition of furans with aldehydes gives mainly *exo* diastereomer of *cis*-fused bicyclic oxetanes **12.12** (Fig. 12.13A).[27] This reversal of diastereoselectivity with furan also arises from the preferred spin inversion geometry of the triplet 1,4-diradical.[28] Here the conformer leading to the *exo* product (cf. **12.9** in Fig. 12.11B) is favoured by secondary orbital interaction of the exocyclic radical with the residual π orbital of furan. This interaction also facilitates ISC with an increase in spin–orbit coupling. The regioselectivity of the cycloaddition is attributed to the bonding of the oxygen atom to the electron-rich 2-position (large HOMO coefficient) in the furan or to the formation of more stable diradical intermediate containing allylic radical moiety.

As shown in Fig. 12.13A, the adduct **12.12** can be unmasked under acidic conditions to afford *syn* aldol products of 1,4-dicarbonyl compounds.[31, 32] This strategy has been exploited in the synthesis of a variety of natural products.

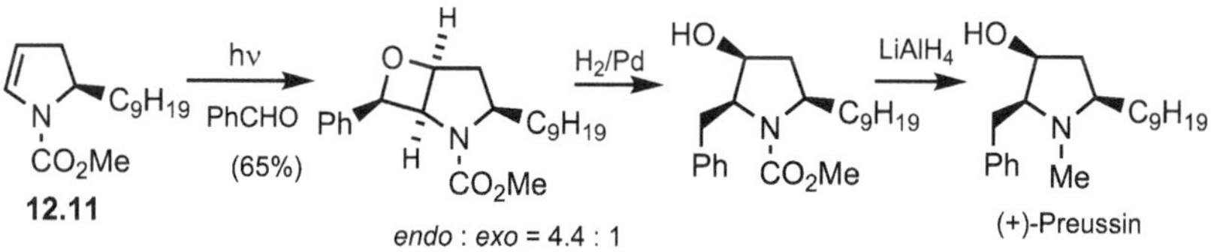


**FIG. 12.12** Application of *endo*-selective Paternò–Büchi reaction in the synthesis of an antifungal agent (+)-preussin.

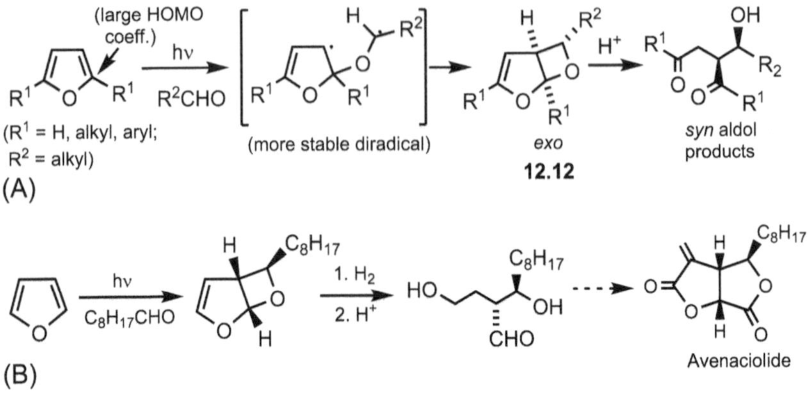


**FIG. 12.13** (A) *exo*-diastereoselectivity in the Paternò–Büchi reaction of furans with aldehydes and (B) application of the *exo*-selective reaction in a synthesis of (±)-avenaciolide.

Fig. 12.13B shows an application of this strategy in a synthesis of the antifungal metabolite avenaciolide.[33] Photocycloaddition of furan with nonanal gives the *exo* product, which is hydrogenated and hydrolyzed to produce an aldehyde intermediate in a synthetic route to avenaciolide.

**Problem 12.4** Predict the product of the following reaction:

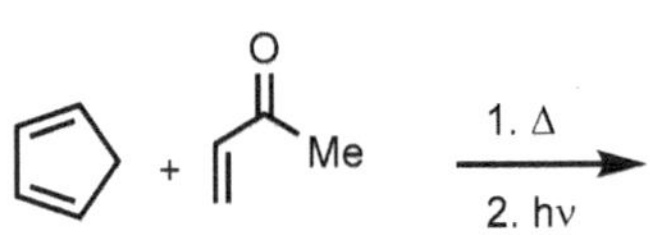


**Problem 12.5** Explain the following observations:

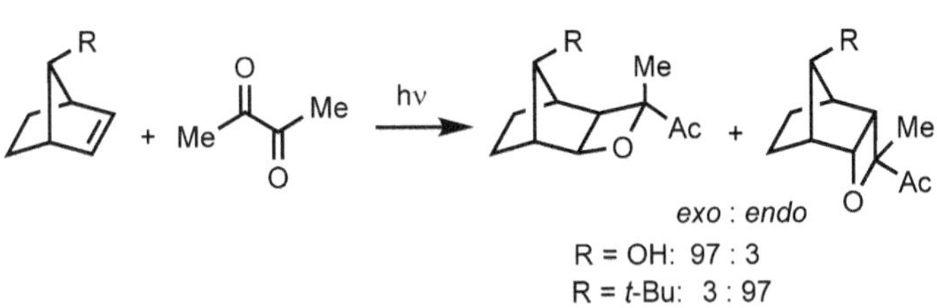


## 12.2 Stereochemistry of photochemical electrocyclic reactions

Photochemical electrocyclic reactions obey the Woodward–Hoffmann selection rules (see Section 4.4.6). The stereochemistry of 4-electron photoelectrocyclization arises from the disrotatory (dis) process, whereas 6-electron photochemical electrocyclic reaction occurs by the conrotatory (con) process. The product stereochemistry can be easily delineated using the stereochemical rule described earlier (see Section 8.8.1).

### 12.2.1 4-Electron photoelectrocyclic process

The photochemical ring closing of (*E*,*E*)-2,4-hexadiene gives a *cis* cyclobutene, which arises from the allowed disrotatory process (Fig. 12.14A).[34] See also the use of the stereochemical rule (Section 8.8.1). The 4-electron photoelectrocyclization has been useful in synthesis.[35] This is illustrated with two examples in Fig. 12.14B. Dewar benzene is synthesized via a photochemical ring closing followed by hydrolysis and bis-decarboxylation with $Pb(OAc)_4$. A synthesis of cyclobutadiene iron tricarbonyl from an unsaturated lactone involves a photochemical ring closure followed by a photodecarboxylation in the presence of iron pentacarbonyl.

### 12.2.2 6-Electron photoelectrocyclic process

The photochemical ring opening of *cis* and *trans* isomers of a tetrasubstituted cyclohexadiene is shown in Fig. 12.15A. The stereochemistry of the products results from the conrotatory ring-opening process. See also the use of the stereochemical rule.

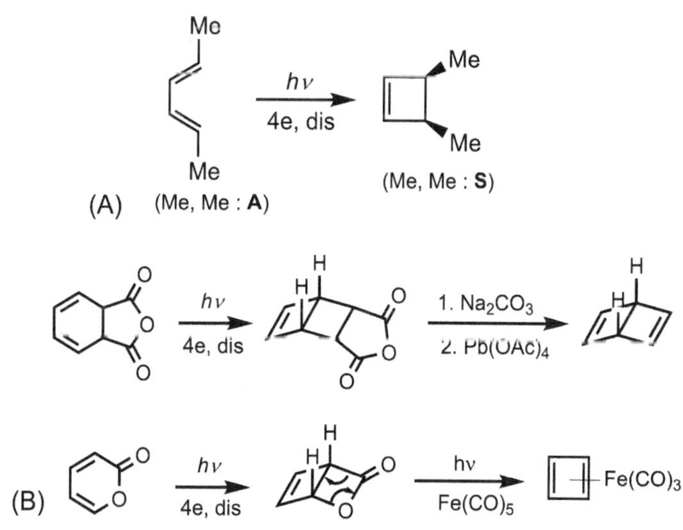


**FIG. 12.14** (A) Photochemical electrocyclic ring closing of (*E*,*E*)-2,4-hexadiene and (B) application of 4-electron photoelectrocyclization in the synthesis of Dewar benzene and cyclobutadiene iron tricarbonyl.

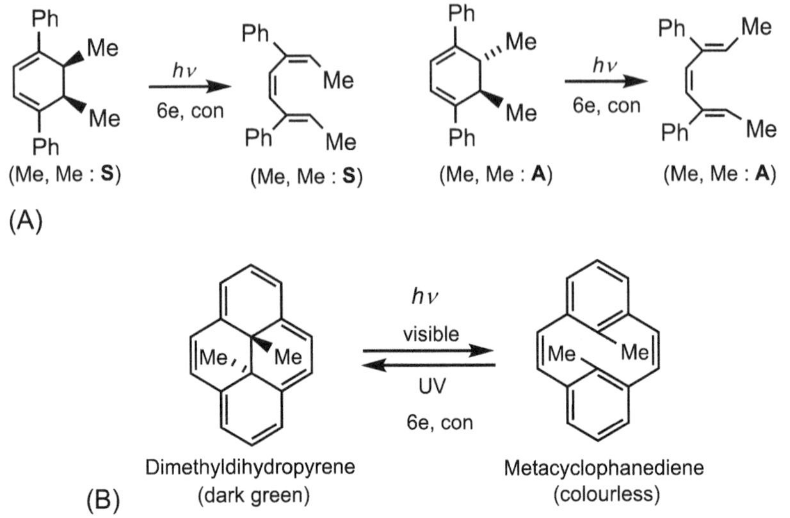


**FIG. 12.15** (A) Stereochemistry of 6-electron photochemical ring-opening reactions and (B) photochromism of dimethyldihydropyrene.

Fig. 12.15B shows an example of photochromism,[36] which is a reversible phototransformation of a species between two states of different electronic absorptions.[37] With visible light, dimethyldihydropyrene (dark green) undergoes an electrocyclic ring opening to form metacyclophanediene (colourless) while the reverse ring closing occurs on UV irradiation.

The vitamin $D_3$ synthesis occurs in the human body upon exposure to sunlight on the skin. A photochemical 6-electron ring opening of 7-dehydrocholesterol produces previtamin $D_3$, which spontaneously isomerizes to vitamin $D_3$ (cholecalciferol) by the [1,7]-sigmatropic hydrogen shift (Fig. 12.16).[38,39] Another form of vitamin D, called vitamin $D_2$, is derived from yeast sterol ergosterol and is found naturally in sun-exposed mushrooms. Vitamin $D_2$ (calciferol) is also produced by photochemical ring opening followed by the [1,7]H shift (Fig. 12.16). Note that human beings do not make vitamin $D_2$; however, vitamins $D_2$ and $D_3$ have similar biological activity in humans.

A useful 6-electron photochemical electrocyclic reaction is the cyclization of 1,2-diayl alkenes.[40, 41] For example, photocyclization of (*Z*)-stilbene gives *trans*-dihydrophenanthrene (Fig. 12.17A). In the reaction, aromaticity of stilbene is lost but the dihydrophenanthrene cannot revert to stilbene by thermal disrotatory ring opening because it would produce a highly strained *trans* cyclohexatriene ring. Oxidation by air or iodine converts the dihydrophenanthrene into phenanthrene.

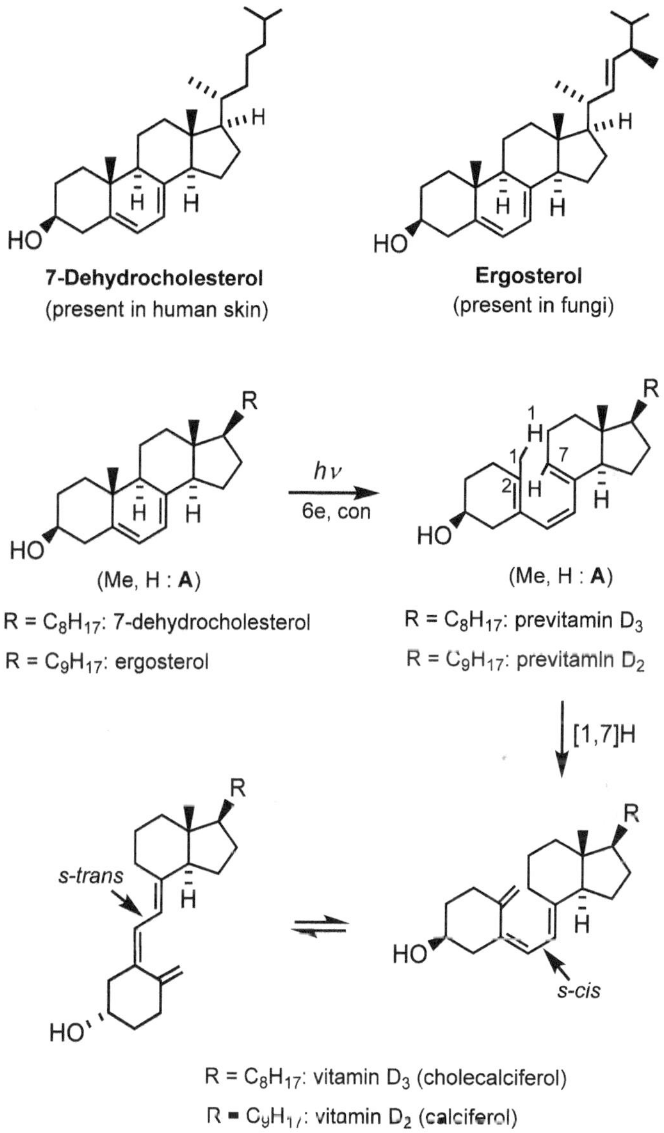


**FIG. 12.16** Vitamin D ($D_3$ or $D_2$) synthesis.

The 6-electron electrocyclization requires that the central double bond must be (*Z*); however, the action of light on stilbenes can promote *E*/*Z* isomerization and thus photocyclization of (*E*)- and/or (*Z*)-diaryl alkenes is possible. Fig. 12.17B shows a photocyclization of an *E*/*Z* mixture of the diaryl alkene **12.13** which gives, after oxidation with $I_2$, the polycyclic aromatic compound **12.14**, in a synthesis of the antibiotic cervinomycin A.[42]

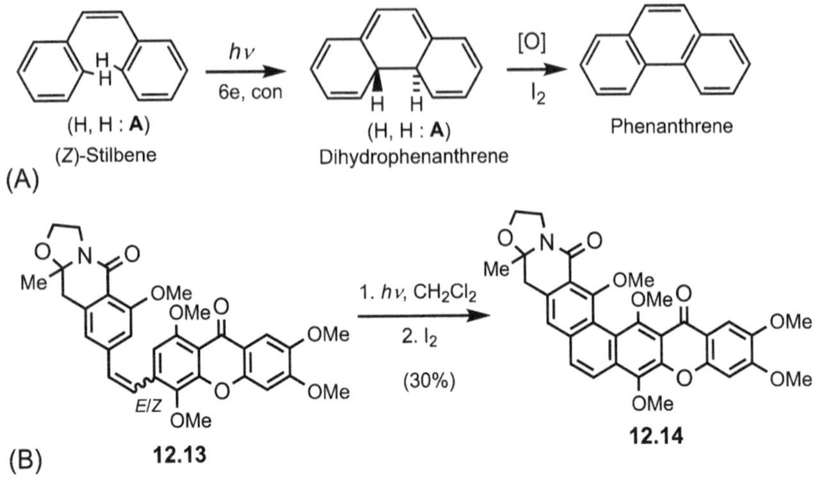


**FIG. 12.17** (A) Conversion of (*Z*)-stilbene to phenanthrene via 6-electron photocyclization and (B) photocyclization of a mixture of (*E*)- and (*Z*)-diaryl alkenes, used in a synthesis of the antibiotic cervinomycin A.

**Problem 12.6** Predict the products of the following reactions:

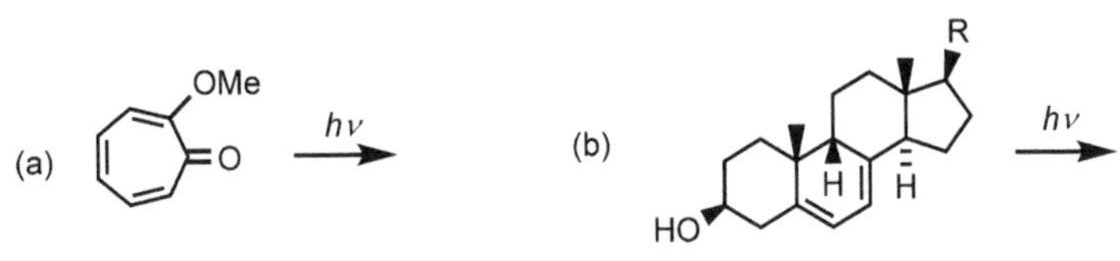


**Problem 12.7** Accomplish the following conversion:

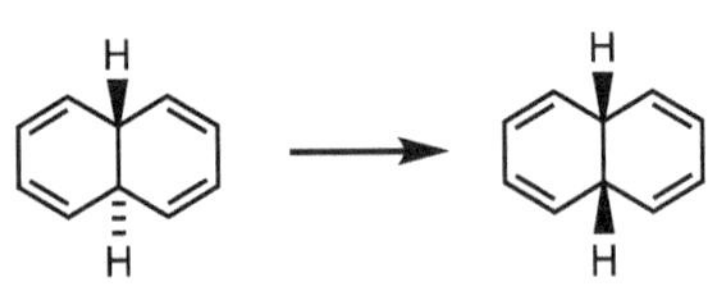

## 12.3 Stereochemistry of photo-induced remote functionalization

Photolytic cleavage of the N—X or O—X bond (X = halogen or other heteroatom) gives a nitrogen- or oxygen-centred radical, which can participate in an intramolecular C—H abstraction. The most common hydrogen transfer is a 1,5-shift via a six-membered cyclic TS, forming a carbon radical. In substrates that possess certain structural and geometrical constraints, the new carbon radical can be formed at an unactivated position, thereby allowing the introduction of a functional group at that position. This is referred to as 'remote functionalization'.[43, 44] Two such processes used successfully in the synthesis of steroidal alkaloids in particular, make use of the Hofmann–Löffler–Freytag reaction and the Barton reaction.

### 12.3.1 The Hofmann–Löffler–Freytag reaction

The Hofmann–Löffler–Freytag reaction provides a method for the construction of a pyrrolidine ring by irradiation of N-halogenated amines in strong acid, followed by basification (Fig. 12.18).[45] The photochemical cleavage of the N—Cl bond of the *N*-chloro-ammonium salt (formed by protonation of the *N*-chloro-amine) produces a nitrogen-centred radical, which undergoes 1,5-hydrogen atom transfer to form a carbon radical. This radical then abstracts a Cl atom from another molecule of the *N*-chloro-ammonium salt to produce a δ-chloro-amine (as a salt) and propagate the chain. Basification of the reaction mixture gives the pyrrolidine by a simple intramolecular $S_N2$ reaction.

An application of the Hofmann–Löffler–Freytag reaction in a synthesis of the steroidal compound dihydroconessine is shown in Fig. 12.19.[46] Remote functionalization of the C18 angular methyl group is possible because the rigid steroid framework and the suitable disposition of the C20 side chain bearing the N-radical allow an easy 1,5-H atom transfer.

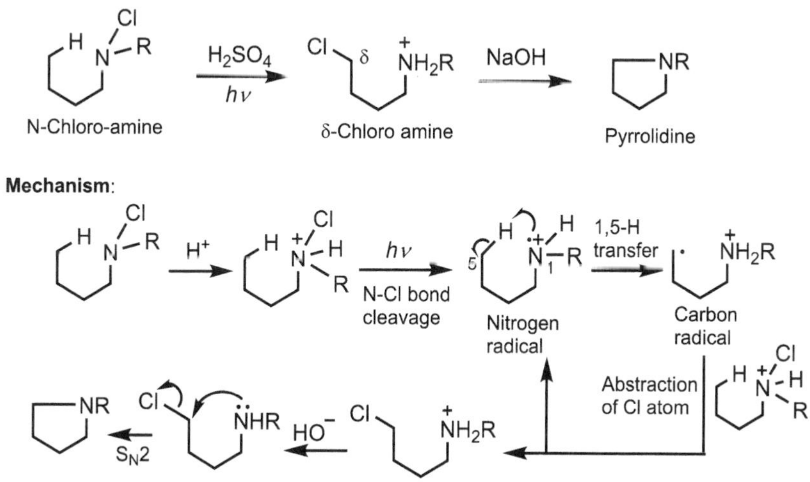


FIG. 12.18 The Hofmann–Löffler–Freytag reaction for the synthesis of pyrrolidines from N-halogenated amines.

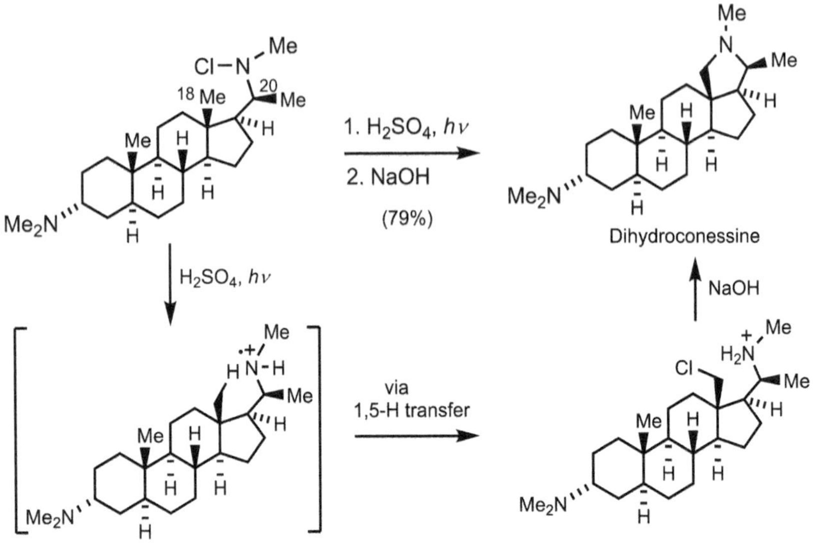


**FIG. 12.19** Application of the Hofmann–Löffler–Freytag reaction in a synthesis of the steroidal alkaloid derivative dihydroconessine.

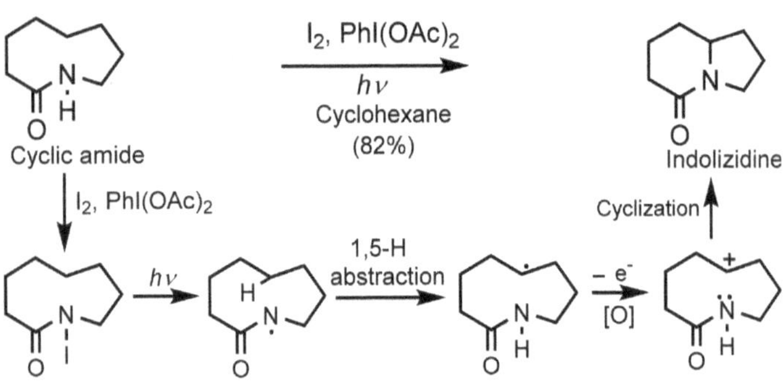


**FIG. 12.20** A modification of the Hofmann–Löffler–Freytag reaction using N-iodo compound in a synthesis of an indolizidine.

A modification of the Hofmann–Löffler–Freytag reaction involves the N-iodo compound to avoid the harshly acidic conditions. The N-iodo compound can be generated by the reaction of an amine derivative with $I_2$ in the presence of a hypervalent iodine(III) compound $PhI(OAc)_2$ (iodobenzene diacetate).[43] Irradiation of the reaction mixture gives the N-radical that promotes remote 1,5-H atom abstraction. The reaction proceeds well with amides, sulphonamides or phosphoramidates.[47] Fig. 12.20 illustrates an example with a cyclic amide substrate. The N-radical generated by the photolytic cleavage of the N—I bond participates in 1,5-transannular H abstraction to give an indolizidine, after one electron oxidation and cyclization.[48]

### 12.3.2 The Barton reaction

Remote functionalizaion is also possible using oxygen-centred radicals.[49] The Barton reaction[50] is the photolytic conversion of organic nitrites into nitroso compounds or oximes via oxygen-centred radicals that promote remote 1,5-H atom transfer. The nitrites (usually prepared from the alcohols) absorb weakly in the region 320–380 nm; thus the reaction can be performed using the wavelengths greater than 300 nm, thereby avoiding the side reactions induced by the more energetic shorter wavelength radiation.

The Barton reaction has found extensive applications in functionalizing the angular methyl groups in steroids.[51] Fig. 12.21A illustrates the use of the Barton

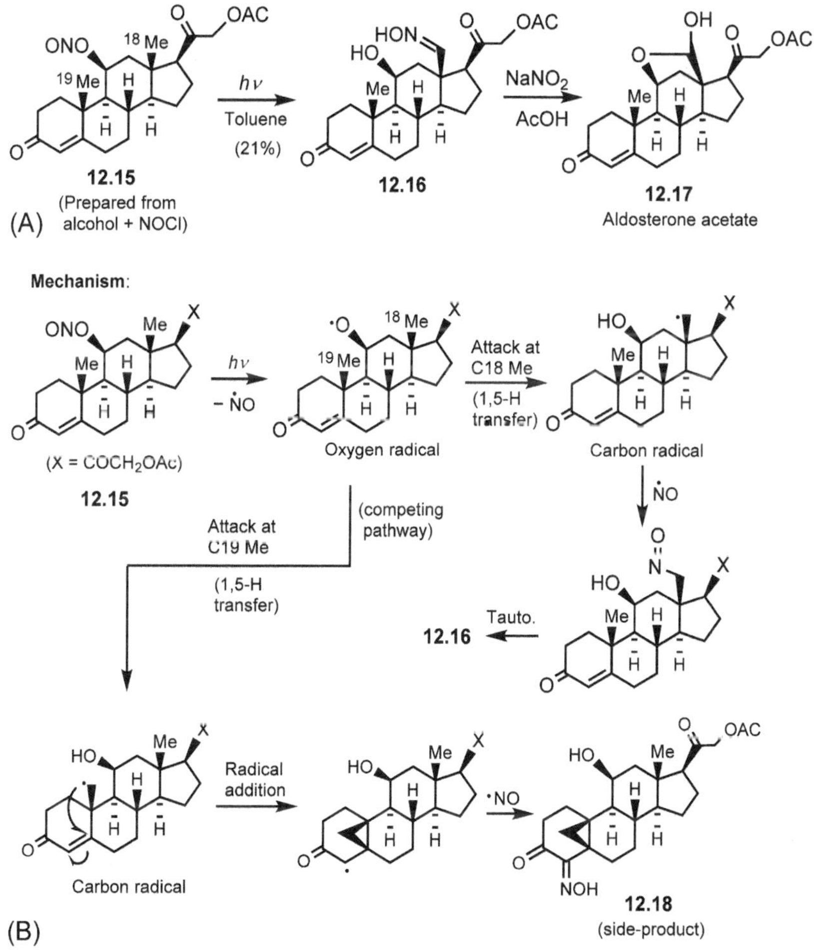


FIG. 12.21 (A) Application of the Barton reaction as a key step in the synthesis of a steroid hormone aldosterone acetate and (B) mechanism of the reaction showing the formation of the desired oxime **12.16** along with a side-product **12.18**.

reaction as a key step in the synthesis of a steroid hormone aldosterone (as acetate) **12.17**.[52] Irradiation of the nitrite **12.15** (derived from the reaction of the corresponding alcohol with nitrosyl chloride) gives the oxime **12.16**, which is hydrolyzed with nitrous acid to furnish aldosterone-21-acetate **12.17** via a rapid cyclization of the intermediate γ-hydroxy ketone (not shown).

The mechanism of the reaction is shown in Fig. 12.21B. Photolysis of the nitrite **12.15** forms an alkoxy radical and nitrogen monoxide. Subsequent attack of the alkoxy radical at C18 methyl group gives a carbon radical by 1,5-H abstraction. The carbon radical then combines with nitrogen monoxide to produce the nitroso compound that tautomerizes to the oxime **12.16**. A side-product **12.18** is also obtained via an alternative 1,5-H abstraction from the C19 methyl group.

The alkoxy radical can also be generated by photolysis of a hypoiodite (obtained by treatment of the corresponding alcohol with $I_2$ and $PhI(OAc)_2$). This is illustrated with the synthesis of a tetrahydrofuran in Fig. 12.22.[53]

If 1,5-H abstraction is unfavourable or not possible, a β-cleavage of the alkoxy radical can occur. Fig. 12.23 shows that the alkoxy radical generated from the compound **12.19** undergoes a β-cleavage (since no H atom is placed at the 5-position for 1,5-H abstraction) to give a carbonyl function and a carbon radical that undergoes β-fragmentation to give the lactone **12.20**, used in a synthesis of 8-deoxyvernolepin.[54]

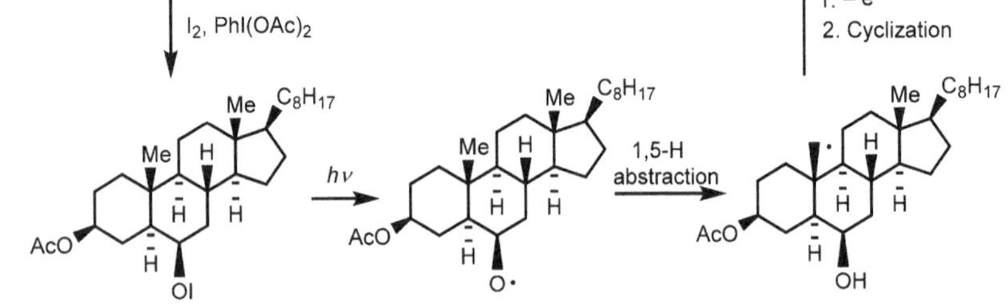


**FIG. 12.22** A method for alkoxy radical generation by photolysis of the hypoiodite, used in the synthesis of a tetrahydrofuran.

**Problem 12.8** Explain the following observation:

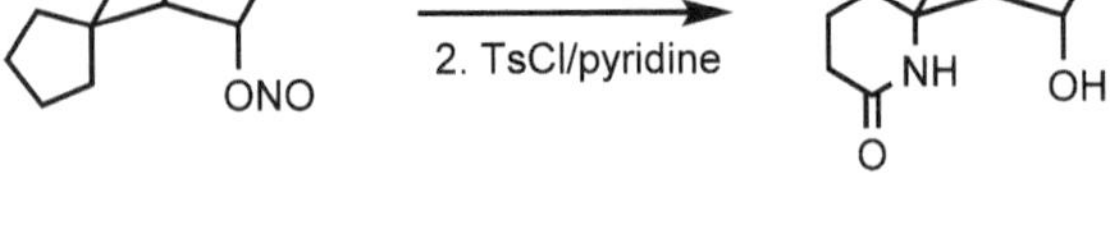

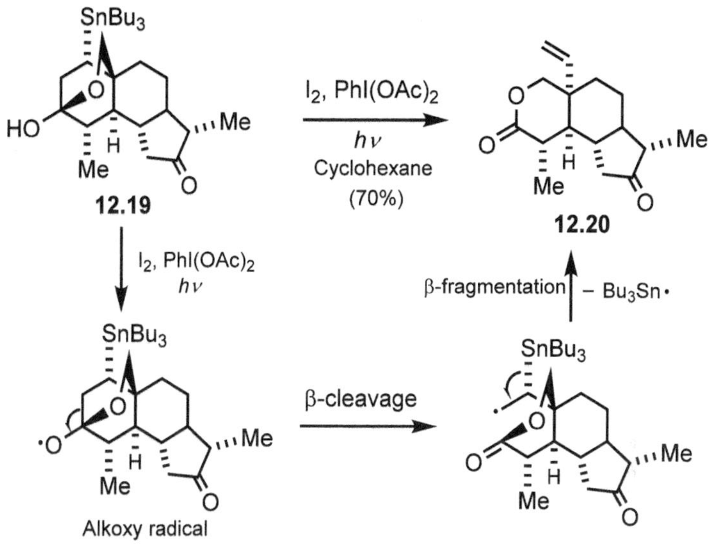


FIG. 12.23 A synthetic route involving β-cleavage of an alkoxy radical.

## 12.4 Stereochemistry of photochemical rearrangements

Two well-studied photochemical rearrangements are the cyclohexadienone rearrangement and the di-π-methane rearrangement. A significant stereochemical feature of these rearrangements is the degree of stereospecificity in their reactions.

### 12.4.1 Cyclohexadienone rearrangement

A classic photochemical cyclohexadienone rearrangement is the rearrangement of the sesquiterpene α-santonin to the cyclopropyl ketone lumisantonin in a nonnucleophilic solvent (Fig. 12.24).[55] The rearrangement involves an inversion of configuration at C4. The product stereochemistry can be drawn with

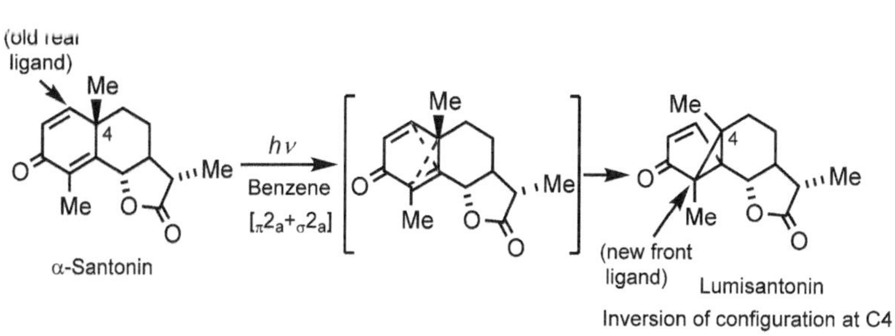


FIG. 12.24 Photochemical cyclohexadienone rearrangement of α-santonin.

the help of the rules in Fig. 2.11. Note that an old rear ligand is replaced by a new front ligand at C4 stereocentre. This result can be attributed formally to concerted bond reorganization by an allowed $[_\pi2_a + _\sigma2_a]$ or $[_\pi2_a + _\sigma2_a + _\pi2_a]$ (taking into account the second double bond) process.[56] It is of note that the other allowed $[_\pi2_s + _\sigma2_s]$ process will lead to a severely strained *trans*-fused 5-3 bicyclic system.

The rearrangement has been extensively studied by Zimmerman, and the Zimmerman mechanism to rationalize the inversion of configuration at the migrating centre is shown in Fig. 12.25A.[57, 58] The reaction is believed to involve a $(n,\pi^*)$ triplet when 3-5 bonding occurs in the excited state. Thereafter, the excitation energy is lost by $\pi^*$–$n$ demotion to give the zwitterion **12.21**, which then rearranges in the ground state by a [1,4]C shift to give the product. The observed stereospecificity is thus explained in terms of the concerted ground state [1,4]C shift that occurs suprafacially with inversion of configuration (see Fig. 4.24B).

An experimental support in favour of the Zimmerman mechanism is shown in Fig. 12.25B.[57, 59] When the zwitterion **12.22** (similar to **12.21**) is generated thermally from a diarylbicyclo[3.1.0]hexanone with a base (KOBu-*t*), it rearranges stereospecifically with inversion of configuration by a suprafacial [1,4]C shift in the allyl cation system. Note that *exo* substituent remains *exo* as a result of inversion of configuration (cf. Fig. 8.50).

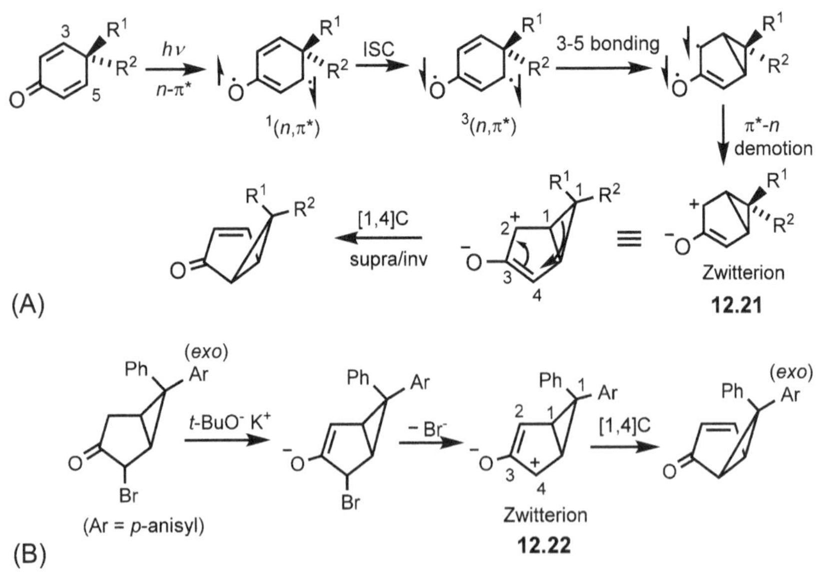


**FIG. 12.25** (A) Zimmerman mechanism for the photochemical cyclohexadienone rearrangement and (B) experimental evidence in favour of stereochemical inversion at the migrating centre.

**Problem 12.9** Suggest a plausible mechanism for the following photochemical rearrangement:

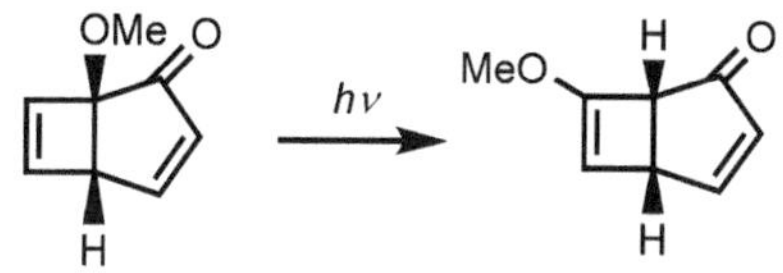


## 12.4.2 Di-π-methane rearrangement

The photochemical rearrangement of 1,4-dienes in which two π systems are separated by an $sp^3$ carbon atom (methane-like) is called the di-π-methane rearrangement.[60–62] The rearrangement produces a vinylcyclopropane. The π systems can be a double bond, an aromatic ring or a carbonyl group. Fig. 12.26A shows a di-π-methane rearrangement of the acyclic substrate **12.23** which gives the vinylcyclopropane **12.24**.[63] Isolated double bonds absorb light below 200 nm; extended conjugation of one of the double bonds by phenyl groups shifts absorption into an experimentally accessible region of the UV spectrum.

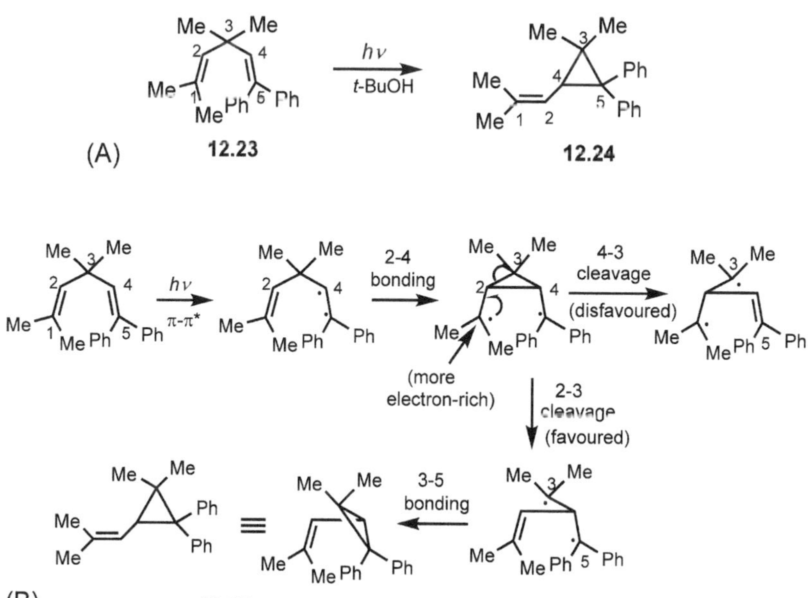


**FIG. 12.26** (A) Di-π-methane rearrangement of an acyclic 1,4-diene and (B) rationalization of regioselectivity of the reaction.

The transformation can be formulated as shown in Fig. 12.26B. After π–π* excitation, an initial 2-4 bonding forms a cyclopropyl diradical, which then fragments by 2-3 cleavage to give a new 1,3-diradical. Finally, 3-5 bonding gives the cyclopropane product. While this scheme is useful in predicting the products, the diradical structures are not necessarily the intermediates in the reaction. The substitution at C3 promotes the di-π-methane rearrangement. The mechanism also explains the regioselectivity of the rearrangement. Cleavage of the cyclopropyl diradical can proceed in two possible pathways for unsymmetrical 1,4-dienes. However, the cleavage involving the more electron-rich radical centre (bearing two Me groups) is favoured. The participation of the other radical centre (delocalized by two Ph groups, less electron-rich) in the bond cleavage is disfavoured.

In general, the di-π-methane rearrangement can proceed via a singlet excited state on direct irradiation or a triplet excited state in a sensitized reaction. The singlet state reaction is presumably a concerted process which gives rise to the stereochemical features of the reaction for acyclic substrates. One stereochemical aspect refers to the double bond stereochemistry; the double bond that remains uncyclized retains the (*E*) or (*Z*) stereochemistry present in the starting material.[64] This is consistent with a concerted mechanism that excludes any cyclopropyl diradical intermediate with a freely rotating terminal radical. The most important element of stereochemistry is observed when the 1,4-diene substrate contains a stereocentre at the $sp^3$ carbon (C3). Fig. 12.27 depicts the

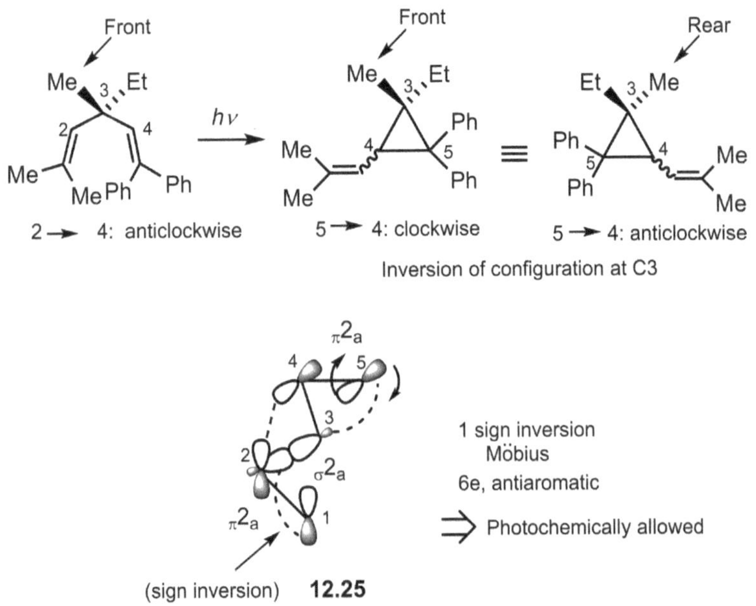


**12.25**

**FIG. 12.27** Stereochemistry of the di-π-methane rearrangement of a chiral substrate showing the inversion of configuration at the stereocentre.

di-π-methane rearrangement of a chiral substrate which proceeds with inversion of configuration at C3.[65] Both diastereomeric cyclopropane products are obtained, each with inversion of configuration at C3. See the use of the rules in Fig. 2.11 to delineate product stereochemistry in which an old ligand C2 is replaced by a new ligand C5 at the stereocentre.

This stereochemical outcome has been rationalized by Zimmerman in terms of a photochemically allowed [${}_{\pi}2_a + {}_{\sigma}2_a + {}_{\pi}2_a$] pericyclic TS (**12.25**). The orbital array involves a single sign inversion between the lobes at C1 and C2 positions, and is therefore of Möbius topology.[66] According to the TS aromaticity approach, a photochemical pericyclic reaction proceeds via an antiaromatic TS. With six electrons, the Möbius TS is antiaromatic indicating an allowed photochemical reaction. As the back lobe at C3 is involved in 3-5 bond formation in the TS (cf. $S_N2$ mechanism), the result is inversion of configuration at C3 in the product.

## 12.5 Asymmetric photochemical reactions[67, 68]

The development of asymmetric photochemistry has been continuing to be a challenge, and the first highly enantioselective photochemical reactions were not developed until recently. Here, we will briefly outline the asymmetric photochemical approaches based on chiral substrates, chiral auxiliaries, chiral photocatalysts and dual catalysis. Asymmetric photochemistry using chiral supramolecular assemblies[69] and chiral crystalline matrices[70] has also been reported.

### 12.5.1 Chiral substrates

Natural products in a chiral pool can be a good source of enantiomerically pure alkene substrates for asymmetric [2+2] photocycloadditions. For example, the enantiopure cyclopentene **12.26** (derived from the natural product (+)-limonene) undergoes asymmetric photocycloaddition with the cyclopentenone **12.27** to give a mixture of head-to-tail (HT) and head-to-head (HH) tricyclic ketones (Fig. 12.28).[71] The reaction is π-facially selective, with the preferential attack of the enone from the face opposite to the isopropyl group on the alkene. The HT isomer was used in a synthesis of the isodaucane sesquiterpene (+)-alphanamol I.

Asymmetric intramolecular enone photocycloadditions are possible when the two alkene units are linked by a short tether. These can provide enantiomerically enriched cyclobutanes as a result of π-facial selectivity involved. Fig. 12.29 shows that the butenolide **12.28** undergoes an intramolecular photocycloaddition to give the tetracyclic lactone **12.29**, used in a synthesis of the tricyclic terpene (+)-stoechospermol.[72]

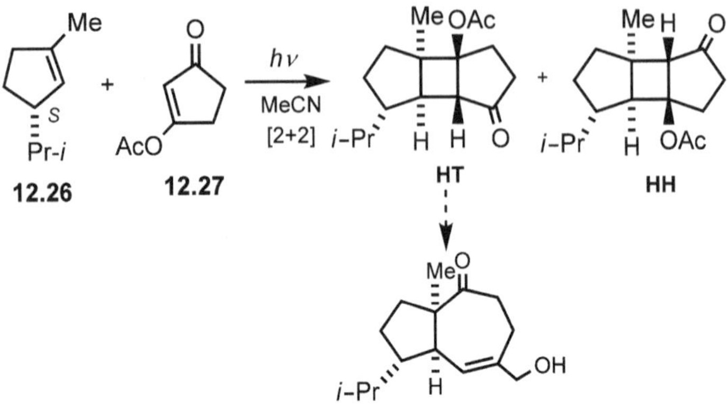


(+)-Alphanamol I

**FIG. 12.28** Asymmetric intermolecular enone photocycloaddition using a chiral cyclopentene, used in a synthesis of the sesquiterpene (+)-alphanamol I.

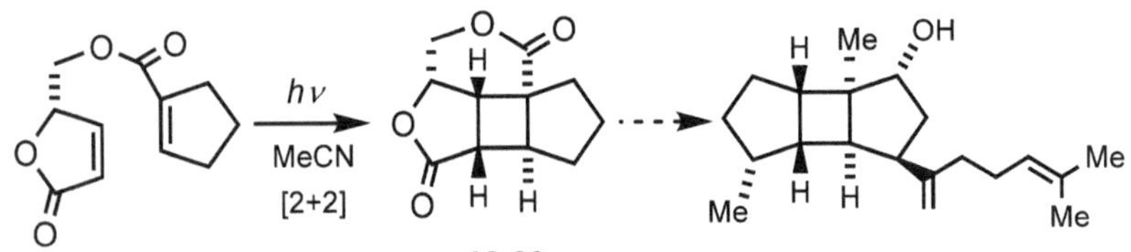


12.28 12.29 (+)-Stoechospermol

**FIG. 12.29** Asymmetric intramolecular enone photocycloaddition.

## 12.5.2 Chiral auxiliaries

In asymmetric enone photocycloadditions, chiral auxiliaries can be introduced in the enone or the alkene component. Fig. 12.30A shows asymmetric photocycloaddition of a cyclopentenone and a cyclohexenone using the chiral auxiliary 8-phenylmenthol.[73] The auxiliary is attached to the enone through ester linkage. The reaction gives a low-to-moderate diastereoselectivity (de 18%–56%) in the formation of cyclobutanes. Attack of the alkene occurs preferentially from the upper face of the enone as the lower face is shielded by a phenyl group on the auxiliary. Removal of the auxiliary by hydrolysis will give the enantiomeric product. The diasteroselectivity can be improved by the attachment of a second chiral auxiliary (e.g. (−)-menthol) to the alkene as a result of double asymmetric induction. The selectivity depends on whether the two auxiliaries are 'matched' or 'mismatched'.[74]

An asymmetric intermolecular photocycloaddition with high diastereoselectivity is shown in Fig. 12.30B.[75] The photocycloaddition of a menthone ketal containing an enone moiety with cyclopentene gives the tricyclic ketone with 90% de.

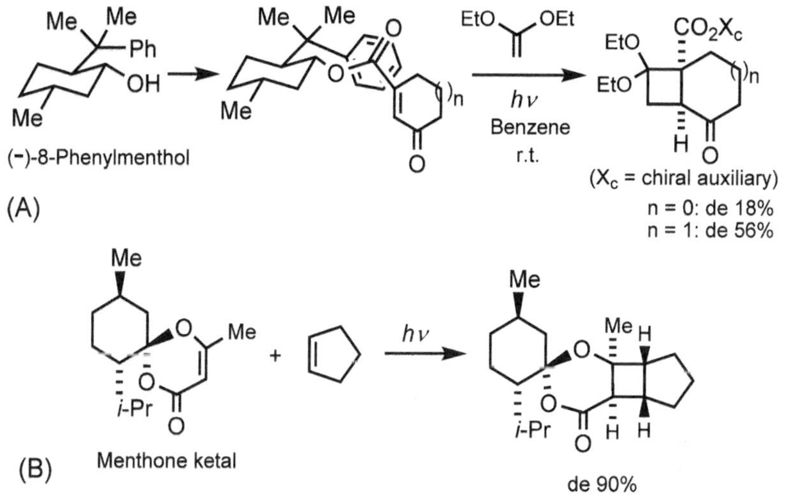


**FIG. 12.30** Asymmetric enone photocycloadditions using (A) 8-phenylmenthol as a chiral auxiliary and (B) a menthone ketal.

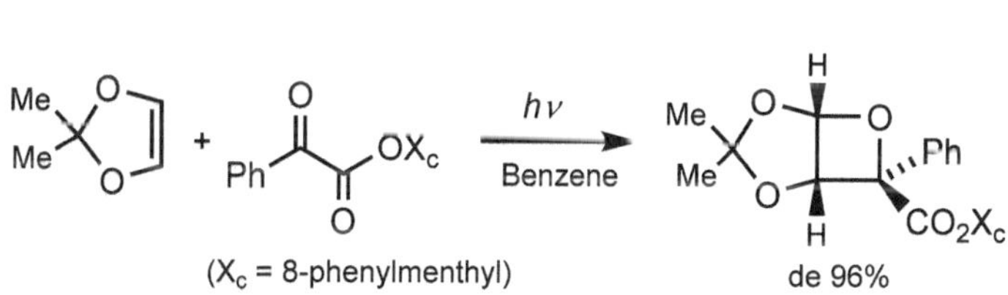


**FIG. 12.31** Asymmetric Paternò–Büchi reaction of phenylglyoxylic ester with a 1,3-dioxole using 8-phenylmenthol as a chiral auxiliary.

Asymmetric induction is also possible in the Paternò–Büchi reaction using the chiral auxiliary approach. Fig. 12.31 shows an asymmetric Paternò–Büchi reaction of phenylglyoxylate attached to the chiral auxiliary 8-phenylmenthol with the alkene 1,3-dioxole.[76, 77] The reaction gives *endo*-phenyl oxetane with excellent diastereoselectivity. Temperature dependence studies show that the selectivity arises presumably from the second C—C bond formation.[78]

### 12.5.3 Chiral photocatalysts

In a photochemical reaction, a catalyst does not need to accelerate the photoreaction itself because the molecule upon excitation gains sufficient energy to undergo a fast subsequent reaction. To achieve a catalysed asymmetric photoreaction,[68, 79–81] it is necessary that the chiral photocatalyst interacts with the substrate during the excitation step. However, a major problem in asymmetric photocatalysis is that the photoreaction can as well proceed as a background racemic reaction at the selected wavelength without involving the catalyst. This unselective background reaction must therefore be suppressed for successful asymmetric catalysis. Extensive screening of chiral catalysts has been made to develop suitable chiral photocatalysts for highly enantioselective photocycloadditions. Chiral Lewis acids, for example, have been found to be an appropriate photocatalyst for the enantioselective photocycloaddition.

Fig. 12.32A depicts an asymmetric intramolecular enone photocycloaddition of a coumarin in the presence of the chiral Lewis acid photocatalyst **12.30**.[82] This catalyst is a chiral oxazaborolidine ligand in complexation with the Lewis acid $AlBr_3$. The reaction gives the product in good yield (84%) and with high ee (82%). Incorporation of a suitable substituent in the tether can improve the selectivity up to 90% ee.

Fig. 12.32B shows another example with a dihydropyridone, catalysed by the chiral Lewis acid photocatalyst **12.30**.[83] The product is obtained with high enantioselectivity (ee 88%), which was used as a key intermediate in a synthesis of the quinolizidine alkaloid (+)-lupinine.

The catalysed asymmetric photocycloaddition of a simple enone substrate is difficult because of the possible uncatalysed background racemic reaction via the ($n$,$\pi$*) triplet excited state at the irradiation wavelength. But the substrate–Lewis acid catalyst complex causes an extensive bathochromic shift due to Lewis acid coordination to the enone and has a very strong absorption for the allowed $\pi$–$\pi$* transition. Thus the light is absorbed exclusively by the complex and the asymmetric reaction will then proceed from this complex.

A model for enantioselectivity of the asymmetric photocycloaddition is shown in Fig. 12.33 using the coumarin substrate.[82, 83] The coordination of carbonyl oxygen to the Lewis acidic boron centre of the catalyst as well as a hydrogen bond between the α-H and the O atom of the oxazaborolidine ring fix the

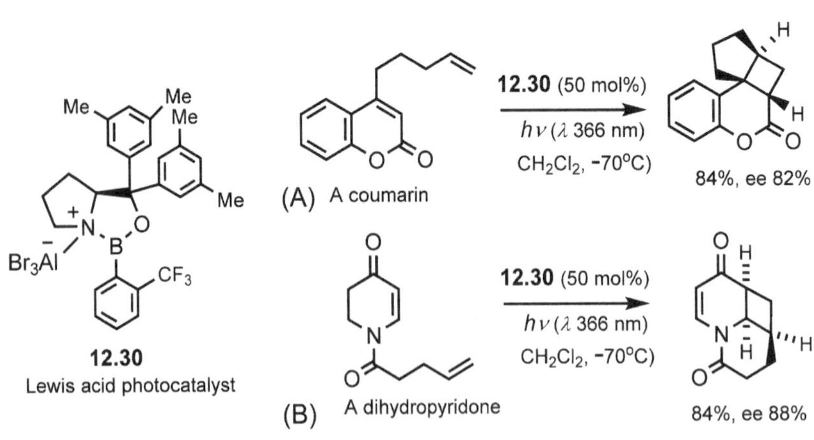


**FIG. 12.32** Asymmetric Lewis acid catalysis of intramolecular enone photocycloaddition of (A) a coumarin and (B) a dihydropyridone.

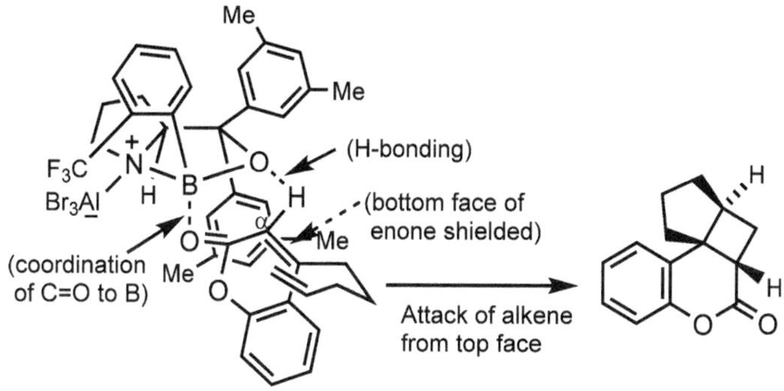


**FIG. 12.33** Model for enantioselectivity in the intramolecular coumarin photocycloaddition in Fig. 12.32A.

substrate in the substrate-catalyst complex in which the bottom face of the coumarin double bond is shielded. The intramolecular cycloaddition of the alkene then takes place preferentially from the top face to give the product.

## 12.5.4 Dual photoredox/chiral Lewis acid catalysis[84]

The problem of racemic background reaction in the asymmetric [2+2] photocycloaddition can be effectively avoided by using a dual-catalyst system comprising a photoredox catalyst and a stereocontrolling Lewis acid cocatalyst. Fig. 12.34 shows an asymmetric [2+2] enone photocycloaddition using a combination of the photoredox catalyst $Ru(bpy)_3Cl_2$ (see Fig. 11.44) and an optimal chiral lanthanide Lewis acid catalyst, a $Eu(OTf)_3$ complex of proline-valine

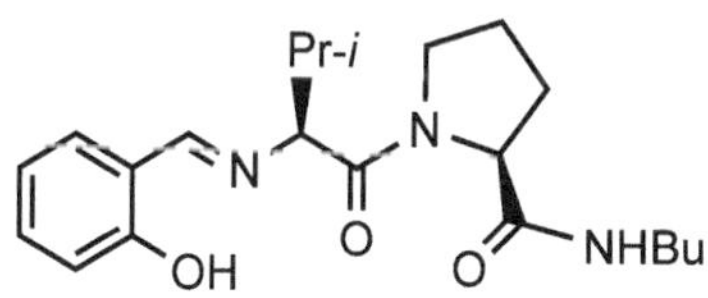


**12.31**

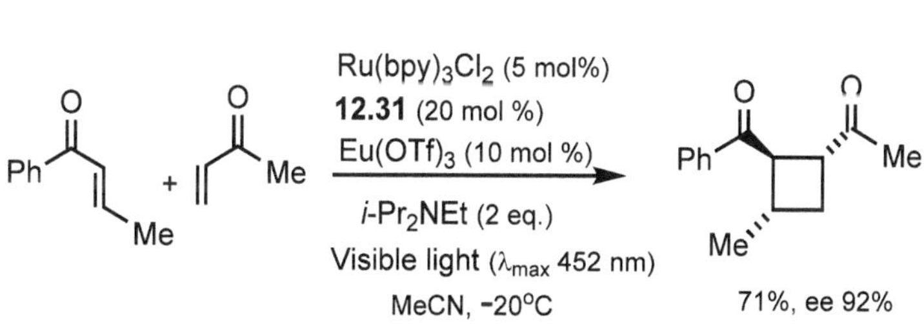


**FIG. 12.34** Asymmetric [2+2] enone photocycloaddition using a dual-catalyst system comprising a photoredox catalyst ($Ru(bpy)_3Cl_2$) and a chiral Lewis acid catalyst ($Eu(OTf)_3$ complex of the chiral ligand **12.31**).

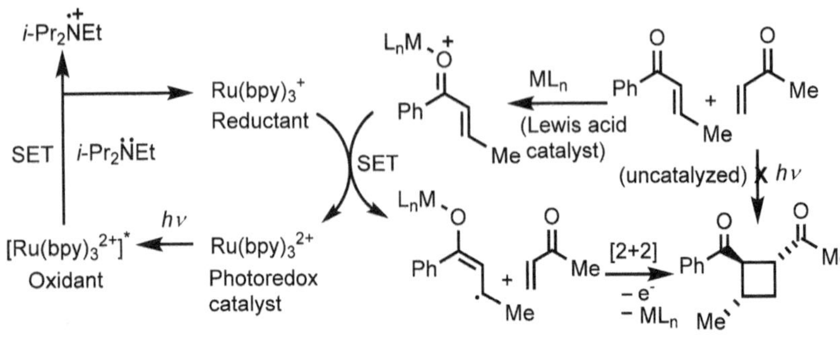


**FIG. 12.35** Dual catalysis mechanism of the asymmetric [2+2] enone photocycloaddition in Fig. 12.34.

Schiff base ligand **12.31**.[85] The reaction gives the cyclobutane product in good yield and with high enantioselectivity.

The dual-catalysis mechanism is depicted in Fig. 12.35.[85] Photoexcitation of the photoredox catalyst by visible light results in a photoexcited state $[Ru(bpy)_3^{2+}]^*$ that removes a single electron from the amine ($i$-$Pr_2NEt$) to form the active reductant $Ru(bpy)_3^+$. Then a single electron transfer (SET) to the Lewis acid-activated aryl enone initiates the [2+2] cycloaddition. One-electron oxidation of the adduct leads to the product and regenerates the Lewis acid catalyst ($ML_n$). Notably, the process prevents the uncatalysed background reaction since the enone does not absorb the visible light ($\lambda_{max}$ 452 nm).

## References


1. Baldwin, S. W. In *Organic Photochemistry*; Padwa, A., Ed.; Vol. 5; Marcel Dekker: New York, 1981; p. 123.
2. Crimmins, M. T. *Chem. Rev.* **1988**, *88*, 1453.
3. Schuster, D. I.; Lem, G.; Kaprinidis, N. A. *Chem. Rev.* **1993**, *93*, 3.
4. Hoffmann, N. *Chem. Rev.* **2008**, *108*, 1052.
5. Yamazaki, H.; Cvetanovic, R. J. *J. Am. Chem. Soc.* **1969**, *91*, 520.
6. Saltiel, J.; Ng Lim, L.-S. *J. Am. Chem. Soc.* **1969**, *91*, 5404.
7. Becker, D.; Haddad, N. *Org. Photochem.* **1989**, *10*, 1.
8. Hammond, G. S.; Turro, N. J.; Fischer, A. *J. Am. Chem. Soc.* **1961**, *83*, 4674.
9. Schenck, G. O.; Steinmetz, R. *Chem. Ber.* **1963**, *96*, 520.
10. Barborak, J. C.; Watts, L.; Pettit, R. *J. Am. Chem. Soc.* **1966**, *88*, 1328.
11. Crimmins, M. T.; Reinhold, T. L. *Org. React.* **1993**, *44*, 296.
12. Bach, T. *Synthesis* **1998**, 683.
13. Corey, E. J.; Mitra, R. B.; Uda, H. *J. Am. Chem. Soc.* **1964**, *86*, 485.
14. Corey, E. J.; Bass, J. D.; LeMahieu, R.; Mitra, R. B. *J. Am. Chem. Soc.* **1964**, *86*, 5570.
15. Wender, P. A.; Lechleiter, J. C. *J. Am. Chem. Soc.* **1978**, *100*, 4321.
16. Hoffman, R.; Wells, P.; Morrison, H. *J. Org. Chem.* **1971**, *36*, 102.
17. Pirrung, M. C. *J. Am. Chem. Soc.* **1981**, *103*, 82.

18. Birch, A. M.; Pattenden, G. *J. Chem. Soc. Chem. Commun.* **1980**, 1195.
19. Kalyanasundaram, K. *Coord. Chem. Rev.* **1982**, *46*, 159.
20. Juris, A.; Balzani, V.; Barigelletti, F.; Campagna, S.; Belser, P.; von Zelewsky, A. *Coord. Chem. Rev.* **1988**, *84*, 85.
21. Ischay, M. A.; Anzovino, M. E.; Du, J.; Yoon, T. P. *J. Am. Chem. Soc.* **2008**, *130*, 12886.
22. Du, J.; Yoon, T. P. *J. Am. Chem. Soc.* **2009**, *131*, 14604.
23. Porco, J. A.; Schreiber, S. L. In *Comprehensive Organic Synthesis*; Trost, B. M., Fleming, I., Eds.; Vol. 5; Pergamon Press: Oxford, 1991; p. 151.
24. Bach, T. *Synthesis* **2000**, 1699.
25. Turro, N. J. *Modern Molecular Photochemistry*; Menlo Park: Benjamin-Cummings, 1978.
26. Kopecky, J. *Organic Photochemistry, A Visual Approach*; VCH Publishers: New York, 1992; p. 126.
27. Griesbeck, A. G.; Stadtmüller, S. *J. Am. Chem. Soc.* **1990**, *112*, 1281.
28. Griesbeck, A. G.; Mauder, H.; Stadtmüller, S. *Acc. Chem. Res.* **1994**, *27*, 70.
29. Bach, T.; Brummerhop, H. *Angew. Chem. Int. Ed. Engl.* **1998**, *37*, 3400.
30. Bach, T.; Brummerhop, H.; Harms, K. *Chem. A Eur. J.* **2000**, *6*, 3838.
31. Schreiber, S. L. *Science* **1985**, *227*, 857.
32. Schreiber, S. L.; Hoveyda, A. H.; Wu, H.-J. *J. Am. Chem. Soc.* **1983**, *105*, 660.
33. Schreiber, S. L.; Hoveyda, A. H. *J. Am. Chem. Soc.* **1984**, *106*, 7200.
34. Srinivasan, R. *J. Am. Chem. Soc.* **1968**, *90*, 4498.
35. Corey, E. J.; Streith, J. *J. Am. Chem. Soc.* **1964**, *86*, 950.
36. Irie, M. *Chem. Rev.* **2000**, *100*, 1683. 100, 1685.
37. Blattmann, H.-R.; Schmidt, W. *Tetrahedron* **1970**, *26*, 5885.
38. Jacobs, H. J. C.; Havinga, E. *Adv. Photochem.* **1979**, *11*, 305.
39. Schlatmann, J. L. M. A.; Pot, J.; Havinga, E. *Rec. Trav. Chim.* **1964**, *83*, 1173.
40. Mallory, F. B.; Mallory, C. W. *Org. React.* **1984**, *30*, 1.
41. Laarhoven, W. H. *Org. Photochem.* **1989**, *10*, 163.
42. Mehta, G.; Shah, S. R.; Venkateswarhu, Y. *Tetrahedron* **1994**, *50*, 11729.
43. Majetich, G.; Wheless, K. *Tetrahedron* **1995**, *51*, 7095.
44. Togo, H.; Katohgi, M. *Synlett* **2001**, 565.
45. Wolff, M. E. *Chem. Rev.* **1963**, *63*, 55.
46. Corey, E. J.; Hertler, W. R. *J. Am. Chem. Soc.* **1959**, *81*, 5209.
47. Francisco, C. G.; Herrera, A. J.; Suárez, E. *Tetrahedron Asymmetry* **2000**, *11*, 3879.
48. Dorta, R. L.; Francisco, C. G.; Suárez, E. *J. Chem. Soc. Chem. Commun.* **1989**, 1168.
49. Cekovic, Z. *Tetrahedron* **2003**, *59*, 8073.
50. Barton, D. H. R. *Pure Appl. Chem.* **1968**, *16*, 1.
51. Coombes, L. G.; Sutherland, I. O., Eds., Vol. 2; *Comprehensive Organic Chemistry*; Pergamon Press: Oxford, 1979; p. 356.
52. Barton, D. H. R.; Beaton, J. M. *J. Am. Chem. Soc.* **1961**, *83*, 4083.
53. de Armas, P.; Concepción, J. I.; Francisco, C. G.; Hernández, R.; Salazar, J. A.; Suárez, E. *J. Chem. Soc. Perkin Trans. I* **1989**, 405.
54. Hernández, R.; Velázquez, S. M.; Suárez, E.; Rodríguez, M. S. *J. Org. Chem.* **1994**, *59*, 6395.
55. Asher, J. D. M.; Sim, G. A. *J. Chem. Soc.* **1965**, 1584.
56. Woodward, R. B.; Hoffmann, R. *The Conservation of Orbital Symmetry*; Verlag Chemie: Weinheim, 1971. Academic Press: New York, 1970.
57. Zimmerman, H. E.; Crumrine, D. S. *J. Am. Chem. Soc.* **1968**, *90*, 5612.
58. Zimmerman, H. E.; Crumrine, D. S.; Dopp, D.; Huyffer, P. S. *J. Am. Chem. Soc.* **1969**, *91*, 434.
59. Brennan, T. M.; Hill, R. K. *J. Am. Chem. Soc.* **1968**, *90*, 5614.

60. Zimmerman, H. E. In *Rearrangements in Ground and Excited States*; de Mayo, P., Ed.; Vol. 3; Academic Press: New York, 1980; p. 131.
61. Zimmerman, H. E. *Acc. Chem. Res.* **1982**, *15*, 312.
62. Zimmerman, H. E.; Armesto, D. *Chem. Rev.* **1996**, *96*, 3065.
63. Zimmerman, H. E.; Pratt, A. C. J. *Am. Chem. Soc.* **1970**, *92*, 1407. 92, 6259.
64. Zimmerman, H. E.; Pratt, A. C. *J. Am. Chem. Soc.* **1970**, *92*, 1410. 92, 6267.
65. Zimmerman, H. E.; Robbins, J. D.; McKelvey, R. D.; Samuel, C. J.; Sousa, L. R. *J. Am. Chem. Soc.* **1974**, *96*, 4630.
66. Zimmerman, H. E. *Acc. Chem. Res.* **1971**, *4*, 272.
67. Inoue, Y. *Chem. Rev.* **1992**, *92*, 741.
68. Brimioulle, R.; Lenhart, D.; Maturi, M. M.; Bach, T. *Angew. Chem. Int. Ed. Engl.* **2015**, *54*, 3872.
69. Nishijima, M.; Wada, T.; Mori, T.; Pace, T. C. S.; Bohne, C.; Inoue, Y. *J. Am. Chem. Soc.* **2007**, *129*, 3478.
70. Addadi, L.; van Mil, J.; Lahav, M. *J. Am. Chem. Soc.* **1982**, *104*, 3422.
71. Hansson, T.; Wickberg, B. *J. Org. Chem.* **1992**, *57*, 5370.
72. Tanaka, M.; Tomioka, K.; Koga, K. *Tetrahedron* **1994**, *50*, 12829. 50, 12843.
73. Herzog, H.; Koch, H.; Scharf, H.-D.; Runsink, J. *Tetrahedron* **1986**, *42*, 3547.
74. Herzog, H.; Koch, H.; Scharf, H.-D.; Runsink, J. *Chem. Ber.* **1987**, *120*, 1737.
75. Denmuth, M.; Palomer, A.; Shima, H.-D.; Dey, A. K.; Krüger, C.; Tsay, Y.-H. *Angew. Chem. Int. Ed. Engl.* **1986**, *25*, 1117.
76. Buschmann, H.; Scharf, H.-D.; Hoffmann, N.; Esser, P. *Angew. Chem. Int. Ed. Engl.* **1991**, *30*, 477.
77. Nehrings, A.; Scharf, H.-D.; Runsik, J. *Angew. Chem. Int. Ed. Engl.* **1985**, *24*, 877.
78. Buschmann, H.; Scharf, H.-D.; Hoffmann, N.; Plath, M. W.; Runsink, J. *J. Am. Chem. Soc.* **1989**, *111*, 5367.
79. Inoue, Y. *Nature* **2005**, *436*, 1099.
80. Wessig, P. *Angew. Chem. Int. Ed. Engl.* **2006**, *45*, 2168.
81. Neier, R. *Science* **2014**, *344*, 368.
82. Guo, H.; Herdtweck, E.; Bach, T. *Angew. Chem. Int. Ed. Engl.* **2010**, *49*, 7782.
83. Brimioulle, R.; Bach, T. *Science* **2013**, *342*, 840.
84. Yoon, T. P. *Acc. Chem. Res.* **2016**, *49*, 2307.
85. Du, J.; Skubi, K. L.; Schultz, D. M.; Yoon, T. P. *Science* **2014**, *344*, 392.

Part IV

# Appendix

# Appendix

# Answers to problems

## 1. Answers to Problems in Chapter 1

**1.1** The effect of increasing size ($F \rightarrow Cl \rightarrow Br \rightarrow I$) is offset by the corresponding increase in the C-halogen bond distance leading to a similar steric effect in the eclipsed conformation of all ethyl halides. Since the staggered conformations do not encounter steric interactions, the energy difference between eclipsed and staggered conformations remains similar.

**1.2** Eclipsing interactions: 1 (H/H) + 2 (Me/H) = $1 + 2 \times 1.4 = 3.8\,\text{kcal mol}^{-1}$.

**1.3** The eclipsing interactions in the conformation at $\phi = 0$ degrees contribute $5.1\,\text{kcal mol}^{-1}$, i.e. 2 (H/H) + 1 (Me/Me) = $5.1\,\text{kcal mol}^{-1}$. Therefore, Me/Me eclipsing $= 5.1 - 2 = 3.1\,\text{kcal mol}^{-1}$.

**1.4** According to von Auwers–Skita's rule,[1] the conformer of higher enthalpy has lower molecular volume. Therefore, gauche conformer has lower molecular volume than anti. This is quite reasonable as gauche is more compact. As a result, the gauche conformer has a higher surface/volume ratio which will lead to greater intermolecular van der Waals attraction. Thus gauche is relatively more stabilized in the liquid phase and has lower energy than that in the vapour phase. This implies that, although anti is more stable even in the liquid phase, the *gauche*–anti enthalpy difference is smaller in the liquid phase compared to that in the gaseous phase.

**1.5** (a) An attractive electrostatic attraction between $Cl^{\delta-}$ and $C^{\delta+}$ of methyl in the gauche conformer is responsible for its preference over the anti slightly.[2]

(b) The electrostatic attraction between $N^+$ and $O^{\delta-}$ strongly stabilizes the gauche conformer.

(c) The gauche conformer is stabilized by n–σ* interaction when X is located antiperiplanar to an oxygen lone pair of $OCH_3$ (see the anomeric effect, Section 3.8).

(d) The two IR bands in each case are due to free OH and intramolecularly H-bonded OH. The intramolecular H-bonding is stronger in the case of chiral (racemic) diastereomer because the increase in torsion angle between two Me's or between Me and OH due to steric effect pushes the two OH groups closer. In contrast, the increase in the torsion angle between two Me's in the *meso* isomer pushes the two OH groups away from each other. The more strongly H-bonded OH for the chiral diastereomer is shifted to lower frequency, thereby causing greater band separation between free OH and H-bonded OH. Note that the *meso* isomer exists as a nonresolvable enantiomeric pair, one enantiomer of which is drawn. Remember that each diastereomer has three staggered conformers (see Fig. 2.32B); here the conformer(s) involving intramolecular H-bonding are only shown.

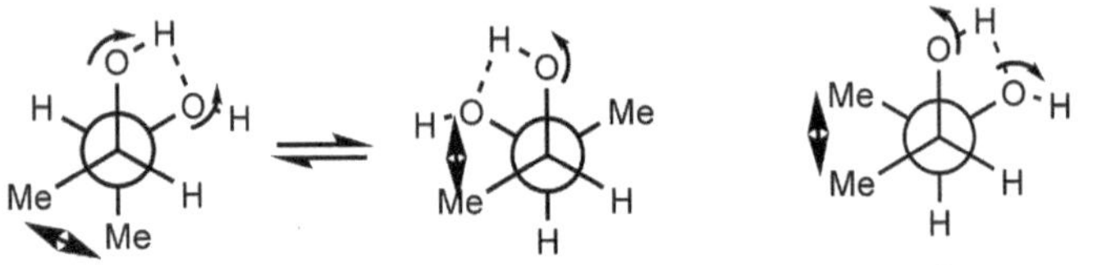

Chiral diastereomer (one enantiomer)

*meso* diastereomer

**1.6** The nine staggered conformers of pentane are

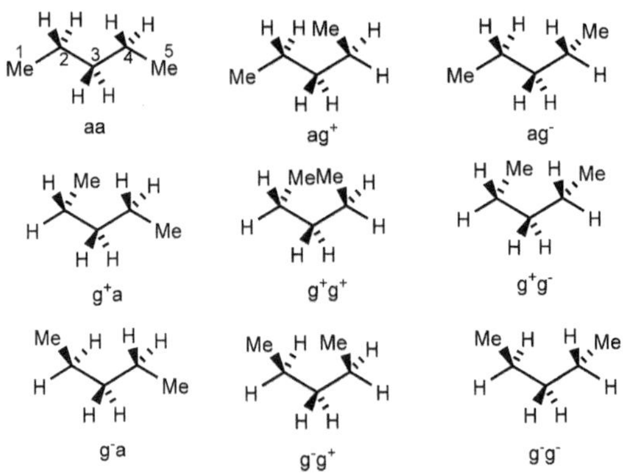


**1.7**

| | Point group | Sym. elements | Sym. operations | Order ($h$) | Sym. no. ($\sigma$) |
|---|---|---|---|---|---|
| (a) | $\mathbf{C_1}$ | $C_1$ | $C_1=E$ | 1 | 1 |
| (b) | $\mathbf{C_2}$ | $C_2$ | $E$, $C_2$ | 2 | 2 |
| (c) | $\mathbf{C_{2v}}$ | $C_2$, $\sigma_v$, $\sigma_v'$ | $E$, $C_2$, $2\sigma_v$ | 4 | 2 |
| (d) | $\mathbf{D_{2h}}$ | $C_2$, $C_2'$, $C_2''$, $\sigma_h$ ($\sigma_v$, $\sigma_v'$, $S_2=i$) | $E$, $C_2$, $2C_2'$, $\sigma_h$, $2\sigma_v$, $i$ | 8 | 4 |
| (e) | $\mathbf{C_{2v}}$ | $C_2$, $\sigma_v$, $\sigma_v'$ | $E$, $C_2$, $2\sigma_v$ | 4 | 2 |
| (f) | $\mathbf{C_{2h}}$ | $C_2$, $\sigma_h$ ($i$) | $E$, $C_2$, $\sigma_h$, $i$ | 4 | 2 |
| (g) | $\mathbf{T_d}$ | 4 $C_3$, 3 $C_2$, 3 $S_4$, 6 $\sigma_d$ | $E$, $8C_3$, $3C_2$, $6S_4$, $6\sigma_d$ | 24 | 12 |
| (h) | $\mathbf{C_{3v}}$ | $C_3$, $\sigma_v$, $\sigma_v'$, $\sigma_v''$ | $E$, $2C_3$, $3\sigma_v$ | 6 | 3 |
| (i) | $\mathbf{D_{3h}}$ | $C_3$, $C_2'$, $C_2''$, $C_2'''$, $\sigma_h$ ($\sigma_v$, $\sigma_v'$, $\sigma_v''$, $S_3$) | $E$, $2C_3$, $3C_2$, $\sigma_h$, $3\sigma_v$, $2S_3$ | 12 | 6 |
| (j) | $\mathbf{D_{3d}}$ | $C_3$, $C_2'$, $C_2''$, $C_2'''$, $\sigma_d$, $\sigma_d'$, $\sigma_d''$ ($S_6$, $i$) | $E$, $2C_3$, $3C_2$, $i$, $2S_6$, $3\sigma_d$ | 12 | 6 |
| (k) | $\mathbf{D_3}$ | $C_3$, $C_2'$, $C_2''$, $C_2'''$ | $E$, $2C_3$, $3C_2$ | 6 | 6 |

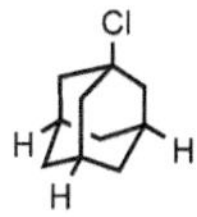


(h) 1-Chloroadamantane

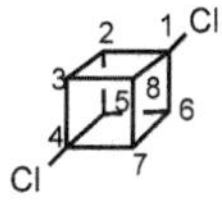


(j) 1,4-Dichlorocubane

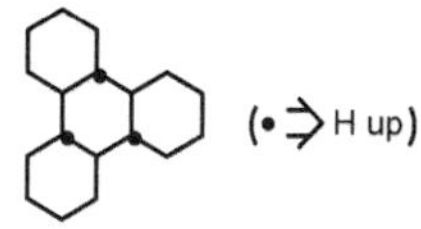


(k) *trans-transoid-trans-transoid-trans*-Perhydrotriphenylene (see Fig. 3.69 for nomenclature)

(see the text for details)

## 2. Answers to Problems in Chapter 2

**2.1** Optical rotation depends on path length and concentration. In polarimetry, optical rotation $\alpha \equiv \alpha \pm 180$. Therefore, measurement at a single concentration cannot differentiate between rotations of +50 degrees and −130 degrees. To make the distinction, the measurement must be made at least at one other concentration (changing concentration is more convenient than varying path length). If the concentration of the solution is made one-half, a compound that is dextrorotatory will give a rotation of +25 degrees but if it is levorotatory, the observed rotation will be −65 degrees. Note that +25 degrees is not equivalent to −65 degrees.

**2.2** Assuming enantiomeric excess (ee) to be the same as optical purity (op), ee = op = (8.11/13.52) × 100% = 60%. Therefore, the dextrorotatory sample has 60% (+) and 40% (±), i.e. 80% (+) and 20% (−).

**2.3** The product is *R*-(−)-PhCHDMe, as shown below.

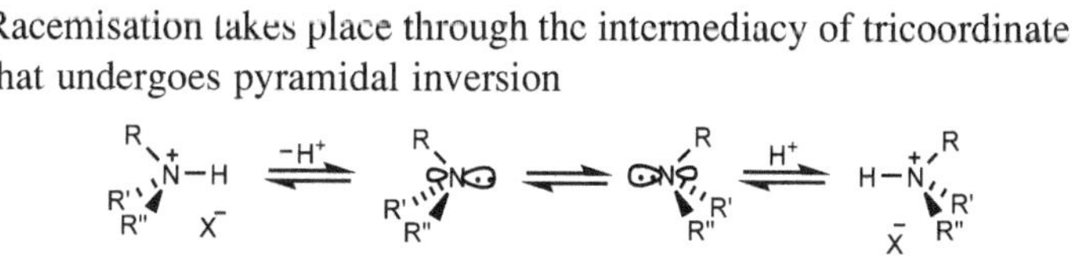


**2.4** Racemisation takes place through the intermediacy of tricoordinate amine that undergoes pyramidal inversion

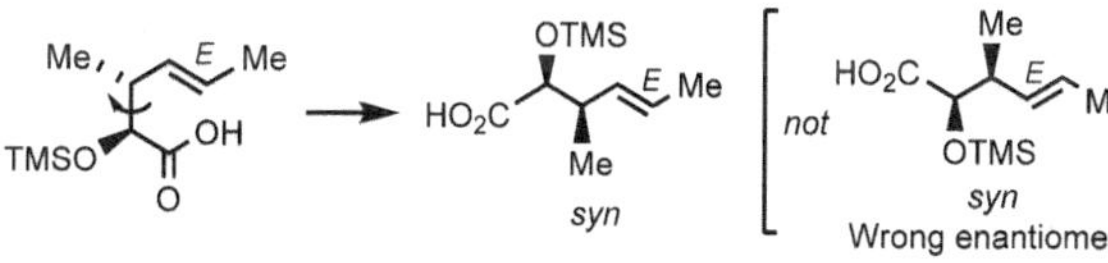


**2.5** (a) *R* (b) *R* (c) *R* (d) 3*S*,4*R* (e) 1*R*,2*S* (f) 3*S*,4*R* (g) *S* (h) *S*

**2.6** The chiral diastereomer is *syn*.

**2.7**

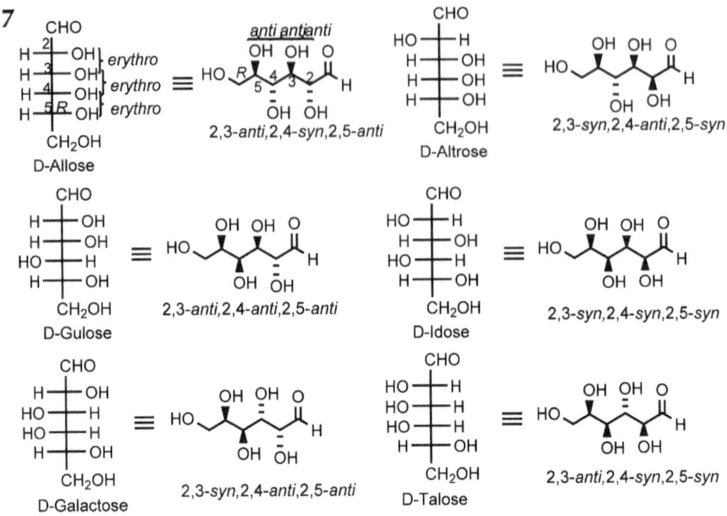


**2.8** In **X**, two rings are noncoplanar due to steric interaction between bulky *ortho* substituents ($NO_2$/$CO_2H$) but **X** is achiral (either ring has a vertical plane of symmetry) and hence the question of resolvability does not arise. In **Y**, two rings are also noncoplanar but **Y** is chiral (each ring lacks a vertical plane of symmetry due to the presence of a *meta* substituent $NO_2$ or $CO_2H$); the steric repulsion between bulky *ortho* substituents ($NO_2$/ $CO_2H$) leads to a high barrier to rotation so that the atropisomers are not interconvertible at room temperature and hence **Y** is resolvable.

**2.9** **A**, **B** and **C** exhibit atropisomerism and are resolvable, whereas **D** is planar and achiral and hence nonresolvable.

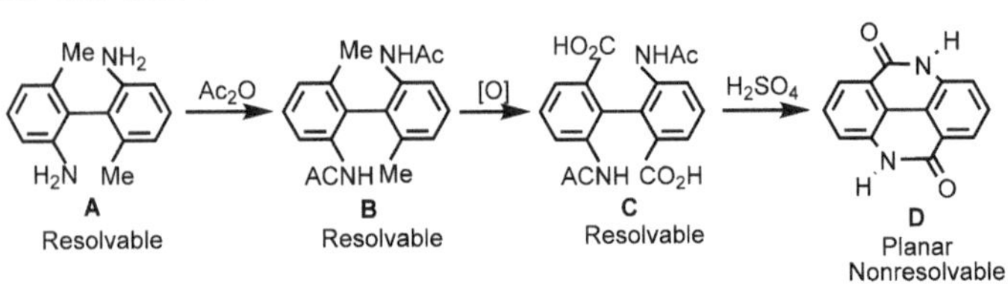


**2.10** The chiral axis passes through C2 and C6 and the geometrical centre of the molecule. The configuration is *R* as shown below.

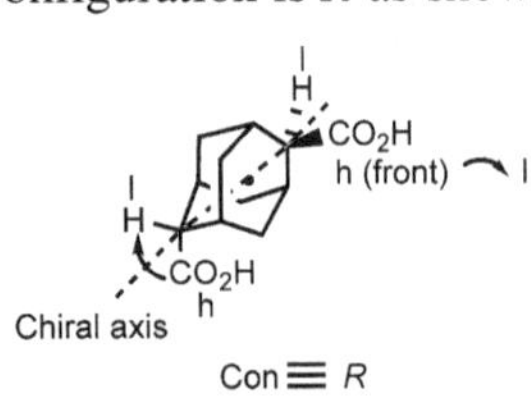

**2.11**

**Two homotopic $CH_2OH$ ligands in D-mannitol:**

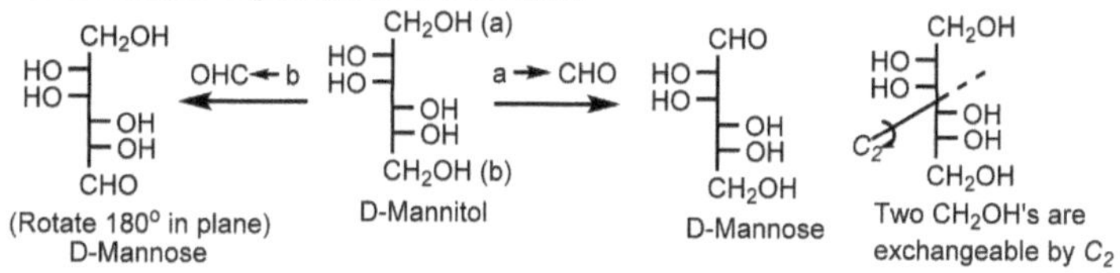


**Two enantiotopic $CH_2OH$ ligands in D-galactitol:**

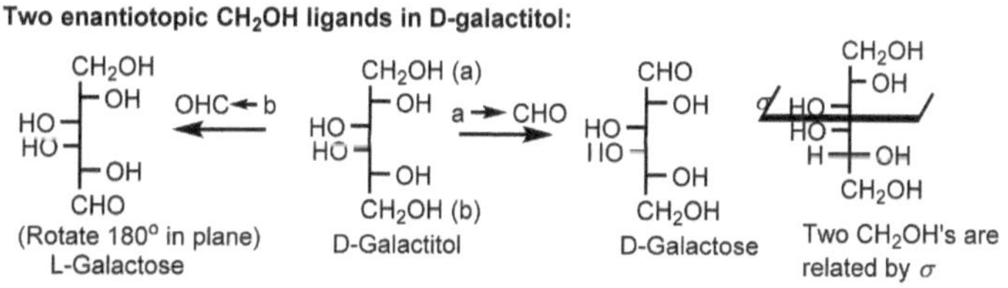


**Two diastereotopic $CH_2OH$ ligands in D-glucitol:**

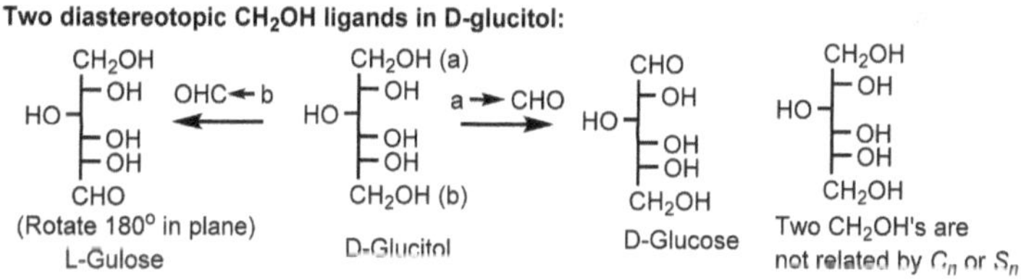


**2.12** In citric acid, C2 and C4 are prochiral. As shown below, $H_a$ and $H_c$ are enantiotopic (related by $\sigma$); similarly, $H_b$ and $H_d$ are enantiotopic. But $H_a$ is diastereotopic with $H_b$ and $H_d$; $H_b$ is diastereotopic with $H_c$. $H_a$ is *pro-S*, $H_b$ is *pro-R*, $H_c$ is *pro-R* and $H_d$ is *pro-S*. The third prochiral centre is C3 which bears two homomorphic $CH_2CO_2H$ ligands, the replacement of either by a new ligand makes C3 a chiral centre. These ligands are enantiotopic (related by $\sigma$). The top $CH_2CO_2H$ is *pro-R* and the bottom $CH_2CO_2H$ is *pro-S*.

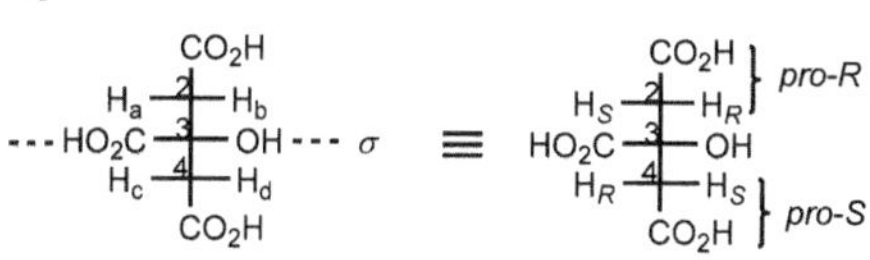


Citric acid

**2.13** This is explained by the substitution criterion as shown below. In the case of the acid, substitution procedure leads to two *meso* diastereomers in which C3 is stereogenic but achirotopic. Thus C3 in the acid is prostereogenic but proachirotopic. However, in the case of the half-ester, substitution gives two chiral diastereomers in which C3 is stereogenic and chirotopic. Thus C3 in the half-ester is prostereogenic and prochirotopic.

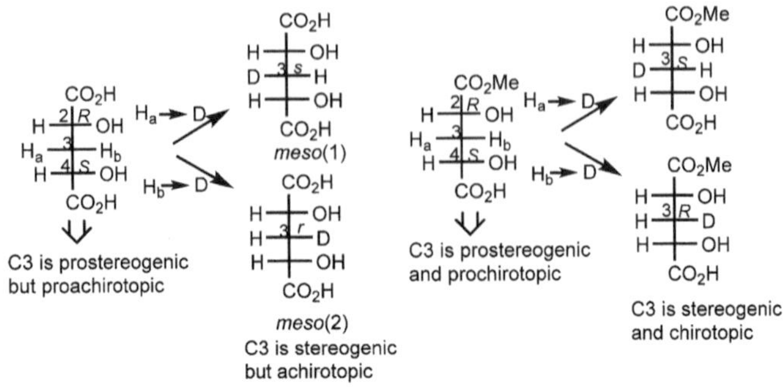


**2.14** In the diester, the alkene protons $H_a$ and $H_b$ are homotopic (exchangeable by $C_2$). They are, therefore, isochronous in [1]H NMR (same chemical shift) and appear as a 2-proton singlet. But in epoxide, $H_a$ and $H_b$ are diastereotopic (not related by $C_n$ or $S_n$). They are anisochronous (different chemical shifts) with the splitting of proton signals. Each proton appears as a doublet due to spin coupling. Note that epoxidation from the front or the rear face gives the same epoxide since two alkene faces in the diester are homotopic.

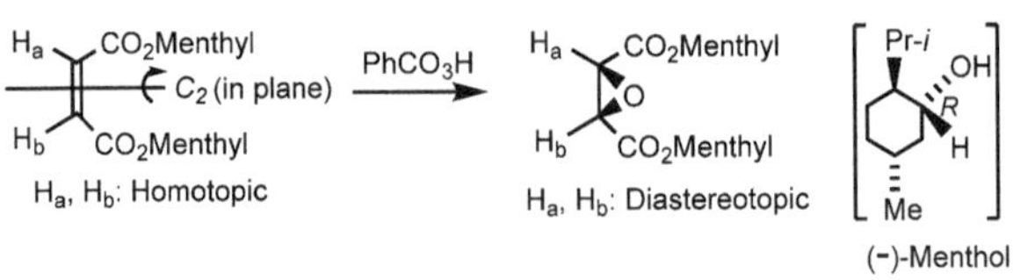


## 3. Answers to Problems in Chapter 3

**3.1** All diastereomers except α-truxilic acid have at least one plane of symmetry (σ) passing through diagonal carbons. The diastereomers are shown below.

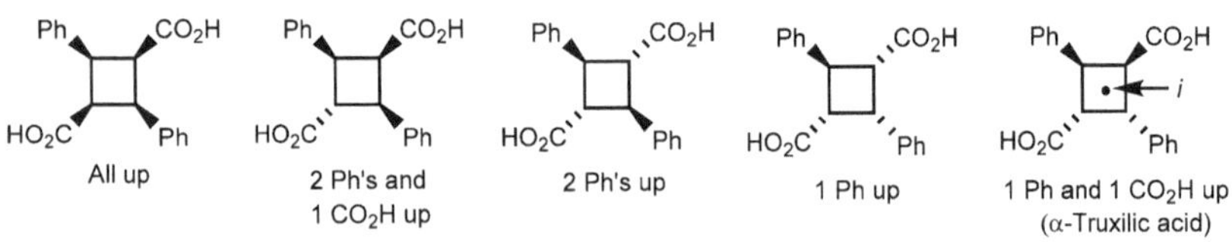


**3.2** (a) 1*s*,4*s* (b) 1*r*,4*r*

**3.3** The 1,3-diaxial interactions in the *cis* and *trans* isomers are shown below.

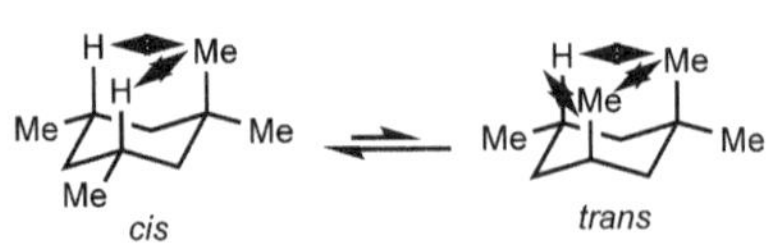


Given: $\Delta H^0_{trans} - \Delta H^0_{cis} = 3.7\,\text{kcal mol}^{-1}$. Therefore, 1,3-diaxial interaction energy:

$$[2\,(\text{Me/H}) + 1\,(\text{Me/Me})] - 2\,(\text{Me/H}) = 3.7\,\text{kcal mol}^{-1}$$
$$\text{or, 1,3-diaxial (Me/Me)} = 3.7\,\text{kcal mol}^{-1}.$$

**3.4** The estimated $A$ value of $i$-Pr is $2.22\,\text{kcal mol}^{-1}$ as shown below.

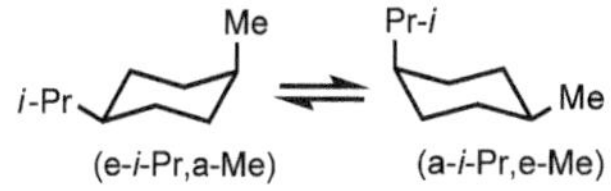


$\Delta G^0 = 0.48$ kcal mol$^{-1}$
or, $A$ ($i$-Pr) - A (Me) = 0.48 kcal mol$^{-1}$
or, $A$ ($i$-Pr) ) = 0.48 + 1.74 = 2.22 kcal mol$^{-1}$

**3.5** The conformational equilibrium is shown below.

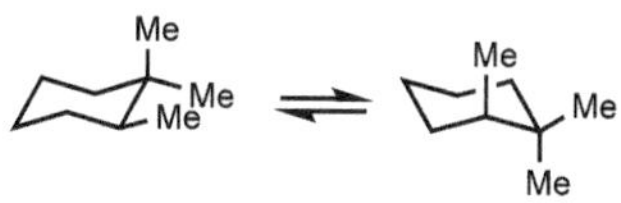


$\Delta G^0 = 1.53$ kcal mol$^{-1}$

By the additivity rule,

$$\Delta G^0 = [2A\,(\text{Me}) + 1\,\text{gb}\,(\text{e}-\text{Me/a}-\text{Me})] \\ - [A\,(\text{Me}) + 1\,\text{gb}\,(\text{e}-\text{Me/a}-\text{Me}) + 1\,\text{gb}\,(\text{e}-\text{Me/e}-\text{Me})]$$

$$= A\,(\text{Me}) - 1\,\text{gb}\,(\text{e}-\text{Me/e}-\text{Me}) = 1.74 - 0.9 = 0.84\,\text{kcal mol}^{-1}$$

Thus the calculated value is much less than the observed value which shows vicinal nonadditivity. This is probably due to the deviation of bond angles (cf. 2,3-dimethylbutane, Section 1.3.6).[3]

**3.6** The preferred conformations of eight inositol diastereomers (one chiral and seven *meso*; see Fig. 3.3B) are shown below. A preferred conformation has maximum possible OH's in the equatorial positions.

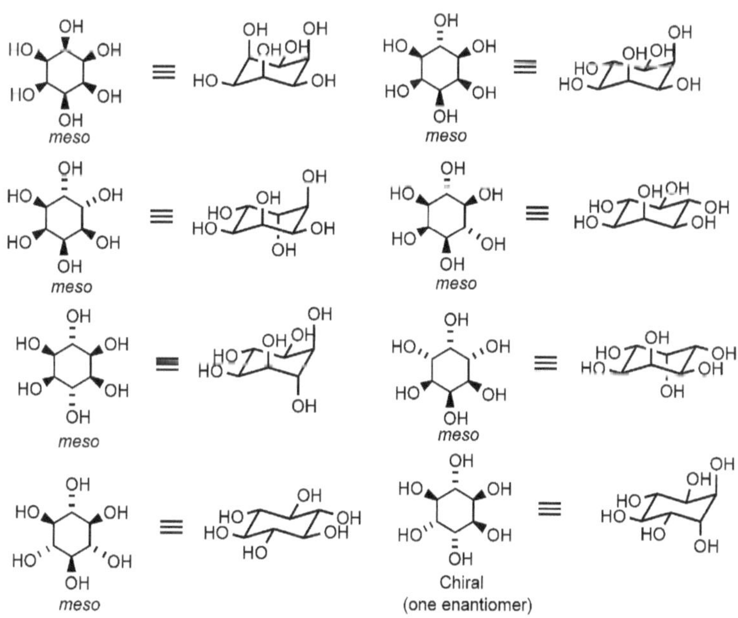

**3.7** In a concentrated solution, the sterically favourable diequatorial conformation is further stabilized by intermolecular H-bonding and is, therefore, preferred. But in a dilute solution, intermolecular H-bonding is not significant but intramolecular H-bonding is possible in the diaxial conformer which is stabilized and favoured at equilibrium.[4]

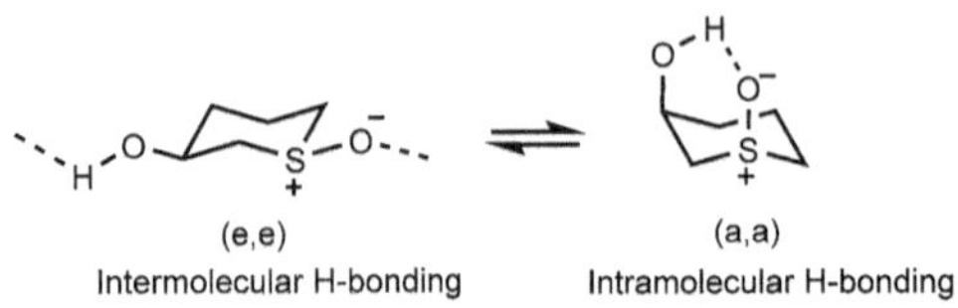


**3.8** Anomeric effect strongly favours the diaxial conformation (see Section 3.8).

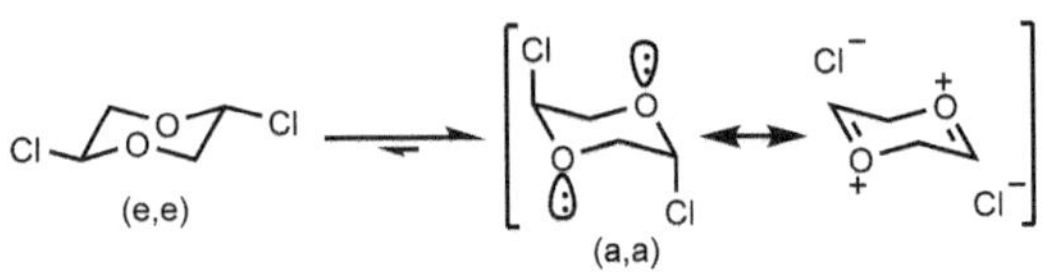


**3.9** The cyclic amide is favoured in the diaxial (a,a) conformation as the diequatorial (e,e) conformation is severely destabilized by $A^{1,3}$ strain. This is in contrast to *cis*-1,3-dimethylcyclohexane which exists practically as (e,e) conformer due to severe steric strain in the (a,a) conformation (see Section 3.5.3).[5]

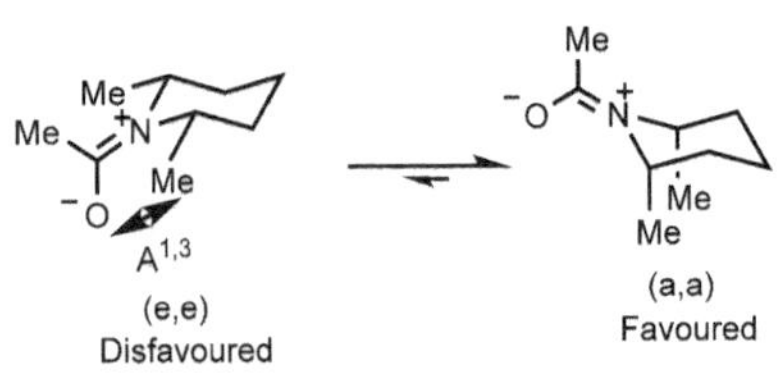


**3.10** The reaction proceeds through a cascade of cyclopropylmethyl-cyclobutyl rearrangement (fivefold) via carbocation intermediates as shown below.[6]

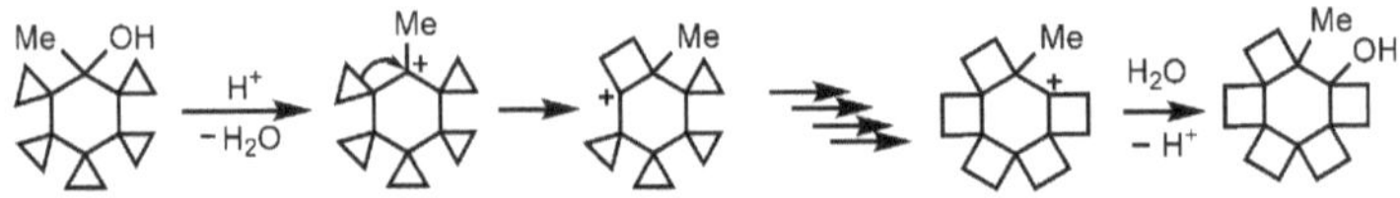

**3.11** (a) In the case of small rings, the change in hybridization from $sp^3$ to $sp^2$ in $S_N1$ hydrolysis leads to an increase in I-strain primarily due to an increase in angle strain. The increase in angle strain is larger for the cyclopropane ring and hence it reacts slower.

(b) I-strain affects both rates and equilibria. The hydrate formation involves a change in hybridization from $sp^2$ to $sp^3$ at the reacting centre. Thus hydrate formation is favoured for cyclopropanone due to some relief of I-strain (angle strain). In contrast, in the acyclic analogue (acetone) such angle strain factor is not operative.

**3.12** (a) The e-Me conformation is sterically more favourable and is preferred.

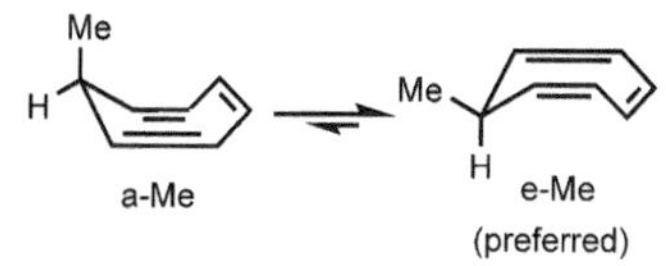


(b) Due to severe steric interaction between Me and *t*-Bu in the equatorial conformation, axial *t*-Bu conformation is preferred.

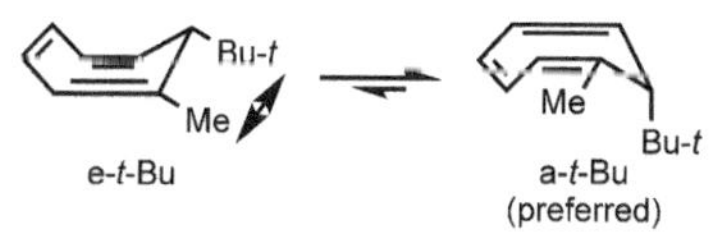


**3.13** Introduction of C=O at C1 creates two unlike chiral centres at bridgehead carbons C9 and C10 and both *trans*- and *cis*-1-decalones are chiral. *trans*-1-Decalone has a rigid chair-chair framework and ring inversion to another chair-chair conformation is not possible. Thus the two enantiomers of the *trans* isomer are not interconvertible and hence resolvable. *cis*-1-Decalone can undergo chair-chair ring inversion but the process does not give the other enantiomer. Therefore, the *cis*-isomer is also resolvable.

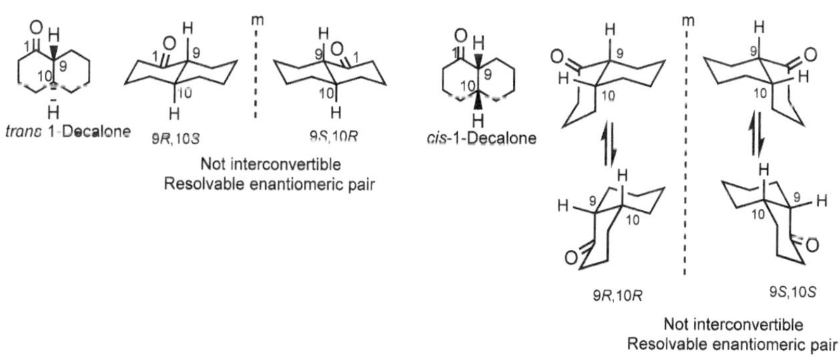


**3.14** In contrast to *cis*-decalin, the two inverted chair conformers of *cis*-decahydroquinoline are not enantiomeric and hence have different energies. The steroid conformation is of higher energy due to a 1,3-diaxial ($CH_2$/H) interaction in comparison to a 1,3-diaxial ($CH_2$/lone pair) interaction in the nonsteroid conformation. Thus the nonsteroid conformation is preferred.[7]

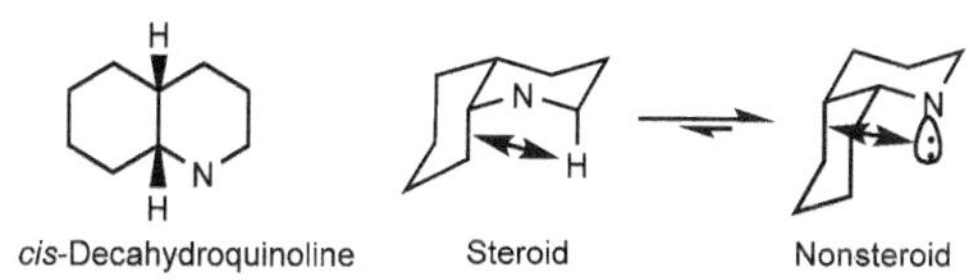

**3.15** Chlorination of *R*-(+)-3-methylcyclohexanone can give *cis* or *trans* isomer of the 2-chloro-5-methyl derivative as shown below. By the axial haloketone rule, the *cis* isomer will show a positive CE whereas the *trans* isomer will show a negative CE. The stereochemistry of the isolated crystalline derivative is, therefore, *trans* in conformity with a negative CE in octane. The preferred conformation of the *trans* isomer in octane has axial Cl because the equatorial Cl conformation is highly destabilized by dipole repulsion. But in a polar solvent ($CH_3OH$), the dipole repulsion is minimized and the sterically favoured diequatorial (e-Cl, e-Me) conformation predominates. Since the equatorial α-halogen has little effect on the sign of the CE, the sign of the CE in $CH_3OH$ is expected to be that of the parent ketone, i.e. positive (as given).

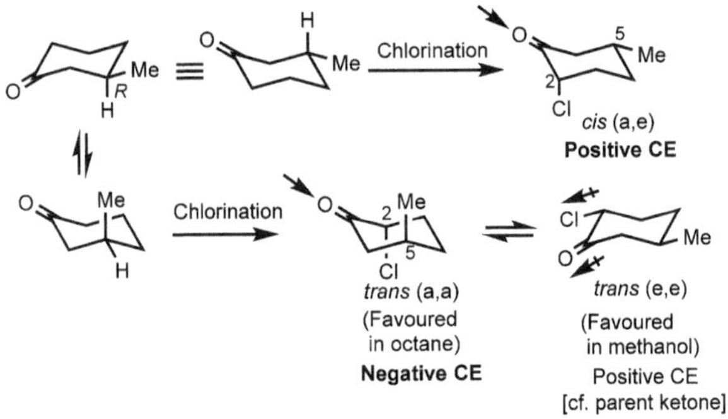


**3.16** The below enantiomer of *cis*-10-methyl-2-decalone is predicted to show a negative CE for the steroidal conformation and a positive CE for the nonsteroidal conformation. Note that the cyclohexanone ring (A ring) is highlighted in the Octant rule projections.

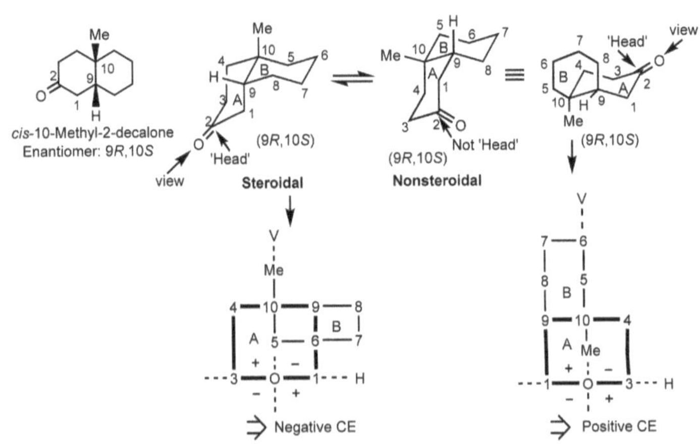

**3.17** The Octant rule projections shown below indicate that 5$\alpha$-cholestan-1-one exhibits a negative CE whereas 5$\alpha$-cholestan-3-one exhibits a positive CE. Note that the cyclohexanone ring (A ring) is highlighted in the Octant rule projections.

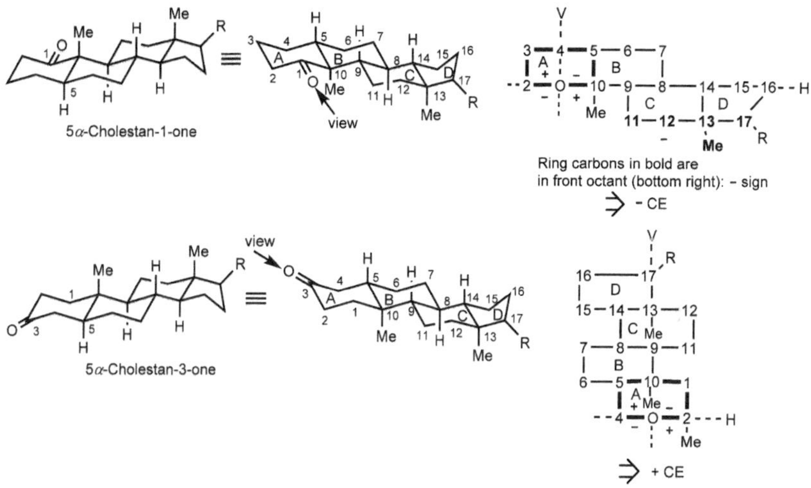


## 4. Answers to Problems in Chapter 4

**4.1** The sketches of the HOMO ($\psi_3$, 2 nodes) and the LUMO ($\psi_4$, 3 nodes) are shown below. Note that the nodes are placed symmetrically in the prospective MO.

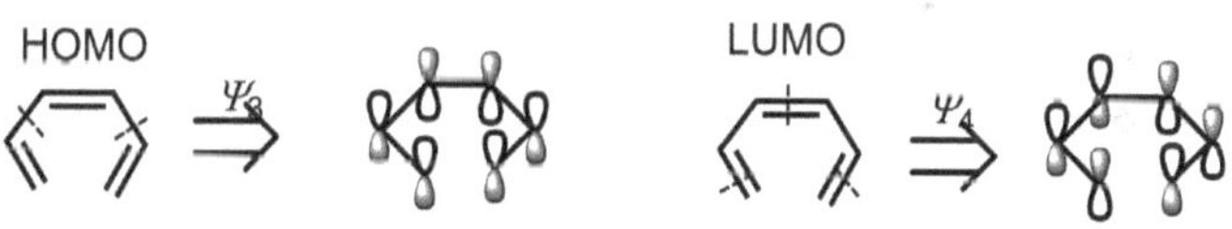


**4.2** (a) 6-*endo-tet*: Disfavoured. Note that no new ring is formed but the breaking Me—O bond is within the ring TS.[8]

(b) 4-*exo-trig* (at the nucleophilic carbon attached to CN) and 4-*exo-tet* (at the electrophilic carbon of C—O bond): Favoured.[9]

**4.3** This is an inverse electron demand Diels-Alder (IEDDA) reaction. The large–large + small–small interactions predict that the major product is 'ortho' regioisomer. (Note that only the interacting lobe of each p orbital is shown.)

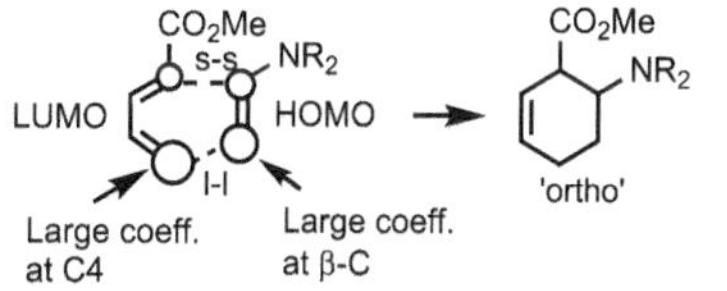


**4.4** Conjugate addition of the oxime to methyl acrylate followed by abstraction of a proton by the enolate produces a nitrone which then undergoes the 1,3-dipolar cycloaddition with a second molecule of methyl acrylate to give the adduct. Here, the regioselectivity is governed by $LUMO_{dipole}/HOMO_{dipolarophile}$ interaction which leads to a 5-substituted isoxazolidine product (see Fig. 4.17B).

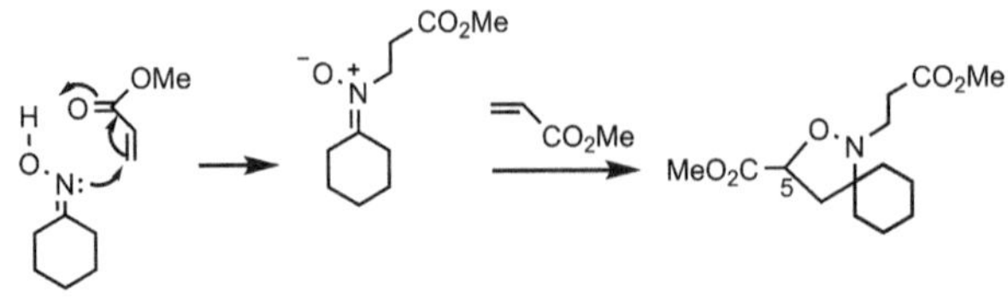


**4.5** The leading bond formed between the large LUMO coefficient on the carbonyl carbon of the ketene and the large HOMO coefficient on the terminal carbon of the conjugated chain gives the major regioisomer.[10]

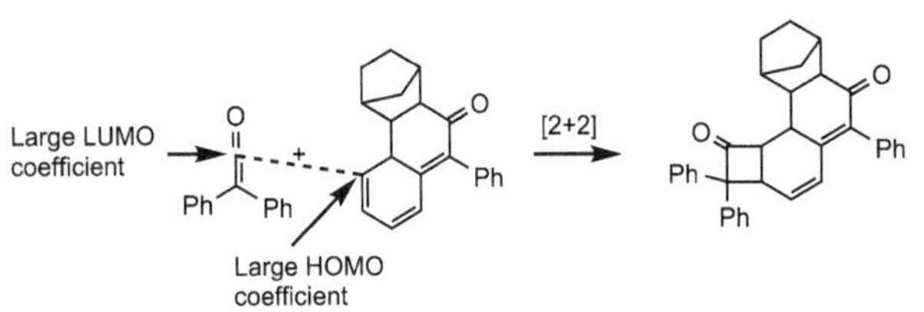


**4.6** Under Lewis acid catalysis, the ene reaction is highly regioselective, and the major product is obtained via preferential formation of C—C bond with the enophile β-carbon having a large LUMO coefficient.

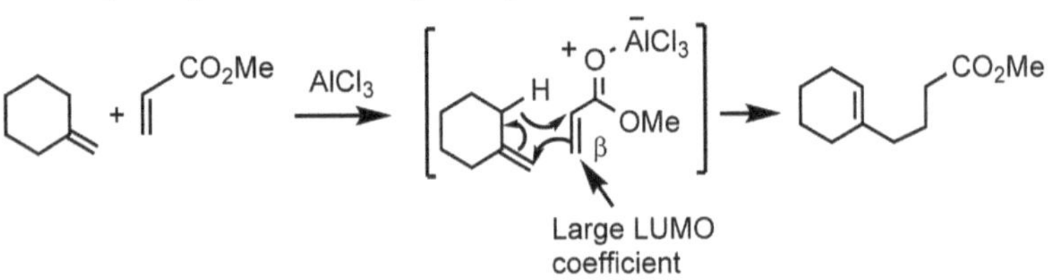


**4.7** In each case, the 6-*endo-trig* cyclization is preferred over the normal 5-*exo-trig* under special situations.

(a) One reason is that the reaction is under thermodynamic control. The radical centre attached to two electron-withdrawing groups in the substrate is highly stabilized. Thus the faster formed 5-membered ring can reopen (the reaction is reversible) and then cyclize again to form the thermodynamically more stable 6-membered ring with a secondary radical centre. Another reason is a favourable frontier orbital coefficient interaction. The radical attached to two electron-withdrawing groups is highly electrophilic with a low-energy SOMO, and the attack of the radical will be preferentially on the unsubstituted terminus having a larger coefficient in the alkene HOMO.[11,12]

(b) The reaction is under kinetic control. But the long Si—C bond reduces the strain in the 6-*endo-trig* TS, and the regioselectivity is determined by the formation of the more stable secondary radical in the 6-*endo* closure.[13]

**4.8** Acrylonitrile is an electron-deficient alkene and has a low-energy LUMO which is also strongly polarized. The regioselectivity is governed by the LUMO/LUMO interaction. The two larger coefficients in the LUMOs are on the carbonyl carbon and on the unsubstituted end of the alkene, which interact preferentially to give the observed oxetane product.[14]

## 6. Answers to Problems in Chapter 6

**6.1** (a) The reaction proceeds with an $S_N2$ inversion at C1 to give (1*s*,3*s*) diastereomer. Remember that inversion at one pseudoasymmetric centre changes descriptors at both pseudoasymmetric centres C1 and C3 (the relative priority of Br and SPh being the same) in 1,3-disubstituted cyclobutanes (see Section 3.1.4).

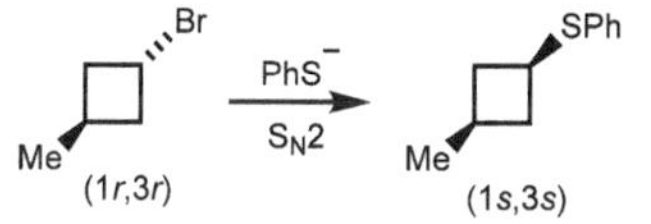


(b) The reaction involves two $S_N2$ inversions in intermolecular and intramolecular steps to give the *cis* product.

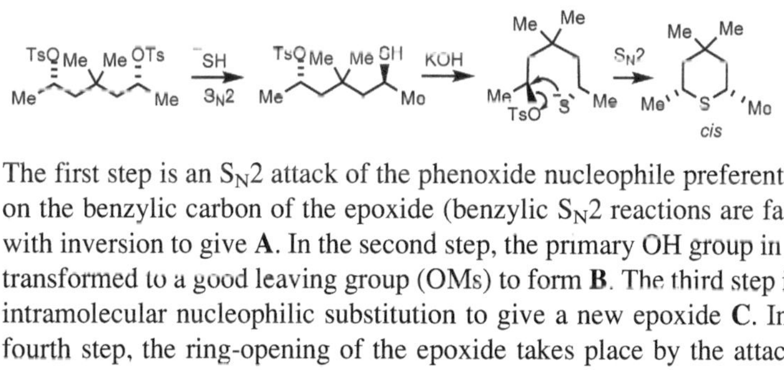


**6.2** The first step is an $S_N2$ attack of the phenoxide nucleophile preferentially on the benzylic carbon of the epoxide (benzylic $S_N2$ reactions are faster) with inversion to give **A**. In the second step, the primary OH group in **A** is transformed to a good leaving group (OMs) to form **B**. The third step is an intramolecular nucleophilic substitution to give a new epoxide **C**. In the fourth step, the ring-opening of the epoxide takes place by the attack of azide nucleophile at the less hindered end, followed by reduction to give an amine **D**. The amine is then converted into an amide **E** in the fifth step. The sixth step involves another intramolecular substitution with the alkoxide as a good nucleophile to give **F**. In the last step, the cyclic amide is reduced with $LiAlH_4$ to an amine, the antidepressant product reboxetine.

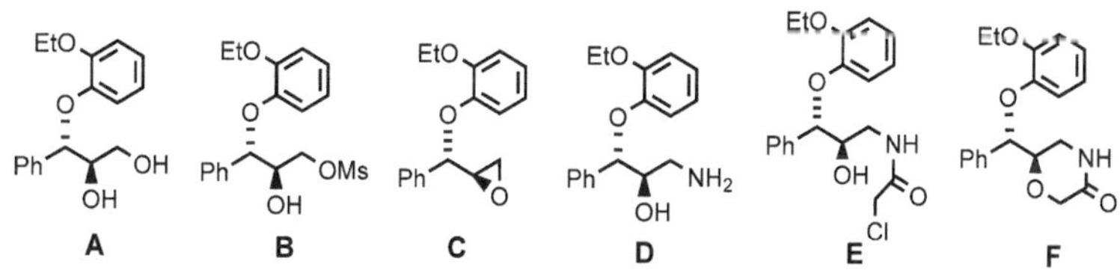


**6.3** Both reactions involve neighbouring-group participation. In reaction (a), the participation of a lone pair on Br atom gives *meso*-2,3-dibromobutane with retention of configuration. In reaction (b), OMe is far away from the leaving group (OTs) in the chair conformation but the molecule can assume a boat conformation for favourable participation by OMe leading to the formation of the *trans* product with the scrambling of the isotopic label (T) between the two substituted positions.

(A)

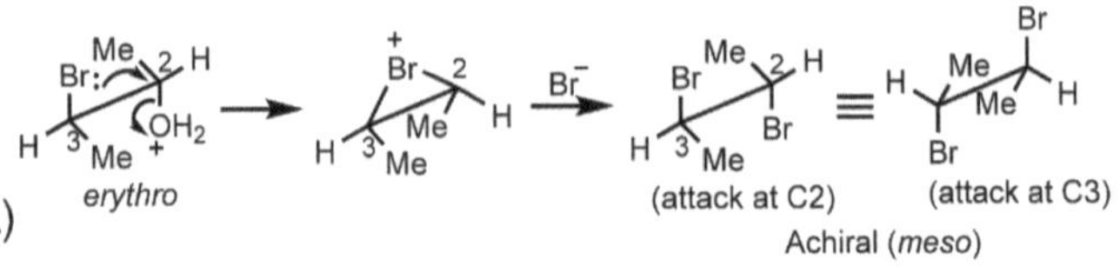


(B)

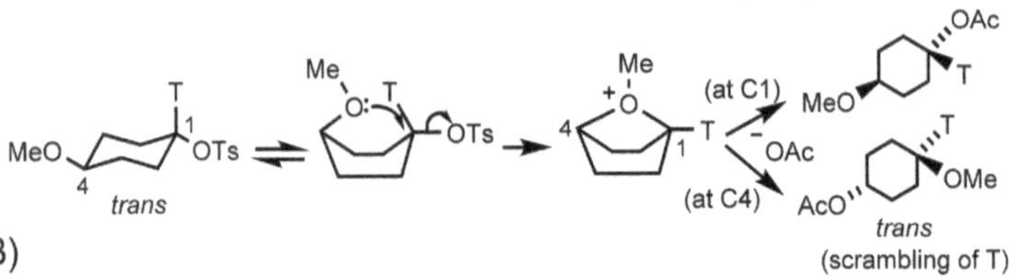


**6.4** (a) The *trans* (a-OH) isomer will react with the relief of a severe 1,3-diaxial (OX/Me) interaction as well as a 1,3-diaxial (OX/H) interaction ($X=CrO_3H$). In comparison, the *cis* (e-OH) isomer will get the relief of only a 1,3-diaxial (Me/H) interaction. Therefore, the *trans* isomer reacts much faster than the *cis* isomer ($k_{trans}/k_{cis}=33.5$).

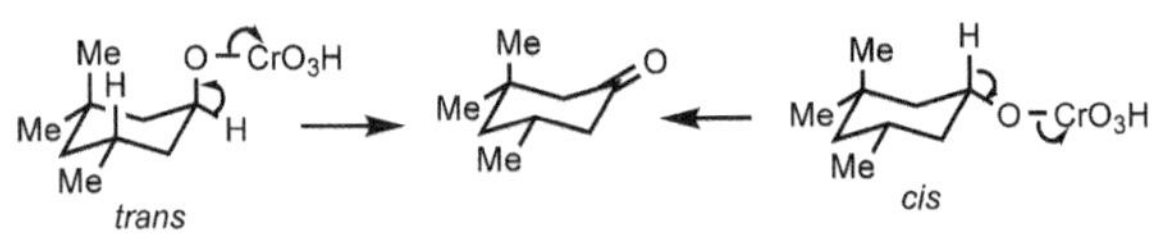


(b) Both cholestanols possess equatorial OH. But 4α-cholestanol reacts faster because of the relief of 1,3-diaxial (Me/H) interaction ($k_{4\alpha}/k_{3\beta}=1.9$).[15]

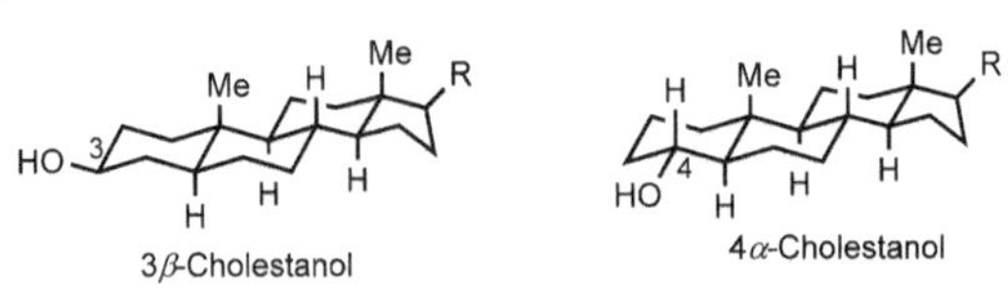


**6.5** The reactive conformer (H, Cl antiperiplanar) of the chiral diastereomer is most populated since the two bulky Ph groups are anti. In contrast, for the *meso* isomer, the reactive conformer is much less populated because the two Ph's are gauche. According to the Winstein-Holness kinetics, the rate constant of the *meso* isomer ($k_{meso}$) will be so small that the reaction does not practically occur at 200°C. Therefore, it is the chiral diastereomer that undergoes E2 dehydrohalogenation to give (Z)-α-chlorostilbene.

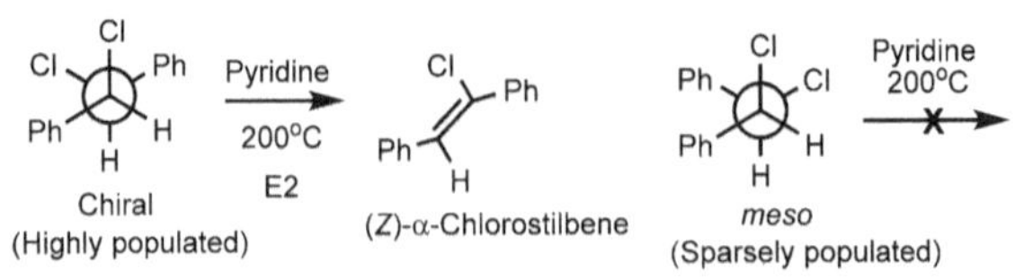

**6.6** Under basic conditions, the Peterson reaction proceeds via the formation of a four-membered ring intermediate, which then fragments by *syn* elimination to give the (*Z*)-alkene. But under acid conditions, an E2 elimination (*anti*) takes place to produce the (*E*)-alkene.[16]

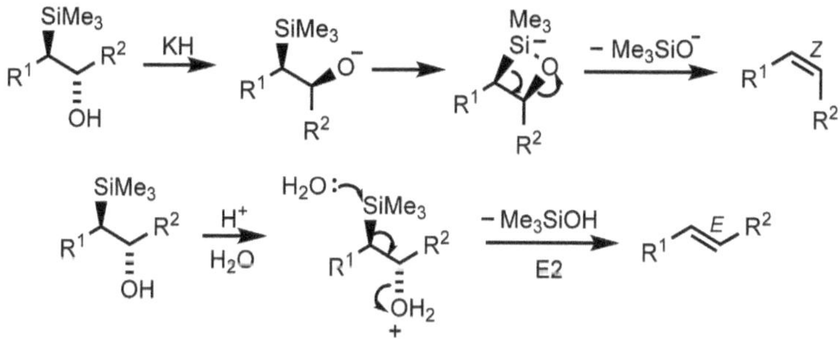

**6.7** In the β isomer, Cl and H must not be antiperiplanar (diaxial) so that facile E2 elimination does not occur. Thus all Cl's will be equatorial in the β isomer as shown below. Ring flipping can give a highly unstable chair in which all Cl's are axial but then all H's will be equatorial.

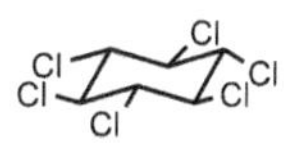

**6.8** In the *cis* bromide, Br is necessarily equatorial in the bridged structure and hence E2 elimination does not occur. But the *trans* bromide has antiperiplanar Br and H, and, therefore, undergoes easy elimination of HBr with base.

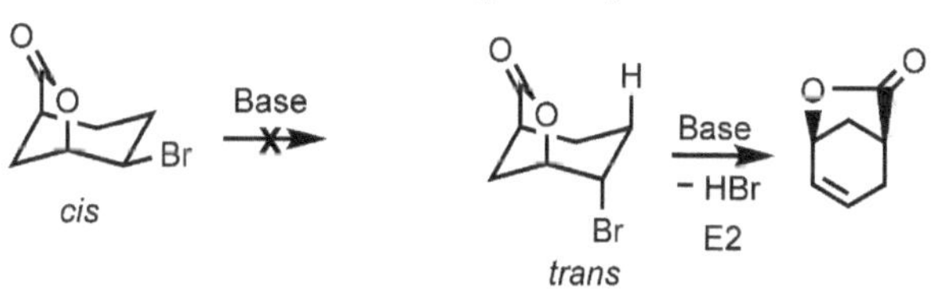

**6.9** (a) In **A**, though Cl and H on adjacent carbons are anti, the dihedral angle is not 180 degrees but is 120 degrees due to the crowding of the rest of the molecule. Thus the TS deviates from being antiperiplanar, and elimination of HCl is much slower.

(b) In **B**, Cl and H on adjacent carbons are syn and the dihedral angle is 0 degrees (synperiplanar). Therefore, syn elimination takes place, and it is faster than anti elimination in **A**. This demonstrates the importance of the planar geometry of the TS in eliminations.[17]

**6.10** Hydroboration on the less hindered face of the alkene produces the borane with a *trans*-decalin structure, which then fragments to the diene product upon the nucleophilic attack of $HO^-$ on boron. Fragmentation occurs because the breaking bonds are antiperiplanar.[18]

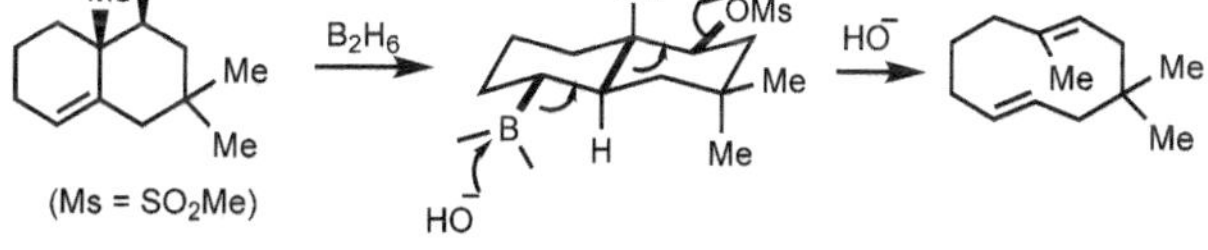

**6.11** The reaction involves two fragmentation steps as shown below. The first fragmentation makes the double bond **1** with (*E*) stereochemistry since H and Me are *trans* during fragmentation. In this step, the oxygen lone pair pushes and OTs pulls. The second fragmentation also leads to (*E*) stereochemistry of the double bond **2** as the two H's at the incipient double bond are *trans*.[19]

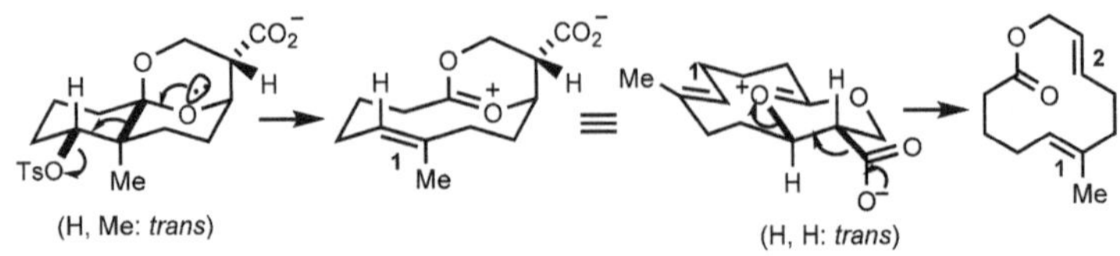


**6.12** (a) NBS (*N*-bromosuccinimide) generates a low concentration of $Br_2$ in solution and thus acts as a source of electrophilic 'Br$^{+}$'. The reaction proceeds through the formation of a bromonium ion intermediate followed by a nucleophilic attack in the second step. Since the concentration of nucleophilic $Br^-$ is always low, $H_2O$ as a nucleophile competes favourably to open the bromonium ion in an $S_N2$ process. The product is, therefore, an *anti* bromohydrin.

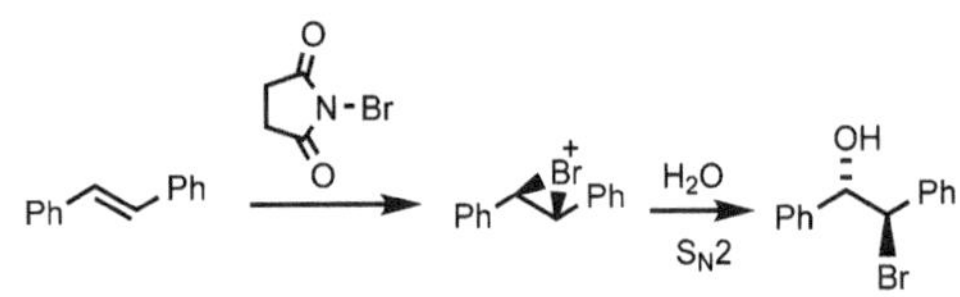


(b) The first dihydroxylation reaction takes place via the preferential addition of $OsO_4$ on the alkene face opposite to the OH group to give the predominant triol product.

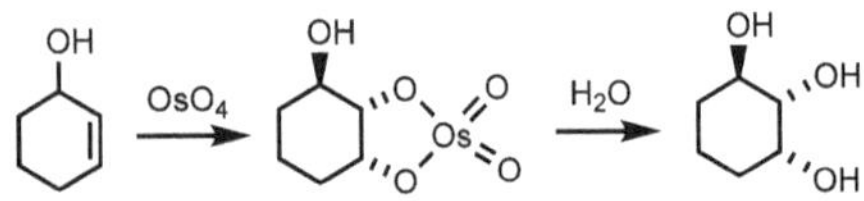


The second dihydroxylation reaction gives the opposite stereoselectivity. Here tetramethylethylenediamine (TMEDA) coordinates to the osmium atom of $OsO_4$ and increases the negative charge on the oxo ligands which favours H-bonding to the allylic OH. This H-bonding controls the direction of dihydroxylation and leads to the preferential addition of $OsO_4$ on the same face of the alkene as the OH group.[20,21]

**6.13** The regio-and stereoselectivity of the reactions result from the epoxide ring-opening by axial nucleophilic attack leading to the chair diaxial product. The other possible axial attack involving twist-boat is unfavourable.[22]

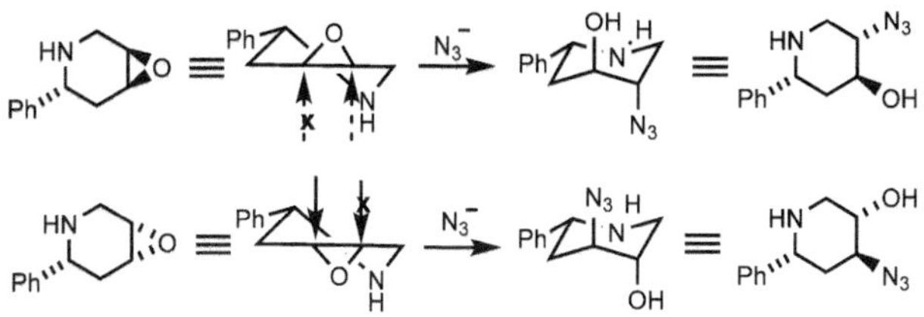

**6.14** This is a Michael addition of methyl acrylate to an enolate of anancomeric cyclohexanone. The reaction proceeds through an axial attack from the top face of the enolate leading directly to the chair product with the axial electrophile. Notably, the large electrophile is on the same side as the *t*-Bu group because the chair-like TS is favoured over the twist-boat TS arising from the bottom side attack.

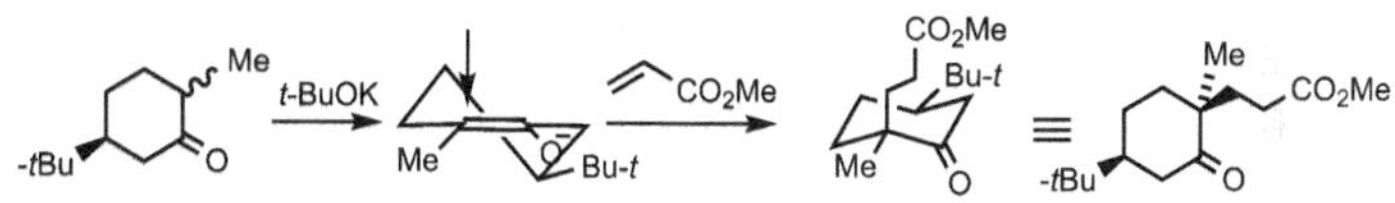


**6.15** The alkyne undergoes *syn* hydroboration to form an alkenylborane. The reaction of the borane with $I_2$ and NaOH proceeds through the formation of an iodonium ion followed by [1,2] shift of a cyclohexyl group from B to C, and then E2 elimination of boron and iodine gives the (*Z*)-alkene.[23]

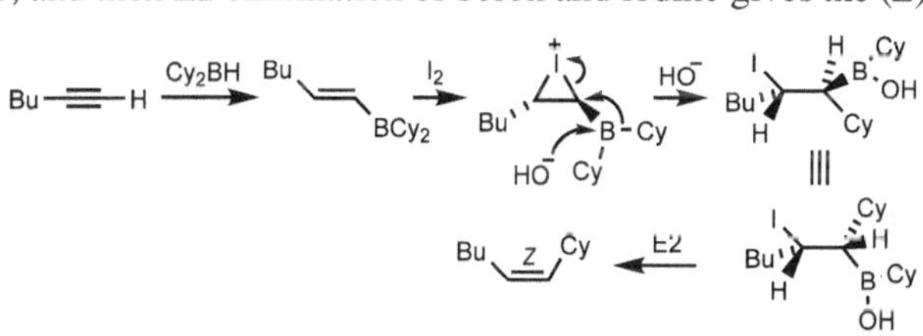


**6.16** The nucleophilic attack on the *syn* face gives the *anti* alcohol while the attack on the *anti* face gives the *syn* alcohol. The *syn* or *anti* attack is determined by the antiperiplanar σ bonds that are more electron-rich (or better donors), which lead to the more effective σ-σ* hyperconjugation in the TS.

(a) The 2,3-substituents are electron-withdrawing ($CO_2Me$) and would cause the depletion of electron density in the 1–2 and 3–4 bonds. Therefore, 1–6 and 4–5 σ bonds (more electron-rich) would make more effective hyperconjugation with the antiperiplanar σ* orbital of the incipient bond formed by the *syn* face nucleophilic attack. The major product is, therefore, the *anti* alcohol.

(b) With electron releasing substituents ($C_2H_5$), the more electron-rich 1–2 and 3–4 σ bonds participate in the hyperconjugation when the attack occurs from the *anti* face. The major product is the *syn* alcohol.

(c) The vinyl substituents, though electron-withdrawing by induction, are conjugating with the 1–2 and 3–4 σ bonds. The conjugation makes these σ bonds relatively better donors since the interaction of σ with π gives the higher energy HOMO (σ+π). The *anti* face attack then leads to the formation of *syn* alcohol as the predominant product.[24]

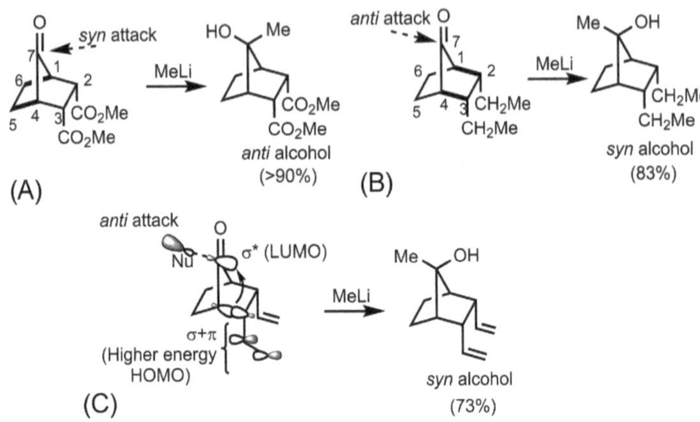


## 7. Answers to Problems in Chapter 7

**7.1** The diastereoselectivity can be explained using the polar Felkin–Anh model as shown below.

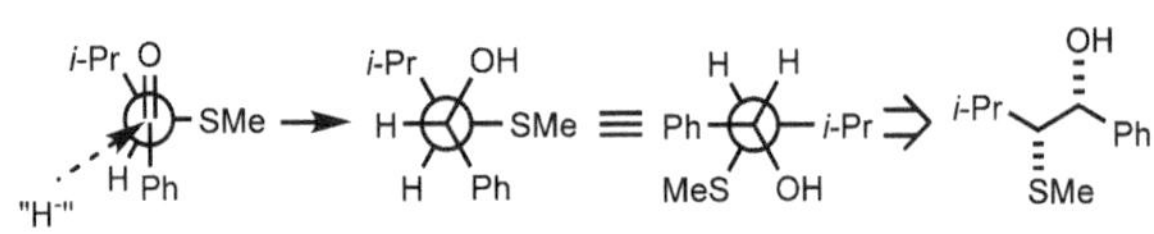


**7.2** The reaction is under chelation control, and the diastereoselectivity can be predicted using the Cram chelate model. The major product is an *anti* isomer. The reaction is also fast and occurs at low temperatures (−78°C).

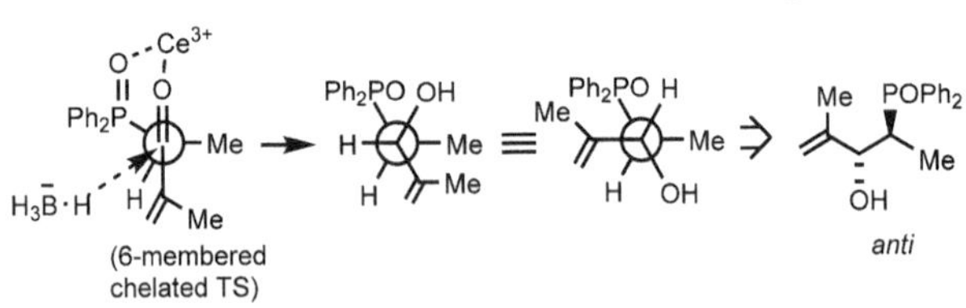


**7.3** (a) When R = $CH_2OBn$, the reaction is mainly under chelation control but if R = $SiPh_2$*t*-Bu (a highly bulky substituent), chelation is sterically disfavoured and the reaction proceeds through a polar Felkin–Anh TS. The chelated TS leads to the *anti* isomer, whereas the polar Felkin–Anh TS gives the *syn* isomer[25]

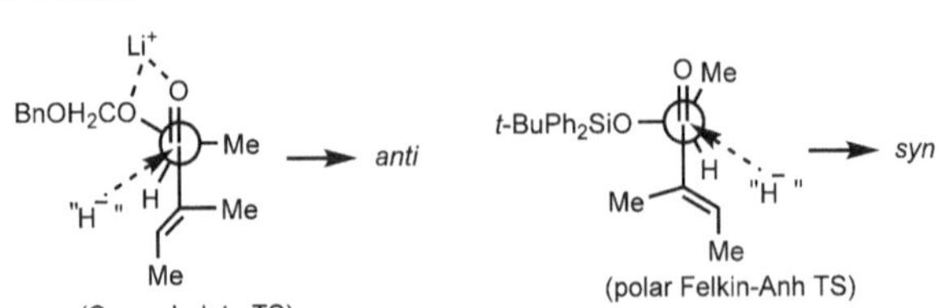

(b) Similar reasoning as in (a). When R = $SiMe_3$, the reaction proceeds through chelated TS. Since chelation is stronger with $Mg^{2+}$, the *anti* (OR/OH) product is obtained almost exclusively. When R = $Si(i\text{-}Pr)_3$, the polar Felkin–Anh TS is favoured, and the major product is the *syn* (OR/OH) isomer.

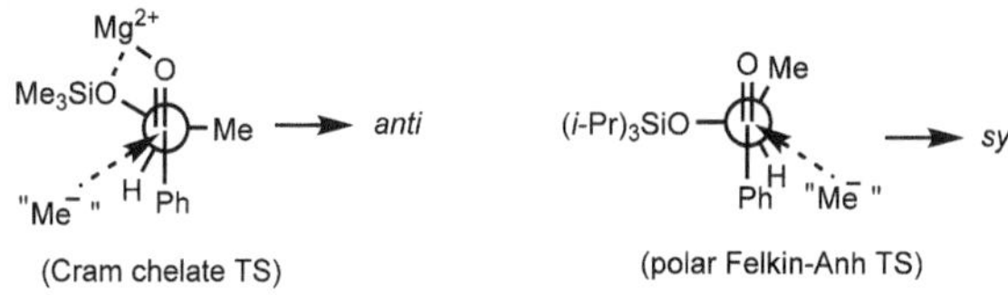


**7.4** The active Lewis acid catalyst for the reduction is not $SmI_2$ but an Sm(III) species generated by pinacol coupling of some aldehyde in the presence of $SmI_2$. The reaction proceeds through the formation of a hemiacetal which, in the presence of Sm(III) species, leads to the *anti*-1,3-diol monoacetate via chair-like TS as shown below.[26]

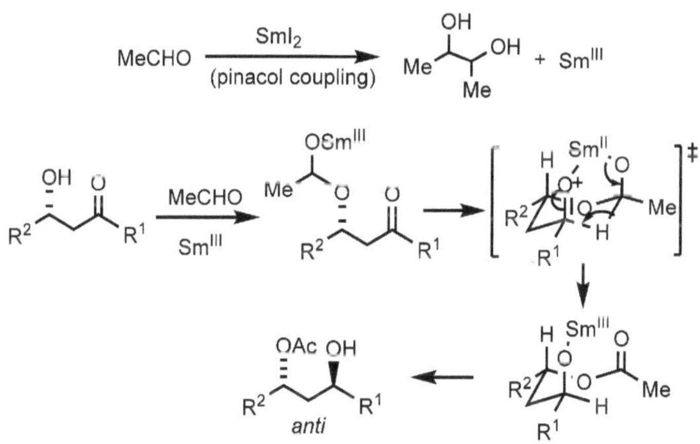


**7.5** The major product is (*R*) enantiomer shown below, which results from the chair TS in which $R_L$ (aromatic ring bearing silyloxy substituent) is equatorial (see Fig. 7.19C). The use of bulky catecholborane instead of $BH_3$ itself gives better enantioselectivity and cleaner products.[27]

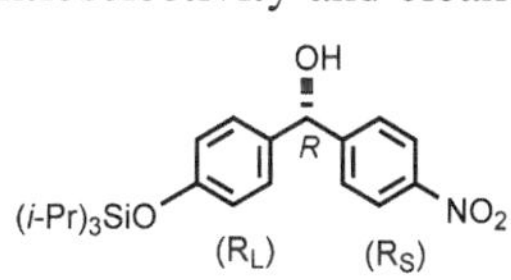


**7.6** Oxidation of the achiral alcohol with PCC (pyridinium chlorochromate) gives the acid which forms *N*-acyloxazolidinone (via the mixed anhydride from the reaction with *t*-BuCOCl). Subsequent *cis*-enolate formation and asymmetric alkylation from the *Si* face followed by removal of the auxiliary gives the (*S*)-alcohol.

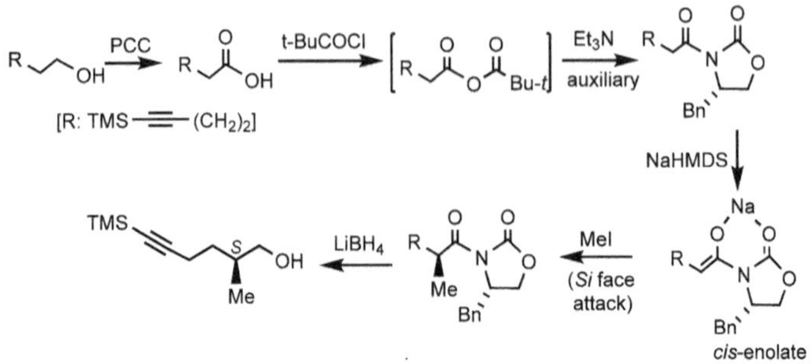


**7.7** (a) The asymmetric bromination of the *cis*-boron enolate with *N*-bromosuccinimide (a source of electrophilic Br), followed by $S_N2$ reaction with azide ion gives the product. The removal of the auxiliary by ester exchange will give α-azido esters which can be readily reduced to α-amino acid derivatives.[28]

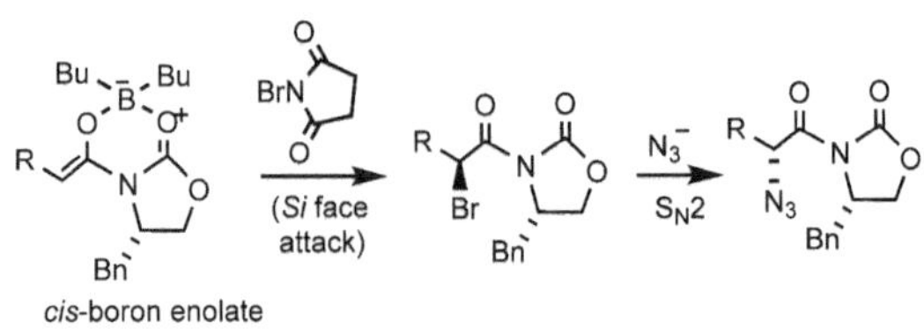


(b) This is a direct asymmetric introduction of α-azido group via the *cis*-enolate using the electrophilic reagent triisopropylbenzenesulphonyl azide. The α-azido product is a precursor to α-amino acid derivatives.[29]

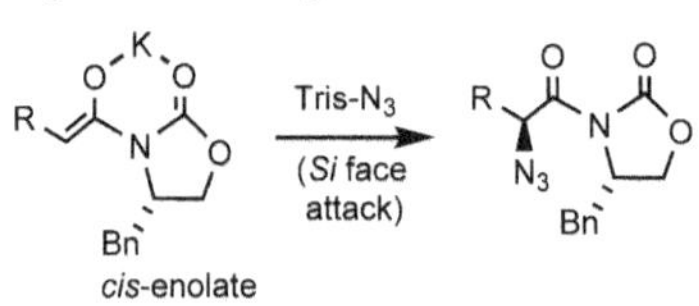


**7.8** Treatment of the acetylenic acid with pivaloyl chloride in the presence of base gives the mixed anhydride **A**. Reaction of **A** with the (+)-camphor sultam gives the N-acylated sultam **B** which on treatment with NaHMDS and 1-chloro-1-nitrosocyclohexane gives the α-hydroxyamino derivative **C** via the chelated *cis*-enolate. The attack of the nitroso compound occurs preferentially from the lower *Si* face of the enolate. Hydrolysis of **C** with aq. HCl gives the (*S*) acid **D**. The mechanism for the formation of **C** via the electrophilic attack on the enolate by nitroso compound is shown.[30]

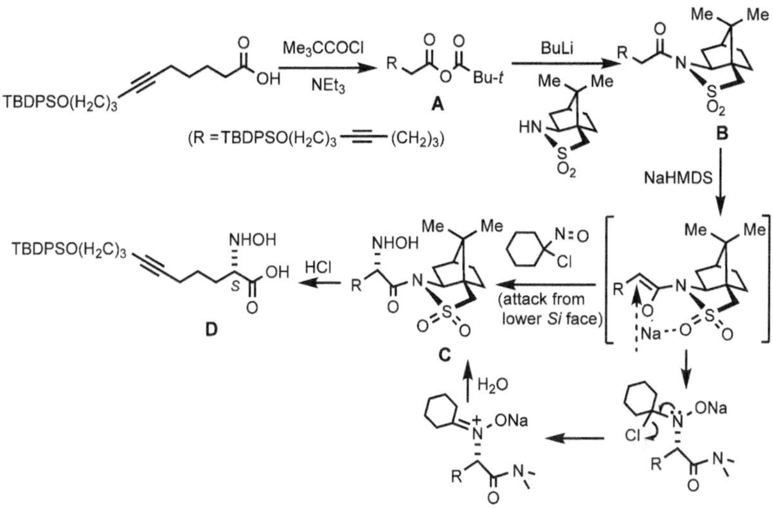

**7.9** The SAMP hydrazone method gives ($R$)-α-silyl ketone (cf. Fig. 7.29B), which then undergoes asymmetric enolate alkylation via the *cis*-enolate to give, after desilylation with tetrabutylammonium fluoride (TBAF), ($R$)-ketone as the final product. In situations where direct asymmetric inductions are not complete, the α-silyl-controlled asymmetric alkylation works better.[31]

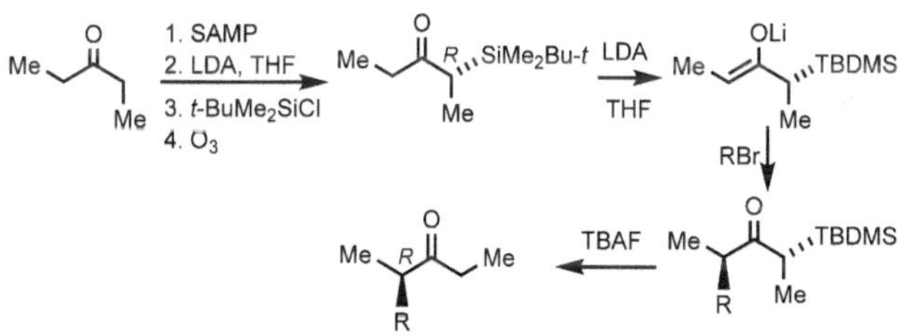

**7.10** The enolate is formed by preferential abstraction of the less hindered proton by the hindered base ($t$-BuLi). Then electrophilic attack by allyl bromide from the upper face of the enolate (opposite to $i$-Pr), followed by hydrolysis gives the product.[32]

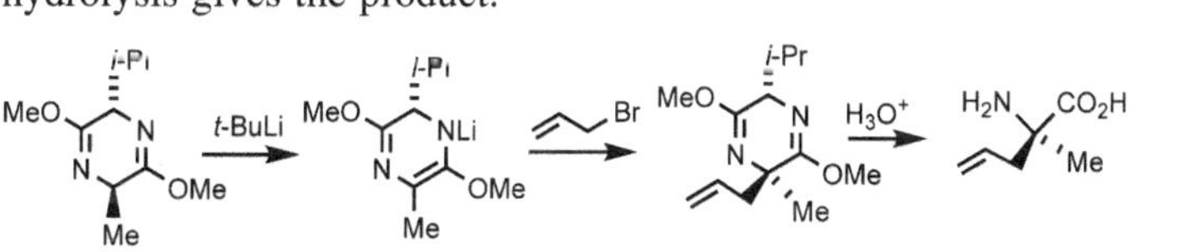

**7.11** The asymmetric alkylation of the glycine derivative (formed in the first step) proceeds through the *cis*-enolate followed by attack of EtI on the opposite face of solvated lithium ions (cf. Fig. 7.31) to give the product with high diastereoselectivity (de 96%). Subsequent hydrolysis provides the α-amino acid essentially as a single enantiomer ($S$).[33]

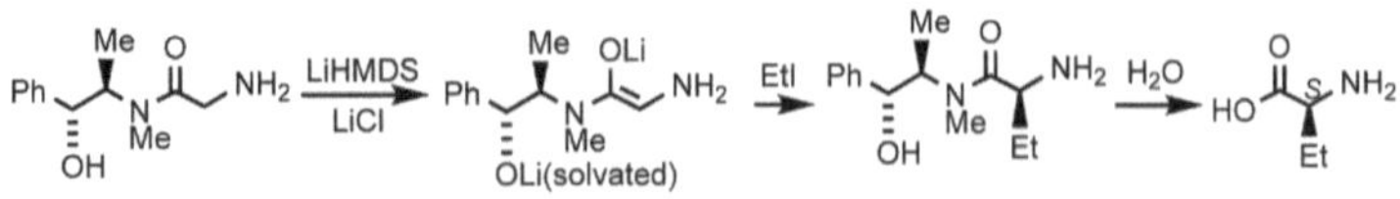

**7.12** The asymmetric aldol reaction proceeds through more stable Zimmerman–Traxler TS via the *cis*-boron enolate to give Evans-*syn* diastereomer (cf. Fig. 7.40) which, after removal of auxiliary by hydrolysis, produces an enantiopure *syn*-aldol product. This *syn* enantiomer (*R*,*R*) is opposite to that obtained using the chiral oxazolidinone in Fig. 7.39.

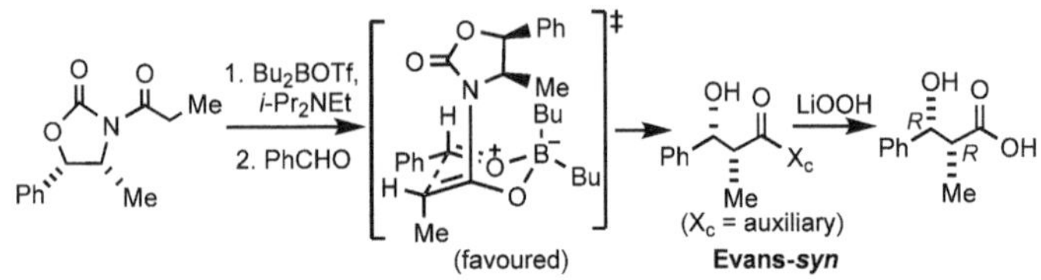


**7.13** The tryptophan-based oxaborazolidinone catalyst induces nucleophilic attack by the enol ether on the *Re* face of the aldehyde carbonyl coordinated to boron in the catalyst-substrate complex and leads to essentially enantiopure *syn* aldol product. The *Si* face of the aldehyde is shielded by the π-stacking interaction between the indole ring and the aldehyde carbonyl group.[34,35] For the model of facial selectivity, see Ref. 36.

**7.14** Each reaction involves an enamine intermediate and proceeds via Zimmerman–Traxler chair-like TS to give the product. The intramolecular reaction (b) is known as the Hajos–Parrish–Eder–Sauer–Wiechert aldol reaction. This was the first report of a direct asymmetric aldol reaction catalysed by a small molecule.[37,38]

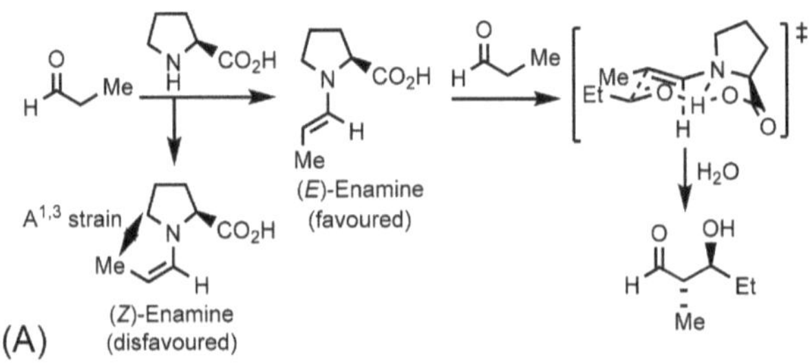

(A)

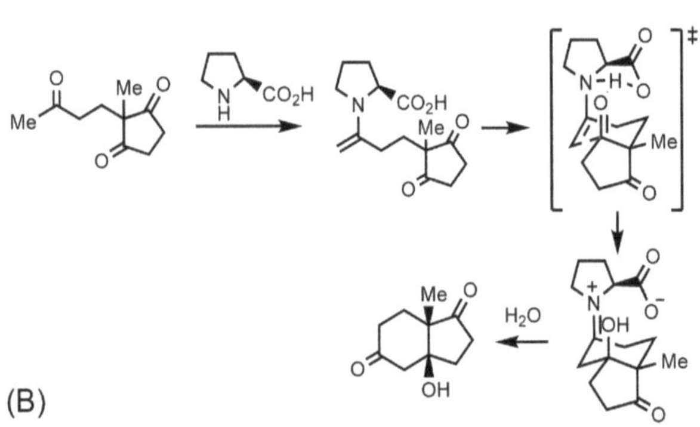

(B)

**7.15** The proline-catalysed asymmetric α-functionalization involves an enamine intermediate and proceeds via chair-like TS to give the product.[39]

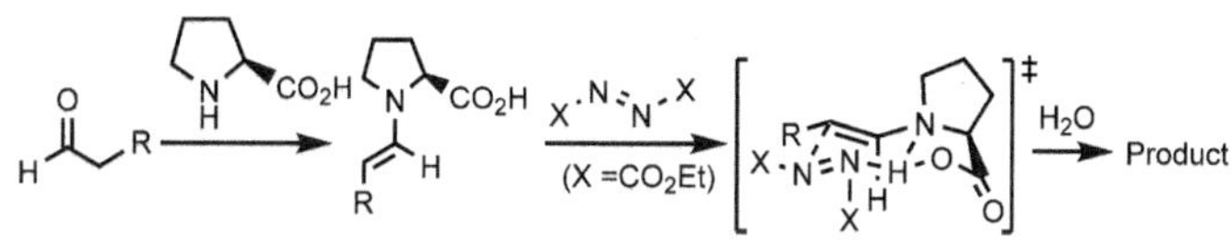


**7.16** (*E*)-2-Butene gives an (*E*)-crotylboronate reagent, and its asymmetric addition to the aldehyde proceeds through more stable chair TS (cf. fig. 7.53B) to give the major enantiomer of the *anti* product (*anti*: *syn* = 98:2, ee 85%), as shown below. (If (*Z*)-2-butene is used in place of (*E*)-2-butene, the product will be a *syn* enantiomer.)[40,41]

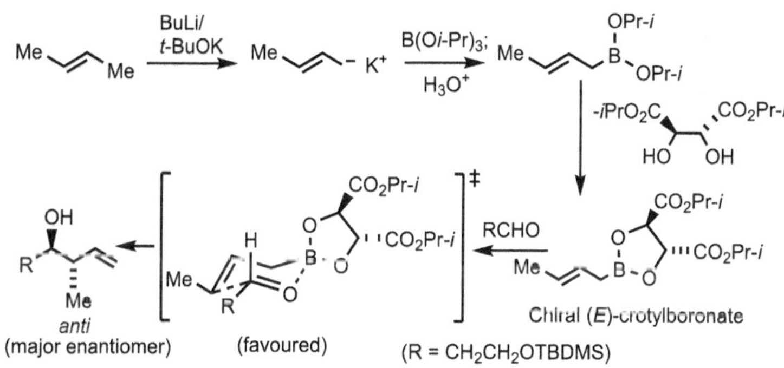


**7.17** The major diastereomer arises from the preferred chair-like TS (equatorial-Ar) involving the coordination of oxygen to Mg atom (cf. Fig. 7.54B).[42]

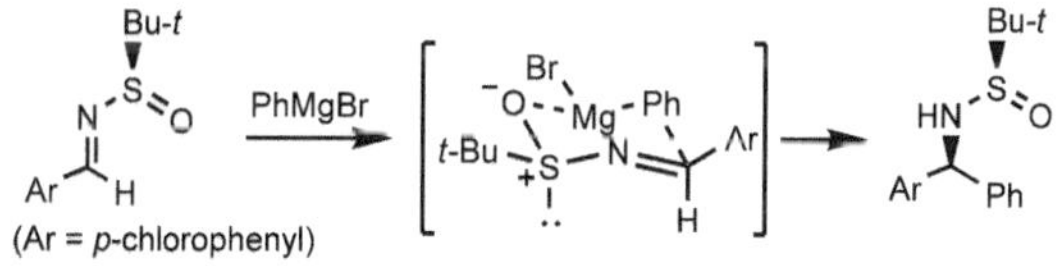


**7.18** (a) The favourable hydrogen bonding possible between the allylic OH and *m*CPBA controls the facial bias of the electrophilic attack. The attack takes place from the same face as the allylic OH to give the major diastereomer (dr = 95:5)

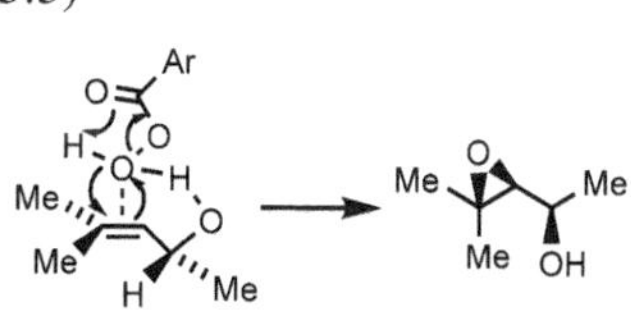

(b) The reaction represents a diastereoselective enolate alkylation. The lowest energy conformation of the intermediate enolate has the hydrogen eclipsing the double bond. Attack by the electrophile (MeI) from the less hindered face, opposite to the silyl group, leads to the major diastereomer (dr = 95:5).[43]

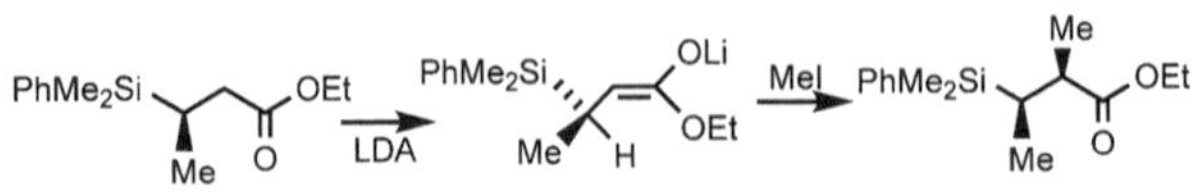


(c) The reaction is diastereoselective hydroboration. The lowest energy conformation of the alkene has the hydrogen eclipsing the double bond. Hydroboration from the less hindered face, opposite to the furyl group (Ar), gives the major borane diastereomer (dr = 8:1), which on oxidation provides the corresponding alcohol with retention of configuration.[44]

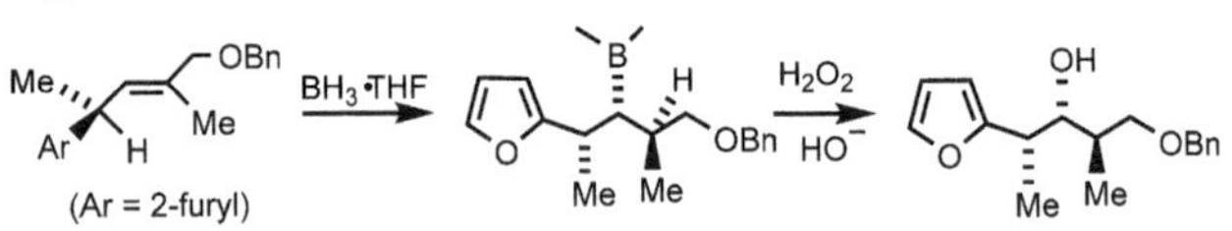


**7.19** With 9-BBN, the hydroboration is *anti*-selective (*anti*:*syn* = 95:5), whereas the reaction is *syn*-selective with $BH_3$ (*syn*:*anti* = 92:8), in terms of Houk models (cf. Fig. 7.60).[45]

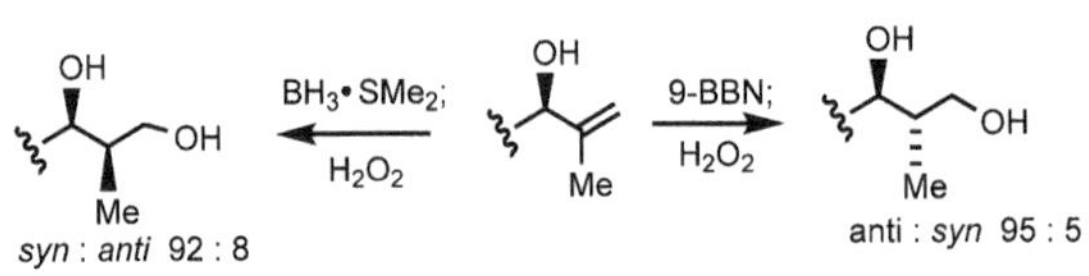


**7.20** The scheme, as shown below, involves the formation of enantiomerically pure boronate ester **D** which is converted to a higher homologue **F**. Oxidation of **F** with alkaline $H_2O_2$ gives the corresponding enantiopure alcohol **G** with retention of configuration. Notably, the homologation procedure (**D** to **F**) is general and can be repeated, if necessary.[46–49]

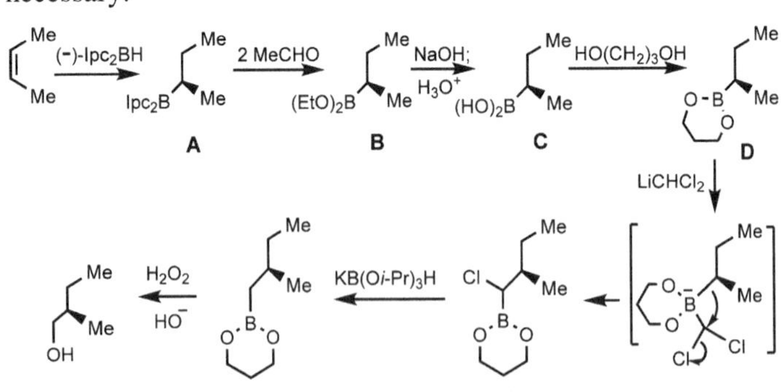


**7.21** The organocatalytic asymmetric conjugate addition proceeds through a favoured iminium ion followed by nucleophilic attack of the indole derivative (via its 3-position) from the exposed rear face (β-*Re*) of C=C as shown below. The resulting conjugate adduct on oxidation gives the inhibitor.[50]

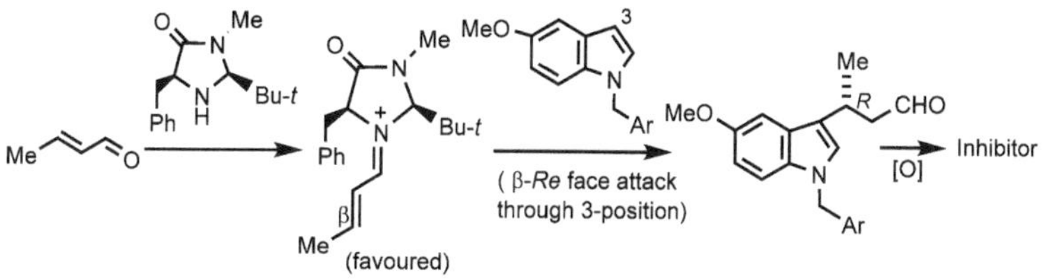


## 8. Answers to Problems in Chapter 8

**8.1** (a) The reaction is regioselective with Me (diene)/$CO_2Me$: 'ortho'. The *endo* and *exo* products can be easily drawn with the help of mnemonics as shown below.

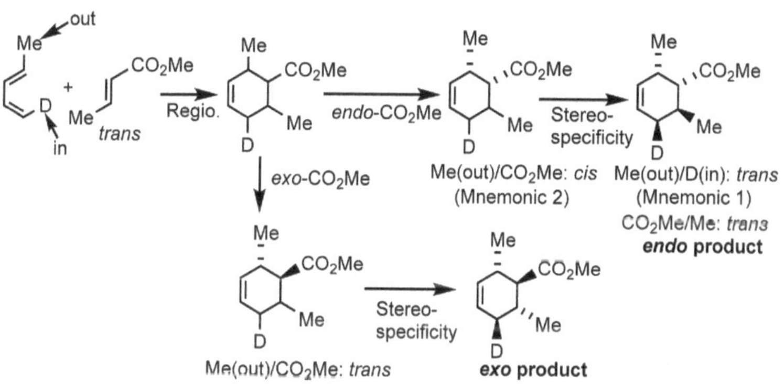


(b) This is an inverse electron demand Diels–Alder (IEDDA) reaction. Using mnemonic 2, the *endo* product can be drawn as shown below.[51]

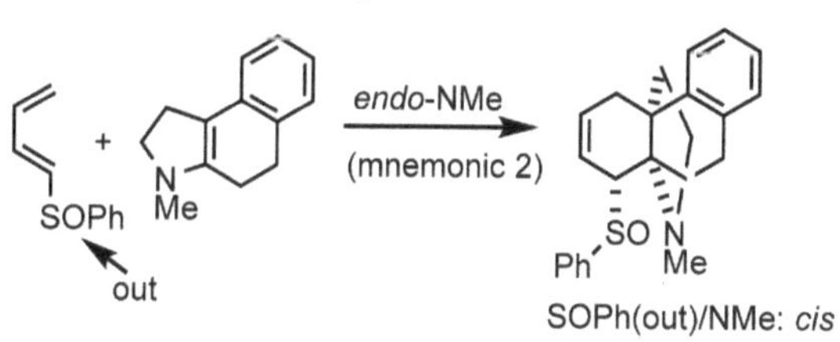


**8.2** Electrocyclic ring-opening of cyclobutene ring would give an *ortho*-quinodimethane as a very reactive diene, with the substituent OH adopting *E* configuration. The observed product results from the preferred *endo* addition. The reaction is also stereospecific concerning the quinone dienophile.

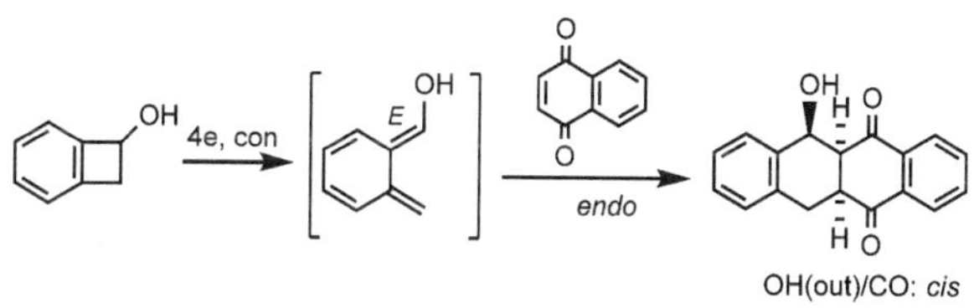

**8.3** Electrocyclic ring-opening of the benzocyclobutene produces an *ortho*-quinodimethane as a diene unit. Intramolecular Diels–Alder (IMDA) reaction with the alkene dienophile moiety lacking an activating carbonyl substituent proceeds preferably through *exo* TS giving the steroid product.[52]

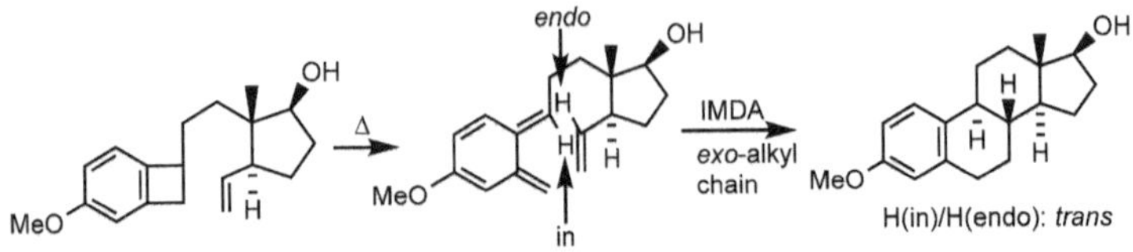


**8.4** The conrotatory ring opening of the oxirane produces a carbonyl ylide dipole which undergoes a 1,3-dipolar cycloaddition with maleic anhydride to give the cycloadduct. The cycloaddition is stereospecific as a result of suprafacial addition to the dipole making Pr/Ph *cis* in the product. The ring junction is also *cis* as a result of suprafacial addition to the dipolarophile.

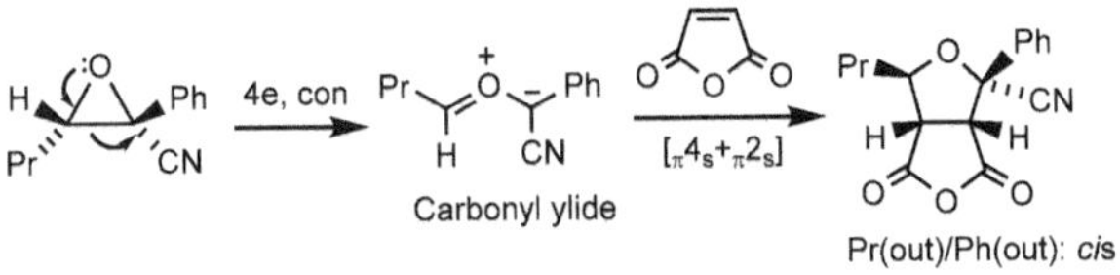


**8.5** The ketene cycloaddition is stereoselective and leads to the thermodynamically less stable cyclobutanone product with Et in the more hindered *endo* position. The lower energy crossed TS has larger Et pointing away from the diene. As the product is reached, Et moves to the *endo* position (cf. Fig. 8.17).[53]

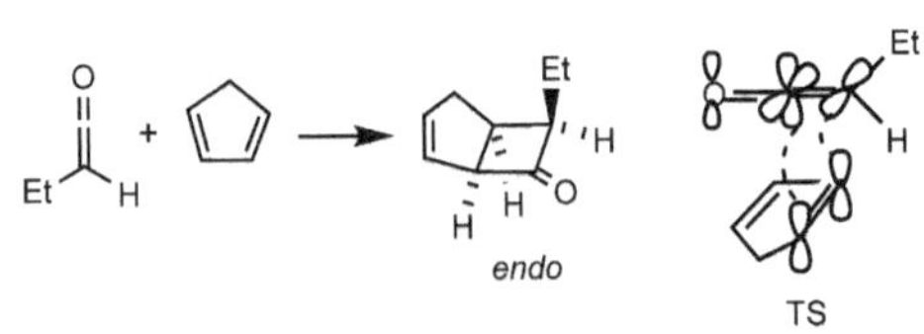


**8.6** The first Wittig reaction involves a stabilized ylide and produces (*E*) double bond. Then a second Wittig reaction with a nonstabilized ylide gives (*Z*) double bond. In the final step, the ester group ($CO_2Me$) is reduced to a primary alcohol ($CH_2OH$). Thus bombykol is an (*E*,*Z*) diene.

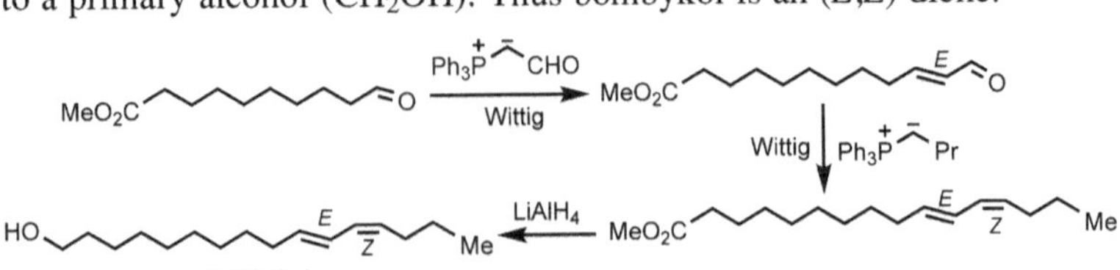

**8.7** This is the ene reaction between β-pinene (the ene component) and benzyne (the enophile) generated from 1-bromo-2-fluorobenzene and Mg. The reaction proceeds through a transfer of allylic *trans* D (not *cis* H) of β-pinene to the benzyne to give the product because the *cis* H transfer would involve severe steric interference by the β-pinene bridge.[54]

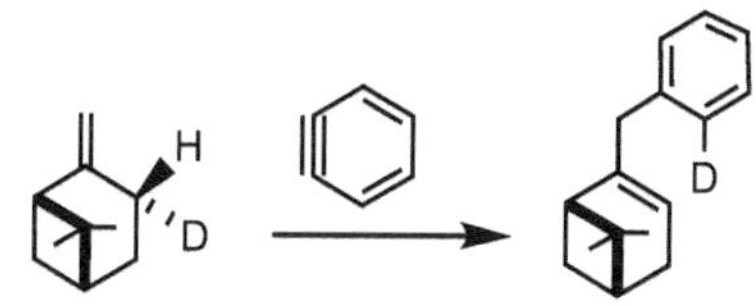


β-Pinene

**8.8** The scheme involves an intramolecular metallo-ene reaction via the *endo* envelope TS followed by a Grignard reaction to give the intermediate **X**.[55]

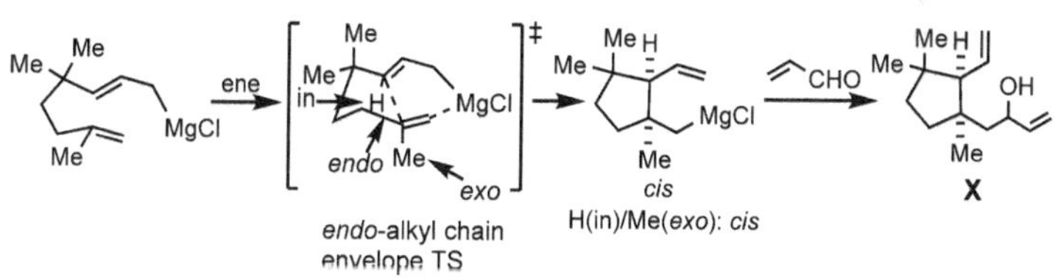


**8.9** Initial protection of the ketone as the acetal with ethylene glycol followed by the formation of ester enolate in the presence of LDA and reaction with dimethyl di sulphide gives, after acetal deprotection, the α-methylthio derivative **A**. Oxidation of **A** with $NaIO_4$ gives the sulphoxide which on heating undergoes *syn* elimination to produce the α,β-unsaturated carbonyl compound **B**.

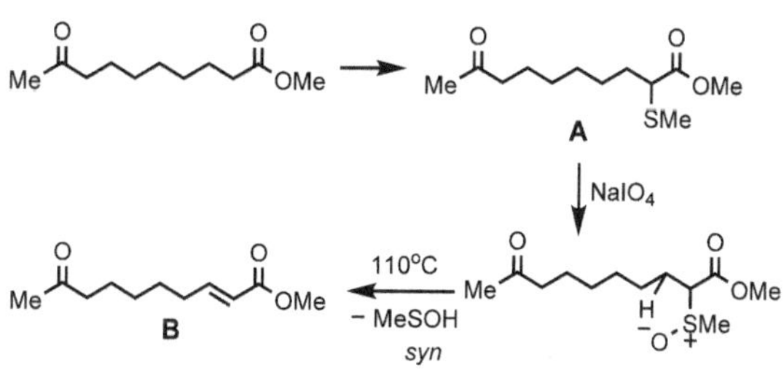


**8.10** Electrocyclic ring-opening of the benzocyclobutene generates an *ortho*-quinodimethane, which undergoes a Diels–Alder (DA) cycloaddition with methyl acrylate to give the predominant *endo* product. See also the use of the stereochemical rule for electrocyclic reactions (Section 8.8.1).

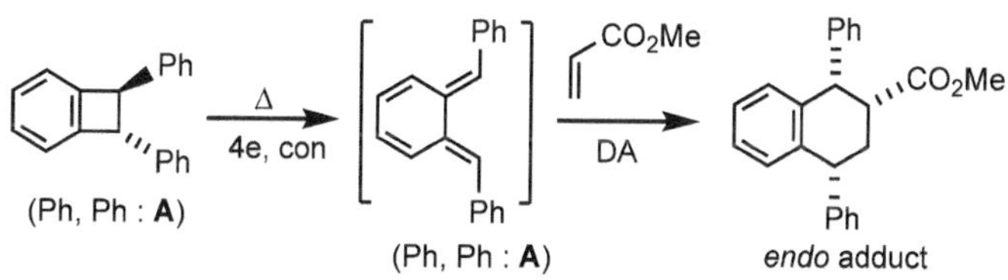

**8.11** The racemization occurs by consecutive electrocyclic ring-opening (ERO) and electrocyclic ring-closing (ERC) processes. The two possible dis modes for the 2-electron ring-opening of cyclopropyl cation give the same sickle-shaped allyl cation, but for ring-closing, the two dis modes give two enantiomers of the *trans* isomer forming a racemate.[56]

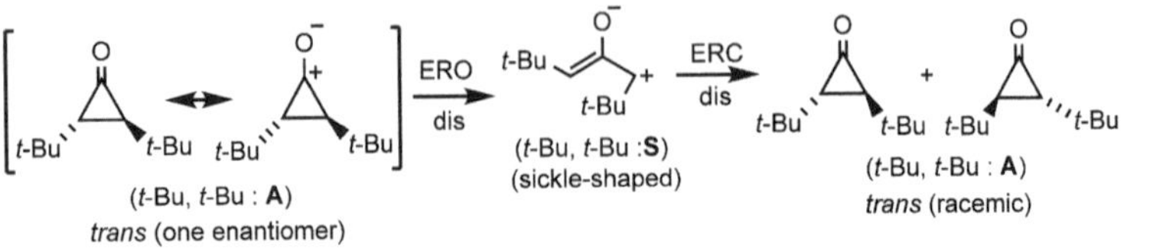


**8.12** The first silicon-directed Nazarov cyclization gives the compound **A**. Conjugate reduction of **A** with the borohydride followed by triflation with the mild triflating reagent gives **B**. Palladium-catalysed Stille coupling of **B** with the stannane in the presence of CO gives the carbonylated product **C** as the substrate for another Nazarov cyclization. The second silicon-directed Nazarov cyclization then gives **D**, which on catalytic hydrogenation and subsequent Wittig reaction produces the sesquiterpene capnellene.[57]

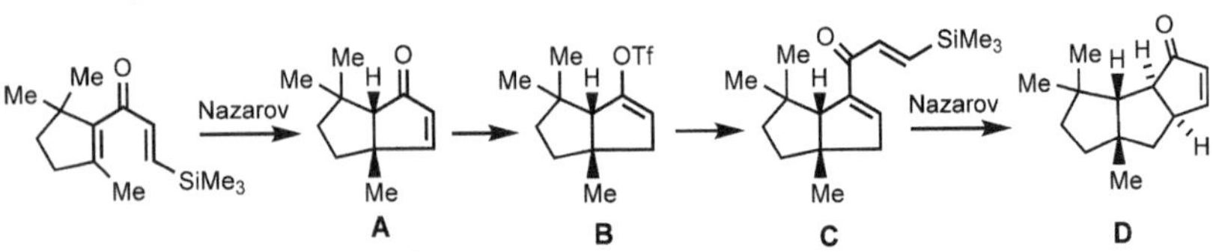


**8.13** This is a quasi Favorskii rearrangement with the anancomeric α-bromoketone that does not possess a hydrogen atom at the α position. The substrate cannot rearrange via a cyclopropanone intermediate and undergoes a base-catalysed rearrangement, sometimes called a semibenzilic rearrangement. The rearrangement leads to the stereochemical inversion at the migration terminus (attached to Br in the substrate) and gives the product.

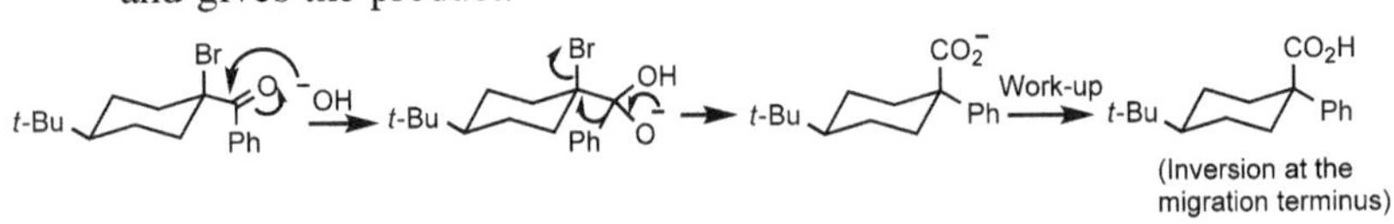


**8.14** The first step involves photochemical [2+2] cycloaddition to form a four-membered ring with *cis* ring junctions. Treatment of the tricyclic adduct with MeLi in the second step gives tertiary alcohol which then undergoes Wagner–Meerwein rearrangement (third step) with ring expansion to give the product α-caryophyllene alcohol.[58]

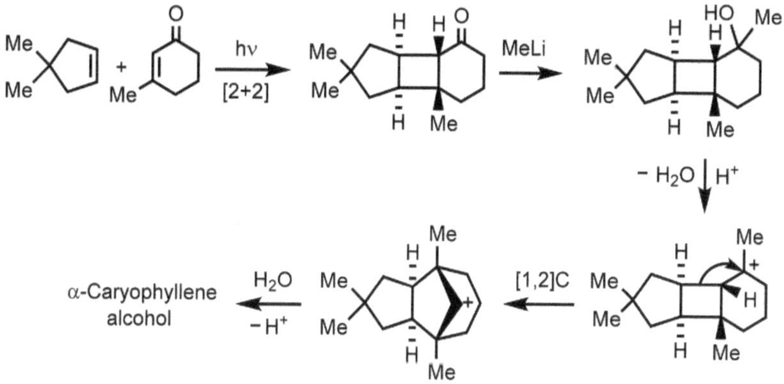


**8.15** (a) Treatment of the ketone with $NH_2OH$ gives the more stable (*E*)-oxime which undergoes Beckmann rearrangement with the migration of the chiral group from carbon (migration origin) to nitrogen (migration terminus). The [1,2] shift proceeds with retention of the configuration of the migrating group

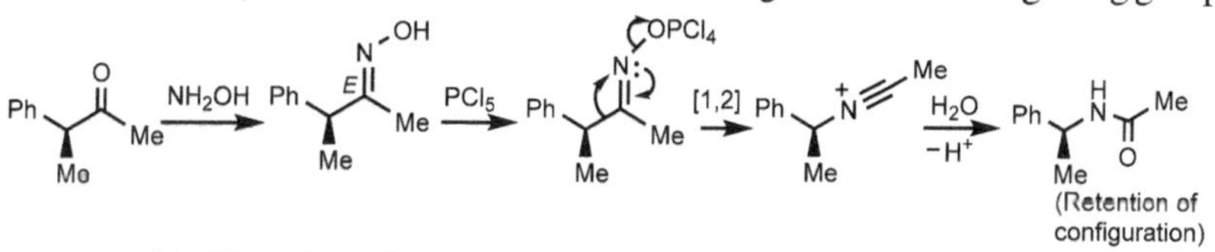


(b) For the diastereomeric 2-bromo-4-phenylcyclohexanols, the relatively large conformational free energy of Ph (2.9 kcal $mol^{-1}$) serves to bias the conformational equilibria to the side in which Ph is equatorial; however, the systems are not anancomeric and ring flipping can give the higher energy conformation with axial Ph. The reactions proceed from the stereoelectronic antiperiplanar conformation of the groups involved. With a base ($Ag_2O$ or $OH^-$), the first diastereomer (OH/Br: *cis*) gives 4-phenylcyclohexanone by a hydride shift. The second diastereomer (OH/Br: *trans*), however, gives different products depending on the base employed. With $Ag_2O$, the departure of the equatorial Br can be induced when the antiperiplanar ring C—C bond migrates to give the ring-contracted aldehyde *cis*-3-phenylcyclopentanecarboxaldehyde. But with hydroxide, the driving force for the ring contraction reaction is not sufficient, and the reaction takes place from the higher energy conformation to give a *trans*-epoxide by an intramolecular $S_N2$ reaction.[59]

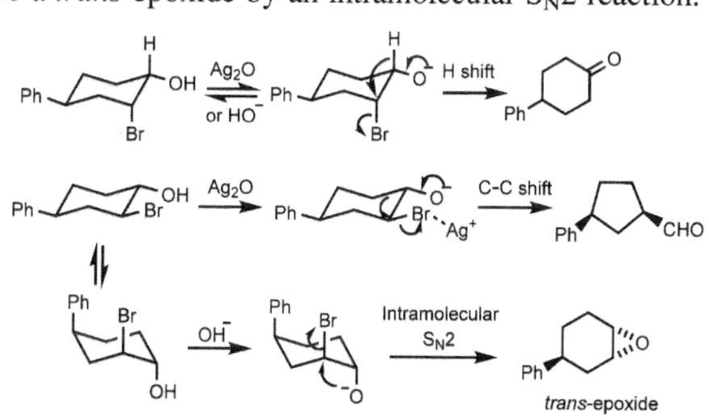

**8.16** The reaction proceeds through a [1,5]C shift that takes place suprafacially with retention of the configuration of the migrating group. As a result of retention of configuration, the relative *cis* stereochemistry of two Me groups is retained in the product. The initial product then undergoes a rapid [1,5]H shift to give the more stable product.[60]

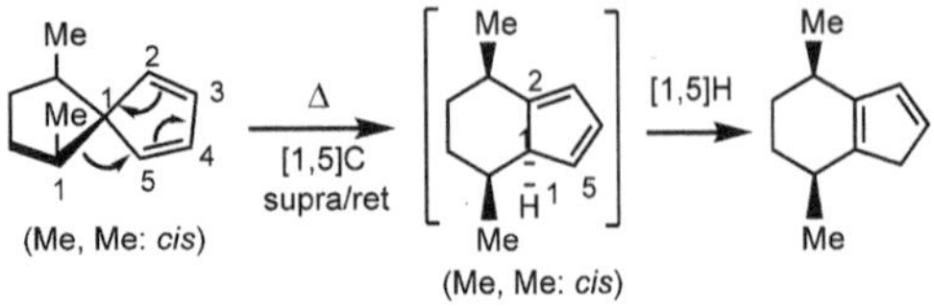


**8.17** The reaction involves two consecutive Cope rearrangements. The first Cope rearrangement occurs through chair TS and the second Cope of *cis*-1,2-divinylcyclobutane proceeds through boat TS.

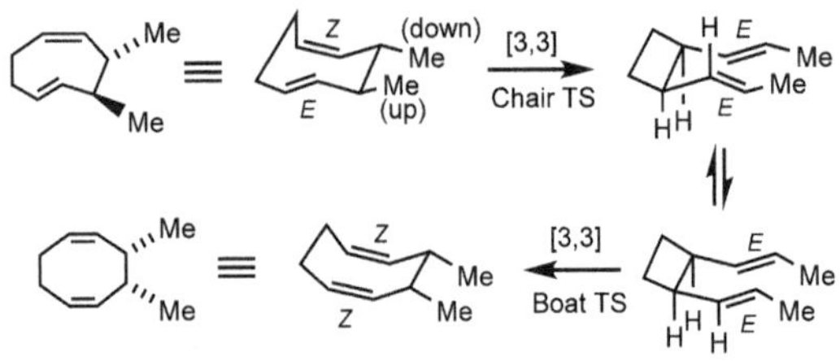


**8.18** The supra/supra pathway of the Eschenmoser-Claisen rearrangement gives the stereochemistry of the product as shown below.[61]

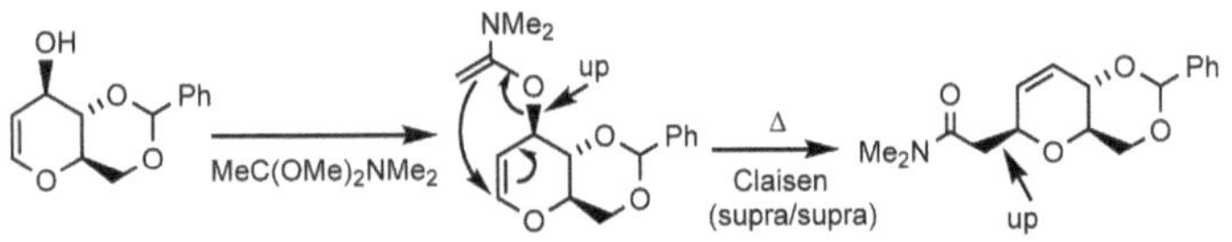


**8.19** Treatment with BuLi generates carbanion of an allyl ether which undergoes a [2,3] Wittig rearrangement. The suprafacial migration across the allylic unit gives the product.

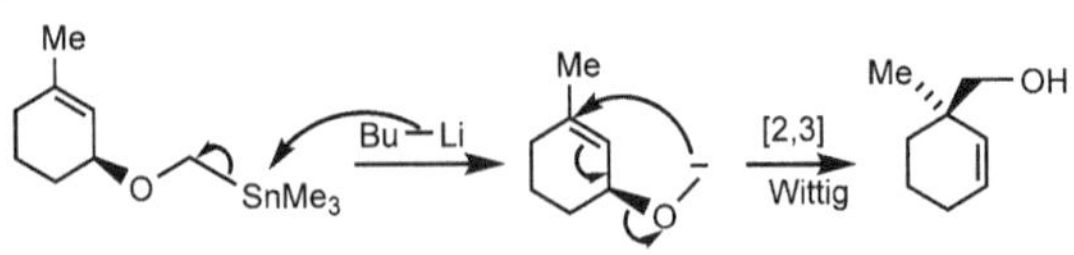


## 9. Answers to Problems in Chapter 9

**9.1** The diene and dienophile units are tethered in the same molecule in which the dienophile is attached with a chiral sultam auxiliary. The Lewis acid-catalysed intramolecular Diels–Alder reaction proceeds through a chelated *endo* TS in which the dienophile adopts the *s-cis* conformation with the Al atom complexed to carbonyl oxygen and one of the two sulphone oxygen atoms, and gives predominantly one diastereomer of *trans*-fused bicyclic *endo* product.[62]

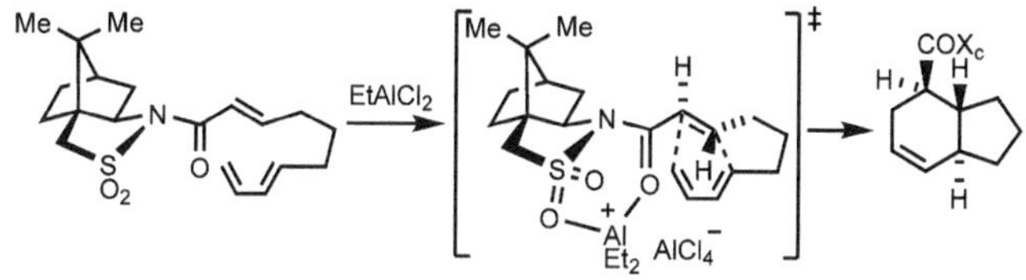


**9.2** Retrosynthetic analysis of enantiomer **1** or enantiomer **2** with the help of mnemonic for asymmetric Diels–Alder reaction gives the oxazolidinone chiral auxiliary as shown below. The reaction involves cyclopentadiene and an acrylic dienophile.

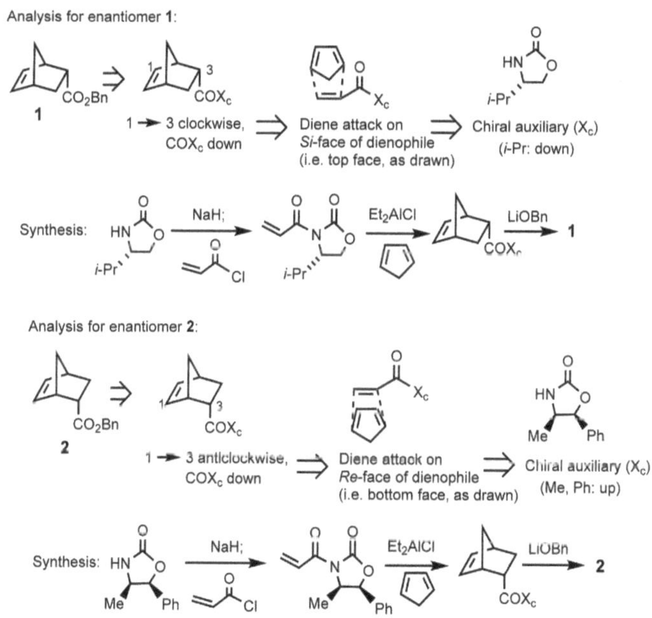


**9.3** Chelation of two carbonyl groups of the hetero-dienophile (methyl pyruvate) to the copper forms a catalyst-dienophile complex. The diene addition then takes place to the lower face of the ketone dienophile (the upper face is shielded by a *t*-Bu group) to give the preferred *endo* adduct. Subsequent treatment with TFA leads to the (*S*) enantiomer of the product.[63,64]

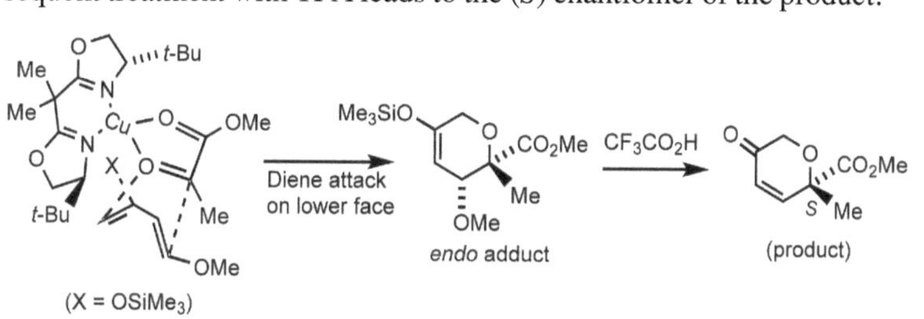

**9.4** PhI=NTs acts as a source of nitrene and forms a complex with the chiral copper catalyst as shown below. The alkene then approaches from the less hindered front side of nitrene (opposite to rear Ph on the catalyst) to give the aziridine with high enantioselectivity.[65–68]

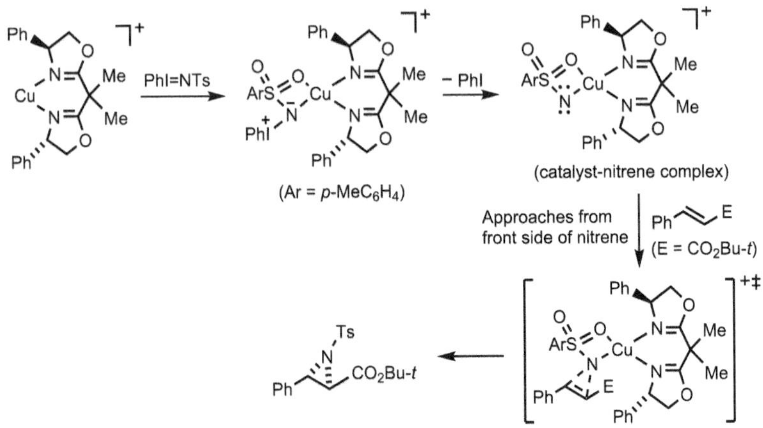


**9.5** (a) The reaction proceeds through a chelated chair TS involving the attack of the alkene ($E$-2-butene) on the upper *Si* face of the aldehyde carbonyl of the glyoxylate ester (the lower face being shielded by the phenyl group on the auxiliary) to give the product.

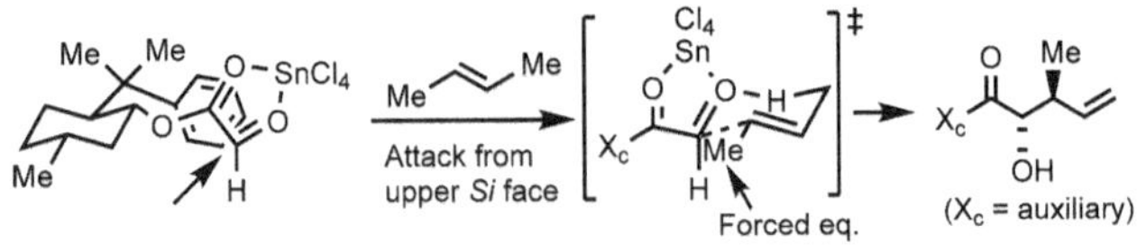


(b) This is an asymmetric desymmetrization using the glyoxylate ester of the chiral auxiliary *trans*-2-phenylcyclohexanol which is available in both enantiomeric forms. The product arises from the approach of the glyoxylate enophile from the less hindered upper face of a cyclopentene ring via an *endo* TS. The product is a key intermediate in a synthesis of (−)-specionin, a potent inhibitor of the feeding of the spruce budworm.[69]

**9.6** Treatment of the chiral starting material with LDA gives the substrate for the aza-Claisen rearrangement which proceeds through a chair TS to give the product **X**.[70,71]

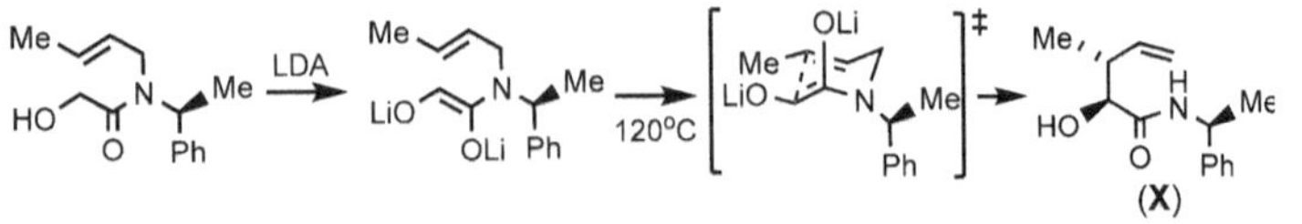


(**X**)

**9.7** This is an asymmetric Johnson–Claisen rearrangement in which ($R$) enantiomer of the substrate gives ($R$) enantiomer of an allene. The supra/supra pathway of the rearrangement establishes the correlation of the absolute configuration of the chiral centre in the substrate with that of the chiral axis of the allene.[72]

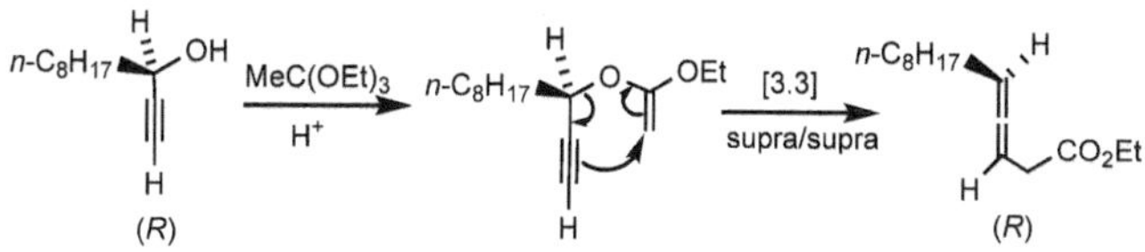


## 10. Answers to Problems in Chapter 10

**10.1** Retrosynthetic analysis gives the starting allylic alcohol and the required Sharpless asymmetric epoxidation with D-(−)-tartrate ligand [(-)-SAE].

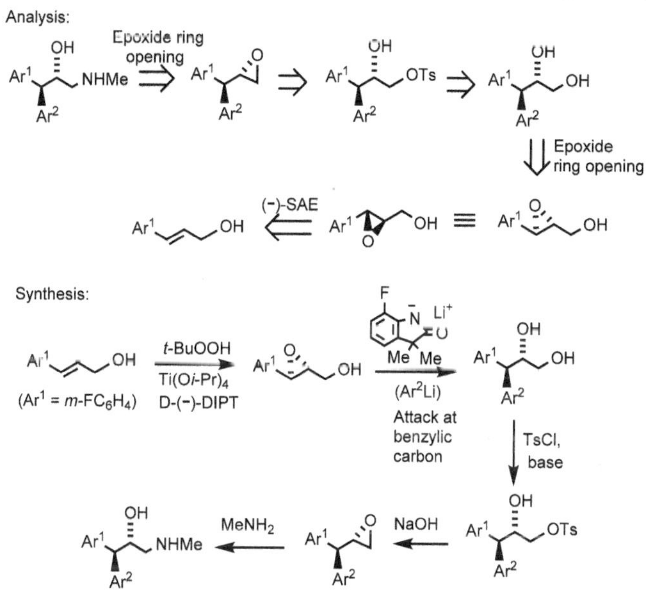


**10.2** The scheme is shown below.[73,74]

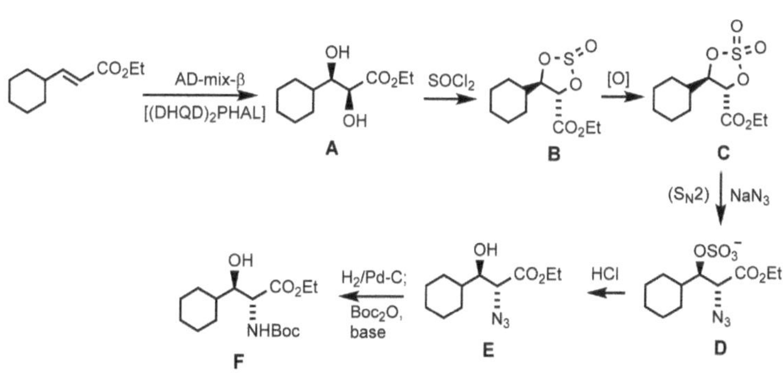


**10.3** Retrosynthetic analysis by disconnection at the amide bonds gives two chiral components (**A** and **B**) and one achiral component (**C**). The achiral aminopyridine **C** can be synthesized by standard methods. The chiral component **B** is an amino acid, (*S*)-glutamic acid which is readily available as a member of the chiral pool. The chiral carboxylic acid **A** requires an asymmetric synthesis which can be accomplished efficiently by using ruthenium-catalysed asymmetric hydrogenation. The synthesis of NC-00637 is then possible involving the protection/deprotection of the side-chain $CO_2H$ group of glutamic acid (**B**).

Analysis:

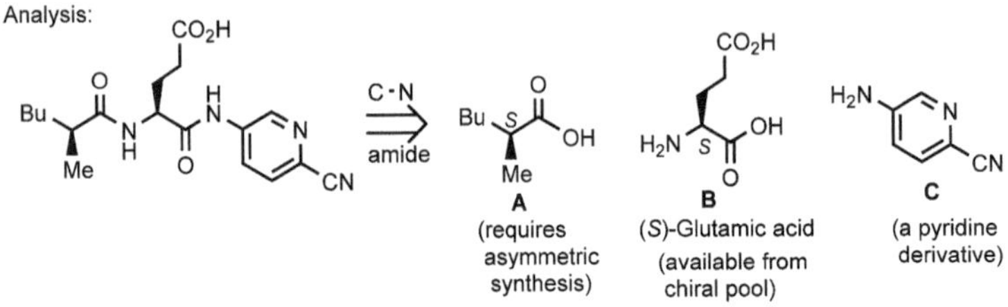


Analysis and Synthesis of **A**:

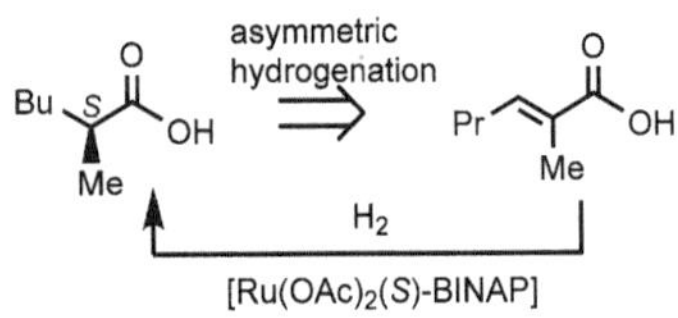


Synthesis of NC-00637:

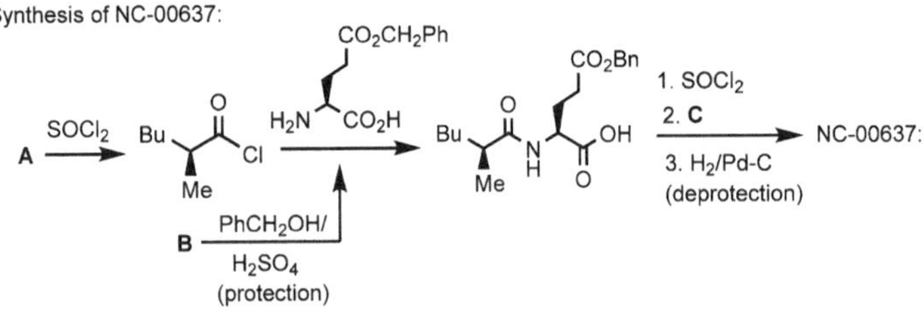


**10.4** The reaction is ruthenium-catalysed asymmetric hydrogenation with a chiral β-keto ester substrate. The direction of stereoselection with (*R*)-BINAP gives the *syn* product whereas (*S*)-BINAP leads to the *anti* product (cf. Fig. 10.31). Due to the inherent *syn* selectivity of the chiral substrate (cf. polar Felkin–Anh model, Fig. 7.7), the (*R*)-BINAP catalysed case represents a stereochemically matched case, and the *syn*-selectivity is almost exclusive. In contrast, the (*S*)-BINAP case is mismatched and has to override the inherent *syn* selectivity of the substrate. Analysis of the results indicate that the catalyst control is >32:1, while the substrate control is 3:1.[75]

## 11. Answers to Problems in Chapter 11

**11.1** The radical reaction is initiated by thermal decomposition of the peroxide to form the radical $C_{11}H_{23}•$ which abstracts I atom to give the cyclic radical. Then radical addition α to the sulphone followed by β-fragmentation gives the product, and the radical $EtSO_2•$ which fragments to $SO_2$ and Et•

to continue the chain process. The major product arises from the approach of the alkene from the less hindered face of the cyclic radical.[76]

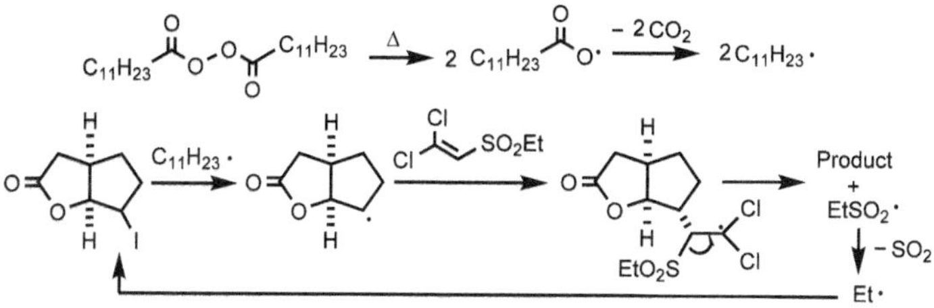


**11.2** (a) Intramolecular amido-mercuration of the alkene, followed by reduction with the borohydride gives an alkylmercury hydride which fragments to give an alkyl radical. Addition of the radical to methyl acrylate gives the product. This radical reaction was used in a synthesis of the alkaloid δ-coniceine.[77]

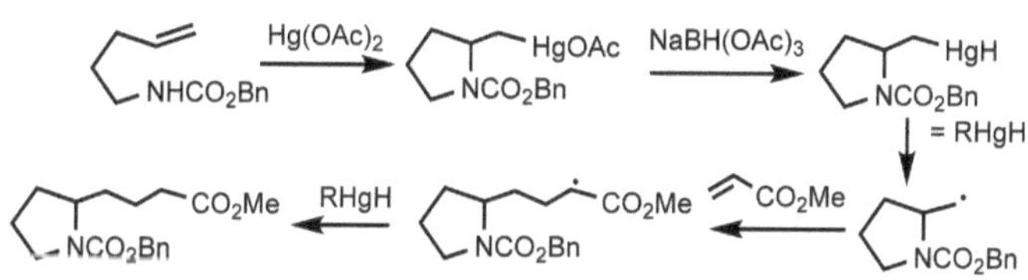


(b) This is a stereoselective radical allylation reaction which involves preferential attack *anti* to the shielding β-substituent to give the predominant product.[78] For initiation step, see Fig. 11.2.

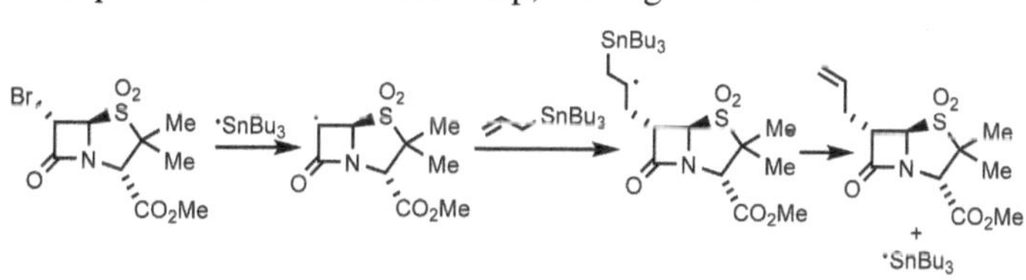


**11.3** The stereoselectivity of the radical reduction in the absence of a Lewis acid arises from the attack of the tin hydride on the less hindered face (*anti* to Ph) of the uncomplexed radical. In the presence of the Lewis acid $MgI_2$, the complexation of the intermediate radical occurs, and the less hindered attack, in this case, results in the reversal of stereoselectivity.[79]

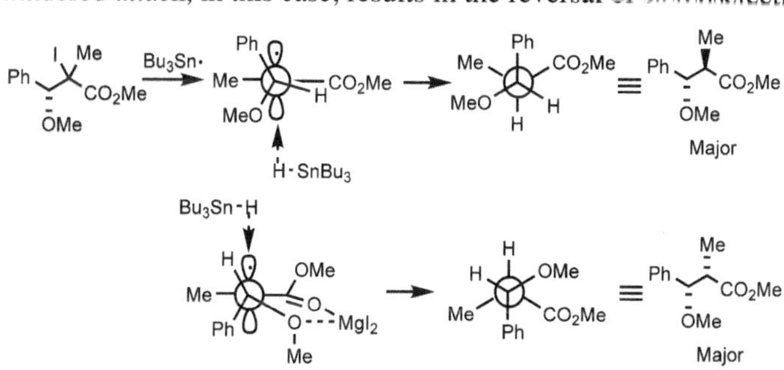

**11.4** $Et_3B/O_2$ generates Et radical to initiate the radical chain reaction (see Fig. 11.13). 5-*exo-trig* cyclization gives selectively the *cis* diastereomer (cf. 3-substituted hexenyl radical). $Bu_3SnH$ acts as the terminal hydride donor.[80]

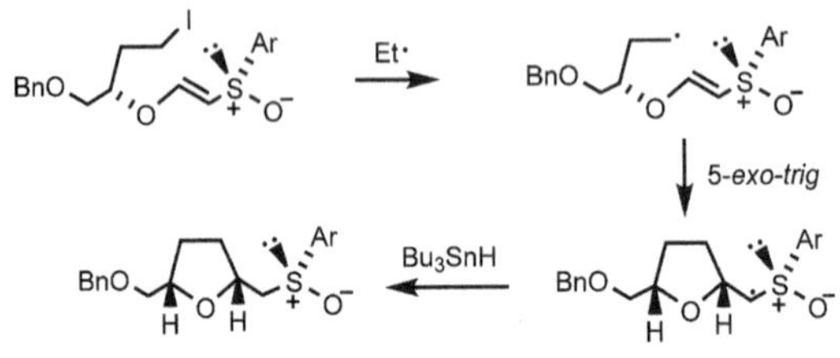


**11.5** Photolysis of the N—O bond followed by decarboxylation produces the carbon radical, which undergoes 5-*exo-trig* closure. The resulting radical reacts with the thiohydroxamic ester to give the product and propagates the cycle (cf. Fig. 11.4).[81]

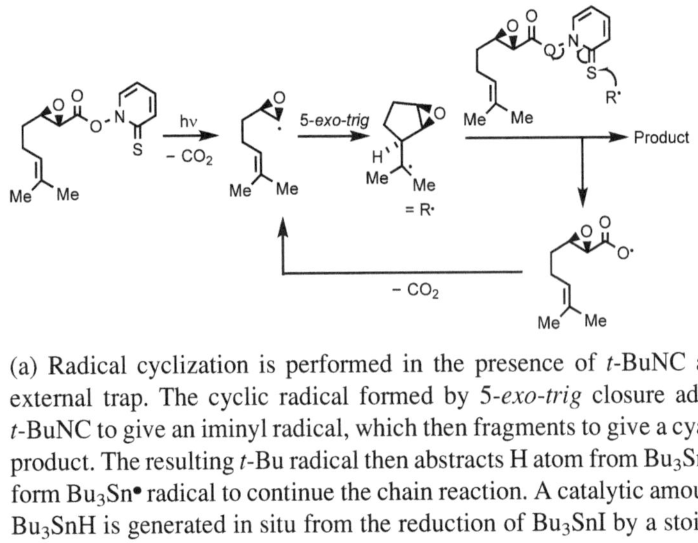


**11.6** (a) Radical cyclization is performed in the presence of *t*-BuNC as an external trap. The cyclic radical formed by 5-*exo-trig* closure adds to *t*-BuNC to give an iminyl radical, which then fragments to give a cyanide product. The resulting *t*-Bu radical then abstracts H atom from $Bu_3SnH$ to form $Bu_3Sn$• radical to continue the chain reaction. A catalytic amount of $Bu_3SnH$ is generated in situ from the reduction of $Bu_3SnI$ by a stoichiometric $NaBH_3CN$.

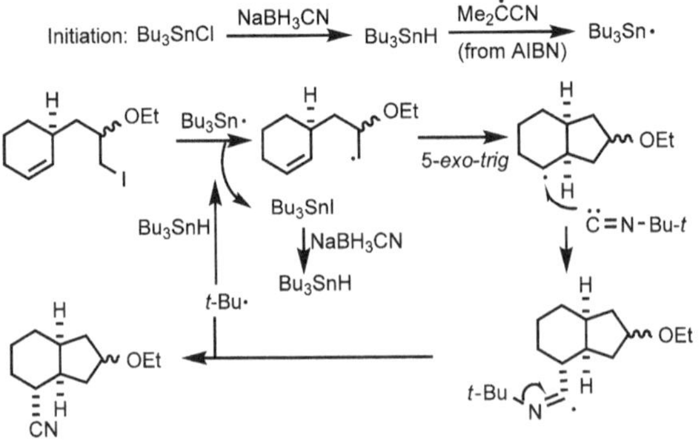

(b) Homolytic C—S bond cleavage by $Bu_3Sn^\bullet$ radical forms a cyclic radical which undergoes 5-*exo-trig* cyclization to give the product. This reaction was used as a key step in a synthesis of the alkaloid gelsemine.[82]

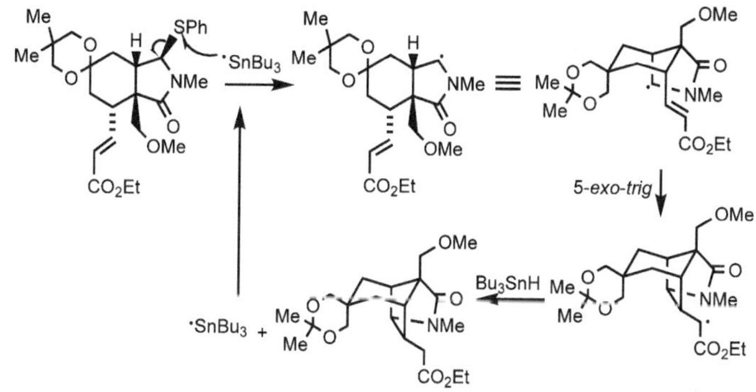


**11.7** The addition of $Bu_3Sn^\bullet$ radical to the alkyne forms an alkenyl radical which undergoes 5-*exo-trig* cyclization to give the observed product after protodestannylation.[83,84]

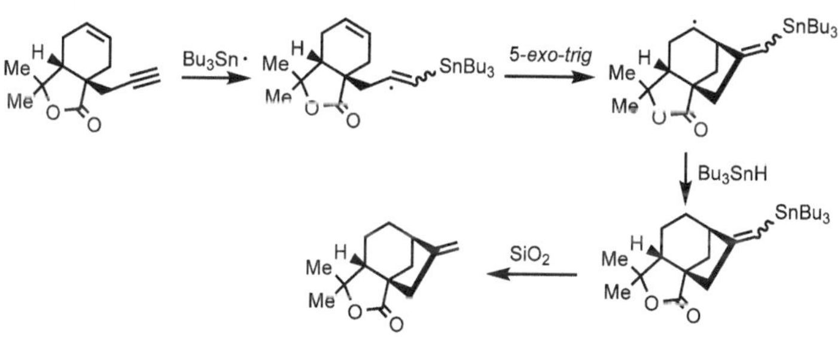


**11.8** (a) The initial alkyl radical undergoes a 5-*exo-dig* cyclization followed by a 5-*exo-trig* closure. The resulting radical intermediate adds to acrylonitrile to form a new radical which then cyclizes by a 5-*exo-trig* process to give the product. The cyclization gives *cis* ring fusion with CN occupying the less hindered position.

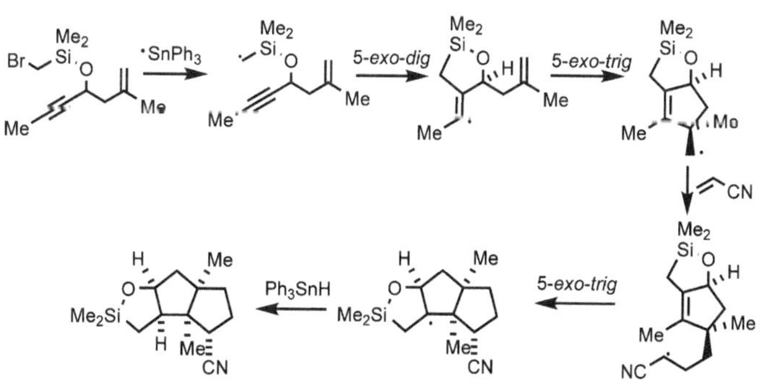

(b) The reaction involves the addition of tributyltin radical to the alkyne to generate an initial alkenyl radical. Then a 5-*exo-trig* cyclization of the alkenyl radical followed by a 5-*exo-dig* addition of the resulting alkyl radical to the nitrile leads to an imine. Hydrolysis of the imine gives the product. Notably, the initiating carbon radical is not generated by the usual atom or group abstraction but by the addition of a tin radical to an alkyne. This addition is normally reversible; however, the alkenyl radical is trapped irreversibly by a 5-*exo-trig* cyclization. This tandem radical cyclization was used in a synthesis of (−)-α-pipitzol.[85]

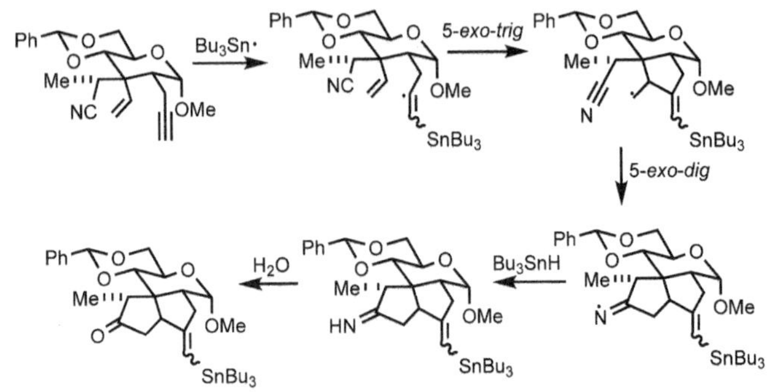


**11.9** (a) As shown below, the $SmI_2$-mediated ketyl radical cyclizes onto the alkyne by a 5-*exo-dig* process to give the product. This reaction was used in a synthesis of isocarbacyclin.[86]

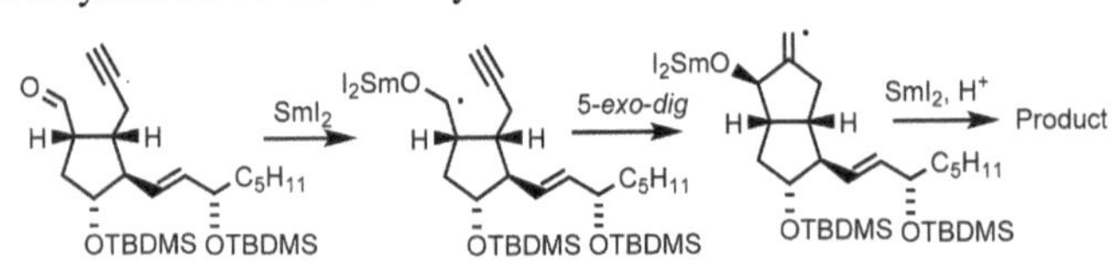


(b) The $SmI_2$-mediated ketyl radical is generated from a 1,3-dicarbonyl substrate. A 5-*exo-trig* cyclization of the chelated ketyl radical favours a *cis* stereochemistry ($CO_2Et$/OH) in the product.[87]

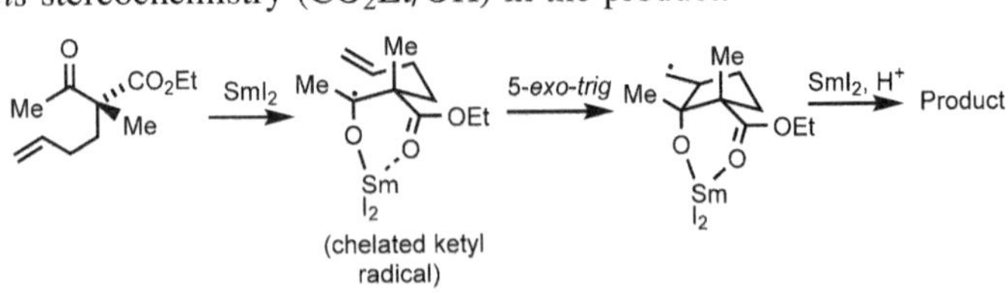


**11.10** Dissolving metal reduction of the terminal alkyne can be prevented by converting it to its sodium salt with $NaNH_2$, and the internal triple bond is then reduced with Na in liquid $NH_3$.[88]

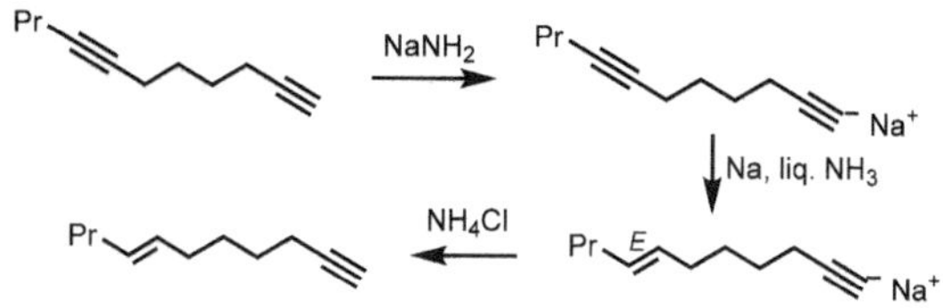


**11.11** The enamine (derived from the aldehyde and chiral imidazolidinone catalyst) is oxidized by the SET reagent (ferrocenium cation, $Cp_2Fe^+$), and the resulting radical is trapped by TEMPO (2,2,6,6-tetramethyl piperidine-1-oxyl) via the *Si*-face attack to give an α-oxyaminated aldehyde product with high enantioselectivity. The product is usually reduced with $NaBH_4$ to the primary alcohol for analysis. Further reduction with $Zn(OAc)_2$ will give the enantioselective 1,2-diol.[89]

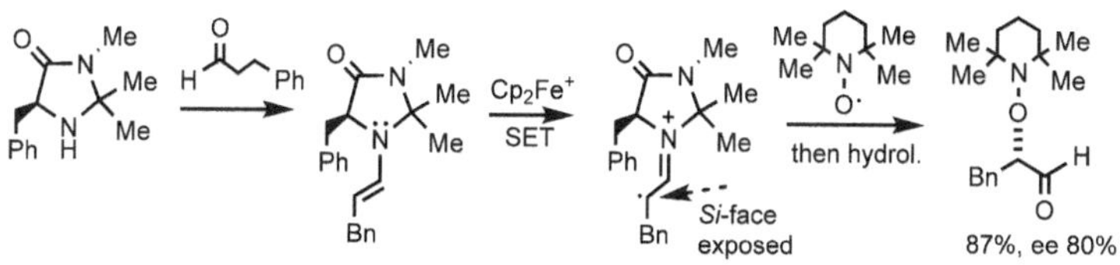


## 12. Answers to Problems in Chapter 12

**12.1** (a) Consecutive Diels–Alder reaction and [2+2] photocycloaddition give the product.

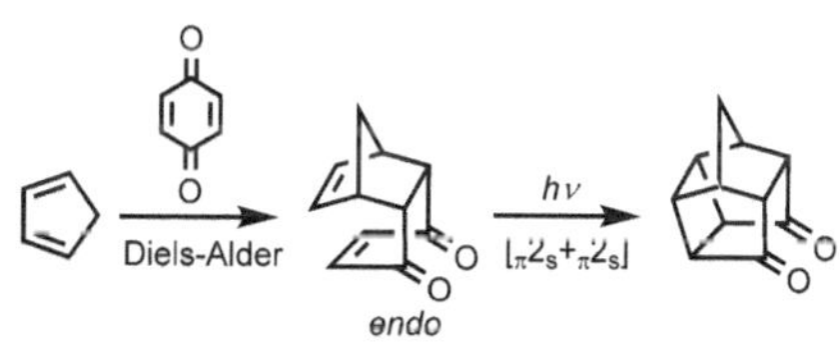


(b) Photodimerization of 9-bromoanthracene via [4+4] cycloaddition forms a head-to-tail dimer, which undergoes Zn-induced transannular debromination (since the two Br's are *anti*) to give the product.[90]

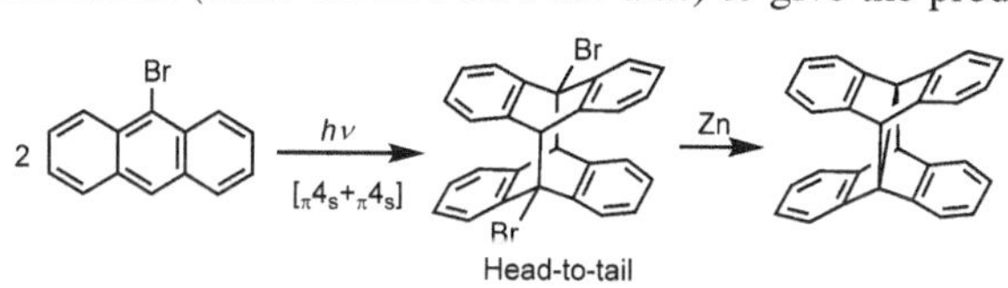


**12.2** The synthesis involves three key pericyclic steps: electrocyclic ring-closing, Diels–Alder reaction and [2+2] photocycloaddition. The photocycloadduct on hydrolysis and bis-decarboxylation gives basketene.[91]

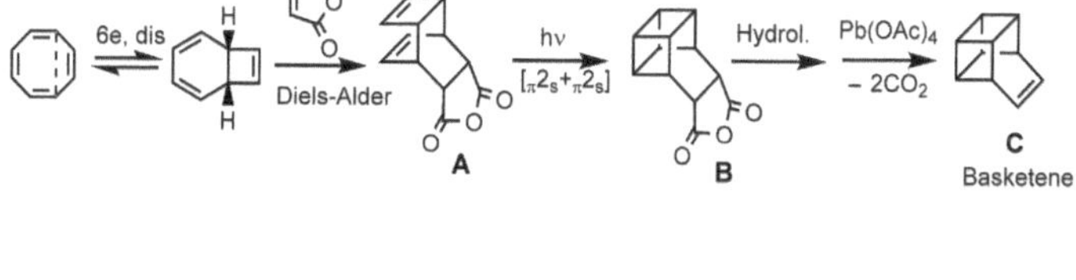

**12.3** This is an intramolecular [2+2] enone photocycloaddition. The conformational constraints dictate the initial C—C bond formation between the nearer ends of the two components, leading to the preferred regioisomer of the product. The *exo* (*anti*) diastereoselectivity in cyclobutane formation is achieved with control of stereochemistry of three new stereocentres by the existing chiral centre.[92]

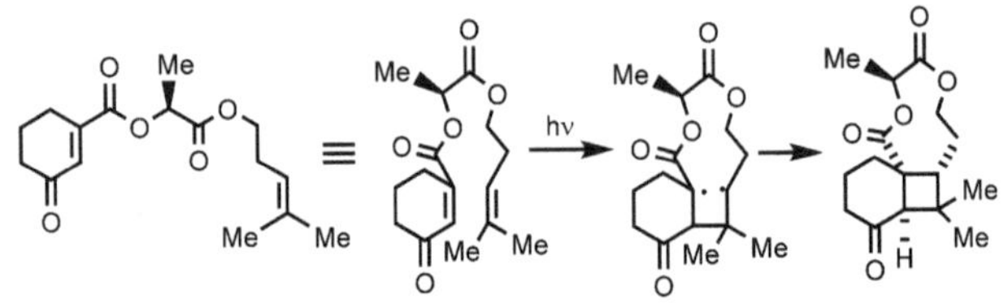


**12.4** Consecutive Diels–Alder reaction and Paternò–Büchi reaction give the product.

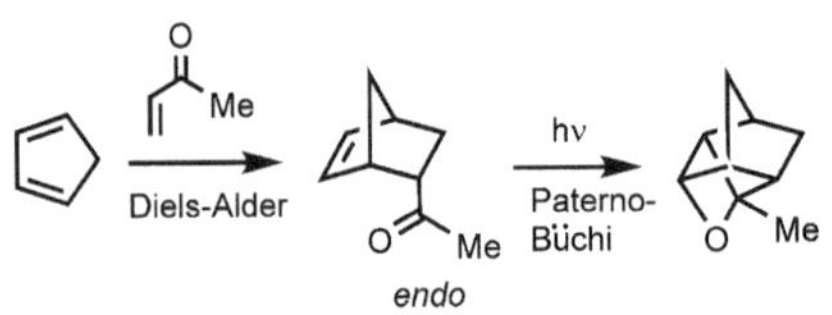


**12.5** When R = OH, the hydrogen bonding interaction of biacetyl with the OH group of the norbornene facilitates its *exo*-face attack in the Paternò–Büchi reaction. In contrast, when R = *t*-Bu, the bulky *t*-Bu group on the norbornene hinders the *exo*-attack of biacetyl leading to the formation of *endo* diastereomer.[93]

**12.6** (a) The cycloheptatrienone undergoes 4-electron disrotatory photoelectrocyclization because photochemical 6-electron ring-closing by the conrotatory process would lead to an impossibly strained *trans* fused 6–3 bicyclic structure. The 4-electron reaction, on the other hand, gives a favourably *cis* fused 5–4 bicyclic system.

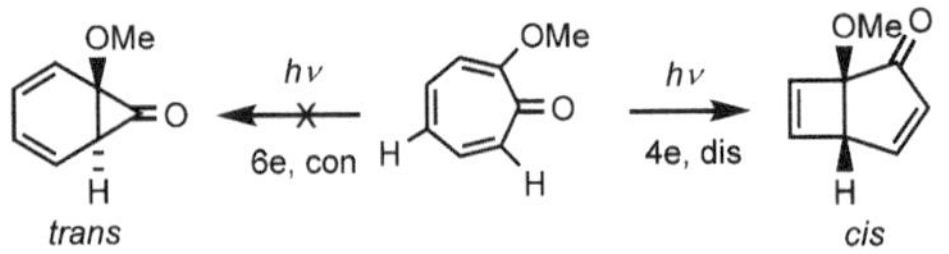


(b) The compound pyrocalciferol (R = $C_9H_{17}$) undergoes 4-electron ring-closing to give the product photopyrocalciferol with favourable *cis* fusion for the four-membered rings. The 6-electron conrotatory ring opening does not occur as the process introduces a *trans* double bond in a six-membered ring.[94]

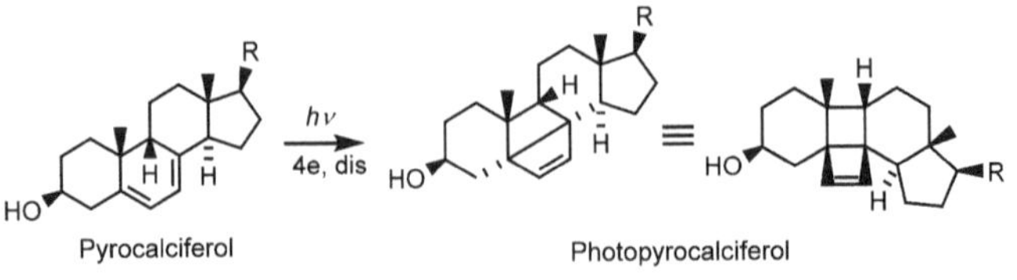

**12.7** Consecutive photochemical ring-opening and thermal ring-closing convert the *trans*-fused bicyclic system to the corresponding *cis*-fused system. See the use of the stereochemical rule.[95]

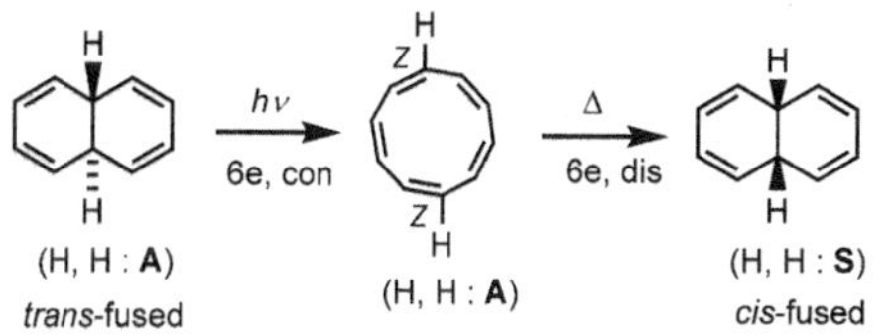


**12.8** The first step is the Burton reaction that produces an (*E*)-oxime, which then undergoes a Beckmann rearrangement in the second step to give the spirocyclic lactam. This reaction was used as a key step in Corey's synthesis of perhydrohistrionicotoxin.[96]

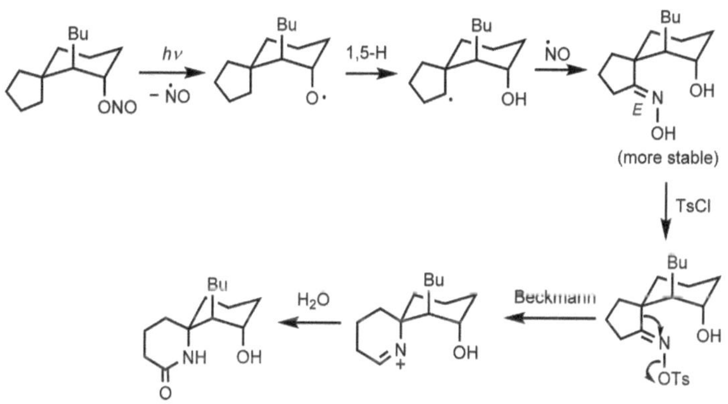


**12.9** The mechanism involves three steps: Norrish Type I cleavage, radical recombination, and [3,3] Cope rearrangement.[97]

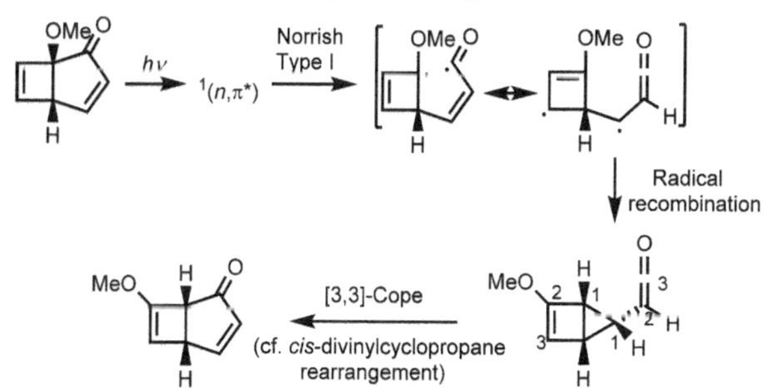


## References


1. Allinger, N. L. *J. Am. Chem. Soc.* **1957**, *79*, 3443.
2. Szasz, G. J. *J. Chem. Phys.* **1955**, *23*, 2449.
3. Mursakulov, I. G.; Ramazanov, E. A.; Guseinov, M. M.; Zefirov, N. S.; Samoshin, V. V.; Eliel, E. L. *Tetrahedron* **1980**, *36*, 1885.

4. Brunet, E.; Eliel, E. L. *J. Org. Chem.* **1986**, *51*, 677.
5. Chow, Y. L.; Colón, C. J.; Tam, J. N. S. *Can. J. Chem.* **1968**, *46*, 2821.
6. Fitjier, L.; Wehle, D. *Angew. Chem. Int. Ed. Engl.* **1979**, *18*, 868.
7. Booth, H.; Griffiths, D. V. *J. Chem. Soc. Perkin Trans. 2* **1973**, 842.
8. Tenud, L.; Farooq, S.; Seibl, J.; Eschenmoser, A. *Helv. Chim. Acta* **1970**, *53*, 2059.
9. Stork, G.; Cohen, J. F. *J. Am. Chem. Soc.* **1974**, *96*, 5270.
10. Feiler, L. A.; Huisgen, R.; Koppitz, P. *J. Am. Chem. Soc.* **1974**, *96*, 2270.
11. Julia, M.; Maumy, M.; Mion, L. *Bull. Soc. Chim. Fr.* **1967**, 2641.
12. Julia, M.; Maumy, M. *Bull. Soc. Chim. Fr.* **1969**, 2427.
13. Chatgilialoglu, C.; Woyner, H.; Ingold, K. U.; Davies, A. G. *J. Chem. Soc. Perkin Trans. 2* **1983**, 555.
14. Barltrop, J. A.; Carless, H. A. J. *J. Am. Chem. Soc.* **1972**, *94*, 1951.
15. Schreiber, J.; Eschenmoser, A. *Helv. Chim. Acta* **1955**, *38*, 1529.
16. Ager, D. *J. Org. React.* **1990**, *38*, 1.
17. Cristol, S. J.; Hause, N. L. *J. Am. Chem. Soc.* **1952**, *74*, 2193.
18. Marshall, J. A. *Synthesis* **1971**, 229.
19. Sternbach, D.; Jaisli, F.; Bonetti, M.; Eschenmoser, A.; Shibuya, M. *Angew. Chem. Int. Ed. Engl.* **1979**, *18*, 634.
20. Donohoe, T. J. *Synlett* **2002**, 1223.
21. Donohoe, T. J.; Blades, K.; Moore, P. R.; et al. *J. Org. Chem.* **2002**, *67*, 7946.
22. Coombs, T. C.; Lushington, G. H.; Douglas, J.; Aubé, J. *Angew. Chem. Int. Ed. Engl.* **2011**, *50*, 2734.
23. Brown, H. C.; Basavaiah, D.; Kulkarni, S. U. *J. Org. Chem.* **1982**, *47*, 171.
24. Mehta, G.; Khan, F. A. *J. Am. Chem. Soc.* **1990**, *112*, 6141.
25. Overman, L. E.; McCready, R. J. *Tetrahedron Lett.* **1982**, *23*, 2355.
26. Evans, D. A.; Hoveyda, A. H. J. *J. Am. Chem. Soc.* **1990**, *112*, 6447.
27. Corey, E. J.; Helal, C. J. *Tetrahedron Lett.* **1995**, *36*, 9153.
28. Evans, D. A.; Britton, T. C.; Ellman, J. A.; Dorow, R. L. *J. Am. Chem. Soc.* **1990**, *112*, 4011.
29. Evans, D. A.; Morrisey, M. M.; Dorow, R. L. *J. Am. Chem. Soc.* **1985**, *107*, 4346.
30. Williams, G. M.; Roughley, S. D.; Davies, J. E.; Holmes, A. B.; Adams, J. P. *J. Am. Chem. Soc.* **1999**, *121*, 4900.
31. Enders, D.; Adam, J.; Klein, D.; Otten, T. *Synlett* **2000**, 1371.
32. Schöllkopf, U. *Tetrahedron* **1983**, *39*, 2085.
33. Myers, A. G.; Schnider, P.; Kwon, S.; Kung, D. W. *J. Org. Chem.* **1999**, *64*, 3322.
34. Corey, E. J.; Cywin, C. L.; Roper, T. D. *Tetrahedron Lett.* **1992**, *33*, 6907.
35. Corey, E. J.; Loh, T.-P.; Roper, T. D.; Azimioara, M. D.; Noe, M. C. *J. Am. Chem. Soc.* **1992**, *114*, 8290.
36. Nelson, S. G. *Tetrahedron Asymmetry* **1998**, *9*, 357.
37. Hajos, Z. G.; Parrish, D. R. German Pat. 1971, DE 2 102 623.
38. Eder, U.; Sauer, G. R.; Wiechert, R. German Pat. 1971, DE 2 014 757.
39. Mukherjee, S.; Yang, J. W.; Hoffmann, S.; List, B. *Chem. Rev.* **2007**, *107*, 5471.
40. Roush, W. R.; Ando, K.; Powers, D. B.; Palkowitz, A. D.; Halterman, R. L. *J. Am. Chem. Soc.* **1990**, *112*, 6339.
41. Roush, W. R.; Palkowitz, A. D.; Palmer, M. A. J. *J. Org. Chem.* **1987**, *52*, 316.
42. Pflum, D. A.; Krishnamurthy, D.; Han, Z.; Wald, S.; Senanayake, C. H. *Tetrahedron Lett.* **2002**, *43*, 923.
43. Bernhard, W.; Fleming, I.; Waterson, D. *J. Chem. Soc., Chem Commun.* **1984**, 28.
44. Schmid, G.; Fukuyama, T.; Akasaka, K.; Kishi, Y. *J. Am. Chem. Soc.* **1979**, *101*, 259.

45. Evans, D. A.; Ratz, A. M.; Huff, B. E.; Sheppard, G. S. *J. Am. Chem. Soc.* **1995**, *117*, 3448.
46. Brown, H. C.; Singaram, B. *J. Am. Chem. Soc.* **1984**, *106*, 1797.
47. Brown, H. C.; Imai, T.; Desai, M. C.; Singaram, B. *J. Am. Chem. Soc.* **1985**, *107*, 4980.
48. Brown, H. C.; Naik, R. G.; Singaram, B.; Pyun, C. *Organometallics* **1985**, *4*, 1925.
49. Brown, H. C.; Naik, R. G.; Bakshi, R. K.; Pyun, C.; Singaram, B. *J. Org. Chem.* **1985**, *50*, 5586.
50. Austin, J. F.; MacMillan, D. W. C. *J. Am. Chem. Soc.* **2002**, *124*, 1172.
51. Evans, D. A.; Bryan, C. A.; Sims, C. L. *J. Am. Chem. Soc.* **1972**, *94*, 2891.
52. Funk, R. L.; Vollhardt, K. P. C. *J. Am. Chem. Soc.* **1979**, *101*, 215. **1980**, *102*, 5253.
53. Rey, M.; Roberts, S. M.; Dreiding, A. S.; et al. *Helv. Chim. Acta* **1982**, *65*, 703.
54. Garsky, V.; Koster, D. F.; Arnold, R. T. *J. Am. Chem. Soc.* **1974**, *96*, 4207.
55. Oppolzer, W. *Angew. Chem. Int. Ed. Engl.* **1989**, *28*, 38.
56. Greene, F. D.; Sclove, D. B.; Pazos, J. F.; Camp, R. L. *J. Am. Chem. Soc.* **1970**, *92*, 7488.
57. Crisp, G. T.; Scott, W. J.; Stille, J. K. *J. Am. Chem. Soc.* **1984**, *106*, 7500.
58. Corey, E. J.; Nozoe, S. *J. Am. Chem. Soc.* **1964**, *86*, 1652.
59. Curtin, D. Y.; Harder, R. J. *J. Am. Chem. Soc.* **1960**, *82*, 2357.
60. Boersma, M. A. M.; Haan, d. J. W.; Kloosterziel, H.; Ven, v.d. L. J. M. *J. Chem. Soc. D, Chem. Commun.* **1970**, 1168.
61. Tulshian, D. B.; Fraser-Reid, B. *J. Org. Chem.* **1984**, *49*, 518.
62. Oppolzer, W.; Dupuis, D. *Tetrahedron Lett.* **1985**, *26*, 5437.
63. Jørgensen, K. A. *Angew. Chem. Int. Ed. Engl.* **2000**, *39*, 3558.
64. Jørgensen, K. A.; Johannsen, M.; Yao, S.; Audrain, H.; Thorhauge, J. *Acc. Chem. Res.* **1999**, *32*, 605.
65. Evans, D. A.; Faul, M. M.; Bilodeau, M. T. *J. Am. Chem. Soc.* **1994**, *116*, 2742.
66. Brandt, P.; Södergren, M. J.; Andersson, P. G.; Norrby, P.-O. *J. Am. Chem. Soc.* **2000**, *122*, 8013.
67. Gillespie, K. M.; Sanders, C. J.; O'Shaughnessy, P.; Westmoreland, I.; Thickitt, C. P.; Scott, P. *J. Org. Chem.* **2002**, *67*, 3450.
68. Müller, P.; Fruit, C. *Chem. Rev.* **2003**, *103*, 2905.
69. Whitesell, J. K.; Allen, D. E. *J. Org. Chem.* **1985**, *50*, 3025. *J. Am. Chem. Soc.* **1988**, *110*, 3585.
70. Tsunoda, T.; Tatsuki, S.; Shiraishi, Y.; Akasaka, M.; Ito, S. *Tetrahedron Lett.* **1993**, *34*, 3297.
71. Tsunoda, T.; Sakai, M.; Sako, Y.; Honda, Y.; Ito, S. *Tetrahedron Lett.* **1992**, *33*, 1651.
72. Evans, R. J. D.; Landor, S. R.; Regan, J. P. *Chem. Commun.* **1965**, 397.
73. Lohray, B. B. *Synthesis* **1992**, 1035.
74. Lohray, B. B.; Sharpless, K. B. *Tetrahedron Lett.* **1989**, *30*, 2623.
75. Nishi, T.; Kitamura, M.; Okhuma, T.; Noyori, R. *Tetrahedron Lett.* **1988**, *29*, 6327.
76. Bertrand, F.; Quiclet-Sire, B.; Zard, S. Z. *Angew. Chem. Int. Ed. Engl.* **1999**, *38*, 1943.
77. Danishefsky, S.; Taniyama, E.; Webb, R. R. *Tetrahedron Lett.* **1983**, *24*, 11.
78. Hanessian, S.; Alpegiani, M. *Tetrahedron* **1989**, *45*, 841.
79. Guindon, Y.; Lavallée, J.-F.; Llinas-Brunet, M.; Horner, G.; Rancourt, J. *J. Am. Chem. Soc.* **1991**, *113*, 9701.
80. Keum, G.; Kang, S. B.; Kim, Y.; Lee, E. *Org. Lett.* **2004**, *6*, 1895.
81. Ziegler, F.; Wang, Y. *Tetrahedron Lett.* **1996**, *37*, 6299.
82. Atarashi, S.; Choi, J.-K.; Ha, D.-C.; et al. *J. Am. Chem. Soc.* **1997**, *119*, 6226.
83. Toyota, M.; Yokota, M.; Ihara, M. *J. Am. Chem. Soc.* **2001**, *123*, 1856.
84. Stork, G.; Mook, R., Jr. *J. Am. Chem. Soc.* **1987**, *109*, 2829.
85. Pak, H.; Canalda, I. I.; Fraser-Reid, B. *J. Org. Chem.* **1990**, *55*, 3009.
86. Bannai, K.; Tanaka, T.; Okamura, N.; et al. *Tetrahedron* **1990**, *46*, 6689.
87. Molander, G. A.; Kenny, C. *J. Am. Chem. Soc.* **1989**, *111*, 8236.

88. Dobson, N. A.; Raphael, R. A. *J. Chem. Soc.* **1955**, 3558.
89. Sibi, M. P.; Hasegawa, M. *J. Am. Chem. Soc.* **2007**, *129*, 4124.
90. Applequist, D. E.; Little, R. L.; Friedrich, E. C.; Wall, R. E. *J. Am. Chem. Soc.* **1959**, *81*, 452.
91. Masamune, S.; Cuts, H.; Hogben, M. G. *Tetrahedron Lett.* **1966**, 7, 1017.
92. Faure, S.; Piva-Le-Blanc, S.; Bertrand, C.; Pete, J.-P.; Faure, R.; Piva, O. *J. Org. Chem.* **2002**, *67*, 1061.
93. Sauers, R. R.; Valenti, P. C.; Tavss, E. *Tetrahedron Lett.* **1975**, 16, 3129.
94. Jacobs, H. J. C.; Havinga, E. *Adv. Photochem.* **1979**, *11*, 305.
95. Masamune, S.; Seidner, R. T. *J. Chem. Soc. D* **1969**, 542.
96. Corey, E. J.; Arnett, J. F.; Widiger, G. N. *J. Am. Chem. Soc.* **1975**, *97*, 430.
97. Calvert, J. G.; Pitts, J. N., Jr. *Photochemistry*; Wiley: New York, 1966.

# Index

Note: Page numbers followed by *f* indicate figures and *t* indicate tables.

## B

# D

## K



## L



## M



## N

# T

FSC
www.fsc.org
MIX
Papier | Fördert gute Waldnutzung
FSC® C083411

Kolibri 360 GmbH
Ferdinand-Jühlke-Straße 7
99095 Erfurt, Deutschland
produktsicherheit@kolibri360.de